SYMMETRIES IN SCIENCE VII

Spectrum-Generating Algebras and Dynamic Symmetries in Physics

SYMMETRIES IN SCIENCE VII

Spectrum-Generating Algebras and Dynamic Symmetries in Physics

Edited by

Bruno Gruber
Southern Illinois University at Carbondale in Niigata
Niigata, Japan

and

Takaharu Otsuka
University of Tokyo
Tokyo, Japan

SPRINGER SCIENCE+BUSINESS MEDIA, LLC

Library of Congress Cataloging-in-Publication Data

Symmetries in science VII : spectrum-generating algebras and dynamic
symmetries in physics / edited by Bruno Gruber and Takaharu Otsuka.
      p.   cm.
   "Proceedings of a Symposium on Symmetries in Science VII: Spectrum
-Generating Algebras and Dynamic Symmetries in Physics, held August
28-31, 1992, in Nakajo, Niigata, Japan."
   Includes bibliographical references (p.   -   ) and index.
   ISBN 978-1-4613-6285-2   ISBN 978-1-4615-2956-9 (eBook)
   DOI 10.1007/978-1-4615-2956-9
   1. Symmetry (Physics)--Congresses. 2. Mathematical physics-
-Congresses. 3. Iachello, F.   I. Gruber, Bruno, 1936-   .
II. Otsuka, Takaharu. III. Symposium on Symmetries in Science VII:
Spectrum-Generating Algebras and Dynamic Symmetries in Physics (1992
: Nakajō-machi, Japan) IV. Title: Symmetries in science 7.
QC793.3.S9S94 1994
539.7'01'512--dc20
                                                         93-47986
                                                              CIP

Proceedings of a Symposium on Symmetries in Science VII:
Spectrum-Generating Algebras and Dynamic Symmetries in Physics,
held August 28–31, 1992, in Nakajo, Niigata, Japan

ISBN 978-1-4613-6285-2

© 1993 Springer Science+Business Media New York
Originally published by Plenum Press, New York in 1993

FOREWORD

The Symposium "Symmetries in Science VII: Spectrum Generating Algebras and Dynamic Symmetries in Physics" was held at the Southern Illinois University at Carbondale in Niigata, Japan Campus, during the period August 28-31, 1992. The Symposium was held in honor of Professor Francesco Iachello on the occasion of his 50th birthday.

We wish to thank the colleagues and friends of Franco for their participation in the Symposium as well as for contributing articles to this volume honoring him. It was their commitment and involvement which made this Symposium a success.

We also wish to thank Dr. Jared H. Dorn, the director of SIUC-N, for his support in the planning and the execution of the Symposium. Moreover we wish to thank Mayor Nobuo Kumakura of Nakajo town and Mr. Kaichi Suzuki of the school entity "The Pacific" for their friendly support.

Bruno Gruber, SIUC-N
Takaharu Otsuka, University of Tokyo

LAUDATIO ON THE OCCASION OF THE
50TH BIRTHDAY OF PROFESSOR FRANCESCO IACHELLO

I first met Franco Iachello in 1974. Driving a smart Alfa-Romeo, he came to meet me at the station at Groningen where I was to spend a summer conducting research.

As soon as I arrived at the Kernfysisch Versneller Instituut we tackled our first problem--the Kaon absorption of nuclei. Our goal was to investigate the influence of nuclear deformation on the absorption. However, in the process of our research, we became intrigued by the discovery of a new model that changed the course of our study. As you all know, this model became known as the Interacting Boson Model.

Pleased with his findings using the d-bosons in his calculations, Franco showed me his results. I realized that his hamiltonian could be precisely diagonalized using the group theoretical method. Franco, too, understood the significance of applying the method to future calculations.

The next step in constructing the Interacting Boson Model was made when we introduced the concept of the "s" boson on the top of the "d" boson. I still remember both of us opening our notes and checking each other's calculations step by step at KVI and Argonne. Naturally, I made many mistakes. Franco, on the other hand, was careful and diligent.

Several years later in Tokyo, my colleague, Takaharu Otsuka and I were discussing how to introduce proton and neutron degrees of freedom to the same model and found a concept like "isospin." We called it "pseudo-isospin."

Shortly afterwards in a letter from the Weizman Institute of Science we learned that Franco and Igal Talmi had independently reached the exact same conclusion! The name they coined for "pseudo-isospin" was "F-spin"--a catchier phrase, we all agreed, "F" being borrowed from the quarks.

The four of us continued to develop a generalization of the IBM by introducing proton bosons as well as neutron bosons. The model was called "np IBM" or "IBM2" a design much closer to the nuclear shell model. We gave a microscopic foundation to the Interacting Boson Model from the shell model point of view.

Franco's imagination and enthusiasm have yielded many contributions to the world of physics. For example, he extended the IBM to odd mass nuclei by adding a Fermion degree of freedom. Furthermore, he tried to unify the boson degree of freedom and that of the Fermion, namely the even-even nuclei and odd nuclei. He succeeded in finding examples of the super-symmetry of nuclei, which has had a strong impact on the study of elementary particle physics.

He's also succeeded in applying the concept of interacting bosons to molecular spectra and the method of group theory to scattering problems. I could never imagine that the group theory would work in a continuum state!

Born in Francofonte, Sicily, Franco was surrounded by Greek temples, Roman villas and Norman-Islamic architecture. The symmetry and beauty of those impressive structures must have inspired him at an early age.

With a degree in nuclear engineering, he graduated from Politecnico di Torino in 1964 and went on to receive his Ph.D. from MIT. Afterwards he spent several years in Europe--Copenhagen, Torino and KVI before moving to Yale in 1978. In 1991 he received the honorable appointment--the Josiah W. Gibbs Professor of Physics. He spent some years in Trento, too, as Professor.

Franco has received many distinguished honors including the AKZO Prize from the Dutch Academy of Sciences, the Wigner medal of the Group Theory and Fundamental Physics Foundation, the Taormina Prize for the Arts and Sciences and an honorary Degree in Physics from the University of Ferrara, to name a few.

Franco is a man of science and the humanities. His knowledge of literature, history and Italian art is encyclopedic. Whenever we meet, I look forward to deepening my appreciation of Italian culture. Another one of his passions is music history, which he dreams of writing about one day. I hope he achieves that goal.

Now, please join me in wishing Franco a very happy and healthy fiftieth birthday. May he continue to contribute to the world of physics.

Akito Arima
President

University of Tokyo

CONTENTS

INTRODUCTORY REMARKS

ON THE OCCASION OF THE PUBLIC LECTURE GIVEN

IN THE NIIGATA-ILLINOIS FRIENDSHIP HALL

Francesco Iachello

Yale University
New Haven, Connecticut 06511, USA

Ladies and Gentlemen, Members of the Community of Nakajo, it is a special pleasure for me to speak to you today in this Niigata-Illinois Friendship Hall. I am particularly happy to contribute to the development of Japanese-American relations and to a better understanding of one another. I am very grateful to Professor Bruno Gruber for organizing this Symposium with speakers from both East and West.

May the objective of the Director, Mr. Jared Dorn, and of the Faculty of Southern Illinois University at Niigata on one side, and of the Mayor, Mr. Nobuo Kumakura and of the entire Community of Nakajo on the other side, be met with success. That knowledge may lead to understanding and understanding to wisdom.

SYMMETRY: THE SEARCH FOR ORDER IN THE UNIVERSE

Francesco Iachello

Yale University

One of the main aspirations of Mankind has always been that of finding an order in the Universe. The idea of order must have come, at least in part, from the observation that many forms of Nature, even the most complex, are often ordered. The first two figures show the order observed

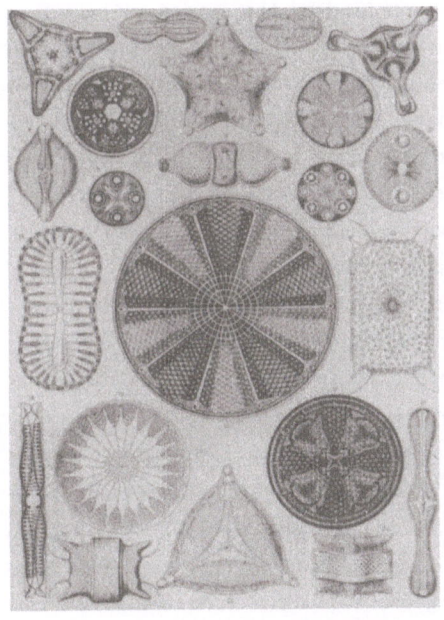

Fig.1. Symmetric forms of Nature (from Ernst Haeckel, <u>Kunstformen der Natur</u>, Leipzig und Wien, 1899).

Symmetries in Science VII, Edited by B. Gruber
and T. Otsuka, Plenum Press, New York, 1994

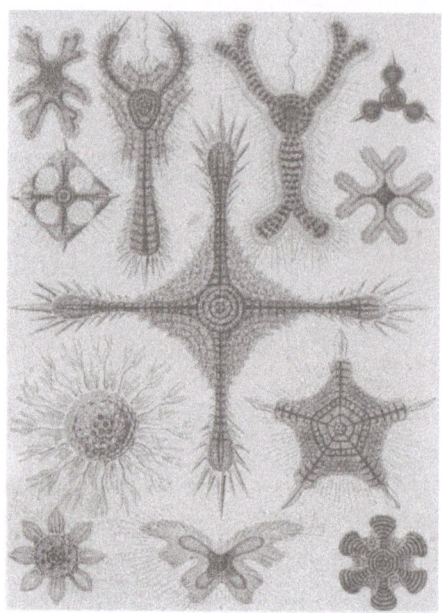

Fig.2. Symmetric forms of
Nature (from Ernst
Haeckel).

in some protoplants and protoanimals, belonging to the classes of
Diatomeas (Fig.1) and Discoides (Fig.2), taken from the illustrations of
the treatise <u>Kunstformen</u> <u>der</u> <u>Natur</u> compiled by the German scientist Ernst
Haeckel in 1899. Order is synonymous with symmetry. The word symmetry,
from the Greek σύμμετρος, describes an object that is well-ordered, well-
organized. Many ancient writers have discussed the meaning of symmetry.
The Greek sculptor Polykleitos in his book on the art of construction
(Πολικλειτος, Περί βελοπιïκῶν) uses the word symmetry extensively. The
Roman architect Vitruvius states that symmetry is "the result of
proportions between different parts". The symmetries that one often
observes in Nature are translations, rotations and reflexions. Figures 1
and 2 clearly display reflexion and rotation symmetries.

All ancient civilizations attempted to imitate the forms of Nature in
art. The Sumerians and all Mesopotamian civilizations that followed them
made extensive use of translation symmetries. Fig.3 shows a portion of a
decoration, dated approximately 2000 B.C., which is symmetric with
respect to translations along the horizontal axis. The Greek civilization
took the concept of symmetry and expanded it even further. Fig.4 shows a
tile from a floor found at the Megaron in Tiryns from the late Helladic
period, about 1200 B.C. The figure is symmetric with respect to reflexion
on a plane perpendicular to it and going through its middle. The Greek

Fig.3. Decoratif motif (Sumerian, circa 2000 B.C.).

Fig.4. Tile found at the Megaron
in Tiryns (Late Helladic,
circa 1200 B.C.).

civilization developed even a Canon, that is a set of rules that works of art must follow.

Together with the development of art, there was also, in the Greek time, a development of mathematics, especially geometry. The discovery of the five regular polyedra, the tetrahedron, the cube, the octahedron, the dodecahedron and the icosahedron, with their symmetric shapes, led the ancient Greeks to think that the entire Universe is built out of these polyhedra. The tetrahedron, octahedron, cube, and icosahedron were associated with fire, air, earth and water, respectively (the building blocks of the Universe), while the (penta)dodecahedron was the image of the Universe itself. (The beginning of the atomistic theory of Nature). The Universe is thus ordered, as the Greek philosopher Plato says in the dialogue _Timaeus_. It is astoninshing to see that, although the Greek construction of the Universe is not correct in a strict sense (the building blocks of Nature are atoms which do not have polyhedric shapes), yet many living forms display to an incredible accuracy the symmetric shapes discovered by the Greeks. Figure 5 shows the skeletons of several

Radiolarias taken from another book of Ernst Haeckel (<u>Report on the Scientific Results of the Voyage of H.M.S. Challenger</u>, Vol. XVIII, pl.117, 1887). Nos. 2,3, and 5 in this figure are perfect examples of octahedric, icosahedric and dodecahedric symmetry.

The study of symmetry took another step forward in the Italian Renaissance, when artists and mathematicians developed further the Greek concept. The five regular polyhedra were complemented with other types, such as the Archimedean polyhedra. The works of Piero della Francesca (<u>Libellus de quinque corporibus regularis</u>, 1482), Luca Pacioli (<u>Divina proportione</u>, 1509) and others describe in full detail the symmetries of

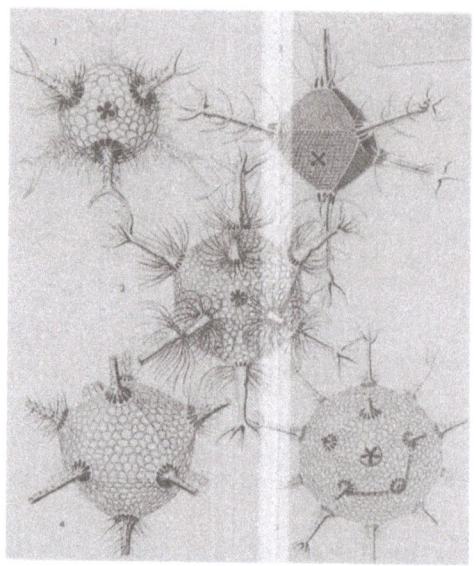

Fig.5. Forms of Nature displaying polyhedric shapes (from Ernst Haeckel, <u>The Challenger Report</u>, London, 1887).

these bodies and begin to introduce what in modern mathematical language is called the theory of group transformations, or, simply, group theory. Once more in these works the basic idea is that the Universe is "proportionate" and thus it follows strict mathematical (geometrical) rules (simmetria). Symmetry is assumed to be the fundamental law of Nature, and is therefore above all (divina).

In 1595, the German astronomer Kepler, in his book <u>Mysterium cosmographicum</u>, noticed that the planetary system known at the time, Saturn, Jupiter, Mars, Earth, Venus and Mercurius, could be reduced to regular bodies which are alternatingly inscribed and circumscribed to spheres (Fig.6). He was so impressed that he concluded his book with the now famous sentence "Credo spatioso numen in orbe", that is "I believe in a geometric order of the Universe". Again, order and symmetry are used in an interchangeable way.

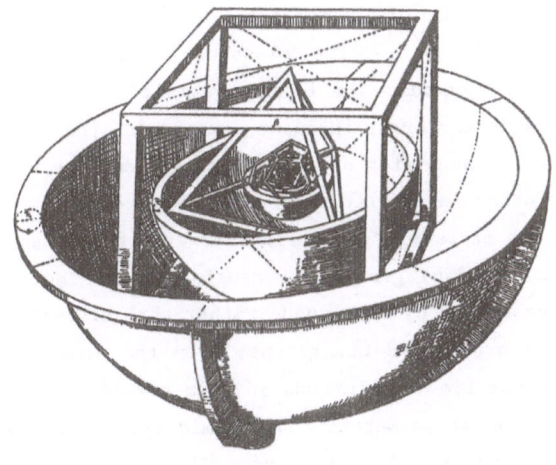

Fig.6. Construction of the Universe according to Kepler. (From the Mysterium cosmographicum, Tübingen, 1595).

The idea of order (and symmetry) began then its entry into physics (the Science that describes Nature through a mathematical language). By the end of the 19th Century, as physics changed more and more from the macroscopic to the microscopic level, it became clearer that symmetries play a fundamental role in physics. Many aspects of Nature are observed to be ordered. The best examples are molecules and crystals. Figure 7 shows the molecule H_3-C-C-Cl_3, whose symmetry is evident (rotations of angles multiple of $120°$, C_3 in mathematical language).   The symmetries

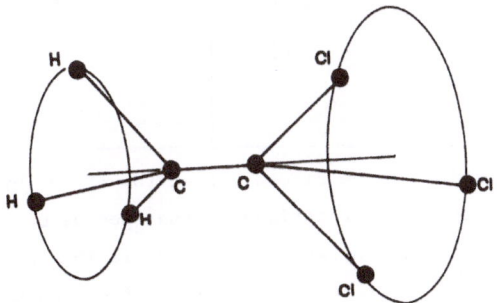

Fig.7. The molecule H_3-C-C-Cl_3.

encountered in molecules and crystals are called "geometric" symmetries. They are similar to those encountered in art (rotations, reflexions and translations). These symmetries were the first ones to be recognized in the microscopic world. But, as time went on, it became clear that other types of symmetry occur in Nature, not related to the geometric arrangement of atoms in a molecule or in a crystal but rather to the dynamic laws of Nature. The role of these other symmetries emerged with the development of quantum mechanics, which occurred around 1920. Contrary to what happens in the classical mechanics of Galileo and Newton, in quantum mechanics the bound states of a physical system are discrete. The search for order in quantum mechanics therefore becomes the search for order in the states of a physical system. The manifestation of symmetry is the observed regularity of the discrete states of the system. As in the previous cases of "geometric" symmetries, "dynamic" symmetries are also characterized by groups of transformations. However, "dynamic" symmetries are described by continuous rather than discrete groups. These continuous groups are called Lie groups, from the name of the Norwegian mathematician Sophus Lie, who introduced them towards the end of the 19th Century. In addition to geometric and dynamic symmetries, other types of symmetry, both discrete and continuous, ("kinematic" symmetries, gauge symmetries, permutation symmetries, ...) have been shown to play an important role in physics. These other types of symmetry will not be discussed in this lecture.

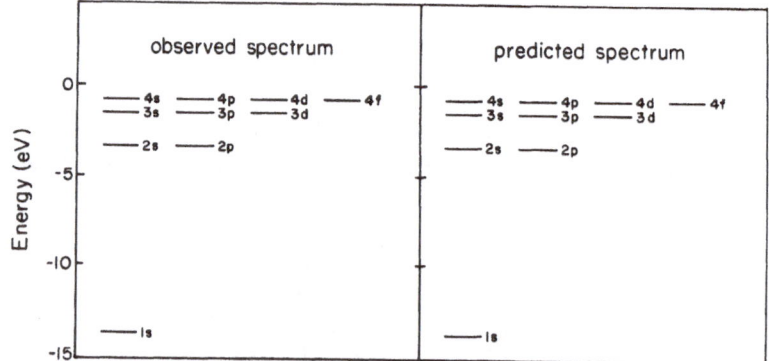

Fig. 8.   A portion of the discrete spectrum of the hydrogen atom. Levels are characterized by the specroscopic notation 1s, 2s, 2p, ... The scale of energy is in electron Volt (eV). On the left the experimental spectrum, on the right the calculated spectrum.

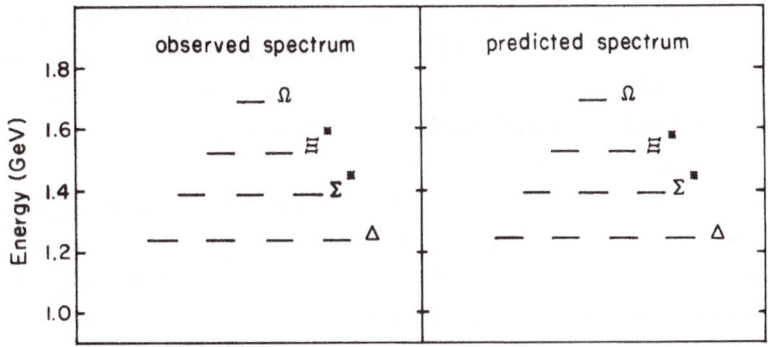

Fig. 9. The energy spectrum of the baryon decuplet (heavy hadrons).
States are characterized by the notation Δ, Σ, ... The
energy scale is in Gev (giga electron Volt) = 10^{12} eV. On the
left the experimental spectrum, on the right the calculated
spectrum.

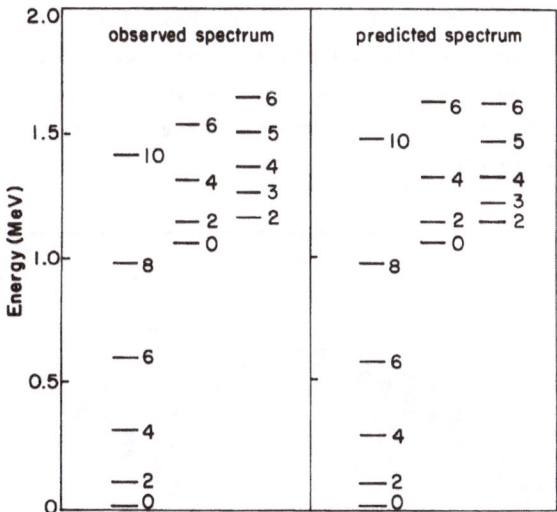

Fig. 10. The energy spectrum of the nucleus ^{156}Gd is shown as an example
of the dynamic symmetry $U(6) \supset SU(3)$. States are characterized by
the angular momenta 0,2,4,.... The energy scale is in MeV (mega
electron Volt) = 10^{6} eV. On the left the experimental spectrum,
on the right the theoretical spectrum. The extent to which the
two parts of the figure agree, is a manifestation of symmetry.

In the course of the last 60 years, dynamic symmetries have been discovered at every level of quantum physics (in molecules that make up the organic and inorganic world, in atoms which make up molecules, in atomic nuclei which are the central part of atoms, in hadrons which are the constituents of nuclei).

In 1926, Wolfgang Pauli ("Über das Wasserstoffspektrum vom Standpunkt der Neuen Quantenmechanik", Z. Physik 36,336 (1926)) noted that the regularity in the discrete spectrum of the hydrogen atom, Fig. 8 (the fundamental atomic structure) is due to the occurrence of a dynamic symmetry, described, in mathematical language, by the group of transformations called SO(4) (special orthogonal transformations in four dimensions).

In 1961, Murray Gell'Mann ("Symmetry of baryons and mesons", Phys. Rev. 125, 1067 (1962)) and Yuval Ne'eman ("Derivation of strong interactions from a gauge invariance", Nucl. Phys. 26, 222 (1961)) discovered that the spectrum of states of the particles called hadrons (from the Greek $\alpha\delta\rho o$, a particle that interacts strongly), Fig. 9, is very regular. It can be described by means of the group SU(3) (special unitary transformations in three dimensions).

In 1974, Akito Arima and myself ("Collective nuclear states as representations of a SU(6) group", Phys. Rev. Lett. 35, 1069 (1975)) discovered that the discrete spectra of many atomic nuclei are regular and proposed that these spectra be described by the group of transformations U(6) (unitary transformations in six dimensions). An example is shown in Fig. 10.

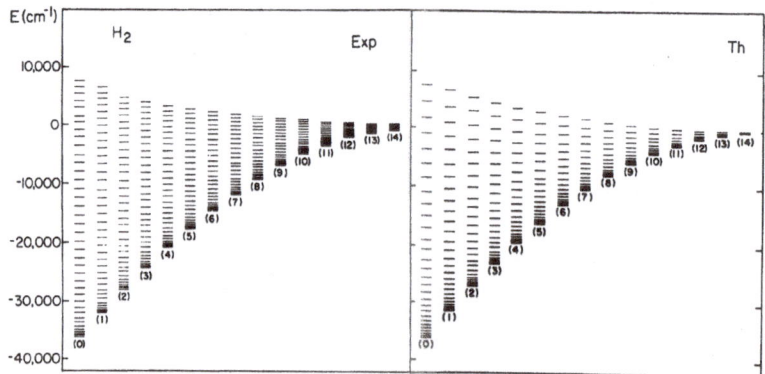

Fig. 11. The energy spectrum of the H_2 molecule is shown as an example of the dynamic symmetry U(4)⊃SO(4). The energy scale is in cm^{-1} = 1.24×10^{-4} eV. On the left the experimental spectrum, on the right the theoretical spectrum.

In 1981, I ("Algebraic methods for molecular rotation-vibration spectra", Chem. Phys. Lett. 78, 581 (1981)) observed that the discrete spectra of many molecules are also regular and suggested that these spectra be described by the group of transformations U(4) (unitary transformations in four dimensions). An example is shown in Fig. 11. This suggestion led to the formulation, together with Raphael D. Levine, of a comprehensive theory of molecules.

One may note in all the cases, shown in the figures 8-11, the arrangement of the energy levels into "patterns" characteristic of the dynamical symmetry which governs the quantum motion of the constituent particles. (Electrons in the case of atoms, Fig. 8, "quarks" in the case of hadrons, Fig.9, protons and neutrons in the case of nuclei, Fig. 10, and atoms in the case of molecules, Fig. 11). The occurrence of these regular "patterns" is the manifestation of dynamic symmetry.

The study of the "patterns" of energy levels of quantum systems, allows us once more to remark on the formal analogy between symmetries in art and in physics. Figure 12 shows a sketch of the energy levels of one of the dynamic symmetries of nuclei (the group SO(6), subgroup of U(6)). The energy spectrum is shown here in a way different from that of Figures 8,9 10 and 11. One may note the regularity of the spectrum and the repetition of the patterns. Fig. 13 shows the pattern of a frieze from Darius' palace in Susa, Mesopotamia. Also here the pattern repeats itself and one can observe the formal analogy between symmetries in art and in physics. One must say, however, to be more precise, that despite this formal analogy, there are important differences between symmetries in art and

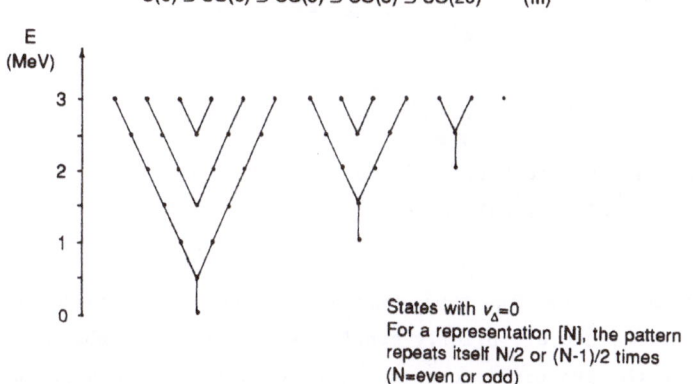

$$U(6) \supset SO(6) \supset SO(5) \supset SO(3) \supset SO(20) \qquad (III)$$

States with $v_\Delta=0$
For a representation [N], the pattern repeats itself N/2 or (N-1)/2 times
(N=even or odd)

Fig.12. Sketch of a portion of the discrete spectrum of the U(6)⊃SO(6) symmetry, composed of a series of "patterns" with a "flower" structure. The pattern repeats itself many times.

physics. The major difference is that while symmetries in art "live" in the usual three dimensional space, symmetries in physics "live" in spaces that include the usual three dimensional space, but often are larger, like in the case of the theory of special relativity in which space

Fig. 13. Frieze from Darius' palace in Susa (Mesopotamian, circa 500 B.C.).

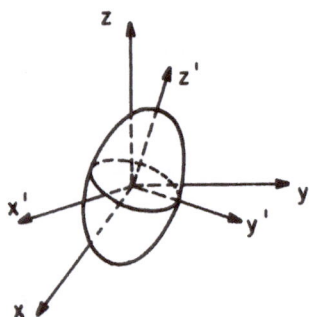

Fig. 14. Representation of the shape of nuclei in the usual three dimensional space.

becomes space-time and thus contains a fourth dimension. The dynamic symmetries of nuclei, briefly mentioned in the paragraphs above, live in a space with five dimensions (one less than that of the group of global transformations, $U(6)$). For this reason, while it is relatively simple to show the symmetries of a work of art, it is much more difficult to show symmetries in physics, unless one represents them in ordinary space, or unless one presents only the energy levels as shown above. In the particular case of atomic nuclei, the representation of the five dimensional space of the group U(6) in ordinary space is an ellipsoid. The possible dynamic symmetries of nuclei correspond to the different shapes of the ellipsoid: sphere, ellipsoid with axial symmetry, etc., Fig. 14.

$$U(5) \supset SO(5) \supset SO(3) \supset SO(2)$$
$$\diagup$$
$$U(6) \quad - \quad SU(3) \supset SO(3) \supset SO(2)$$
$$\diagdown$$
$$SO(6) \supset SO(5) \supset SO(3) \supset SO(2)$$

Fig. 15. Dynamic symmetries in nuclei.

Another difference between symmetries in art and physics, is that symmetries in physics are related to groups of transformations which are more complicated than those describing symmetries in art, such as the reflexion of Fig.4. Therefore, in order to obtain results that can be compared with the observation, one must make use of a much more complicated mathematics. Fig. 15 shows the complete classification scheme of dynamic symmetries in nuclei, an example of which was shown in Fig. 10. Fig.16 shows the complete classification scheme of dynamic symmetries in in molecules, an example of which was shown in Fig. 11. These classification schemes describe in a mathematical language, the "order" of the Universe at the level of atomic nuclei and molecules. They summarize in few lines the beauty and simplicity of Nature, much in the same way in which the three

$$U(3) \supset SO(3) \supset SO(2)$$
$$U(4) \diagup$$
$$\diagdown$$
$$SO(4) \supset SO(3) \supset SO(2)$$

Fig. 16. Dynamic symmetries in molecules.

lines of a "Haiku" with their symmetric structure in terms of 5-7-5 syllables summarize the complex feelings of Man, or in the way in which the simple (and symmetric) lines of a pagoda express the Harmony of art.

In conclusion, Nature (and physics) appear to display at all levels an ordered structure. This is particular evident at the quantum level in the ordered structure of the spectra of molecules, atoms, nuclei and hadrons. "Nature loves symmetry", as the German mathematician Hermann Weyl used to say. The deeper and deeper we go into the mysteries of Nature, the more and more we find symmetry. Despite the widening experience which has repeatedly tested our ideas of symmetry, we still find a mathematical Harmony in the Universe.

U(15)⊃ SU(6) DESCRIPTION OF THE SDG BOSON MODEL

Yoshimi Akiyama

Physics Department, College of Humanities & Sciences
Nihon University, Setagaya-ku, Tokyo 156, Japan

INTRODUCTION

The most prominent feature of the interacting boson model[1] proposed by Arima and Iachello is that it can describe vibrational, rotational and γ-unstable nuclei in one model space, spanned by s- and d-bosons, in terms of the subgroup-reduction chains, U(6)⊃SU(5)⊃O(5)⊃O(3), U(6)⊃SU(3)⊃O(3) and U(6)⊃O(6)⊃O(5)⊃O(3). To describe detailed properties of various nuclei, however, the model space is sometimes too small, and one introduces other bosons, g-boson for example. The sdg boson model has been studied[2] in the U(15)⊃SU(3)⊃O(3) reduction chain to improve the sd description of deformed nuclei. It is interesting to investigate how the introduction of the g-boson changes other reduction chains. In this paper a counterpart of the U(6)⊃O(6)⊃O(5)⊃O(3) reduction chain will be considered. One of its obvious extension to the sdg model space is the U(15)⊃O(15)⊃O(14)⊃O(3) chain. However, the U(15)⊃SU(6)⊃Sp(6)⊃O(3) chain is found to be another sdg extension. Essential features of the O(6) description are easily seen to be carried over into the O(15) one, but details of the SU(6) description are not well known. Their investigation is the main purpose of this paper.

Intrinsic nuclear deformation is studied in terms of the coherent state given by the U(15)⊃SU(6) chain. The group Sp(6) no longer specifies the coherent state. The nuclear shape predicted is very unstable: the result not only shows competition between the prolate and oblate shapes but even suggests possible breaking up of the whole nucleus.

Symmetries in Science VII, Edited by B. Gruber
and T. Otsuka, Plenum Press, New York, 1994

U(15)⊃SU(6)⊃Sp(6)⊃O(3) AS AN EXTENSION OF U(6)⊃O(6)⊃O(5)⊃O(3)

Let us consider an "unphysical fermion" carrying angular momentum j with its projection m. Wave functions of this fermion provide the basis for the irreducible representation of the group chain SU($2j+1$)⊃Sp($2j+1$)⊃O(3). A boson of physical interest consists of two unphysical fermions. The physical boson constructed in this way has its angular momentum $L=0,2,\cdots,2j-1$. The wave functions of the bosons provide the basis for the representation of the group U($j(2j+1)$). Thus the reduction chain U($j(2j+1)$)⊃SU($2j+1$)⊃Sp($2j+1$)⊃O(3) is obtained. If $j=3/2$, one has $L=0$ and 2; and the sd IBM of Arima and Iachello emerges. Since the groups SU(4) and Sp(4) are isomorphic to O(6) and O(5), respectively, the U(6)⊃SU(4)⊃Sp(4)⊃O(3) chain is just another statement of the O(6) chain. If $j=5/2$, one has $L=0$, 2 and 4. The sdg IBM with the group reduction chain U(15)⊃SU(6)⊃Sp(6)⊃O(3) is now obtained. This is clearly an sdg counterpart of the O(6) chain in the sd IBM as seen from its mathematical construction. This fact is also pointed out by Kota et al.[3] Before going to boson description, the SU(6)⊃Sp(6) reduction is briefly summarized.

Creation and destruction operators of unphysical fermion with angular momentum j are denoted as

$$a^\dagger_m = a^\dagger_{jm}, \quad \tilde{a}_m = (-1)^{j+m} a_{j-m}. \tag{1}$$

They satisfy the anti–commutation relation

$$\{\tilde{a}_{m'}, a^\dagger_m\} = (-1)^{j+m} \delta_{-mm'}. \tag{2}$$

The generators of SU($2j+1$) are

$$g^{LM} = (a^\dagger \tilde{a})^{LM} \tag{3}$$

with $L=1,2,\cdots,2j$. Their commutation relation reads

$$[g^{LM}, g^{L'M'}] = \sum C_{LM\,L'M'}^{\quad\quad L''M''} g^{L''M''}, \tag{4}$$

where

$$C_{LM\,L'M'}^{\quad\quad L''M''}$$

$$= (-1)^{2j}(1-(-1)^{L+L'-L''})\sqrt{(2L+1)(2L'+1)}\ W(jLjL';jL'')\ (LL'MM'|L''M'') \tag{5}$$

is the structure constant. It vanishes unless $L+L'-L''$ is odd. Therefore, the commutation relation (4) closes among the g^{LM}'s with odd L. They constitute the generators of the subgroup Sp($2j+1$).

For $j=5/2$, the Killing forms for SU(6) and Sp(6) are constructed in the standard way:

$$K_{SU(6)} = 1/12 \{g^1 \cdot g^1 + g^2 \cdot g^2 + g^3 \cdot g^3 + g^4 \cdot g^4 + g^5 \cdot g^5\} \qquad (6)$$

and

$$K_{Sp(6)} = 1/8 \{g^1 \cdot g^1 + g^3 \cdot g^3 + g^5 \cdot g^5\}. \qquad (7)$$

It is convenient to define the Casimir operators of SU(6) and Sp(6) by

$$C_{SU(6)} = 12 \ K_{SU(6)} = \{g^1 \cdot g^1 + g^2 \cdot g^2 + g^3 \cdot g^3 + g^4 \cdot g^4 + g^5 \cdot g^5\} \qquad (8)$$

and

$$C_{Sp(6)} = 8 \ K_{Sp(6)} = \{g^1 \cdot g^1 + g^3 \cdot g^3 + g^5 \cdot g^5\}. \qquad (9)$$

The irreducible representations of SU(6) and Sp(6) are characterized by $(\mu_1\mu_2\mu_3\mu_4\mu_5)$ and $(\tau_1\tau_2\tau_3)$, respectively. The eigenvalue of Casimir operator for SU(6) is

$$C_{SU(6)}(\mu_1\mu_2\mu_3\mu_4\mu_5) = 1/6 \{ 5 \mu_1^2 + 8 \mu_2^2 + 9 \mu_3^2 + 8 \mu_4^2 + 5 \mu_5^2$$
$$+ 2 (4 \mu_1\mu_2 + 3 \mu_1\mu_3 + 2 \mu_1\mu_4 + \mu_1\mu_5 + 6 \mu_2\mu_3$$
$$+ 4 \mu_2\mu_4 + 2 \mu_2\mu_5 + 6 \mu_3\mu_4 + 3 \mu_3\mu_5 + 4 \mu_4\mu_5) \qquad (10)$$
$$+ 6 (5 \mu_1 + 8 \mu_2 + 9 \mu_3 + 8 \mu_4 + 5 \mu_5)\}$$

and that for Sp(6) is

$$C_{Sp(6)}(\tau_1\tau_2\tau_3) = 1/2 \{\tau_1(\tau_1 + 6) + \tau_2(\tau_2 + 4) + \tau_3(\tau_3 + 2)\}. \qquad (11)$$

State classification relevant to the sdg IBM1 is given in the next section.

BOSON MODEL DESCRIPTION ACCORDING TO U(15)⊃SU(6)⊃Sp(6)

The U(15)⊃SU(6) reduction in the boson space is accomplished by mapping the fermion generators g^{LM} onto the boson generators,

$$T^{(L)}{}_M = \Sigma \ (\tfrac{5}{2} lm | g^{LM} | \tfrac{5}{2} l'm') \ b^\dagger{}_{lm} \ b_{l'm'}, \qquad (12)$$

where the boson creation and destruction operators $b^\dagger{}_{l'm'}$ and b_{lm} satisfy the commutation relation,

$$[b_{lm}, b^{\dagger}_{l'm'}] = \delta_{ll'} \delta_{mm'} . \tag{13}$$

Denoting $b^{\dagger}_{lm}$ for $l=0$, 2 and 4 by $s^{\dagger}_m$, $d^{\dagger}_m$ and $g^{\dagger}_m$ and b_{lm} correspondingly, explicit forms of the SU(6) generators are

$$T^{(1)} = \frac{2}{\sqrt{7}} (d^{\dagger}\tilde{d})^1 + \frac{2\sqrt{6}}{\sqrt{7}} (g^{\dagger}\tilde{g})^1, \tag{14}$$

$$T^{(2)} = \frac{\sqrt{2}}{\sqrt{3}} ((s^{\dagger}\tilde{d}) + (d^{\dagger}\tilde{s})) - \frac{5\sqrt{2}}{7\sqrt{3}} (d^{\dagger}\tilde{d})^2 + \frac{9}{7\sqrt{2}} ((d^{\dagger}\tilde{g})^2 + (g^{\dagger}\tilde{d})^2) + \frac{\sqrt{33}}{7} (g^{\dagger}\tilde{g})^2 \tag{15}$$

$$T^{(3)} = -\frac{9}{7\sqrt{2}} (d^{\dagger}\tilde{d})^3 + \frac{5\sqrt{3}}{7} ((d^{\dagger}\tilde{g})^3 + (g^{\dagger}\tilde{d})^3) - \frac{\sqrt{11}}{7\sqrt{2}} (g^{\dagger}\tilde{g})^3, \tag{16}$$

$$T^{(4)} = \frac{\sqrt{2}}{\sqrt{3}} ((s^{\dagger}\tilde{g}) + (g^{\dagger}\tilde{s})) + \frac{3\sqrt{5}}{7\sqrt{2}} (d^{\dagger}\tilde{d})^4 + \frac{\sqrt{55}}{7\sqrt{3}} ((d^{\dagger}\tilde{g})^4 + (g^{\dagger}\tilde{d})^4) - \frac{\sqrt{143}}{7\sqrt{2}} (g^{\dagger}\tilde{g})^4 \tag{17}$$

and

$$T^{(5)} = -\frac{\sqrt{15}}{\sqrt{14}} ((d^{\dagger}\tilde{g})^5 + (g^{\dagger}\tilde{d})^5) - \frac{\sqrt{13}}{\sqrt{7}} (g^{\dagger}\tilde{g})^5. \tag{18}$$

These operators satisfy the same commutation relation as that for g^{LM}, Eq.(4). Explicit forms for the Casimir operators in this boson description are obtained by replacing the g^{LM} in Eqs.(8,9) by the $T^{(L)}_M$.

For N bosons, possible SU(6) labels $(\mu_1\mu_2\mu_3\mu_4\mu_5)$ are those with $\mu_1=\mu_3=\mu_5=0$,

Table 1. SU(6) decomposition of N boson system.

N	$(\mu_1\mu_2\mu_3\mu_4\mu_5)$
0	(00000)
1	(01000)
2	(02000), (00010)
3	(03000), (01010), (00000)
4	(04000), (02010), (00020), (01000)
5	(05000), (03010), (01020), (02000), (00010)
6	(06000), (04010), (02020), (03000), (00030), (01010), (00000)

and two numbers, μ_2 and μ_4, specify the SU(6) symmetry with $\mu_2+2\mu_4 \le N$. Examples of the SU(6) decomposition are shown in Table 1. Low lying states belong to the representation with the highest symmetry given by $(0,N,0,0,0)$. The Sp(6) representations contained in $(0,N,0,0,0)$ are (000), (110), $\cdots$, $(\tau\tau0)$, $\cdots$, $(NN0)$. We restrict ourselves to these representations in the following considerations. The angular momentum decomposition of the $(\tau\tau0)$ representation is shown in Table 2. Note that the single boson state with $L=0$ belongs to the (000) representation and that with $L=2$ or 4 to the (110).

Table 2. Angular momentum decomposition of the Sp(6) representation $(\tau\tau0)$.

$(\tau\tau0)$	L
(000)	0
(110)	2,4
(220)	$0,2^2,3,4^2,5,6^2,8$
(330)	$0^2,1,2^3,3^3,4^5,5^3,6^5,7^3,8^3,9^2,10^2,12$
(440)	$0^2,1^2,2^7,3^5,4^{10},5^8,6^{10},7^8,8^{10},9^6,10^7,11^4,12^4,13^2,14^2,16$

The hamiltonian in this approach is written as

$$H = a\ C_{SU(6)} + b\ C_{Sp(6)} + c\ L(L+1) \qquad (19)$$

with $a<0$ and $b>0$. By the help of Eqs.(10,11), its eigenvalue for the states belonging to the $(0,N,0,0,0)$, $(\tau\tau0)$ representation is

$$E = a\ \{4/3\ N(N+6)\} + b\ \tau(\tau+5) + c\ L(L+1). \qquad (20)$$

Although this formula is similar to that of the O(6) description, the level structure predicted is rather peculiar. If we set $c=0$, all the states belonging to the same $(\tau\tau0)$ are degenerate. The first excited state is doubly degenerate with $L=2$ and 4. With increasing excitation energy, degeneracy of levels increases much stronger than the O(6) case, as seen from Table 2. With a finite value of c, of course, the states with different L can be separated, but still there are many states with a given L in one $(\tau\tau0)$ representation. The level structure is quite different from that of the O(6) case.

E2 and E4 transition properties can be discussed in terms of the generators $T^{(2)}$ and $T^{(4)}$. Selection rules for the transition among the symplectic states of our interest, $(\tau\tau0)$ to $(\tau+\Delta\tau,\tau+\Delta\tau,0)$, are

$$\Delta\tau=\pm1,0 \qquad (21)$$

in both E2 and E4 transitions. Unlike the O(6) case, $\Delta\tau=0$ is allowed. The symplectic states can have static quadrupole and hexadecapole moments, in contrast to the O(6) description.

COHERENT STATE

Since the operators $T^{(1)}_0$, $T^{(2)}_0$ and $T^{(4)}_0$ commute each other, a coherent state $|N\rangle$ can be defined in terms of the eigen mode operator $b^\dagger_{K=0}$ as

$$|N\rangle = A\,(b^\dagger_{K=0})^N\,|0\rangle, \tag{22}$$

where A is a normalization constant and $|0\rangle$ is the boson vacuum. The $b^\dagger_K$ is a simultaneous solution of the eigen-mode equations,

$$[T^{(1)}_0, b^\dagger_K] = K\,b^\dagger_K \tag{23}$$

$$[T^{(2)}_0, b^\dagger_K] = \xi\,b^\dagger_K \tag{24}$$

$$[T^{(4)}_0, b^\dagger_K] = \eta\,b^\dagger_K . \tag{25}$$

The eigen-mode operator $b^\dagger_0$ for $K=0$ is expressed as

$$b^\dagger_0 = \alpha_0\,s^\dagger + \alpha_2\,d^\dagger_0 + \alpha_4\,g^\dagger_0 . \tag{26}$$

We have three solutions for the eigen-mode equation. They are shown in Table 3. If one adopts the ordinary interpretation that the $T^{(2)}_0$ and $T^{(4)}_0$ give the quadrupole and hexadecapole moments, one can say that the first solution gives a prolate shape with a positive hexadecapole deformation, the second an oblate shape with a negative one, and the third an oblate shape with a positive one. There arises, however, a problem concerning geometric interpretation of our results, as given shortly. Notice that the Sp(6) character of $b^\dagger_0$ is a mixture of (000) and (110). As a result, the coherent state defined by Eq.(21) is no longer a pure Sp(6) state.

Table 3. Three solutions of eigen mode equations for $K=0$.

ξ	η	$b^\dagger_0$
$5/\sqrt{21}$	$1/\sqrt{7}$	$1/\sqrt{3}\,s^\dagger_0 + 5/\sqrt{42}\,d^\dagger_0 + 1/\sqrt{14}\,g^\dagger_0$
$-1/\sqrt{21}$	$-3/\sqrt{7}$	$1/\sqrt{3}\,s^\dagger_0 - 1/\sqrt{42}\,d^\dagger_0 - 3/\sqrt{14}\,g^\dagger_0$
$-4/\sqrt{21}$	$2/\sqrt{7}$	$1/\sqrt{3}\,s^\dagger_0 - 2\sqrt{2}/\sqrt{21}\,d^\dagger_0 + \sqrt{2}/\sqrt{7}\,g^\dagger_0$

To obtain the ground state band, one eigen-mode operator should be selected out among the three solutions of Table 3. For this purpose the interaction,

$$V = T^{(2)} \cdot T^{(2)} + T^{(4)} \cdot T^{(4)} = C_{SU(6)} - C_{Sp(6)},$$ (27)

is considered, which plays the same role as the quadrupole–quadrupole force does in the case of the SU(3) limit. All the three solutions, however, give the same expectation value, $(4/3)N^2$ (in the large N limit), for this interaction V, and we cannot determine which one gives the ground state band. Thus we cannot predict whether the nuclear shape is prolate or oblate. Rather, the result may be interpreted as explaining coexistence of prolate and oblate bands within a single nucleus. It should be remembered that nuclei, having prolate deformations in the low-lying states, have large oblate deformations in the "high spin states". Similarly, two solutions of the eigen-mode equation for the $K=2$ case give the same kind of degeneracy, a stable γ band being not obtained.

The nuclear surface is written as

$$R(\theta,\phi) = R_0 (1 + \beta_2 \, Y_{20}(\theta,\phi) + \beta_4 \, Y_{40}(\theta,\phi)).$$ (28)

If one assumes, following Devi and Kota,[4] $\beta_2 = \alpha_2/\alpha_0$ and $\beta_4 = \alpha_4/\alpha_0$, although this is not a good approximation for large β's, resulting nuclear shape given by each eigen-mode solution is very much deformed. In particular, the second solution has an unphysical shape: for $0° \leq \theta \leq 14°$ and $166° \leq \theta \leq 180°$, $R(\theta,\phi)$ becomes negative. This point raises a problem concerning geometric interpretation of the interacting boson model with hexadecapole degree of freedom. If one takes the "unphysical shape" seriously, the solution should be abandoned or might be understood to suggest break-up of the whole nucleus. The interacting boson model, however, has been originally developed as an approximation to the nuclear shell model, and it has no direct connection with the nuclear shape. From this standpoint, there is no need for excluding our second solution. Whether the geometric interpretation of the interacting boson model has really physical significance is an open problem, particularly if the hexadecapole degree of freedom is included.

CONCLUDING REMARKS

The level structures predicted by this description do not seem to be realized in real nuclei. For applying it to nuclear structure problem, two points should be taken into consideration. First, the single particle energy of the g-boson should be higher than that of d-boson. Second, the hexadecapole-hexadecapole interaction should not be so strong as the model implies.

Finally we point out that the current study opens its possible application to "super-symmetry" theory. It is clear from our derivation of the reduction chain $U(15) \supset SU(6) \supset Sp(6)$ that the boson with $SU(6)$ symmetry and the real fermion with $j=5/2$ are both subject to the group structure $SU(6) \supset Sp(6)$. Therefore, an IBFM described in terms of the $SU(6) \supset Sp(6)$ chain can exist.

REFERENCES

1. Akito Arima and Francesco Iachello, The interacting boson model, Ann.Rev.Nucl. Part.Sci.,31:75(1981).

2. Y.Akiyama, sdg boson model in the $SU(3)$ scheme, Nucl.Phys.A, 433:369(1985); Y.Akiyama, A.Gelberg and P.von Brentano, Experimental consequences of $SU(3)$ symmetry in an sdg boson model, Z.Phys.A326:517(1987); N.Yoshinaga, Y.Akiyama and A.Arima, sdg interacting-boson model in the $SU(3)$ scheme and its application to ^{168}Er, Phys. Rev.C, 38:419(1988).

3. V.K.B.Kota, J.Van der Jeugt, H.De Meyer, and G.Vanden Berghe, Group theoretical aspects of the extended interacting boson model, J.Math.Phys.,28:1644(1987).

4. Y.D.Devi and V.K.B.Kota, Geometric shapes with g-bosons in the interacting boson model, Z.Phys.A,337:15(1990).

DYNAMICAL SYMMETRY BREAKING AND THE ONSET OF CHAOS IN THE INTERACTING BOSON MODEL OF NUCLEI

Y. Alhassid

Center for Theoretical Physics
Sloane Physics Laboratory and
A.W. Wright Nuclear Structure Laboratory
Yale University
New Haven, Connecticut 06511

1. INTRODUCTION

The interacting boson model introduced by Iachello and Arima[1],[2], has been very successful in describing the low lying collective levels and electromagnetic transition intensities of heavy nuclei. In its simplest version, the IBM-1, the nuclear Hamiltonian is described with few parameters and possesses three physical limits where it is exactly solvable. Even the general IBM-1 Hamiltonian is relatively easy to solve due to the moderate size of its basis.

These same features make the model very attractive for the study of the onset of chaos in the nuclear collective motion[3]. One can investigate both the statistical fluctuation properties of the quantal Hamiltonian and the dynamics of its classical limit. The quantal analysis requires an accurate and complete set of energy levels and transition intensities. The finite Hilbert space of the model allows an exact solution with no truncation errors. The classical (mean-field) limit is easily obtained through the use of coherent boson states in the limit of large numbers of bosons. Moreover, the completely integrable limits of the model are easily indentified without any detailed calculations.

On the experimental side, the neutron and proton resonances data collected from various nuclei in the so called nuclear data ensemble[4], are still the best example of a "chaotic" spectrum. Their fluctuation properties were explained by the use of a random matrix theory (RMT)[5] which is consistent with the time reversal symmetry, i.e. the gaussian orthogonal ensemble (GOE). The use of RMT[6] was justified by the complexity of the compound nucleus Hamiltonian[6]. It was however more recently conjectured that the GOE is the proper ensemble to describe systems whose classical limit is chaotic[7]. Following that, much attention was paid to analysis of experimental levels in the low-lying collective regime of the nuclear spectrum. Weidenmuller et al[8] analyzed such levels by grouping together levels of similar properties in different nuclei. Shriner et al[9] analyzed the statistical properties of low lying levels in ^{26}Al and Raman et al[9] analyzed a complete set of levels in ^{116}Sn. Levels from a large number of nuclei were also studied in Ref. 10.

The advantage in a theoretical analysis is that one can analyze a given Hamiltonian and states with given spin/parity and still obtain reasonable statistics. The model has to be realistic and tractable at the same time. For that purpose the IBM is very suitable.

2. MODEL

The degrees of freedom on the IBM are one monopole s boson and five quadrupole bosons d_μ. The total number of bosons $N = s^\dagger s + \sum_\mu d_\mu^\dagger d_\mu$ is conserved. The Hamiltonian is a rotational scalar built for one- and two-body terms in the bosons. To study the dynamics of the IBM throughout its parameter space, it is best to use its most economical parametrization, the consistent Q formalism[11].

The Hamiltonian is

$$H = E_0 + c_0 \hat{n}_d + c_2 \; Q^\chi \cdot Q^\chi + c_1 \; \vec{L}^2 \quad , \tag{1}$$

where $n_d = d^\dagger \cdot \tilde{d}$ is the number of d bosons, Q^χ is a quadrupole operator

$$Q_\chi = (d^\dagger \times \tilde{s} + s^\dagger \times \tilde{d})^{(2)} + \chi (d^\dagger \times \tilde{d})^{(2)} \quad , \tag{2}$$

and $\vec{L}$ is the angular momentum. We have used the notation $\tilde{d}_\mu = (-)^\mu d_{-\mu}$ so that $\tilde{d}_\mu$ transforms like $d_\mu^\dagger$ under rotations. The important terms in (1) are $\hat{n}_d$ which take into account the energy difference between the d boson (pair of fermions coupled to spin 2) and the s boson (pair coupled to spin 0), and the attractive quadrupole-quadrupole interaction. The same operator as in (2) is used to describe the E2 transitions

$$T(E2) = \alpha_2 \; Q^\chi \quad . \tag{3}$$

3. CLASSICAL LIMIT

In the IBM, $1/N$ plays the role of $\hbar$ so that the classical limit is obtained for $N \to \infty$ through a mean field approximation. A mean-field approximation of a bosonic model can be obtained by using coherent states[12]. In our case we define for any six complex parameters $\vec{\alpha} = \{\alpha_s, \alpha_\mu; \; \mu = -2, \ldots, 2\}$

$$|\vec{\alpha}> = \exp(-|\alpha|^2/2) \; \exp(\alpha_s s^\dagger + \sum_\mu \alpha_\mu d_\mu^\dagger) |0> \quad . \tag{4}$$

Using the time-dependent variational principle $\delta \int <\psi | i\partial/\partial t - H | \psi> = 0$ for a trial function $|\psi> = |\alpha(t)>$ we obtain equations of motion which have an Hamiltonian form. The Hamiltonian is $<\alpha | H | \alpha>$ and $\alpha_j, i\alpha_j^*$ are canonically conjugate variables. Rescaling $\alpha \to \alpha/\sqrt{N}$ and defining $h = <\alpha | H | \alpha>/N$, we obtain in the $N \to \infty$ limit[13]

$$h(\alpha, \alpha^*) = \epsilon_0 + \bar{c} \, [\eta n_d - (1-\eta) q^\chi \cdot q^\chi] + \bar{c}_1 \vec{\ell}^2 \qquad (5)$$

n_d, q^χ_μ and $\vec{\ell}$ are c numbers obtained from $\hat{n}_d$, Q^χ_μ and $\vec{L}$, respectively, by the substitution

$$s^\dagger, \, d^\dagger_\mu \to \alpha^*_s, \, \alpha^*_\mu$$

$$ \qquad (6)$$

$$s, \, d_\mu \to \alpha_s, \, \alpha_\mu .$$

The parameters in (5) are given by

$$\epsilon_0 = E_0/N \quad , \quad \bar{c}_1 = Nc_1$$

$$\bar{c} = c_0/\eta \quad , \quad \frac{\eta}{1-\eta} = - \frac{c_0}{Nc_2} \qquad . \qquad (7)$$

$\vec{\ell}$ in (5) is the angular momentum per boson. Since $\vec{\ell}^2$ is a constant we can set $\bar{c}_1 = 0$ without loss of generality. Furthermore with proper scaling of the Hamiltonian we can take $\bar{c} = 1$. This leaves us with two parameters: η ($0 \leq \eta \leq 1$) and χ ($-\sqrt{7}/2 \leq \chi \leq 0$).

The Hamiltonian (5) is in six degrees of freedom. It is possible to eliminate one degree of freedom by factoring out a total phase $\exp(i\theta)$ from all six α_j. The phase is chosen such that α_s becomes real. The action variable conjugate to θ is the number of bosons N. After the scaling $\alpha \to \alpha/\sqrt{N}$, the constraint on the boson number becomes

$$\alpha^*_s \alpha_s + \Sigma_\mu \, \alpha^*_\mu \alpha_\mu = 1 . \qquad (8)$$

By substituting $\alpha_s = \sqrt{1 - \Sigma_\mu \alpha^*_\mu \alpha_\mu}$ in (5) we obtain a problem in five degrees of freedom. The remaining α_μ (after factoring out $\exp(i\theta)$) can be rewritten as

$$\alpha_\mu = [(-)^\mu q_{-\mu} + i p_\mu]/\sqrt{2}$$

$$ \qquad (9)$$

$$\alpha^*_\mu = [q_\mu - (-)^\mu i p_{-\mu}]/\sqrt{2} \qquad .$$

The real tensors q_μ, p_μ play the role of the quadrupole deformation variables and its conjugate momenta, respectively. It is possible to transform q_μ to the intrinsic variables β, γ and Euler angles Ω, and p_μ to the respective conjugate momenta.

For the special case $\ell=0$, the β,γ variables decouple from the Euler angles and it is possible to rewrite (5) in the intrinsic variables alone

$$h = \eta n_d + (1-\eta)\ q^\chi \cdot q^\chi$$

$$= \frac{1}{2}\ [\eta - 2(1-\eta)\beta^2]\ (\beta^2 + T) + 2(1-\eta)\beta^2$$

$$+ \frac{\chi(1-\eta)}{\sqrt{7}/2}\ \sqrt{1-(\beta^2+T)/2}\ \ [(p_\gamma^2/\beta - \beta p_\beta^2 - \beta^3)\cos 3\gamma + 2p_\beta p_\gamma \sin 3\gamma]$$

$$+ \frac{\chi^2}{7/4}\ (1-\eta)\ [(\beta^2+T)^2/8 - p_\gamma^2/2] \tag{10}$$

where $T = p_\beta^2 + p_\gamma^2/\beta^2$.

4. DYNAMICAL SYMMETRIES AND INTEGRABILITY

A dynamical symmetry occurs in an algebraic model with algebra G when the Hamiltonian can be written as a function of the Casimir invariants of a chain of subalgebras of G

$$G \supset G' \supset G'' \supset \ldots \qquad . \tag{11}$$

The eigenstates of H can then be labeled by the eigenvalues of the Casimir invariants $C(G)$, $C(G')$, $C(G'')$, ... and the energy levels are given analytically. A dynamical symmetry implies integrability[3,14] since the above set of Casimir invariant are constants of motion in involution (i.e. commute among themselves).

$$[H,\ C(G^{(i)})] = 0$$

$$[C(G^{(i)}),\ C(G^{(i)})] = 0\ . \tag{12}$$

If the set is not complete, there are missing labels. If a missing label occurs for example in the reduction $G' \supset G''$ then it is possible[18] to construct from the generators of G' an invariant of G''. With these additional invariants we obtain a complete set of constants and hence integrability. One can therefore identify apriori the completely integrable limits of an algebraic model by simply identifying its dynamical symmetries.

It is well known that the IBM has three such limits

$$U(6) \supset \left\{ \begin{array}{c} U(5) \supset O(5) \\ SU(3) \\ O(6) \supset O(5) \end{array} \right\} \supset O(3)\ , \tag{13}$$

which are denoted by (I), (II), and (III), respectively[1]. (I) describes vibrational nuclei, (II) rotational nuclei and (III) γ-unstable nuclei. (I) is obtained in (5) for $\eta=1$, (II) for $\eta=0$ and $\chi=-\sqrt{7}/2$ and (III) for $\eta=0$ and $\chi=0$.

The corresponding classical limits of (13) are also completely integrable. The substitution (6) provides the classical constants which satisfy the same relations as in (12) but with the Poisson brackets replacing the commutator.

We remark that $\{N, L^2, L_z\}$ are always constants in involution. To get a complete set we need three more (since the problem is in six degrees of freedom). In the above three limits they are the following:

(i) U(5); two are the number of d bosons, $\hat{n}_d$, and the quadratic Casimir invariant of O(5). The third is a O(3) invariant built from the O(5) generators.

(ii) SU(3); the quadratic and cubic Casimir invariants of SU(3) and the O(3) invariant $(\vec{L} \times Q)^{(1)} \cdot \vec{L}$ where Q is as in (2) with $\chi=-\sqrt{7}/2$.

(iii) O(6); two constants are the quadratic O(6) and O(5) Casimir invariants and the third is as discussed in (i).

We note that in all the cases above the Hamiltonian does not depend on the invariants which are associated with the missing labels. This will cause exact degeneracies which are beyond what the Poisson statistics predicts. These situations will be termed "overintegrable".

The family of Hamiltonians can be described by a triangle whose three vertices corresponds to the three dynamical symmetry limits. A point inside the triangle is parametrized by coordinates (χ, η) such that η is the height above the base and χ is the intersection with the base of a line originating at the upper vertex and passing through the given point. The base describes the transition between the SU(3) $(\chi=-\sqrt{7}/2)$ and O(6) $(\chi=0)$.

5. CLASSICAL CHAOS

In the limit of dynamical symmetry the classical dynamics is completely integrable. As we break the symmetry chaotic dynamics may set in part of the phase space. This can be detected by examining the Lyapunov exponents. A Lyapunov exponent is defined as the rate of exponential separation D(t) between two neighboring trajectories $\alpha_j(t)$ and $\alpha_j(t) + \Delta\alpha_j(t)$

$$\lambda = \lim_{t \to \infty} \frac{1}{t} \ln \left[\frac{D(t)}{D(0)} \right] \tag{14}$$

Depending on the choice of $\Delta\alpha_j(0)$ there are several such exponents whose number is equal to the dimension of phase space. Since the time evolution has a symplectic structure, they come in pairs $\pm\lambda$. Since N, L^2, L_z and the Hamiltonian H are always constants of the motion there are at most two non-zero pairs of Lyapunov exponents. This is shown in Fig. 1 for the Hamiltonian (5) with $\eta=0$ and $\chi = -0.7$. The motion is thus effectively in three degrees of freedom so that the method of Poincare sections to demonstrate chaos is impractical. Instead we observe the maximal λ and call a point in phase space chaotic when this λ is positive. For a regular trajectory $\lambda=0$.

We define two measures of classical chaos in the subspace of given energy and angular momentum: the fraction σ of chaotic volume and the average maximal Lyapunov exponent. Instead of the latter we may calculate an average of all the non-negative Lyapunov exponents which is equal to the Kolmogrov entropy. In practice we solve Hamilton's equations in the full twelve-dimensional phase space. The initial points are chosen randomly on a twelve-dimensional sphere $\Sigma_j |\alpha_j|^2 = 1$ and their maximal Lyapunov exponent is determined by solving the classical equations of motion for two neighboring trajectories. By choosing initial points which are within narrow energy and angular momentum windows, we can determine σ and $\bar{\lambda}$ as a function of energy and angular momentum as well.

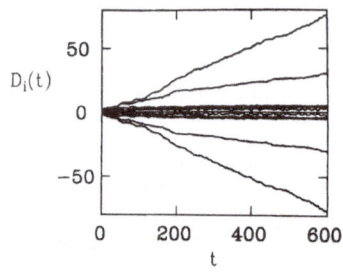

Fig. 1. Example of $\ell n(D_i(t)/D_i(0))$ versus time t where $D_i(t)$ is the distance between neighboring trajectories. The slopes are the Lyapunov exponents. Only two pairs are non-zero.

Fig. 2. shows the classical results in the Casten triangle where σ and $\bar{\lambda}$ are averaged over energy and over angular momentum (per boson) in the range $0.16 \leq \ell \leq 0.80$. The unshaded regions ($0 \leq \sigma \leq 0.2$) are highly regular while the dotted areas ($0.8 \leq \sigma \leq 1$) are highly chaotic. In addition to the expected regularity of the three vertices notice that the transition between U(5) and O(6) is regular. It is in fact completely integrable since O(5) is a common subalgebra. The quadratic Casimir invariant of O(5) and the invariant which corresponds to the O(5) ⊃ O(3) reduction, together with N, L^2, L_z and H form a complete set of constants in involution. In the transitions from SU(3) to O(6) and from SU(3) to U(5) we see highly chaotic behavior in the intermediate regions. An unexpected result is the existence of an almost regular narrow region inside the triangle connecting the SU(3) and U(5) vertices.

To demonstrate the energy dependence we show in Fig. 3 an energy (ϵ)-χ diagram at constant angular momentum and for three values of η. ϵ_{min} denotes the lowest possible energy for a given χ. The immediate region above ϵ_{min} is always regular so that for each χ (and η) there is a critical energy for the onset of chaos.

As mentioned earlier, for $\ell=0$ the β,γ degrees of freedom decouple from the Euler angle and there is only one non-zero pair of Lyapunov exponents. In that case we can construct two dimensional Poincare sections. An example is shown in Fig. 4 where Poincare sections of (10) for $\eta=0$ $\chi=-0$. Notice that chaos first sets in at the vicinity of the unstable hyperbolic 3-cycles. The onset of chaos is very rapid in this example[17].

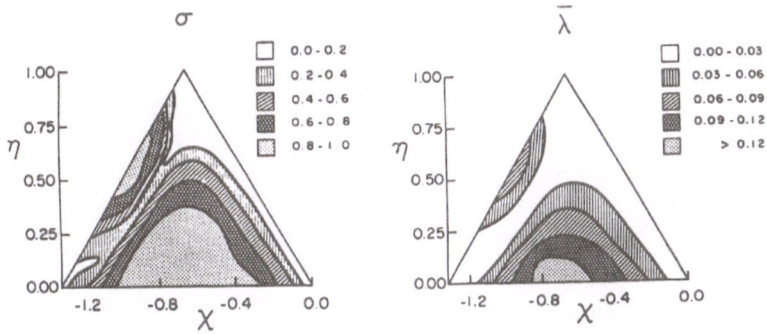

Fig. 2. Classical diagrams of Casten's triangle. Left: the chaotic fraction σ of phase space for $0.16 < \ell < 0.81$. Right: the average maximal Lyapunov exponent $\overline{\lambda}$.

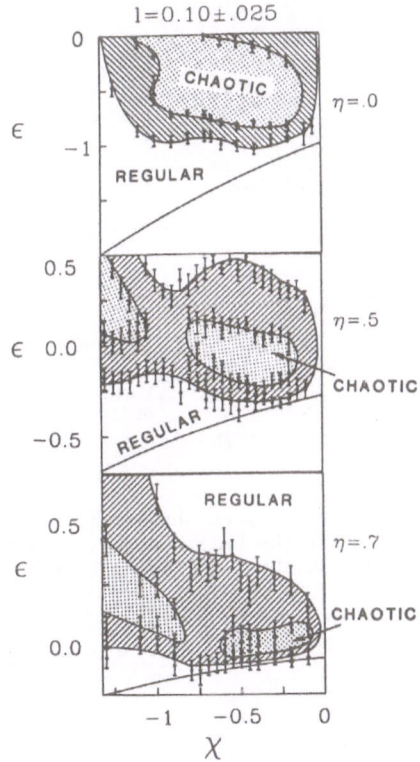

Fig. 3. Classical phase diagram at constant $\ell=0.1$ in the ϵ(energy per boson)-χ plane and for several values of η. The regular, intermediate and chaotic regions correspond to $\sigma < 0.3$, $0.3 < \sigma < 0.7$ and $0.7 < \sigma < 1$, respectively.

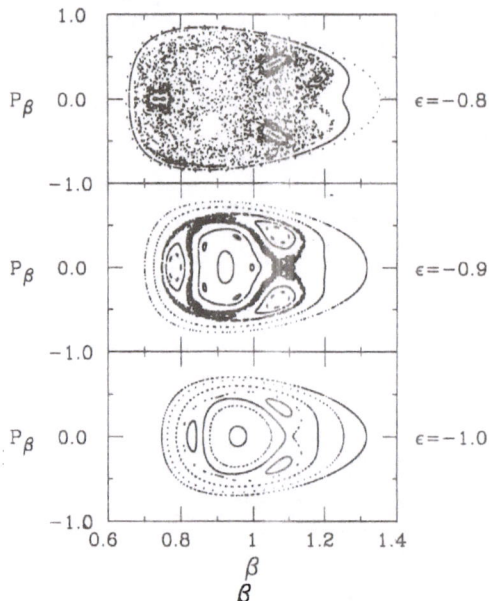

Fig. 4. Poincare Sections $\gamma=0$ in the P_β-β plane for the Hamiltonian (10) with $\eta=0$, $\chi=-0.7$ at three different energies.

The dependence of chaos on spin is demonstrated in Fig. 5, where a χ-ϵ diagram (η=0) is shown for four values of the spin (the highest possible spin per boson is ℓ=2). We see that at low spins there is only a weak dependence of chaoticity on spin. However at higher spins the chaotic region shrinks. We also show (top part of Fig. 5) the classical level density as a function of energy. It is seen that for the chaotic Hamiltonians (intermediate values of χ) only a small fraction of the states are regular.

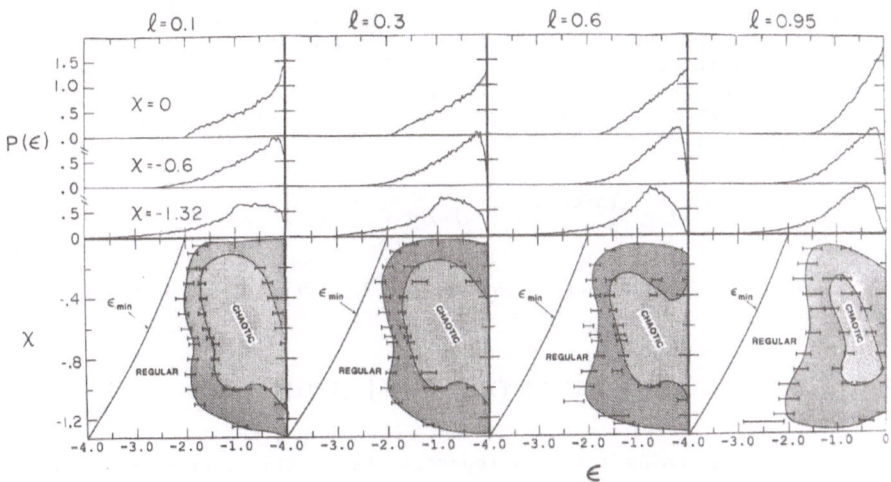

Fig. 5. Bottom: Phase diagram in the ϵ-χ plane for ℓ=0.1, 0.3, 0.6 and 0.95. Top: classical level densities P(ϵ).

6. QUANTAL CHAOS

To study the signatures of classical chaos in the quantal IBM Hamiltonians we have investigated fluctuation properties of the spectrum and the E2 transition intensities[18]. A study of spectral fluctuations alone was also carried out in Ref. 19 and 20.

To separate the average part of the spectrum from its fluctuating part we calculate the staircase function N(E) and find $N_{av}(E)$ by fitting to it a sixth order polynomial. The unfolded energy levels $\tilde{E}_i$ are then defined by $\tilde{E}_i$=N(E_i). The nearest neighbor level spacing distribution is calculated and fitted to a Brody distribution

$$P_\omega(S) = AS^\omega \exp(-\alpha S^{1+\omega}) \tag{15}$$

where $\alpha=\Gamma[(2+\omega)/(1+\omega)]^{1/2}$ and A=(1+ω)α are determined by the conditions that <S>=1 and P is normalized to 1. The Brody distribution interpolates between the Poisson distribution (ω=0) and Wigner distribution (ω=1). The latter is very close to the one obtained from the GOE.

Another statistical measure of the spectrum is the Δ_3 statistics of Dyson and Metha

$$\Delta_3(\alpha,L) = \min_{A,B} \frac{1}{L} \int_{\alpha}^{\alpha+L} [N(\tilde{E}) - (A\tilde{E}+B)]^2 d\tilde{E} \quad . \tag{16}$$

It measures the deviation of the staircase function (of the unfolded spectrum) from a straight line. To obtain a smoother function $\bar{\Delta}_3(L)$, we average $\Delta_3(\alpha,L)$ over n_α intervals $(\alpha,\alpha+L)$, $\bar{\Delta}_3(L)=\Sigma_\alpha \Delta_3(\alpha,L)/n_\alpha$. The successive intervals are taken to overlap by $L/2$. A more rigid spectrum has smaller values of Δ_3, so we except $\bar{\Delta}_3$ in the GOE limit to be smaller than $\bar{\Delta}_3$ for a Poisson spectrum. For a Poisson statistics $\bar{\Delta}_3(L)=L/15$ and the asymptotic result for the GOE is $\bar{\Delta}_3(L)\approx \ln L/\pi^2 - 0.007$ $(L\gg1)$. In general, we fit $\bar{\Delta}_3$ to a function which depends on a parameter q and interpolates between the Poisson statistics (q=0) and the GOE (q=1)

$$\Delta_3^q(L) = \Delta_3^{Poisson}((1-q)L) + \Delta_3^{GOE}(qL) \quad . \tag{17}$$

We also analyze the distribution of the E2 transition intensities y, where

$$y = B(E2; i\to f) = \frac{1}{2J_i+1} | (f\|T(E2)\|i)|^2, \tag{18}$$

and $T(E2)$ is given by (3). To separate the secular variation of $B(E2)$ with energy we first divide (18) by an average intensity calculated by broadening each level with a Gaussian of width γ. The choice for γ is discussed in Ref. 18. We then construct the distribution $P(y)$ such that $P(y)dy$ is the probability to find an intensity y in the interval dy aroung y.

The histogram $P(y)$ is fitted to the following distribution[21]

$$P_\nu(y) = \frac{(\nu/2<y>)^{\nu/2}}{\Gamma(\nu/2)} y^{\nu/2-1} \exp(-\nu y/2<y>) , \tag{19}$$

which is a χ^2 distribution in ν degrees of freedom. For $\nu=1$, this is the Porter-Thomas distribution obtained from GOE in the limit of large matrices[6]. This should correspond to a chaotic behavior. When the dynamics becomes more regular, we expect selection rules to set in so that ν will decrease[21].

Fig. 6 shows an example of the three quantal measures described above $(P(S)$, $\Delta_3(L)$ and $P(y))$ for the J=6 states of the IBM Hamiltonian with $\eta=0$ and $-\sqrt{7}/2 \leq \chi \leq 0$ describing the SU(3)$\to$O(6) transition. The quantal results are strongly correlated to the classical ones. Near the dynamical symmetry limits the spectral fluctuations are close to Poisson and ν is small. For intermediate values of χ (where the classical dynamics is chaotic), the spectral fluctuations are similar to those of the GOE and ν is largest.

The negative values of ω at the dynamical symmetry limits reflect the overintegrability explained in Section 4. Due to missing labels we have exact degeneracies beyond what the Poisson statistics predicts. It is possible to restore a generic behavior without breaking the dynamical symmetry by adding to the Hamiltonian invariant terms which are associated with the missing labels.

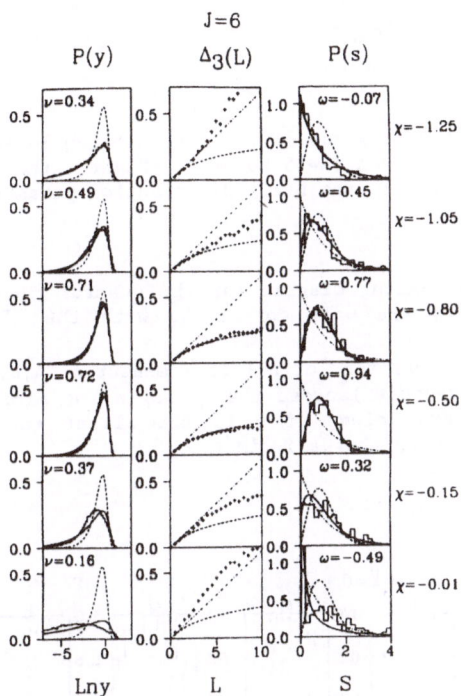

Fig. 6. Spectral and intensity fluctuations
for the $J=6^+$ levels of the Hamiltonian (1)
with $N=25$ bosons, $\eta=0$ and various values of χ
between $-\sqrt{7}/2$ (SU(3)) and 0 (O(6)).
Right: Level spacing distribution P(S) where
the solid line is the best fitted Brody distribution.
Middle: Δ_3 statistics denoted by the + symbols.
Left: The E2 intensity distribution P(y) where
the solid line is a χ^2 distribution in ν degrees of
freedom. In all columns the dotted-dashed lines are the
Poisson statistics and the dashed lines are the GOE
statistics.

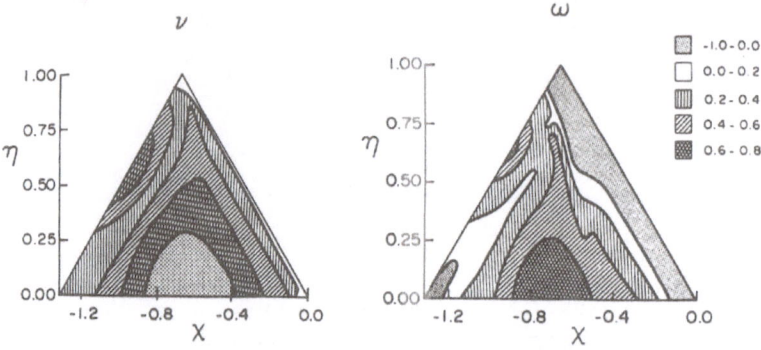

Fig. 7. Quantal diagrams of the Casten triangle for states
with spin $4 \leq J \leq 20\hbar$ (N=25 bosons). Right: the level spacing
parameter ω. Left: the B(E2) distribution parameter ν.

Fig. 7 shows a contour diagram for the quantal measures ω and ν in
the triangle. They are well correlated with the classical measures
shown in Fig. 2.

Fig. 8 is a comparison of all four measures: λ, σ, ω and ν versus χ
for several cuts (η=const.) across the triangle. The minimum which occurs
for the intermediate values of η is the almost regular region that
connects the SU(3) with the U(5) limits.

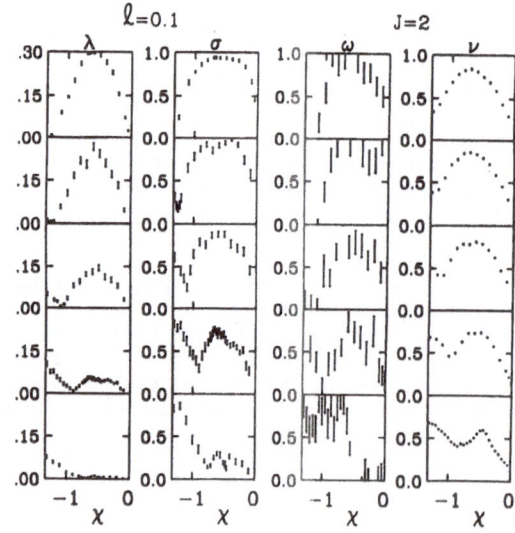

Fig. 8. Measures of chaos λ, σ, ω and ν versus χ
for several η; η=0, 0.1, 0.3, 0.5 and 0.7 (from
top to bottom). The number of bosons is N=25.

7. Spectral Autocorrelation Function

The measures of quantal chaos discussed in Section 6 require an accurate knowledge of a complete set of energy levels and/or transition intensities. Experimentally this requires high revolution data. It is desirable to have a quantity which can be directly extracted from the observed spectrum, yet is less sensitive to finite resolution. The spectral autocorrelation function has been suggested[22],[23] to be such an observable:

$$G(\omega) = \int_{-\infty}^{\infty} S(\omega') \, S(\omega+\omega') d\omega' \tag{20}$$

where $S(\omega) = \Sigma_n p_n \delta(\omega-\omega_n)$ is the spectrum. The Fourier trasform of (20) is the survival probability $P(t) = |<\phi(0)|\phi(t)>|^2$ of an initial non-stationary state $\phi(0) = \Sigma a_n |n>$ (such that $p_n = a_n^2$). Assuming the initial state is randomly distributed over the sphere $\Sigma a_n^2 = 1$ we have

$$P_{av}(t) = \frac{3}{M+2} \left[1 + \frac{1}{3M} \sum_{n \neq m} e^{i(\omega_n - \omega_m)t} \right] \tag{21}$$

where M is the number of eigenstates in the expansion for $\phi(0)$. Using random matrix theory we can evaluate[23],[24] the ensemble average of (21)

$$\overline{P_{av}(t)} = \frac{3}{M+2} \left\{ 1 + \frac{1}{3} \Delta_M * \left[\delta(t) - b_{2\beta}(t) \right] \right\} \quad, \tag{22}$$

where $\tau = t/2\pi\bar{\rho}$ ($\bar{\rho}$ is the average level density) and $\beta = 1,2,4$ for the Gaussian orthogonal, unitary and symplectic ensembles, respectively. $b_{2\beta}(t)$, the two level form factor is the Fourier transform of the two point cluster function $Y_{2\beta}(\omega)$. The asterik denotes a convolution $\Delta_M * f(t) = \int dt' \Delta_M(t') f(t-t')$ and $\Delta_M(t) = M^{-1}[\sin(\pi M t)/\pi t]^2$. For a chaotic system with time reversal symmetry $\beta = 1$. The δ function in (22) describes a rapid dephazing of the survival probability while the form factor describes a correlation hole where the probability descreases below its asymptotic value[25],[26].

When the dynamics are intermediate between chaotic and regular, only a fraction of the classical phase space is chaotic. As in Ref. 27 we assume that a corresponding fraction of the levels β are GOE while a fraction $1-\beta$ are Poisson. Using the rule for calculating b_2 for the superposition of spectra we find

$$b_{2\beta}(t) = \beta b_2(t)/\beta + \left[(1-\beta)/M \right] \delta(t) \quad, \tag{23}$$

where b_2 on the righ hand side is the GOE one. Eqs. (22) and (23) provide $\overline{P_{av}(t)}$ for intermediate statistics.

We have investigated the measure (22) for the IBM. We chose M=10 and to obtain good statistics we averaged over successive groups of ten levels between the 20th and 200th levels of spin J=10. We calculated $P_{av}(t)$ for three cases shown in Fig. 9: regular, intermediate and chaotic. The dashed lines are the theoretical prediction (22) with β given by the chaotic fraction in each case (β=0.24,0.80 and 0.46 from top to bottom). Notice the correlation hole whose width and depth are both of order β. As the system becomes more regular the hole shrinks and eventually disappears in the regular case. We note that when each line in the spectrum is replaced by a Lorentzian with typical width Γ (to take into account finite resolution), Eq. (22) is multiplied by an envelope exp(-Γt). As long as $\Gamma \lesssim D$ (D is an average level spacing), the decay of P(t) will not mask the effects seen in Fig. 9.

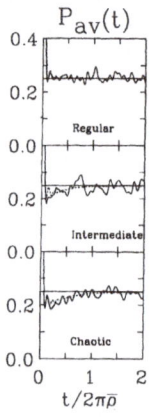

Fig. 9. The survival probability $P_{av}(t)$ for a random initial state with M=10 levels. Top to bottom: regular (χ=-1,η=0.3), intermediate (χ=-0.9, η=0.3) and chaotic (χ=-0.65), η=0) cases. The dashed lines are $\overline{P_{av}(t)}$ of Eqs.(21) and (23) with β=0.24, 0.80 and 0.96 (top to bottom, respectively).

8. Conclusions

We have studied the onset of chaos in the collective dynamics of nuclei, using the framework of the interacting boson model. Strong correlations are found between spectral and transition intensity statistics on one hand and the character of the classical dynamics on the other hand. At higher spin and/or energy, the bosons break into pairs of fermions and it is then important to include these additional fermionic degrees of freedom to obtain a realistic description. A first step towards such a study can be found in Ref. 28. Another interesting question is the role played by the neutron-proton interaction in the onset of chaotic collective motion.

I would like to thank A. Novoselsky and N. Whelan for their collaboration on various parts of this work. This work was supported in part by DOE Grant No. DE-FG-0291ER-40608.

References

1. F. Iachello and A. Arima, The Interacting Boson Model,
   Cambridge University Press, (1987).
2. A. Arima and F. Iachello, Ann. Phys $\underline{99}$, 253 (1976); $\underline{111}$, 201
   (1978); $\underline{123}$, 468 (1979).
3. Y. Alhassid, A. Novoselsky and N. Whelan, Phys. Rev. Lett. $\underline{65}$,
   2971 (1990).
4. R.V. Haq, A. Pandey and O. Bohigas, Phys. Rev. Lett. $\underline{48}$, 1086
   (1982).
5. M.L. Metha, Random Matrices, Academic Press (1990).
6. C.E. Porter, Statistical Theories of Spectra: Fluctuations,
   Academic Press (1965).
7. O. Bohigas, M.J. Giannoni and C. Schmit, Phys. Rev. Lett. $\underline{52}$,
   1 (1984).
8. A.Y. Abul-Magd and H.A. Weidenmüller, Phys. Lett. B$\underline{162}$, 223 (1985).
9. J.F. Shriner, Jr. et al., Z. Phys. A$\underline{335}$, 393 (1990);
   S. Raman et al, Phys. Rev. C$\underline{43}$, 521 (1991).
10. J.F. Shriner, Jr. et al., Z. Phys. A$\underline{338}$, 309 (1991).
11. R.F. Casten and D.D. Warner, Rev. Mod. Phys. $\underline{60}$, 389 (1988).
12. R.L. Hatch and S. Levit, Phys. Rev. C$\underline{25}$, 614 (1982).
13. Y. Alhassid and N. Whelan, Phys. Rev. C$\underline{67}$, 816 (1991).
14. W. Zhang, C.C. Martens, D.H. Feng and J.Yuan, Phys. Rev. Lett $\underline{61}$,
    2167 (1988).
15. G. Draayer and R. Gilmore, J. Math. Phys. $\underline{26}$, 3053 (1986).
16. Y. Alhassid and N. Whelan, Phys. Rev. Lett. $\underline{67}$, 816 (1991).
17. N. Whelan, Ph.D. thesis, Yale University (1993).
18. Y. Alhassid and A. Novoselsky, Phys. Rev. C$\underline{45}$, 1677 (1992);
    N. Whelan and Y. Alhassid, Nucl. Phys. A in press (1993).
19. V. Paar and V. Vorkapic, Phys. Lett. B$\underline{205}$, 7 (1988);
    Phys. Rev. C$\underline{41}$, 2379 (1990).
20. T. Mizusaki, N. Yoshinaga, T. Shigenhara and T. Cheon,
    Phys. Lett. B$\underline{269}$, 6 (1991).
21. Y. Alhassid and R.D. Levine, Phys. Rev. Lett. $\underline{57}$, 2789 (1986);
    Y. Alhassid and M. Feingold, Phys. Rev. A$\underline{34}$, 374 (1989).
22. L. Leviander, M. Lombardi, R. Just and J.P. Pique,
    Phys. Rev. Lett. $\underline{56}$, 2449 (1986).
23. Y. Alhassid and R.D. Levine, Phys. Rev. A$\underline{46}$, 4650 (1992).
24. Y. Alhassid and N. Whelan, Phys. Rev. Lett. $\underline{70}$, 572 (1993).
25. T. Guhr and H.A. Weidenmüller, Chem. Phys. $\underline{146}$, 21 (1990).
26. J. Wilkie and P. Brumer, Phys. Rev. Lett. $\underline{67}$, 1185 (1991).
27. M.V. Berry and M. Robnik, J. Phys. A$\underline{17}$, 2413 (1984).
28. Y. Alhassid and D. Vretenar, Phys. Rev. C$\underline{46}$, 1334 (1992).

GENERATING THE SPECTRUM OF NONLINEAR HAMILTONIANS

L. Ya. Baranov and R.D. Levine

The Fritz Haber Research Center for Molecular Dynamics
The Hebrew University of Jerusalem
Jerusalem 91904, Israel

INTRODUCTION

Much of the current interest in the spectroscopy and dynamics of few/many body systems, be they nuclei, atoms or molecules is due to the nonlinearities of their Hamiltonians. Specifically, for the purpose of this discussion we mean by 'nonlinearity' that even in a reasonable but zeroth order approximation the energy eigenvalues are nonlinear functions of the quantum numbers. As is by now well recognized[1-4] this is at the heart of the quantum manifestations of the behavior that, in the classical limit, we call 'chaotic'. There are however a multitude of other reasons for being interested not the least of which is the ability to obtain a quantitative agreement with experiment.

In an algebraic approach[5] the dynamics of systems described by a Hamiltonian linear in the generators of its spectrum generating algebra[6] can be reduced to a closed form solution in terms of a finite number of time dependent parameters.[7-10] For such systems the time independent generators can be interpreted as either constants of the motion or as lowering and raising operators[6]

$$[H, L_k^{\pm}] = L_k^{\pm}\omega_k \quad . \tag{1}$$

Unfortunately, (1) also implies that the spectrum of the Hamiltonian is restricted to be linear in the quantum numbers[6]

$$E(n_1 \cdots, n_N) = E_0 + \sum_{k=1}^{N} n_k \omega_k \quad . \tag{2}$$

Symmetries in Science VII, Edited by B. Gruber
and T. Otsuka, Plenum Press, New York, 1994

In both nuclear[11-13] and molecular[14-17] physics it has been shown possible to obtain very realistic spectra using Hamiltonians which are bilinear in the generators of a Lie group. In such an approach, the frequencies ω_k are, even in zeroth order, dependent on the quantum numbers. It would be very useful to extend this remarkably successful approach to the dynamics of such systems. In this contribution we explore one possible approach based on allowing the frequencies ω_k in equation (1) to be not numbers but operators which are constants of the motion $[H, \omega_k] = 0$. This requires that the raising and lowering operators be linearly closed under commutation in the generalized sense

$$[L_i, L_j] = L_k c_{ij}^k \tag{3}$$

where the structure 'constants' c_{ij}^k are not numerical constants but constants of the motion.

A procedure for obtaining spectral generators which satisfy equation (3) has been discussed in ref. [18]. In the applications use has been made of the form of recurrence relations known from the theory of special functions.[19,20] Here we rederive the generators using a different technique and demonstrate a more systematic approach to obtaining Hamiltonians that admit such symmetries. We demonstrate the approach by obtaining the first and second Pöschl-Teller potentials,[21] the Morse potential[22] and a third, unnamed potential. All these potentials have a quadratic spectrum. It is our intention to extend the examples to cover the family of Natanson[23,24] potentials. These are obtained from the hypergeometric differential equation using an energy independent change of variables, (see also ref. [25]). These potentials have been discussed using the factorization[26] and the potential algebra[27,28] methods (see also refs. [29,30]).

The essential new point in our approach is that we obtain energy dependent lowering and raising operators for the spectrum of a given Hamiltonian. We moreover do so in a way that suggests the possible applications to dynamics. On the technical level our procedure has the advantage that computing the normalization of wavefunctions, (cf. refs. [23a,29,30]) or of matrix elements can be done in a simple fashion, without the use of special functions.

Elsewhere we shall discuss the time dependent constants of the motion of the potentials considered here, and show that they have a clear interpretation in classical mechanics. These constants of the motion depend on a finite (small) number of time dependent parameters and hence are particularly convenient for discussing both the coherence and the dephasing[31] of wavepackets.

SYMMETRIES AND LADDER OPERATORS

We look for ladder operators, equation (1), for the one dimensional Hamiltonian

$$H = -\partial^2/\partial x^2 + V(x) \quad . \tag{4}$$

Following[18,20,32] we look for an ansatz for L and to a relation between L and H. In particular, there are several reasons suggesting a realization of L as a first order differential operator in x

$$L \sim a(x, n)(\partial/\partial x) + b(x, n)$$

with coefficients which are functions of x and of the quantum number n (or, equivalently, of the energy E). The reasons include the feasibility of solving (1) by equating powers of $\partial/\partial x$ on both sides.[32] Furthermore, this ansatz generalizes the known ladder operators for the harmonic oscillator and angular momentum.[33] Finally, operators of this form can be found in the theory of special functions,[19] factorization method[26] and potential algebra approach[27,28] where they generate a solution for one potential from a solution from another potential of the same family.

A naive way to introduce the n dependence into L or the frequency ω would be by the formal definition $L \sim a(x, H)(\partial/\partial x) + b(x, H)$ where a and b are defined by a power series expansion. Equation (1) could then be understood as a definition of L and the operator equality becomes a functional one when applied to a suitable function of x. However, any non-trivial dependence of L on H makes the commutator $[H, L]$ rather awkward to evaluate.

An alternative, discussed in[18] is to substitute another operator for H, with the choice

$$\lambda \equiv i\partial/\partial t \tag{5}$$

being explicitly considered therein. This choice corresponds to the determination of the symmetries I of the time dependent Schrödinger equation

$$(i(\partial/\partial t) - H)\psi(x, t) = 0 \tag{6}$$

The required L's are then of the form

$$I(x, \lambda, t) = \exp(-it\omega(\lambda))L(x, \lambda)$$

$$L(x, \lambda) = a(x, \lambda)(\partial/\partial x) + b(x, \lambda) \tag{7}$$

Such a symmetry transforms any solution $\psi(x, t)$ of (6)

$$\psi(x, t) = \int dE C(E) \exp(-iEt)\psi(x; E)$$

where $H\psi(x; E) = E\psi(x; E)$, into a solution of the same equation

$$I\psi(x, t) = \int dE C(E) \exp(-i(E + \omega(E))t)L(x, E)\psi(x; E)$$

$$= \int dE C'(E) \exp(-i(E + \omega(E))t)\psi(x; E + \omega(E)) . \tag{9}$$

Here L is defined as

$$L(x, E)\psi(x; E) \equiv (a(x, E)(\partial/\partial x) + b(x, E))\psi(x; E) = c(E)\psi(x; E + \omega(E)) . \tag{10}$$

Therefore, starting from the equation for the symmetry I[18,32]

$$[i(\partial/\partial t) - H, I] = R(i(\partial/\partial t) - H) \tag{11}$$

an equation for L has been obtained which no longer requires the cumbersome evaluation of commutators.

We take our cue from this successful reduction of the problem. We wish however, if possible, to start directly with the simple result (10). This can be done if we interpret equation (1) not as an operator equation but as an equality when both sides act on a solution $\psi(x; E)$ of the time independent Schrödinger equation

$$[H, L]\psi(x; E) = L\omega\psi(x; E) \ . \tag{12}$$

Then

$$L\omega\psi(x; E) = \omega(E)(a(x, E)(\partial/\partial x) + b(x, E))\psi(x; E) = [H, L]\psi(x; E)$$

$$= (-2a'(\partial^2/\partial x^2) - a''(\partial/\partial x) - 2b'(\partial/\partial x) - b'' - aV')\psi(x; E) \tag{13}$$

$$= (a'2(E - V) - (a'' + 2b')(\partial/\partial x) - (b'' + aV'))\psi(x; E)$$

where the prime denotes a derivative *wrt* x. Sufficient conditions for equation (13) to be satisfied are

$$a'' + 2b' + \omega a = 0 \tag{14}$$

$$b'' + aV' + a'2(V - E) + \omega b = 0 \tag{15}$$

Up to now E is a parameter in $\psi(x; E)$. Equations (14) and (15) can also be derived by an alternative route, where we regard all operators as acting in a space of two variables x and E. Then, with L defined by equation (10), we argue that I

$$I \equiv TL \ , \tag{16}$$

where T is an energy shift operator,

$$Tf(E) = f(E + k(E)) \ , \tag{17}$$

is a symmetry of the time independent Schrödinger equation and converts solutions to solutions. To show this we define the function $k(E)$ in (17) by

$$E + k(E) + \omega(E + k(E)) = E \ . \tag{18}$$

With the definitions (16)-(18), and using (10),

$$I\psi(x, E) \equiv TL\psi(x, E)$$

$$= Tc(E)\psi(x, E + \omega(E))$$

$$= c(E + k(E))\psi(x, E + k(E) + \omega(E + k(E)))$$

$$= c(E + k(E))\psi(x, E) \tag{19}$$

CONSTRUCTING LADDER OPERATORS

The set of two coupled differential equations

$$\begin{pmatrix} (\partial^2/\partial x^2) + \omega & 2(\partial/\partial x) \\ 2(V - E)(\partial/\partial x) + V' & (\partial^2/\partial x^2) + \omega \end{pmatrix} \begin{pmatrix} a \\ b \end{pmatrix} = 0 \tag{20}$$

or, equivalently, the fourth order equation for $a(x, E)$

$$((\partial^4/\partial x^4) + 2(\omega + 2E - 2V)(\partial^2/\partial x^2) - 6V'(\partial/\partial x) + \omega^2 - 2V'')a = 0 \tag{21}$$

can be solved (given $V(x)$, $\omega(E)$ and boundary conditions) to determine $a(x, E)$, $b(x, E)$ and hence the ladder operator obeying equation (10). In general, the solution of equation (21) will be more difficult than that of the Schrödinger equation itself. Hence we following what 'symmetry folks' sometimes do and change the problem. Instead of solving (21) for a given $V(x)$ we ask 'what should $V(x)$ be so that (21) has a simple solution for $a(x, E)$'.

Equation (21) does not contain derivatives *wrt* E and the coefficients depend on only one variable, either x or E. This suggests some 'separation-of-variables-like' ansatz for a. The simplest such guess is for a to depend only on x. Differentiating (21) *wrt* E for $a = a(x)$ leads to

$$2(\partial(\omega(E) + 2E)/\partial E)a'' + (\partial\omega^2(E)/\partial E))a = 0 \ , \tag{22}$$

or

$$a'' = \pm\alpha^2 a \tag{23}$$

where α,

$$\alpha^2 = -\omega(d\omega/dE)/((d\omega/dE) + 2) \ , \tag{24}$$

is a real, non negative, constant so that

$$\omega^2 \pm 2\alpha^2(\omega + 2E) = const \tag{25}$$

The system of equations (20) and (22) admits, among others, the following solutions: Family (I): the harmonic oscillator/centrifugal potential. This corresponds to $\alpha^2 = 0$ and so ω is E-independent. Then $\alpha'' = 0$ and $a = a_1 x + a_0$ with a_0, a_1 constants. After some straightforward manipulations one arrives for $a = x - x_0$ to

$$H = -(\partial^2/\partial x^2) + \omega^2(x - x_0)^2/16 + \lambda(x - x_0)^{-2} + V_0 \tag{26}$$

and

$$L(x, E) = a(x)(\partial/\partial x) - \omega(x - x_0)^2/4 + 2(E - V_0)/\omega + 1/2 \ . \tag{27}$$

Other cases are $a_1 = 0$ so that $a(x, E) = 1$ corresponding to the harmonic oscillator. This family is equivalent to the symmetries previously discussed by Miller.[31]

Family (II) is the case of $\alpha^2 \neq 0$. Since α in (23) can be regarded as the scale of x we write it as $a'' = \eta a$ with $\eta = \pm 1$ so that

$$(a')^2 = \eta a^2 + \sigma \qquad (28)$$

For $\sigma = 0$, $a(x)$ is an exponential function of x leading to the Morse potential.[22] We do not discuss this familiar potential in detail but proceed to the other members of this family for which $\sigma = \pm 1$. The final result is

$$H = -(\partial^2/\partial x^2) + (C_1 + C_2 a')/a^2$$

$$L = a(\partial/\partial x) - \eta(\eta + \omega)a'/2 - C_2\eta/\omega \qquad (29)$$

$$(\omega + \eta)^2 + 4\eta E = 0 \qquad (30)$$

where $a'(x)$ is given by (28).

Figure 1 is a plot of $V(x)$ vs. x for the different potentials with $\eta = 1$ of this family. The case $\eta = -1$, $\sigma = +1$ corresponds to the first Pöschl-Teller potential.

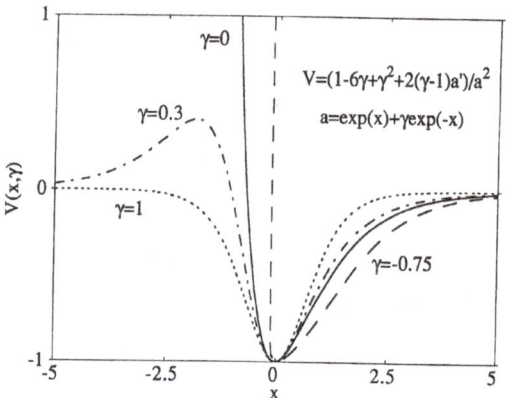

Figure 1. Different possible potentials vs. the reduced coordinate x. The potentials and $a(x)$ are defined in the insert. The reduced variables are such that the minimum is at $x=0$. $V(0)=-1$ and the curvature at $x=0$ is unity. All these potentials have a spectrum quadratic in the quantum number. $\gamma > 0$ is a potential not previously discussed in the literature except for $\gamma = 1$ which is the usual Pöschl-Teller potential. $\gamma = 0$ is the Morse potential. $\gamma < 0$ is the Pöschl-Teller potential of the second kind.

SPECTRUM AND MATRIX ELEMENTS

In this section we discuss the explicit computation of the spectrum, wavefunctions and their normalization and matrix elements. Length limitations preclude giving detailed derivations and these will be presented elsewhere.

Equation (30) admits two solutions $\omega_{\pm}$ corresponding to lowering or raising ladder operators $[H, L_{\pm}] = L_{\pm}\omega_{\pm}$. In the following we choose $\omega(E) = \omega_{+} = -\eta + 2(-\eta E)^{1/2}$. One thus obtains $\omega(E_{n+1}) = \omega(E_n) - 2\eta$ or

$$\omega(E_n) = \omega(E_0) - 2\eta n$$

$$2(-\eta E_n)^{1/2} = 2(-\eta E_0)^{1/2} - 2\eta n \quad . \tag{31}$$

$L_{+}L_{-}$ and $L_{-}L_{+}$ should be equivalent to a function of H. This allows the determination of the 'lowest weight vector' ψ_0 along with its energy E_0

$$0 = L_{-}\psi_0 = L_{+}L_{-}\psi_0 = f(H)\psi_0 = f(E_0)\psi_0 \tag{32}$$

where $f(E_0) = 0$. After some algebra one obtains that

$$\left.\begin{array}{c} L_{-}L_{+} \\ L_{+}L_{-} \end{array}\right\} = C_1 - (C_2^2/\omega_{\pm}^2) + (\sigma/4)(1 - \omega_{\pm}^2) \tag{33}$$

here, equality is when acting on solutions. Taking the least familiar case, $\eta = 1$, $\sigma = -1$ in equations (28)-(30), (or $\gamma > 0$ in figure 1), and putting $C_1 = (\alpha^2 + 1 - \beta^2)/4$, $C_2 = \alpha\beta/2$, we have

$$V(x) = ((\alpha^2 - \beta^2 + 1) + 2\alpha\beta \sin hx)/4\cos h^2 x$$

$$\omega_{-}(E_0) = -\beta$$

$$E_n = -(n - (\beta - 1)/2)^2$$

$$\psi_0 \propto (\cos hx)^{(1-\beta)/2} \exp(-\alpha tn^{-1}(\exp(x))) \tag{34}$$

where α is arbitrary and $\beta > 1$. The similar results for the other potentials will be published elsewhere.

Excited states can be generated as

$$|n> = \frac{(L_{+})^n}{\Pi\lambda_k} |0> \tag{35}$$

or, explicitly in the coordinate representation

$$< x|n> = \psi_n(x) = (a(\partial/\partial x) + b(x, n-1)) \cdot \cdot (a(\partial/\partial x) + b(x, 0))\psi_0(x)/\lambda_{n-1} \cdot \cdot \lambda_0 \quad . \tag{36}$$

The λ_k's are defined by

$$L_{+}|k> = \lambda_k|k+1> \tag{37}$$

where the $|k>$'s are normalized. L, a and b are given in (20) and (28)-(30). The general result is

$$|\lambda_n|^2 = |<n+1|L_{+}|n>|^2 = (C_1 - (C_2^2/\omega_{-}^2(n+1)) + (\sigma/4)(1 - \omega_{-}^2(n+1))) \cdot$$

$$(4\eta + \omega_{+}(n+1) + \omega_{+}(n))/(4\eta + \omega_{-}(n+1) + \omega_{-}(n)) \quad . \tag{38}$$

The highest value of n is determined by the condition that $|n>$ is normalizable, i.e. that λ_n^2 is a finite positive number. For the special case considered in (34), n_{max} is the smaller integer between $(\beta - 1)/2 - 1$ and $(\beta - 1)/2$. This is, of course, consistent with the result for E_n given in (34).

CONCLUDING REMARKS

A general approach to the construction of ladder operators for nonlinear spectra was discussed. It is found easier to reverse the problem and to construct potentials corresponding to special choices for the general functional form of the ladder operators. The simplest such choice led to a family of potentials the spectra of which are linear and/or quadratic in the quantum number. These include the Morse potential,[22] the two Pöschl-Teller potentials[21] and a potential not previously known to us. The four mentioned potentials all have exactly the same spectrum. All four potentials have closed form analytic wavefunctions and, in classical mechanics, their trajectories (i.e., x as a function of time) can be explicitly solved for using known elementary functions.

As far as we know, it was first pointed out by Pöschl and Teller[21] (and, by many others, since) that the energy spectrum cannot define a unique potential. Indeed, the four mentioned potentials have the same spectrum but considerably differ in their shape, figure 1. Of course, this difference in shape will be manifested in other observables such as the scattering amplitude.

REFERENCES

1. (a) L.E. Reichl, "The Transition to Chaos in Conservative Classical Systems: Quantum Manifestations", Spinger-Verlag, New York (1992); (b) S.N. Rasband, "Chaotic Dynamics of Nonlinear Systems", Wiley, New York (1990); (c) M.C. Gutzweiler, "Chaos in Classical and Quantum Mechanics", Springer, N.Y. (1990); (d) M. Tabor, "Chaos and Integrability in Nonlinear Dynamics", Wiley, New York (1989).

2. R.Z. Sagdeev, D.A. Usikov and G.M. Zaslavsky, "Nonlinear Physics, From the Pendulum to Turbulence and Chaos", Harwood, Chur. (1988).

3. E.L. Sibert, Variational and Perturbative Descriptions of Highly Vibrationally Excited Molecules, *Int'l Rev. Phys. Chem.* 9:1 (1990).

4. O. Bohigas and H.A. Weidenmüller, Aspects of Chaos in Nuclear Physics, *Ann. Rev. Nucl. Part. Sci.* 38:421 (1988); D.W. Noid, M.L. Koszykowski and R.A. Marcus, Quasiperiodic and Stochastic Behavior in Molecules, *Ann. Rev. Phys. Chem.* 32:267 (1981).

5. A. Bohm, Y. Ne'eman and A.O. Barut, "Dynamical Groups and Spectrum Generating Algebras", World Scientific, Singapore (1988).

6. Y. Dothan, Finite-Dimensional Spectrum-Generating Algebras, *Phys. Rev.* D 2:2944 (1970).

7. Y. Alhassid and R.D. Levine, Connection Between the Maximal Entropy and the Scattering Theoretic Analyses of Collision Processes, *Phys. Rev.* A 18:89 (1978).

8. R.D. Levine, Lie Algebraic Approach to Molecular Structure and Dynamics, *in*: "Mathematical Frontiers in Computational Chemical Physics", D.G. Truhlar, Ed., Springer-Verlag, Berlin (1988).

9. R.D. Levine, Dynamical Symmetries, *J. Phys. Chem.* 89:2122 (1985).

10. A. Perelemov, "Generalized Coherent States and Their Applications", Springer-Verlag, Berlin (1986).

11. A. Arima and F. Iachello, Interacting Boson Model of Collective Nuclear States, *Ann. Phys.* 123:468 (1979).

12. F. Iachello, Algebraic and Geometric Properties of Interacting Boson Model-1, *Lecture Notes in Physics*, 161:1 (1982).

13. F. Iachello and A. Arima, "The Interacting Boson Model", Cambridge University Press (1987).

14. (a) R.D. Levine and C.E. Wulfman, Energy Transfer to A Morse Oscillator, *Chem. Phys. Lett.* 60:372 (1979). (b) F. Iachello, Algebraic Methods for Molecular Rotation-Vibration Spectra, *Chem. Phys. Letts.* 78:581 (1981); (c) F. Iachello and R.D. Levine, Algebraic Approach to Molecular Rotation-Vibration Spectra. I. Diatomic Molecules, *J. Chem. Phys.* 77:3046 (1982); (d) C.E. Wulfman and R.D. Levine, Isotopic Substitution As A Symmetry Operation in Molecular Vibrational Spectroscopy, *Chem. Phys. Letts.* 104:9 (1984).

15. (a) O.S. van Roosmalen, F. Iachello, R.D. Levine and A.E.L. Dieperink, Algebraic Approach to Molecular Rotation-Vibration Spectra. II. Triatomic Molecules, *J. Chem. Phys.* 79:2515 (1983); (b) O.S. van Roosmalen, I. Benjamin and R.D. Levine, A Unified Algebraic Model Description for Interacting Vibrational Modes in ABA Molecules, *J. Chem. Phys.* 81:5986 (1984).

16. (a) F. Iachello, A. Leviatan and A. Mengoni, Algebraic Approach to Molecular Rotation-Vibration Spectra. III. Infrared Intensities, *J. Chem. Phys.* 95:1449 (1991); (b) F. Iachello and S. Oss, Vibrational Spectra of Linear Triatomic Molecules in the Vibron Model, *J. Mol. Spectr.* 146:56 (1991); (c) I.L. Cooper and R.D. Levine, Computed Overtone Spectra of Linear Triatomic Molecules by Dynamical Symmetry, *J. Mol. Spectr.* 148:391 (1991).

17. F. Iachello and R.D. Levine, "Algebraic Theory of Molecules", Oxford University Press, N.Y. (1993).

18. E.G. Kalnins, R.D. Levine and W. Miller, Jr., Conformal Symmetries and Generalized Recurrences for Heat and Schrödinger Equations in One Spatial Dimension, *in*:

"Mechanics, Analysis and Geometry: 200 years after Lagrange", M. Francaviglia, Ed., North Holland, Amsterdam (1991).

19. E.T. Whittaker and G.N. Watson, "Modern Analysis", Cambridge University Press, London (1946).

20. W. Miller, "Lie Theory and Special Functions", Academic Press, New York (1968).

21. Von G. Pöschl and E. Teller, Bemerkungen zur Quantenmechanik des anharmonischen Oszillators, *Z. Phys.* 83:143 (1933).

22. P.M. Morse, Diatomic Molecules According to the Wave Mechanics. II. Vibrational Levels, *Phys. Rev.* 34:57 (1929).

23. (a) G.A. Natanson, Investigation of One-Dimensional Schrödinger's Equation Generated by the Hypergeometric Equation, (Russian) *Vestn. Leningrad Univ.* 10:22 (1970); (b) G.A. Natanson, General Properties of Potentials for which the Schrödinger Equation can be Solved by Means of Hypergeometric Functions, *Teoret. Mat. Fiz.* (Engl. Transl.), 38:146 (1979); (c) G.A. Natanson, Comment on "Four-parameter Exactly Solvable Potential for Diatomic Molecules", *Phys. Rev.* A 44:3377 (1991).

24. J.N. Ginocchio, A Class of Exactly Solvable Potentials. I. One-Dimensional Schrödinger Equation, *Ann. Phys.* 152:203 (1984).

25. A.K. Bose, A Class of Solvable Potentials, *Nuov. Cim.* 32:679 (1964).

26. L. Infeld and T.E. Hull, The Factorization Method, *Rev. Mod. Phys.* 23:21 (1951).

27. Y. Alhassid, F. Gürsey and F. Iachello, Group Theory Approach to Scattering, *Ann. Phys.* 148:346 (1983).

28. J. Wu, Y. Alhassid and G. Gürsey, Group Theory Approach to Scattering. IV. Solvable Potentials Associated with $SO(2,2)$, *Ann. of Physics* 196:163 (1989).

29. M.M. Nieto, Exact Wave-Function Normalization Constants for the $B_0 \tan hz - U_0 \cos h^{-2}z$ and Pöschl-Teller Potentials, *Phys. Rev.* A 17:1273 (1978).

30. A.O. Barut, A. Inomata and R. Wilson, Algebraic Treatment of Second Pöschl-Teller, Morse-Rosen and Eckart Equations, *J. Phys.* A 20:4083 (1987).

31. (a) M.M. Nieto and L.M. Simmons, Jr., Coherent States for General Potentials, *Phys. Rev. Letts.* 41:207 (1978); (b) M.M. Nieto, L.M. Simmons, Jr. and V.P. Gutschick, Coherent States for General Potentials. VI. Conclusions About the Classical Motion and the WKB Approximation, *Phys. Rev.* D 23:927 (1981).

32. W. Miller, Jr., "Symmetry and Separation of Variables", Addison-Wesley, Reading (1977).

33. H.S. Green, "Matrix Mechanics", P. Noordhoff Ltd., Groningen (1965).

IDENTIFYING MIXED-SYMMETRY STATES IN THE O(6) LIMIT

OF THE PROTON-NEUTRON INTERACTING BOSON MODEL: E0

DECAYS OF BETA-VIBRATIONS

Bruce R. Barrett[1] and Takaharu Otsuka[2]

[1]Department of Physics, University Arizona, Tucson AZ 85721 USA
[2]Department of Physics, University of Tokyo
Hongo, Bunkyo-ku, Tokyo 113, Japan

INTRODUCTION

In going from the original Interacting Boson Model[1] (or IBM-1) of only $J = 0$ s bosons and $J = 2$ d bosons to the neutron-proton IBM[1,2] (or IBM-2) of proton and neutron s and d bosons, one introduces an important new feature to the model, namely the combined proton and neutron boson states of mixed symmetry or states partly antisymmetric under the interchange of the proton and neutron boson degrees of freedom.

The classification of the lowest-lying proton-neutron configurations in the three dynamical symmetry limits of the IBM-2 was given by Iachello.[3] The formula for the large M1 strength from the low-lying 1^+ mixed-symmetry state in the SU(3) limit was presented by Dieperink.[4] The existence of such strong M1 decays, representing a kind of scissor or wobble excitation mode between the prolate deformed protons and the prolate deformed neutrons was first observed in ^{156}Gd by Bohle et al.[5]

In the U(5) limit[1] the low-lying mixed-symmetry state corresponds to proton and neutron surface vibrations about a spherical minimum, where the protons and neutrons oscillate out of phase.[3,6,7] The O(6) limit[1] of the IBM, being the least studied and understood of the three symmetry limits, also presents special challenges in trying to determine and observe its antisymmetric configurations. As in the U(5) limit, the O(6) limit possesses reasonably low-lying mixed-symmetry states corresponding to surface or β-vibrations, but in this case about a deformed minimum.[3,8] Consequently, in the O(6) limit one wants to determine the structure and properties of these symmetric and antisymmetric β-vibrational states.

In the O(6) limit these symmetric and antisymmetric β-vibrations do not belong to the lowest O(6) representations, $\langle N, 0 \rangle$ and $\langle N - 1, 1 \rangle$, respectively ($N = N_\pi + N_\nu$, N_ρ = number of ρ bosons, where $\rho = \pi$ or ν), but to the next higher-lying

Symmetries in Science VII, Edited by B. Gruber
and T. Otsuka, Plenum Press, New York, 1994

O(6) representation, being $\langle N-2,0 \rangle$ in both cases.[8] Because of O(6) selection rules,[1] these β-vibrations cannot decay to the 0^+ ground state (gs) or to other ground-band configurations with O(6) symmetry $\langle N,0 \rangle$ by M1 or E2 electromagnetic transitions. Because of all of these restrictions, we found in earlier investigations[9] no strong signature for the M1 and E2 decay modes of the lowest-lying antisymmetric β-vibration, i.e., $0^+(M')$. Consequently, we would now like to recommend E0 decay strengths as a signature for these different β-vibrational modes.

In this paper we report on the strengths of E0 decays of symmetric and antisymmetric β-vibrations in the O(6) limit of the IBM-2. Because of its one-body nature, E0 decays will connect only to the 0^+ gs. Hence, the observation of strong E0 transitions would provide a signature for these β-vibrational configurations.

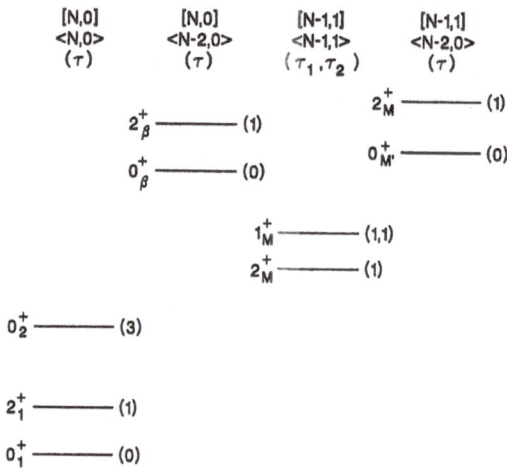

Fig. 1. "Typical" spectrum of a nucleus with IBM-2 O(6) dynamical symmetry.

FORMALISM

The O(6) limit of the IBM-2 corresponds to the group chain[1]

$$U_\pi(6) \times U_\nu(6) \supset U_{\pi+\nu}(6) \supset O_{\pi+\nu}(6) \supset O_{\pi+\nu}(5) \supset O_{\pi+\nu}(3) \, ,$$
$$[N-f,f] \quad (\sigma_1,\sigma_2) \quad \langle \tau_1, \tau_2 \rangle \quad L \tag{1}$$

where the quantum numbers for each step in the subgroup chain are listed below each group classification. Detailed studies[1] have been made of the low-lying properties of O(6) nuclei within the IBM-2 for the totally symmetric states $[N]$, (i.e., the IBM-1 configurations or equivalently the states of maximal F-spin,[2] $F_{max} = N/2$) and the first set of mixed-symmetry states $[N-1,1]$ (i.e., those with an F-spin value of $F_{max}-1$). A possible spectrum of a "typical" nucleus with the dynamical symmetry chain (1) is shown in Fig. 1.

We consider only the decay of the symmetric β-vibrational bandhead state, which we denote as (using the notation of Van Isacker *et al.*[10])

$$|0^+(\beta)\rangle \equiv |[N_\pi][N_\nu]; [N]\langle N-2\rangle(0)0^+\rangle \qquad (2)$$

and the mixed-symmetry β-vibrational bandhead state, denoted as

$$|0^+(M')\rangle \equiv |[N_\pi][N_\nu]; [N-1,1]\langle N-2\rangle(0)0^+\rangle \qquad (3)$$

to the ground state

$$|0^+(1)\rangle \equiv |[N_\pi][N_\nu]; [N]\langle N\rangle(0)0^+\rangle . \qquad (4)$$

Hereafter we omit second quantum numbers (e.g., f, σ_2, or τ_2) which are zero. We use M' to distinguish the $\langle N-2\rangle$ mixed-symmetry states from the $\langle N-1,1\rangle$ mixed-symmetry states which are usually denoted by a subscript M.

Within the IBM-2 the "effective" $E0$ operator is of the form[1]

$$T(E0) = f_\pi \hat{n}_{d_\pi} + \gamma_{0_\pi} N_\pi + f_\nu \hat{n}_{d_\nu} + \gamma_{0_\nu} N_\nu , \qquad (5)$$

where $f_\rho =$ the monopole effective charge of a ρ boson and $\hat{n}_{d_\rho} =$ the number operator for a d boson of type ρ, where $\rho = \pi$ or ν. The terms $\gamma_{0_\rho} N_\rho$ are constant for a given nucleus and do not give rise to transitions. We choose the parameters f_ρ and γ_{0_ρ} to have the units of fm^2, since, in configuration space, that part of the $E0$ operator that gives rise to transitions is proportional to r^2. Although the terms in the $E0$ operator depending upon N_ρ do not contribute to transitions, they definitely do contribute to the calculated values of the nuclear radii.

As Van Isacker *et al.*,[10] have shown, it is straightforward to calculate the reduced matrix elements of transition operators, such as (5), once one has constructed the states $|0^+(1)\rangle$, $|0^+(\beta)\rangle$, and $|0^+(M')\rangle$ for $N_\nu = N-1$ and $N_\pi = 1$. The present situation, however, is a little tricky,[11,12] because $\hat{n}_{d_\rho}$ is not a $[2,1,1,1,1]$ tensor under U(6), which is the condition for using the procedure of Van Isacker *et al.*[10] Instead, $\hat{n}_{d_\rho}$ is a combination of a $[2,1,1,1,1]$ tensor, namely

$$\tilde{N}_\rho = \hat{n}_{s_\rho} - \hat{n}_{d_\rho} \quad , \qquad (6a)$$

and a $[1,1,1,1,1,1] = [0]$ tensor (*i.e.*, a scalar), namely the number operator N_ρ, given by

$$\hat{N}_\rho = \hat{n}_{s_\rho} + \hat{n}_{d_\rho} . \qquad (6b)$$

The operator $\tilde{N}_\rho$ has $\sigma = 2$, while $\hat{N}_\rho$ has $\sigma = 0$, so it is the $\tilde{N}_\rho$ part of $T(E0)$ which produces the $E0$ decays of the β-vibrational bandhead states. The operator $\hat{N}_\rho$ does not contribute to transitions. Taking these points into consideration, we obtain for the reduced matrix elements of $T(E0)$

$$\langle 0^+(1)||T(E0)||0^+(\beta)\rangle = [f_\pi N_\pi + f_\nu N_\nu] \times \frac{\sqrt{(N+2)(N+3)(N-1)}}{[2N(N+1)]} \qquad (7a)$$

$$\langle 0^+(1)||T(E0)||0^+(M')\rangle = [f_\pi - f_\nu] \times \frac{\sqrt{(N-2)(N+3)N_\pi N_\nu/(N+1)}}{2N} \qquad (7b)$$

$$\langle 0^+(\beta)||T(E0)||0^+(M')\rangle = 2[f_\pi - f_\nu] \times \frac{\sqrt{(N-2)(N+2)N_\pi N_\nu/[(N-1)(N+1)]}}{[N(N-1)]} . \qquad (7c)$$

So, depending upon the values of f_π and f_ν, (7a) and (7b) are of the same order in N, while (7c) is smaller by a factor of $N^{(-3/2)}$. We have checked formulas (7a) - (7c) by explicit IBM-2 numerical calculations using the computer code NPBOS[13] and obtain exact agreement in all cases. The expressions for $\rho(E0)$ are obtained by multiplying the reduced matrix elements (7a)-(7c) by Z/R^2, where R is the nuclear radius, given by $r_0 \times A^{1/3}$ and $r_0 = 1.2$ fm.

If $f_\pi = f_\nu$, then the antisymmetric β-vibrational bandhead state $0^+(M')$ has no E0 strength in the O(6) limit.

RESULTS

As a "maximal" approximation, we will take $f_\nu = 0$ and $f_\pi \neq 0$ and determine the E0 strength, given by the square of the reduced matrix elements in (7a)-(7c), for the nucleus ^{132}Ba(Z=56), considered in our previous study.[9] Because it has $N_\pi = N_\nu = 3$, the $0^+(M')$ state in ^{132}Ba cannot decay by either M1 or E2 transitions.[9] Then in units of $[f_\pi]^2$,

$$B(E0, 0^+(1) \to 0^+(\beta))^2 = 0.4592 \tag{8a}$$

$$B(E0, 0^+(1) \to 0^+(M'))^2 = 0.3214 \tag{8b}$$

$$B(E0, 0^+(\beta) \to 0^+(\dot{M}'))^2 = 0.0366 \tag{8c}$$

So the E0 transitions to the gs of ^{132}Ba from the symmetric and antisymmetric β-vibrational bandhead states have similar strengths, while the E0 transition between these two β-vibrational states is an order of magnitude smaller.

We have also performed IBM-2 calculations for the Ba isotopes using the computer code NPBOS[13] and the standard IBM-2 Hamiltonian,[1,14] i.e., not possessing the form appropriate for any of the three symmetry limits of the IBM. In the case of ^{132}Ba we find that the second 0^+ state mixes 50-50 with the third 0^+ state, where both of these states are predominantly symmetry. The B(E0) strengths from the $0^+(2)$ and $0^+(3)$ states to the gs are 66% and 12%, respectively, of the O(6) limit value predicted by (7a); hence, we conclude that the second 0^+ configuration corresponds to what we have called the $0^+(\beta)$ configuration.

The remaining E0 strength to the gs is clustered in a group of mixed symmetry 0^+ states around 3MeV. The summed E0 strength for these levels is about 30% of the O(6)-limit value predicted by (7b), with the maximum B(E0) strength for an individual level being about 16%. Thus, there is considerable fragmentation of the E0 strength for the mixed-symmetry states.

In doing the IBM-2 calculations we noted: 1) that the B(E0) strength becomes larger if we force the Hamiltonian to be more O(6)-like and 2) that the B(E0) strength from the mixed-symmetry states to the gs becomes larger, if we make the d-boson single-particle energy smaller, i.e, we make the nucleus more deformed by increasing the amount of mixing between s and d bosons.

CONCLUSIONS

We have derived the expressions for the reduced matrix elements giving the B(E0) transition strengths from the symmetric and antisymmetric β-vibrational 0^+ bandhead configurations to the gs, as well as the E0 decay matrix element governing the transition between these two bandhead states. To obtain numerical results, we assume that the Ba isotopes are near to O(6) symmetry and then predict that the two E0 decays from the symmetric and antisymmetric β-vibrational bandhead states to the gs should be fairly large and similar in magnitude, while the E0 transition between these same two configurations should be an order of magnitude weaker.

Our IBM-2 calculations for the Ba isotopes indicate that, even though the O(6) symmetry is broken for these nuclei, there is a symmetric 0^+ configuration (generally the second 0^+) which has a large E0 strength to the gs (66% of the O(6)-limit value for $0^+(\beta)$) and, hence, should be detectable. Although the E0 strength is fragmented for the mixed-symmetry 0^+ states, this is always true in real nuclei, whether one is considering the U(5), SU(3) or O(6) limit. Our IBM-2 results also predict increased B(E0) strength for more deformed nuclei (*i.e.*, smaller d-boson single-particle energy), so Ba isotopes nearer to the center of the neutron major shell should demonstrate a larger signal experimentally. Because the M1 and E2 transitions from the symmetric and antisymmetric β-vibrational states to the gs correspond to two-step processes, we believe that experimental measurements of E0 transition strengths in the mass-130 region are still the best possibility for locating and identifying these configurations.

ACKNOWLEDGMENTS

We would like to thank J. Ciezewski, J. N. Ginocchio, P. Halse, P. van Isacker, and R. F. Casten for helpful discussions. We would like to thank H. A. Weidenmüller for his hospitality and that of the Max Planck Institut für Kernphysik, Heidelberg, Germany, where part of this work took place. One of us (T. O.) is grateful for the support by the Grant-in-Aid for General Scientific Research (No. 01540231) by the Ministry of Education, Science and Culture. This work was also supported in part by the National Science Foundation, Grant No. PHY87-23182 and the Max Planck Institut für Kernphysik.

REFERENCES

1. See, for example, F. Iachello and A. Arima, "The Interacting Boson Model," Cambridge University Press, Cambridge (1987), and references therein.
2. A. Arima, T. Otsuka, F. Iachello, and I. Talmi, Collective nuclear states as symmetric couplings of proton and neutron excitations, *Phys. Lett.* 66B:205 (1977).
3. F. Iachello, New class of low-lying collective modes in nuclei, *Phys. Rev. Lett.* 53:1427 (1984).
4. A. E. L. Dieperink, Geometrical analysis of the interacting boson model, *Prog. Part. Nucl. Phys.* 9:121 (1983).
5. D. Bohle, A. Richter, W. Steffen, A. E. L. Dieperink, N. Lo Iudice, F. Palumbo, and O. Scholten, New magnetic dipole excitation mode studied in the heavy deformed nucleus ^{156}Gd by inelastic electron scattering, *Phys. Lett.* 137B:27 (1984).
6. A. B. Balantekin and B. R. Barrett, Collective 2^+ states in the U(5) classical limit of the proton-neutron interacting boson model, *Phys, Rev. C* 35:1878 (1987).
7. T. Otsuka, Ph.D. dissertation, University of Tokyo (1979), unpublished.
8. A. B. Balantekin, B. R. Barrett, and P. Halse, β vibrations in the O(6) limit of the proton-neutron interacting boson model, *Phys. Rev. C* 38:1392 (1988).

9. B. R. Barrett and T. Otsuka, Structure and decay modes of antisymmetric β vibrations in the O(6) limit of the neutron-proton interacting boson model, *Phys. Rev. C* 42:2438 (1990).

10. P. van Isacker, K. Heyde, J. Jolie, and A. Sevrin, The F-spin symmetric limits of the neutron-proton interacting boson model, *Ann. Phys. (N.Y.)* 171:253 (1986).

11. Private communications by J. N. Ginocchio, P. Halse, and P. van Isacker.

12. F. Iachello, Group theory and nuclear spectroscopy, in: "Nuclear Spectroscopy: Lecture Notes in Physics, vol. 119," G. F. Bertsch and D. Kurath, eds., Springer-Verlag, Berlin (1980).

13. T. Otsuka and N. Yoshida, Computer code NPBOS, Japan Atomic Energy Research Institute Report JAERI-M85-094 (1985).

14. G. Puddu, O. Scholten, and T. Otsuka, Collective quadrupole states of Xe, Ba and Ce in the interacting boson model, *Nucl. Phys. A* 348 (1980) 109.

DYNAMICAL SYMMETRY AND STRING THEORY

Itzhak Bars [1]

Department of Physics and Astronomy
University of Southern California
Los Angeles, CA 90089-0484, USA

1. INTRODUCTION

Dynamical symmetries and spectrum genererating algebras have found applications in many branches of physics. Franco Iachello has championed their applications in Nuclear Physics and has explored several other areas of physics by using ingenious schemes. One of the most effective methods for constructing solvable models of string theory can be viewed in the same light. The so called coset method for affine (Kac-Moody) algebras is, in fact, a "dynamical symmetry" in the spirit of Iachello's work. After describing the method, I show how to use it to construct models of strings propagating in curved spacetime. Furthermore, as is known from Iachello's work in Nuclear Physics, there is a relation between the geometry of the nucleus and the dynamical symmetry chain. Similarly in string theory there is an explicit relation between the geometry of spacetime and the algebraic coset chain, as I describe in this paper.

According to the interacting boson model (IBM) of Iachello and Arima [1], and later work on supersymmetry in nuclear physics (IBM-SUSY) [2], it is possible to describe the spectrum, decay rates and other properties of large complex nuclei in various parts of the periodic table by making simple algebraic models of their Hamiltonian. Similar methods have been successful in molecular physics as well. Dynamical symmetries, such as Heisenberg's isospin, Gell-Mann and Nee'man's SU(3) and Gürsey and Radicati's SU(6) were also very successful in the early development of particle physics and they eventually led to the underlying quark model and quantum chromodynamics.

The algebraic scheme is as follows. One assumes a symmetry limit described by a group G. The actual physical system does not respect the symmetry, and the symmetry is broken according to a chain of subgroups $G \in H_1 \in \cdots \in H_n$. The Hamiltonian and/or transition operator is taken as a function of the generators of the symmetry. For example in the IBM or IBM-SUSY case the Hamiltonian is a

---

[1] Research supported by DOE grant DE-FG03-84ER-40168.

function of the quadratic Casimir operators of the chain. Then the spectrum of the physical system must be described by representations of the group G, decomposed according to the subgroup chain. For IBM's six effective bosons (pairs of protons or neutrons) the physically relevant group is $SU(6)$. The bosons are placed in the fundamental representation **6**. The angular momentum assignment ($l = 0$ and $l = 2$) permits only the pure chains $SU(6) \in SU(5) \in SO(5) \in SO(3)$ or $SU(6) \in SO(6) \in SO(5) \in SO(3)$ or $S(6) \in SU(3) \in SO(3)$ where the last $SO(3)$ is the rotation group. By adjusting a few parameters in this Hamiltonian it is remarkable that hundreds of low lying levels of very complex nuclei are successfully described qualitatively as well as quantitatively as multi-boson states that fill large irreducible representations of $SU(6)$. From the droplet model of nuclei it is also possible to compute some of the same properties of these nuclei. By relating the two approaches it is then possible to understand that the shape of the nucleus (spherical, ellipsoid, deformed ellipsoid) is directly related to the algebraic chain that describes it successfully. More complicated geometrical shapes may not be described by a pure chain, but may correspond to a more complicated pattern of symmetry breaking.

The so called coset models [3] of string theory are very similar. Here there is an infinite dimensional symmetry whose generators form an affine algebra, or an "affine current algebra". There is a chain of subalgebras, usually taken in one step $G \in H$, although it could be more complex. There is a quadratic form in the generators (called the Sugawara form) which plays a role similar to the Casimir operator (but not quite an invariant). Actually because of the infinite dimensional nature of the algebra there are an infinite number of such quadratic forms which, taken together close among themselves under commutation and, form the Virasoro algebra. Instead of the Hamiltonian the relevant operator is the energy-momentum tensor (or stress tensor) whose Fourier coefficients are precisely the Virasoro generators. The stress tensor for the G/H coset model for string theory is constructed by subracting the quadratic Sugawara form for H from the quadratic Sugawara form for G. This is analogous to the IBM or IBM-SUSY schemes for the Hamiltonian that is constructed from a linear combination of quadratic Casimir operators in the chain. However, while the relative coefficients in the IBM or IBM-SUSY models are determined phenomenogically, in the G/H coset model the relative coefficient is fixed by gauge invariance of the stress tensor under H. The G/H coset scheme is very tight because of a requirement of conformal invariance that underlies string theory. Conformal invariance is successfully incorporated by demanding the closure of the Virasoro algebra mentioned above, and this is achieved by the scheme automatically. This feature makes the G/H model an exactly solvable string theory by using the representation theory of the symmetry, in much the same way that the IBM or IBM-SUSY models correspond to exactly solvable models of nuclei.

During the past couple of years it has been understood that, by appropriately choosing G and H, one can describe strings propagating in gravitational backgrounds that correspond to curved spacetime with a single time coordinate [4]. All possible cosets that have the single time property have been found and classified [5][6]. Furthermore, by using the equivalence of the G/H scheme to the gauged Wess-Zumino-Witten model (GWZW) it has been possible to connect the geometry of spacetime to the algebraic scheme [4] [7]. The first example $SL(2, \mathbb{R})/SO(1, 1)$ which was worked out explicitly was interpreted as a string propagating in the gravitational background of a black hole in two dimensions [7]. By now there are examples in three and four dimensions as well as with supersymmetry. These new geometries solve the Einstein equations in dilaton gravity and have singularities that are more intricate than black holes. Such geometrical studies has a parallel in the IBM and related models.

2. THE COSET SCHEME IN STRING THEORY

2.1. An introduction to affine algebras and WZW models

Let us consider a affine algebra associated with a group G which may be compact or non-compact [8]

$$[J_{nA}, J_{mB}] = i f_{AB}^C J_{n+m,C} - n\frac{k}{2}\eta_{AB}\delta_{n+m}, \tag{2.1}$$

where η_{AB} is taken proportional to the Killing metric and is defined by $g_{AB} = f_{AC}^D f_{BD}^C = g\eta_{AB}$. The number g is the Coxeter number, e.g. for $SU(N)$, $g = N$; k is positive and larger than g when G is non-compact and a negative constant when G is compact. The generators of the affine algebra are Fourier components of a local current defined on the string worldsheet $J_A(z) = \sum_{n=-\infty}^{+\infty} J_{nA} z^{-n-1}$ ($z = exp(\tau + i\sigma)$ and (τ, σ) parametrize the Euclidean continuation of the worldsheet). It may be constructed from a group element $g(z, \bar{z}) \in G$ as $t^A J_A = -i g^{-1}\partial_z g$, where t^A is a basis of matrices for the Lie algebra, say in the fundamental representation. After writing an appropriate action, which is the Wess-Zumino-Witten (WZW) action, and quantizing it, one finds that the commutation rules above correspond to the canonical quantization of the model [9][10]. Actually one finds that there is another current, $t^A \bar{J}_A(\bar{z}) = -i\partial_{\bar{z}} g g^{-1}$, whose Fourier components $\bar{J}_{nA}$ are independent canonical degrees of freedom than J_{nA}, and form another affine algebra with the same structure as above. The WZW action that gives rise to these structures may be written as (in the Minkowski version of the worldsheet)

$$S_0(g) = \frac{k}{8\pi} \int_M d^2\sigma \ Tr(g^{-1}\partial_+ g \ g^{-1}\partial_- g) - \frac{k}{24\pi} \int_B Tr(g^{-1}dg \ g^{-1}dg \ g^{-1}dg). \tag{2.2}$$

In order to gain a bit more insight into the meaning of these equations it is useful to consider the large k limit. Large k is equivalent to small $\hbar$ and hence it corresponds to the semi-classical limit of the theory. In this limit it is useful to rescale the generators by defining $\alpha_{nA} = J_{nA}/\sqrt{k}$ and then taking the $k \to \infty$ limit in eq.(2.1). The result is

$$[\alpha_{nA}, \alpha_{mB}] = -n\eta_{AB}\delta_{n+m,0} \tag{2.3}$$

which is equivalent to harmonic oscillator commutation rules. A similar result holds for $\bar{\alpha}_{nA} = \bar{J}_{nA}/\sqrt{k}$. Thus, the generators of the affine algebra may be thought of as a non-Abelian version of harmonic oscillators. This can be made more explicit as follows. With an appropriate parametrization of the group element by coordinates X_A, say $g = exp(it^A X_A/\sqrt{k})$, the large k limit of the action reduces to an action for the free field $X_A(z, \bar{z})$. The $X_A(z, \bar{z})$ will be interpreted as the string coordinates defined on the Euclidean worldsheet $(z, \bar{z})$, with $z = exp(\tau + i\sigma)$. In fact, in the large k limit, the free field can be thought of as a free string propagating in flat spacetime. For finite k the spacetime will be curved. At this point the signature of spacetime is not yet correct (i.e. one needs only one time coordinate and additional space coordinates), but we will see later how to construct a good model. The Fourier components of the free field are, in fact, the $\alpha_{nA}, \bar{\alpha}_{nA}$ oscillators. The perturbative expansion of the non-linear theory in powers of $1/k$ can be worked out by starting with the free field Hilbert space defined by the Fock space of the α_{nA}, α_{nA}. This will correspond to an expansion around flat spacetime. However, the theory can also be solved non-perturbatively for any k, in curved spacetime, by using the representation theory of the affine algebra, and this is the principal advantage of considering such a theory.

The conserved energy-momentum tensor has components $T_{zz}^G, T_{\bar{z}\bar{z}}^G, T_{z\bar{z}}^G$. After taking quantum ordering problems into account they take the form $T_{zz}^G =: J_A(z)J_B(z):$ $\eta^{AB}/(-k+g)$ with a similar form for $T_{\bar{z}\bar{z}}^G$ in terms of $\bar{J}_A$, while $T_{z\bar{z}}$ vanishes. The colon ":" is used to indicate normal ordering. Note that the form of T^G has the same structure of the quadratic Casimir operator, and it is called the Sugawara form. It has the Fourier expansion $T_{zz}^G = \sum_n L_n^G z^{-n-2}$. The L_n^G are called Virasoro generators and they could be considered to be a kind of generalization of the quadratic Casimir operator. However, unlike the usual Casimir operator, the Virasoro operators do not commute with the currents or with each other, rather

$$[L_n^G, L_m^G] = (n-m)L_{n+m}^G + \frac{c_G}{12}n(n^2-1)\delta_{n+m,0}, \qquad [L_n^G, J_{mA}] = -mJ_{n+m,A}. \quad (2.4)$$

where the so called Virasoro central charge is $c_G = kdim(G)/(k-g)$. The Hamiltonian of the theory is constructed from the zero mode Virasoro operators and is given by $H = L_0^G + \bar{L}_0^G$.

The eigenstates of the Hamiltonian are formed by constructing the representations of the affine algebra as follows. First one defines the so called ground states, or level zero states, $|R>_G^0$ which correspond to the states in any unitary representation R of the group G. One assumes that these states form the vacuum states for the "non-Abelian oscillators". For $n \geq 1$ the $J_{-n,A}, J_{n,A}$ are considered creators and annihilators repectively, while for $n = 0$ the $J_{0,A}$ act with a matrix representation t_A^R appropriate for representation R

$$J_{0,A}|R>_G^0 = t_A^R|R>_G^0, \qquad J_{nA}|R>_G^0 = 0, \quad n \geq 1. \quad (2.5)$$

The excited higher level states are obtained by applying integer powers p_i of the creators J_{-n_i,A_i}, $n_i \geq 1$ on the ground state, thus constructing the Hilbert space

$$|R>_G^l = \prod_i (J_{-n_i,A_i})^{p_i})|R>_G^0 . \quad (2.6)$$

The level is given by $l = \sum_i n_i p_i$. This space is somewhat analogous to Fock space. In fact, for large k the states (2.6) reduce to ordinary Fock space constructed from the α_{nA} oscillators introduced above. Of course, the non-Abelian nature of the creators provide for a richer structure. The eigenvalue of the Hamiltonian is then obtained from

$$L_0^G|R>_G^l = (\frac{C(R)}{-k+g} + l)|R>_G^l \ . \tag{2.7}$$

where $C(R)$ is the eigenvalue of the usual quadratic Casimir operator for representation R. A similar result is obtained for $\bar{L}_0^G$. Thus, for any value of k *the spectrum is calculated from the quadratic Casimir*. Furthermore, the non-linearity of the theory has been shoved entirely into the Casimir operator through the representation content R of the state. The excitations contribute to the energy through the level, the integer l, in a way that is quite analogous to excitations in a Fock space. As we shall see, the geometry of spacetime is also embedded in the Casimir operator.

2.2. Introduction to Cosets and Gauged WZW models

The action (2.2) may be modified by gauging a subgroup H of G. The new action takes the form $S = S_0 + S_1$ with [11]

$$S_1(g,A) = -\frac{k}{4\pi}\int_M d^2\sigma \ Tr(A_-\partial_+ gg^{-1} - \tilde{A}_+ g^{-1}\partial_- g + A_- g\tilde{A}_+ g^{-1} - A_- A_+) \ . \tag{2.8}$$

where the gauge fields A_+, A_- are matrices in the Lie algebra of H. (A generalization of this action with A_+ twisted relative to A_- exists [12], and this is important for a phenomenon called "duality", but we will not be concerned with it in this paper.) The gauge invariance leads to constraints that require the subgroup currents $J_{n,i}, \bar{J}_{n,i}$, with $i \in H$, to vanish. In the quantum theory the constraints are applied on the states by demanding the subgroup currents with $n \geq 1$ to vanish on kets $|R>_G^l$, and those with $n \leq 1$ to vanish on bras, while for $n = 0$ one must take only the combination $J_{0,i} + \bar{J}_{0,i}$ to vanish. The physical states are only those states that satisfy the constraint equations.

With these constraints one is effectively removing gauge degrees of freedom that correspond to the subgroup H, thus dealing with a theory whose degrees of freedom are in some sense associated with the coset G/H. Again, it is useful to consider the large k limit for which the constraints are solved very simply: take only the Fock space constructed from the oscillators $\alpha_{n,\mu}, \bar{\alpha}_{n,\mu}$ with $\mu \in G/H$ and ignore the subgroup oscillators (this is equivalent to a free string $X_\mu(\tau,\sigma)$ in flat spacetime). The removal of degrees of freedom is a consequence of the subgroup gauge invariance: one may imagine choosing a gauge in which the H-gauge fields "eat up" degrees of freedom from the group element $g \in G$. The gauge fields are not dynamical since they appear without derivatives, thus they are just Lagrange multipliers which, through the equations of motion are functions of the remaining degrees of freedom. Therefore, indeed the gauged WZW model depends only G/H degrees of freedom. For finite k the explicit dependence of the theory on the G/H degrees of freedom is highly non-trivial and is different than the old G/H sigma models. Furthermore it contains the information on the geometry as will be seen later.

The energy-momentum tensor for the gauged WZW model is constructed by subtracting the Sugawara form for the subgroup H from the corresponding form for G [3]

$$T_{z\bar{z}}^{G/H} = \frac{:J_A J_B : \eta^{AB}}{-k+g} - \frac{:J_i J_j : \eta^{ij}}{-k+h} \ , \tag{2.9}$$

where g, h are the Coxeter numbers for G, H respectively. Therefore the Virasoro generators for the gauged WZW model is given by $L_n^{G/H} = L_n^G - L_n^H$, and therefore the Hamiltonian is $H = L_0^{G/H} + \bar{L}_0^{G/H}$. The eigenvalues of this Hamiltonian are computed as follows: First decompose the representation R into representations r_h of the subgroup H, i.e. $|R>_G^l = \sum_h \oplus |r_h>_H^{l_h}$, then compute the difference between the Casimir eigenvalues

$$L_0^{G/H}|r_h>_H^{l_h} = \left[\frac{C^G(R)}{-k+g} - \frac{C^H(r_h)}{-k+h} + (l - l_h) \right]|r_h>_H^{l_h}. \tag{2.10}$$

At this stage it is interesting to *note the similarity between this spectrum and the kind of spectrum that emerges from the dynamical symmetry approach of the IBM, IBM-SUSY or other similar models.* Note however that the relative coefficients between the group and subgroup Casimirs are fixed, unlike the IBM schemes. The reason for this fact is the gauge invariance of the theory that demands that any observable operator, such as the Hamiltonian or the energy-momentum tensor, should commute with the constraints. That is, the subgroup currents $J_{n,a}, \bar{J}_{n,a}$ must commute with the Hamiltonian H, and indeed with all the Virasoro generators $L_n^{G/H}$. This requirement fixes the relative coefficients as above.

In the full string theory that we will consider below we also need to impose the gauge constraints that follows from the reparametrization invariance, or conformal invariance of the string theory. This corresponds to the vanishing of the Virasoro generators $L_n^{G/H}, \bar{L}_0^{G/H}$, for $n \geq 1$, on physical kets, and demanding $L_0^{G/H} = bar L_0^{G/H}$ as well. Therefore, the eigenvalue of the Hamiltonian on a physical state is $H = 2L_0^{G/H}$. A physical state involves both sectors of the Hilbert space constructed from the J_{nA} and $\bar{J}_{nA}$ in "modular invariant" combinations (not to be explained here). These constraints, as well as the G/H constraints outlined above are all automatically satisfied for the ground states ($l = 0 = l_h$) that are gauge invariant under the subgroup H. These are called the tachyonic states (although they are not necessarily tachyons in a supersymmetric theory, or in curved spacetime). Thus, a physical state of the gauged WZW model at the tachyonic level (i.e. ground state level) is automatically obtained for any state that satisfies

$$(J_{0,a} + \bar{J}_{0,a})|Tachyon, R, r_h >= 0. \tag{2.11}$$

This is nothing but the requirement of gauge invariance. The energy eigenvalue follows from (2.10). We will examine the states that satisfy these conditions in order to extract the spacetime geometry.

2.3. Time Coordinate

Let me review how the single time coordinate condition restricts the possible cosets G/H. Let us consider a WZW model based on a non-compact group. Let us parametrize the group element by $X^A(\tau, \sigma)$, where A is an index in the adjoint representation. The left or right moving currents take the form $J^A = \partial X^A + \cdots$, where the dots stand for non-linear terms in an expansion in powers of X. The Fourier components of these currents J_n^A satisfy a affine algebra as in eq.(2.1) where k is the central extension and η^{AB} is proportional to the Killing metric. In an appropriate basis one can choose a diagonal $\eta^{AB} = diag(1, \cdots, 1, -1, \cdots - 1)$ with $+1$ entries corresponding to compact generators and -1 entries to non-compact ones. For example, for $SL(2, \mathbb{R})$ with currents (J^0, J^1, J^2), one has the Minkowski metric in $2 + 1$ dimensions: $\eta^{AB} = diag(1, -1, -1)$.

As explained above, when $k \to \infty$ the currents behave like the free field oscillators of the flat string theory as in (2.3). Examining the signature of the oscillators as given by the Killing metric (with the extra sign in front) we interpret the free fields $X^A \sim \sum_n \frac{1}{n} \alpha_n^A z^n + \cdots$ as time coordinates when A corresponds to compact generators and as space coordinates when A corresponds to non-compact generators. The signature of the coordinates are the same for finite positive k. This is seen by specializing the commutation rules (2.1) to $A = B$ for which the structure constant of the Lie algebra drops out. The timelike oscillators create negative norm states that ruin the unitarity of the theory. The negative norm states must be eliminated from the theory by a consistent set of constraints.

For ordinary string theory in flat spacetime the necessary constraint emerges automatically from the conformal invariance of the theory. Conformal invariance requires the vanishing of the Virasoro generators, and it can be shown that indeed these conditions remove all negative norms. This result goes under the name of the no-ghost theorem.

In a string theory one can tolerate only one time coordinate. This is because, by naive counting, the Virasoro constraints $L_n^{G/H} \sim 0$ can eliminate only the ghosts generated by the negative norm of one time-like oscillator α_n^0, just like string theory in flat spacetime. Therefore, one must put constraints that set to zero the unwanted time-like compact generators $J_{n,i}$, except for one of them. However, first class constraints of this type must close to form an algebra. Therefore, the currents that are set equal to zero ($J^i \sim 0$ weakly on states) must form a subalgebra corresponding to a subgroup of the non-compact group $H \in G$. The subalgebra may include compact and non-compact generators. The remaining currents J^μ, $\mu = 0, 1, 2, \cdots (d-1)$ stand in one-to-one correspondance to the coset coordinates X^μ that include just one time coordinate. Thus, one must choose a subgroup H such that the coset G/H has the signature of Minkowski space $\eta_{\mu\nu}$ in d dimensions. As seen in the previous section this set of constraints defines an exact conformal field theory that fits the algebraic framework of the gauged WZW model. The important ingredient is that one must take an appropriate non-compact coset G/H. The only simple groups that give a single time coordinate were classified in [5]

$$
\begin{array}{cc}
SO(d-1, 2)/SO(d-1, 1) & SO(d, 1)/SO(d-1, 1) \\
SU(n, m)/SU(n) \times SU(m) & SO(n, 2)/SO(n) \\
SO(2n)^*/SU(n) & Sp(2n)^*/SU(n) \\
E_6^*/SO(10) & E_7^*/E_6
\end{array}
\tag{2.12}
$$

This list, which contains only simple groups, may be extended with direct products of simple groups $G_1 \times G_2 \times \cdots$ including $U(1)$ or $\mathbb{R}$ factors, or their cosets, so long as the additional factors do not introduce additional time coordinates [4][5][6].

There is another way to see the same result by using a Lagrangian method at the classical level rather than the algebraic Hamiltonian argument given above at the quantum level. As discussed in the previous section, a coset theory correponds to a gauged WZW model with the subgroup H local. Using the gauge invariance one can eat-up $dim(H)$ degrees of freedom, leaving behind $dim(G/H)$ group parameters that contain just one timelike coordinate. Since the gauge fields are non-dynamical they can be integrated out. This leaves behind a sigma model type theory with the desired signature. The large k limit of this theory has free field quantum oscillators with a single time coordinate.

Both the Hamiltonian and Lagrangian arguments were first given by Bars and Nemeschansky [4]. The Hamiltonian approach was given more weight in [4] where several examples, including $SL(2, \mathbb{R})/\mathbb{R}$ at $k = 9/4$, were investigated. The Lagrangian method was explicitly carried out for $SL(2, \mathbb{R})/\mathbb{R}$ by Witten [7] who interpreted the sigma model metric as a black hole. With the realization that non-compact group coset methods generate singular geometries there has been a flurry of activity to determine the geometries of higher dimensional cosets [12][13]. While these models represent only a small subset of all possible curved spacetime models described by the general sigma model, they have the advantage of being solvable in principle thanks to the algebraic algebra formulation. Thus a lot more can be said about the spectrum, correlation functions, etc. of the quantum string theory based on these models. Furthermore, it has been realized that the special geometries described by these non-compact groups are relevant to gravitational singularities such as black holes and cosmological Big Bang. For these reasons this class of models has received considerable attention during the past couple of years. It is hoped that through such solvable models new light will be shed on unresolved gravitational issues, in string theory as well as general relativity, such as singularities, quantization and finiteness or renormalizability in curved spacetime, the question of Euclidean-Minkowski continuation, spectrum of low energy particles and excited string states in the presence of curvature, etc..

2.4. Heterotic Superstring in Curved Spacetime

The other important question for string theory is the nature and content of the low energy matter it is supposed to predict in the form of quarks, leptons, gauge bosons, etc.. The new models have opened up the possibility of heterotic superstring theories in four spacetime dimensions (with or without additional compactified dimensions) [14]. This is possible because the Virasoro central charge $c = 26$ (or $c = 15$ with supersymmetry) condition can be satisfied in fewer dimensions provided the space is curved. For example it has been possible to construct consistent purely four dimensional heterotic string theories based on non-compact current algebra cosets [15][16]. We can now ask, what are the heterotic models that can be constructed with the non-compact group method? Due to the lack of space I will only give a couple of tables that are self explanatory. For details see the original literature. In these tables only the models in purely four dimensions are given. The left movers are fully described in Table-1 in the form of a supersymmetric Kazama-Suzuki coset, where $SO(3, 1)_1$ represents four free Minkowski fermions. The right movers are described in both tables. Table-1 contains the 4-dimensional coset for right movers (connected with the left movers) while Table-2 gives the gauge group that is needed in order to satisfy the $c_R = 26$ condition. The contribution of the gauge group to the central charge is given by $c_R(int)$ in Table-2.

Table 1. Current algebraic description of left movers and right movers.

#	left movers with N=1 SUSY	right movers
1	$SO(3,2)_{-k} \times SO(3,1)_1/SO(3,1)_{-k+1}$	$SO(3,2)_{-k}/SO(3,1)_{-k}$
2	$\frac{SL(2,\mathbf{R})_{-k_1} \times SL(2,\mathbf{R})_{-k_2} \times SO(3,1)_1}{SL(2,\mathbf{R})_{-k_1-k_2+2}} \times \mathbb{R}$	$\frac{SL(2,\mathbf{R})_{-k_1} \times SL(2,\mathbf{R})_{-k_2}}{SL(2,\mathbf{R})_{-k_1-k_2}} \times \mathbb{R}$
3	$(SO(2,2)_{-k} \times SO(3,1)_1/SO(2,1)_{-k+2}) \times \mathbb{R}$	$(SO(2,2)_{-k}/SO(2,1)_{-k}) \times \mathbb{R}$
4	$SL(2,\mathbb{R})_{-k} \times SO(3,1)_1 \times \mathbb{R}$	$SL(2,\mathbb{R})_{-k} \times \mathbb{R}$
5	$\frac{SL(2,\mathbf{R})_{-k_1} \times SL(2,\mathbf{R})_{-k_2} \times SO(3,1)_1}{\mathbb{R}^2}$	$SL(2,\mathbb{R})_{-k_1} \times SL(2,\mathbb{R})_{-k_2}/\mathbb{R}^2$
6	$SL(2,\mathbb{R})_{-k_1} \times SU(2)_{k_2} \times SO(3,1)_1/\mathbb{R}^2$	$SL(2,\mathbb{R})_{-k_1} \times SU(2)_{k_2}/\mathbb{R}^2$
7	$(SL(2,\mathbb{R})_{-k} \times \mathbb{R}^2 \times SO(3,1)_1)/\mathbb{R}$	$(SL(2,\mathbb{R})_{-k} \times \mathbb{R}^2)/\mathbb{R}$
8	$\mathbb{R}^3 \times \mathbb{R}_Q \times SO(3,1)_1$	$\mathbb{R}^3 \times \mathbb{R}_Q$

Table 2. Conditions for $c_L = 15$ and examples of symmetries that give $c_R = 26$.

#	conditions for $c_L = 15$	$c_R(int)$	gauge group, right movers
1	$k = 5$	11	$(E_7)_1 \times SU(5)_1$
2	$k_1 - 2 = \frac{k_2-2}{2}(-1 + \sqrt{\frac{3k_2}{3k_2-8}})$	$13 - \delta$	$\delta = \frac{12}{(k_1+k_2-4)(k_1+k_2-2)}$
3	$k = 3$	$11\frac{1}{2}$	$(E_7)_1 \times SU(3)_1 \times SU(2)_2 \times U(1)_1$
4	$k = 8/3$	13	$(E_8)_1 \times SO(10)_1$
5	$k_1 = \frac{8k_2-20}{3k_2-8}, \quad k_1, k_2 > \frac{8}{3}$	13	$(E_8)_1 \times SO(10)_1$
6	$k_1 = \frac{8k_2+20}{3k_2+8}, \quad k_2 = 1,2,3,\cdots$	13	$(E_8)_1 \times SO(10)_1$
7	$k = 8/3$	13	$(E_8)_1 \times SO(10)_1$
8	$Q^2 = \frac{3}{4}$	13	$(E_8)_1 \times SO(10)_1$

It is encouraging to note that *the desirable low energy symmetries, including $SU(3) \times SU(2) \times U(1)$, are contained in these curved spacetime string models that have only four dimensions.* Also, the grand unified gauge groups that emerge are the familiar and desirable ones. The quark and lepton states, which come in color triplets and $SU(2)$ doublets, are expected to emerge in several families. Compared to the popular approach of four flat dimensions plus compactified dimensions, the gauge groups are either the same or closely related. This gives the hope that the quark/lepton spectrum of a curved purely four dimensional heterotic superstring that describes the very early universe may be closely related to the quarks and leptons that survive to the present times.

In trying to solve the puzzles of gravitational singularities and cosmology, and those of the Standard Model with respect to the spectrum of matter (i.e. quark/lepton families) and gauge bosons, we may hope that a complete string theory in curved spacetime may guide us. For this reason I believe that it is valuable to study in great detail the models presented in Table-1. These are solvable models that should direct us toward a realistic unified theory.

3. GEOMETRY OF THE MANIFOLD

A gauged WZW model given by (2.2)(2.8) can be rewritten in the form of a non-linear sigma model by choosing a unitary gauge that eliminates some of the degrees of freedom from the group element, and then integrating out the non-propagating gauge fields [4][7]. The remaining degrees of freedom are identified with the string coordinates $X^\mu(\tau, \sigma)$. The resulting action exhibits a gravitational metric $G_{\mu\nu}(X)$ and an antisymmetric tensor $B_{\mu\nu}(X)$ at the classical level.

$$S_{eff} = \int d^2\sigma (G^{\mu\nu}(X)\partial_\alpha X_\mu \partial^\alpha X_\nu + \cdots) . \tag{3.1}$$

At the one loop level there is also a dilaton $\Phi(X)$. These fields govern the spacetime geometry of the manifold on which the string propagates. Conformal invariance at one loop level demands that they satisfy coupled Einstein's equations. Thanks to the exact conformal properties of the gauged WZW model *the Einstein equations are automatically satisfied*. Therefore, any of our non-compact gauged WZW models can be viewed as generating automatically a solution of these rather unyielding equations. One only needs to do some straightforward algebra to extract the explicit forms of $G_{\mu\nu}, B_{\mu\nu}, \Phi$.

This algebra can be carried out by starting from the Lagrangian, such as in (2.2), and has been done for all the models in four dimensions listed in Table-1. The first case was $SL(2, \mathbb{R})/\mathbb{R}$ which was interpreted by Witten [7] as the geometry of a 2D black hole. The higher dimensional cases yield more intricate but singular geometries [12][13][14][6] . Although the Lagrangian method is straightforward, it has a number of drawbacks. First, it yields the geometry only in a patch that is closely connected to a particular choice of a unitary gauge. The ramaining patches of the global geometry can be recovered only in other unitary gauges and may have no resemblance to the analytic form of the metric, dilaton, etc. in another unitary gauge. To overcome this problem we have introduced global coordinates [17] on the complete geometry. The global coordinates are gauge invariant. The second problem with the Lagrangian method is that it yields the semi- classical geometry up to one loop in an expansion in powers of $1/k$. However, since the gauged WZW model is conformally exact one would rather obtain the conformally exact geometry by using alternative methods. It turns out that the Hamiltonian method that utilizes the GKO construction solves both of these problems simultaneously and yields an exact metric and dilaton to all orders in $1/k$ [18][19][20]. Therefore in this paper we concentrate on the Hamiltonian approach.

With the Hamiltonian approach one can compute the gravitational metric and dilaton backgrounds to all orders in the quantum theory (all orders in the central extension k). We have managed to obtain these quantities for bosonic, type-II supersymmetric, and heterotic string theories in $d \leq 4$. It turns out that the geometry of the heterotic and type-II superstrings are obtained by deforming the geometry of the purely bosonic string by definite shifts in the exact k-dependence. Therefore, it is sufficient to first concentrate on the purely bosonic string. The following relations have been proven for $G/H = SO(d-1,2)/SO(d-1,1)$ which is relevant to string theory [18]: (i) For type-II superstrings the conformally *exact* metric and dilaton are identical to those of the non-supersymmetric *semi-classical* bosonic model except for an overall renormalization of the metric obtained by $k \rightarrow k - g$. (ii) The exact expressions for the heterotic superstring are derived from their exact bosonic string counterparts by shifting the central extension $k \rightarrow 2k - h$ (but an overall factor $(k-g)$ remains unshifted). (iii) The combination $e^\Phi \sqrt{-G}$ is independent of k and therefore can be computed in lowest order perturbation theory. Cases 2,5,6 in Table-1 are a bit more complicated because of the two central extensions, but the results that relate the bosonic string to superstrings are analogous. Case 6 is explicitly discussed in [20], and the others are just analytic continuations of this one.

The main idea is the following. For the bosonic string the conformally exact Hamitonian is the sum of left and right Virasoro generators $L_0^L + L_0^R$. They may be written purely in terms of Casimir operators of G and H when acting on a state $T(X)$ at the tachyon level. The exact dependence on the central extension k is included in this form by using the GKO formalism in terms of currents. For example for the left-movers (instead of $J, \bar{J}$ notation we will use J^L, J^R respectively) [2]

$$
L_0^L T = \left(\frac{\Delta_G^L}{k-g} - \frac{\Delta_H^L}{k-h} \right) T
$$
$$
\Delta_G^L \equiv Tr(J_G^L)^2, \qquad \Delta_H^L \equiv Tr(J_H^L)^2 ,
$$

(3.2)

The exact quantum eigenstate $T(X) =< X|Tachyon >$ can be analyzed in X-space. Then the Casimir operators become Laplacians constructed as differential operators in group parameter space $(dimG)$. Consider a state $T(X)$ which is a singlet under the gauge group H (acting simultaneously on left and right movers)

$$
(J_H^L + J_H^R) T = 0 .
$$

(3.3)

Because of the $dimH$ conditions $T(X)$ can depend only on $d = dim(G/H)$ parameters, X^μ (string coordinates), which are H-invariants constructed from group parameters (see below). The fact that there are exactly $dim(G/H)$ such independent invariants is not immediately obvious but it should become apparent to the reader by considering a few specific examples. As discussed in [17] these are in fact the coordinates that globally describe the sigma model geometry. Consequently, using the chain rule, we reduce the derivatives in (3.2) to only derivatives with respect to the d string coordinates X^μ. In this way we can write the conformally exact Hamiltonian $L_0^L + L_0^R$ as a Laplacian differential operator in the global curved space-time manifold involving only the string coordinates X^μ. By comparing to the expected general form

$$
(L_0^L + L_0^R)T = \frac{-1}{e^\Phi \sqrt{-G}} \partial_\mu (e^\Phi \sqrt{-G} G^{\mu\nu} \partial_\nu T)
$$

(3.4)

for the singlet T, we read off the exact global metric and dilaton.

We have applied this program to all the models in Table-1 and obtained the exact geometry to all orders in $1/k$. The large k limit of our results agree with the semi-classical computations of the Lagrangian method. In the special case of two dimensions we also agree with another previous derivation of the exact metric and dilaton for the $SL(2, \mathbb{R})/\mathbb{R}$ bosonic string [21]. We summarize here the global and conformally exact results for the metric and dilaton in the case of $SO(d-1,2)_{-k}/SO(d-1,1)_{-k}$ for d=2,3,4 [18]. Due to the more complex expressions we refer the reader to

---

[2] Here J_G^L and J_H^L are *antihermitian* group and subgroup generators obeying the appropriate Lie algebras, and g, h are the Coxeter numbers for the group and the subgroup. For the cases of interest in this paper $g = d-1$, $h = d-2$ for $d \geq 3$, and $g = 2$, $h = 0$ for $d = 2$.

the original literature for the remaining cases [19][20]. The group element g for $SO(d-1,2)/SO(d-1,1)$ can be parametrized as a $(d+1) \times (d+1)$ matrix in the form

$$g = \begin{pmatrix} 1 & 0 \\ 0 & (\frac{1+a}{1-a})_\alpha{}^\beta \end{pmatrix} \begin{pmatrix} b & (b+1)x^\beta \\ -(b+1)x_\alpha & (\eta_\alpha{}^\beta - (b+1)x_\alpha x^\beta) \end{pmatrix}, \qquad (3.5)$$

where $b = \frac{1-x^2}{1+x^2}$. The d parameters x_α and $d(d-1)/2$ parameters $a_{\alpha\beta}$ transform as vector and antisymmetric tensor respectively under the Lorentz subgroup $H = SO(d-1,1)$ which acts on both sides of the matrix as $g \to hgh^{-1}$. By considering the infinitesimal left transformations $\delta_L g = \epsilon_L g$ we can read off the generators that form an $SO(d-1,2)$ algebra for left transformations.

$$J_{\alpha\beta}^L = \frac{1}{2}(1+a)_{\alpha\alpha'}(1+a)_{\beta\beta'}\frac{\partial}{\partial a_{\alpha'\beta'}}$$

$$J_\alpha^L = -\frac{1}{2}(1+x^2)(\frac{1+a}{1-a})_\alpha{}^\beta\frac{\partial}{\partial x^\beta} + \frac{1}{2}(1+a)_{\alpha\alpha'}(1+a)_{\beta'\gamma}x^\gamma\frac{\partial}{\partial a_{\alpha'\beta'}}. \qquad (3.6)$$

If we consider instead, the infinitesimal right transformations $\delta_R g = g\epsilon_R$ we find the following expressions for the generators of right transformations

$$J_{\alpha\beta}^R = -\frac{1}{2}(1-a)_{\alpha\alpha'}(1-a)_{\beta\beta'}\frac{\partial}{\partial a_{\alpha'\beta'}} - x_{[\alpha}\frac{\partial}{\partial x^{\beta]}}$$

$$J_\alpha^R = \frac{1}{2}(x^2-1)\frac{\partial}{\partial x^\alpha} - x_\alpha x^\beta\frac{\partial}{\partial x^\beta} - \frac{1}{2}(1-a)_{\alpha\alpha'}(1-a)_{\gamma\beta'}x^\gamma\frac{\partial}{\partial a_{\alpha'\beta'}}. \qquad (3.7)$$

The J^R currents obey the same commutation rules as J^L and moreover commute with each other $[J^L, J^R] = 0$. The quadratic Casimirs for the group and subgroup on either the left or the right are obtained by squaring these currents. For the explicit expressions see [18].

As argued above the global parametrization of the manifold is given in terms of H-invariants, i.e. Lorentz invariants in the present case. In order to obtain a diagonal metric on the manifold one must find d convenient combinations of these Lorentz invariants in d dimensions. We give here the basis that diagonalizes the semi-classical metric at large k. One of the natural invariants already occurs in the construction of the group element for every d, namely $b = \frac{1-x^2}{1+x^2}$.

3.1. Two dimensions

For $d = 2$ the antisymmetric tensor is Lorentz invariant $a_{\alpha\beta} = a\epsilon_{\alpha\beta}$, and it is convenient to parametrize $a = \tanh(t)$ or $\coth(t)$. Then the global string coordinates can be taken as $X^\mu = (t, b)$. Given all possible values for (a, x^α) the ranges of the two invariants cover the entire plane $-\infty < t, b < +\infty$. The metric is given by the line element

$$ds^2 = 2(k-2)(\frac{db^2}{4(b^2-1)} - \beta(b)\frac{b-1}{b+1}dt^2), \qquad \beta^{-1}(b) = 1 - \frac{2}{k}\frac{b-1}{b+1}. \qquad (3.8)$$

For the dilaton the corresponding expression is

$$\Phi = \ln\left(\frac{b+1}{\sqrt{\beta(b)}}\right) + const .$$ (3.9)

The scalar curvature for this metric is

$$R = \frac{2k}{k-2} \frac{(k-2)b + k - 4}{((k-2)b + k + 2)^2} .$$ (3.10)

The curvature is singular at $b = -(k+2)/(k-2)$, which is also where $\beta(b) = \infty$. These are the properties of the exact 2d metric. The semi-classical metric is obtained by taking the large k limit, for which $\beta = 1$. Then the singularity is at $b = -1$. Following Witten this singularity is interpreted as a black hole while the horizon is at $b = 1$. The signature of the space is $(+-)$ or $(-+)$ depending on the region in the (t, b) plane as indicated in Fig-2 of [17]. The signature is understood by examining the semi-classical metric. To see the connection to the Kruskal coordinates used by Witten let $b = 1 - 2uv$ and $u^2 = e^{2t}|b - 1|/2$, $v^2 = e^{-2t}|b - 1|/2$.

There are asymptotically flat regions which are displayed by the change of co-ordinates $b = \pm\cosh\frac{2z_1}{\sqrt{2(k-2)}}$, $t = \frac{z_0}{\sqrt{2k}}$. For large $z_1 \to \pm\infty$ and any z_0 the exact metric and dilaton have the asymptotic forms

$$ds^2 = dz_1^2 - dz_0^2, \qquad \Phi = \sqrt{\frac{2}{k-2}}|z_1|,$$ (3.11)

displaying a dilaton which is asymptotically linear in the space direction, just like a Liouville field in 2d quantum gravity with a background charge. Despite the flat metric there is no Poincaré invariance due to the linear dilaton. Note that both the region outside the horizon ($b \to +\infty$) and the naked singularity region ($b \to -\infty$) are asymptoticaly flat.

3.2. Three dimensions

For $d = 3$ the antisymmetric tensor is equivalent to a pseudo-vector $a_{\alpha\beta} = \epsilon_{\alpha\beta\lambda}y^\lambda$, from which we construct two convenient invariants $v = 2/(1 + y^2)$ and $u = -v(x \cdot y)^2/x^2$, which together with b provide a basis for the string coordinates $X^\mu = (v, u, b)$. Given all possible values taken by (x^α, y^α) the allowed ranges for the invariants are

$$(+-+) \ or \ (-++) \quad \{b^2 > 1 \ \& \ uv > 0\},$$
$$(++-) \quad \{b^2 < 1 \ \& \ uv < 0\}, \quad except \quad 0 < v < u + 2 < 2.$$ (3.12)

The 3d conformally exact metric is given by the line element [18]

$$ds^2 = 2(k - 2)(G_{bb}db^2 + G_{vv}dv^2 + G_{uu}du^2 + 2G_{vu}dvdu) .$$ (3.13)

where

$$G_{bb} = \frac{1}{4(b^2 - 1)}$$

$$G_{vv} = -\frac{\beta(v, u, b)}{4v(v - u - 2)}\left(\frac{b+1}{b-1} + \frac{1}{k-1}\frac{u+2}{v-u-2}\right)$$

$$G_{uu} = \frac{\beta(v, u, b)}{4u(v - u - 2)}\left(\frac{b-1}{b+1} - \frac{1}{k-1}\frac{v-2}{v-u-2}\right) \qquad (3.14)$$

$$G_{vu} = \frac{1}{4(k-1)}\frac{\beta(v, u, b)}{(v - u - 2)^2} ,$$

and

$$\beta^{-1}(v, u, b) = 1 + \frac{1}{k-1}\frac{1}{v-u-2}\left(\frac{b-1}{b+1}(u+2) - \frac{b+1}{b-1}(v-2) - \frac{2}{k-1}\right) . \qquad (3.15)$$

The exact dilaton is

$$\Phi = \ln\left(\frac{(b^2-1)(v-u-2)}{\sqrt{\beta(v, u, b)}}\right) + \Phi_0 , \qquad (3.16)$$

In the large k limit one obtains the global version of a semi-classical metric derived in [17] with Lagrangian methods

$$\frac{ds^2}{2(k-2)}\Big|_{k\to\infty} = \frac{db^2}{4(b^2-1)} - \frac{1}{v-u-2}\left(\frac{b+1}{b-1}\frac{dv^2}{4v} - \frac{b-1}{b+1}\frac{du^2}{4u}\right) \qquad (3.17)$$

The signature $(+-+)$, or $(-++)$, or $(++-)$ depends on the region and is indicated in Fig-1 of [17]. A three dimensional view of this metric is given in Figs-4 of [17]. The surface is where the scalar curvature blows up. This coincides with the location where the dilaton blows up in the large k limit as seen from the above expression. The space has two topological sectors denoted by the sign of a conserved "charge" $\pm = sign(v(b+1)) = sign(u(b-1))$. The sign never changes along geodesics. A more intuitive view of the space is obtained in another set of coordinates for the plus sector (b, λ_+, σ_+) and the minus sector (b, λ_-, σ_-), which are given by $\lambda_{\pm}^2 = \pm v(b+1)$ and $\sigma_{\pm}^2 = \pm u(b-1)$. Then the singularity surface is shown in Figs-3 of [17]. In the plus region the singularity surface has the topology of the double trousers with pinches in the legs. In the minus region we have the topology of two sheets that divide the space into three regions.

There are asymptotically flat regions that may be displayed by a change of variables to $b = \pm\cosh\frac{1}{\sqrt{3(k-2)}}(2z_1 - z_0)$, $u = (\pm)/\cosh\frac{1}{\sqrt{3(k-2)}}(-z_1 + 2z_0)\cosh^2 z_2$, $v = (\pm)/\cosh\frac{1}{\sqrt{3(k-2)}}(-z_1 + 2z_0)\sinh^2 z_2$. For large values of $z_1 \to \pm\infty$, and finite values of (z_0, z_2), the semiclassical metric and dilaton take the form

$$ds^2 = -dz_0^2 + dz_1^2 + dz_2^2, \qquad \Phi = \sqrt{\frac{6}{k-2}}\left|\frac{5}{3}z_1 - \frac{4}{3}z_0\right|, \qquad (3.18)$$

showing that the dilaton is linear in a space-like direction $z_1' = \frac{5}{3}z_1 - \frac{4}{3}z_0$ in the asymptotically flat region. Then z_1' behaves just like a Liouville field, while the

Lorentz transformed $z_0' = \frac{5}{3}z_0 - \frac{4}{3}z_1$ is a time coordinate, and the diagonal metric is rewritten as $d\dot{s}^2 = -(dz_0')^2 + (dz_1')^2 + dz_2^2$. The exact metric is not flat when only $|z_1|$ is large. To display its asymptotically flat region one requires somewhat different coordinates.

3.3. Four dimensions

For $d = 4$ one can construct the Lorentz invariants

$$x^2 , \qquad z_1 = \frac{1}{4}Tr(a^2) , \qquad z_2 = \frac{1}{4}Tr(a^*a) , \qquad z_3 = xa^2x/x^2 , \qquad (3.19)$$

where $a_{\alpha\beta}^* = \frac{1}{2}\epsilon_{\alpha\beta\alpha'\beta'}a^{\alpha'\beta'}$ is the dual of $a_{\alpha\beta}$. However, the semi-classical metric is diagonal for a different set of four invariants $X^\mu = (v, u, w, b)$ given by

$$b = \frac{1 - x^2}{1 + x^2} , \qquad u = \frac{1 + z_2^2 + 2(z_1 - z_3)}{1 - 2z_1 - z_2^2}$$

$$v = \frac{1 + z_1 + \sqrt{z_1^2 + z_2^2}}{1 - z_1 - \sqrt{z_1^2 + z_2^2}} , \qquad w = \frac{1 + z_1 - \sqrt{z_1^2 + z_2^2}}{1 - z_1 + \sqrt{z_1^2 + z_2^2}} . \qquad (3.20)$$

To find the ranges in which the above global coordinates take their values we consider a Lorentz frame that can cover all possibilities without loss of generality. First we notice that by Lorentz transformations the antisymmetric matrix $a_{\alpha\beta}$ can always be transformed to a block diagonal matrix with the non-zero elements

$$a_{01} = \tanh t \text{ or } \coth t , \qquad a_{23} = \tan\phi . \qquad (3.21)$$

Then using (3.20) one can deduce the form of the global variables: $v = \pm\cosh 2t$, $w = \cos 2\phi$, and $u = \frac{1}{x^2}\left(w(x_0^2 - x_1^2) - v(x_2^2 + x_3^2)\right)$. Therefore the string variables can take values in the following regions with the signature in the (v, u, w, b) basis

$(-+++):$ $\quad b^2 > 1,$ $\quad \{-1 < w < u < 1 < v$ or $v < -1 < u < w < 1$

$\text{or } -1 < w < 1 < u < v\},$

$(+-++):$ $\quad b^2 > 1,$ $\quad \{-1 < w < 1 < v < u$ or $u < v < -1 < w < 1\}$

$(+++-):$ $\quad b^2 < 1,$ $\quad \{u < w < 1 1 < v \text{ or } v < -1 < w < u \text{ or } v < u < -1 < w < 1\}.$

$$(3.22)$$

With this set of coordinates we compute the conformally exact dilaton and metric as before. The dilaton field is

$$\Phi = \ln\left(\frac{(b^2 - 1)(b - 1)(v - u)(w - u)}{\sqrt{\beta(b, u, v, w)}}\right) + \Phi_0 . \qquad (3.23)$$

and the metric is given by

$$ds^2 = 2(k - 3)\left(G_{bb}db^2 + G_{uu}du^2 + G_{vv}dv^2 + G_{ww}dw^2\right.$$
$$\left. + 2G_{uv}dudv + 2G_{uw}dudw + 2G_{vw}dvdw\right) , \qquad (3.24)$$

where

$$G_{bb} = \frac{1}{4(b^2 - 1)}$$

$$G_{uu} = \frac{\beta(b,u,v,w)}{4(u-w)(v-u)}\left(\frac{b-1}{b+1} - \frac{1}{k-2}\frac{(v-w)^2}{(v-u)(u-w)}\left(1 - \frac{1}{k-2}\frac{b+1}{b-1}\right)\right)$$

$$G_{vv} = -\frac{(v-w)\beta(b,u,v,w)}{4(v^2-1)(v-u)}\left(\frac{b+1}{b-1} - \frac{1}{k-2}\frac{1}{(v-u)(u-w)}\left[1 - u^2 +\right.\right.$$
$$\left.\left. + (\frac{b+1}{b-1})^2(v-u)(v-w) + \frac{1}{k-2}\frac{b+1}{b-1}\frac{(1+v^2)(u+w) - 2v(1+uw)}{v-w}\right]\right)$$

$$G_{ww} = \frac{(v-w)\beta(b,u,v,w)}{4(1-w^2)(u-w)}\left(\frac{b+1}{b-1} - \frac{1}{k-2}\frac{1}{(v-u)(u-w)}\left[1 - u^2 +\right.\right. \qquad (3.25)$$
$$\left.\left. + (\frac{b+1}{b-1})^2(u-w)(v-w) - \frac{1}{k-2}\frac{b+1}{b-1}\frac{(1+w^2)(u+v) - 2w(1+uv)}{v-w}\right]\right)$$

$$G_{uv} = \frac{\beta(b,u,v,w)}{4(k-2)(v-u)^2}\left(1 - \frac{1}{k-2}\frac{b+1}{b-1}\frac{v-w}{u-w}\right)$$

$$G_{uw} = \frac{\beta(b,u,v,w)}{4(k-2)(u-w)^2}\left(1 - \frac{1}{k-2}\frac{b+1}{b-1}\frac{v-w}{v-u}\right)$$

$$G_{vw} = \frac{1}{(k-2)^2}\frac{b+1}{b-1}\frac{\beta(b,u,v,w)}{4(v-u)(u-w)} \, ,$$

and the function $\beta(b,u,v,w)$ is defined by

$$\beta^{-1}(b,u,v,w) = 1 + \frac{1}{k-2}\frac{(v-w)^2}{(v-u)(w-u)}\left(\frac{b+1}{b-1} + \frac{b-1}{b+1}\frac{1-u^2}{(v-w)^2}\right)$$
$$+ \frac{1}{k-2}\left(\frac{vw + u(v+w) - 3}{(v-w)^2} - (\frac{b+1}{b-1})^2\right) + \frac{2}{(k-2)^3}\frac{b+1}{b-1}\frac{vw-1}{(v-u)(u-w)} \, .$$
$$(3.26)$$

The large k limit of these expressions reduce to the semiclassical dilaton and metric that follow from the Lagrangian approach

$$\frac{ds^2}{2(k-2)}\Big|_{k\to\infty} = \frac{db^2}{4(b^2-1)} + \frac{b-1}{b+1}\frac{du^2}{4(v-u)(u-w)}$$
$$+ \frac{b+1}{b-1}(v-w)\left(\frac{dw^2}{4(1-w^2)(u-w)} - \frac{dv^2}{4(v^2-1)(v-u)}\right) . \qquad (3.27)$$

We can see that the signature of the semiclassical metric for different ranges of the parameters (3.22) is precisely as required by the group parameter space which led to (3.22). However, for the exact metric $\beta(u,v,w,b)$ must remain positive to keep $-det(G)$ positive. This implies that part of the regions in (3.22) are screened out by quantum effects for the exact geometry. This screening phenomenon is true for every dimension $d = 2,3,4$ and the screened regions must be interpreted in the quantum theory as tunneling or decay regions for probability amplitudes (such as the tachyon wavefunction). Under any circumstances the manifold cannot go outside of the range (3.22) dictated by the group theory.

As in the previous $d = 2, 3$ cases, we can check that our explicit expressions for the dilaton and metric give the k-independent combination $\sqrt{-G}e^\Phi$. Therefore this quantity takes the same value for either the exact metric and dilaton or the semiclassical metric and dilaton. Since it is unrenormalized by quantum effects (other than one loop), it may be computed in lowest order perturbation theory. This combination appears in the Dalambertian and is also closely related to the integration measure in the path integral. Through group theoretical arguments given in [12][14] it was possible to guess that this combination should remain unrenormalized by quantum effects. Similar to the $d = 2, 3$ cases the 4d manifold has an asymptotically flat region, but it will not be discussed here.

3.4. Particle and String Geodesics

Having global coordinates and a global geometry is not sufficient to get a feeling of the geometry, one also needs to know the behavior of the geodesics. However, for the complicated metrics that are displayed above the geodesic equation seems to be completely unmanageable. Fortunately, we have developed a procedure that relies on group theory and managed to solve for all particle geodesics. The trick is to take advantage of the fact that the global coordinates are gauge invariant under H-transformations. Then we may solve the equations of motion for the group element g in any gauge, and use *the solution* to construct the H-invariant combinations that form the global coordinates of the geometry. In fact, there is an axial gauge in which g is solved easily [17]. For a point particle (string shrunk to a point) it is given as a function of proper time

$$g(\tau) = e^{\alpha \tau} g_0 e^{(p-\alpha)\tau}, \tag{3.28}$$

where g_0 is a constant group element at initial proper time τ, and α, p are constant matrices in the Lie algebras of H and G/H respectively. The equations of motion require that these constants satisfy a constraint

$$(g_0(p - \alpha)g_0^{-1})_H + \alpha = 0, \tag{3.29}$$

where the subscript H implies a projection to the Lie algebra of H. This solution applies to any group and subgroup. As shown in [17] the standard geodesics equations for the geometries displayed above are automatically solved when the H-invariants are constructed from the solution (3.28)(3.29). In this way all light-like, space-like and time-like geodesic solutions are obtained.

With the point geodesics at hand we have learned a number of additional interesting properties about the $d = 2, 3, 4$ manifolds [17] which generalize to other non-compact gauged WZW models as well. The most striking feature is that the manifolds that are pictured in the figures have many copies and the complete manifold must include all the copies. The gauge invariant coordinates (e.g. (b, t) for $d = 2$) are not sufficient to fully describe the structure. There are additional *discrete* gauge invariants constructed from the group element g that label the copies of the manifold. This can be seen easily in our examples since the gauge subgroup is just the Lorentz group and its properties are well known. In this case the invariants are Lorentz dot products constructed from a vector x^α and a tensor $a^{\alpha\beta}$. Let us consider the invariant $b = (1 - x^2)/(1 + x^2)$, say in the region $x^2 > 0$. It is known that the time component x^0 could be either positive or negative and that a Lorentz transformation cannot change this sign. Therefore, the sign of x^0 is a discrete gauge invariant which does not show up in the metric or dilaton that characterized the manifolds discussed above. However, the model as whole knows about this sign through the group element g. Such discrete invariants are present in every *non-compact* gauged WZW model and they label copies of the manifolds described above. We may then ask whether these copies communicate with each other? The answer is yes, they do, and this can be seen by following the behaviour of a particle geodesic. The full information about the particle geodesic is contained in the solution for g in (3.28)(3.29). From this it can be verified that at the proper time that a particle touches a curvature singularity the discrete invariant switches sign and then the particle continues its journey smoothly from one copy of the manifold to the next. For example, in the 2d black hole case this happens for a time-like geodesic (i.e. massive particle) in a finite amount of proper time (on the other hand, a light-like geodesic takes an infinite amount of proper time to reach the singularity and therefore ends its journey without changing copies of the manifold). This behavior is present in all non-compact models in this paper as well as other models (e.g. we have verified it in the $SL(2, \mathbb{R}) \times SU(2)/\mathbb{R}^2$ model). It is reminiscent of the Reissner-Nordtsrom black hole in which geodesics move on to other worlds. The difference is that in our case this happens at the singularity itself. When quantum corrections are included and the exact metric considered, then the singularity and the transition to other worlds no longer seem to be at the same place, at least this is the case for the 2d black hole. The spectrum of the discrete invariant depends on the group representation and therefore one expects different numbers of copies in different quantum states. The number of copies is infinite for quantum states with non-fractional quantum numbers, which is typical in unitary non-holomorphic representations of non-compact groups. When the number of copies is infinite the particle can never come back to the same world, but for a finite number of copies the particle returns to the original world by emerging from a white singularity.

So far we have discussed particle geodesics that correspond to a string collapsed to a single point. We may also investigate string geodesics in the same manifolds. That is we are also interested in solutions for the strings moving in curved spacetime, just like one has a complete solution in flat spacetime in terms of harmonic oscillator normal modes. This problem has been solved in principle for the non-compact gauged WZW models in [12]. There the solution for the group element $g(\tau, \sigma)$ has been obtained explicitly in terms of normal modes. This is the analog of (3.28) above. There remains to construct the appropriate dot products to form the invariants, which in turn are the solutions to the string geodesics. This last part has not yet been performed explicitly, but it is only a matter of straightforward algebra of the kind performed for the particle geodesics in [17]. This procedure gives all the solutions in curved spacetime and can answer questions of the type "what happens when a string falls into a black hole ?"

3.5. Duality

Due to the lack of space we have not covered other interesting topics such as duality properties of these manifolds. It was shown in [12][17] that there is a dynamical duality that generalizes the $R \to 1/R$ duality properties of conformal field theories based on tori. This is related to the axial/vector duality that is present in the 2d black hole. It was shown in [17] that the duality transformation is equivalent to an inversion in group parameter space $(x_\alpha, a_{\alpha\beta})$ given in (3.5). This inversion generates discrete leaps for the group parameter that corresponds to interchanging different regions of the geometrical manifold. For details the reader is refered to [17]. This duality property is closely related to mirror symmetry of the kind discussed for Calabi-Yau manifolds, as will be explained elsewhere. The duality symmetry mentioned here is different than the one discussed in recent months by Verlinde, Giveon, Rocek and others.

4. CONCLUSIONS

We have only scratched the surface of the subject of non-compact gauged WZW models. We have shown that this approach is very useful for learning about strings in curved spacetime that may be relevant for the early part of the Universe. It is during this era that string theory should be relevant and it is during this era that the matter we know was formed. Therefore, in trying to solve the puzzles of the Standard Model with respect to the spectrum of matter and gauge bosons we may hope that a string theory in curved spacetime may guide us. For this reason I believe that it is valuable to study in great detail the models presented in Table-1. These are solvable models that should direct us toward a realistic unified theory.

We have pointed out a number of similarities between the coset string method and the IBM-like dynamical symmetry method used by Iachello in nuclear physics and molecular physics. The geometrical picture that corresponds to the dynamical symmetry has, to my knowledge, not been well developed. It may be possible to expand on the analogies presented in this paper to better understand more precisely the geometrical aspects of dynamical symmetries in nuclear and molecular physics.

References

[1] F. Iachello and Arima, ; see also articles in this volume.

[2] F. Iachello, Phys. Rev. Lett. ;
A. B. Balantekin, I. Bars and F. Iachello.

[3] K. Bardakci and M.B. Halpern, Phys. Rev. D3 (1971) 2493; ;
P. Goddard, A. Kent and D. Olive, Phys. Lett. 152B (1985) 88.

[4] I. Bars and D. Nemeschansky, Nucl. Phys. B348 (1991) 89.

[5] I. Bars, "Curved Spacetime Strings and Black Holes", in Proc. *XXth Int. Conf. on Diff. Geometrical Methods in Physics*, eds. S. Catto and A. Rocha, Vol. 2, p. 695, (World Scientific, 1992).

[6] P. Ginsparg and F. Quevedo, Nucl. Phys. B385 (1992) 527.

[7] E. Witten, Phys. Rev. D44 (1991) 314.

[8] V.G. Kac, Func. Anal. App. 1 (1967) 238; ;
R.V. Moody, Bull. Am. Math. Soc. 73 (1967) 217.

[9] E. Witten, Comm. Math. Phys. 92 (1984) 455.

[10] I. Bars, in *Vertex Operators in Math and Physics*, Eds. Lepowsky, Mandelstam, Singer, (Springer Verlag NY, 1985), page 373.

[11] E. Witten, Nucl. Phys. **B223** (1983)422; ;
K. Bardakci, E. Rabinovici andB. Säring, Nucl. Phys. **B299** (1988) 157; ;
K. Gawedzki and A. Kupiainen, Phys. Lett. **215B** (1988) 119; Nucl. Phys. **B320** (1989) 625.

[12] I. Bars and K. Sfetsos, Mod. Phys. Lett. **A7** (1992) 1091.

[13] M. Crescimanno, Mod. Phys. Lett. **A7** (1992) 489;;
J. B. Horne and G. T. Horowitz, Nucl. Phys. **B368** (1992) 444;;
E. S. Fradkin and V. Ya. Linetsky, Phys. Lett. **277B** (1992) 73;;
P. Horava, Phys. Lett. **278B** (1992) 101;;
E. Raiten, "Perturbations of a Stringy Black Hole", Fermilab-Pub 91-338-T; ;
D. Gershon, "Exact Solutions of Four-Dimensional Black Holes in String Theory", TAUP-1937-91.

[14] I. Bars and K. Sfetsos, Phys. Lett. **277B** (1992) 269.

[15] I. Bars, Phys. Lett **293B** (1992) 315.

[16] I. Bars, "Superstrings on Curved Spacetimes", USC-92/HEP-B5 (hep-th 9210079).

[17] I. Bars and K. Sfetsos, Phys. Rev. **D46** (1992) 4495.

[18] I. Bars and K. Sfetsos, Phys. Rev. **D46** (1992) 4510.

[19] K. Sfetsos, "Conformally Exact Results for $SL(2,\mathbb{R}) \otimes SO(1,1)^{d-2}/SO(1,1)$ Coset Models", USC-92/HEP-S1 (hep-th/9206048), to appear in Nucl. Phys.

[20] I. Bars and K. Sfetsos $SL(2,\mathbb{R}) \times SU(2)/R^2$ String Model in Curved Spacetime and Exact Conformal Results", USC-92/HEP-B3 (hep-th/ 9208001), to appear in Phys. Lett.

[21] R. Dijgraaf, E. Verlinde and H. Verlinde, Nucl. Phys. **B371** (1992) 269.

U(7) SPECTRUM GENERATING ALGEBRA FOR ROTATIONS AND VIBRATIONS IN TRIATOMIC MOLECULES

R. Bijker[1], A.E.L. Dieperink[2], and A. Leviatan[3]

[1] R.J. Van de Graaff Laboratory, University of Utrecht
P.O. Box 80000, 3508 TA Utrecht, The Netherlands

[2] K.V.I., Zernikelaan 25, 9747 AA Groningen, The Netherlands

[3] Racah Institute of Physics, The Hebrew University
Jerusalem 91904, Israel

INTRODUCTION

Algebraic methods have proved to be very useful in providing a unified description of rotational and vibrational spectra, both in collective nuclei [1, 2] and in molecules [3, 4]. The first version of the vibron model [3] was introduced for diatomic molecules. It describes the spectra in terms of the three components of a dipole (or p-) boson and a scalar (or s-) boson, under the restriction that the total number of bosons $N = n_s + n_p$ is conserved by the hamiltonian. The three dipole degrees of freedom are associated with the three components of the radius vector connecting the two atoms. The vibron model thus provides a unified description of rotational and vibrational excitations of diatomic molecules in terms of the spectrum generating algebra (SGA) of $U(4)$. An interesting aspect is that one of the dynamic symmetries corresponds in lowest approximation to the Morse oscillator which has been used widely in the study of diatomic molecules. Other examples are the $U(6)$ Interacting Boson Model (IBM) [1] for the description of quadrupole rotations and vibrations in collective nuclei and the $U(4)$ vibron model for the relative coordinate between the quark and anti-quark in mesons [5].

The molecular vibron model becomes especially useful in its extension to triatomic and polyatomic molecules. For these more complex systems conventional methods become increasingly hard to implement, despite the rapid development of computing power. It is here that the vibron model offers an attractive alternative. The main computational effort involves a finite-dimensional matrix diagonalization. In addition the vibron model has several dynamical symmetries in which the hamiltonian can be solved in closed analytic form. In 1982 it was proposed [4], in close analogy with the neutron-proton version of the IBM [2], to describe triatomic molecules in terms of two coupled $U(4)$ groups, one for each intrinsic dipole degree of freedom: $U_1(4) \otimes U_2(4)$. The extension to n-atomic molecules involves $n-1$ coupled $U(4)$ groups [6]. In these models the number of bosons of each type $N_i = n_{s_i} + n_{p_i}$ is conserved separately. In [7] a simplified version of the vibron model was introduced in terms a set of coupled one-dimensional anharmonic oscillators to describe the stretching vibrations in polyatomic molecules, such as XY_6 and X_6Y_6.

Despite its success at a phenomenological level, a closer examination of the geometric properties of the $U_1(4) \otimes U_2(4)$ vibron model shows that there are a few open questions concerning its application to nonlinear molecules.

(i) In a recent study [8] of the intrinsic and collective structure of the vibron hamiltonian it was found that whereas for an arbitrary two-body hamiltonian for a linear molecule a full resolution of the vibron hamiltonian into intrinsic (i.e. vibrational) and collective (i.e. rotational) parts can be obtained, for bent molecules this is only possible for a restricted class of hamiltonians.

(ii) A normal mode analysis of the intrinsic vibron hamiltonian (with two-body interactions) for bent molecules shows that whenever in the equilibrium shape the coordinates that are associated with the dipole bosons are perpendicular, the corresponding radial modes are decoupled from each other as well as from the angular mode [9]. Examples are symmetric X_2Y and X_3 molecules, where the two dipole bosons are associated with the Jacobi coordinates. In a conventional description such radial modes are necessarily coupled and the resulting normal modes can be interpreted in terms of symmetric and antisymmetric stretching vibrations. In the $U_1(4) \otimes U_2(4)$ model such a coupling can only be obtained by introducing higher order (four-body) interactions in the intrinsic hamiltonian [9].

(iii) In the description of a symmetric X_3 molecule it is important to preserve the permutation symmetry of the three identical atoms. In the $U_1(4) \otimes U_2(4)$ model this can only be achieved in the large N_1, N_2 limit. For finite values of N_1 and N_2 the permutation symmetry is broken. Moreover a normal mode analysis of the intrinsic hamiltonian shows that the characteristic pattern of the normal vibrations for a X_3 shape cannot be reproduced with two-body interactions only.

In this contribution we suggest an alternative description of triatomic molecules in terms of an $U(7)$ spectrum generating algebra which is realized by adding a single scalar boson to the two dipole bosons. The only restriction is that the total number of quanta is conserved. We show that for this reason the $U(7)$ model is better suited to incorporate the point group symmetries of nonlinear triatomic molecules and that it solves the above mentioned problems with the application of the $U_1(4) \otimes U_2(4)$ model to bent molecules.

U(7) SGA FOR TRIATOMIC MOLECULES

In this section we introduce a new algebraic model based on a $U(7)$ spectrum generating algebra to describe the rotations and vibrations of triatomic molecules. For a system of k degrees of freedom supporting a finite number of bound states, a convenient choice of SGA is that associated with the unitary group in $k+1$ dimensions: $U(k+1)$. For triatomic molecules the six degrees of freedom of the two relative coordinates, lead to a spectrum generating algebra of $U(7)$.

The building blocks of the $U(7)$ model are the six components of two dipole bosons with $L^\pi = 1^-$, denoted in second quantization by $p_{1,m}^\dagger$ and $p_{2,m}^\dagger$, and a scalar boson with

$L^\pi = 0^+$, denoted by $s^\dagger$. All many-body states are classified according to the totally symmetric representation $[N]$ of $U(7)$, where $N = n_s + n_{p_1} + n_{p_2}$ is the total number of bosons. The most general one- and two-body rotationally and parity invariant $U(7)$ hamiltonian that conserves the total number of bosons, N, can be expressed in second quantized form as

$$H = \sum_i \epsilon_i \sum_m a^\dagger_{i,m} a_{i,m} + \sum_L \sum_{ijkl} v^{(L)}_{ijkl} (a^\dagger_i a^\dagger_j)^{(L)} \cdot (\tilde{a}_l \tilde{a}_k)^{(L)} , \qquad (1)$$

where $a_0 = s$, $a_1 = p_{1,}$, $a_2 = p_2$. Since only the total boson number is conserved, the $U(7)$ hamiltonian contains several two-body terms, which are not present in the $U_1(4) \otimes U_2(4)$ model (e.g. $(p^\dagger_1 \cdot p^\dagger_1)(\tilde{p}_2 \cdot \tilde{p}_2) + \text{h.c.}$). Here $\tilde{p}_m = (-1)^{1-m} p_{-m}$ for each type of dipole bosons and h.c. stands for hermitian conjugate. Thus in the $U(7)$ model the two types of dipole degrees of freedom are treated on equal footing in the sense that the distribution of quanta among them is determined dynamically by the hamiltonian.

In the next section we apply the $U(7)$ model to triatomic molecules and study the connection between its normal modes and the fundamental vibrations. Another application of this model is to baryon resonances, in which the two dipole degrees of freedom are associated with the intrinsic coordinates of the three quarks in baryons. This is the subject of a separate contribution to these proceedings [10].

VIBRATIONAL SPECTRA AND NORMAL MODES

The vibrational spectrum of an arbitrary $U(7)$ hamiltonian can be obtained from its intrinsic part. The procedure is as follows. First a coherent state [11] is introduced, which allows one to study the minima of the classical energy surface. Next the coherent state representing the equilibrium shape of the molecule, is used to decompose the hamiltonian into intrinsic and collective parts and to extract the normal modes from the intrinsic hamiltonian [8].

The ground state band is approximated by a condensate wave function, which for a system of coupled dipole degrees of freedom can be taken as

$$| N; c \rangle = \frac{1}{\sqrt{N!}} \left(b^\dagger_c \right)^N | 0 \rangle ,$$

$$b^\dagger_c = (1 + R^2)^{-1/2} \left[s^\dagger + r_1 p^\dagger_{1,0} + r_2 \sum_m d^{(1)}_{m,0}(\theta) p^\dagger_{2,m} \right] . \qquad (2)$$

Here $R = \sqrt{r_1^2 + r_2^2}$ and $d^{(1)}_{m,m'}(\theta)$ is a reduced Wigner d-function. The two vectors $\vec{r}_1$ and $\vec{r}_2$ in the ground state condensate of eq. (2) span the xz-plane. We have chosen the z-axis along the direction of $\vec{r}_1 = r_1 \hat{z}$, and $\vec{r}_2$ is rotated by an angle θ about the y-axis, $\vec{r}_1 \cdot \vec{r}_2 = r_1 r_2 \cos \theta$.

The classical energy surface is obtained by evaluating the expectation value of the hamiltonian in the coherent state, $E(r_1 r_2 \theta) = \langle N; c | H | N; c \rangle$, and therefore it depends on the two coordinates, r_1 and r_2, and the relative angle, θ. The equilibrium shape, characterized by $r_1 = \bar{r}_1$, $r_2 = \bar{r}_2$ and $\theta = \bar{\theta}$, is determined by minimizing the energy surface. In the following we only consider hamiltonians for which the equilibrium value of the two coordinates $\bar{r}_1$, $\bar{r}_2 \neq 0$.

Both for linear and nonlinear molecules it is possible to achieve an *exact* resolution of an arbitrary two-body $U(7)$ hamiltonian into an intrinsic and a collective part. The intrinsic part annihilates the ground state condensate and has the same energy surface as the original hamiltonian, whereas the collective part has a completely flat energy surface, independent of the two coordinates, r_1 and r_2, and the angle θ. For non-linear shapes in the $U_1(4) \otimes U_2(4)$ model with two-body interactions, such a separation can only be achieved for a restricted class of hamiltonians. This is due to the fact that, as mentioned in the previous section, the $U(7)$ hamiltonian contains several additional interaction terms.

The vibrational spectrum can be studied by analyzing the intrinsic part of the hamiltonian in more detail. To this end we introduce a set of orthonormal deformed vibrons, consisting of the ground state condensate boson $b_c^\dagger$ of eq. (2), and the following six fluctuation bosons

$$b_u^\dagger(\theta) = (1+R^2)^{-1/2}\left[-R\,s^\dagger + R^{-1}\left(r_1\,p_{1,0}^\dagger + r_2\sum_m d_{m,0}^{(1)}(\theta)\,p_{2,m}^\dagger\right)\right],$$

$$b_v^\dagger(\theta) = R^{-1}\left[-r_2\,p_{1,0}^\dagger + r_1\sum_m d_{m,0}^{(1)}(\theta)\,p_{2,m}^\dagger\right],$$

$$b_w^\dagger(\theta) = R^{-1}\left[-r_2\frac{1}{\sqrt{2}}\left(p_{1,1}^\dagger - p_{1,-1}^\dagger\right) + r_1\sum_m\frac{1}{\sqrt{2}}\left(d_{m,1}^{(1)}(\theta) - d_{m,-1}^{(1)}(\theta)\right)p_{2,m}^\dagger\right],$$

$$b_1^\dagger(\theta) = R^{-1}\left[r_1\frac{1}{\sqrt{2}}\left(p_{1,1}^\dagger + p_{1,-1}^\dagger\right) + r_2\sum_m\frac{1}{\sqrt{2}}\left(d_{m,1}^{(1)}(\theta) + d_{m,-1}^{(1)}(\theta)\right)p_{2,m}^\dagger\right],$$

$$b_2^\dagger(\theta) = R^{-1}\left[r_1\frac{1}{\sqrt{2}}\left(p_{1,1}^\dagger - p_{1,-1}^\dagger\right) + r_2\sum_m\frac{1}{\sqrt{2}}\left(d_{m,1}^{(1)}(\theta) - d_{m,-1}^{(1)}(\theta)\right)p_{2,m}^\dagger\right],$$

$$b_3^\dagger(\theta) = R^{-1}\left[-r_2\frac{1}{\sqrt{2}}\left(p_{1,1}^\dagger + p_{1,-1}^\dagger\right) + r_1\sum_m\frac{1}{\sqrt{2}}\left(d_{m,1}^{(1)}(\theta) + d_{m,-1}^{(1)}(\theta)\right)p_{2,m}^\dagger\right]. \tag{3}$$

The bosons in eq. (3) represent excitations of the condensate which involve radial modes ($b_u^\dagger$, $b_v^\dagger$), angular modes ($b_w^\dagger$, $b_3^\dagger$) and rotational modes ($b_1^\dagger$, $b_2^\dagger$). The physical interpretation of these excitation modes for linear and nonlinear molecules will be discussed in more detail in the next sections.

LINEAR TRIATOMIC MOLECULES

We first discuss the vibrational spectrum of linear triatomic molecules. With the choice of the geometry of eq. (2) the symmetry axis is along the z-axis. The intrinsic part of an arbitrary two-body $U(7)$ hamiltonian for linear molecules that annihilates the condensate with $r_1 = \bar{r}_1 \neq 0$, $r_2 = \bar{r}_2 \neq 0$ and $\bar{\theta} = 0$, can be written as (to simplify the notation, we omit the bars from the equilibrium shape parameters)

$$
\begin{aligned}
H_{\text{int}} = \ & A_1[p_1^\dagger \cdot p_1^\dagger - r_1^2\,s^\dagger s^\dagger][\text{h.c.}] + A_2[p_2^\dagger \cdot p_2^\dagger - r_2^2\,s^\dagger s^\dagger][\text{h.c.}] \\
& + A_3[r_1^2\,p_2^\dagger \cdot p_2^\dagger - r_2^2\,p_1^\dagger \cdot p_1^\dagger][\text{h.c.}] + B_0[p_1^\dagger \cdot p_1^\dagger - r_1 r_2\,s^\dagger s^\dagger][\text{h.c.}] \\
& + B_1[r_1\,p_1^\dagger \cdot p_2^\dagger - r_2\,p_1^\dagger \cdot p_1^\dagger][\text{h.c.}] + B_2[r_2\,p_1^\dagger \cdot p_2^\dagger - r_1\,p_2^\dagger \cdot p_2^\dagger][\text{h.c.}] \\
& + D(p_1^\dagger p_2^\dagger)^{(1)} \cdot (\tilde{p}_2 \tilde{p}_1)^{(1)}.
\end{aligned} \tag{4}
$$

The normal modes can be found by rewriting the above hamiltonian in terms of the deformed vibrons of eqs. (2) and (3) (with $\theta = 0$), replacing the condensate bosons, b_c and $b_c^\dagger$, by their classical mean field value $\sqrt{N}$ and examining the Bogoliubov image of H_{int} [8, 9]. As a result we find

$$
\begin{aligned}
\frac{1}{N}H_{\text{int}} = \ & \epsilon_u\,b_u^\dagger b_u + \epsilon_v\,b_v^\dagger b_v + \epsilon_{uv}\left(b_u^\dagger b_v + b_v^\dagger b_u\right) \\
& + \epsilon_\theta\left[b_{\theta,1}^\dagger b_{\theta,1} + b_{\theta,-1}^\dagger b_{\theta,-1}\right] + \mathcal{O}(1/\sqrt{N}),
\end{aligned} \tag{5}
$$

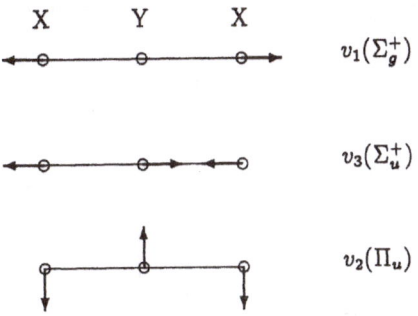

Figure 1. Schematic representation of the normal vibrations of a linear XYX molecule. The transformation character under the $D_{\infty h}$ point group is shown in parenthesis.

with

$$\epsilon_u = 4[A_1 r_1^4 + A_2 r_2^4 + B_0 r_1^2 r_2^2]R^{-2} ,$$

$$\epsilon_v = \left\{[4(A_1 + A_2)r_1^2 r_2^2 + B_0(r_1^2 - r_2^2)^2]R^{-2} \right.$$
$$\left. + [4A_3 r_1^2 r_2^2 + B_1 r_1^4 + B_2 r_2^4]R^2\right\}(1 + R^2)^{-1} ,$$

$$\epsilon_{uv} = 2r_1 r_2[-2A_1 r_1^2 + 2A_2 r_2^2 + B_0(r_1^2 - r_2^2)]R^{-2}(1 + R^2)^{-1/2} ,$$

$$\epsilon_\theta = \frac{1}{2}DR^2(1 + R^2)^{-1} . \tag{6}$$

For a linear triatomic molecule the b_u- and b_v- vibrons with angular momentum projection $K = 0$ on the symmetry (or z-) axis, represent coupled radial excitations. The normal modes in the radial direction can easily be obtained from eq. (5) by diagonalizing a two-dimensional matrix. The two orthogonal linear combinations of $b_u^\dagger$ and $b_v^\dagger$ represent the symmetric and antisymmetric stretching vibrations. The corresponding eigen frequencies are

$$\epsilon_\pm = \frac{1}{2}\left(\epsilon_u + \epsilon_v \pm \sqrt{(\epsilon_u - \epsilon_v)^2 + 4\epsilon_{uv}^2}\right) . \tag{7}$$

The b_θ-vibron operators

$$b_{\theta \pm 1}^\dagger = \frac{1}{\sqrt{2}}\left[b_3^\dagger(0) \pm b_w^\dagger(0)\right] = R^{-1}\left(r_1 p_{2,\pm 1}^\dagger - r_2 p_{1,\pm 1}^\dagger\right) , \tag{8}$$

with $K = \pm 1$ correspond to the two-dimensional bending mode. These assignments of intrinsic modes are in agreement with the usual point group classification [12]. Eq. (5) shows that the bending modes are decoupled from the radial modes. The $b_1^\dagger$ and $b_2^\dagger$ bosons missing from eq. (5) are Goldstone modes associated with rotations of the linear shape about the x- and y- directions perpendicular to the symmetry (z-) axis.

For a symmetric linear molecule of the type XYX additional constraints have to be imposed on the intrinsic hamiltonian to reflect the permutation symmetry. If the dipole bosons, p_1 and p_2, are associated with the two XY bonds, it is natural to take $A_1 = A_2$, $B_1 = B_2$ and $r_1 = r_2$ in eq. (4). Eq. (6) shows that in this case the two radial bosons ($b_u^\dagger$ and $b_v^\dagger$) are uncoupled and correspond to the symmetric stretching and antisymmetric stretching vibrations (see figure 1).

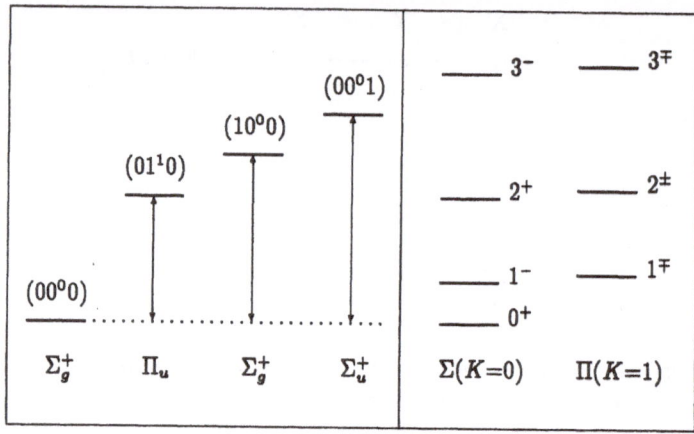

Figure 2. Schematic representation of the vibrational levels (left) of a linear X_2Y molecule in the $U(7)$ vibron model. The vibrational excitations are labeled by the spectroscopic notation $(v_1 v_2^K v_3)$ and by their transformation character under the $D_{\infty h}$ point group. The rotational structure is shown on the right (for $N \to \infty$). The levels are labeled by angular momentum and parity L^π.

The eigenstates of a linear X_2Y molecule are labeled in the standard spectroscopic notation by $(v_1 v_2^{l_2} v_3) L^\pi M$, where v_1 and v_3 represent the number of quanta in the symmetric and antisymmetric stretching vibration, respectively, and v_2 that in the bending vibration. The l_2 label denotes the projection of the vibrational angular momentum on the symmetry (or z-) axis: $K = l_2 = v_2, v_2 - 2, \ldots, 1$ or 0. In addition, each vibration is either symmetric (gerade) or antisymmetric (ungerade) with respect to exchange of the identical nuclei. The rotational spectrum of a linear X_2Y molecule is that of a prolate top, since the moments of inertia satisfy $I_x = I_y > I_z = 0$. The angular momentum and parity content is $L = K, K + 1, K + 2, \ldots$ with parity $\pi = (-)^L$ for $K = 0$ and $\pi = \pm$ for $K \neq 0$. In the $U_1(4) \otimes U_2(4)$ model these degenerate parity doublets can arise as a consequence of the $SO(4)$ symmetry for any value of N_1 and N_2. Furthermore, in that case, the vibrational quantum numbers $(v_1 v_2^K v_3)$ are in one-to-one correspondence with the $SO(4)$ labels [4]. In the $U(7)$ model no such symmetry is present and although degenerate parity doublets occur in the large N limit, they are not guaranteed to occur for finite values of N.

In figure 2 we show the vibrational excitations of a linear X_2Y molecule as well as the rotational structure of each of them.

NONLINEAR TRIATOMIC MOLECULES

Next we turn to a discussion of vibrational excitations of nonlinear molecules in the $U(7)$ model. The intrinsic part of an arbitrary two-body $U(7)$ hamiltonian for bent molecules that annihilates the condensate with $r_1 \neq 0$, $r_2 \neq 0$ and $0 < \theta < \pi$, can be written as

$$
\begin{aligned}
H_{\text{int}} = \; & A_1 [p_1^\dagger \cdot p_1^\dagger - r_1^2 \, s^\dagger s^\dagger][\text{h.c.}] + A_2 [p_2^\dagger \cdot p_2^\dagger - r_2^2 \, s^\dagger s^\dagger][\text{h.c.}] \\
& + A_3 [r_1^2 \, p_2^\dagger \cdot p_2^\dagger - r_2^2 \, p_1^\dagger \cdot p_1^\dagger][\text{h.c.}] + B_0 [p_1^\dagger \cdot p_2^\dagger - r_1 r_2 \cos\theta \, s^\dagger s^\dagger][\text{h.c.}]
\end{aligned}
$$

$$+B_1[r_1\, p_1^\dagger \cdot p_2^\dagger - r_2 \cos\theta\, p_1^\dagger \cdot p_1^\dagger][\text{h.c.}]$$
$$+B_2[r_2\, p_1^\dagger \cdot p_2^\dagger - r_1 \cos\theta\, p_2^\dagger \cdot p_2^\dagger][\text{h.c.}] \ . \tag{9}$$

A normal mode analysis of this intrinsic hamiltonian yields the following results

$$\frac{1}{N}H_{\text{int}} = \epsilon_u\, b_u^\dagger b_u + \epsilon_v\, b_v^\dagger b_v + \epsilon_w\, b_w^\dagger b_w + \epsilon_{uv}\,(b_u^\dagger b_v + b_v^\dagger b_u)$$
$$+\epsilon_{vw}\,(b_v^\dagger b_w + b_w^\dagger b_v) + \epsilon_{uw}\,(b_u^\dagger b_w + b_w^\dagger b_u) + \mathcal{O}(1/\sqrt{N}) \ . \tag{10}$$

The eigenfrequencies can be determined from

$$\begin{aligned}
\epsilon_u &= 4[A_1 r_1^4 + A_2 r_2^4 + B_0 r_1^2 r_2^2 \cos^2\theta]R^{-2} \ , \\
\epsilon_v &= \left\{[4(A_1 + A_2)r_1^2 r_2^2 + B_0(r_1^2 - r_2^2)^2 \cos^2\theta]R^{-2}\right. \\
&\quad \left.+[4A_3 r_1^2 r_2^2 + (B_1 r_1^2 + B_2 r_2^2)\cos^2\theta]R^2\right\}(1 + R^2)^{-1} \ , \\
\epsilon_w &= \sin^2\theta[B_0 + B_1 r_1^2 + B_2 r_2^2]R^2(1 + R^2)^{-1} \ , \\
\epsilon_{uv} &= 2r_1 r_2[-2A_1 r_1^2 + 2A_2 r_2^2 + B_0(r_1^2 - r_2^2)\cos^2\theta]R^{-2}(1 + R^2)^{-1/2} \ , \\
\epsilon_{vw} &= \sin\theta\cos\theta[B_0(r_1^2 - r_2^2) + (B_1 r_1^2 - B_2 r_2^2)R^2](1 + R^2)^{-1} \ , \\
\epsilon_{uw} &= 2B_0 r_1 r_2 \sin\theta\cos\theta(1 + R^2)^{-1/2} \ .
\end{aligned} \tag{11}$$

The b_u- and b_v- vibrons represent radial excitations and the b_w-vibron an angular excitation. For a bent molecule of the type XYZ no further symmetry is present and we have three non-degenerate fundamental vibrations, which can be obtained simply by diagonalizing a three-dimensional matrix. The resulting normal modes involve three orthogonal combinations of $b_u^\dagger$, $b_v^\dagger$ and $b_w^\dagger$. The bosons $b_i^\dagger$, $(i = 1, 2, 3)$ missing from eq. (10) are Goldstone modes associated with rotations of the condensate. For triatomic molecules of the type X_2Y or X_3 it is essential to preserve the permutation symmetry between the identical atoms. In the next section we show how this can be achieved in the context of the $U(7)$ model.

PERMUTATION SYMMETRY

For bent X_2Y and X_3 molecules it is convenient to associate the dipole bosons, p_1 and p_2, with the two Jacobi coordinates. The equilibrium value of the angle θ then corresponds to the angle between the two Jacobi coordinates, which for X_2Y and X_3 molecules is $\theta = \pi/2$. Eq. (10) shows that, since in this case $\epsilon_{vw} = \epsilon_{uw} = 0$, the angular mode $(b_w^\dagger)$ is decoupled from the two radial modes. It corresponds to oscillations in the angle between the two Jacobi coordinates (the v_2-vibration in figure 3). The radial normal modes are composed of two orthogonal combinations of $b_u^\dagger$ and $b_v^\dagger$ and correspond to the symmetric and antisymmetric stretching vibrations for bent molecules, in which the radial motions in the direction of the two Jacobi coordinates are coupled and move in and out of phase, respectively (the v_1- and v_3- vibrations in figure 3). In contrast, in the $U_1(4) \otimes U_2(4)$ model with two-body interactions, the radial modes in the two Jacobi coordinates, are decoupled for $\theta = \pi/2$. This precludes an interpretation in terms of symmetric and antisymmetric stretching vibrations. Such a coupling can be achieved only at the expense of introducing higher order (four-body) interactions in the intrinsic hamiltonian [9].

A nonlinear X_3 molecule has a threefold symmetry axis and is therefore an oblate top [12]. For a proper description of a triatomic oblate top molecule one has to preserve the permutation symmetry of the three identical atoms. The transformation properties

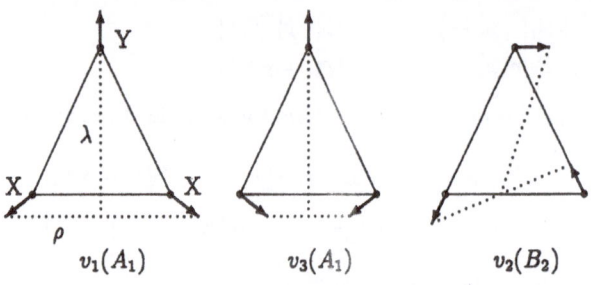

Figure 3. Schematic representation of the normal vibrations of a nonlinear X_2Y molecule. The Jacobi coordinates are indicated by the dotted lines. The transformation character under the C_{2v} point group is shown in parenthesis.

under the S_3 permutation group are most easily studied when we associate the dipole bosons with the Jacobi coordinates, $\vec{\rho} = (\vec{x}_1 - \vec{x}_2)/\sqrt{2}$ and $\vec{\lambda} = (\vec{x}_1 + \vec{x}_2 - 2\vec{x}_3)/\sqrt{6}$. The permutation symmetry of the dipole bosons is determined by the transformation properties under the two basic operations: the interchange $P(12)$ and the cyclic permutation $P(123)$,

$$P(12)\begin{pmatrix} p^\dagger_{\rho,m} \\ p^\dagger_{\lambda,m} \end{pmatrix} = \begin{pmatrix} -1 & 0 \\ 0 & 1 \end{pmatrix} \begin{pmatrix} p^\dagger_{\rho,m} \\ p^\dagger_{\lambda,m} \end{pmatrix},$$

$$P(123)\begin{pmatrix} p^\dagger_{\rho,m} \\ p^\dagger_{\lambda,m} \end{pmatrix} = \begin{pmatrix} \cos 2\pi/3 & \sin 2\pi/3 \\ -\sin 2\pi/3 & \cos 2\pi/3 \end{pmatrix} \begin{pmatrix} p^\dagger_{\rho,m} \\ p^\dagger_{\lambda,m} \end{pmatrix}. \tag{12}$$

The s-boson is a scalar under the permutation group. It is straightforward to show that the intrinsic part of an arbitrary S_3 invariant hamiltonian can be written as

$$\begin{aligned} H_{\text{int}} = & \ A_1[p^\dagger_\rho \cdot p^\dagger_\rho + p^\dagger_\lambda \cdot p^\dagger_\lambda - R^2\, s^\dagger s^\dagger][\text{h.c.}] \\ & + A_2 \left\{ [p^\dagger_\rho \cdot p^\dagger_\rho - p^\dagger_\lambda \cdot p^\dagger_\lambda][\text{h.c.}] + 4[p^\dagger_\rho \cdot p^\dagger_\lambda][\tilde{p}_\lambda \cdot \tilde{p}_\rho] \right\}. \end{aligned} \tag{13}$$

H_{int} annihilates the condensate of eq. (2) with $r_1 = r_2$ ($= R/\sqrt{2}$) and $\theta = \pi/2$.

Whereas in the $U(7)$ model the permutation symmetry can be preserved *exactly* for any value of N, in the $U_1(4) \otimes U_2(4)$ model this can only be achieved in the limit of $N_1 \to \infty$ and $N_2 \to \infty$. The difference is due to the fact that in the $U(7)$ model we deal with a single s- boson and with a full oscillator basis, where only the total number of dipole quanta, $n_p = n_{p_\rho} + n_{p_\lambda}$, is limited.

In the large N limit the normal modes of the hamiltonian in eq. (13) are diagonal in the deformed vibron basis of eq. (3) (with $r_1 = r_2$ and $\theta = \pi/2$)

$$\frac{1}{N} H_{\text{int}} = \epsilon_u\, b^\dagger_u b_u + \epsilon_v\, b^\dagger_v b_v + \epsilon_w\, b^\dagger_w b_w + \mathcal{O}(1/\sqrt{N}). \tag{14}$$

The three fundamental vibrations of the oblate top have the eigen frequencies

$$\begin{aligned} \epsilon_u &= 4A_1 R^2, \\ \epsilon_v &= 4A_2 R^2 (1 + R^2)^{-1}, \\ \epsilon_w &= \epsilon_v, \end{aligned} \tag{15}$$

with $R^2 = 2r_1^2 = 2r_2^2$. The intrinsic vibrational modes (see figure 3) consist of two radial modes, a symmetric stretching and an antisymmetric stretching (represented by $b_u^\dagger$ and $b_v^\dagger$ respectively) and an angular bending mode (represented by $b_w^\dagger$). The antisymmetric stretching mode ($b_v^\dagger$) and the bending mode ($b_w^\dagger$) are degenerate, which is in agreement with the point group classification of the fundamental vibrations for a symmetric X_3 molecule [12]. It should be noted that although the two radial modes (b_u and b_v) in eq. (14) are uncoupled, each of them is a linear combination of p_ρ and p_λ (see eq. (3)) and hence already contains the required coupling between the Jacobi coordinates.

A similar analysis of the S_3 invariant intrinsic (two-body) hamiltonian in the large N_1, N_2 limit of the $U_1(4) \otimes U_2(4)$ model yields that in this case, the two radial modes associated with the Jacobi coordinates are uncoupled and have the same frequency. These features do not agree with the expected pattern of normal vibrations for a symmetric X_3 shape, and thus pose a problem with an interpretation in terms of Jacobi coordinates. This problem may be overcome by introducing higher-order terms (four-body) in the $U_1(4) \otimes U_2(4)$ hamiltonian [9].

The collective (or rotational) hamiltonian for the oblate symmetric top is given by

$$H_{\rm rot} = \kappa \, \hat{L} \cdot \hat{L} - 4\eta \, \hat{F}_y^2 \,, \tag{16}$$

where $\hat{L}$ is the total angular momentum operator

$$\hat{L}_m = \sqrt{2} \, (p_\rho^\dagger \tilde{p}_\rho + p_\lambda^\dagger \tilde{p}_\lambda)_m^{(1)} \,, \tag{17}$$

and $\hat{F}_y$ is given by

$$\hat{F}_y = \frac{1}{2i} \sum_m (p_{\rho,m}^\dagger p_{\lambda,m} - p_{\lambda,m}^\dagger p_{\rho,m}) \,. \tag{18}$$

It is straightforward to show that the rotational hamiltonian of eq. (16) is a scalar under the permutation group, as is required for an oblate symmetric top.

A geometrical interpretation of the second term in $H_{\rm rot}$ of eq. (16) can be obtained in the mean-field limit by rewriting it in terms of the deformed vibrons (with $r_1 = r_2 = R/\sqrt{2}$ and $\theta = \pi/2$) and replacing the condensate bosons by $\sqrt{N}$. The result for the three components of the angular momentum and $\hat{F}_y$ is to leading order in N,

$$
\begin{aligned}
\hat{L}_x/\sqrt{N} &= \frac{1}{2}\left(b_1^\dagger - b_3^\dagger + b_1 - b_3\right) R(1+R^2)^{-1/2} + \mathcal{O}(1/\sqrt{N}) \,, \\
\hat{L}_y/\sqrt{N} &= -i\left(b_2^\dagger - b_2\right) R(1+R^2)^{-1/2} + \mathcal{O}(1/\sqrt{N}) \,, \\
\hat{L}_z/\sqrt{N} &= -\frac{1}{2}\left(b_1^\dagger + b_3^\dagger + b_1 + b_3\right) R(1+R^2)^{-1/2} + \mathcal{O}(1/\sqrt{N}) \,, \\
\hat{F}_y/\sqrt{N} &= i\frac{1}{2}\left(b_2^\dagger - b_2\right) R(1+R^2)^{-1/2} + \mathcal{O}(1/\sqrt{N}) \,.
\end{aligned}
\tag{19}
$$

This shows that for large N, $2\hat{F}_y$ is related to the projection of the angular momentum on the threefold symmetry (or y-) axis. The rotational spectrum of the oblate top is then given by

$$E_{\rm rot} = \kappa \, L(L+1) - \eta \, K_y^2 \,, \tag{20}$$

where $K_y = 2F_y$. It is interesting to note that in this case we have an interaction that splits but does not admix the different K_y-bands. In fact, the eigenstates of any S_3 invariant $U(7)$ hamiltonian have in addition to the total angular momentum L and its projection M, also K_y as a good quantum number.

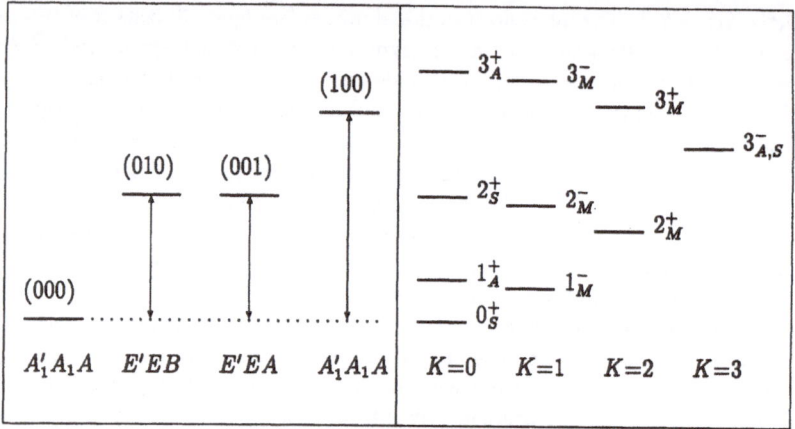

Figure 4. Schematic representation of the vibrational levels (left) of a nonlinear X_3 molecule. The vibrational excitations are labeled by the spectroscopic notation $(v_1 v_2 v_3)$ and by their transformation character under the $D_{3h} \supset D_3 \supset C_2$ point groups. The rotational levels are labeled by angular momentum, its projection on the symmetry axis and parity (L^π, K) and the transformation character under the S_3 permutation group.

The eigenstates of the oblate symmetric top are labeled in the standard spectroscopic notation by $(v_1 v_2 v_3) K L^\pi M$, where v_1 and v_3 represent the number of quanta in the symmetric and antisymmetric stretching vibration, respectively, and v_2 that in the bending vibration. On top of each vibrational excitation $(v_1 v_2 v_3)$ there is a whole series of rotational levels with angular momentum $L = K, K+1, \ldots$. The label K denotes the projection of the angular momentum on the symmetry (y-) axis and can have values $K = 0, 1, 2, \ldots$. For $K > 0$ the angular momentum states are doubly degenerate and can only be distinguished by their transformation character under the permutation group. In the $U(7)$ model this degeneracy is present for any value of N and is a consequence of the permutation symmetry of the hamiltonian, whereas in the $U_1(4) \otimes U_2(4)$ model this occurs only in the large N_1, N_2 limit.

In figure 4 we show the vibrational excitations of an oblate symmetric top as well as the rotational structure.

SUMMARY AND CONCLUSIONS

In this contribution we proposed the $U(7)$ as a spectrum generating algebra for triatomic molecules. We studied the normal modes for both linear and bent molecules and wherever possible made a comparison with a description in terms of a $U_1(4) \otimes U_2(4)$ model.

For linear molecules both models are, as far as the vibrational excitations are concerned, in complete agreement with the conventional point group classification of the fundamental vibrations. In the $U_1(4) \otimes U_2(4)$ model it is possible to establish a one-to-one correspondence between the vibrational quantum numbers $(v_1 v_2^{l_2} v_3)$ of a linear molecule and the labels of the $SO(4)$ dynamical symmetry. For such a molecule the bending vibrations consist of a series of parity doublets, which arise naturally as a consequence of the $SO(4)$ symmetry. In the $U(7)$ model however no such symmetry is present and exact parity doublets occur in the large N limit.

For nonlinear molecules the situation is completely different. Although in the large N limit the two models give the same results for ground state properties, this does *not* always hold for excitations. As an example we showed explicitly how the textbook example of the oblate symmetric top arises in the $U(7)$ vibron model. Since the $U(7)$ model uses a full oscillator basis for the two coupled dipole degrees of freedom, the permutation symmetry of the three identical atoms in a symmetric X_3 molecule can be preserved *exactly* for any value of N. A normal mode analysis of the intrinsic oblate top hamiltonian showed that the characteristic pattern of the normal vibrations of a symmetric X_3 molecule is recovered in the $U(7)$ model with two-body interactions, whereas in the $U_1(4) \otimes U_2(4)$ model this can only be be achieved by the inclusion of higher order terms (four-body) in the hamiltonian.

The main difference between the two models is in the choice of the model space. In the $U_1(4) \otimes U_2(4)$ model the number of vibrons in each mode is conserved separately. In the $U(7)$ model only the total number of vibrons is conserved and the distribution of quanta among the two dipole degrees of freedom is determined dynamically by the hamiltonian.

In conclusion, we have shown that linear triatomic molecules are most conveniently described in terms of a $U_1(4) \otimes U_2(4)$ SGA in which each bond is treated as a Morse oscillator (via the $SO(4)$ symmetry). Rotational-vibrational states in larger linear molecules with n atoms would then be amenable to a description in terms of $n-1$ coupled $U(4)$ groups. On the other hand, for nonlinear triatomic molecules, we found that a description in terms of a $U(7)$ SGA is favored, since one has to use a full oscillator basis to incorporate the relevant point groups and permutation symmetry. This is most conveniently realized in an interpretation of the dipole bosons in terms of Jacobi coordinates. The merits of $U(7)$ for nonlinear triatomic molecules suggest to use a $U(3n-2)$ SGA for larger nonlinear molecules with n atoms.

ACKNOWLEDGEMENTS

It is a great pleasure to dedicate this contribution to the 50th birthday of F. Iachello. This work was inspired by his pioneering work on the interacting boson model and the vibron model and has benefited greatly from his continuous interest and unlimited enthusiasm. Happy birthday, Franco!

REFERENCES

[1] A. Arima and F. Iachello, Phys. Rev. Lett. **35**, 1069 (1975); F. Iachello and A. Arima, 'The Interacting Boson Model' (Cambridge Univ. Press, Cambridge, 1987).

[2] A. Arima, T. Otsuka, F. Iachello and I. Talmi, Phys. Lett. **B66**, 205 (1977); T. Otsuka, A. Arima, F. Iachello and I. Talmi, Phys. Lett. **B76**, 139 (1978).

[3] F. Iachello, Chem. Phys. Lett. **78**, 581 (1981); F. Iachello and R.D. Levine, J. Chem. Phys. **77**, 3046 (1982).

[4] O.S. van Roosmalen, A.E.L. Dieperink and F. Iachello, Chem. Phys. Lett. **85**, 32 (1982); O.S. van Roosmalen, F. Iachello, R.D. Levine and A.E.L. Dieperink, J. Chem. Phys. **79**, 2515 (1983).

[5] F. Iachello and D. Kusnezov, Phys. Lett. **B255**, 493 (1991) and Phys. Rev. **D45**, 4156 (1992); F. Iachello, N.C. Mukhopadhyay and L. Zhang, Phys. Lett. **B256**, 295 (1991) and Phys. Rev. **D44**, 898 (1991).

[6] J. Hornos and F. Iachello, J. Chem. Phys. **90**, 5284 (1989); F. Iachello, S. Oss and R. Lemus, J. Mol. Spectry. **146**, 56 (1991) and **149**, 132 (1991).

[7] F. Iachello and S. Oss, Phys. Rev. Lett. **66**, 2976 (1991) and J. Mol. Spectry. **153**, 225 (1992); A. Frank and R. Lemus, Phys. Rev. Lett. **68**, 413 (1992).

[8] A. Leviatan and M.W. Kirson, Ann. Phys. **188**, 142 (1988).

[9] A. Leviatan, J. Chem. Phys. **91**, 1706 (1989).

[10] R. Bijker and A. Leviatan, these proceedings.

[11] See e.g. R. Gilmore and D.H. Feng, Nucl. Phys. **A301**, 189 (1978); J.N. Ginocchio and M.W. Kirson, Phys. Rev. Lett. **44**, 1744 (1980); A.E.L. Dieperink, O. Scholten and F. Iachello, Phys. Rev. Lett. **44**, 1747 (1980).

[12] G. Herzberg, 'Molecular Spectra and Molecular Structure II: Infrared and Raman Spectra of Polyatomic Molecules' (Van Nostrand Rheinhold, Princeton, NJ 1945).

ALGEBRAIC TREATMENT OF COLLECTIVE
EXCITATIONS IN BARYON SPECTROSCOPY

R. Bijker[1] and A. Leviatan[2]

[1] R.J. Van de Graaff Laboratory, University of Utrecht
P.O. Box 80000, 3508 TA Utrecht, The Netherlands

[2] Racah Institute of Physics, The Hebrew University
Jerusalem 91904, Israel

INTRODUCTION

Algebraic methods have been used extensively in hadronic physics for the description of the internal degrees of freedom (flavor-spin-color) [1]. Spectrum generating algebras and dynamic symmetries have been very instrumental in the classification of hadronic states and the construction of mass formulas, such as the Gell-Mann-Okubo mass formula [2].

Recently Iachello suggested [3] to use algebraic methods for the geometric structure of hadrons as well, and thus, by combining the degrees of freedom of the internal and relative motion, to obtain a fully algebraic description of hadrons. The key ingredient in such an approach is the choice of a spectrum generating algebra (SGA) for the relevant spatial degrees of freedom, *e.g.* relative quark coordinates or geometric shape variables. The operators representing physical observables, such as masses, are then expressed in terms of elements of the algebra. All matrix elements of interest can be calculated exactly without having to make further approximations. As a first application of this approach the string-like configuration of a quark (q) and an antiquark ($\bar{q}$) in a meson was described in terms of a $U(4)$ SGA [4, 5]. The $U(4)$ algebra is realized in terms of bosons whose mutual interactions simulate the dynamics of the radius vector connecting the quark and antiquark. This model corresponds to a collective string-like description of mesons.

The situation for baryons is more complex. In a valence quark model, baryons are made up of three quarks (qqq), some of which may be identical. Therefore, unlike the situation in mesons, for baryons one has to construct states with good permutation symmetry. To fulfill the Pauli principle only particular combinations of space, spin-flavor and color representations are allowed. In quark potential models, either nonrelativistic

Symmetries in Science VII, Edited by B. Gruber
and T. Otsuka, Plenum Press, New York, 1994

[6] or in a relativized form [7], baryons are described as a system of quarks interacting through two-body interactions. Most calculations are done in a harmonic oscillator basis and usually only a few oscillator shells are taken into account exactly. As an alternative to such a single-particle type of description, it is of interest to consider the possibility of a collective description of baryons. The quarks participating in a collective motion are strongly correlated and move coherently. In terms of a harmonic oscillator basis collectivity requires the mixing of many different oscillator shells. This is in analogy to the case encountered in nuclei for which a description of α-clustering requires mixing of many shells in the nuclear shell model. In any collective description of baryons one has to mix different oscillator shells in such a way that the permutation symmetry of the quarks is preserved. Furthermore, in order to be able to compare the single-particle with the geometric-collective description of baryons, it would be beneficial to have a framework that encompasses both points of view.

In this contribution we introduce a $U(7)$ spectrum generating algebra as a new algebraic string-like model of baryons. This model satisfies all the above mentioned requirements for a collective description. We show explicitly that in the $U(7)$ model it is possible to incorporate the permutation symmetry of the three quarks exactly and to accommodate both single-particle and collective types of motion. The model space consists of many oscillator shells, but since it is finite dimensional, the $U(7)$ SGA provides a tractable computational scheme, that involves only the diagonalization of finite dimensional matrices. A great advantage is that within the assumptions of the model all calculations can be done exactly. We discuss some special cases in which we derive closed analytic solutions that give a clear insight into the properties of the model. Finally we note that the same $U(7)$ SGA can be used to describe rotations and vibrations in triatomic molecules [8].

THE U(7) QUARK MODEL

The hamiltonian in quark potential models contains a kinetic energy term, a confining potential and the hyperfine interaction. The confining potential is written as the sum of two uncoupled harmonic oscillators in the Jacobi coordinates and a residual two-body interaction. The two relative Jacobi coordinates are

$$\vec{\rho} = (\vec{r}_1 - \vec{r}_2)/\sqrt{2} ,$$
$$\vec{\lambda} = (\vec{r}_1 + \vec{r}_2 - 2\vec{r}_3)/\sqrt{6} , \qquad (1)$$

where $\vec{r}_i$ is the coordinate of the i-th quark. The group structure associated with harmonic oscillator part is given by the chain,

$$U(6) \supset U_\rho(3) \otimes U_\lambda(3) \supset SO_\rho(3) \otimes SO_\lambda(3) \supset SO_{\rho\lambda}(3) . \qquad (2)$$

The unperturbed eigenvalues are given in terms of $n = n_\rho + n_\lambda$, the total number of quanta in the ρ and λ oscillators. The group chain in eq. (2) provides a convenient basis to diagonalize the full hamiltonian. Typical calculations involve only a few oscillator shells. For a given harmonic oscillator shell, n, there exist well known techniques to construct states of good permutation symmetry, angular momentum and parity [9].

Collective behavior of quarks inside baryons corresponds to a coherent motion of the quarks. A description of such correlated motion in a harmonic oscillator basis requires strong coupling of many different oscillator shells. In order to accommodate such mixing we discuss an extension of the $U(6)$ symmetry group of the harmonic oscillator. From experience gained with algebraic models in nuclear [10] and molecular physics [11], it is known that this can be achieved by embedding the $U(6)$ algebra in the compact SGA of $U(7)$. This is done by adding a scalar boson under the restriction that the total number of bosons is conserved. The last condition guarantees that the $U(7)$ SGA still describes only six degrees of freedom, which can be associated with the two relative Jacobi coordinates.

The building blocks of the $U(7)$ model are the six components of two dipole bosons with $L^\pi = 1^-$, denoted in second quantization by $p^\dagger_\rho$ and $p^\dagger_\lambda$, and a scalar boson with $L^\pi = 0^+$, denoted by $s^\dagger$. A mass operator that conserves the total number of quanta $N = n_s + n_\rho + n_\lambda$ can be expressed in terms of the 49 generators of $U(7)$,

$$
\begin{aligned}
D_\lambda &= (p^\dagger_\lambda s - s^\dagger \tilde{p}_\lambda)^{(1)}, & A_\lambda &= i\,(p^\dagger_\lambda s + s^\dagger \tilde{p}_\lambda)^{(1)}, & G^{(l)}_\lambda &= (p^\dagger_\rho \tilde{p}_\rho - p^\dagger_\lambda \tilde{p}_\lambda)^{(l)}, \\
D_\rho &= (p^\dagger_\rho s - s^\dagger \tilde{p}_\rho)^{(1)}, & A_\rho &= i\,(p^\dagger_\rho s + s^\dagger \tilde{p}_\rho)^{(1)}, & G^{(l)}_\rho &= (p^\dagger_\rho \tilde{p}_\lambda + p^\dagger_\lambda \tilde{p}_\rho)^{(l)}, \\
G^{(l)}_S &= (p^\dagger_\rho \tilde{p}_\rho + p^\dagger_\lambda \tilde{p}_\lambda)^{(l)}, & G^{(l)}_A &= i\,(p^\dagger_\rho \tilde{p}_\lambda - p^\dagger_\lambda \tilde{p}_\rho)^{(l)}, & \hat{n}_s &= s^\dagger s,
\end{aligned}
\tag{3}
$$

with $l = 0, 1, 2$ and $\tilde{p}_\mu = (-1)^{1-\mu} p_{-\mu}$. All many-boson states are classified according to the totally symmetric representation $[N]$ of $U(7)$. The value of N determines the number of states in the model space. In view of confinement we expect N to be large. The model space contains several oscillator shells. For a given value of N, it consists of oscillator shells with $n = n_\rho + n_\lambda = 0, 1, \ldots, N$. In order to construct a mass operator that properly takes into account the permutation symmetry among the three quarks inside a baryon, we study the transformation properties of the generators of eq. (3) under the S_3 permutation group.

PERMUTATION SYMMETRY IN THE U(7) MODEL

The harmonic oscillator basis is very well suited to construct a set of basis states that in addition to good angular momentum and parity also have good permutation symmetry under the interchange of any of the three quarks. For baryons with strangeness -1 or -2 the only permutation that is involved is the interchange of the two identical quarks, $P(12)$, which without loss of generality are taken to be the first two. For baryons with strangeness 0 or -3 the three quarks are indistinguishable and therefore we have to use in addition to $P(12)$ also the cyclic permutation $P(123)$. All other permutations can be expressed in terms of these two elementary ones. Although the $U(7)$ model can describe both strange and nonstrange baryons, in the present contribution we discuss only the nonstrange sector, namely the nucleon and the delta resonances. The spatial wave functions then have to be combined with a spin-flavor and a color part so that the full wave function of the baryon is antisymmetric.

In the current approach the spatial part of the wave function is described in terms of the $U(7)$ SGA. The eigenstates are required to have well defined transformation properties under the permutation group, or equivalently, the mass operator for nonstrange baryon resonances has to be invariant under S_3. The transformation properties under S_3 of all operators in the model follow from those of $s^\dagger$, $p^\dagger_\rho$ and $p^\dagger_\lambda$,

$$
P(12) \begin{pmatrix} s^\dagger \\ p^\dagger_{\rho,m} \\ p^\dagger_{\lambda,m} \end{pmatrix} = \begin{pmatrix} 1 & 0 & 0 \\ 0 & -1 & 0 \\ 0 & 0 & 1 \end{pmatrix} \begin{pmatrix} s^\dagger \\ p^\dagger_{\rho,m} \\ p^\dagger_{\lambda,m} \end{pmatrix},
$$

$$
P(123) \begin{pmatrix} s^\dagger \\ p^\dagger_{\rho,m} \\ p^\dagger_{\lambda,m} \end{pmatrix} = \begin{pmatrix} 1 & 0 & 0 \\ 0 & \cos 2\pi/3 & \sin 2\pi/3 \\ 0 & -\sin 2\pi/3 & \cos 2\pi/3 \end{pmatrix} \begin{pmatrix} s^\dagger \\ p^\dagger_{\rho,m} \\ p^\dagger_{\lambda,m} \end{pmatrix}.
\tag{4}
$$

The s-boson is a scalar under the permutation group. There are three different symmetry classes for the permutation of three objects: a symmetric one, S, an antisymmetric

Table 1. Transformation properties of linear and bilinear operators under the S_3 permutation group. Here $l = 0, 1, 2$ and $l' = 0, 2$.

Operator	$P(12)$	$P(123)$	S_3	Young tableau
$s^\dagger$, $s^\dagger s^\dagger$, $(p_\rho^\dagger p_\rho^\dagger + p_\lambda^\dagger p_\lambda^\dagger)^{(l')}$, $\hat{n}_s$, $G_S^{(l)}$	1	1	S	▦ $\boxed{1}\boxed{2}\boxed{3}$
$p_\lambda^\dagger$, $s^\dagger p_\lambda^\dagger$, $(p_\rho^\dagger p_\rho^\dagger - p_\lambda^\dagger p_\lambda^\dagger)^{(l')}$, D_λ, A_λ, $G_\lambda^{(l)}$	1	λ	M_λ	$\boxed{1}\boxed{2}\atop\boxed{3}$
$p_\rho^\dagger$, $s^\dagger p_\rho^\dagger$, $(p_\rho^\dagger p_\lambda^\dagger + p_\lambda^\dagger p_\rho^\dagger)^{(l')}$, D_ρ, A_ρ, $G_\rho^{(l)}$	-1	ρ	M_ρ	$\boxed{1}\boxed{3}\atop\boxed{2}$
$(p_\rho^\dagger p_\lambda^\dagger)^{(1)}$, $G_A^{(l)}$	-1	1	A	$\boxed{1}\atop\boxed{2}\atop\boxed{3}$

one, A, and a two-dimensional one of mixed symmetry type, denoted by M_ρ and M_λ. The latter have the same transformation properties as the creation operators, $p_\rho^\dagger$ and $p_\lambda^\dagger$. Alternatively, the three symmetry classes can be labeled by the irreducible representations of the point group D_3 (which is isomorphic to S_3) as A_1, A_2 and E, respectively. It is now straightforward to find the bilinear combinations of creation and annihilation operators that transform irreducibly under the permutation group. The results are shown in table 1.

Next one can use the multiplication rules for S_3 to construct all rotationally invariant interactions that preserve parity and that are scalars under S_3. As a result we find two terms linear in the $U(7)$ generators of eq. (3),

$$\hat{n}_s, \quad \hat{n} = \sqrt{3}\, G_S^{(0)}, \tag{5}$$

where $\hat{n}_s = \hat{N} - \hat{n}$ is the number operator for scalar bosons, and $\hat{n} = \hat{n}_\rho + \hat{n}_\lambda$ counts the total number of ρ and λ bosons. In addition we have the following quadratic multipole operators,

$$\hat{n}_s\hat{n}_s, \quad \hat{n}_s\hat{n}, \quad D_\rho \cdot D_\rho + D_\lambda \cdot D_\lambda, \quad A_\rho \cdot A_\rho + A_\lambda \cdot A_\lambda,$$
$$G_S^{(l)} \cdot G_S^{(l)}, \quad G_A^{(l)} \cdot G_A^{(l)}, \quad G_\rho^{(l)} \cdot G_\rho^{(l)} + G_\lambda^{(l)} \cdot G_\lambda^{(l)}. \tag{6}$$

We note that not all terms in eqs. (5,6) are independent. A set of independent two-body interactions that are scalars under the S_3 permutation group can be obtained by transforming to normal ordered form,

$$s^\dagger s^\dagger s s, \quad s^\dagger (p_\rho^\dagger \cdot \tilde{p}_\rho + p_\lambda^\dagger \cdot \tilde{p}_\lambda)s, \quad (p_\rho^\dagger p_\lambda^\dagger)^{(1)} \cdot (\tilde{p}_\lambda\tilde{p}_\rho)^{(1)},$$
$$(p_\rho^\dagger \cdot p_\rho^\dagger + p_\lambda^\dagger \cdot p_\lambda^\dagger)ss + s^\dagger s^\dagger (\tilde{p}_\rho \cdot \tilde{p}_\rho + \tilde{p}_\lambda \cdot \tilde{p}_\lambda),$$

$$(p_\rho^\dagger p_\rho^\dagger - p_\lambda^\dagger p_\lambda^\dagger)^{(l')} \cdot (\tilde{p}_\rho \tilde{p}_\rho - \tilde{p}_\lambda \tilde{p}_\rho)^{(l')} + 4\,(p_\rho^\dagger p_\rho^\dagger)^{(l')} \cdot (\tilde{p}_\lambda \tilde{p}_\rho)^{(l')}\,,$$

$$(p_\rho^\dagger p_\rho^\dagger + p_\lambda^\dagger p_\lambda^\dagger)^{(l')} \cdot (\tilde{p}_\rho \tilde{p}_\rho + \tilde{p}_\lambda \tilde{p}_\lambda)^{(l')}\,, \tag{7}$$

with $l' = 0, 2$.

Diagonalization of a general S_3-invariant mass operator composed of the terms in eqs. (5-7) yields wave functions of good permutation symmetry. The transformation properties of these wave functions under S_3 can be determined by taking the overlap with harmonic oscillator wave functions that by construction have good (and known) permutation symmetry [9]. We used a more direct method which is also based on the transformation properties under $P(12)$ and $P(123)$. The symmetry classes (S, M_λ) are distinguished from (A, M_ρ) by a choice of basis states with the number of quanta in the ρ oscillator, n_ρ, even or odd, respectively. The classes (S, A) are distinguished from (M_λ, M_ρ) by the expectation value K_y^2 (with K_y integer) of the operator, $\hat{K}_y^2 = 3\,G_A^{(0)}G_A^{(0)}$, namely, $|K_y| = 0, 3, 6, \ldots$ for the former and $|K_y| = 1, 2, 4, 5, \ldots$ for the latter. Formally K_y corresponds to the quantum number, m, that was used in ref. [12] to classify the states in the quark potential model. It is important to note that our method is valid for any oscillator shell.

GEOMETRIC SHAPE AND ITS EXCITATIONS

The previous analysis shows that already at the level of one- and two-body interactions, there are many possible terms that can be included in a S_3-invariant mass operator in the $U(7)$ model. In this section we analyze the model in terms of geometric variables and study its elementary excitations to gain a better understanding of the physical content of each of the allowed interaction terms. This will provide a selection criterion which terms to include in the mass operator.

A geometric interpretation of the algebraic model can be obtained by means of a coherent state [13], which for the $U(7)$ model takes the form of a condensate of N bosons,

$$|N; c\rangle \;=\; \frac{1}{\sqrt{N!}}\left(b_c^\dagger\right)^N |0\rangle\,, \tag{8}$$

with

$$b_c^\dagger \;=\; (1 + R^2)^{-1/2}\!\left[s^\dagger + r_\rho\, p_{\rho,0}^\dagger + r_\lambda \sum_m d_{m,0}^{(1)}(\theta)\, p_{\lambda,m}^\dagger\right]. \tag{9}$$

Here $R = \sqrt{r_\rho^2 + r_\lambda^2}$. The two vectors, $\vec{r}_\rho$ and $\vec{r}_\lambda$, in the condensate span the xz-plane. We have chosen the z-axis along the direction of $\vec{r}_\rho = r_\rho \hat{z}$, and $\vec{r}_\lambda$ is rotated by an angle θ about the out-of-plane y-axis, $\vec{r}_\rho \cdot \vec{r}_\lambda = r_\rho r_\lambda \cos\theta$.

The expectation value of the mass operator in the condensate defines a classical potential function, $V(r_\rho r_\lambda \theta) = \langle N; c \mid \hat{M}^2 \mid N; c\rangle$. The equilibrium shape is determined by minimizing the potential function with respect to the coordinates, r_ρ and r_λ, and the relative angle θ. The equilibrium values of these shape parameters are denoted by $\bar{r}_\rho$, $\bar{r}_\lambda$ and $\bar{\theta}$, respectively. For the problem at hand, three identical quarks inside a baryon, the most general S_3-invariant $U(7)$ mass operator with one- and two-body terms yields a rigid nonlinear equilibrium shape characterized by

$$\bar{r}_\rho = \bar{r}_\lambda\,, \qquad \bar{\theta} = \frac{\pi}{2}\,. \tag{10}$$

These are precisely the conditions satisfied by the Jacobi coordinates of eq. (1) for an equilateral triangular shape. This strongly suggests to associate these coordinates with the algebraic shape parameters in eq. (9). The nonnegative equilibrium value $\bar{r}_\rho = \bar{r}_\lambda = \bar{r}$ is a measure of the dipole deformation in the ground state condensate. For the special case of $\bar{r} = 0$ we recover the spherical condensate of the quark potential model.

The equilibrium configuration of three identical quarks in a baryon is represented in the $U(7)$ model by an intrinsic state, eq. (8), composed of N condensate bosons with $r_\rho = r_\lambda$ and $\theta = \pi/2$:

$$b_c^\dagger = (1 + R^2)^{-1/2} \left\{ s^\dagger + R \frac{1}{\sqrt{2}} \left[p_{\rho,0}^\dagger - \frac{1}{\sqrt{2}} (p_{\lambda,1}^\dagger - p_{\lambda,-1}^\dagger) \right] \right\} . \tag{11}$$

It corresponds geometrically to an equilateral triangular shape with the y-axis as a threefold symmetry axis. Excitations of the equilibrium shape are represented by the following six deformed bosons which, together with $b_c^\dagger$, form a complete orthonormal basis,

$$b_u^\dagger = (1 + R^2)^{-1/2} \left\{ -R\, s^\dagger + \frac{1}{\sqrt{2}} \left[p_{\rho,0}^\dagger - \frac{1}{\sqrt{2}} (p_{\lambda,1}^\dagger - p_{\lambda,-1}^\dagger) \right] \right\} ,$$

$$b_v^\dagger = \frac{1}{\sqrt{2}} \left[p_{\rho,0}^\dagger + \frac{1}{\sqrt{2}} (p_{\lambda,1}^\dagger - p_{\lambda,-1}^\dagger) \right] ,$$

$$b_w^\dagger = \frac{1}{\sqrt{2}} \left[-\frac{1}{\sqrt{2}} (p_{\rho,1}^\dagger - p_{\rho,-1}^\dagger) + p_{\lambda,0}^\dagger \right] ,$$

$$b_x^\dagger = \frac{1}{\sqrt{2}} \left(p_{\rho,1}^\dagger + p_{\rho,-1}^\dagger \right) ,$$

$$b_y^\dagger = \frac{1}{\sqrt{2}} \left[\frac{1}{\sqrt{2}} (p_{\rho,1}^\dagger - p_{\rho,-1}^\dagger) + p_{\lambda,0}^\dagger \right] ,$$

$$b_z^\dagger = \frac{1}{\sqrt{2}} \left(p_{\lambda,1}^\dagger + p_{\lambda,-1}^\dagger \right) . \tag{12}$$

The above boson operators represent excitations of the condensate which involve three vibrational intrinsic modes and three rotational Goldstone modes. The vibrational modes are a symmetric radial mode ($b_u^\dagger$), an antisymmetric radial mode ($b_v^\dagger$) and an angular mode ($b_w^\dagger$). The boson operators $b_i^\dagger$ ($i = x, y, z$) are associated with rotational modes. They are obtained by applying the three components of the angular momentum operator, $\hat{L}_i$ ($i = x, y, z$), on the condensate (8) with the shape parameters of eq. (10). The physical interpretation of these excitation modes are discussed in more detail in the following sections.

The vibrational and rotational excitations of a S_3-invariant $U(7)$ mass operator can be studied by decomposing it into an intrinsic (vibrational) and a collective (rotational) part,

$$\hat{M}_{U(7)}^2 = \hat{M}_{int}^2 + \hat{M}_{coll}^2 . \tag{13}$$

This resolution is obtained [14] by requiring that the intrinsic part annihilates the ground state condensate and has the same shape of the potential function as the original mass operator. The collective part of the mass operator has by construction a completely flat potential function. We now discuss each of these parts separately.

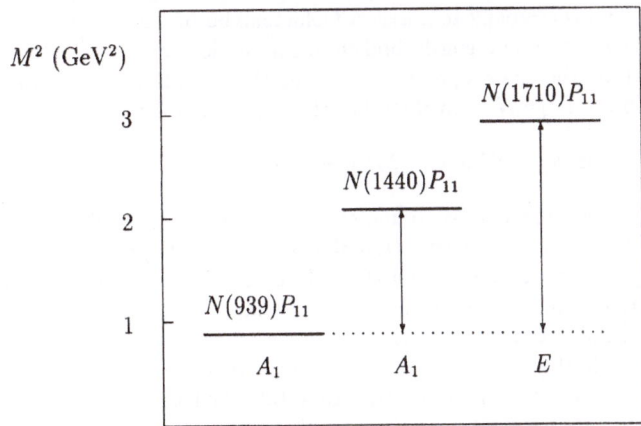

Figure 1. Schematic representation of the vibrational spectrum of nucleon resonances. The resonances are labeled by the usual spectroscopic notation [16] and their vibrational permutation symmetry.

VIBRATIONS

The intrinsic part of a S_3-invariant $U(7)$ mass operator that annihilates the condensate with $r_\rho = r_\lambda = \bar{r} \neq 0$ and $\theta = \bar{\theta} = \pi/2$, consists two terms only,

$$
\begin{aligned}
\hat{M}^2_{\text{int}} &= \xi_1 \left(R^2 s^\dagger s^\dagger - p^\dagger_\rho \cdot p^\dagger_\rho - p^\dagger_\lambda \cdot p^\dagger_\lambda \right) \left(R^2 ss - \tilde{p}_\rho \cdot \tilde{p}_\rho - \tilde{p}_\lambda \cdot \tilde{p}_\lambda \right) \\
&+ \xi_2 \left[\left(p^\dagger_\rho \cdot p^\dagger_\rho - p^\dagger_\lambda \cdot p^\dagger_\lambda \right) \left(\tilde{p}_\rho \cdot \tilde{p}_\rho - \tilde{p}_\lambda \cdot \tilde{p}_\lambda \right) + 4 \left(p^\dagger_\rho \cdot p^\dagger_\lambda \right) \left(\tilde{p}_\lambda \cdot \tilde{p}_\rho \right) \right] , \quad (14)
\end{aligned}
$$

with $R^2 = 2\bar{r}^2$. The vibrational excitations can be obtained from a normal mode analysis of the intrinsic part of the mass operator. This is done [14] by rewriting the mass operator in terms of the deformed bosons of eqs. (11,12), replacing the condensate bosons, b_c and $b^\dagger_c$, by their classical mean field value, $\sqrt{N}$, and keeping terms of leading order in N. In the large N limit the normal modes are diagonal in the deformed boson basis and the mass operator reduces to a harmonic oscillator form,

$$
\frac{1}{N} \hat{M}^2_{\text{int}} = \lambda_1 b^\dagger_u b_u + \lambda_2 \left(b^\dagger_v b_v + b^\dagger_w b_w \right) + \mathcal{O}(1/\sqrt{N}) , \quad (15)
$$

with eigenvalues

$$
\begin{aligned}
\lambda_1 &= 4\xi_1 R^2 , \\
\lambda_2 &= 4\xi_2 R^2/(1 + R^2) . \quad (16)
\end{aligned}
$$

Eq. (15) identifies the deformed bosons that correspond to the three fundamental vibrations: a symmetric stretching (u), an antisymmetric stretching (v) and a bending vibration (w). The first two are radial excitations, whereas the third is an angular mode which corresponds to oscillations in the angle θ between the two Jacobi coordinates. The angular mode is degenerate with the antisymmetric radial mode (with eigenvalue λ_2). This is in agreement with the point group classification of the fundamental vibrations for a symmetric X_3 configuration [15]. This shows that $\hat{M}^2_{\text{int}}$ describes the vibrational excitations of an oblate symmetric top. The rotational bosons, $b^\dagger_i$ ($i = x, y, z$), are missing from eq. (15) and correspond to massless Goldstone bosons associated with the rotation symmetry which is spontaneously broken in the condensate.

The intrinsic part of the mass operator describes the vibrational excitations of baryon resonances. In the large N limit the vibrational spectrum is harmonic,

$$M^2_{\text{vib}} = N \left[\lambda_1 \, n_u + \lambda_2 \, (n_v + n_w) \right] . \tag{17}$$

In the nucleon sector, the vibrationless ground state ($n_u = n_v = n_w = 0$) is identified with the nucleon $N(939)$. The symmetric stretching vibration with frequency, λ_1, has the same symmetry properties as the ground state. It is therefore possible to associate the vibration ($n_u = 1, n_v + n_w = 0$) with the Roper resonance $N(1440)$. The $N(1710)$ resonance could be a candidate for the other (two-dimensional) vibrational mode with ($n_u = 0, n_v + n_w = 1$). Similar considerations apply to the delta sector.

A great advantage of this analysis is that we have established a relation between the parameters of otherwise abstract 'algebraic' interactions, and measured experimental quantities, in this case the mass of specific resonances. The vibrational contribution to the masses, eq. (17), is obtained in the large N approximation which is expected to be valid in view of confinement. The extracted value of the parameters, ξ_1 and ξ_2, can be used as a good starting value in a more detailed fitting procedure in an exact numerical calculation. In figure 1 we show a schematic representation of the vibrational spectrum of nonstrange baryon resonances.

ROTATIONS

On top of each vibrational excitation there is a whole series of rotational states. In a geometric description the rotational excitations are labeled by the angular momentum, L, its projection on the threefold symmetry axis, $K = 0, 1, \ldots$, parity, and the transformation character under the permutation group. For a given value of K, the states have angular momentum $L = K, K+1, \ldots$, and parity $\pi = (-)^K$. For rotational states built on vibrations of type A_1 (symmetric under S_3) each L state is single for $K = 0$ and twofold degenerate for $K \neq 0$. For rotational states built on vibrations of type E (mixed S_3 symmetry) each L state is twofold degenerate for $K = 0$ and fourfold degenerate for $K \neq 0$. The transformation property of the states under the permutation group is found by multiplying the symmetry character of the vibrational and the rotational wave functions. The results are shown schematically in figure 2.

The collective part of the full mass operator contains the contribution of rotations to the mass. By construction $\hat{M}^2_{\text{coll}}$ consists of interaction terms which do not affect the shape of the potential function. Apart from the one-body term, $\hat{n}$, whose contribution to the mass is negligible for large N, the collective part of the mass operator can be expressed in terms of the Casimir operators of the following group chain,

$$U(7) \supset SO(7) \supset SO(6) \supset SO(3) \otimes SO(2) . \tag{18}$$

The bosonic Casimir invariants of $U(7)$ involve only the total number operator, $\hat{N} = \hat{n}_s + \hat{n}$, which is a conserved quantity. The only terms in the collective part of the mass operator that contribute to the excitation spectrum are therefore,

$$\hat{M}^2_{\text{coll}} = \kappa_1 \, \hat{C}_{SO(7)} + \kappa_2 \, \hat{C}_{SO(6)} + \kappa_3 \, \hat{C}_{SO(3)} + \kappa_4 \, [\hat{C}_{SO(2)}]^2 . \tag{19}$$

Explicit expressions for the Casimir operators ($\hat{C}_G$) in terms of the generators of the groups (G) are shown in table 2.

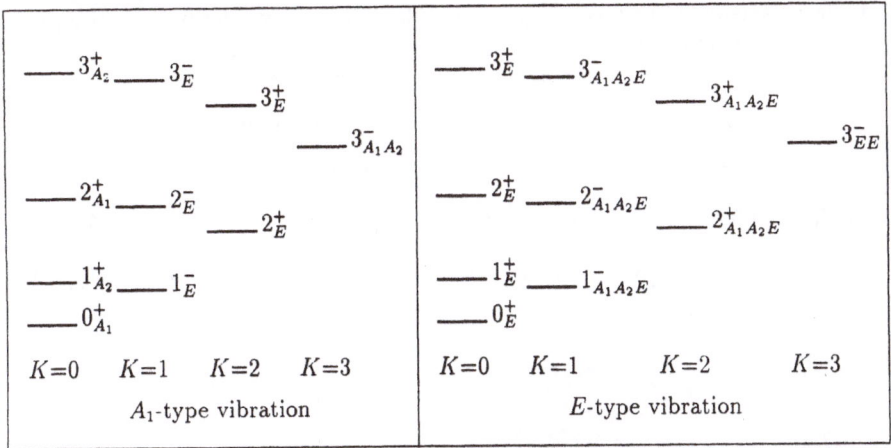

Figure 2. Schematic representation of the rotational structure built on vibrations of type A_1 (lefthand side) and vibrations of type E (righthand side). The levels are labeled by K, L_t^τ, where t denotes the overall (vibrational plus rotational) permutation symmetry. Each E state is doubly degenerate.

The last two terms in the collective part of the mass operator (19) commute with any S_3-invariant mass operator and thus correspond to *exact* symmetries. Their eigenvalues are similar in form to those of a symmetric top, $\kappa_3 L(L+1) + \kappa_4 K_y^2$. For the lowest eigenstates of each L the value of K_y coincides with that of the projection of the angular momentum on the threefold symmetry axis, denoted previously by K. Since L and K_y are always good quantum numbers, it is straightforward to include in the collective part of the mass operator higher orders of the corresponding Casimir operators. Generalizing the mass operator to contain arbitrary functions, $f(\hat{L} \cdot \hat{L})$ and $g(\hat{K}_y^2)$, the expression for the resulting rotational spectrum becomes $f[L(L+1)] + g[K_y^2]$.

In general the first two terms in eq. (19) do not correspond to exact symmetries of a S_3-invariant mass operator and do not commute with the intrinsic part of the mass operator, eq. (14). Therefore, in addition to shifting and splitting the bands generated by $\hat{M}_{\rm int}^2$, they can also mix them. Their effect on the spectrum can be studied numerically. However, just as was done for the intrinsic part of the mass operator, we can gain physical insight of the different rotational terms in the collective part of the mass operator, by studying the large N limit,

$$\frac{1}{N} M_{\rm coll}^2 = -\eta_1 (b_u^\dagger - b_u)^2 - \eta_2 \left[(b_v^\dagger - b_v)^2 + (b_w^\dagger - b_w)^2 \right]$$
$$+\eta_3 \left[(b_x^\dagger + b_x)^2 + (b_z^\dagger + b_z)^2 \right] - \eta_4 (b_y^\dagger - b_y)^2 + \mathcal{O}(1/\sqrt{N}) , \quad (20)$$

with

$$\begin{aligned}
\eta_1 &= \kappa_1 , \\
\eta_2 &= (1 + R^2)^{-1}[\kappa_1 + \kappa_2 R^2] , \\
\eta_3 &= (1 + R^2)^{-1}[\kappa_1 + (\kappa_2 + \tfrac{1}{2}\kappa_3)R^2] , \\
\eta_4 &= (1 + R^2)^{-1}[\kappa_1 + (\kappa_2 + \kappa_3 + \kappa_4)R^2] .
\end{aligned} \quad (21)$$

The b_x, b_y and b_z bosons are the Goldstone bosons connected with rotations in configuration space, whereas the b_u, b_v and b_w bosons correspond to generalized rotations in

Table 2. Generators and Casimir operators of the groups relevant for the collective part of the mass operator.

Group G	Generators	Casimir operator $\hat{C}_G$
$SO(7)$	$G_S^{(1)}, G_\rho^{(1)}, G_\lambda^{(1)}, G_A^{(0)}, G_A^{(2)}, A_\rho, A_\lambda$	$\hat{C}_{SO(6)} + A_\rho \cdot A_\rho + A_\lambda \cdot A_\lambda$
$SO(6)$	$G_S^{(1)}, G_\rho^{(1)}, G_\lambda^{(1)}, G_A^{(0)}, G_A^{(2)}$	$G_S^{(1)} \cdot G_S^{(1)} + G_\rho^{(1)} \cdot G_\rho^{(1)} + G_\lambda^{(1)} \cdot G_\lambda^{(1)}$ $+ G_A^{(0)} \cdot G_A^{(0)} + G_A^{(2)} \cdot G_A^{(2)}$
$SO(3)$	$\hat{L} = \sqrt{2}\, G_S^{(1)}$	$\hat{L} \cdot \hat{L}$
$SO(2)$	$\hat{K}_y = -\sqrt{3}\, G_A^{(0)}$	$\hat{K}_y$

a higher dimensional space. It is seen from eqs. (20,21) that the κ_1-term of eq. (19) is associated with rotations in a six-dimensional space (u, v, w, x, y, z) space and the κ_2-term with rotations in a five-dimensional subspace (v, w, x, y, z). The κ_3-term, $\hat{L} \cdot \hat{L}$, corresponds to ordinary three-dimensional (x, y, z) rotations and the κ_4-term, $\hat{K}_y^2$, is associated with a rotation about the threefold symmetry y-axis. The equality of the coefficients of the (v, w) and the (x, z) terms in eq. (20) is a reflection of the oblate-top nature of the S_3-invariant mass operator.

Summarizing, we have shown that a S_3-invariant $U(7)$ mass operator corresponds geometrically to rotations and vibrations of an equilateral equilibrium configuration of three quarks in a baryon. There are only six independent terms in the mass operator that determine the excitation spectrum: two for the vibrational excitations and four for the rotational excitations. The two vibrational terms cluster the states into bands. Two of the rotational terms correspond to exact symmetries of the mass operator and their eigenvalues are obtained in closed form. Whereas these terms are diagonal and hence only cause band splitting, the remaining two rotational terms may in addition cause band mixing.

BARYON RESONANCES

In the previous sections we introduced a $U(7)$ spectrum generating algebra for the spatial part of the baryon wave function. The total wave function for baryon resonances consists of a spatial, a spin-flavor and a color part,

$$\psi = \psi_L\, \phi_{sf}\, \psi_c \ . \tag{22}$$

To satisfy the Pauli principle for a system of identical quarks these parts have to be combined so that the total wave function is antisymmetric.

Both the spatial and the spin-flavor degrees of freedom contribute to the mass operator. In principle there could be also spin-flavor dependence in the coefficients of the interaction terms in $\hat{M}_{\text{space}}^2$. Although these effects can be included without difficulty in an algebraic approach, in the present study we do not take them into account and write

$$\hat{M}^2 = \hat{M}^2_{\text{space}} + \hat{M}^2_{\text{spin-flavor}} \cdot \tag{23}$$

For the spin-flavor part we consider the $SU(6) \supset SU(3) \otimes SU(2)$ dynamic symmetry of Gürsey and Radicati [17]. In general, this symmetry may be broken (as is the case with the hyperfine interaction in the quark potential model), which would lead to coupling terms in eq. (23). Here, for simplicity, we limit ourselves to a diagonal breaking for which the eigenvalues are given in closed form,

$$M^2_{\text{spin-flavor}} = a\left[\langle\hat{C}_{SU(6)}\rangle - 45\right] + b\left[\langle\hat{C}_{SU(3)}\rangle - 9\right] + c\,S(S+1) \,. \tag{24}$$

The first term involves the Casimir operator of the $SU(6)$ spin-flavor group with eigenvalues 45, 33 and 21 for the representations $56 \leftrightarrow A_1$, $70 \leftrightarrow E$ and $20 \leftrightarrow A_2$, respectively. The second term involves the Casimir invariant of the $SU(3)$ flavor group with eigenvalues 9 and 18 for the octet and decuplet, respectively. The last term contains the eigenvalues $S(S+1)$ of the total spin operator.

For the spatial part of the mass operator we take the S_3 invariant $U(7)$ mass operator discussed in the previous section which is decomposed into an intrinsic and a collective part

$$\hat{M}^2_{\text{space}} = \hat{M}^2_{U(7)} = \hat{M}^2_{\text{int}} + \hat{M}^2_{\text{coll}} \,. \tag{25}$$

Here, for simplicity, we take a simplified form for the collective part of the mass operator, containing only spatial rotations

$$\hat{M}^2_{\text{coll}} \rightarrow \hat{M}^2_{\text{rot}} = \alpha\sqrt{\hat{L}\cdot\hat{L} + \frac{1}{4}} \,, \tag{26}$$

with eigenvalues

$$M^2_{\text{rot}} = \alpha\,(L + 1/2) \,. \tag{27}$$

The total mass operator can now be diagonalized numerically to get a fit for the masses of nonstrange baryon resonances. Instead, for the vibrational part we use the large N expression in eq. (17) with $R^2 = 1$, and obtain the following analytic expression for the masses,

$$M^2 = M^2_0 + M^2_{\text{vib}} + M^2_{\text{rot}} + M^2_{\text{spin-flavor}} \cdot \tag{28}$$

Here M^2_0 is a constant and the other contributions are given by eqs. (17,24,27).

For the nucleon which is a member of the flavor octet the relevant mass formulas read

$$\begin{aligned}
M^2_N(A_1, L, S = 1/2) - M^2_{N(939)} &= M^2_{\text{vib}} + \alpha\,L \,, \\
M^2_N(E, L, S = 1/2) - M^2_{N(939)} &= M^2_{\text{vib}} + \alpha\,L - 12a \,, \\
M^2_N(E, L, S = 3/2) - M^2_{N(939)} &= M^2_{\text{vib}} + \alpha\,L - 12a + 3c \,, \\
M^2_N(A_2, L, S = 1/2) - M^2_{N(939)} &= M^2_{\text{vib}} + \alpha\,L - 24a \,.
\end{aligned} \tag{29}$$

Here $M^2_{N(939)} = 0.882$ GeV$^2 = M^2_0 + \alpha/2 + 3c/4$. For the delta which is a member of the flavor decuplet the relevant mass formulas read

$$\begin{aligned}
M^2_\Delta(A_1, L, S = 3/2) - M^2_{\Delta(1232)} &= M^2_{\text{vib}} + \alpha\,L \,, \\
M^2_\Delta(E, L, S = 1/2) - M^2_{\Delta(1232)} &= M^2_{\text{vib}} + \alpha\,L - 12a - 3c \,,
\end{aligned} \tag{30}$$

Table 3. Oblate top classification of nonstrange baryons of the N family with $I = 1/2$. Here t denotes the overall permutation symmetry. The last column lists the dominant representation of the quark model assignment in a $SU_{sf}(6) \otimes O(3)$ basis [16]. M^2 is given in GeV^2. The experimental values are taken from [16].

Mass	Status	M^2_{exp}	J^π	L^π, K	S	t	M^2_{calc}	% Error	$(D, L^\pi_n)S$
$N(939)P_{11}$	****	0.882	$\frac{1}{2}^+$	$0^+, 0$	$\frac{1}{2}$	A_1	0.882	0	$(56, 0^+_0)\frac{1}{2}$
$N(1440)P_{11}$	****	2.074	$\frac{1}{2}^+$	$0^+, 0$	$\frac{1}{2}$	A_1	2.074	0	$(56, 0^+_2)\frac{1}{2}$
$N(1520)D_{13}$	****	2.310	$\frac{3}{2}^-$	$1^-, 1$	$\frac{1}{2}$	E	2.442	-5.7	$(70, 1^-_1)\frac{1}{2}$
$N(1535)S_{11}$	****	2.356	$\frac{1}{2}^-$	$1^-, 1$	$\frac{1}{2}$	E	2.442	-3.6	$(70, 1^-_1)\frac{1}{2}$
$N(1650)S_{11}$	****	2.772	$\frac{1}{2}^-$	$1^-, 1$	$\frac{3}{2}$	E	2.817	-1.6	$(70, 1^-_1)\frac{3}{2}$
$N(1675)D_{15}$	****	2.806	$\frac{5}{2}^-$	$1^-, 1$	$\frac{3}{2}$	E	2.817	-0.4	$(70, 1^-_1)\frac{3}{2}$
$N(1680)F_{15}$	****	2.822	$\frac{5}{2}^+$	$2^+, 0$	$\frac{1}{2}$	A_1	2.994	-6.0	$(56, 2^+_2)\frac{1}{2}$
$N(1700)D_{13}$	***	2.890	$\frac{3}{2}^-$	$1^-, 1$	$\frac{3}{2}$	E	2.817	-2.5	$(70, 1^-_1)\frac{3}{2}$
$N(1710)P_{11}$	***	2.924	$\frac{1}{2}^+$	$0^+, 0$	$\frac{1}{2}$	E	2.924	0	$(70, 0^+_2)\frac{1}{2}$
$N(1720)P_{13}$	****	2.958	$\frac{3}{2}^+$	$2^+, 0$	$\frac{1}{2}$	A_1	2.994	-1.2	$(56, 2^+_2)\frac{1}{2}$
$N(2190)G_{17}$	****	4.796	$\frac{7}{2}^-$	$3^-, 1$	$\frac{1}{2}$	E	4.554	5.0	$(70, 3^-_3)\frac{1}{2}$
					$\frac{3}{2}$	E	4.929	-2.8	
$N(2220)H_{19}$	****	4.928	$\frac{9}{2}^+$	$4^+, 0$	$\frac{1}{2}$	A_1	5.106	-3.6	$(56, 4^+_4)\frac{1}{2}$
$N(2250)G_{19}$	****	5.063	$\frac{9}{2}^-$	$3^-, 1$	$\frac{3}{2}$	E	4.929	2.6	$(70, 3^-_3)\frac{3}{2}$
$N(2600)I_{1,11}$	***	6.760	$\frac{11}{2}^-$	$5^-, 1/5$	$\frac{1}{2}$	E	6.666	1.4	

and $M^2_{\Delta(1232)} = 1.518$ GeV^2 = $M^2_{N(939)} + 9b + 3c$. The parameters M^2_0, b, $\xi_1 N$ and $\xi_2 N$ are determined from the mass squared of $N(939)$, $\Delta(1232)$, $N(1440)$ and $N(1710)$ respectively. The remaining parameters a, c, and α are determined by fitting the mass of the baryon resonances with *** or **** status. The values of the parameters (in GeV^2) extracted in a least square fit are

$$M^2_0 = 0.260 , \quad \xi_1 N = 0.298 , \quad \xi_2 N = 0.769 , \quad \alpha = 1.056 ,$$
$$a = -0.042 , \quad b = 0.029 , \quad c = 0.125 . \tag{31}$$

Table 4. Oblate top classification of nonstrange baryons of the Δ family with $I = 3/2$. For further information see table 3.

Mass	Status	$M^2_{\rm exp}$	J^π	L^π, K	S	t	$M^2_{\rm calc}$	% Error	$(D, L^\pi_n)S$
$\Delta(1232)P_{33}$	****	1.518	$\frac{3}{2}^+$	$0^+, 0$	$\frac{3}{2}$	A_1	1.518	0	$(56, 0^+_0)\frac{3}{2}$
$\Delta(1620)S_{31}$	****	2.624	$\frac{1}{2}^-$	$1^-, 1$	$\frac{1}{2}$	E	2.703	-3.0	$(70, 1^-_1)\frac{1}{2}$
$\Delta(1700)D_{33}$	****	2.890	$\frac{3}{2}^-$	$1^-, 1$	$\frac{1}{2}$	E	2.703	6.5	$(70, 1^-_1)\frac{1}{2}$
$\Delta(1900)S_{31}$	***	3.610	$\frac{1}{2}^-$	$1^-, 1$	$\frac{1}{2}$	E	3.895	-7.9	
$\Delta(1905)F_{35}$	****	3.629	$\frac{5}{2}^+$	$2^+, 0$	$\frac{3}{2}$	A_1	3.630	-0.03	$(56, 2^+_2)\frac{3}{2}$
$\Delta(1910)P_{31}$	****	3.648	$\frac{1}{2}^+$	$2^+, 0$	$\frac{3}{2}$	A_1	3.630	0.5	$(56, 2^+_2)\frac{3}{2}$
$\Delta(1920)P_{33}$	***	3.686	$\frac{3}{2}^+$	$2^+, 0$	$\frac{3}{2}$	A_1	3.630	1.5	$(56, 2^+_2)\frac{3}{2}$
$\Delta(1930)D_{35}$	***	3.725	$\frac{5}{2}^-$	$2^-, 1$	$\frac{1}{2}$	E	3.759	-0.9	
$\Delta(1950)F_{37}$	****	3.803	$\frac{7}{2}^+$	$2^+, 0$	$\frac{3}{2}$	A_1	3.630	4.5	$(56, 2^+_2)\frac{3}{2}$
$\Delta(2420)H_{3,11}$	****	5.856	$\frac{11}{2}^+$	$4^+, 0$	$\frac{3}{2}$	A_1	5.742	1.9	$(56, 4^+_4)\frac{3}{2}$

In table 3 and 4 we show the fit to the nonstrange baryon masses of the nucleon and the delta family, respectively. These results obtained in the large N limit are substantiated by exact numerical calculations. We find a reasonable overall fit for the *** and **** resonances with an r.m.s. deviation of $\delta_{\rm rms} = 0.14$ GeV^2. Some characteristic features of the fit are:

(i) All experimentally well established resonances are reproduced by the calculations. Experimentally known but uncertain resonances with ** status can be accommodated in the fit as well. For example, for the $\Delta(1600)P_{33}$ resonance with $M^2_{\rm exp} = 2.56$ GeV^2 we find $M^2_{\rm calc} = 2.71$ GeV^2.

(ii) In the present calculation most states listed in the tables are rotational members of the ground band ($n_u = 0, n_v + n_w = 0$). We have associated $N(1440)$ and $\Delta(1900)$ with A_1 vibrational bands ($n_u = 1, n_v + n_w = 0$), and $N(1710)$ with the E vibrational band ($n_u = 0, n_v + n_w = 1$).

(iii) The oblate top assignments of orbital angular momentum, spin and permutation symmetry are similar to the quark model assignment.

(iv) The low-lying nucleon resonances, P_{11}, D_{13}, S_{11}, the cluster of $I = 1/2$ resonances in the mass range 1.6-1.7 GeV and the cluster of $I = 3/2$ resonances near 1.9 GeV are well reproduced. Above this mass range there are many more resonances predicted than observed experimentally. This is due to the fact that we have associated the low-lying $N(1440)$ and $N(1710)$ resonances with vibrational bandheads. Consequently the rotational states built on top of them occur low in the mass spectrum. This problem of missing resonances is known to exist in quark potential models as well.

One possible explanation is that the 'missing' resonances indeed do exist, but that they cannot be resolved individually since there are many overlapping resonances in that mass region. Another explanation could be that these resonances are decoupled from the πN channel [18]. Since at present most experimental information is from pion scattering, they could simply not be excited in this type of experiments. Future experiments with electromagnetic probes may shed more light on this question. It has also been suggested that the Roper and the $N(1710)$ are hybrid states [19]. If that were the case these resonances are outside the $U(7)$ model space and therefore the coefficients, ξ_1 and ξ_2, in the intrinsic part of the mass operator cannot be determined from the present data. A large value of ξ_1, ξ_2 would shift the vibrational bands up in mass, without affecting the rotational members of the ground band.

Finally, we note that in the $U(7)$ model we have obtained a fit to the nonstrange baryon masses which is comparable to that in quark potential models [6, 7], although the underlying quark dynamics is quite different. This shows that the masses alone are not sufficient to distinguish between different forms of quark dynamics, *e.g.* single-particle *vs.* collective motion.

SUMMARY

In this contribution we have proposed to use $U(7)$ as a spectrum generating algebra for a geometry-oriented description of baryons. It is combined with the spin-flavor and color parts into a $U(7) \otimes SU_{sf}(6) \otimes SU_c(3)$ SGA for baryon spectroscopy. Although we have limited the discussion to nonstrange baryons, the $U(7)$ model can accommodate strange baryons as well. The present model allows one to study both collective-like and single-particle-like motion in a single algebraic framework. Collectivity corresponds to coherent and strongly correlated motion of quarks which requires a strong coupling of harmonic oscillator shells. We have shown explicitly that there exists $U(7)$ mass operators that strongly mix states with different oscillator quanta, but still preserve the permutation symmetry.

We have applied the model to the family of nucleon and delta resonances and found good overall agreement with the observed masses. The fit is of comparable quality to that obtained in quark potential models. In addition to mass spectra, the model provides wave functions which can be used to calculate other observables such as helicity amplitudes and (transition) form factors. These quantities provide a far more sensitive test to details in the wave functions than the mass spectrum. We are in the process of examining a variety of such observables in the $U(7)$ model. Our goal is to identify signatures which may distinguish single-particle from collective aspects of quark dynamics in baryons, as well as to provide guidance to future experiments.

ACKNOWLEDGEMENTS

On the occasion of his 50th birthday, it is gratifying and most proper to dedicate this contribution to F. Iachello who is actively involved in advancing the methods and ideas reported in this work.

REFERENCES

[1] See *e.g.* M. Gell-Mann and Y. Ne'eman, 'The eightfold way', W.A. Benjamin (1964).

[2] S. Okubo, Progr. Theor. Phys. **27**, 949 (1962).

[3] F. Iachello, Nucl. Phys. **A497**, 23c (1989); *ibid.* **A518**, 173 (1990); *ibid.* Proc. of the '4th workshop on perspectives in nuclear physics at intermediate energies', S. Boffi, C. Ciofi degli Atti and M. Giannini, eds., World Scientific, Singapore (1989), page 17.

[4] F. Iachello and D. Kusnezov, Phys. Lett. **255B**, 493 (1991); *ibid.* Phys. Rev. **D45**, 4156 (1992).

[5] F. Iachello, N.C. Mukhopadhyay and L. Zhang, Phys. Lett. **256B**, 295 (1991); *ibid.* Phys. Rev. **D44**, 898 (1991).

[6] N. Isgur and G. Karl, Phys. Rev. **18**, 4187 (1978); *ibid.* **D19**, 2653 (1979); *ibid.* **D20**, 1191 (1979); K.-T. Chao, N. Isgur and G. Karl, Phys. Rev. **D23**, 155 (1981).

[7] S. Capstick and N. Isgur, Phys. Rev. **D34**, 2809 (1986).

[8] R. Bijker, A.E.L. Dieperink and A. Leviatan, these proceedings.

[9] M. Moshinsky, in 'The Harmonic Oscillator in Modern Physics: From Atoms to Quarks', Gordon and Breach, 1969 and references therein.

[10] A. Arima and F. Iachello, Phys. Rev. Lett. **35**, 1069 (1975); F. Iachello and A. Arima, 'The Interacting Boson Model' (Cambridge Univ. Press, Cambridge, 1987).

[11] F. Iachello, Chem. Phys. Lett. **78**, 581 (1981).

[12] K.C. Bowler, P.J. Corvi, A.J.G. Hey, P.D. Jarvis and R.C. King, Phys. Rev. **D24**, 197 (1981); A.J.G. Hey and R.L. Kelly, Phys. Rep. **96**, 71 (1983).

[13] R. Gilmore and D.H. Feng, Nucl. Phys. **A301**, 189 (1978).

[14] A. Leviatan and M.W. Kirson, Ann. Phys. **188**, 142 (1988); A. Leviatan, J. Chem. Phys. **91**, 1706 (1989).

[15] G. Herzberg, 'Molecular Spectra and Molecular Structure II: Infrared and Raman Spectra of Polyatomic Molecules' (Van Nostrand Rheinhold, Princeton, NJ 1945).

[16] Particle Data Group, Phys. Rev. **D45**, S1 (1992).

[17] F. Gürsey and L.A. Radicati, Phys. Rev. Lett. **13**, 173 (1964).

[18] R. Koniuk and N. Isgur, Phys. Rev. **D21**, 1868 (1980).

[19] Z. Li, Phys. Rev. **D44**, 2841 (1991); Z. Li, V. Burkert and Z. Li, Phys. Rev. **D46**, 70 (1992).

CAN IACHELLO'S IDEA OF A SPECTRAL SUPERSYMMETRY BE EXTENDED INTO THE RELATIVISTIC DOMAIN?

Arno Bohm and L.C. Biedenharn

Center for Particle Physics
Department of Physics
University of Texas at Austin
Austin, TX 78712

Dedicated to Francesco Iachello

INTRODUCTION

Supersymmetry is a symmetry structure relating fermions and bosons, but this general structure can take different forms. *Spacetime supersymmetry*, realized by Z_2-graded Lie algebras, is a fundamentally new, and fruitful, symmetry that has been introduced in quantum field theory[1,2]. Because this new symmetry structure could resolve a number of basic field-theoretic difficulties (tachyons, convergence, anomalies,...) it soon became the basis for present supergravity theory and quantum string theory. Spacetime supersymmetry implies strong constraints on particle spectra: *every fermionic state must have a bosonic partner.* To date there is unfortunately no experimental evidence whatsoever that such a supersymmetric structure is realized in the elementary particle domain for which it was devised.

For composite systems, for which the elementary constituents are particles with half-integer spin, a less tightly constrained form of supersymmetry[3] can be realized, since the spectra of such composite fermionic (bosonic) systems with odd (even) number of constituents are related. There are many physical systems that occur in 'pairs' in which one partner differs from the other by an additional fermion. If each of the partners is described by an irrep of a spectrum generating group (SGG)-also called dynamical symmetry-then the spectrum of the pair combines into an irrep of a spectrum generating supergroup (SGSG), or dynamical supersymmetry. This observation led Francesco Iachello in 1980 to the introduction of *spectral supergroups* into nuclear physics[4]: the additional fermion (nucleon) of an odd A nucleus which is added to an even-even nuclear core is expected to couple weakly. This results in similar level structure and level spacing for this pair of odd A and even A nuclei. If one has a collective model for the structure of such pairs of nuclei then the spectra can be described by the spectral supergroups based[5] on U(6/N). There appears to be experimental evidence for the occurrence of these spectral supergroups in nuclear physics[6].

We wish to investigate, in our contribution to this colloquium dedicated to Francesco Iachello, whether or not similar ideas can be applied more generally. The occurrence of

Symmetries in Science VII, Edited by B. Gruber
and T. Otsuka, Plenum Press, New York, 1994

spectral supersymmetries in atomic and molecular physics is discussed in another article in this colloquium volume[7]. We shall address the question whether or not these ideas can be extended into the *relativistic domain* and applied to hadrons.

There is a strong structural similarity between nuclei and hadrons: both can be considered as composite systems built up from more elementary constituents. The nucleus consists mainly of neutrons and protons for which non-relativistic dynamics is a reasonable first approximation. Even with this simplification the solution of the many body problem is complicated; one therefore resorts to collective models, principally oscillators and rotators. In the collective model of Bohr-Mottelson[8] the elementary modes are rotations, β-vibrations and γ-vibrations. Thus one has on the one extreme the constituents (neutrons and protons) as 'parts' and on the other extreme the collective motions as 'parts'. For many nuclei the truth lies between these two extremes and is a combination of the constituent picture and the collective motion picture. The 'parts' can in general be described by different subspaces of an irreducible representation space of a spectrum generating group. These different subspaces are provided by the three different reduction chains of the SGG U(6) in the interacting boson model[9] (IBM), which is, in consequence, the bosonic part of the spectrum supergroup U(6/N) mentioned above.

For hadrons the basic structure is similarly composite: the constituents are quarks and the microscopic theory is QCD. Perturbative QCD has had impressive success for hard processes (for example, jets in e+e- annihilation) but in most cases the results are uncertain due to renormalization scheme dependence and the presence of higher twists at present energies. Lattice QCD provides a computational scheme for the hadron spectrum, but the results are so far semi-quantitative (and dependent upon the quark masses). The direct use of QCD to determine hadronic spectra is a theoretical task of daunting complexity – the already difficult many-body problem of nuclear physics is for hadrons inherently relativistic (and thus in effect having unlimitedly many particles). In such a case it appears even more reasonable to model the problem as a collective one, in which the elementary constituents are replaced by the lowest collective modes of an extended relativistic system.

One can use standard models to devise an intuitive classical picture which can serve as the starting point for such a collective quantum theoretical model of hadrons. In QCD hadrons are understood as color singlet bound states of quarks with the force between the quarks arising from the exchange of gluons. The classical collective model for hadrons is accordingly a flux tube connecting an antiquark q̄ and quark q for *mesons*, or a diquark[10] (qq) and a quark q for baryons. (The diquark-quark structure for hadrons can be deduced from the observed spectra: there is only a *single* orbital angular momentum evident in these spectra, not *two* as a three body q-q-q picture would entail.)

This string of flux can perform various kinds of intrinsic motions of which a rotating rigid rod, or one-dimensional oscillations are useful special cases. These particular cases are specified by the choice of the relativistic collective Hamiltonian in the same way as the choice of Hamiltonian in the IBM gives the three reduction chains.

Motions are connected with groups; the motion of the hadron as a whole (center of mass motion) is given by the Poincaré symmetry group, and the intrinsic collective motion is given by the spectrum generating group. We thus have a similar situation for the meson and baryon as we had for the pair of even and odd A nuclei which were combined into a spectrum supermultiplet by Iachello[4]. There is, however, an important difference: for non-relativistic objects the center of mass motion given by the Galilei group can be separated off trivially (direct product) and then ignored. For relativistic extended objects the Poincaré group and the SGG are more intimately connected and the representation space is not given by the direct product. There is good experimental evidence in favor of a relativistic spectral supersymmetry for hadrons: the similarity between the meson and baryon spectra (of the same flavor) determined by the equality of the slope for the experimental meson and baryon Regge trajectories.

RELATIVISTIC SPECTRAL SUPERALGEBRAS

The construction of relativistic spectrum generating superalgebras begins with the prior construction of a relativistic spectrum generating group model of hadrons obeying a mass-spin (Regge trajectory) constraint[11]. If realized in Minkowski space (as opposed to quasi-Newtonian coordinates), with the Poincaré group as the center of mass motion group, then the SGG must contain the group $SO(3,1)_{S_{\mu\nu}}$, generated by the *intrinsic* Lorentz generators $S_{\mu\nu}$ ($\mu, \nu = 0,1,2,3$). Together with the orbital Lorentz generators $L_{\mu\nu}$ of the center of mass motion one forms the *physical* Lorentz generators: $J_{\mu\nu} \equiv L_{\mu\nu} + S_{\mu\nu}$.

Spectrum generating algebras connect states having different spins (aside from the uninteresting case where the spectrum has but a single spin). There are two ways in which this may occur: either (a) by *spinorial* generators which connect integer with half-integer states or (b) by *vectorial* generators which connect integer to integer (or half-integer to half-integer) states.

An example of case (a) occurs if we adjoin to the $SO(3,1)$ generators $S_{\mu\nu}$ the Majorana spinorial operators, Q:

$$Q \equiv \begin{pmatrix} Q_\alpha \\ \bar{Q}^{\dot{\alpha}} \end{pmatrix} = \begin{pmatrix} Q_1 \\ Q_2 \\ Q_2^+ \\ -Q_1^+ \end{pmatrix}. \tag{1}$$

Under the action of the generators $S_{\mu\nu}$, Q transforms as a standard four-dimensional $(0, \frac{1}{2}) \oplus (\frac{1}{2}, 0)$ irrep (spinorial tensor operator).

An example of case (b) occurs if we adjoin to the $SO(3,1)$ generators $S_{\mu\nu}$ the four-vector operator Γ_μ. Under the action of the generators $S_{\mu\nu}$ the operator Γ_μ transforms as a four-dimensional Lorentz irrep $(\frac{1}{2}, \frac{1}{2})$, that is, as a four-vector tensor operator.

To complete the algebraic structure one must specify the commutation relations for the adjoined operators. For the vectorial operator, Γ_μ, one chooses, for example, the commutation relations:

$$[\Gamma_\mu, \Gamma_\nu] = -iS_{\mu\nu}, \tag{2}$$

then $S_{\mu\nu}$ and Γ_μ together form the ten (Hermitian) generators of the non-compact spectrum generating group SO(3,2) with infinite-dimensional irreps.

For spinorial operators there are two fundamentally different choices: one can choose either *commutation* or *anti-commutation* relations for the spinorial operators. If one chooses commutation relations, one obtains a Lie algebra structure (and thus a group), even though the symmetry directly relates integer and half-integer states[11].

In sharp contrast to this, when one chooses anti-commutation relations for the spinorial generators, one obtains a Lie superalgebra. In this case it is proper to speak of *fermionic generators* (spinorial generators with anti-commutation relations) and *bosonic generators* (vectorial operators with commutation relations), which fit together in a Z_2-graded Lie (super) algebra, with graded commutators and a graded Jacobi relation.

(Note that there is no algebraic spin-statistics theorem for operators which would link commutation properties with spinorial/vectorial properties.)

Supersymmetries can have finite or infinite (super) multiplets, depending on the choice of superalgebra. If, for example, one adjoins both Q and Γ_μ to $S_{\mu\nu}$ and chooses the anti-commutation relation:

$$\left\{ Q_\alpha, \bar{Q}^\beta \right\} = i(\sigma^{\mu\gamma})_\alpha{}^\beta S_{\mu\gamma} + (\gamma^\mu)_\alpha{}^\beta \Gamma_\mu, \tag{3}$$

then these fourteen generators form the superalgebra $Osp(1,4) \supset SO(3,2) \supset SO(3,1)$, which describes infinite supermultiplets. (The additional commutation relation $[\Gamma_\mu, Q]$ $\subset Q$, is a consequence of the other defining relations.) We remark for completeness that for *space-time supersymmetry* the fermionic anti-commutation relations yield bosonic Poincaré generators, and, in fact, space-time super-symmetry contains the Poincaré group as a proper sub-(super) group structure, rather than the "intrinsic" $SO(3,2)_{S_{\mu\gamma}\Gamma_\mu}$ in our case.

THE REPRESENTATION OF THE RELATIVISTIC SSG AND THE PREDICTED SPECTRUM

In this brief survey we cannot give a detailed construction of the representations of the "relativistic spectrum supergroups". This requires an explanation of the interplay between the Poincaré group of the center of mass motion and the SG of the intrinsic (collective) motion, which are connected by a constraint equation that follows from a relativistic Hamiltonian (constrained relativistic quantum mechanics) as in the relativistic string theory. Such detailed discussions are given in refs 11 and 15. Here we will just describe the spectrum of the vibration and rotation quantum number (ν, j) which follows from the choice (6) for the reduction chain (5) and quote the mass formula (7) which follows from the relativistic Hamiltonian.

The spectrum supergroup $Osp(1,4)$ is the relativistic analog of the supergroup $Osp(1,2)$. Non-relativistic supersymmetric quantum mechanics is usually formulated in terms of the group chain:[12]

$$SU(1,1/1) \supset Osp(1,2) \supset SO(2,1)_{H,K,D} \supset SO(2)_{\Gamma_0=(H+K)}. \tag{4}$$

It is used for the radial excitations, for example, the radial part of the harmonic oscillator.

The representations $D_S(q_0)$ of $SU(1,1/1)$ which are usually used are the so-called "non-typical" representations which remain irreducible when restricted to $Osp(1,2)$. Thus the reduction chain (4) really starts with $Osp(1,2)$, though one usually likes to adjoin an extra spinor operator S. These irreps of $Osp(1,2)$ (or $SU(1,1/1)$, reduce into the direct sum of $SO(2,1)$ irreps: $D_S(q_0) \Rightarrow D_+(q_0) \oplus D_+(q_0 + \frac{1}{2})$. Each $D_+(q)$ describes a tower of vibrational levels with $\mu =$ eigenvalue $\Gamma_0 = q, q+1, q+2, \cdots$ and $D_S(q_0)$ describes the spectrum of the one-dimensional (radial) super-symmetric oscillator with lowest energy value q_0.

The relativistic spectrum supergroup $Osp(1,4)$ is obtained by enlarging $Osp(1,2)$ with an $SO(3)_{S_{ij}}$ to take the rotational degrees of freedom into account and making it relativistic. This leads to the following supersubgroup chain which is the relativistic analogue of (4):

$$SU(2,2/1) \supset Osp(1,4) \supset SO(3,2)_{\Gamma_\mu S_{\mu\nu}} \supset SO(3)_{S_{ij}} \times SO(2)_{\Gamma_0} \tag{5}$$

(We have written here an SU(2, 2/1) of which one usually uses the so-called "massless" "positive energy" representations which are characterised by one number s_0. But as these irreps remain irreducible when restricted to the supersubgroup $Osp(1,4)$, the spectrum is already obtained from $Osp(1,4)$. We want to ignore the additional spinor operator S and vector and tensor operators and restrict ourselves to the chain (5) starting with $Osp(1,4)$.

The irreps of the superalgebra $Osp(1,4)$ that we are interested in are denoted by $D_S(s_0 + 1, s_0)$[13] and reduce with respect to $SO(3,2)$ into the sum of two irreducible representations:

$$D_S(s_0 + 1, s_0) = D(\mu_{min} = s_0 + 1, j_{min} = s_0) \oplus D(\mu_{min} = s_0 + \tfrac{3}{2}, j_{min} = s_0 + \tfrac{1}{2}). \tag{6}$$

The weight diagram of $D_S(s_0 + 1, s_0)$ is given in Figure 1 below for the case $s_0 = \tfrac{1}{2}$. Here j is the angular momentum (hadron spin) and μ is again the vibrational quantum number[14] = eigenvalue Γ_0. Each dot represents a hadron level with quantum numbers (μ, j).

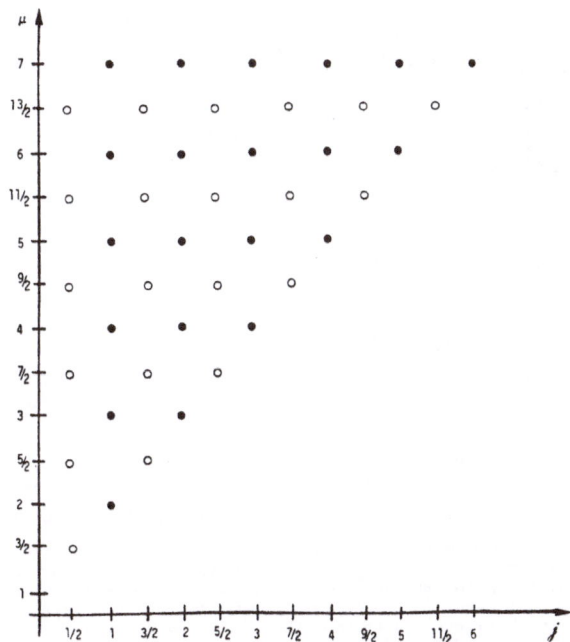

Fig. 1 Weight diagram of the representation
$D_S(3/2, 1/2) = D(3/2, 1/2) \oplus D(2, 1)$. The o represent the weights of the representation $D(3/2, 1/2)$ of $SO(3,2)$ and • represent the weights of the representation $D(2, 1)$ of $S, O(3, 2)$.

The physical interpretation of the spectrum supergroup $Osp(1,4)$ and of it bosonic part $SO(3,2)$ is the following: μ = eigenvalue (Γ_0) and j—with $j(j+1)$ = eigenvalue (S^2)— are the discrete quantum numbers that label the hadron resonances. In the full representation of the quantum relativistic oscillator and rotator away from rest[11,15] μ = eigenvalue $(\Gamma_\mu \hat{P}^\mu)$ and $j(j + 1)$ = eigenvalue $(-\hat{W}_\mu \hat{W}^\mu)$, where $\hat{W}_\mu = \tfrac{1}{2}\varepsilon_{\mu\nu\rho\sigma} \hat{P}^\nu J^{\rho\sigma} = \tfrac{1}{2}\varepsilon_{\mu\nu\rho\sigma} \hat{P}^\nu S^{\rho\sigma}$ and $\hat{P}_\mu = P_\mu/M$. The operators P_μ and $J_{\mu\nu} = M_{\mu\nu} + S_{\mu\nu}$ are the generators of physical Poincaré transformations; the mass operator is $M = (P_\mu P^\mu)^{\frac{1}{2}}$. Therefore the physical interpretation of j is the hadron spin, m^2 = eigenvalue(M^2) is the hadron mass squared, μ is a new principal quantum number, which in the non-relativistic limit goes into the vibrational quantum number.[14] The quantum numbers

μ_{min}, the lowest value of μ and $s = j_{min}$ the lowest value of j characterize the $SO(3,2)$ representation. The irreducible representation chosen in (6) has the property that the maximal compact subgroup $SO(2)_{\Gamma_0} \times SO(3)_{S_{ij}}$ characterized by the pair (μ, j) occurs at most once. Physically this means that μ and j are a sufficient set of quantum numbers to label the hadrons in one super-tower. The quantum number $s = j_{min}$ can in the non-relativistic correspondence be interpreted as the total quark spin. As $s = 1$ for ρ; ω etc. and $s = 1/2$ for N the representation $D(2,1)$ in Fig.1 should describe the vector mesons and the representation $D(3/2, 1/2)$ in Fig.1 should describe the nucleon tower.

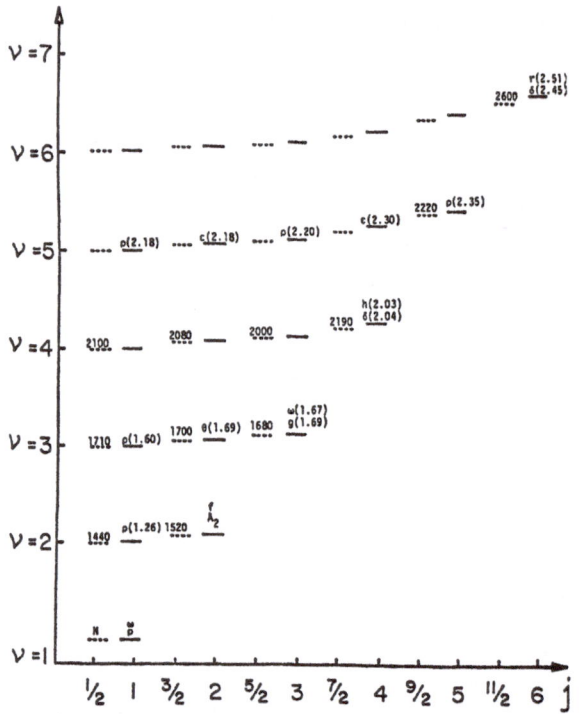

FIG. 2. The mass level diagram as obtained from a fit of the nucleon- and of the $(Y = 0, CP = +1, j^P =$normal) meson- resonances to the mass formula (7). On the horizontal axis is plotted the spin j of the resonance. Vertically is plotted $m^2 - \tilde{m}_0^2 - \beta(2 - s^2)$, where m^2 is the value calculated from (7) with the parameters (8), so that the baryon and meson ground-state levels coincide. The values of the parameters have been obtained in a χ^2 fit of the experimental data for the resonances which are shown [by particle symbol and mass (in MeV for nucleons and GeV for mesons)] near the levels to which they have been assigned. These are only the resonances whose existence and whose spin are fairly well established.

In Figure 2 above we have drawn a level for each pair of numbers $(\nu = \mu - 1, j)$ that occurs in $D(2,1)$ and we have also drawn another level for each pair $(\nu = \mu - 1/2, j)$ that occurs in $D(3/2, 1/2)$. If the levels would have been drawn all with ν as y-coordinate, then Fig. 2 would be identical with Fig.1. But in Fig. 2 the y-coordinate of the levels is not exactly ν but:

$$m^2 - \tilde{m}_0^2 = \frac{1}{\alpha'}\mu + \lambda^2 j(j+1), \qquad (7)$$

where m^2 is the mass of the hadron that has been assigned to the (μ, j) level. The empirical parameters $1/\alpha'$, λ^2 and $\tilde{m}_0^2$ have been determined by fits to the meson and nucleon spectrum separately and it is found that the values of $1/\alpha'$ and λ^2 for the meson and nucleon towers agree within error. Therefore a joint fit to all resonances in the $\omega - \rho-$ and $N-$towers was performed and gives the values

$$\frac{1}{\alpha'} = (1.03 \pm 0.036)\text{GeV}^2, \quad \lambda^2 = (0.015 \pm 0.008)\text{GeV}^2 \text{ with } \frac{\chi^2}{n_0} = \frac{9.9}{28}. \quad (8)$$

For the y-coordinate of the level with half integer j we have chosen $m^2 - \tilde{m}_0{}^2$(baryon) and for the y-coordinate of the level with integer j we have chosen $(m^2 - \tilde{m}_0{}^2$(meson)). In this way the ground state masses have been adjusted to the same y-coordinate (with separately fitted values for $\tilde{m}_0{}^2$ (meson) and $\tilde{m}_0{}^2$ (baryon). (For direct comparison we have also drawn the experimental values of $m^2 - \tilde{m}_0{}^2$ with their error bars). Figure 2 illustrates the evidence for the supersymmetry. The resonances on the vector meson trajectory and on the nucleon Regge trajectory are the "yrast" states with $\nu = j$. The states with $j = \nu - 1, j = \nu - 2 \cdots$ are the places for the daughters. The slopes for the nucleon trajectory with daughters and for the vector meson trajectory with daughters are the same and given by $\frac{1}{\alpha'}$ of (8).

There is no experimentally well-established resonance (with $j > 0$) which cannot be accommodated in the supermultiplet of Fig. 2, and there are no predictions of the model which contradict experimental facts. But then the value of the new quantum number ν or (μ) can be arbitrarily assigned to the resonances and have been assigned such that eq. (7) gives the best fit. Thus the experimental evidence for this supermultiplet of hadrons is good but not overwhelming.

Our fits to the predicted mass formulas show that the mass splittings between the meson and baryon levels are approximately the same which we take as an indication for supersymmetry, in analogy to the dynamical supersymmetries in nuclear physics.

So we conclude that the idea put forward by Francesco Iachello in nuclear physics can also be extended successfully to the relativistic domain.

REFERENCES

1. Y.A. Gol'fand, and E.P. Likhtman, "Extension of the algebra of Poincaré group generators and violation of P invariance", *JETP Lett.* **13**, 323 (1971).

2. J. Wess, and B. Zumino, "Supergauge transformations in four dimensions", Nucl. Phys. **B70**, 39 (1974).

3. A. Bohr, and B.R. Mottelson, "Nuclear Structure" Vol I *Single-Particle Motion*, W.A. Benjamin, Reading (1969).

4. F. Iachello, "Dynamical supersymmetries in nuclei", *Phys. Rev. Lett.* **44**, 772 (1980).

5. A.B. Balantekin, I. Bars, and F. Iachello, "U(6/4) Dynamical supersymmetry in nuclei", *Phys Rev. Lett* **47**, 19 (1981).

6. D.D. Warner, R.F. Casten, M.L. Stelts, H.G. Börner, and G. Barreau, "Nuclear structure of ^{195} Pt", *Phys. Rev.* **C26**, 1921 (1982).

7. V.A. Kostelecký, this volume.

8. A. Bohr, and B.R. Mottelson, "Nuclear Structure" Vol II *Nuclear Deformations*, W.A. Benjamin, Inc., Reading (1969).

9. A. Arima, F. Iachello, Ann Rev. Nucl. Part. Sci. **31**, 75 (1981).

q-DEFORMED INTERACTING BOSON MODELS
IN NUCLEAR AND MOLECULAR PHYSICS

Dennis Bonatsos

Institute of Nuclear Physics, N.C.S.R. "Demokritos"
GR-15310 Aghia Paraskevi, Attiki, Greece

1. INTRODUCTION

It is a particular honor for me to present this paper on the occassion of the 50th birthday of Professor F. Iachello. As it will be realized by the end of this paper, not only the present work, but in addition large parts of my past work could not have been done without the pathbreaking contributions of Professor F. Iachello.

Quantum algebras (also called quantum groups) [1-4], which from the mathematical point of view are q-deformations of the universal enveloping algebras of the corresponding Lie algebras, are recently receiving much attention in physics. They are concrete examples of Hopf algebras [3,4]. When the deformation parameter q is set equal to 1, the corresponding usual Lie algebras are obtained. Initially used for solving the quantum Yang Baxter equation [5], quantum algebras are now finding applications in several branches of physics, especially after the introduction of the q-deformed harmonic oscillator [6,7]. Applications in conformal field theory [8,9], quantum gravity [10], quantum optics [11,12], as well as in the study of spin chains [13,14] have already appeared. Rotational spectra and electric transition probabilities of deformed nuclei [15-17], superdeformed nuclei [18], and diatomic molecules [19,20] have been described in terms of the q-deformed rotator having the symmetry $SU_q(2)$, the deformation parameter τ^2 (where $q = e^{i\tau}$) found [16] to correspond to the softness parameter of the Variable Moment of Inertia (VMI) model. Vibrational spectra of diatomic molecules have been described in terms of q-deformed harmonic [21] and anharmonic oscillators [22,23] having the $SU_q(1,1)$ symmetry, as well as in terms of suitable generalized deformed oscillators [24-26] (see [27] for a review). Classical potentials giving the same spectrum as the q-deformed harmonic [28] and anharmonic [29-31] oscillators have also been constructed. The pairing correlations in a nuclear single-j shell can be described in terms of the usual q-deformed harmonic oscillators approximately [32], or in terms of a generalized q-deformed harmonic oscillator exactly [33,34], the deformation parameter τ (with $q = e^{i\tau}$), which is serving as a small parameter, being related to the inverse of the size of the shell [32].

The applications of quantum algebras in nuclear physics have been limited so far to the relatively simple mathematical constructions of the q-deformed harmonic oscillator and the $SU_q(2)$ algebra. In terms of usual Lie algebras, however, several successful phenomenological models exist for the description of collective nuclear spectra, like Elliott's SU(3) model [35], the Interacting Boson Model (IBM) ([36], see [37,38] for recent overviews), the Fermion Dynamical Symmetry Model (FDSM) [39], the Interacting two-Vector Boson Model (IVBM) [40]. The q-deformed generalization of these models could possibly open the way for an improved description of collective nuclear spectra in medium and heavy mass nuclei, where shell model calculations are not yet possible.

In this paper we take a first step in this direction, encouraged by the above mentioned facts that for rotational spectra the $SU_q(2)$ symmetry gives a better description than the usual SU(2) symmetry using a deformation parameter with a well defined physical meaning [16], giving successful predictions for the electric transition probabilities [17] as well, while for the pairing correlations in a single-j shell the description in terms of q-bosons is much simpler than in terms of usual bosons, with a deformation parameter again having a well defined physical meaning [32-34]. The q-deformed generalization of the full IBM, which is built out of s (J=0) and d (J=2) bosons, having a U(6) overall symmetry and the three limiting symmetries U(5), SU(3) and O(6), is technically quite demanding. Therefore we first focus attention on a two-dimensional toy version of the Interacting Boson Model [41], having an overall SU(3) symmetry and containing two limiting symmetries: the vibrational SU(2) symmetry and the rotational SO(3) symmetry. We focus attention on both the q-deformation of the vibrational limit, which has the algebraic structure SU(3) ⊃ SU(2), and the q-deformation of the rotational limit, which has the algebraic structure SU(3) ⊃ SO(3), present in the rotational limit of all the above mentioned nuclear collective models.

In section 2 of this paper a brief description of the the q-deformed version of the toy IBM will be given. In sections 3 and 4 the vibrational and rotational limits of the model will be studied respectively, while section 5 will contain discussion of the present results and plans for further work.

2. A TOY INTERACTING BOSON MODEL WITH $U_q(3)$ SYMMETRY

In the classical version of the toy IBM [41] one introduces bosons with angular momentum $m = 0, \pm 2$, represented by the creation (annihilation) operators a_0^+, a_+^+, a_-^+ (a_0, a_+, a_-) and satisfying usual boson commutation relations

$$[a_i, a_j^+] = \delta_{ij}, \quad [a_i, a_j] = [a_i^+, a_j^+] = 0. \tag{1}$$

The 9 bilinear operators

$$\Lambda_{ij} = a_i^+ a_j \tag{2}$$

satisfy then the commutation relations

$$[\Lambda_{ij}, \Lambda_{kl}] = \delta_{jk}\Lambda_{il} - \delta_{il}\Lambda_{jk}, \tag{3}$$

which are the standard U(3) commutation relations. The total number of bosons

$$N = \Sigma_i \Lambda_{ii} = a_0^+ a_0 + a_+^+ a_+ + a_-^+ a_- \tag{4}$$

Table 1. $U_q(3)$ commutation relations [41], given in the form $[A, B]_a = C$. A is given in the first column, B in the first row. C is given at the intersection of the row containing A with the column containing B. a, when different from 1, follows C, enclosed in parentheses.

	A_{11}	A_{22}	A_{33}	A_{12}	A_{23}
A_{11}	0	0	0	A_{12}	0
A_{22}	0	0	0	$-A_{12}$	A_{23}
A_{33}	0	0	0	0	$-A_{23}$
A_{12}	$-A_{12}$	A_{12}	0	0	$A_{13}\,(q)$
A_{23}	0	$-A_{23}$	A_{23}	$-q^{-1}A_{13}(q^{-1})$	0
A_{13}	$-A_{13}$	0	A_{13}	$0\,(q)$	$0\,(q^{-1})$
A_{21}	A_{21}	$-A_{21}$	0	$-[A_{11} - A_{22}]$	0
A_{32}	0	A_{32}	$-A_{32}$	0	$-[A_{22} - A_{33}]$
A_{31}	A_{31}	0	$-A_{31}$	$-q^{-A_{11}+A_{22}}A_{32}$	$-A_{21}q^{A_{22}-A_{33}}$

	A_{13}	A_{21}	A_{32}	A_{31}
A_{11}	A_{13}	$-A_{21}$	0	$-A_{31}$
A_{22}	0	A_{21}	$-A_{32}$	0
A_{33}	$-A_{13}$	0	A_{32}	A_{31}
A_{12}	$0\,(q^{-1})$	$[A_{11} - A_{22}]$	0	$-q^{-A_{11}+A_{22}}A_{32}$
A_{23}	$0\,(q)$	0	$[A_{22} - A_{33}]$	$A_{21}q^{A_{22}-A_{33}}$
A_{13}	0	$-A_{23}q^{A_{11}-A_{22}}$	$q^{-A_{22}+A_{33}}A_{12}$	$[A_{11} - A_{33}]$
A_{21}	$A_{23}q^{A_{11}-A_{22}}$	0	$-qA_{31}(q)$	$0\,(q^{-1})$
A_{32}	$-q^{-A_{22}+A_{33}}A_{12}$	$A_{31}(q^{-1})$	0	$0\,(q)$
A_{31}	$-[A_{11} - A_{33}]$	$0\,(q)$	$0\,(q^{-1})$	0

is kept constant. Since we are dealing with a system of bosons, only the totally symmetric irreducible representations (irreps) $\{N, 0, 0\}$ of U(3) occur.

In the quantum case one has the $U_q(3)$ commutation relations given in Table 1 [42], where A_{ij} are the generators of $U_q(3)$ and the q-commutator is defined as

$$[A, B]_q = AB - qBA. \tag{5}$$

One then needs a realization of $U_q(3)$ in terms of q-bosons, satisfying the commutation relations [6,7]

$$a_i a_i^\dagger - q a_i^\dagger a_i = q^{-N_i}, \quad a_i a_i^\dagger - q^{-1} a_i^\dagger a_i = q^N, \tag{6}$$

and satisfying

$$a_i^\dagger a_i = [N_i], \quad a_i a_i^\dagger = [N_i + 1], \tag{7}$$

where

$$[x] = \frac{q^x - q^{-x}}{q - q^{-1}} \tag{8}$$

is the definition of q-numbers. Starting with [43,44]

$$A_{12} = a_1^\dagger a_2, \quad A_{21} = a_2^\dagger a_1, \tag{9}$$

113

$$A_{23} = a_2^+ a_3, \quad A_{32} = a_3^+ a_2, \tag{10}$$

one can easily verify that the $U_q(3)$ commutation relations involving these generators are satisfied. One can now determine the boson realizations of A_{13} and A_{31} from other commutation relations, as follows

$$A_{13} = [A_{12}, A_{23}]_q = a_1^+ a_3 q^{-N_2}, \tag{11}$$

$$A_{31} = [A_{32}, A_{21}]_{q^{-1}} = a_3^+ a_1 q^{N_2}. \tag{12}$$

It is by now a straightforward task to verify that all commutation relations of Table 1 are satisfied. For example, one has

$$[A_{12}, A_{21}] = [N_1 - N_2], \tag{13}$$

$$[A_{23}, A_{32}] = [N_2 - N_3], \tag{14}$$

$$[A_{13}, A_{31}] = [N_1 - N_3], \tag{15}$$

using the identifications

$$N_1 = A_{11}, \quad N_2 = A_{22}, \quad N_3 = A_{33}, \tag{16}$$

and the identity

$$[N_i][N_j + 1] - [N_j][N_i + 1] = [N_i - N_j]. \tag{17}$$

3. THE VIBRATIONAL LIMIT

So far we have managed to write a boson realization of $U_q(3)$ in terms of 3 q-bosons, namely a_1, a_2, a_3. Omitting the generators involving one of the bosons, one is left with an $SU_q(2)$ subalgebra. Omitting the generators involving a_3, for example, one is left with A_{12}, A_{21}, N_1, N_2, which satisfy the $SU_q(2)$ commutation relations [6,7]

$$[J_+, J_-] = [2J_0], \quad [J_0, J_\pm] = \pm J_\pm, \tag{18}$$

where the identifications

$$J_+ = A_{12}, \quad J_- = A_{21}, \quad J_0 = \frac{1}{2}(N_1 - N_2) \tag{19}$$

have been made. J_0 alone forms an $SO_q(2)$ subalgebra. Therefore the relevant chain of subalgebras is

$$SU_q(3) \supset SU_q(2) \supset SO_q(2). \tag{20}$$

The second order Casimir operator of $SU_q(2)$ is known to have the form [6,7]

$$C_2(SU_q(2)) = J^2 = J_- J_+ + [J_0][J_0 + 1]. \tag{21}$$

Substituting the above expressions for the generators one finds

$$C_2(SU_q(2)) = \left[\frac{N_1 + N_2}{2}\right]\left[\frac{N_1 + N_2}{2} + 1\right]. \tag{22}$$

All of the above equations go to their classical counterparts by allowing $q \to 1$, for which $[x] \to x$, i.e. q-numbers become usual numbers. In the classical case [41] out of the three bosons (a_0, a_+, a_-) forming SU(3), one chooses to leave out a_0, the boson with zero angular momentum, in order to be left with the SU(2) subalgebra formed by a_+ and a_-, the two bosons of angular momentum two. The choice of the $SU_q(2)$ subalgebra made above is then consistent with the following correspondence between classical bosons and q-bosons

$$a_+ \to a_1, \quad a_- \to a_2, \quad a_0 \to a_3. \tag{23}$$

(We have opted in using different indices for usual bosons and q-bosons in order to avoid confusion.)

In the classical case [41] the states of the system are characterized by the quantum numbers characterizing the irreducible representations (irreps) of the algebras appearing in the classical counterpart of the chain of eq. (20). For SU(3) the total number of bosons N is used. For SU(2) and SO(2) one can use the eigenvalues of J^2 and J_0, or, equivalently, the eigenvalues of $a_+^{\dagger} a_+ + a_-^{\dagger} a_-$ and $L_3 = 4J_0$, for which we use the symbols n_d (the number of bosons with angular momentum 2) and M. Then the basis in the classical case can be written as [41]

$$|N, n_d, M> = \frac{(a_0^+)^{N-n_d}}{(N-n_d)!} \frac{(a_+^+)^{n_d/2+M/4}}{(n_d/2+M/4)!} \frac{(a_-^+)^{n_d/2-M/4}}{(n_d/2-M/4)!} |0>. \tag{24}$$

In the quantum case for each oscillator one defines the basis as [6,7]

$$|N_i> = \frac{(a_i^+)^{N_i}}{[N_i]!} |0>, \tag{25}$$

where the q-factorial is defined as

$$[N]! = [N][N-1][N-2]\ldots[2][1]. \tag{26}$$

Then one has [6,7]

$$N_i|N_i> = N_i|N_i>, \tag{27}$$

$$a_i^+|N_i> = \sqrt{[N_i+1]}|N_i+1>, \tag{28}$$

$$a_i|N_i> = \sqrt{[N_i]}|N_i-1>, \tag{29}$$

so that the full basis in the q-deformed case is

$$|N, n_d, M>_q = \frac{(a_3^+)^{N-n_d}}{[N-n_d]!} \frac{(a_1^+)^{n_d/2+M/4}}{[n_d/2+M/4]!} \frac{(a_2^+)^{n_d/2-M/4}}{[n_d/2-M/4]!} |0>, \tag{30}$$

where $N = N_1 + N_2 + N_3$ is the total number of bosons, $n_d = N_1 + N_2$ is the number of bosons with angular momentum 2, and M is the eigenvalue of $L = 4J_0$. n_d takes values from 0 up to N, while for a given value of n_d, M takes the values $\pm 2n_d, \pm 2(n_d - 2), \ldots, \pm 2$ or 0, depending on whether n_d is odd or even [41]. In this basis the eigenvalues of the second order Casimir operator of $SU_q(2)$ are then

$$C_2(SU_q(2))|N, n_d, M>_q = [\frac{n_d}{2}][\frac{n_d}{2}+1]|N, n_d, M>_q. \tag{31}$$

In the case of $N = 5$ one can easily see that the spectrum will be composed by the ground state band, consisting of states with $M = 0, 2, 4, 6, 8, 10$ and $n_d = M/2$, the first excited band with states characterized by $M = 0, 2, 4, 6$ and $n_d = M/2+2$, and the second excited band, containing states with $M = 0, 2$ and $n_d = M/2 + 4$.

In the case that the Hamiltonian has the $SU_q(2)$ dynamical symmetry, it can be written in terms of the Casimir operators of the chain (20). Then one has

$$H = E_0 + AC_2(SU_q(2)) + BC_2(SO_q(2)), \qquad (32)$$

where E_0, A, B are constants. Its eigenvalues are

$$E = E_0 + A[\frac{n_d}{2}][\frac{n_d}{2} + 1] + BM^2. \qquad (33)$$

Realistic nuclear spectra are characterized by strong elecric quadrupole transitions among the levels of the same band, as well as by interband transitions. In the framework of the present toy model one can define, by analogy to the classical case [41], quadrupole transition operators

$$Q_+ = a_1^+ a_3 + a_3^+ a_2, \qquad (34)$$

$$Q_- = a_2^+ a_3 + a_3^+ a_1. \qquad (35)$$

In order to calculate transition matrix elements of these operators one only needs eqs. (28), (29), i.e. the action of the q-boson operators on the q-deformed basis. The selection rules, as in the classical case, are $\Delta M = \pm 2$, $\Delta n_d = \pm 1$, while the corresponding matrix elements are

$$_q < N, n_d + 1, M \pm 2 |Q_\pm| N, n_d, M >_q = \sqrt{[N - n_d][\frac{n_d}{2} \pm \frac{M}{4} + 1]}, \qquad (36)$$

$$_q < N, n_d - 1, M \pm 2 |Q_\pm| N, n_d, M >_q = \sqrt{[N - n_d + 1][\frac{n_d}{2} \mp \frac{M}{4}]}. \qquad (37)$$

From these equations it is clear that both intraband and interband transitions are possible.

In order to get a feeling of the qualitative changes in the spectrum and the transition matrix elements resulting from the q-deformation of the model, one can make a simple calculation for a system of 20 bosons ($N = 20$). We distinguish two cases:

i) q can be real ($q = e^\tau$, with τ real), in which case q-numbers can be written as

$$[x] = \frac{\sinh \tau x}{\sinh \tau}. \qquad (38)$$

ii) q can be a phase ($q = e^{i\tau}$, with τ real), in which case q-numbers can be put in the form

$$[x] = \frac{\sin \tau x}{\sin \tau}. \qquad (39)$$

In order to isolate the effects of q-deformation in the spectrum, we consider a Hamiltonian (32) with $E_0 = 0$, $A = 1$, $B = 0$. As we have already remarked, the ground state band contains states characterized by $M = 2n_d$. Results for the lowest 10 members of the ground state band are reported in [44] for the classical case

$(\tau = 0)$, as well as for the two q-deformed cases (q real, q a phase) for two different values of the deformation parameter ($\tau = 0.05, 0.1$). When q is real the spectrum is increasing more rapidly than in the classical case, while when q is a phase the spectrum increases more slowly than in the classical case. This is in agreement with the findings of the q-rotator model [15-19], which is equivalent to the VMI model for q being a phase, with τ having a well defined physical meaning (τ^2 is equivalent to the softness parameter of the VMI model [16]).

In [44] we also report for the same cases results for the transition matrix element

$$_q < N, n_d + 1, M + 2|Q_+|N, n_d, M >_q, \tag{40}$$

which within the ground state band (where $M = 2n_d$) takes the form (see eq. (36))

$$_q < N, n_d + 1, 2n_d + 2|Q_+|N, n_d, 2n_d >_q = \sqrt{[N - n_d][n_d + 1]}. \tag{41}$$

Results up to $n_d = 9$ are reported, since the matrix elements for n_d and $N - n_d - 1$ are equal, as it is easily seen from (41). The transition matrix elements in the case that q is real increase more rapidly than in the classical case, while they increase less rapidly than the classical case when q is a phase. We also remark that transition matrix elements are much more sensitive to q-deformation than energy spectra. This is an interesting feature, showing that q-deformed algebraic models can be much more flexible in the description of transition probabilities than their classical counterparts.

In summary, we have developed the q-deformed generalization of the vibrational limit of a two-dimensional toy IBM model with $SU_q(3) \supset SU_q(2) \supset SO_q(2)$ symmetry. Spectra and transition matrix elements are influenced in different ways depending on whether q is real or a phase. Transition matrix elements are much more sensitive to q-deformation than energy levels.

4. THE ROTATIONAL LIMIT

We turn now to the study of the rotational limit of the q-deformed toy model. Our task is then to determine the $SO_q(3)$ subalgebra. For the totally symmetric representations of $U_q(3)$, which are the only ones needed in the present model, since we are dealing with a system of bosons, this problem has been solved [45]. Using the notation

$$a_+ \to a_1, \quad a_- \to a_2, \quad a_0 \to a_3, \tag{42}$$

the basis states are of the form [45]

$$|n_+ n_0 n_- > = \frac{(a_+^\dagger)^{n_+}(a_0^\dagger)^{n_0}(a_-^\dagger)^{n_-}}{\sqrt{[n_+]![n_0]![n_-]!}}|0 >, \tag{43}$$

with $a_i|0 > = 0$ and $N_i|n_+ n_0 n_- > = n_i|n_+ n_0 n_- >$, where $i = +, 0, -$. The principal subalgebra $SO_q(3)$ is then generated by [45]

$$L_0 = N_+ - N_-, \tag{44}$$

$$L_+ = q^{N_- - \frac{1}{2}N_0}\sqrt{q^{N_+} + q^{-N_+}}\ a_+^\dagger a_0 + a_0^\dagger a_-\ q^{N_+ - \frac{1}{2}N_0}\sqrt{q^{N_-} + q^{-N_-}}, \tag{45}$$

$$L_- = a_0^\dagger a_+ q^{N_- - \frac{1}{2}N_0}\sqrt{q^{N_+} + q^{-N_+}} + q^{N_+ - \frac{1}{2}N_0}\sqrt{q^{N_-} + q^{-N_-}}\ a_-^\dagger a_0, \tag{46}$$

satisfying the commutation relations

$$[L_0, L_\pm] = \pm L_\pm, \quad [L_+, L_-] = [2L_0]. \tag{47}$$

L_0 alone generates then the $SO_q(2)$ subalgebra. Therefore the relevant chain of subalgebras is

$$SU_q(3) \supset SO_q(3) \supset SO_q(2). \tag{48}$$

The $SO_q(3)$ basis vectors can be written in terms of the vectors of eq. (43) as [45]

$$|v(N, L, M)> = q^{-[(L+M)(L+M-1)]/4} \sqrt{\frac{[N+L]!![2L+1][L+M]![L-M]!}{[N-L]!![N+L+1]!}}$$

$$\sum_x q^{(2L-1)x/2} \; s^{(N-L)/2} \; \frac{|x, L+M-2x, x-M>}{\sqrt{[2x]!![L+M-2x]![2x-2M]!!}}, \tag{49}$$

where

$$s = (a_0^+)^2 q^{N_+ + N_- + 1} - \sqrt{\frac{[2N_+][2N_-]}{[N_+][N_-]}} a_+^+ a_-^+ q^{-N_0 - \frac{1}{2}}, \tag{50}$$

x takes values from $\max(0, M)$ to $[(L+M)/2]$ in steps of 1, $L = N, N-2, \ldots, 1$ or 0, $M = -L, -L+1, \ldots, +L$, and $[2x]!! = [2x][2x-2]\ldots[2]$. The action of the generators of $SO_q(3)$ on these states is given by

$$L_0 |v(N, L, M)> = M \quad |v(N, L, M)>, \tag{51}$$

$$L_\pm \quad |v(N, L, M)> = \sqrt{[L \mp M][L \pm M + 1]} \quad |v(N, L, M \pm 1)>. \tag{52}$$

The second order Casimir operator of $SO_q(3)$ has the form

$$C_2(SO_q(3)) = L^2 = L_- L_+ + [L_0][L_0 + 1]. \tag{53}$$

Its eigenvalues in the above basis are given by

$$C_2(SO_q(3)) \quad |v(N, L, M)> = [L][L+1] \quad |v(N, L, M)>. \tag{54}$$

All of the above equations go to their classical counterparts by allowing $q \to 1$, for which $[x] \to x$, i.e. q-numbers become usual numbers. In the classical case [41] the states of the system are characterized by the quantum numbers characterizing the irreducible representations (irreps) of the algebras appearing in the classical counterpart of the chain of eq. (48). For SU(3) the total number of bosons N is used. For SO(3) and SO(2) one can use L and M, respectively. In [41], however, the eigenvalue of $L_0' = 2L_0$ is used, which is $M' = 2M$.

Since the rules for the decomposition of the totally symmetric $U_q(3)$ irreps into $SO_q(3)$ irreps are the same as in the classical case, it is easy to verify that for a system with $N = 6$ the spectrum will be composed by the ground state band, consisting of states with $M' = 0, 2, 4, 6, 8, 10, 12$ and $L = N$, the first excited band with states characterized by $M' = 0, 2, 4, 6, 8$ and $L = N - 2$, the second excited band, containing states with $M = 0, 2, 4$ and $L = N - 4$, and the third excited band, containing a state with $M' = 0$ and $L = N - 6$.

In the case that the Hamiltonian has the $SO_q(3)$ dynamical symmetry, it can be written in terms of the Casimir operators of the chain (48). Then one has

$$H = E_0 + AC_2(SO_q(3)) + BC_2(SO_q(2)), \qquad (55)$$

where E_0, A, B are constants. Its eigenvalues are

$$E = E_0 + A[L][L+1] + BM^2. \qquad (56)$$

It is then clear that in this simple model the internal structure of the rotational bands is not influenced by q-deformation. What is changed is the position of the bandheads.

Realistic nuclear spectra are characterized by strong electric quadrupole transitions among the levels of the same band, as well as by interband transitions. In the framework of the present toy model one can define, by analogy to the classical case [41], quadrupole transition operators $Q_\pm$ proportional to the $SO_q(3)$ generators $L_\pm$

$$Q_\pm = L_\pm. \qquad (57)$$

In order to calculate transition matrix elements of these operators one only needs eq. (52). The selection rules, as in the classical case, are $\Delta M' = \pm 2$, $\Delta L = 0$, i.e. only intraband transitions are allowed. The relevant matrix elements are

$$< v(N, L, M'+2)|Q_+|v(N, L, M') > = \sqrt{[L - \frac{M'}{2}][L + \frac{M'}{2} + 1]}, \qquad (58)$$

$$< v(N, L, M'-2)|Q_-|v(N, L, M') > = \sqrt{[L + \frac{M'}{2}][L - \frac{M'}{2} + 1]}. \qquad (59)$$

In order to get a feeling of the qualitative changes in the spectrum and the transition matrix elements resulting from the q-deformation of the model, one can make a simple calculation. As in the case of the vibrational limit, we distinguish two cases:

i) q can be real ($q = e^\tau$, with τ real), in which case q-numbers can be written as in eq. (38).

ii) q can be a phase ($q = e^{i\tau}$, with τ real), in which case q-numbers can be put in the form of eq. (39).

In order to isolate the effects of q-deformation in the spectrum, we consider a Hamiltonian (55) with $E_0 = 0$, $A = 1$, $B = 0$. As we have already remarked, q-deformation influences only the position of bandheads, while it leaves the internal structure of the bands intact. Making a calculation for the classical case ($\tau = 0$), as well as for the two q-deformed cases (q real, q a phase) for two different values of the deformation parameter ($\tau = 0.05, 0.1$), we remark that when q is real the bandheads are increasing more rapidly than in the classical case, while when q is a phase the bandheads increase more slowly than in the classical case. This result is in qualitative agreement with the findings of the $SU_q(2)$ model [15-19].

Furthermore, the transition matrix element

$$< v(N, L, M+2)|Q_+|v(N, L, M) > \qquad (60)$$

can be calculated for a system of 10 bosons ($N = 10$) in its ground state band, in which $L = 10$, too. One can then see that the transition matrix elements in the case that q is real have values higher than in the classical case, while they have values lower than in the classical case when q is a phase.

In summary, we have developed the q-deformed generalization of the rotational limit of a two-dimensional toy IBM model with $SU_q(3) \supset SO_q(3) \supset SO_q(2)$ symmetry. Bandheads and intraband transition matrix elements are influenced in different ways depending on whether q is real or a phase, while the spacing of the levels within each band remains intact.

5. DISCUSSION

In this paper we have studied the q-deformed generalization of both the vibrational and the rotational limit of a toy IBM. In the vibrational limit the extra flexibility of the model lies mainly in the transition probabilities and to a lesser extend in the energy levels, while in the rotational limit it lies in the transition probabilities and the bandheads, the spacing of the levels within each band remaining intact. Such an extra flexibility added to an exactly soluble model, like the Interacting Boson Model (IBM) [36] by the q-deformation might be useful in overcoming certain difficulties appearing in the classical version of the model. It should be mentioned at this point that recent work on the complementarity between $SU_q(3)$ and $U_q(2)$ [46,47] opens the way for the q-deformation of the rotational limit of the Interacting two-Vector Boson Model (IVBM) [40]. The q-deformation of the vibron model [48,49] (see [38] for a list of references), widely used in molecular physics, is treated through these techniques in the contribution of Yu. F. Smirnov to this volume.

REFERENCES

1. P. P. Kulish and N. Yu. Reshetikhin, *Zapiski Semenarov LOMI* 101:101 (1981) (in Russian).
2. E. K. Sklyanin, *Funkts. Anal. Prilozh.* 16:27 (1982) (in Russian).
3. V. G. Drinfeld, Quantum groups, in: "Proceedings of the International Congress of Mathematicians", A. M. Gleason, ed., American Mathematical Society, Providence, RI (1986).
4. E. Abe. "Hopf Algebras", Cambridge University Press, Cambridge (1980).
5. M. Jimbo, Introduction to the Yang-Baxter equation, in: "Braid Group, Knot Theory and Statistical Mechanics," C. N. Yang and M. L. Ge, eds, World Scientific, Singapore (1989).
6. L. C. Biedenharn, The quantum group $SU_q(2)$ and a q-analogue of the boson operators, *J. Phys. A* 22:L873 (1989).
7. A. J. Macfarlane, On q-analogues of the quantum harmonic oscillator and the quantum group $SU_q(2)$, *J. Phys. A* 22:4581 (1989).
8. L. Alvarez-Gaumé, C. Gomez and G. Sierra, Duality and quantum groups, *Nucl. Phys. B* 330:347 (1990).
9. V. Pasquier and H. Saleur, Common structures between finite systems and conformal field theories through quantum groups, *Nucl. Phys. B* 330:523 (1990).
10. S. Mizoguchi and T. Tada, Three-dimensional gravity from the Turaev–Viro invariant, *Phys. Rev. Lett.* 68:1795 (1992).
11. E. Celeghini, M. Rasetti and G. Vitiello, Squeezing and quantum groups, *Phys. Rev. Lett.* 66:2056 (1991).
12. V. Buzek, Dynamics of a q-analogue of the quantum harmonic oscillator, *J. Mod. Optics* 38:801 (1991).
13. M. T. Batchelor, L. Mezincescu, R. I. Nepomechie and V. Rittenberg, q-deformations of the O(3) symmetric spin-1 Heisenberg chain, *J. Phys. A* 23:L141 (1990).

14. P. P. Kulish and E. K. Sklyanin, The general $U_q[sl(2)]$ invariant XXZ integrable quantum spin chain, *J. Phys. A* 24:L435 (1991).

15. P. P. Raychev, R. P. Roussev and Yu. F. Smirnov, The quantum algebra $SU_q(2)$ and rotational spectra of deformed nuclei, *J. Phys. G* 16:L137 (1990).

16. D. Bonatsos, E. N. Argyres, S. B. Drenska, P. P. Raychev, R. P. Roussev and Yu. F. Smirnov, $SU_q(2)$ description of rotational spectra and its relation to the Variable Moment of Inertia model, *Phys. Lett. B* 251:477 (1990).

17. D. Bonatsos, A. Faessler, P. P. Raychev, R. P. Roussev and Yu. F. Smirnov, B(E2) transition probabilities in the q-rotator model with $SU_q(2)$ symmetry, *J. Phys. A* 25:3275 (1992).

18. D. Bonatsos, S. B. Drenska, P. P. Raychev, R. P. Roussev and Yu. F. Smirnov, Description of superdeformed bands by the quantum algebra $SU_q(2)$, *J. Phys. G* 17:L67 (1991).

19. D. Bonatsos, P. P. Raychev, R. P. Roussev and Yu. F. Smirnov, Description of rotational molecular spectra by the quantum algebra $SU_q(2)$, *Chem. Phys. Lett.* 175:300 (1990).

20. Z. Chang and H. Yan, The $SU_q(2)$ quantum group symmetry and diatomic molecules, *Phys. Lett. A* 154:254 (1991).

21. Z. Chang and H. Yan, $H_q(4)$ and $SU_q(2)$ quantum group symmetries in diatomic molecules, *Phys. Rev. A* 43:6043 (1991).

22. D. Bonatsos, P. P. Raychev and A. Faessler, Quantum algebraic description of vibrational molecular spectra, *Chem. Phys. Lett.* 178:221 (1991).

23. D. Bonatsos, E. N. Argyres and P. P. Raychev, $SU_q(1,1)$ description of vibrational molecular spectra, *J. Phys. A* 24:L403 (1991).

24. C. Daskaloyannis, Generalized deformed oscillator and nonlinear algebras, *J. Phys. A* 24:L789 (1991).

25. C. Daskaloyannis, Generalized deformed oscillator corresponding to the modified Pöschl-Teller energy spectrum, *J. Phys. A* 25:2261 (1992).

26. D. Bonatsos and C. Daskaloyannis, Generalized deformed oscillators for vibrational spectra of diatomic molecules, *Phys. Rev. A* 46:75 (1992).

27. D. Bonatsos, E. N. Argyres, S. B. Drenska, P. P. Raychev, R. P. Roussev, Yu. F. Smirnov and A. Faessler, Quantum algebraic symmetries in nuclear and molecular physics, *in:* "2nd Hellenic Symposium in Nuclear Physics", G. S. Anagnostatos, D. Bonatsos and E. Mavrommatis, eds, University of Athens, Athens (1992) p. 168.

28. D. Bonatsos, C. Daskaloyannis and K. Kokkotas, WKB equivalent potentials for the q-deformed harmonic oscillator, *J. Phys. A* 24:L795 (1991).

29. D. Bonatsos, C. Daskaloyannis and K. Kokkotas, Classical potentials for q-deformed anharmonic oscillators, *Phys. Rev. A* 45:R6153 (1992).

30. D. Bonatsos, C. Daskaloyannis and K. Kokkotas, WKB equivalent potentials for q-deformed anharmonic oscillators, *Chem. Phys. Lett.* 193:191 (1992).

31. D. Bonatsos, C. Daskaloyannis and K. Kokkotas, WKB equivalent potentials for q-deformed harmonic and anharmonic oscillators, *J. Math. Phys.* (in press).

32. D. Bonatsos, Are q-bosons suitable for the description of correlated fermion pairs?, *J. Phys. A* 25:L101 (1992).

33. D. Bonatsos and C. Daskaloyannis, Generalized deformed oscillator for the pairing correlations in a single-j shell, *Phys. Lett. B* 278:1 (1992).

34. D. Bonatsos and C. Daskaloyannis, q-deformed oscillators in nuclear and molecular physics, *in:* "Symmetries in Science VI: From the Rotation Group to Quantum Algebras", H. D. Doebner, B. Gruber and F. Iachello, eds, Plenum, New York (in press).

35. J. P. Elliott, Collective motion in the nuclear shell model I. Classification schemes for states of mixed configurations, *Proc. R. Soc. London Ser. A* 245:128 (1958).

36. A. Arima and F. Iachello, Collective nuclear states as representations of a SU(6) group, *Phys. Rev. Lett.* 35:1069 (1975).

37. F. Iachello and A. Arima. "The Interacting Boson Model", Cambridge University Press, Cambridge (1987).

38. D. Bonatsos. "Interacting Boson Models of Nuclear Structure", Clarendon, Oxford (1988).

39. C. L. Wu, D. H. Feng, X. G. Chen, J. Q. Chen and M. W. Guidry, A fermion dynamical symmetry model of nuclei I. Basis, Hamiltonian and symmetries, *Phys. Rev. C* 36:1157 (1987).

121

40. A. Georgieva, P. Raychev and R. Roussev, Interacting two-vector-boson model of collective motions in nuclei, *J. Phys. G* 8:1377 (1982).

41. D. Bhaumik, S. Sen and B. Dutta-Roy, A toy version of the interacting boson model, *Am. J. Phys.* 59:719 (1991).

42. Yu. F. Smirnov, V. N. Tolstoy and Yu. I. Kharitonov, The quantum algebra $U_q(3)$, *Sov. J. Nucl. Phys.* 54:437 (1991).

43. P. P. Kulish, Quantum algebras and symmetries of dynamical systems, *in*: "Group Theoretical Methods in Physics", V. V. Dodonov and V. I. Man'ko, eds, Springer Verlag, Heidelberg (1991) p. 195.

44. D. Bonatsos, A. Faessler, P. P. Raychev, R. P. Roussev and Yu. F. Smirnov, An exactly soluble nuclear model with $SU_q(3) \supset SU_q(2) \supset SO_q(2)$ symmetry, *J. Phys. A* 25:L267 (1992).

45. J. Van der Jeugt, On the principal subalgebra of quantum enveloping algebras $gl_q(1+1)$, *J. Phys. A* 25:L213 (1992).

46. Yu. F. Smirnov and V. N. Tolstoy, On complementary relations between the unitary quantum algebras, *in*: "Group Theory and Special Symmetries in Nuclear Physics", J. P. Draayer and J. W. Jänecke, eds, World Scientific, Singapore (1992).

47. C. Quesne, Complementarity of $SU_q(3)$ and $U_q(2)$ and q-boson realization of the $SU_q(3)$ irreducible representations, U. Libre de Bruxelles preprint PNT/15/91.

48. F. Iachello and R. D. Levine, Algebraic approach to molecular rotation-vibration spectra. I. Diatomic molecules, *J. Chem. Phys.* 77:3046 (1982).

49. O. S. van Roosmalen, F. Iachello, R. D. Levine and A. E. L. Dieperink, Algebraic approach to molecular rotation-vibration spectra. II. Triatomic molecules, *J. Chem. Phys.* 79:2515 (1983).

LOW LYING ELECTRIC DIPOLE EXCITATIONS
AND THE INTERACTING BOSON MODEL

P. von Brentano,[1] A. Zilges,[1] N. V. Zamfir,[2] and R.-D. Herzberg[1]

[1]Institut für Kernphysik
Universität zu Köln
W-5000 Köln 41, Germany
[2]Central Institute of Physics
Bucharest, Romania

INTRODUCTION

The idea of the Interacting Boson Approximation (IBA) by Arima and Iachello simplified the description of low lying collective states enormously [1, 2]. Now it became relatively easy to reproduce the experimental observables in a wide variety of nuclei ranging from spherical to well deformed [3]. Further improvement was achieved when proton– and neutron–bosons were distinguished explicitly in the IBA–2 [4]. In the early eighties Iachello used the IBA–2 to predict a new class of states which are not fully symmetric under the exchange of protons and neutrons. In axially deformed nuclei the lowest lying member of these states should be a collective 1^+–state around 2 – 4 MeV excitation energy [5]. The experimental finding of this so called "scissors mode" in electron scattering experiments by a group around A. Richter from Darmstadt [6] was an impressive proof for the power of the IBA–2. Whereas the original IBA–models with s– and d–bosons (representing bosons with $J^\pi=0^+$ and $J^\pi=2^+$) were only capable of describing positive parity states, the model can be extended in a natural way to allow the description of negative parity states by including a p–boson with $J^\pi=1^-$ and a f–boson with $J^\pi=3^-$ [7–10] The investigation of negative parity states will be the subject of this contribution.

In a collective picture E1–transitions are forbidden if a homogeneous charge distribution and reflection symmetry are assumed. But a small number of 1^-–states in deformed as well as in spherical nuclei exhibit E1 transition strengths which are orders

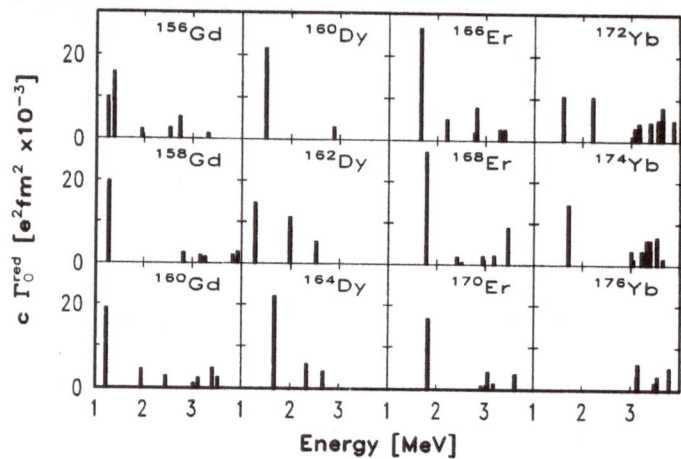

Figure 1. Distribution of E1–strength in rare earth nuclei. The numerical value of the parity independent unit $c \cdot \Gamma_0^{red}$ can be identified with the B(E1)↑–strength in $10^{-3} e^2 fm^2$ (from ref. [12]).

of magnitudes larger than the values usually found in these nuclei [11, 12]. One assumes that a dynamic electric dipole moment is induced by an octupole vibration of the nucleus which allows the enhanced E1–transitions [13].

In the present contribution we will investigate octupole induced 1^-–states in well deformed rare earth nuclei. In Nuclear Resonance Fluorescence (NRF) experiments we detected in nearly all well deformed nuclei enhanced E1–groundstate transitions. We will show that the sdf–IBA with an improved E1–operator and an Alaga rule constraint is able to predict the branching ratios and transition strengths of these states very precisely.

ELECTRIC DIPOLE STRENGTHS IN WELL DEFORMED NUCLEI

The distribution of E1–strength between 1 and 4 MeV in well deformed rare earth nuclei has been obtained in photon scattering or Nuclear Resonance Fluorescence experiments of the Stuttgart–Giessen–Cologne collaboration [12]. Due to its high selectivity in spin and strength, NRF is a very powerful tool for the investigation of dipole excitations. The experimental observables are the spins, energies, decay branching ratios, absolute transition strengths (i.e. lifetimes), and cross sections of the excited states [14]. For several states the parities have been measured in additional experiments [15, 16].

Figure 1 shows the E1–systematics for different Gd–, Dy–, Er–, and Yb–isotopes. The parities of most of the higher lying, weaker excitations have not yet been determined, but is assumed to be negative due to the K–quantum number K=0 of these states. Therefore the strengths are given in the parity independent unit $c \cdot \Gamma_0^{red}$ where

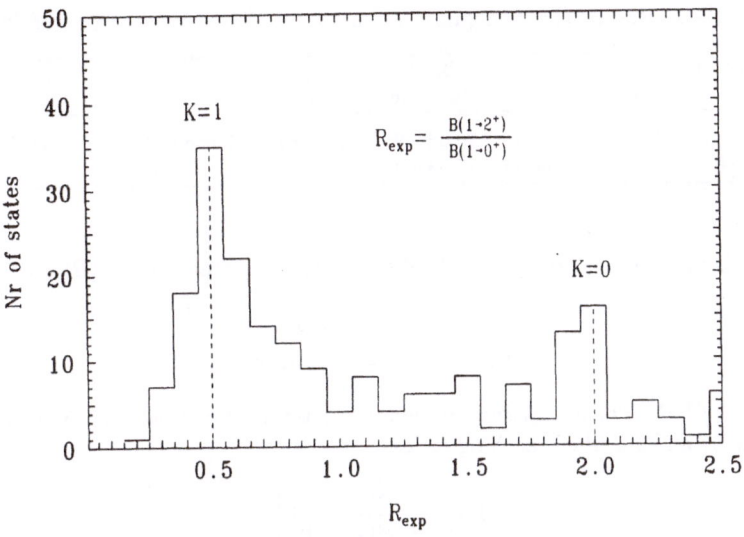

Figure 2. Distribution of branching ratios in rare earth nuclei.

Γ_0^{red} is the energy reduced groundstate decay width ($\Gamma_0^{red} = \Gamma_0/E_x^3$). The factor c is chosen in such a way that the product $c \cdot \Gamma_0^{red}$ gives the B(E1)↑–strength in $10^{-3}e^2fm^2$ if the transition in question is an E1–transition. In the examined nuclei the dipole strength distribution can roughly be divided into two groups: For all nuclei we observe one and sometimes two strong E1–transitions around 1.5 MeV excitation energy. Here we disregard ^{176}Yb where we could not detect any transitions below 2 MeV due to experimental reasons. A second group of weaker $\Delta K=0$ transitions is distributed between 2 and 4 MeV. In the following we will focus on the lower group of electric dipole transitions.

It is interesting to study the distribution of the observed decay branching ratios for states with spin $J = 1$ in the rare earth region. In figure 2 the branching ratios of 171 states from several nuclei are included [17, 18, 19]. The two maxima at $R = 0.5$ and $R = 2.0$ clearly indicate that for most states the Alaga rules for pure K states are well fulfilled. However, a small number of states exhibit branching ratios which are incompatible with either $K = 1$ or $K = 0$. As will be shown below, these branching ratios can be explained if one assumes K mixing to occur. These results are consistent with theoretical calculations in a microscopic mean-field model [20].

The lowest lying 1^-–states in the deformed nuclei are interpreted as states with an octupole component. The strong coupling between the octupole vibration and the rotational motion of the quadrupole deformed core leads to different bands characterized by their K–quantum number [13]. However, in most of the examined nuclei the photon scattering experiments detected only the 1^-–band-head of the K=0 band which points to the interesting feature, that $J^\pi=1^-$–states with K=0 are much stronger populated than $J^\pi=1^-$–states with K=1. This is in agreement with previous experimental studies [21, 22, 23] and theoretical predictions [24, 25]. There are a few cases where one finds two relatively strong E1–transitions at low energies. We assume that in these cases the $K^\pi=1^-$ and $K^\pi=0^-$–states mix strongly and that the E1–strength is shared between them due to width attraction.

Table 1. Results of a K mixing calculation for ^{156}Gd. Values from ref. [26].

Level	E_x [keV]	R_{exp}	$B(E1)\uparrow$ $[10^{-3}e^2fm^2]$	V_{mix} [keV]	$|Z|$	γ^2
$|A\rangle$	1367	2.34±0.14	16.0±5.9	52.6±7.5	0.066±0.018	0.31±0.09
$|B\rangle$	1243	1.23±0.14	10.0±4.0			

In this picture, the experimentally observed mixed states $|A\rangle$ and $|B\rangle$ are

$$|A\rangle = \alpha|K=0\rangle + \beta|K=1\rangle \tag{1}$$
$$|B\rangle = \beta|K=0\rangle - \alpha|K=1\rangle \tag{2}$$

with $\alpha^2 + \beta^2 = 1$. The observables in photon scattering experiments are the excitation energies E_A and E_B, the branching ratios

$$R_A = \frac{B(1_A \rightarrow 2_1^+)}{B(1_A \rightarrow 0_1^+)} \tag{3}$$

$$R_B = \frac{B(1_B \rightarrow 2_1^+)}{B(1_B \rightarrow 0_1^+)}, \tag{4}$$

and the absolute transition strengths, $B(1_A \rightarrow 0_1^+)$ and $B(1_B \rightarrow 0_1^+)$. Figure 3 shows this two level system. For pure K-states the branching ratios R_A and R_B follow the Alaga rules leading to $R_{K=0} \approx 2.0$ and $R_{K=1} = 0.5$. For the mixed states we derive the branching ratios

$$R_A = 5 \cdot \left(\frac{\langle 2010|10\rangle + \left(\frac{\beta}{\alpha}\right) \cdot \left(\frac{\langle K_f=1|\mathcal{M}|0_1^+\rangle}{\langle K_f=0|\mathcal{M}|0_1^+\rangle}\right) \cdot \langle 2011|11\rangle \cdot \sqrt{2}}{\langle 0010|10\rangle + \left(\frac{\beta}{\alpha}\right) \cdot \left(\frac{\langle K_f=1|\mathcal{M}|0_1^+\rangle}{\langle K_f=0|\mathcal{M}|0_1^+\rangle}\right) \cdot \langle 0011|11\rangle \cdot \sqrt{2}} \right)^2 \tag{5}$$

and

$$R_B = 5 \cdot \left(\frac{\left(\frac{\beta}{\alpha}\right) \cdot \left(\frac{\langle K_f=0|\mathcal{M}|0_1^+\rangle}{\langle K_f=1|\mathcal{M}|0_1^+\rangle}\right) \cdot \langle 2010|10\rangle - \langle 2011|11\rangle \cdot \sqrt{2}}{\left(\frac{\beta}{\alpha}\right) \cdot \left(\frac{\langle K_f=0|\mathcal{M}|0_1^+\rangle}{\langle K_f=1|\mathcal{M}|0_1^+\rangle}\right) \cdot \langle 0010|10\rangle - \langle 0011|11\rangle \cdot \sqrt{2}} \right)^2. \tag{6}$$

We define

$$\gamma := (\beta/\alpha) \tag{7}$$

and

$$Z := \langle K_f=1|\mathcal{M}|0_1^+\rangle / \langle K_f=0|\mathcal{M}|0_1^+\rangle. \tag{8}$$

The variable Z denotes the ratio between the E1-matrix elements with $\Delta K=1$ and $\Delta K=0$.

The mixing matrix element V_{mix} can be expressed as

$$V_{mix} = \alpha \cdot \beta \cdot (E_A - E_B) = \frac{\gamma}{1+\gamma^2} \cdot (E_A - E_B) \tag{9}$$

From the formulas it can be seen that even a very small K-admixture may change the branching ratio for the K=1 state drastically if Z is very small. We performed the calculation for the two close lying levels at 1243 and 1367 keV in ^{156}Gd which could be seen in fig. 1.

Within the error bars the results shown in table 1 are in agreement with previous Coriolis–mixing calculations by Bäcklin et al. [23], but their method needs additional information about other members of the octupole–vibrational band. Our method instead utilizes the newly measured absolute B(E1) values. The agreement between the two methods thus gives rise to a consistent picture.

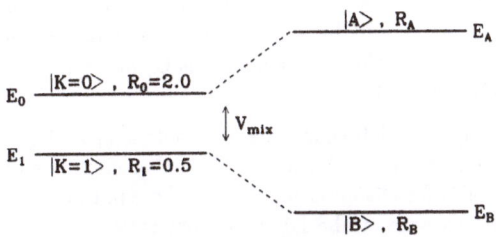

Figure 3. K–mixing of octupole bands.

We summarize that the observables of the photon–scattering experiments allow the determination of K–mixing between two bands if one assumes a simple two–level picture. The idea of K–mixing will be used again in the next chapter to simplify calculations in the sdf–Interacting Boson Model.

E1-TRANSITIONS IN WELL DEFORMED NUCLEI IN THE FRAMEWORK OF THE SDF-IBA

To describe the E1-transitions in the framework of the IBA we first write the T(E1) operator in the full spdf-space [7, 8, 9, 10]:

$$T(E1)_\mu^{spdf} = \alpha_{fd}(d^\dagger \tilde{f})_\mu^{(1)} + \alpha_{dp}(d^\dagger \tilde{p})_\mu^{(1)} + \alpha_{sp}(s^\dagger \tilde{p})_\mu^{(1)} + h.c. \tag{10}$$

It is important that we assume the parameters α_{fd}, α_{dp} and α_{sp} to depend only on the total number of bosons, N.

If one assumes that the p-boson represents the Giant Dipole Resonance (GDR) thus having a very high excitation energy of $\simeq 15\,\text{MeV}$, it is possible to project the E1–operator from the full spdf-space onto the sdf-space, i.e. the p–boson is eliminated via perturbation theory. The T(E1)–operator then takes the form [25, 27]:

$$T(E1)_\mu^{sdf} = \left[\alpha_{fd}(d^\dagger \tilde{f})_\mu^{(1)} + \alpha_{ps}O_{1,\mu} + \alpha_{dp}O'_{1,\mu}\right] + h.c. \tag{11}$$

$$= \alpha_{fd}\left[(d^\dagger \tilde{f})_\mu^{(1)} + \frac{\alpha_{ps}}{\alpha_{fd}}O_{1,\mu} + \frac{\alpha_{dp}}{\alpha_{fd}}O'_{1,\mu}\right] + h.c. \tag{12}$$

where $O_{1,\mu}$ and $O'_{1,\mu}$ are the two–body terms

$$O_{1,\mu}(\chi_2) = [Q_{sd}^{(2)}(\chi_2) \times (s^\dagger \tilde{f} + f^\dagger s)^{(3)}]_\mu^{(1)} \tag{13}$$

$$O'_{1,\mu}(\chi_2) = \sum \sqrt{2l+1}(-1)^{m+1} \left\{ \begin{matrix} 2 & 1 & 1 \\ 2 & 3 & m \end{matrix} \right\} [Q^{(2)}_{sd}(\chi_2) \times (d^\dagger \tilde{f} + f^\dagger \tilde{d})^{(m)}]^{(1)}_\mu \qquad (14)$$

with

$$Q^{(2)}_{sd}(\chi_2) = (s^\dagger \tilde{d} + d^\dagger s)^{(2)} + \chi_2 (d^\dagger \tilde{d})^{(2)}. \qquad (15)$$

Introducing $e_1 := \alpha_{fd}$, $\chi_1 := \alpha_{ps}/\alpha_{fd}$ and $\chi'_1 := \alpha_{dp}/\alpha_{fd}$, the T(E1)–operator is finally written as:

$$T(E1)^{sdf} = e_1 \left[d^\dagger \tilde{f} + \chi_1 O_1(\chi_2) + \chi'_1 O'_1(\chi_2) \right] + h.c. \qquad (16)$$

The parameter e_1 is the "effective charge". The structural parameters χ_1 and χ'_1 weighing the different two–body terms depend (from our assumption) only on the total number of bosons. One should be aware of the implicit χ_2–dependence (due to the quadrupole–operator Q_{sd}) of the two–body terms O_1 and O'_1. This parameter χ_2 varies considerably for different nuclei.

One would now start to determine the parameters χ_1 and χ'_1 in the usual way, i.e. through a fit to experimental data. But we will rather employ a different method using an Alaga rule constraint for the determination of these parameters. This method is very simple and it allows to predict the E1–branching ratios from the Hamiltonian alone. In order to use the Alaga rule constraint we consider a well deformed nucleus with a dynamical SU(3) symmetry. Here the states have K as a good quantum number. From experimental data as well as from the geometrical model we expect, that the Alaga rules mentioned in the previous chapter hold for the decay branching ratios of these states, namely that:

$$R_{K=0} = \frac{B(E1; 1^-_{K=0^-} \to 2^+_1)}{B(E1; 1^-_{K=0^-} \to 0^+_1)} = 2.0 \qquad (17)$$

$$R_{K=1} = \frac{B(E1; 1^-_{K=1^-} \to 2^+_1)}{B(E1; 1^-_{K=1^-} \to 0^+_1)} = 0.5 \ . \qquad (18)$$

In order to calculate these branching ratios we first generate wave functions with good K by increasing the quadrupole force parameter A_2 in the Hamiltonian:

$$H_{sdf} = H_{sd}(SU(3)) + \epsilon_f n_f + A_2 (Q_{sd} \cdot Q_f) \ . \qquad (19)$$

If one now calculates the branching ratios of the states with good K with the T(E1)–operator of eq. 16 one finds that the ratios $R_{K=0}$ and $R_{K=1}$ depend strongly on the parameters χ_1 and χ'_1. Therefore it is crucial to have good parameters which can be obtained by solving the equations:

$$R_{K=0}(\chi_1, \chi'_1) = 2.00 \qquad (20)$$
$$R_{K=1}(\chi_1, \chi'_1) = 0.50 \ . \qquad (21)$$

Once the parameters are fixed (and they are assumed to be the same for nuclei with the same number of valence nucleons) it is possible to calculate all B(E1)–ratios without any free parameters. Moreover, the values for χ_1 and χ'_1 obtained in this special case can be used in any IBA Hamiltonian regardless of its structure.

Table 2 summarizes the results of the sdf–IBA–calculations for low lying 1^-–states in the nuclei ^{156}Gd, ^{158}Gd, ^{168}Er and ^{172}Yb. The third column compares the measured experimental branching ratios (data from different experiments are given) with the prediction of the sdf–IBA. One can see that the agreement is very good, even for

Table 2. Comparison of experimental data with the predictions of the sdf–IBA–Model, see explanation in the text.

Nucleus	E_x [keV]		B(1→2)/B(1→0)		B(E1)↑ $[10^{-3}e^2fm^2]$		Z (eq. 8)	e_1 [e·fm]
	Exp	IBA	Exp	IBA	Exp	IBA	IBA	IBA
^{156}Gd	1243	1208	1.5±0.4[a] 1.3±0.1[b]	1.5	10±4	12.6	0.11	0.27
	1367	1347	2.5±0.4[a] 2.3±0.1[b]	2.3	16±6	11.7		
^{158}Gd	977	1026	1.0±0.1[c]	1.0	–	4.7	0.06	0.25
	1264	1207	1.9±0.2[a] 1.8±0.2[c]	2.1	20±5	19.8		
^{168}Er	1359	1344	2.9±0.8[d]	4.0	–	0.7	0.06	0.27
	1786	1798	2.0±0.1[a] 2.0±0.1[d]	2.1	27±2	24.6		
^{172}Yb	1155	1142	4.4±0.3[e]	4.7	–	0.3	0.05	0.15
	1599	1582	1.9±0.4[a] 1.8±0.2[e]	1.9	11±3	11.1		

[a] Data from ref. [12].
[b] Data from ref. [23].
[c] Data from ref. [22].
[d] Data from ref. [28].
[e] Data from ref. [29].

those states with branching ratios strongly deviating from the Alaga–rule values. In the next column the B(E1)↑–strength measured in our photon–scattering experiments is compared with the values obtained from the model calculations. The IBA–B(E1)↑–strength depends on the effective charge e_1 given in the last column. We fitted this charge for each nucleus to all known E1–transitions, reproducing the experimental data rather well. One can see that e_1 is nearly constant for the lighter nuclei.

As mentioned above the ratio Z of the E1–matrix elements with $\Delta K=1$ and $\Delta K=0$ defined in eq. 8 is a very interesting observable. The fifth column of table 2 gives this ratio Z. It depends only on the total number of bosons. It is predicted to be very small for the examined nuclei which is in agreement with the experimental results. The fact that Z is small in the deformed rare earth region and that it can be reproduced by the IBA without free parameters is very interesting. It is a challenge to gain a deeper understanding of this property.

We conclude that the sdf–Interacting Boson Model is able to reproduce E1–transitions in a wide variety of nuclei. In the well deformed rare earth nuclei it was possible to predict decay branching ratios successfully without any free parameter and absolute transition strengths with only a single parameter. It has been demonstrated that the sdf–IBA is a quite powerful tool to investigate electric dipole excitations.

ACKNOWLEDGEMENTS

We thank our colleagues from the NRF–collaboration in Stuttgart and Giessen, especially U. Kneissl, R. D. Heil, H. H. Pitz, C. Wesselborg, H. Friedrichs, S. Lindenstruth, and B. Schlitt for their outstanding engagement to obtain and discuss the experimental data on dipole excitations. The authors gratefully acknowledge valuable discussions with A. Barfield, R. F. Casten, A. Gelberg, and A. Richter. This work was partially supported by the Deutsche Forschungsgemeinschaft (DFG) under contract Br 799/33, by the BMFT under contract number 06OK143, and by the US Department of Energy under contract No DE-AC02-76CH00016.

REFERENCES

[1] A. Arima and F. Iachello, Collective nuclear representations of a SU(6) group, Phys. Rev. Lett. **35**:1069 (1975).

[2] A. Arima and F. Iachello, Interacting boson model of collective states, Ann. of Phys. **99**:253 (1976).

[3] R. F. Casten and D. D. Warner, The interacting boson approximation, Rev. Mod. Phys. **60**:389 (1988).

[4] A. Arima, T. Otsuka, F. Iachello, and I. Talmi, Collective nuclear states as symmetric couplings of proton and neutron excitations, Phys. Lett. **B66**:205 (1977).

[5] F. Iachello, New class of low–lying collective modes in nuclei, Phys. Rev. Lett. **53**:1427 (1984).

[6] D. Bohle, A. Richter, W. Steffen, A. E. L. Dieperink, N. Lo Iudice, F. Palumbo, and O. Scholten, New magnetic dipole excitation mode studied in the heavy deformed nucleus ^{156}Gd by inelastic electron scattering, Phys. Lett. **B137**:27 (1984).

[7] F. Iachello and A. Arima, "The interacting boson model", Cambridge University press, Cambridge (1987).

[8] J. Engel and F. Iachello, Interacting boson model of collective octupole states, Nucl. Phys. **A472**:61 (1987).

[9] D. Kusnezov and F. Iachello, A study of collective octupole states in barium in the interacting boson model, Phys. Lett. **B209**:420 (1988).

[10] T. Otsuka and M. Sugita, Unified description of quadrupole–octupole collective states in nuclei, Phys. Lett. **B209**:140 (1988).

[11] F. R. Metzger, Low–lying E1–transitions in the stable even Sm isotopes, Phys. Rev. **C14**:543 (1976).

[12] A. Zilges, P. von Brentano, H. Friedrichs, R. D. Heil, U. Kneissl, S. Lindenstruth, H. H. Pitz, and C. Wesselborg, A survey of ΔK=0 dipole transitions from low lying J=1 states in rare earth nuclei, Z. Phys. A – Hadrons and Nuclei **340**:155 (1991).

[13] W. Donner and W. Greiner, Octupole vibrations of deformed nuclei, Z. Phys. **197**:440 (1966).

[14] H. H. Pitz, R. D. Heil, U. Kneissl, S. Lindenstruth, U. Seemann, R. Stock, C. Wesselborg, A. Zilges, P. von Brentano, S. D. Hoblit, and A. M. Nathan, Low energy photon scattering off ^{142,146,148,150}Nd: An investigation in the mass region of a nuclear shape transition, Nucl. Phys. **A509**:587 (1990).

[15] R. D. Heil, B. Kasten, W. Scharfe, P. A. Butler, H. Friedrichs, S. D. Hoblit, U. Kneissl, S. Lindenstruth, M. Ludwig, G. Müller, H. H. Pitz, K. W. Rose, M. Schumacher, U. Seemann, J. Simpson, P. von Brentano, Th. Weber, C. Wesselborg, and A. Zilges, Parity assignments in Nuclear Resonance Fluorescence experiments using Compton polarimeters, Nucl. Phys. **A506**:223 (1990).

[16] U. Kneissl, Parity assignments in photon scattering using Compton polarimeters, Prog. Part. and Nucl. Phys. **28**:331 (1992).

[17] A. Zilges, P. von Brentano, A. Richter, R. D. Heil, U. Kneissl, H. H. Pitz, C. Wesselborg, Uncommon decay branching ratios of spin-one states in the rare-earth region and evidence for K mixing, Phys. Rev. **C42**:1945 (1990).

[18] A. Richter, Electron scattering and elementary excitations, Nucl. Phys. **A522**:139c (1991).

[19] P. von Brentano, A. Zilges, R. Jolos, A. Richter, R. D. Heil, U. Kneissl, H. H. Pitz, C. Wesselborg, in: AIP conference proceedings **238**, "Capture Gamma-Ray Spectroscopy", p. 234, Editor R. W. Hoff, American Institute of Physics, New York, 1991.

[20] E. Hammarén, P. Heikkinen, K. W. Schmid, A. Faessler, Microscopic and phenomenological analysis of the Alaga rule for dipole states, Nucl. Phys. **A541**:226 (1992).

[21] L. Kocbach and P. Vogel, Branching ratios for E1 transitions deexciting the octupole states in even–even deformed nuclei, Phys. Lett. **B32**:434 (1970).

[22] R. C. Greenwood, C. W. Reich, H. A. Baader, H. R. Koch, D. Breitig, O. W. B. Schult, B. Fogelberg, A. Bäcklin, W. Mamper, T. von Egidy, and K. Schreckenbach, Collective and two–quasiparticle states in ^{158}Gd observed through study of radiative neutron capture in ^{157}Gd, Nucl. Phys. **A304**:327 (1978).

[23] A. Bäcklin, G. Hedin, B. Fogelberg, M. Saraceno, R. C. Greenwood, H. A. Baader, H. D. Breitig, O. W. B. Schult, K. Schreckenbach, T. von Egidy, and W. Mampe, Levels in ^{156}Gd studied in the (n,γ) reaction, Nucl. Phys. **A380**:189 (1982).

[24] V. G. Soloviev and V. A. Sushkov, Electric–dipole transitions in doubly even deformed nuclei, Phys. Lett. **B262**:189 (1991).

[25] P. von Brentano, N. V. Zamfir, and A. Zilges, E1 operator in the sdf-Interacting Boson Model from an Alaga rule constraint, Phys. Lett. **B278**:221 (1992).

[26] H. H. Pitz, U. E. P. Berg, R. D. Heil, U. Kneissl, R. Stock, C. Wesselborg, and P. von Brentano, Systematic study of low–lying dipole excitations in ^{156,158,160}Gd by photon scattering, Nucl. Phys. **A492**:411 (1989).

[27] N. V. Zamfir, O. Scholten, and P. von Brentano, E1 calculations in the sdf–interacting boson model, Z. Phys. A – Atomic Nuclei **337**:293 (1990).

[28] J. K. Tuli, R. R. Kinsey, and M. J. Martin, Nuclear Data Sheets **53**:223 (1988).

[29] R. C. Greenwood, C. W. Reich, and S. H. Vegors, Level structure of ^{172}Yb form the ^{171}Yb(n,γ) reaction, Nucl. Phys. **A252**:260 (1975).

EFFECTIVE CHARGES, THE VALENCE p-n INTERACTION, AND THE IBM

R. F. Casten[1] and A. Wolf[1,2]

[1]Brookhaven National Laboratory, Upton, New York, 11973, USA
[2]Nuclear Research Centre, Negev, Beer Sheba, Israel

INTRODUCTION

There are three recent themes in nuclear structure that come together in an interesting and useful way via the concept of effective charges and the framework of the IBM. These three concepts are the importance of dynamical symmetries in describing nuclear structure and the benefits that accrue from their exploitation, secondly, the critical role of the p-n interaction in the onset and development of collectivity in nuclei, and, thirdly, the importance of the valence nucleons in determining structure and its evolution.

We will illustrate this by showing that the interpretation of measured B(E2) values in the context of the dynamical symmetries of the IBM leads to new insights into the meaning of effective charges and offers new avenues to understand the role of the proton-neutron (p-n) interaction in modulating the nature of the valence space and the growth of collectivity. In particular, we will show that effective charges in *valence* models, such as the IBM, can be interpreted in terms of *derivatives of the collectivity of the low lying levels*, that is, as measures of the rate of change of collectivity as the proton and neutron numbers vary. This paper is based on recent work[1] by the authors.

EFFECTIVE CHARGES IN VALENCE MODELS

In most nuclear models, effective charges are simply multiplicative factors on calculated transition rates. They are employed to ensure agreement with experiment and to compensate for simplifications or approximations in the model such as truncations of the basis space (in which case the effective charges would mock up effects such as core polarization), simplified forms of the transition operator, or the use of effective interactions amongst the nucleons. [Of course, some microscopic models can estimate the values of these effects, in which case the effective charges may not be free parameters or may be partially constrained. Multi-$\hbar\omega$ symplectic models[2] are a case in point.]

Symmetries in Science VII, Edited by B. Gruber
and T. Otsuka, Plenum Press, New York, 1994

In valence models, however, that is, in models that consider only the valence nucleons in the determination of structure, effective charges take on an additional, and more physically intuitive, role. To see this, we exploit the simplicities of the Interacting Boson Model (IBM)[3].

Let us first consider the three limiting symmetries of this model, the U(5), SU(3), and O(6) limits. We can then write expressions[4] for the key observable $M(E2:2_1^+ \rightarrow 0_1^+)$, where M(E2) is just the square root of the corresponding B(E2) value, as follows:

$$\text{U(5)} \qquad M(E2:2_1^+ \rightarrow 0_1^+) = \sqrt{1/N} \ (e_\pi N_\pi + e_\nu N_\nu) \qquad (1)$$

$$\text{O(6)} \qquad M(E2:2_1^+ \rightarrow 0_1^+) = \left(\frac{N+4}{5N}\right)^{1/2} (e_\pi N_\pi + e_\nu N_\nu) \qquad (2)$$

$$\text{SU(3)} \qquad M(E2:2_1^+ \rightarrow 0_1^+) = \left(\frac{2N+3}{5N}\right)^{1/2} (e_\pi N_\pi + e_\nu N_\nu) \qquad (3)$$

Note that all three expressions have the same form, namely, a function of the total boson number N and a separate factor of the form $(e_\pi N_\pi + e_\nu N_\nu)$, incorporating the boson effective charges e_π, e_ν and the proton and neutron boson numbers N_π, N_ν.

Most nuclei, of course, do not satisfy the exact strictures of a specific symmetry, but are intermediate in structure. Exact analytic expressions cannot be written down for such cases, but very good approximate expressions are available: these incorporate parameters from the IBM-1 Hamiltonian. For example, for nuclei near U(5), Ginocchio and Van Isacker[4] have developed the expression:

$$M(E2:2_1^+ \rightarrow 0_1^+) = \sqrt{\frac{1}{N}} \left(1 - \frac{\kappa(N-1)}{\varepsilon}\right)^2 (e_\pi N_\pi + e_\nu N_\nu) \qquad (4)$$

where ε and κ are the coefficients of the $\hat{n}_d$ and Q·Q terms in the IBM Hamiltonian, respectively, and where the latter gives the scale of U(5) symmetry breaking. For nuclei between O(6) and SU(3) [this category includes the vast majority of deformed nuclei such as the rare earths or the actinides], the following expression has been developed[5]

$$M(E2:2_1^+ \rightarrow 0_1^+) = 0.5 \left(\frac{N+1}{N}\right) (1-0.1\chi) (e_\pi N_\pi + e_\nu N_\nu) \qquad (5)$$

Here, χ is an internal parameter in the IBM quadrupole operator, Q. χ varies from 0 [for O(6)] to -1.32 [for SU(3)] and is typically near -0.5 for deformed nuclei. [Equation 5, it will be noted, has only a weak dependence on χ.] A more general (and more complicated) formula exists[6] which spans the entire IBM symmetry "triangle." It need not be given here. These approximate formulas, of course, differ in detail from those applicable to the limiting symmetries (eqs. 1-3), but they again share the factorizable form of an expression in N and one in the effective charges.

Thus, rather generally, in the IBM, we can write:

$$M(E2:2_1^+ \rightarrow 0_1^+) = f(N) \ (e_\pi N_\pi + e_\nu N_\nu) \qquad (6)$$

Similar expressions may apply to other valence models.

There are two interesting aspects of eq. 6. One is that it allows the easy extraction of the effective charges from sets of B(E2) values for a sequence of nuclei (e.g., an isotopic series). The second is that it reveals a new interpretation of the effective charges.

We can see the first point simply by re-writing eq. 6 as

$$\frac{M(E2:2^+_1 \to 0^+_1)}{f(N)\,N_\nu} = \left(e_\nu + e_\pi \frac{N_\pi}{N_\nu}\right) \tag{7}$$

The left hand side (lhs) is linear in the ratio N_π/N_ν with slope e_π and intercept e_ν. Since N_π and N_ν are known from counting the numbers of valence nucleons (more on this later), $f(N)$ is known from the structure in a given region, and the M(E2) values are measured quantities, it is possible to plot the lhs of eq. 7 against N_π/N_ν for a series of nuclei and, if a linear behavior results, to extract the boson effective charges. (We will illustrate this process momentarily.) It is important to stress here that the effective charges thus extracted refer to, and indeed are only defined in the context of, a *series* of nuclei, not an individual nuclide.

The second point is evident if one inspects the forms of eqs. 1-5 and notes that $f(N)$ is nearly always more slowly varying than $(e_\pi N_\pi + e_\nu N_\nu)$. If we therefore make the approximation that $f(N)$ is constant, and differentiate eq. 6, we obtain the fascinating results that

$$\frac{\partial M(E2)}{\partial N_\nu} \approx f(N)e_\nu \quad \text{and} \quad \frac{\partial M(E2)}{\partial N_\pi} \approx f(N)e_\pi \tag{8}$$

These equations have the immediate physical interpretation that e_π and e_ν are proportional to the rate of change of collectivity (as reflected in the $M(E2:2^+_1 \to 0^+_1)$ values), with respect to N_π and N_ν, respectively. Thus, extracting these effective charges by means of eq. 7 immediately discloses the evolution of collective structure for a series of nuclei.

Of course, a key to this procedure is the assumption that $f(N)$ varies slowly enough that it is reasonable to treat it as constant for a set of nuclides. Except for U(5), $f(N)$ typically changes nearly an order of magnitude slower than $(e_\pi N_\pi + e_\nu N_\nu)$, especially for large N_π and N_ν. This is clear, for example, in eqs. 2 and 3, where the $f(N)$ functions $((N+4/5N))^{1/2}$ and $(2N+3/5N))^{1/2}$ are asymptotically constant for large N. Even in the case of U(5) the leading order term in the derivative of eq. 1 is independent of $f(N)$. A more rigorous derivation of this result is given in the second article of ref. 7.

EMPIRICAL EXTRACTION AND INTERPRETATION OF EFFECTIVE CHARGES

We now illustrate the extraction of the boson effective charges and their interpretation in light of eq. 8 and the discussion above. In fig. 1 we show plots of the lhs eq. 7 against N_π/N_ν for the Pd and Sm isotopes. We note that the plots are indeed linear, in fact, often to higher accuracy than might be expected from the experimental uncertainties. In each case we show more than one fit, using different expressions of the type of eqs. 1-6 to illustrate that the extracted effective

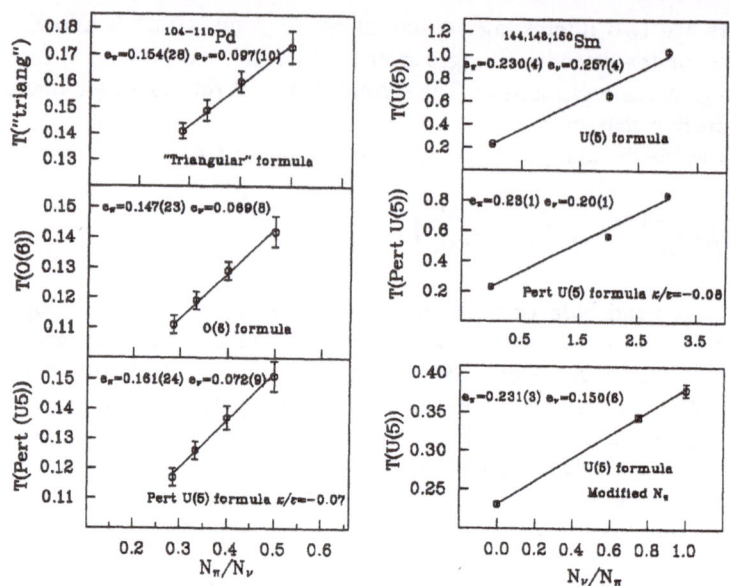

Figure 1. Examples of the linear fits described in the text. $T=M(E2)/f(N)N_\rho$, where $\rho=\nu$ for Pd and π for Sm. Left: Fits, with different formulas, for Pd. Top right: two fits to Sm. (Note that $e_\nu > e_\pi$.) In the "perturbed" U(5) formula, κ and ε are IBM parameters. Bottom right: fit to Sm using effective N_π values, yielding "normal" e_π and e_ν values. (See text.)

charges are not very sensitive to uncertainties in which forms of f(N) are used. (We will discuss the lower right panel of fig. 1 later.)

We have carried out fits like those in fig. 1 for isotopic chains from Mo to the actinides. The extracted effective charges reveal some interesting features, especially in light of the natural expectation that e_ν should be substantially less than e_π. We show plots of e_π, e_ν and e_ν/e_π against Z in fig. 2. The e_π values, with few disruptions, exhibit a nearly linear, monotonically increasing, behavior with Z, from values near 0.1 for A~100 to 0.35 for the actinides. More striking, e_ν shows two remarkable features, namely, first, a repeated undulation or oscillating behavior in each shell, with large values near magic numbers and in shape transitional regions and very small values near midshell, and, secondly, values in the former regions that can actually *exceed* e_π. This is highlighted in the ratio plot on the top of the figure which shows at least two regions where $e_\nu/e_\pi > 1$. (Incidentally, these plots dramatically point to the need for further data in the Pb region which has all the earmarks of another region where e_ν may be greater than e_π.)

The "derivative" interpretation of these effective charges allows a simple physical interpretation of the large e_ν values which have, heretofore, been extremely puzzling. At the most superficial level, large e_ν values merely imply that collectivity [M(E2)] changes faster with neutron than proton number in a given region. But the real issue is *why* should this be so? Disregarding minor orbit dependencies of the p-n interaction, the success of the N_pN_n scheme[7]

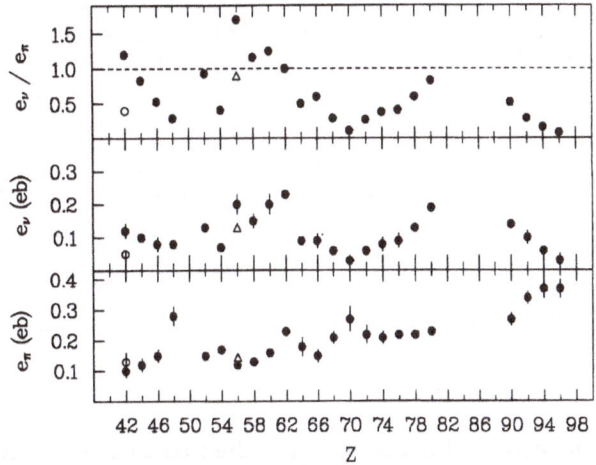

Figure 2. Empirical e_π, e_ν and e_ν/e_π values for isotopic sequences of all non-magic even-even nuclei with Z>40. The uncertainties reflect the fit quality and the variations in e_π and e_ν with different analytical formulas. Open symbols for Mo and Ba give values for the rotational isotopes of these elements following the spherical-deformed phase transitional regions.

suggests that changes in collectivity should vary symmetrically with N_π and N_ν. However, the regions where $e_\nu > e_\pi$ are exactly those where this is not the case because, here, the neutrons play a dual role[8,9]. Of course, additional neutrons add to the total number of valence nucleons and to the $N_p N_n$ product, and thus increase the collectivity, primarily via the quadrupole component of the p-n interaction[10]. However, they *also* affect the single particle energies of the *protons* via the attractive monopole p-n interaction[10]. For example, in the A = 100 (150) region, the addition of neutrons in the $1g_{7/2}$ ($1h_{9/2}$) orbit, lowers the effective single particle energy of the proton $1g_{9/2}$ ($1h_{11/2}$) orbit. This effect can be large, can grow as a function of neutron number, and can actually alter (create or destroy) important sub-shell gaps in the proton energy levels. Again, in the A = 100 (150) regions, near N = 60 (90), the addition of neutrons obliterates the proton sub-shell gaps at Z = 40 (64). This in turn affects the way one counts the number of valence proton bosons, N_π, used in eqs. 1-7. However, since N_π is *defined* to be constant in applying these equations for an iso*topic* series, this effect shows up as an enhanced multiplier, e_ν, of N_ν. Thus, we see that the effective charges can, in a formalism of valence models, give important information on subtle yet critical aspects of the p-n interaction which are key to understanding the onset of collectivity and shape/phase transitions in nuclei.

With this explanation in hand, one could, of course, turn the procedures we have used around by making a *model* of the changing proton shell structure (that is, a model for how N_π changes with neutron number) in some region, insert these N_π values in eq. 7, re-do the fits and the extraction of the effective charges

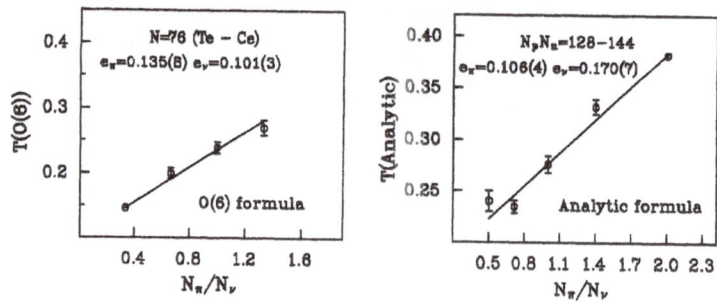

Figure 3. As in Fig. 1, except for the isotonic and iso-N_pN_n sequences indicated.

and determine if "normal" (that is, $e_\nu < e_\pi$) values result as expected. The lower right panel in fig. 1, for Sm, illustrates this procedure (we took $N_\pi = 1,2,4$ for ^{144,148,150}Sm, respectively) and shows that more traditional values of e_ν do result.

We noted before that the e_π and e_ν values extracted were inherently defined with respect to a specifically chosen series of nuclei. (In the cases above, these were isotopic chains.) They should not necessarily be expected to be identical to values extracted, for example, from several transitions in a single nucleus. More significantly, e_π and e_ν values for a nucleus in, say, an isotopic chain, should not be expected to coincide with those for the same nucleus extracted from a *different* (say, isotonic) sequence. For example, the e_π and e_ν values for ^{106}Pd obtained from the Pd isotopes in fig. 1 will not necessarily be the same as those extracted from the N = 60 isotonic series. This is evident from our derivative interpretations. Collectivity does not vary isotropically in the N-Z plane: since e_π and e_ν reflect these rates of change, they will, in general, be different for different "cuts" through the plane. Note also that one is not limited to isotopic or isotonic sequences. One can inspect isobaric sequences or more exotic sets such as iso-N_pN_n, iso-P, iso-F-spin, and so on. We illustrate these ideas in fig. 3 which shows an isotonic and an iso-N_pN_n sequence. In isotopic (isotonic) series, if the B(E2) values were constant, the e_ν (e_π) values extracted would be zero and, hence, values of e_π (e_ν) of the order of 0.1-0.3 eb [see plots in ref. 1 and, here, in figs. 1, 2 and 3 (left)] point to significant rates of change of collectivity with N(Z) in these chains of nuclei. In contrast, for an iso-N_pN_n sequence, inspection of the form of eq. 6 shows that constant B(E2) values lead to equal (and non-zero) e_π and e_ν. For the iso-N_pN_n nuclei in fig. 3 (right) a constant B(E2) at the average value observed for these nuclei would give $e_\pi = e_\nu = 0.14$. It is then to this value that the extracted $e_\pi = 0.106$, $e_\nu = 0.170$ values should be compared: we see that the small deviations of about 0.03 eb for both e_π and e_ν from the ideal average value do indeed reflect the relative constancy of structure in this N_pN_n multiplet.

Extension of these results to other nuclides is an on-going project. Clearly, with extensive mapping in this way, and with the availability of sufficient data, the full 2-dimensional surface of the derivatives of collectivity could be mapped out. Such a result could be highly interesting in itself and a challenge to both algebraic and microscopic models of nuclear structure.

ACKNOWLEDGEMENTS

We would like to express our deep appreciation to Franco Iachello, in whose honor this volume is dedicated. He, and his seminal work with Akito Arima, has been the inspiration for much of our own research over a decade and a half. Beyond this, we are particularly grateful for innumerable discussions, suggestions, guidance, and encouragement over this time period. We would also like to acknowledge many fruitful discussions, on the specific subject matter of this article, with J. N. Ginocchio, O. Schölten, A. Leviatan, I. Talmi, K. Heyde, S. Pittel, B. Barrett, P. von Brentano, and D. D. Warner. Research has been performed in part under contract No. DE-AC02-76CH00016 with the United States Department of Energy.

REFERENCES

1. R.F. Casten and A. Wolf, *Phys. Lett.* B280:1 (1992); A. Wolf and R.F. Casten, *Phys. Rev.*, to be published.
2. O. Castanos et al., *Nucl. Phys.* A524:469 (1991).
3. A. Arima and F. Iachello, *Phys. Ref. Lett.* 35:1069 (1975).
4. J.N. Ginocchio and P. Van Isacker, *Phys. Rev.* C33:365 (1986).
5. R.F. Casten and A. Wolf, *Phys. Rev.* C35:1156 (1987).
6. A. Wolf, O. Scholten and R.F. Casten, *Phys. Rev.* C43:2279, (1991).
7. R. F. Casten, *Phys. Rev. Lett.* 54:1991 (1985).
8. P. Federman and S. Pittel, *Phys. Lett.* B69:385 (1977); B77:29 (1977).
9. R.F. Casten et al., *Phys. Rev. Lett.* 47:1433 (1981).
10. K. Heyde et al., *Nucl. Phys.* A466:189 (1987).

DYNAMICAL SUPERSYMMETRIES IN
SUPERDEFORMED NUCLEI

Jolie A. Cizewski

Department of Physics and Astronomy
Rutgers University
New Brunswick, New Jersey 08903 USA

INTRODUCTION

Since the initial observation of superdeformed rotational bands in A≈190 nuclei in 1989, there has been an explosion of data in the Hg-Tl-Pb isotopes with as many as 35 superdeformed (SD) rotational bands reported in 15 isotopes in this region.[1-11] The current status is summarized in Fig. 1, which also indicates the number of bands seen in each isotope.

What was totally unexpected was the observation of SD band transitions with identical or simply related energies. The initial observation[12] was in pairs of A≈150 nuclei (^{151}Tb*-^{152}Dy) and (^{150}Gd*-^{151}Tb) where the γ-ray energies in the pairs were identical within 1-3 keV. This was followed shortly by the observation[13] of excited superdeformed bands in ^{194}Hg with γ-ray energies related to those of the only known SD band in ^{192}Hg. Since these initial observations, 6 pairs of "identical" SD bands have been observed in A≈150 nuclei[12] and 14 cases[8,13] have been observed in A≈190 SD bands. In this paper the A≈190 data and the theoretical symmetry implications will be discussed.

SD BAND MEASUREMENTS AND THE DETERMINATION OF SPINS

The SD excitations have been found in a variety of reactions, with beams ranging from light heavy-ions to ^{48}Ca. Typically they represent 1-2% of the channel cross section, are populated at 30-40ℏ of angular momentum, and depopulate to normal yrast structures in one or two transitions, in many cases with the lowest SD transition energy ≈250 keV. While in many cases the isotopic identification of the SD bands is confirmed by the observation of prompt coincidences with known low-lying yrast transitions, in no case has a transition which links the superdeformed band to normal excitations been observed.

Symmetries in Science VII, Edited by B. Gruber
and T. Otsuka, Plenum Press, New York, 1994

Therefore, no definitive knowledge of the excitation energy of these structures can be determined, although the second minima in the potential energy surfaces are predicted[14] to occur at 4-6 MeV in excitation at zero angular momentum. Although searches for superdeformed excitations have been made in Pt, Au, Bi, and Po isotopes, to date, no additional element in this region has been observed to have an SD band.

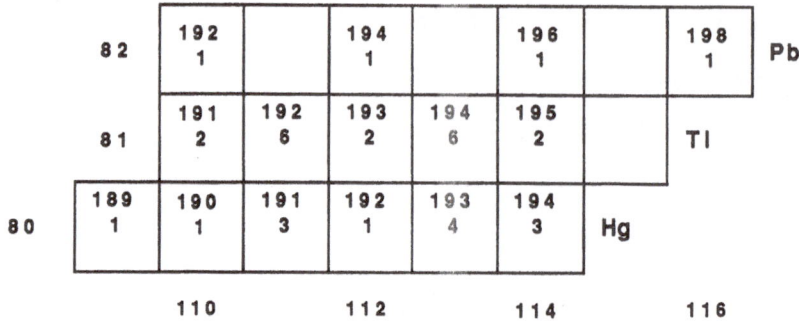

Figure 1   SD bands in the A=190 nuclei.

That connecting transitions between the SD bands and the normal low-lying states have not been observed also means that the angular momenta of the lowest states in these SD bands cannot be determined in a model-independent manner. Fortunately, these SD cascades extend to low γ-ray energy, as low as 169 keV in [194]Pb. Also, in many cases the γ rays have been confirmed to be of $\Delta L = \Delta J = 2$ character, and this assumption has been extended to all SD band cascades. Therefore, spins of individual band members can be determined[15] using simple expectations of quantum rotors.

In a nuclear rotor the excitation energies can be fit by an expansion in J(J+1):

$$E_x(J) = \hbar^2/2\vartheta \, J(J+1) + B \, [J(J+1)]^2 + C \, [J(J+1)]^3 + \ldots \tag{1}$$

The γ-ray transition energy is the difference in excitation energies, $E_x(J+1) - E_x(J-1)$. In a typical superdeformed cascade, 9-17 γ rays are measured, each of which carries 2 units of angular momentum. Therefore, the spin, J_f, of the level at the bottom of the cascade, and the moment of inertia parameters, $A = \hbar^2/2\vartheta$, B, and possibly C, can be determined from a least-squares fit of the empirical transition energies to the theoretical expectations determined from eq. 1. For the A$\approx$190 SD bands, J_f is determined to be integer with high confidence for all even-even nuclei; for most odd-A nuclei J_f is determined to be half-integer. Alternative formulae have also been used to extract angular momenta of SD bands;[15] within errors all methods give the same values for J_f.

"IDENTICAL" BAND STRUCTURES

As displayed in Fig. 2, one of the [194]Hg* bands has γ-ray energies at the arithmetic mean or "midpoint" values compared to the values for the strongly populated SD band in

these curves are parallel, offset by an angular momentum or alignment of exactly $1.0\hbar$. The alignment is quantized for at least 8 SD bands with respect to ^{192}Hg, and all six SD bands in ^{194}Tl with respect to ^{193}Tl, as was shown in refs. 13 and 8, respectively.

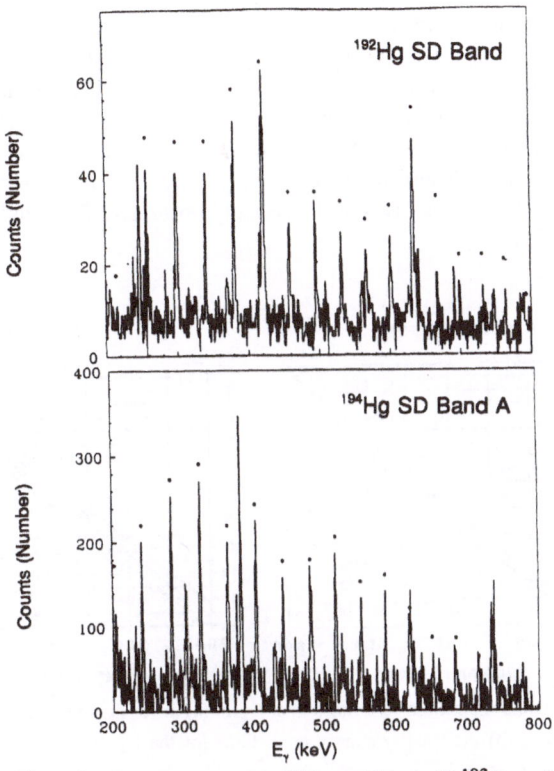

Figure 2   Gamma-ray spectra of SD transitions in (a) ^{192}Hg and (b) the even-spin excited SD band in ^{194}Hg.

There are then two aspects of the experimental results which are unexpected and need to be understood. First are the identical or related γ-ray energies in a large number of nuclei when compared to ^{192}Hg or ^{193}Tl. Second is the observed integer alignment, which sets in at moderate rotational frequencies. The remainder of this paper will present a possible theoretical interpretation of these observations in terms of dynamical supersymmetries.

In Fig. 3a the angular momenta as a function of γ-ray energy are plotted for the excited SD bands in ^{194}Hg and the ^{192}Hg SD band. Especially at the higher γ-ray energies, ^{192}Hg. From naive expectations one expects the moment of inertia to be proportional to $A^{5/3}$, or a 1.7% difference between the moments of inertia for ^{192}Hg and ^{194}Hg; the γ-ray energies indicate a difference of <<1%. However, this example is not isolated. For A≈190 nuclei almost all of the SD bands which involve odd neutron configurations, as well as ^{194}Pb, exhibit "identical" bands. The main exceptions are the only SD band in ^{189}Hg and the strongly populated SD band in ^{191}Hg. Both exceptions probably involve a valence neutron in a single-particle "intruder" configuration[1,3] which has high-j and small projection on the symmetry axis, such as the 3/2-[751] $j_{15/2}$ orbital. The other exception occurs in ^{193}Hg: two of the four bands display anomalous behavior[4] of their dynamical moments of inertia as a function of γ-ray energy, a behavior which has been interpreted as evidence for the interaction between two separate intrinsic excitations. However, the other 2 bands in ^{193}Hg are part of the systematic pattern observed also for the two weaker bands in ^{191}Hg, the two weaker bands in ^{194}Hg, the only band in ^{194}Pb, and all six SD bands in ^{194}Tl (related to ^{193}Tl). Also, two of the recently observed SD bands[11] in ^{192}Tl can be related those in ^{191}Tl.

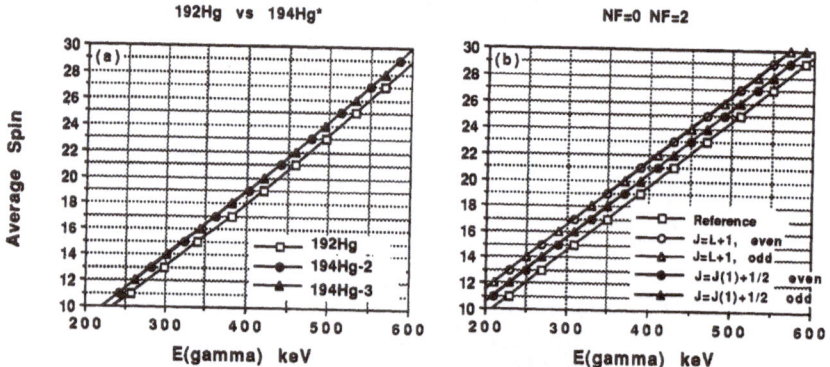

Figure 3   Average angular momentum as a function of E_γ for (a) the excited SD bands in ^{194}Hg compared to the ^{192}Hg reference; (b) the supersymmetry predictions for the $N_F=2$ predictions from eq. 2 and 5 compared to the $N_F=0$ reference.

194 Pb*		
NF=2		
193 Tl	194 Tl	
NF=1	NF=2	
192 Hg	193 Hg	194 Hg*
NF=0	NF=1	NF=2

Figure 4   Supermultiplets in A≈194 nuclei.

DYNAMICAL SUPERSYMMETRIES IN DEFORMED NUCLEI

It has been suggested[16,17] that the identical structures in superdeformed nuclei are examples of the dynamical supersymmetries that occur in the framework of the Interacting Boson Fermion Approximation (IBFA) model. The beauty of a supersymmetry is that it predicts the γ-ray energies in all members of a supermultiplet to be related. The supermultiplets that could be appropriate for the A≈190 SD nuclei are shown in Fig. 4. The nuclei are labeled by N_B and N_F, where N_B is the number of bosons, N_F is the number of fermions, and $N_B + N_F$ = constant for all members of a supermultiplet. For a supersymmetry the yrast SD band in ^{192}Hg would be related to N_F=1 one-particle bands in ^{193}Hg, and N_F=2 two-particle bands in ^{194}Hg. The yrast SD band in ^{194}Hg could be the N_F=0 member of another supermultiplet and, in general, will have no relation to ^{192}Hg.

For a supersymmetry to occur three criteria must be met. First, the core structure must be an example of a boson symmetry; SU(3) symmetry is appropriate for these highly deformed nuclei. While most tests of boson-fermion symmetries have assumed that the core structure is dominated by s and d bosons, and that the valence particle occupies a limited number of orbitals, the present work requires no such restrictions. The boson SU(3) symmetry is realized for many boson systems. Second, the valence nucleons must be in specific orbitals, which are part of a complete major shell, e.g., j=1/2,3/2,5/2, or j=1/2,3/2,5/2,7/2, etc. In particular, the valence nucleon cannot occupy a unique-parity orbital, isolated from other orbitals of its major shell. The observed dynamic moments of inertia of the nuclei with structures related to ^{192}Hg suggest that the valence particle is not in a unique-parity configuration. The requirement of a complete shell is met if one exploits the pseudo-harmonic oscillator.[18] For all major shells with principal quantum number N>3, the highest j=ℓ+1/2, ℓ=N, orbital has been lowered into the N-1 shell. However, the normal parity orbitals which are left in that shell have exactly the j values of the N-1 shell. Since j is a good quantum number, rather than ℓ and s, the structure of the normal parity orbitals in a realistic N shell can be approximated by a complete pseudo-harmonic oscillator Ñ=N-1 shell. This pseudo-harmonic oscillator approximation is also valid for finite nuclear deformations, as long as the j=ℓ+1/2 spherical state does not contribute significantly to the deformed wave function. The specific set of orbitals for the odd particle need not be specified, since the fermion excitations take on one of a number of generic types of spectra.

In a deformed supersymmetric system the excitation energies can be given by:[17]

$$E = E_0 + A\,S(S+1) + B\,L(L+1) + C\,J(J+1) \qquad (2)$$

where the parameter E_0 gives the band-head energies, and the excitation energies depend upon the intrinsic angular momentum S, an integer orbital angular momentum L, and total angular momentum J=L+S. The third requirement for a supersymmetry is satisfied if the same parameters apply to all nuclei in the supermultiplet.

For the even-even nucleus with N_F=0, S=0; hence, J=L, and the expression for the γ-ray energies, $E_x(J+1)-E_x(J-1)$, becomes:

$$E_\gamma(N_F=0) = (B+C)\,(4L+2) \qquad (3)$$

For the members of the supermultiplet with N_F=1, the expression for the γ-ray energies can depend upon B and C, not just B+C. The parameters A and E_0 only affect the band-head energies, not the transition energies, so only 2 parameters are needed to fit γ-ray energies for all excitations in a supermultiplet.

Three generic types of spectra characterize the S=1/2, N_F=1 nuclei. Two of these involve decoupled structures, which would not give rise to the signature pairs that characterize the A≈190 related structures, although such decoupled, identical structures are observed in A≈150 SD nuclei. The third generic level diagram expected in a supersymmetry description of an odd-A nucleus with N_F=1, S=1/2 is shown in Fig. 5, and will be compared to the data for [193]Hg.

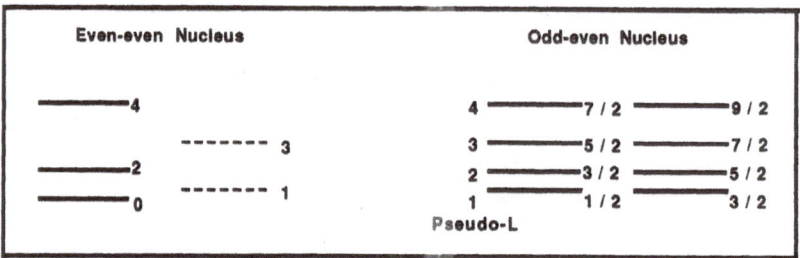

Figure 5   Generic spectrum for N_F=1, S=1/2 (with strong coupling) dynamical supersymmetry.

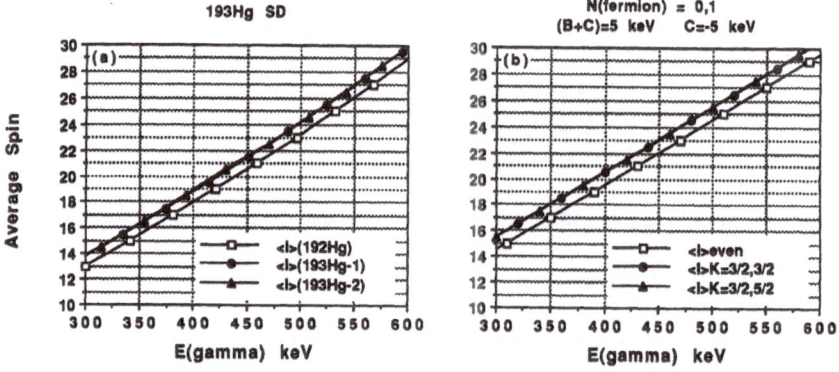

Figure 6   Average angular momentum as a function of E_γ for (a) the excited SD bands in [193]Hg compared to the [192]Hg reference; (b) the supersymmetry predictions for the N_F=1 spectrum shown in Fig.5 and compared to the N_F=0 reference.

A comparison of the plots of the angular momenta as a function of E_γ is presented in Fig. 6 for (a) the [193]Hg data, with [192]Hg as reference, and (b) the expectations of a N_F=1 supersymmetry, with (B+C) = 5 keV, and C = -5 keV. This choice of parameters reproduces all of the trends in the data: the identical moments of inertia, the observation of signature partners, and the critical test of alignment i=1ℏ. The same agreement between experiment and theory would also be obtained for the bands in [191]Hg related to [192]Hg.

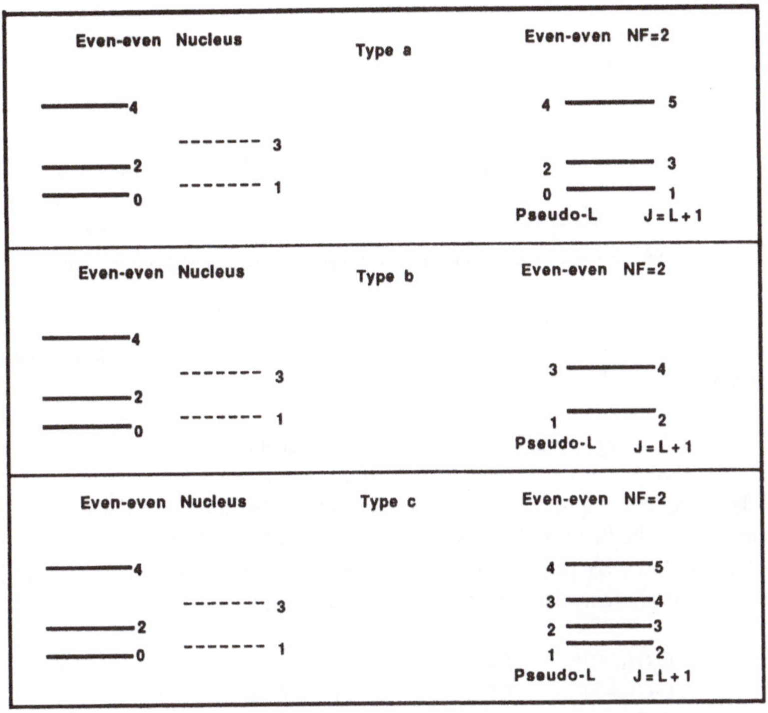

Even-even Nucleus   Type a   Even-even NF=2

```
————— 4                                      4 ————— 5
            ------- 3
————— 2                                      2 ————— 3
            ------ 1
————— 0                                      0 ————— 1
                                            Pseudo-L      J=L+1
```

Even-even Nucleus   Type b   Even-even NF=2

```
————— 4
            ------- 3                         3 ————— 4
————— 2
            ------ 1                          1 ————— 2
————— 0                                      Pseudo-L      J=L+1
```

Even-even Nucleus   Type c   Even-even NF=2

```
————— 4                                      4 ————— 5
            ------- 3                         3 ————— 4
————— 2                                      2 ————— 3
            ------ 1                          1 ————— 2
————— 0                                      Pseudo-L      J=L+1
```

Figure 7   Generic spectra for $N_F=2$, $J=L+1$ (eq. 2) dynamical supersymmetry.

For a multi-fermion system with the particles in the same oscillator shell eq.2 with the same parameters will predict spectra for nuclei in which N_B+N_F = constant, which includes two-particle, as well as many-particle, excitations. The spectra associated with the $N_F=2$ systems take on a large number of forms, since both S=0 and 1 are allowed, as well as the variety of ways that $J=L+S$ can add vectorially. In the case of $^{194}Hg^*$, two bands are observed and the alignment with respect to ^{192}Hg is $1\hbar$. This suggests that a supersymmetry in which the spins of the fermions are aligned, S=1, could be appropriate. The spectra for S=1, and $J=L+1$ $N_F=2$ supersymmetries are given in Fig. 7. In Fig. 3 we compare (a) the $^{194}Hg^*$ data, with ^{192}Hg as reference, and (b) the expectations of the $N_F=2$, S=1 strongly coupled (Fig. 7c, eq. 2) supersymmetry configuration. Unfortunately, the strongly coupled spectrum, type c, gives $i=2\hbar$ with the same choice of parameters as was used to fit the ^{192}Hg-^{193}Hg behavior.

However, there is no guarantee that both fermions in $^{194}Hg^*$ will come from the same oscillator shell as the single fermion in ^{193}Hg, for example. Therefore, the comparison between theory and experiment is not as straightforward for the $N_F>1$ systems. The level diagram for single-particle configurations at large deformations is a complicated mixture of orbitals from many shells. For example, the neutron orbitals for Hg nuclei would usually come from the N=5 oscillator shell; at large deformations one also finds many orbitals from the N=6 shell, as well as "intruder" $j_{15/2}$ configurations from the N=7 shell. Only the isolated high-j N=7 orbitals need be considered as outside of the framework of a dynamical symmetry. Therefore, the two-fermion system can either have two particles in the same $\tilde{N}$ shell, or each fermion can come from an orbital from different $\tilde{N}$ shells, which would give a structure similar to that of dynamical supersymmetries in odd-odd nuclei.

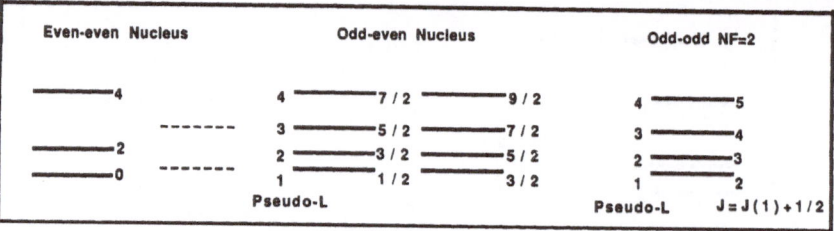

Figure 8   Generic spectrum for $N_F=2$, $J=J_1+1/2$ (eq. 4) dynamical supersymmetry.

For two fermions, from different oscillator (or pseudo-oscillator shells) one appropriate form for the excitation spectrum is:[19]

$$E = E_0 + B\,L(L+1) + C\,J_1(J_1+1) + D\,J(J+1) \qquad (4)$$

where J_1 is the total angular momentum of one of the fermions and J is the total angular momentum. Again, there will be a plethora of bands arising from the different ways that $J_1=L+S_1$ and $J=J_1+S_2$ can couple. For the generic spectra illustrated in Fig. 8, with $J=J_1+1/2$ for the $N_F=2$ nucleus, the transition energies are:

$$
\begin{aligned}
E_\gamma(N_F=0) &= (B+C+D)\ (4L+2) \\
E_\gamma(N_F=1) &= (B+C+D)\ (4L+2) + 2(C+D) \\
E_\gamma(N_F=2) &= (B+C+D)\ (4L+2) + 2C + 4D
\end{aligned}
\qquad (5)
$$

The parameters $(B+C+D)=5$ keV, $(C+D)=-5$ keV, $C=-5$ keV, and $D=0$, will now reproduce all of the observed alignments, as well as the identical moments of inertia. A comparison between experiment and these predictions for ^{192}Hg and ^{194}Hg* transitions are shown in Fig. 3b. The generic spectrum with $J=J_1-1/2$ can also reproduce the data since the D term does not contribute to the γ-ray energies.

The other system with 2 fermions that exhibits "identical" bands is that of ^{194}Tl, especially when related to ^{193}Tl. The odd-odd nucleus ^{194}Tl has 2 fermions which are clearly not identical, so that eq.4 with the same parameter choices for ^{192}Hg, ^{193}Tl, and ^{194}Tl will reproduce the observed pattern of γ-ray transition energies.[20]

In the present analysis we have not attempted to superimpose the supersymmetry predictions on the data. The main reason has been that while the moments of inertia are identical for these nuclei, they are not constant as a function of spin. Rather the dynamical moments of inertia increase by ≈50% over the measured range of γ-ray energies. This could be incorporated by allowing $(B+C)$ or $(B+C+D)$ to have a dependence on J. Also, the quantized alignment sets in at moderate γ-ray energies. This observation can be readily accommodated in the supersymmetry framework. Now it is no longer the ground-state excitations in the second minima which exhibit the symmetry behavior, but rather the excitations which involve broken pairs, and the alignment of these broken pairs gives rise to the dramatic change in the moments of inertia as a function of γ-ray energy. Therefore, the reference in ^{192}Hg is the $N_F=2$ system, which is then compared to the $N_F=3$ excitations in ^{193}Hg, and the $N_F=4$ excitations in ^{194}Hg*, etc. The same generic types of spectra illustrated in Fig. 7 and 5 occur in supersymmetries with N_F even or N_F odd, respectively.

The supersymmetry description can then explain all of the observables: the identical moments of inertia, the identical or related γ-ray energies, the observed quantized alignments, and why the identical behavior does not set in until there is sufficient excitation energy in the second minimum for at least one pair of fermions to be broken.

CONCLUSIONS

A large number of superdeformed bands in A≈190 nuclei show related structures. In even-even nuclei, the γ-ray energies are identical to, or at the mid-point values of, the γ-ray energies of SD ^{192}Hg. In odd-A nuclei the γ-ray energies occur at the "quarter points." The observed pattern of γ-ray energies and the extracted spins of these excitations indicate that the alignment with respect to ^{192}Hg is quantized, frequently with i=1.0ℏ. The simplicity of these data is a signature of a symmetry, and most probably a dynamical supersymmetry which involves many-fermion excitations.

The study of superdeformed excitations and their possible symmetries is a frontier area of nuclear structure study. The limits of the N and Z values of superdeformed shapes in A≈190 nuclei have not been established. Very few of the excited structures in the superdeformed minima have been observed; the present measurements are at the limit of current experimental techniques. Also, it is important to confirm the model-dependent spin assignments which yield the deduced alignments. The present theoretical ideas would be further tested by the identification of additional supersymmetric excitations in these superdeformed nuclei. We look forward to the next generation of gamma-ray detector arrays, which will allow a more detailed study of excitations in superdeformed nuclei and further tests of the supersymmetry predictions.

ACKNOWLEDGMENTS

The author would like to thank her colleagues at Lawrence Livermore and Lawrence Berkeley National Laboratories for their extensive efforts in the acquisition and analysis of the data on superdeformed excitations in A≈190 nuclei. In particular, I would like to thank J. A. Becker, for his work on the spin determinations, F. S. Stephens, for recognizing the importance of alignment, and M. J. Brinkman, for his analysis in ^{193}Hg and ^{194}Pb, as well as his work on the systematical behavior. I would also like to thank M. P Carpenter for providing results prior to publication and R. Bijker and A. B. Balantekin for discussions on the nature of supersymmetries in highly deformed nuclei.

Finally, I am very happy to have been given the opportunity to contribute to this volume in honor of the 50th birthday of Professor Francesco Iachello. Throughout my career I have been fortunate to have been able to learn from Franco the beauty of dynamical symmetries in the spectra of heavy nuclei. From the first days when we recognized the O(6) symmetry in ^{196}Pt (I can still remember his exclamation: "0-2-2! 0-2-2!") to our recent efforts to understand the details of the simple structures in superdeformed nuclei, I have enjoyed our fruitful interactions. I wish him the very best of birthdays.

This work was supported in part by the National Science Foundation.

REFERENCES

1. E. F. Moore, et al., *Phys. Rev. Lett.* **63** (1989) 360.
2. J. A. Becker, et al., *Phys. Rev.* C **41** (1990) R9; D. Ye, et al., *Phys. Rev.* C **41** (1990) R13.
3. C. W. Beausang, et al., *Z. Phys.* **A335** (1990) 325; M. A. Riley, et al., *Nucl. Phys.* **A512** (1990) 178.
4. E. A. Henry, et al., *Z. Phys.* **A335** (1990) 361; C. M. Cullen, et al., *Phys. Rev. Lett.* **65** (1990) 1547.
5. M. P. Carpenter, et al., *Phys. Lett.* **240B** (1990) 44; M. W. Drigert, et al., *Nucl. Phys.* **A530** (1991) 452.
6. M. J. Brinkman, et al., *Z. Phys.* **A336** (1990) 115; K. Theine, et al., *Z. Phys.* **A336** (1990) 113.
7. P. B. Fernandez, et al., *Nucl. Phys.* **A517** (1990) 386.
8. F. Azaiez, et al., *Z. Phys.* **A336** (1990) 243; *Phys. Rev. Lett.* **66** (1991) 1030.
9. E. A. Henry, et al., *Z. Phys.* **A338** (1991) 469; T.F.Wang, et al., *Phys. Rev.* C **43** (1991) R2465.
10. F. Azaiez, et al., *Z. Phys.* **A338** (1991) 471.
11. S. Pilotte, et al., in "Nuclear Structure at High Angular Momentum," to be published; Y. Liang, et al., *Phys. Rev. C* (in press).
12. T. Byrski, et al., *Phys. Rev. Lett.* **60** (1990) 1503.
13. F. S. Stephens, et al., *Phys. Rev. Lett.* **64** (1990) 2623; *Phys. Rev. Lett.* **65** (1990) 301.
14. e.g., P. Bonche, et al., *Nucl. Phys.* **A500** (1989) 309.
15. J. A. Becker, et al., *Phys. Rev.* C **46** (1992) 889.
16. A. Gelberg, P.von Brentano, and R. F. Casten, *J. Phys.* **G16** (1990) L143.
17. F. Iachello, *Nucl. Phys.* **A522** (1991) 83c and private communication.
18. K. T. Hecht and A. Adler, *Nucl.Phys.* **A137** (1969) 129; R. D. Ratna Raju, J. P. Draayer, and K. T. Hecht, *Nucl.Phys.* **A202** (1973) 433; A. Arima, M. Harvey, and K. Shimizu, *Phys. Lett.* **30B** (1969) 517.
19. R. Bijker, private communication.
20. J. A. Cizewski, et al., to be published.

DYNAMICAL ALGEBRAS AND SUPERALGEBRAS FOR INTERACTING ITINERANT MANY-ELECTRON SYSTEMS

Andrea Danani and Mario Rasetti

Dipartimento di Fisica, and Unitá INFM
Politecnico di Torino, I-10129 Torino, Italy

Introduction

The role of dynamical algebras, both (semi-simple) Lie and graded Lie, in the self-consistent solution of many itinerant interacting electron mean field problems is thoroughly reviewed, with particular attention to models of the form of the Hubbard model and generalizations thereof.

1 Dynamical groups and coexistence phenomena

The transition of a system from one (thermodynamic) phase to another is usually accompanied by a spontaneous breaking of symmetry: in its disordered state (above a critical temperature T_C), the system is described by a Hamiltonian H having symmetry group G, whereas in the ordered state (below T_C), it is described by a reduced hamiltonian H_{red}, invariant under a smaller symmetry group $G_0 \subset G$. The appearance of order parameters is thus associated with the disappearance of the more extended symmetry.

Well-known cooperative many-body effects such as superconductivity, charge- and spin-density waves, ferromagnetism, etc. can be investigated as broken symmetries. For a superconductor, for example, above T_C the fields satisfy angle-phase ($U(1)$) invariance (equivalent, by Noether's theorem, to number conservation). In the superconducting state, such $U(1)$ invariance is broken: the order parameters are the fields or their expectation values if quantum theory is used.

On the other hand, the discovery of the possible existence of coexisting phases at low temperatures in many-fermion systems can be suitably handled in a compact way within a unified group theoretical approach. In this context was introduced [25],[1] the concept of *dynamical group*.

Dynamical groups, which describe both the symmetry of the system and its spectrum, arise in the following way: the reduced hamiltonian, typically a mean-field approximation, is a representation of an element of some Lie algebra, the so-called Spectrum Generating Algebra (SGA). The dynamical group is the Lie group of such SGA.

1.1 The linearization procedure

We shall be primarly concerned here with hamiltonians describing systems of interacting fermions:

$$H = \sum_{i,j} \varepsilon_{i,j} a_i^\dagger a_j + \frac{1}{2} \sum_{ijkl} < i,j \mid V \mid k,l > a_i^\dagger a_j^\dagger a_k a_l .$$ (1)

where the label i is a multi-index including both momentum (or position) and spin indices ($i \equiv (\vec{k}, \sigma)$). As usual, $a_i^\dagger, a_i$ are fermion creation and annihilation operators satisfying $\{a_i, a_j\} = 0$, $\{a_i, a_j^\dagger\} = \delta_{ij}$.

The methods of solution which we shall refer to as linearization methods involve either the reduction of the four-fermion terms in the interaction to a subset of multilinear terms (of equal or lower order) or the linearization of the kinetic term, leading to some reduced Hamiltonian H_{MF}. It is the closure, under Lie bracket, of these multilinears that generates the SGA of H_{MF}. Linearization can be achieved, from a general point of view, in the following way. Consider the product of two operators A and B, and the identity

$$AB = (A - < A >)(B - < B >) + < A > B + A < B > - < A >< B > .$$ (2)

If $< A >, < B >$ are assumed to be the expectation values of the respective operators A and B in the most probable state (equilibrium, quantum or thermodynamic), the terms $(A - < A >)$ and $(B - < B >)$ are "small", and their product can be neglected. The above identity is then replaced by the approximate expression

$$AB \simeq < A > B + A < B > - < A >< B > .$$ (3)

At this point, in order to guarantee consistency, we have to consider two possibilities, which will discussed in the following two subsections.

1.1.1 Bosonic linearization

The operators A and B in (3) commute: then BA leads to the same linear approximation of AB with $< A >$ and $< B >$ ordinary c-numbers.

In this case, typically, A and B are identified successively with all different pairs of commuting operators entering the four-fermions interaction and use of the approximate identity (3) transforms H into a bilinear form, which belongs to an ordinary Lie algebra.

Indeed, considered a fermion system with N single particle states, labelled by indices $\alpha, \beta, \gamma, \ldots$, etc., define the pair of fermions operators

$$
\begin{aligned}
E_\beta^\alpha &= a_\alpha^\dagger a_\beta - \frac{1}{2} \delta_{\alpha\beta} , \\
E_{\alpha\beta} &= a_\alpha a_\beta , \quad E^{\alpha\beta} = a_\alpha^\dagger a_\beta^\dagger ,
\end{aligned}
$$ (4)

(with the obvious properties $E_\beta^{\alpha\dagger} = E_\alpha^\beta, E^{\alpha\beta} = E_{\beta\alpha}^\dagger, E_{\alpha\beta} = -E_{\beta\alpha}$). The set consisting of all single and pair operators generates the Lie algebra B_N isomorphic to $so(2N + 1)$. The commutation relations are given by:

$$
\begin{aligned}
[E_\beta^\alpha, E_\delta^\gamma] &= \delta_{\gamma\beta} E_\delta^\alpha - \delta_{\alpha\delta} E_\beta^\gamma , \\
[E_\beta^\alpha, E_{\gamma\delta}] &= \delta_{\alpha\delta} E^{\beta\gamma} - \delta_{\alpha\gamma} E_{\beta\delta} , \\
[E^{\alpha\beta}, E_{\gamma\delta}] &= \delta_{\alpha\delta} E_\gamma^\beta + \delta_{\beta\gamma} E_\delta^\alpha - \delta_{\alpha\gamma} E^{\beta\delta} - \delta_{\beta\delta} E_\gamma^\alpha , \\
[E_{\alpha\beta}, E_{\gamma\delta}] &= 0 ,
\end{aligned}
$$ (5)

$$[\,a_\alpha^\dagger, a_\beta\,] \;=\; 2\,E_\beta^\alpha, \quad [\,a_\alpha, a_\beta\,] = 2\,E_{\alpha\beta},$$

$$[\,a_\alpha, E_\delta^\beta\,] \;=\; \delta_{\alpha\beta} a_\gamma, \quad [\,a_\alpha, E_{\beta\delta}\,] = 0 \quad,$$

$$[\,a_\alpha, E^{\beta\delta}\,] \;=\; \delta_{\alpha\beta} a_\gamma^\dagger - \delta_{\alpha\gamma} a_\beta^\dagger \quad.$$

(All other commutators are straightforwardly obtained by hermitian conjugation.) From (5) one can check that the set of all pair operators (4) is closed under commutation and generates the algebra $D_N \sim so(2N)$, while the subset $\{E_\beta^\alpha\}$ forms a $u(N)$ Lie algebra.

1.1.2 Fermionic linearization

The operators A and B anticommute: then $< A >, < B >$ should anticommute with the operators A, B and among themselves.[18]

In this approach, we linearize the non-diagonal bilinear terms in the following way:

$$\sum_{<i,j>} a_{i,\sigma}^\dagger a_{j,\sigma} \rightarrow q \sum_i (< a_{i\sigma}^\dagger > a_{j\sigma} + a_{i,\sigma}^\dagger < a_{j\sigma} > - < a_{i\sigma}^\dagger >< a_{j\sigma} >) \quad, \tag{6}$$

where q denotes, e.g., the number of nearest neighbours per site in the lattice, whereas for the non-diagonal four-fermion interaction terms, the linearization will consist in replacing

$$a_i^\dagger a_j^\dagger a_\ell a_k \rightarrow < a_i^\dagger a_j^\dagger a_\ell > a_k + a_i^\dagger a_j^\dagger a_\ell < a_k > - < a_i^\dagger a_j^\dagger a_\ell >< a_k > \quad. \tag{7}$$

One should notice that if this linearization procedure is applied to the hamiltonian (1), it can never happen that A equals B or $B^\dagger$ and therefore we have no constraints on the relations between $< A >$ and $< A >^*$ and $< B >, < B >^*$. This implies that there are various possible choices for the algebraic structure $< A >$ and $< B >$ should belong to in order to consistently realize the above linearization condition. Let us define $\Theta_i \equiv< A_i >$, where the A_i's are either linear or trilinear in the fermion operators; then the Θ_i may belong to the field $\mathcal{G}$, which is a:

1. **Grassmann algebra:** $\mathcal{G}$ is the $\mathbb{Z}_2$-graded Grassmann algebra the two non-intersecting subsets of which are products of even ($\mathcal{G}_0$) and odd ($\mathcal{G}_1$) numbers of anticommuting generators $\{\Theta_i\}$, respectively: $\mathcal{G} = \mathcal{G}_0 \oplus \mathcal{G}_1$, with $[\mathcal{G}_0, \mathcal{G}_1] = 0$, realized by the (anti)commutation relations

$$\{\Theta_i, \Theta_j\} = 0 \quad, \quad \{\Theta_i, \bar{\Theta}_j\} = 0 \quad. \tag{8}$$

   Due to associativity, quantities like $(\bar{\Theta}_i \Theta_j)^2$ vanish and this makes it difficult to give to quantities such as $\bar{\Theta}_i \Theta_j$ in the algebra a sound physical interpretation.

2. **Non-associative Banach-Grassmann algebra:** $\mathcal{G}$ has the same structure and algebraic relations (8) of the Grassmann algebra plus a set of new relations requiring that bilinears of the form $\Theta_i \bar{\Theta}_j = -\bar{\Theta}_j \Theta_i$ are c-numbers, which imply non-associativity: $(\bar{\Theta}\Theta) = c \, (c \in \mathbb{R}) \Rightarrow (\bar{\Theta}\Theta)^2 = c^2$. In the reduced hamiltonian we shall then have coefficients at most linear in the variables $\Theta, \bar{\Theta}$. A possible realization consists in interpreting $\Theta_i, \bar{\Theta}_j$ as 1-*forms* over a non-flat super-manifold, each associated with a phase ($\varphi_i, -\varphi_j$, respectively), with (wedge) product $\Theta_i \Theta_j \sim \sin(\varphi_i - \varphi_j)$.

3. **Clifford algebra:** the (anti)commutation relations defining $\mathcal{G}$ are now [17]

$$\{\Theta_i, \Theta_j\} = 0 \quad, \quad \{\Theta_i, \bar{\Theta}_j\} = \delta_{ij} \quad. \tag{9}$$

In the following, we shall generically denote by Grassmann-like algebras the set of three algebras introduced above.

1.2 The generalized Bogolubov rotation

In general, for a rank-l, d-dimensional semi-simple Lie algebra g, one introduces the Cartan-Weyl (CW) basis

$$\{h_1, \ldots, h_l; e_{\pm 1}, \ldots, e_{\pm m}\}, \quad l + 2m = d, \tag{10}$$

with commutation relations

$$[h_i, h_j] = 0, \quad [h_i, e_{\pm \alpha}] = \pm \alpha_i e_\alpha,$$

$$[e_\alpha, e_\beta] = N_{\alpha\beta} e_{\alpha+\beta} \quad (\alpha + \beta \neq 0), \tag{11}$$

$$[e_\alpha, e_{-\alpha}] = \sum_{i=1}^{l} \alpha_i h_i,$$

where $\{\alpha_i\}$ are the roots of g.

If g is the SGA, we can rewrite H_{MF} in the CW-basis as:

$$H_{MF} = \sum_{j=1}^{l} \beta_j \, h_j + \sum_{\alpha=1}^{m} \mu_\alpha (e_\alpha + e_{-\alpha}). \tag{12}$$

H_{MF} can then be rotated [25] to diagonal form $\mathcal{H}_D$ by means of an inner automorphism[1] $\Phi : g \to g$ implemented by the adjoint action of the operator $Z = \sum_{\alpha=1}^{m} \phi_\alpha (e_\alpha - e_{-\alpha})$. We say that $\mathcal{H}_D = \exp(\text{ad } Z)(H_{MF})$ is diagonal if – through an appropriate choice of the coefficents $\{\phi_\alpha\}$ – it can be made to belong to the commuting Cartan subalgebra generated by the set $\{h_i\}$.

The diagonalization proceeds therefore in the following way. First notice that

$$\mathcal{H} = \exp(\text{ad } Z)(H_{MF}) \equiv \sum_{n=0}^{\infty} \frac{1}{n!} \underbrace{[Z, \ldots, [Z, H_{MF}] \ldots]}_{n-\text{times}} = \sum_{n=0}^{\infty} \frac{1}{n!} H^{(n)}, \tag{13}$$

where, as $[Z, H_{MF}] \in g$, we can write, in general,

$$H^{(k)} = [Z, H^{(k-1)}] = \sum_{j=1}^{l} \beta_j^{(k)} \, h_j + \sum_{\alpha=1}^{m} \mu_\alpha^{(k)} (e_\alpha + e_{-\alpha}), \quad k \geq 1, \tag{14}$$

with $H^{(0)} \equiv H_{MF}$. The commutator in (14) defines a recursive relation for the coefficients $\{\beta_j^{(k)}, \mu_\alpha^{(k)}\}$.

Upon introducing the vector $| \, v >^{(k)} = | \, \mu_1^{(k)}, \ldots, \mu_m^{(k)}, \beta_1^{(k)}, \ldots, \beta_1^{(l)} >$, of dimension $p = l + m$, as well as the operator valued vector $| \, \mathcal{G} > = | \, \hat{e}_1, \ldots, \hat{e}_m, h_1, \ldots, h_l >$, where $\hat{e}_\alpha \equiv e_\alpha + e_{-\alpha}$, we write $H^{(k)} = < v^{(k)} | \, \mathcal{G} >$. (14) allows us to define a $p \times p$ matrix A such that $| \, v >^{(k+1)} = A | \, v >^{(k)}$. Clearly $A = A(\{\phi_\alpha\})$ and we set

$$A_d = T^{-1} A T = \text{diag} (\lambda_1, \ldots, \lambda_p). \tag{15}$$

where T is the unitary matrix which diagonalizes A. Upon denoting, in vector form, by $| \, \omega > \equiv | \, \nu_1, \ldots, \nu_m, \gamma_1, \ldots, \gamma_l >$ the set of coefficients relative to $\mathcal{H}$, (so that one has $\mathcal{H} = < \omega | \, \mathcal{G} >$), one obtains, from (13) and (15),

$$| \, \omega > = \sum_{n=0}^{\infty} \frac{1}{n!} | \, v >^{(n)} = \sum_{n=0}^{\infty} \frac{1}{n!} A^n | \, v >^{(0)} = T \left(\sum_{n=0}^{\infty} \frac{1}{n!} A_d^n \right) T^{-1} | \, v >^{(0)}, \tag{16}$$

---

[1]Strictly speaking, if Φ turns out to be an outer automorphism of g, then the smallest algebra which has g as its maximal normal subalgebra should be referred to as dynamical algebra.

namely

$$\omega_i = \sum_{k,l=1}^{p} \exp(\lambda_k) \, T_{ik} T_{kl}^{-1} \, v_l^{(0)} . \tag{17}$$

The m nonlinear coupled equations $\{\nu_\alpha(\{\phi_\beta\}) = 0; \ \alpha, \beta = 1, \ldots, m\}$ (manifestly, $\omega_\alpha \equiv \nu_\alpha$, for $\alpha = 1, \ldots, m$) give the required solutions $\tilde{\phi}_\alpha$ and we are left with

$$\mathcal{H}_D = \sum_{i=1}^{l} \tilde{\gamma}_i \, h_i, \quad \tilde{\gamma}_i = \gamma_i(\{\tilde{\phi}_\alpha\}). \tag{18}$$

1.3 The order parameters

We shall show now how it is possible to associate the appearance of order parameters with the dynamical group, as well as the symmetry group of the system in question. We consider first the case of broken abelian symmetries [1][26].

The mean-field approximation to the system, represented by the hamiltonian H_{MF}, has – in the ordered state – a reduced hamiltonian H_{red}, that is an element of the SGA. The hamiltonian H of the system in the disordered phase is not, in general, an element of the spectrum generating algebra; however, there may be such an element, let us call it H_{sym}, that effectively recovers the symmetry of the original hamiltonian H. When this happens one can use the $H_{sym} \rightarrow H_{red}$ transition to mimic the $H \rightarrow H_{red}$ phase transition, as far as symmetry is concerned. In other words, by means of the former, we can define the order parameters associated with the latter.

Assuming that the reduced hamiltonian is a representation of an element x_{red} of some semisimple complex Lie algebra g, we shall be able to identify the algebra g_s of the (abelian) symmetry group G with a subalgebra of the Cartan subalgebra h of g. Similarly, let g_0 be the algebra of the subgroup $G_0 \subset G$ characterizing the symmetry of H_{red}. We have

$$g_0 \subset g_s \subset h \subset g, \quad [\, H_{red}, \, g_0\,] = 0, \quad \text{and} \quad [\, H_{red}, \, g_s\,] \neq 0 \ . \tag{19}$$

If H_{red} is diagonalizable, x_{red} belongs to some Cartan subalgebra $\bar{h}$ of g. Since, in the semi-simple complex case, all Cartan subalgebras are conjugate under the adjoint group of g, there exists an automorphism $\mathcal{I} : g \rightarrow g$, such that $h = \mathcal{I}(\bar{h})$. This automorphism enables us to define a new element $x_{sym} \in g$, by

$$x_{sym} = \mathcal{I}(x_{red}) \in h \ . \tag{20}$$

Since $[\, x_{sym} \, , \, g_s\,] = 0$, the Hilbert space representative H_{sym} of x_{sym} has the full symmetry of the original hamiltonian H and the same spectrum as H_{red}. Therefore we may label eigenstates of H_{sym} with symmetry labels appropriate to H. These are the eigenstates one can use to mimic those of H and for which we require the order parameters to vanish. For example, the eigenstates of $\mathcal{H}_D$ in (18) are labelled by the eigenvalues λ_j of h_j:

$$h_j |\, \{\lambda_j\} \, > = \lambda_j |\, \{\lambda_j\} \, > . \tag{21}$$

We define order parameter, say η_A, the expectation value of an order operator $\mathcal{O}_A$ such that $\eta_A^d = 0$ in the disordered state $|\, d >$ but $\eta_A^g \neq 0$ in the ordered or broken symmetry state $|\, g >$. The state $|\, d >$ will be identified with a state in which the l mutually commuting operators $h_1, \ldots, h_l$ are all diagonal, and represent therefore conserved quantities.

It is easy to see that the operators in Hilbert space corresponding to the non-Cartan elements e_α of the SGA behave as order operators. This follows from

$$< d \,|\, e_\alpha \,|\, d > = \frac{1}{\alpha_i} < d \,|\, [h_i, \, e_\alpha] \,|\, d > = 0 \tag{22}$$

(provided that $\alpha_i \neq 0$). In that case, e_α will be an order parameter for the phase described by the non-conservation of those h_i which form the subset $\bar{h}$ of $\{h_1, \ldots, h_l\}$. For a system with SGA of rank $l \geq 2$, the $e_\alpha, e_\beta, \ldots$ can correspond physically to coexisting order operators with order parameters $\eta_A^g, \eta_B^g, \ldots$, and a phase boundary can be defined in the space of coupling parameters $\{\beta_j, \mu_\alpha\}$ in H_{MF} by $\eta_A^g = \eta_B^g$, etc.

This approach manifestly fails in the case of non-Abelian broken symmetries. In the latter case, one resorts [16] to the following scheme. Denote by g_s the algebra of the symmetry group of H, and by g_d the dynamical algebra of the linearized hamiltonian H_ℓ. Naturally, the process of linearization affects only part of the hamiltonian. We denote by $H^{(0)}$ the part of H not affected by the approximation. In other words, $H = H^{(0)} + H^{(I)}$, and $H_\ell = H^{(0)} + H_\ell^{(I)}$. $H^{(0)}$ and $H_\ell^{(I)}$ have themselves dynamical algebras which are subalgebras of g_d, and we denote them by $g_d^{(0)}$ and $g_d^{(I)}$ respectively.

The order parameters associated with breaking of abelian symmetries were defined above as the expectations of those operators in g_d generating its root space. This identification holds when, besides being abelian, g_s is a subalgebra of the Cartan subalgebra h_d of g_d: the order parameters correspond then to breaking of those symmetries induced by the operators in $g_s \cap h_d$. A natural generalization of such scheme, also holding for non-abelian symmetries, is the following[2] :

- construct first the commutator set $g \doteq [H_\ell, g_d]$;

- define moreover the set $g' \doteq g_d/g$; as well as

$$
g_p = \begin{cases} g'/g_d^{(0)}, & \text{if } g_d = g_d^{(0)} \oplus g_d^{(I)} \; ; \\ g' \setminus g_d^{(0)}, & \text{otherwise} \end{cases} \tag{23}
$$

- one can then finally identify the order parameters as expectation values of the operators $\mathcal{O}_A \in g_p$.

Both the above procedures imply that $< \mathcal{O}_A >$ is non identically vanishing, where $< \bullet >$ denotes some average implied by H_ℓ (for instance, it can represent a statistical average over the corresponding Gibbs ensemble or the expectation value in its ground state). Any mean field strategy can thus be made explicit by constructing suitable consistency equations for $< \mathcal{O}_A >$. In general, one should expect that in the disordered phase (corresponding to vanishing order parameters) the whole symmetry g_s were restored for the linearized model, and in fact this happens for some of the bosonic and for the fermionic linearization schemes. However, this is not always the case and the phase described by the vanishing of the order parameter may still be an ordered phase.

2 Linearization of Hubbard-like models

2.1 The Yang-Zhang SO(4) symmetry

The Hubbard model [10] is defined, in the grand-canonical ensemble, by the hamiltonian

$$
H = -\mu \sum_i (n_{i\uparrow} + n_{i\downarrow}) - t \sum_{<i,j>} \sum_\sigma a_{i,\sigma}^\dagger a_{j,\sigma} + U \sum_i n_{i\uparrow} n_{i\downarrow} \quad , \tag{24}
$$

---

[2]When the operation is not defined with algebras, we denote by the symbol / subtraction of the common generators in the basis fixed by g. On the other hand the symbol \ denotes subtraction element by element of common operators

where μ is the chemical potential, and i, j denote sites in some lattice Λ ($< i, j >$ are nearest-neighbours in Λ).

In [28], Yang and Zhang have shown that the model at half filling is endowed with a global $SO_4 = (SU_2 \otimes SU_2)/\mathbb{Z}_2$ symmetry. Recall that filling n is the average occupation number $n \doteq \dfrac{< N_e >}{N_\Lambda}$; $N_e \doteq \sum_i (n_{i,\uparrow} + n_{i,\downarrow})$ and N_Λ denoting, respectively, the total number of electrons operator and the number of sites of Λ. Half filling means $n = 1$ and implies $\mu = \frac{1}{2}U$. One can check indeed that at half filling hamiltonian (24) as well as the total momentum operator commute with two mutually orthogonal $su(2)$ algebras. The first (*magnetic*) $su(2)$ is generated by the operators

$$S_+ = \sum_i a_{i\uparrow}^\dagger a_{i\downarrow} \equiv \sum_k a_{k\uparrow}^\dagger a_{k\downarrow} \quad , \quad S_- = S_+^\dagger$$

$$S_0 = \frac{1}{2}\sum_i (n_{i\uparrow} - n_{i\downarrow}) \equiv \frac{1}{2}\sum_k (n_{k\uparrow} - n_{k\downarrow}) \quad , \tag{25}$$

with the usual $su(2)$ commutation relations $[S_0, S_\pm] = \pm S_\pm$, $[S_+, S_-] = 2S_0$, and derives from the fact that the interaction is isotropic in spin space. Besides this "obvious" symmetry, there is a "hidden" (*pseudospin* or *superconductive*) $su(2)$ symmetry, which is generated by

$$J_+ = \sum_j e^{i\mathbf{p}\cdot\mathbf{j}} a_{j\downarrow}^\dagger a_{j\uparrow}^\dagger \equiv \sum_k a_{k\uparrow}^\dagger a_{p-k\downarrow}^\dagger \quad , \quad J_- = J_+^\dagger$$

$$J_0 = \frac{1}{2}\sum_i (n_{i\uparrow} + n_{i\downarrow} - 1) \equiv \frac{1}{2}\sum_k (n_{k\uparrow} + n_{k\downarrow} - 1) \quad , \tag{26}$$

where $\mathbf{p}$ is the vector (π, π, π). By $a_{k\sigma}$, we denote the Fourier transforms of $a_{i\sigma}$.

When the condition of half filling is not met, J_0 still commutes with the hamiltonian, due to the conservation of the particle number but J_+ and J_- do not, and the global symmetry of the model is $u(1) \oplus su(2)$ generated by $\{J_0; S_+, S_-, S_0\}$.

The $\mathbb{Z}_2$ present in the full global symmetry is a duality transformation generated by a unitary operator $\hat{U}$ such that

$$\hat{U} a_{i\uparrow} \hat{U}^{-1} = a_{i\uparrow}^\dagger \quad , \quad \hat{U} a_{i\downarrow} \hat{U}^{-1} = (-1)^i a_{i\downarrow} \quad , \tag{27}$$

which exactly interchanges the spin with the pseudospin algebra.

It is crucial to observe that $J^2 = \frac{1}{2}\{J_+, J_-\} + J_0^2$, being a Casimir operator, is an exactly conserved quantity. Therefore, the four quantum number associated to S^2, S_0, J^2, J_0 can be used to classify all the eigenstates of the Hubbard model. An important consequence of this is the exact one-to-one correspondence of the states at half-filling ($\mu = U/2, J_0 = 0$) and the states away from half-filling [13][14].

In the superconducting phase, the $U(1)$ symmetry associated with J_0 (which describes a coherent resonant excitation of the system) is spontaneously broken.

2.2 Bosonic Linearization Schemes

Three different bosonic linearization schemes have been adopted for the Hubbard model, all of which approximate the on-site interaction U term, quadrilinear in the fermionic operators, by a combination of bilinear terms. Such schemes lead therefore to a description suitable for the weak coupling regime in which the electrons exhibit a band-like behavior (*i.e.* they are almost delocalized). The dynamical algebra for the resulting hamiltonian is easily recognizable in all cases if one represents the site-dependent operators in their Fourier transformed form. In particular, in the latter representation one immediately verifies that $g_d^{(0)} \equiv \oplus_k u(1)_k$, each $u(1)_k$ generated by

$$L_z^{(k)} \doteq \frac{1}{2}(n_{k\uparrow} + n_{k\downarrow} - 1) \quad , \tag{28}$$

as the tight-binding model (corresponding to $U = 0$) is diagonal in wave-vector space.

2.2.1 The Hartree Approximation

In this case ([9]), the hamiltonian (24) is approximated by setting

$$n_{i\uparrow}n_{i\downarrow} \mapsto \left(\alpha - \frac{1}{2}\right)(n_{i\uparrow} + n_{i\downarrow}) - \frac{1}{2}\alpha^2 \quad , \tag{29}$$

where $\alpha = <n_{i\uparrow} + n_{i\downarrow}>$. The resulting linearized hamiltonian H_ℓ' reads

$$H_\ell' = \sum_k \left[\varepsilon_k + U\left(\alpha - \frac{1}{2}\right)\right](2L_z^{(k)} + 1) + \frac{1}{2}N_\Lambda U\alpha^2 \tag{30}$$

where $\varepsilon_k \doteq -\mu + 2t\sum_{r=1}^{d}\cos(k_r)$. H_ℓ' is manifestly diagonal in its wave-vector space representation, and its dynamical algebra g_d coincides with $g_d^{(0)}$. Since H_ℓ' is invariant with respect to all of the transformations generated by $g_s^{(n)}$ ($g_s^{(n)}$ denotes the algebra of the symmetry group corresponding to filling n), the linearization does not provide any non-trivial order parameter.

2.2.2 The Hartree-Fock Approximation

This approximation ([27]) consists in substituting

$$n_{i\uparrow}n_{i\downarrow} \mapsto \frac{1}{2}(n_{i\uparrow} + n_{i\downarrow}) + \gamma a_{i\downarrow}^\dagger a_{i\downarrow} + \gamma^* a_{i\downarrow}a_{i\uparrow}^\dagger + |\gamma|^2 \quad ; \tag{31}$$

where $\gamma = <a_{i\uparrow}^\dagger a_{i\downarrow}> \equiv |\gamma|e^{i\varphi_\gamma}$. The linearized hamiltonian H_ℓ'' has the form

$$H_\ell'' = \sum_k \left(\varepsilon_k + \frac{1}{2}U\right)(2L_z^{(k)} + 1) - U\left(\gamma S_+ + \gamma^* K_- + |\gamma|^2\right) . \tag{32}$$

One can check from eq. (32) that the symmetry algebra of H_ℓ'' is $su(2) \oplus u(1)$ at half filling ($u(1) \oplus u(1)$ for $n \neq 1$), whereas the dynamical algebra is $g_d = g_d^{(0)} \oplus su(2)$. In view of the remaining steps of our scheme, it is convenient to think of the latter $su(2)$ as generated by $\{e^{-i\varphi_\gamma}S_+ + e^{i\varphi_\gamma}S_-, e^{-i\varphi_\gamma}S_+ - e^{i\varphi_\gamma}S_-, S_0\}$. The set g is then given by $g = \{S_0, e^{-i\varphi_\gamma}S_+ - e^{i\varphi_\gamma}S_-\}$, whence $g' = g_d^{(0)} \cup \{e^{-i\varphi_\gamma}S_+ + e^{i\varphi_\gamma}S_-\}$, and $g_p \sim u(1)$ has a unique generator $e^{-i\varphi_\gamma}S_+ + e^{-i\varphi_\gamma}S_-$.

The only order parameter of the Hartree-Fock approach is therefore $|\gamma|$, in that $< e^{-i\varphi_\gamma}S_+ + e^{i\varphi_\gamma}S_- > = 2|\gamma|$. $|\gamma| \neq 0$ describes a phase for which the symmetries induced by S_0 and $e^{-i\varphi_\gamma}S_+ - e^{i\varphi_\gamma}S_-$ are broken, namely a phase endowed with magnetic order. $|\gamma| = 0$ on the other hand provides a good description of the disordered phase, since equation (32) implies that the corresponding hamiltonian commutes with the whole $g_s^{(n)}$.

2.2.3 The BCS Approximation

The BCS bosonic linearization consists in introducing the "pairing" parameter $\delta \doteq <a_{i\downarrow}a_{i\uparrow}> \equiv |\delta|e^{i\varphi_\delta}$, whereby one replaces

$$n_{i\uparrow}n_{i\downarrow} \mapsto \delta a_{i\uparrow}^\dagger a_{i\downarrow}^\dagger + \delta^* a_{i\downarrow}a_{i\uparrow} - |\delta|^2 \quad . \tag{33}$$

The resulting linearized hamiltonian is

$$H_l''' = \sum_k \left[\varepsilon_k \left(2J_z^{(k)} + 1 \right) + U \left(\delta J_+^{(k)} + \delta^* J_-^{(k)} - |\delta|^2 \right) \right] \quad , \tag{34}$$

with

$$J_+^{(k)} \doteq a_{k\uparrow}^\dagger a_{-k\downarrow}^\dagger \; , \; J_-^{(k)} \doteq a_{-k\downarrow} a_{k\uparrow} \; , \; J_z^{(k)} \doteq \frac{1}{2} \left(n_{k\uparrow} + n_{-k\downarrow} - 1 \right) . \tag{35}$$

The dynamical algebra is $g_d = \bigoplus_k su(2)_k$, where each $su(2)_k$ is generated by $\{ J_+^{(k)}, J_-^{(k)}, J_z^{(k)} \}$. We once more refer the latter $su(2)_k$ to a "rotated" basis: $\{ G_1^{(k)} \doteq i[\gamma_k J_z^{(k)} + \sigma_k K_+^{(k)}]$, $G_2^{(k)} \doteq [\sigma_k J_z^{(k)} + \gamma_k K_+^{(k)}]$, $G_3^{(k)} \doteq iK_-^{(k)} \}$, where we defined $K_\pm^{(k)} \equiv \frac{1}{2} \left(e^{i\varphi_\delta} J_+^{(k)} \pm e^{-i\varphi_\delta} J_-^{(k)} \right)$, with $\gamma_k = \cosh z_k$, $\sigma_k = \sinh z_k$, and $z_k = \tanh^{-1} \left(\frac{\varepsilon_k}{U|\delta|} \right)$. In such basis hamiltonian (34) simply reads $H_l''' = 2U|\delta| \sum_k \left\{ \frac{1}{\gamma_k} G_2^{(k)} + \frac{1}{2} (\zeta_k - |\delta|) \right\}$. The set g is thus the collection of mutually commuting elements $g = \bigcup_k \{ G_1^{(k)}, G_3^{(k)} \}$, and hence $g' = \bigcup_k \{ G_2^{(k)} \}$. Since $g_d^{(0)}$ can be thought of as generated by $\bigcup_k \{ J_z^{(k)} \}$, in this case it is manifestly a subalgebra of $g_d^{(I)}$, so that g_d coincides with $g_d^{(I)}$. Thus, finally, $g_p = \bigcup_k \{ K_+^{(k)} \}$. The appropriate order parameter is hence just

$$\frac{1}{N_\Lambda} \sum_k < K_+^{(k)} > = \frac{1}{2N_\Lambda} \sum_i < e^{-i\varphi_\delta} a_{i\downarrow} a_{i\uparrow} + e^{i\varphi_\delta} a_{i\uparrow}^\dagger a_{i\downarrow}^\dagger > \equiv |\delta| \quad . \tag{36}$$

The phase described by $|\delta| \neq 0$ has the "superconductive" symmetry completely broken, whereas the whole "magnetic" symmetry survives.

On the other hand, when the order parameter $|\delta|$ equals zero, the linearized hamiltonian reduces to $H^{(0)}$, which is invariant with respect to the entire $g_s^{(n)}$ only for $n \neq 1$. At $n = 1$, $H^{(0)}$ is invariant with respect to the subalgebra $u(1) \oplus su(2)$ of $g_s^{(1)}$, the whole symmetry $su(2) \oplus su(2)$ being recovered only for the unphysical case $\mu = 0$. Thus at half filling the vanishing of the order parameter $|\delta|$ restores only partially the "superconductive" symmetry, and describes therefore an *order-order* (and not an *order-disorder*) transition, corresponding to the onset of non-vanishing pairing between couples of electrons. One should also point out that the self-consistent implementation of the present linearization scheme implies $U < 0$, in that the self-consistency equation

$$|\delta| = \frac{1}{N_\Lambda} \sum_k \frac{\text{Tr} \{ K_+^{(k)} \exp(-\beta H_l''') \}}{\text{Tr} \{ \exp(-\beta H_l''') \}} \quad , \tag{37}$$

has non-vanishing solutions for $|\delta|$ only if U is < 0. This latter feature is at the basis of the major interest devoted to the *negative-U* Hubbard model in the frame of high T_c superconductivity [24].

2.3 Fermionic Linearization Scheme

In this approach, it is the hopping term in (24) to be linearized instead of the interaction term. Thus the approximate mean-field hamiltonian $H_l^{(F)}$ is particularly suited to describe the system in the strong coupling limit, in which the electrons exhibit an almost atomic behavior (i.e. they are strongly localized).

$H_l^{(F)}$ has a dynamical algebra which is a superalgebra, and can be diagonalized by a straighforward extension to the case of spectrum generating *superalgebras* of the (super)group space Bogolubov rotation. Rotated $H_l^{(F)}$ is diagonal in configuration space, not in the wave vector Fock space.

Linearization proceeds by replacing

$$\sum_{<i,j>} a_{i,\sigma}^\dagger a_{j,\sigma} \mapsto q \sum_i \left(\bar{\vartheta}_\sigma i\sigma + a_{i,\sigma}^\dagger \vartheta_\sigma - \bar{\vartheta}_\sigma \vartheta_\sigma \right) \quad , \tag{38}$$

where q denotes the number of nearest neighbours per site in Λ, whereas the site independent average field $\vartheta_\sigma = < a_{i\sigma} >$ belongs to the odd sector of a Grassmann-like algebra.

The dynamical algebra is the direct sum of N_Λ copies of the same superalgebra $g_d = \oplus_i u(2|2)_i$. The order parameters turn out to be proportional to $(\vartheta_\uparrow \bar{\vartheta}_\uparrow + \vartheta_\downarrow \bar{\vartheta}_\downarrow)$, and to $(\vartheta_\downarrow \bar{\vartheta}_\uparrow + \vartheta_\uparrow \bar{\vartheta}_\downarrow)$.

In the case when also spin-exchange invariance is assumed (i.e. $\vartheta_\uparrow \equiv \vartheta_\downarrow$), g_d reduces [19] to $\oplus_i(u(1|1) \oplus u(1|1))_i$ and the two parameters coalesce into a single one, say $\vartheta\bar{\vartheta}$. In this case the corresponding self-consistency equation has solutions $\vartheta\bar{\vartheta} = 0$, describing a disordered phase in which the whole $g_s^{(n)}$ symmetry is restored, as well as $\vartheta\bar{\vartheta} \neq 0$. In the latter case the original symmetry is completely broken (both the "superconducting" and the "magnetic" $su(2)$). As the order parameter is unique, it will describe, in general, a superposition of the two ordered phases ("mixed" phase).

The global $SO_4 = (SU_2 \otimes SU_2)/\mathbb{Z}_2$ symmetry discovered by Yang and Zhang may be thought of – as pointed out by the same authors – as related to the possible coexistence of "superconductive" and "magnetic" phase. We have described above how the standard bosonic linearization allows one to define order parameters describing the breaking of one symmetry only at a time. On the contrary, the fermionic linearization scheme leads to introducing order parameters which break both symmetries simultaneously, and are therefore able to represent "mixed" phases. In particular, such a scheme is the only one permitting to describe a phase with $U > 0$, characterized by an order parameter associated with the breaking of the superconductive symmetry. It is worth pointing out here that even though states non conserving the local average number of electrons may thus be generated, such states do not correspond to non-vanishing pairing, which can indeed be obtained only from extended Hubbard models [19],[3].

An important characteristic feature which emerges from this analysis is the property that even within a linearized scheme the transition from an ordered to a disordered phase can be consistently described only if the whole original symmetry is restored for the linearized hamiltonian by the vanishing of the order parameters. This is not always the case in the bosonic mean field approximations: when it happens, it leads to situations which are of very little physical relevance (the self-energy term becomes identically zero). The fermionic mean field approach on the other hand allows us to restore the complete symmetry in the disordered phase in a non-trivial dynamical way.

2.3.1 The n-cluster Fermi-linearized Hubbard model

Let us consider now the variant of the Hubbard model (24) defined by

$$H_H = \sum_{i \in \Lambda} \sum_\sigma \varepsilon_i n_{i\sigma} + \sum_{i \in \Lambda} U_i n_{i\uparrow} n_{i\downarrow} + \sum_{<i,j>} \sum_{\sigma,\sigma'} t_{\sigma,\sigma'} a_{i\sigma}^\dagger a_{j\sigma'} . \tag{39}$$

Let moreover $\{H_n/n \leq N\}$ be a sequence of hamiltonians which approximate H as $H = \sum_{\{c_n\}\in\Lambda} H_n$, where $\{c_n\}$ denotes any cluster covering of Λ. The cluster linearization method consists in taking into account exactly the mutual interactions of the particles within each cluster and using fermionic linearization to describe interactions between neighbouring clusters [21].

In the model described by (39), one can recover the standard Hubbard model by choosing the spin-flip hopping amplitude $t_{\sigma,-\sigma} = 0$. On the contrary, we select $t_{\sigma,-\sigma} = t_{\sigma,\sigma} = t$ and this naturally leads to introducing the new set of local spinless fermionic operators

$$
\begin{aligned}
A_i &= \frac{1}{\sqrt{2}}(a_{i\uparrow} + a_{i\downarrow}) \quad, \quad N_i = A_i^\dagger A_i \quad, \\
D_i &= \frac{1}{2}(n_{i\uparrow} + n_{i\downarrow} + a_{i\uparrow}a_{i\downarrow}^\dagger + a_{i\downarrow}a_{i\uparrow}^\dagger) \quad, \quad D_i^2 = D_i \, .
\end{aligned}
\tag{40}
$$

Operators $A_i, A_i^\dagger$ and N_i satisfy the same commutation and anticommutation relations as $a_i, a_i^\dagger$ and n_i, while D_i commutes both $A_j, A_j^\dagger$ and N_j. Moreover, we have the relations $N_i + D_i = \sum_\sigma n_{i\sigma}$, $N_i D_i = n_{i\uparrow}n_{i\downarrow}$, and $N_i = \frac{1}{2}\sum_\sigma(n_{i\sigma} + a_{i\sigma}^\dagger a_{i,-\sigma})$, whereby hamiltonian (39) can be rewritten in terms of the new operators (40) as

$$
H \equiv H_{\{\Lambda\}} = \sum_{i\in\Lambda}(\varepsilon_i + UD_i)N_i + t\sum_{<i,j>}(A_iA_j^\dagger + A_jA_i^\dagger) + \sum_{i\in\Lambda}\varepsilon_i D_i \, ,
\tag{41}
$$

where explicit reference to the spin index has manifestly disappeared.

In this case, fermionic linearization acts only on the hopping interactions between sites on the cluster boundary and sites outside the cluster, by replacing bilinear by linear forms:

$$
A_i^\dagger A_j \rightarrow \bar{\eta}_i A_j + A_i^\dagger \eta_j - \bar{\eta}_i \eta_j \, .
\tag{42}
$$

The coefficients $\bar{\eta}_i = <A_i^\dagger>$ and $\eta_j = <A_j>$ belong to the odd sector $\mathcal{G}_0$ of the non-associative $\mathbb{Z}_2$-graded Grassmann-like algebra $\mathcal{G}$. We adopt here the non-associative form of $\mathcal{G}$, amounting to the physical requirement that $\eta_i^2 = 0$ but $(\eta_i\bar{\eta}_j)^2 \neq 0$. H_n is then given by ($\tau := t(n-1)$):

$$
H_n = \sum_{\alpha=1}^n H_\alpha^{(l)} + \tau \sum_{\substack{\alpha=1 \\ (\text{mod } n)}}^n (A_\alpha^\dagger A_{\alpha+1} + A_{\alpha+1}^\dagger A_\alpha) \, ,
\tag{43}
$$

with

$$
H_\alpha^{(l)} = \epsilon_\alpha(N_\alpha + D_\alpha) + U_\alpha D_\alpha N_\alpha + \sqrt{2}(\vartheta_\alpha A_\alpha + A_\alpha^\dagger \bar{\vartheta}_\alpha) - C_\alpha \, ,
\tag{44}
$$

and

$$
\vartheta_\alpha := t\sum_{\kappa\in\Lambda/c_n}\bar{\eta}_\alpha - t\sum_{\gamma=1}^n\bar{\eta}_\gamma \, ,
\tag{45}
$$

$$
C_\alpha := \frac{\tau}{2}\sum_{\kappa\in\Lambda/c_n}(\bar{\eta}_\alpha\eta_\kappa + \bar{\eta}_\kappa\eta_\alpha) - \tau\sum_\gamma^n\bar{\eta}_\alpha\eta_\gamma \, ,
\tag{46}
$$

where $\kappa = \text{n.n.}(\alpha)$.

It is straighforward to show that the dynamical algebra associated with the n-cluster model hamiltonian H is $\mathcal{S}_n = \bigoplus_1^{2^n} u(n|1)$ where the superalgebra $u(n|1)$ is generated by $(n+1)^2$ elements, (n^2+1) of which form the bosonic sector $\mathcal{B}(u(n|1))$ and $2n$ are in the fermionic sector $\mathcal{F}(u(n|1))$. Explicitly:

$$
\mathcal{B}(u(n|1)) = \{\mathbb{I}, A_i^\dagger A_j; \, i,j\in c_n\} \sim u(n)\oplus u(1) \quad,
\tag{47}
$$

$$
\mathcal{F}(u(n|1)) = \{A_i, A_i^\dagger; \, i\in c_n\} \, .
\tag{48}
$$

As the D_i's are projectors, the 2^n orthogonal copies of $u(n|1)$ which generate $\mathcal{S}_n$ are each labelled by one of the 2^n possible different combinations of eigenvalues $d_i\in\{0,1\}$ of D_i ($i\in$

c_n). We indicate with a multiindex λ one of such combinations and denote by $\mathcal{L}$ the set of all possible λ's, whereas P_λ is the projection operator on the Fock subspace associated with a given λ, $u(n|1)_\lambda = S_n P_\lambda$. Therefore $H_n = \sum_{\lambda \in \mathcal{L}} H_\lambda \equiv \sum_{\lambda \in \mathcal{L}} H_n P_\lambda$. In order to obtain the spectrum of H_n, we apply once again a generalized Bogolubov rotation in the (super)group space, which can be implemented by the adjoint action of some antihermitian operator $Z \in S_n$.

With no loss of generality, we write

$$\exp Z = \exp Z^{(b)} \exp Z^{(f)} , \tag{49}$$

where, due to the direct sum structure of S_n, we can set

$$Z^{(b)} = \sum_{\lambda \in \mathcal{L}} Z_\lambda^{(b)} \equiv \sum_{\lambda \in \mathcal{L}} \sum_{\substack{\alpha=1 \\ (\mathrm{mod}\ n)}}^{n} z_\alpha^{(\lambda)} (A_\alpha^\dagger A_{\alpha+1} - A_{\alpha+1}^\dagger A_\alpha) P_\lambda , \tag{50}$$

$$Z^{(f)} = \sum_{\lambda \in \mathcal{L}} Z_\lambda^{(f)} \equiv \sum_{\lambda \in \mathcal{L}} \sum_{\substack{\alpha=1 \\ (\mathrm{mod}\ n)}}^{n} (\xi_\alpha^{(\lambda)} A_\alpha + \bar{\xi}_\alpha^\lambda A_\alpha^\dagger) P_\lambda , \tag{51}$$

with $\{z_\alpha^\lambda | \lambda \in \mathcal{L}, \alpha = 1, \ldots, n\} \in \mathcal{G}_e$, $\{\xi_\alpha^\lambda | \lambda \in \mathcal{L}, \alpha = 1, \ldots, n\} \in \mathcal{G}_o$. The diagonal form $\mathcal{H}$ of H_n can then be obtained from

$$\mathcal{H} = \bigoplus_{\lambda \in \mathcal{L}} \exp\ (\mathrm{ad}\ Z_\lambda^{(b)})(\exp\ (\mathrm{ad}\ Z_\lambda^{(f)})(H_\lambda) = \bigoplus_{\lambda \in \mathcal{L}} \exp\ (\mathrm{ad}\ Z_\lambda^{(b)})(\mathcal{H}_\lambda^{(f)}) , \tag{52}$$

by suitably choosing the coefficients $z_\alpha^{(\lambda)}$ and $\xi_\alpha^{(\lambda)}$ in such a way that the resulting operator belongs to the Cartan subalgebra of $\mathcal{B}(S_n)$. The first step consists in selecting the variables $\{\xi_\alpha^{(\lambda)}\}$ so that $\mathcal{H}_\lambda^{(f)}$ lives in $\mathcal{B}(S_n)$ and this leads in general to a system of linear equations (homogeneous over the odd sector $\mathcal{G}_0$ of $\mathcal{G}$ and hence equivalent to a regular linear system over $\mathbb{R}$).

Notice that due to the presence of the anticommuting Grassmann-like coefficients for the fermionic operators, we have

$$[\ \eta_\alpha F_\alpha\ ,\ \eta_\beta F_\beta\ ] = \eta_\beta \eta_\alpha \{\ F_\alpha\ ,\ F_\beta\ \} \ , \quad \eta_\alpha, \eta_\beta \in \mathcal{G}_1 \ , \tag{53}$$

$$[\ \eta F\ ,\ \alpha B\ ] = \alpha \eta [\ F\ , B\ ] \ , \quad \alpha \in \mathcal{G}_0, \eta \in \mathcal{G}_1 \ , \tag{54}$$

and therefore we remain – as expected – into the superalgebra.

Since the bosonic sector of a superalgebra is a Lie algebra, we can perform a Bogolubov rotation in $\mathcal{B}(S_n) \equiv u(n) \oplus u(1)$ with the adjoint action of $Z^{(b)}$ in the customary way.

We define now a supercoherent state [22] $| \psi_f >$ as

$$|\psi_f> = \mathcal{U} | \omega >_n, \quad \mathcal{U} := \exp(Z), \quad Z = -Z^\dagger \in S_n \tag{55}$$

where $| \omega >_n$ is the *highest weight vector* of S_n. If $\exp(\mathrm{ad} Z)(H_n)$ is diagonal, then $| \omega >_n$ coincides with the vacuum of H_n. In this case, we may identify the cluster ground state $| \psi_G >$ with that particular supercoherent state which minimizes the system internal energy (namely the expectation value of H_c), for the appropriate value of the chemical potential μ at fixed site occupancy.

Here we sketch the main steps of the computation for the dimer case [19]. Using formulas (49),(50) and (51) with $n = 2$, the $T = 0$ ground state of the model will correspond to the particular choice of the thirteen parameters $\{ z^{(\lambda)}, \xi_\alpha^{(\lambda)} | \lambda = 1, \ldots, 4 ; \alpha = 1, 2\}$ and μ for which

1. all the derivatives of $\mathcal{H} \equiv \langle\, \psi_I|\, H_n - \mu N_n|\, \psi_I\, \rangle$ with respect to the parameters $\{z^{(\lambda)}, \xi_\alpha^{(\lambda)}\}$ are equal zero, where N_n is the total number of particles operator for the dimer.

2. $\langle\, \psi_I|\, N_n\, |\, \psi_I\, \rangle >= n_0$, where n_0 is the fixed expected dimer occupation number.

3. η_α is identified with the average $\langle\, a_{\alpha\sigma}\, \rangle$, in order to guarantee self-consistency of the linearization.

Noticing that $\mathcal{H}$ contains only bilinear products of elements of $\mathcal{G}_0$, since the only elements of $\mathcal{G}_0$ entering the model are ϑ_1 and ϑ_2, the parameters $\xi_\alpha^{(\lambda)}$ have to be linear combinations of the ϑ_β's. Upon resorting to the reparametrization $\xi_\alpha^{(\lambda)} = \kappa_\alpha^{(\lambda)} \vartheta_{\alpha+1}$, which allows transforming the derivatives with respect to $\xi_\alpha^{(\lambda)}$ into usual derivatives with respect to $\kappa_\alpha^{(\lambda)}$, $\mathcal{H}$ does contain (linearly) only elements of $\mathcal{G}_e$ of the form $\vartheta_\alpha \bar{\vartheta}_\alpha$ and $\vartheta_\alpha \bar{\vartheta}_{\alpha+1}$. Adopting the representation defined in sect. **1.1.2** we assume

$$\eta_i \eta_j \equiv \sin(\varphi_i - \varphi_j) \quad , \tag{56}$$

where φ_i has to be determined consistently with the relation $\eta_i \bar{\eta}_i \equiv \sin(2\varphi_i)$.

From (56), we deduce that $\bar{\vartheta}_\alpha \vartheta_\alpha \sim \sin(2\varphi_{\alpha+1})$ and $\bar{\vartheta}_\alpha \vartheta_{\alpha+1} \sim \sin(\varphi_\alpha + \varphi_{\alpha+1})$. Including the self-consistency equations related to this fundamental bilinears in the minimization scheme as constraints with the introduction of two Lagrange multipliers ρ_1, ρ_2, there finally results a system of seventeen equations in the seventeen unknowns $\{z^{(\lambda)},\ \kappa_\alpha^{(\lambda)},\ \mu,\ \varphi_\alpha,\ \rho_\alpha\ |\ \lambda = 1, \ldots, 4\,;\ \alpha = 1, 2\}$. The solutions of this system allows us to evaluate the rotation paramaters $\{\bar{\kappa}_\alpha^{(\lambda)} \bar{z}^{(\lambda)}\}$ and in particular the pairing order parameter

$$\mathcal{P} \equiv 2\,\mathrm{Re}\,\left(\langle\, A_1 A_2\, \rangle - \langle\, A_1\, \rangle \langle\, A_2\, \rangle\right). \tag{57}$$

in the ground state. Besides particular solutions corresponding to zero pairing [19], the system has another set of solutions which exhibit non-vanishing pairing.

3 The extended Falikov-Kimball model

The *Falikov-Kimball* model (FK) [5],[11] is a special case of the Hubbard model in which only one species of electrons (say with spin up) is allowed to move, whereas the other is fixed. The FK model has been interpreted as a good model for crystallization as well as for mixed-valence states in rare-earth compounds, where the moving particles play the role of s-band electrons and the localized ones stand for f-electrons with sharp energy levels.

We consider here an extension of the FK model (derived from the extended Hubbard model[10]) characterized by an additional nearest neighbour Coulomb interaction term $\frac{1}{2} V \sum_{\langle i,j \rangle} \sum_{\sigma, \sigma'} n_{i\sigma} n_{j\sigma'}$. After relabelling the fermion operators as $A_i \doteq a_{i\uparrow}$, $B_i \doteq a_{i\downarrow}$, $N_i \doteq A_i^\dagger A_i$, and $D_i \doteq B_i^\dagger B_i$, FK hamiltonian can be written (as usual in the grand-canonical ensemble) as

$$\begin{aligned}
H = \ & - \ \mu_A \sum_i N_i - \mu_B \sum_i D_i + \frac{t}{2} \sum_{\langle i,j \rangle} \left(A_i^\dagger A_j + A_j^\dagger A_i \right) + U \sum_i N_i D_i \\
& + \ \frac{1}{2} V \sum_{\langle i,j \rangle} (N_i + D_i)(N_j + D_j) \quad .
\end{aligned} \tag{58}$$

where μ_A and μ_B are the chemical potentials for A- and B-particles respectively, to be determined in such a way that the numbers of up and down spins are conserved (as the corresponding operators commute with H). Again, the fact that there is no hopping term between electrons of type B_i has the important consequence that the idempotent number operators D_i commute with each term in (58), and can be dealt with as a classical Ising-like variable.

The exact statistical mechanical solution for the FK model is known only for large number of dimensions [2]. However, a few general theorems are known [11] for the symmetric (or neutral) case $\mu_A = \mu_B = U/2$, and in particular an Ising-like phase transition is expected for dimension $d \geq 2$ at some critical temperature, whose value should vanish both for small and large U. Moreover, there are a number of investigations of the ground state phase diagram [8]. Also a strong-coupling mean-field theory, based on the $d = \infty$ exact solution, was proposed [6]. We proceed here to fermionic linearization of the hopping terms in (58), setting

$$A_i^\dagger A_j \sim \theta_i A_j + A_i^\dagger \bar{\theta}_j - \theta_i \bar{\theta}_j \quad , \tag{59}$$

where $\theta_i = < A_i^\dagger >$. As usual, $\bar{\theta}, \theta \in \mathcal{G}$ are nilpotent variables anticommuting with both the fermion operators $A_i^\dagger, A_i, B_i^\dagger, B_i$ and among themselves.

Furthermore, we perform standard Hartree linearization over the intersite Coulomb interaction terms, i.e.

$$(N_i + D_i)(N_j + D_j) \sim n_0(N_i + D_i + N_j + D_j) - n_0^2 \quad , \tag{60}$$

with $n_0 = < N_i + D_i > = n_A + n_B$.

Both approximations (59) and (60) will be implemented in the following within a dimer-cluster scheme. H reduces thus to an effective hamiltonian denoted here by H_Λ, which is a sum – over an arbitrary set of dimer coverings of Λ – of commuting dimer hamiltonians H_d,

$$H_d = \sum_{i=1}^{2} \left[\varepsilon_i N_i + t(q-1)\left(\theta_{\bar{\imath}} A_i - \bar{\theta}_{\bar{\imath}} A_i^\dagger \right) \right] + V N_1 N_2 + t \left(A_1^\dagger A_2 + A_2^\dagger A_1 \right) + \mathcal{C} \quad , \tag{61}$$

where $\varepsilon_i \doteq -\mu_A + U D_i + V[D_{\bar{\imath}} + (q-1)n_0]$, and

$$\mathcal{C} \doteq \sum_{i=1}^{2} \left[(-\mu_B + V(q-1)n_0)D_i + t(q-1)\bar{\theta}_i \theta_i \right] + V \left(D_1 D_2 - (q-1)n_0^2 \right) \tag{62}$$

is a central term. q denotes once more the number of nearest neighbours per site in Λ, $\bar{\imath}$ the site in the dimer which is not i. Furthermore, we assumed that the mean field sensed by site i is the same as for each of its neighbours.

H_d as given in (61) has a dynamical algebra $\mathcal{A}$ which is the direct sum of 4 copies of a superalgebra $\mathcal{S}$, isomorphic with the Cartan extension by $B_0 \doteq N_1 N_2$ of $su(2|2)^3$. The latter has a bosonic subalgebra $\mathcal{B}$ isomorphic with $su(2) \oplus su(2)$, generated by the two sets of operators $\{B_1, B_5, B_2\}$ and $\{B_3, B_6, B_4\}$:

$$B_1 \doteq A_1 A_2 + A_2^\dagger A_1^\dagger, \ B_5 \doteq A_1 A_2 - A_2^\dagger A_1^\dagger, \ B_2 \doteq \mathbb{I} - (N_1 + N_2) \quad , \tag{63}$$

$$B_3 \doteq A_1^\dagger A_2 + A_2^\dagger A_1, \ B_6 \doteq A_1^\dagger A_2 - A_2^\dagger A_1, \ B_4 \doteq N_1 - N_2 \quad , \tag{64}$$

and a fermionic sector $\mathcal{F}$ with eight generators:

$$F_1 = A_1(1 - N_2), \ F_2 = A_1 N_2, \ F_3 = A_2 N_1, \ F_4 = A_2(1 - N_1) \quad ;$$

$$F_{\kappa+4} = F_\kappa^\dagger \quad ; \quad \kappa = 1, \ldots, 4 \quad . \tag{65}$$

[3]It is interesting to point out that recently Korepin and others [15] have proposed a model for high-T_c superconductivity – which differs from the extended Hubbard model in that it contains state dependent hopping terms, identical with those adopted in the version of the KSSH model [12] solved in ref. [20] – which has this as an exact symmetry

B_2 and B_4 are the two Cartan elements of $\mathcal{B}$. Each copy of $\mathcal{S}$ is characterized by a different distribution of the eigenvalues $(0 , 1)$ for the operators D_i entering (61).

Let us write the commutation (anticommutation) relations of $\mathcal{S}$ as

$$[B_m, B_n] = b^p_{mn} B_p \quad , \tag{66}$$

$$[F_\alpha, B_m] = c^\beta_{\alpha m} F_\beta \quad , \tag{67}$$

$$\{F_\alpha, F_\beta\} = f^m_{\alpha\beta} B_m \quad , \tag{68}$$

where $\{b^p_{mn}, c^\beta_{\alpha m}, f^m_{\alpha\beta}\}$ are the structure constants.

In order to implement the Bogolubov automorphism of $\mathcal{S}$ rotating H_d into the Cartan sector of $\mathcal{B}$, we first act on (61) by the adjoint action of the antihermitian rotation operator $\mathcal{Z}_F \in \mathcal{F}$,

$$\mathcal{Z}_F = \sum_{\kappa=1}^{4}(\varphi_\kappa F_\kappa + \bar\varphi_\kappa F_{\kappa+4}) \quad , \tag{69}$$

which maps H_d into $H' \doteq \exp\left(\mathrm{ad}\mathcal{Z}_F\right)(H_d)$. The coefficients $\varphi_\kappa \in \mathcal{G}_o$ are to be eventually chosen in such a way that H' turns out to be an element of $\mathcal{B}$.

The evaluation of H' for generic φ_ν's gives

$$H' = V B_0 + \sum_{\mu=1}^{4} b_\mu B_\mu + \sum_{\kappa=1}^{4} \left(f_\kappa F_\kappa - \bar f_\kappa F_{\kappa+4}\right) + C' \quad , \quad f_\kappa \in \mathcal{G}_o \quad , \tag{70}$$

where, as H' should naturally be hermitian, the operators B_5 and B_6 do not appear: $b_5 = 0$, and $b_6 = 0$.

The coefficients b_μ and f_κ in (70) can be expressed in terms of the coefficients $X^{(m)}_\mu \in \mathbb{R}$, and $Y^{(m)}_\nu \in \mathcal{G}_o$, defined by the recursive relation

$$(\mathcal{Z}_F)_m \circ H_d \doteq [\mathcal{Z}_F,(\mathcal{Z}_F)_{m-1} \circ H_d]$$
$$\equiv \sum_{\mu=1}^{4} X^{(m)}_\mu B_\mu + \sum_{\kappa=1}^{4} \left(Y^{(m)}_\kappa F_\kappa - \bar Y^{(m)}_\kappa F_{\kappa+4}\right) , \quad m \geq 1 , \tag{71}$$

with $(\mathcal{Z}_F)_0 \circ H_d \equiv H_d$. From the formula

$$\exp(\mathrm{ad}\mathcal{Z}_F) H_d = \sum_{m=0}^{\infty} \frac{1}{m!}((\mathcal{Z}_F)_m \circ H_d) , \tag{72}$$

it follows that

$$b_\mu = \sum_{m=0}^{\infty} \frac{1}{m!}X^{(m)}_\mu \quad , \quad f_\kappa = \sum_{m=0}^{\infty} \frac{1}{m!} Y^{(m)}_\kappa \quad . \tag{73}$$

In order ti deduce the explicit form of the coefficients b_μ's and f_κ's different from zero, we split H_d in two parts

$$H_d = H_B^{(0)} + H_F^{(0)} \tag{74}$$

and define, for $m \geq 1$

$$H_B^{(m)} = [\mathcal{Z}_F, H_F^{(m-1)}] = X^{(m+1)}_0 \mathbb{I} + \sum_{\mu=1}^{4} X^{(m)}_\mu B_\mu \quad , \tag{75}$$

$$H_F^{(m)} = [\mathcal{Z}_B, H_B^{(m-1)}] \equiv \sum_{\kappa=1}^{4}(Y^{(m)}_\kappa F_\kappa - \bar Y^{(m)}_\kappa F_{\kappa+4}) + \delta_{m,1} V(\varphi_2 F_2 + \varphi_3 F_3). \tag{76}$$

Notice that $H_B^{(0)}$ contains the Cartan extension term $V N_1 N_2$ giving rise in $H_F^{(1)}$ to the extra term explicitly written in $H_F^{(m)}$. One gets

$$
\begin{aligned}
H_B^{(m+1)} &= \sum_{\alpha,\kappa=1}^{4} \left(-\phi_\alpha Y_\kappa^{(m)} \{F_\alpha,\, F_\kappa\} + \phi_\alpha \bar{Y}_\kappa^{(m)} \{F_\alpha,\, F_{\kappa+4}\} \right) + \text{h.c} \\
&= X_0^{(m+1)} \mathbb{I} + \sum_{\mu=1}^{4} \underbrace{\left(\left(-\phi_\alpha f_{\alpha\kappa}^\mu Y_\kappa^{(m)} + \phi_\alpha f_{\alpha,\kappa+4}^\mu \bar{Y}_\kappa^{(m)} \right) + \text{h.c.} \right)}_{X_\mu^{(m+1)}} B_\mu\ ;
\end{aligned}
\tag{77}
$$

$$
\begin{aligned}
H_F^{(m+1)} &= \sum_{\mu=1}^{4} X_\mu^{(m)} \sum_{\alpha=1}^{4} \left(\phi_\alpha [\, F_\alpha,\, B_\mu] + \bar{\phi}_\alpha [\, F_{\alpha+4},\, B_\mu] \right) \\
&= \sum_{\kappa=1}^{4} \underbrace{\left(\sum_{\mu=1}^{4} X_\mu^{(m)} \sum_{\alpha=1}^{4} c_{\alpha\mu}^\kappa \phi_\alpha \right)}_{Y_\kappa^{(m+1)}} F_\kappa + \text{h.c.}\ .
\end{aligned}
\tag{78}
$$

The $\{\pm\}$-commutators in (77) and (78) lead to the definition of two matrices transferring fermionic coefficients to bosonic and viceversa, at each step. The more convenient way to solve the problem is to restrict first the attention to the set of coefficients $\{X_\mu^{(m)}, Y_\kappa^{(m)}$; $\mu, \kappa = 1, \dots, 4\}$. The operator B_0 appears only in $H_B^{(0)}$ while the contributions to the constant term arising from the commutators $[\, \mathcal{Z}_F,\, H_F^{(m)}]$, $m \geq 0$ will be calculated separately.

We introduce then the kets $\mid X^{(m)} >$ and $\mid Y^{(m)} >$, denoting 4-vectors whose components are $X_\mu^{(m)}, Y_\kappa^{(m)}$; $\mu, \kappa = 1, \dots, 4$ respectively, and the "reduced" 4×4 matrices $\mathbf{R}, \mathbf{S}$ implicitly defined by (77) and (78), such that $\mid Y^{(m+1)} >= \mathbf{S} \mid X^{(m)} >$, $\mid X^{(m+1)} >= \mathbf{R} \mid Y^{(m)} >$. Merging together the last two relations, the following recursive relation holds as well:

$$
\mid X^{(m+2)} >= \mathbf{P} \mid X^{(m)} >\ ,\qquad \mathbf{P} = \mathbf{RS}\ .
\tag{79}
$$

$\mathbf{P}$ is a 4×4 matrix whose elements belong to $\mathcal{G}_e$.

As the matrix $\mathbf{P}^2$ turns out to be block diagonal (two 2×2 blocks), it is convenient to express the first of equations (73), upon implementing (79), as a series of even powers of $\mathbf{P}$, assuming the first four $\mid X^{(m)} >$'s, namely $m = 0 \dots, 3$, as primitives. We hence obtain

$$
\mid b > = \sum_{l=0}^{3} \sum_{k=0}^{\infty} \frac{\mathbf{P}^{2k}}{(4k+l)!} \mid X^{(l)} >\ ,
\tag{80}
$$

$$
\mid f > = \mid Y^{(0)} > +\mid \Gamma > + \sum_{l=1}^{4} \sum_{k=0}^{\infty} \frac{\mathbf{P}^{2k}}{(4k+l)!} \mid X^{(l-1)} >\ ,
\tag{81}
$$

where $Y_1^{(0)} = Y_2^{(0)} \doteq \tau \theta_1$, $Y_3^{(0)} = Y_4^{(0)} \doteq \tau \theta_2$, and $\Gamma_\nu \in \mathcal{G}_o$ are the coefficients of the fermionic operators obtained by the commutation of $\mathcal{Z}_F$ with the operator $N_1 N_2$.

Upon denoting by $\mathbf{T}$ the orthogonal matrix which diagonalizes $\mathbf{P}$ (i.e. such that $\mathbf{T}^{-1} \mathbf{P}^2 \mathbf{T} = \text{diag}\,(z_\alpha^4)$) and setting $\kappa_{\mu,\mu'}^{(\alpha)} = (\mathbf{T}^{-1})_{\mu,\alpha}\,(\mathbf{T})_{\alpha,\mu'}$, one finally obtains the b_μ's in (80) as

$$
b_\mu = \sum_{\mu'=1}^{4} \sum_{l=0}^{3} \mathcal{L}_{\mu,\mu'}^{(l)} X_{\mu'}^{(l)}\ ,
\tag{82}
$$

with

$$
\mathcal{L}_{\mu,\mu'}^{(l)} = \sum_{\alpha=1}^{4} \kappa_{\mu,\mu'}^{(\alpha)} z_\alpha^{-l} \frac{d^{4-l} \mathcal{Z}_\alpha}{d\, z_\alpha^{4-l}},
\tag{83}
$$

$$
\mathcal{Z}_\alpha = (\cosh(z_\alpha) + \cos(z_\alpha)) \Theta(z_\alpha) - \cosh\left(\frac{z_\alpha}{\sqrt{2}}\right) \cos\left(\frac{z_\alpha}{\sqrt{2}}\right) \Theta(-z_\alpha),
$$

where $\Theta(x)$ is the usual Heaviside step-function: $\Theta(x) = 0$ when $x < 0$ and $\Theta(x) = 1$ when $x \geq 0$. For the coefficients of (81), one obtains, in analogous way,

$$f_\nu = Y_\nu^{(0)} + \Gamma_\nu + \sum_{\mu=1}^{4} S_{\nu,\mu} T_\mu \quad , \tag{84}$$

where

$$T_\mu = \sum_{\mu'=1}^{4} \sum_{\ell=1}^{4} \left(\mathcal{L}_{\mu,\mu'}^{(\ell)} X_{\mu'}^{(\ell-1)} - X_{\mu'}^{(2)} \kappa_{\mu,\mu'}^{(\ell)} z_\ell^{-4} \right) \quad . \tag{85}$$

As for the term proportional to the identity, we find

$$C' = C + \sum_{\alpha=1}^{4} (Y_\alpha^{(0)} \bar{\varphi}_\alpha) + \frac{1}{2} V (\varphi_2 \bar{\varphi}_2 - \varphi_3 \bar{\varphi}_3)$$
$$+ \sum_{\alpha=1}^{4} q_\alpha \sum_{\gamma,l=1}^{4} \left(\mathcal{L}_{\alpha\gamma}^{(l+1)} b_\gamma^{(l-1)} - \kappa_{\alpha\gamma}^{(l)} z_l^{-4} (b_\gamma^{(2)} + b_\gamma^{(3)}) \right) , \tag{86}$$

where the coefficient q_α are bilinear forms in the $\{\varphi_\beta\}$'s.

H' as given by (70) is an element of $\mathcal{B}$, if the coefficients f_ν given by (84) equal zero. The latter requirement gives rise to a system of four equations in the unknown $\varphi_\nu \in \mathcal{G}_o$. In order to solve it, it is convenient to express both the θ's and the φ's as linear combinations with coefficients in $\mathbb{R}$ of two anticommuting units ξ and $\bar{\xi} \in \mathcal{G}_o$, with $\xi \bar{\xi} = 1$. More precisely we choose

$$\theta_i = \rho_i \xi + \sigma_i \bar{\xi} \quad , \quad \varphi_\nu = r_\nu \xi + s_\nu \bar{\xi} \quad . \tag{87}$$

Once the above representation has been inserted in the system (84), by separately setting equal to zero the coefficients of the ξ and the $\bar{\xi}$ part of each equation, one obtains a system of eight equations in the real unknown r_ν and s_ν ($\nu = 1, \ldots, 4$).

We shall denote by $\{\tilde{r}_\nu, \tilde{s}_\nu\}$ the solutions of such system, and by $\tilde{b}_\mu$ the expressions (82) evaluated at $r_\nu = \tilde{r}_\nu$, $s_\nu = \tilde{s}_\nu$. We also denote by $\tilde{H}'$ the rotated bosonic hamiltonian

$$\tilde{H}' = V B_0 + \sum_{\mu=1}^{4} \tilde{b}_\mu B_\mu + \tilde{C}' \quad . \tag{88}$$

It is worth pointing out how $\tilde{H}'$ contains the off-diagonal pairing operator B_1, even though H_d doesn't. As mentioned earlier, this makes the model considered particularly interesting in view of high T_c superconductivity, as such operators, being intrinsically generated by the dynamical algebra of the system, may be expected to give rise to non-vanishing order parameters.

We can now proceed to the rotation of $\tilde{H}'$ into H'', by means of the adjoint action of $\mathcal{Z}_B = z\, B_5 + w\, B_6 \in \mathcal{B}$; $z, w \in \mathbb{R}$. Of course, as the rotation is an automorphism in $\mathcal{B}$, H'' will still be of the form

$$H'' = V B_0 + \sum_{\mu=1}^{4} h_\mu B_\mu + \tilde{C}' \quad . \tag{89}$$

Let us define $\tilde{H}'^{(m+1)} = [\, \mathcal{Z}_B, \; \tilde{H}'^{(m)}]\, , m \geq 0$. Then $|\, \tilde{b}^{(m)} > = \Omega^m \, |\, \tilde{b} >$, with

$$\Omega = \begin{pmatrix} 0 & -2z & 0 & 0 & z \\ 2z & 0 & 0 & 0 & 0 \\ 0 & 0 & 0 & -2w & 0 \\ 0 & 0 & 2w & 0 & 0 \\ z & 0 & 0 & 0 & 0 \end{pmatrix} \quad . \tag{90}$$

where (66) has been used, and $| h > = \exp(\Omega) | \tilde{b} >$. Exponentiation of the matrix Ω leads to

$$
\begin{aligned}
h_1 &= \tilde{b}_1 \cos 2w + (\frac{V}{2} - \tilde{b}_2) \sin 2w \quad , \\
h_2 &= (\tilde{b}_2 - \frac{V}{2}) \cos 2w + \tilde{b}_1 \sin 2w + \frac{1}{2} V \quad , \\
h_3 &= \tilde{b}_3 \cos 2z - \tilde{b}_4 \sin 2z \quad , \\
h_4 &= \tilde{b}_4 \cos 2z + \tilde{b}_3 \sin 2z \quad .
\end{aligned}
\tag{91}
$$

In order for $\mathcal{H}$ to be diagonal, the equations $h_1 = 0 = h_3$ must be satisfied. These can be easily solved in z, w. Denoting the solution by $\tilde{z}, \tilde{w}$, one has

$$
\tilde{w} = \frac{1}{2} \arctan \left(\frac{2\tilde{b}_1}{2\tilde{b}_3 - V} \right) \quad , \quad \tilde{z} = \frac{1}{2} \arctan \left(\frac{\tilde{b}_3}{\tilde{b}_4} \right) \quad .
\tag{92}
$$

With the above choices $\mathcal{H}$ finally reads

$$
\mathcal{H} = (\tilde{h}_4 - \tilde{h}_2) N_1 - (\tilde{h}_4 + \tilde{h}_2) N_2 + V N_1 N_2 + C'' \quad ,
\tag{93}
$$

where

$$
\begin{aligned}
\tilde{h}_2 &= \sqrt{\tilde{b}_1^2 + (\tilde{b}_2 - \frac{V}{2})^2} + \frac{V}{2} \quad , \\
\tilde{h}_4 &= \sqrt{\tilde{b}_3^2 + \tilde{b}_4^2} \quad , \\
C'' &= \tilde{C}' + \tilde{h}_2 \quad .
\end{aligned}
\tag{94}
$$

Of course, the spectrum is given by (93) setting $N_i = 0, 1$ for $i = 1, 2$.

The spectrum of hamiltonian (93) still depends (through $\tilde{h}_2, \tilde{h}_4$) on the set of classical Ising-like configuration variables, D_i's. The expectation value of the latter cannot evolve dynamically under the action of the linearized hamiltonian (61), since their commutator with it (as well as with (58)) vanishes. Nevertheless, since we aim to know which is the configuration of the D_i's most favourable from the point of view of the free energy, we should average over the distribution of the D_i's in Λ. We do this by simply summing within the partition function Z over all possible configurations of the D_i's in the dimer, recalling that a chemical potential μ_B fixing their average number (specified by the magnetization m) was introduced in the hamiltonian when constructing the model (see(58)): in other words, we let μ_B incorporate all necessary information on the distribution of the D_i's over the lattice, with a weight which is a Gibbs probability. Explicitly

$$
\begin{aligned}
Z &\doteq \sum_{D_1, D_2 = 0,1} \sum_{N_1, N_2 = 0,1} \exp -\beta \mathcal{H} \\
&= e^{-\beta C''} \left[1 + e^{\beta \tilde{h}_2} \left(2 \cosh \beta \tilde{h}_4 + e^{\beta(\tilde{h}_2 - V)} \right) \right] \quad .
\end{aligned}
\tag{95}
$$

In order to obtain quantitative predictions from equation (95), we must first evaluate the θ_i's as well as filling n_0 consistently with their definitions (59) and (60). Moreover, the magnetization $m \doteq < N_i - D_i > = n_A - n_B$ has to be fixed. This requires us to compute the expectation values $< A_i^\dagger >$ and $< N_i \pm D_i >$ respectively, which is straightforwardly achieved upon recalling that, for any operator $\Omega \in \mathcal{A}$ (different from B_0), we have

$$
\begin{aligned}
< \Omega > &= \frac{1}{Z} \text{Tr} \left[\Omega \, e^{-\beta H_d} \right] \\
&= \frac{1}{Z} \text{Tr} \left\{ \exp(\text{ad}\tilde{z}_B)(\exp(\text{ad}\tilde{z}_F)(\Omega)) \, e^{-\beta \mathcal{H}} \right\} \quad ,
\end{aligned}
\tag{96}
$$

where $\tilde{Z}_B$ and $\tilde{Z}_F$ are of course those antihermitian rotation operators which diagonalize H_d.

Upon denoting by Ω' the rotated operator $\exp(\mathrm{ad}\tilde{Z}_F)(\Omega)$, we observe that Ω' is given by a formula completely analogous to (70) (with $V = 0$), in which one has to replace the b_μ's, f_κ's and C' by suitable $b_\mu^{(\Omega)}$'s, $f_\kappa^{(\Omega)}$'s and C'_Ω. The latter are still expressed by equations (73) to (86), in which only the initial vectors $\mid X^{(\ell)} >$ ($\ell = 0, 1, 2, 3$) and $\mid Y^{(0)} >$ have been replaced by $\mid X^{(\ell)} >^\Omega$ and $\mid Y^{(0)} >^\Omega$, and are obtained by (71) upon substituting H_d with Ω.

Also, when evaluating $\Omega'' \doteq \exp(\mathrm{ad}\tilde{Z}_B)(\Omega')$ in (96), we can disregard the contribution of the fermionic operators in Ω', as they remain fermionic after bosonic rotation, and hence have vanishing expectation value in the basis in which $\mathcal{H}$ is diagonal. This implies that formula (93)-(94) hold even for Ω'' (once more setting $V = 0$), where of course the appropriate $\tilde{b}_\mu^{(\Omega)}$'s (same expression as in (82) with $|X^{(\ell)}>^\Omega$ instead of $|X^{(\ell)}>$) have to be used instead of $\tilde{b}_\mu$'s in the definition of the corresponding $h_\mu^{(\Omega)}$'s. Recalling once more that only the operators diagonal in the Cartan basis contribute to the trace in (96), we finally get

$$
\begin{aligned}
< \Omega > \;=\; & \frac{1}{Z} \sum_{\{D_i\}} \sum_{\{N_j\}} (h_2^{(\Omega)}(\{D_i\})(1 - N_1 - N_2) + h_4^{(\Omega)}(\{D_i\})(N_1 - N_2) \\
& + C^{(\Omega)}(\{D_i\})) \, e^{-\beta \mathcal{H}(\{D_i\}, \{N_j\})} \quad,
\end{aligned}
\tag{97}
$$

with

$$
h_2^{(\Omega)} = b_1^{(\Omega)} \sin 2\tilde{w} + b_2^{(\Omega)} \cos 2\tilde{w} \quad, \qquad h_4^{(\Omega)} = b_3^{(\Omega)} \sin 2\tilde{z} + b_4^{(\Omega)} \cos 2\tilde{z} \quad.
\tag{98}
$$

By applying (97), successively setting $\Omega = \frac{1}{2}(\theta_1 A_1 - \bar{\theta}_1 A_1^\dagger)$, $\frac{1}{2}(\theta_2 A_2 - \bar{\theta}_2 A_2^\dagger)$, and $\frac{1}{2}\{N_1 + N_2 \pm (D_1 + D_2)\}$, we obtain the four consistency equations to be satisfied, by setting at the l.h.s. of (97) $\theta_1 \bar{\theta}_1$, $\theta_2 \bar{\theta}_2$, n_0 and m respectively. The latter equations of course are to be solved in the four unknowns $\theta_1 \bar{\theta}_1$, $\theta_2 \bar{\theta}_2$, μ_A and μ_B. While the first three of these equations are in general highly non-linear, and must be handled numerically, the fourth can be solved analytically: upon defining $p = < D_i > = (n_0 - m)$, we obtain from it μ_B as a function of the other unknowns:

$$
\exp \beta \mu_B = \frac{1}{(p-2)Y_2} \left\{ Y_1(1 - p) + \sqrt{(p-1)^2 Y_1^2 - p(p-2)Y_0 Y_2} \right\} \quad,
\tag{99}
$$

where

$$
Y_\kappa \doteq \mathcal{H}|_{D_1 + D_2 = \kappa} + \kappa \mu_B \quad, \qquad \kappa = 0, 1, 2 \quad,
\tag{100}
$$

are, in fact, independent of μ_B.

Obviously, any expectation value of operators in S can be evaluated by means of (97)-(98), once the vectors $\mid X^{(\ell)} >^\Omega$ and $\mid Y^{(0)} >^\Omega$ have been specified.

In particular we have, for the pairing operator B_1,

$$
\mid X^{(0)} >^{B_1} = \begin{pmatrix} 1 \\ 0 \\ 0 \\ 0 \end{pmatrix} \quad, \qquad \mid X^{(2)} >^{B_1} = \begin{pmatrix} a \\ 0 \\ -b \\ -c \end{pmatrix} \quad,
$$

$$
\mid X^{(1)} >^{B_1} = \mid X^{(3)} >^{B_1} = \mid Y^{(0)} >^{B_1} = 0 \quad,
\tag{101}
$$

with a, b and c non-zero. It is important noticing that, depending on the physical parameters U, t, and V, the expression (96) implemented with $\Omega \equiv B_1$, together with (101), give a non-vanishing expectation value for the pairing operator. On the contrary, when $V = 0$ the latter turns out to be identically zero. This suggests once more that the intersite Coulomb repulsion could play a possible role in the onset of superconductivity.

4 The Clifford Mean-Field Approximation

We shall discuss, in this section, how fermionic linearization scheme is implemented when the mean-field amplitudes are treated as Θ_i of a Grassmann-Clifford algebra:

$$\{\Theta_i, \Theta_j\} = 0 \quad , \quad \{\Theta_i, \bar{\Theta}_j\} = c_i \, \delta_{ij} \quad , \quad c_i \in \mathbb{R} \quad , \tag{102}$$

where c_i are undeterminates to be defined for each specific problem.

We observe that (102) implies that the dynamical algebra of the linearized model is no longer graded, but simply a Lie algebra. Moreover the values of the c_i's can be determined either self-consistently or in a variational way.

We shall analyze, in this case, the algebraic structure of the Hubbard as well as of the Falikov-Kimball and the extended Falikov-Kimball models.

4.1 The Hubbard model

We fermi-linearize the hopping term of the Hubbard hamiltonian by writing

$$\sum_{<i,j>} A_i^\dagger A_j \doteq \frac{q}{2} \sum_i (\vartheta_i^\dagger A_i + A_i^\dagger \vartheta_i) \quad , \tag{103}$$

with $\vartheta_i = q^{-1} \sum_{j \, n.n.i} A_j$ (q has the usual meaning). As mentioned, we shall implement the fermionic linearization scheme by replacing the ϑ_i's by variables Θ_i still anticommuting with the fermion operators which are locally Clifford:

$$\Theta_i^2 = \bar{\Theta}_i^2 = 0 \quad , \quad \{\Theta_i, \bar{\Theta}_j\} = c \, \delta_{ij} \quad . \tag{104}$$

One thus obtains for the Hubbard model a reduced hamiltonian $\mathcal{H}$ which is a sum over all lattice sites of single-particle hamiltonians $H^{(i)}$, commuting with each other. Defining the new fermionic variables

$$A_1 := a_{i\uparrow}, \quad A_2 := a_{i\downarrow}, \quad A_3 = \Theta_i/\sqrt{c} \tag{105}$$

so that $\{A_\ell, A_\ell^\dagger\} = 1$, $\ell = 1, 2, 3$, we obtain

$$H^{(i)} = -\mu(N_1 + N_2) + U N_1 N_2 - tq\sqrt{c}\,[(A_1^\dagger + A_2^\dagger)A_3 + \text{h.c}] \tag{106}$$

where for simplicity the index i has been omitted at the r.h.s..

Closure under Lie bracket of the operators appearing in (106) leads to the dynamical algebra $\mathcal{A}_H$,

$$\mathcal{A}_H = \left(\bigoplus_{k=1}^4 u_k(1) \right) \oplus su(3) \oplus su(\bar{3}) \,. \tag{107}$$

The four central elements C_l generating the $u(1)$'s are given by:

$$\begin{aligned}
C_0 &= \mathbb{I} \quad , \quad C_1 = N_1 + N_2 + N_3 \quad , \\
C_2 &= N_1 N_2 + N_1 N_3 + N_2 N_3 \quad , \quad C_3 = N_1 N_2 N_3 \quad .
\end{aligned} \tag{108}$$

The Cartan-Weyl basis $\{H_i, E_\alpha\}$ of the first $su(3)$ subalgebra of $\mathcal{A}_H$ is given by:

$$\begin{aligned}
E_1 &= N_3 A_1^\dagger A_2 \quad , \quad E_2 = N_1 A_2^\dagger A_3 \quad , \quad E_3 = N_2 A_1^\dagger A_3 \quad , \\
E_{-1} &= N_3 A_2^\dagger A_1 \quad , \quad E_{-2} = N_1 A_3^\dagger A_2 \quad , \quad E_{-3} = N_2 A_3^\dagger A_1 \quad , \\
H_1 &= \frac{1}{\sqrt{2}} N_3 (N_1 - N_2) \,, \; H_2 = \frac{1}{\sqrt{6}} (N_1(N_2 - N_3) + N_2(N_1 - N_3)) \,,
\end{aligned} \tag{109}$$

with root vectors

$$\alpha_1 = \frac{1}{\sqrt{2}}(2,0) \quad , \quad \alpha_2 = \frac{1}{\sqrt{2}}(-1,\sqrt{3}) \quad , \quad \alpha_3 = \frac{1}{\sqrt{2}}(1,\sqrt{3}) \quad . \tag{110}$$

The remaining $su(3)$ algebra ($su(\tilde{3})$) is defined by the "orthogonal" set of operators $\{\tilde{H}_i, \tilde{E}_\alpha\}$ obtained by the substitution $N_i \rightarrow (1 - N_i)$ for the N_i multiplying the fermionic bilinears in the set $\{E_\alpha\}$ and the differences $N_i - N_j$ in the Cartan subset $\{H_1, H_2\}$.

In terms of the operators define above $H^{(i)}$ can be rewritten as

$$H^{(i)} = \mathcal{H}_{u(1)} \oplus \mathcal{H}_{su(3)} \oplus \mathcal{H}_{su(\tilde{3})} \quad . \tag{111}$$

with

$$\begin{aligned}
\mathcal{H}_{u(1)} &= -\frac{1}{3}(2\mu + tc)\mathcal{C}_1 + \frac{1}{3}U\mathcal{C}_2 \quad , \\
\mathcal{H}_{su(3)} &= \frac{2}{\sqrt{6}}(-\mu + U)H_2 - qt\sqrt{c}(E_2 + E_{-2} + E_3 + E_{-3}) \quad , \\
\mathcal{H}_{su(\tilde{3})} &= -\frac{2}{\sqrt{6}}\mu\tilde{H}_2 - qt\sqrt{c}(\tilde{E}_2 + \tilde{E}_{-2} + \tilde{E}_3 + \tilde{E}_{-3}) \quad .
\end{aligned} \tag{112}$$

$\mathcal{H}_{su(3)}$ and $\mathcal{H}_{su(\tilde{3})}$ have, in the $su(3)$ CW-basis, similar forms:

$$H = \varepsilon_2 H_2 + \tau(E_2 + E_{-2} + E_3 + E_{-3}). \tag{113}$$

Rotation by $Z_1 = -\frac{\pi}{4}(E_1 - E_{-1})$ reduces H to

$$H' = \exp(\mathrm{ad}Z)(H) = \varepsilon H_2 + \sqrt{2}\tau(E_2 + E_{-2}) \quad , \tag{114}$$

with $H' \in u(2) = \{H_1, H_2, E_2, E_{-2}\}$. Now, we can use the general procedure to diagonalize H'. We define the kets $\mid \Phi^{(k)} >= \mid \beta_1^{(k)}, \beta_2^{(k)}, \mu^{(k)} >$ with $\beta_1^{(0)} = 0$, $\beta_2^{(0)} = \varepsilon_2$, $\mu^{(0)} = \sqrt{2}\tau$ and $\mid H' >= \mid H_1, H_2, E_+ >$, where $E_+ \equiv E_2 + E_{-2}$, such that $H' =< \Phi^{(0)}\mid H' >$: the recursive relation $H'^{(k)} = [Z_2, H'^{(k-1)}] =< \Phi^{(k)}\mid H' >$ holds then with $Z_2 = \phi(E_2 - E_{-2})$. This leads to the definition of the 3×3 matrix

$$\mathbf{A} = \begin{pmatrix} 0 & 0 & -2a \\ 0 & 0 & 2b \\ a & -b & 0 \end{pmatrix} \quad , \quad a = \frac{\phi}{\sqrt{2}} \quad , \quad b = \sqrt{\frac{3}{2}}\phi \quad , \tag{115}$$

such that $\mid \Phi^{(k+1)} >= \mathbf{A} \mid \Phi^{(k)} >$. The rotated hamiltonian H'' can then be written as

$$H'' = \exp(\mathrm{ad}Z_2)(H') =< \Phi^{(0)}\mid e^{\mathbf{A}} \mid H' > \quad , \tag{116}$$

where

$$e^{\mathbf{A}} = \frac{1}{4}\begin{pmatrix} \cos 2\phi + 3 & -\sqrt{3}(\cos 2\phi - 1) & -2\sqrt{2}\sin 2\phi \\ -\sqrt{3}(\cos 2\phi - 1) & 3\cos 2\phi + 1 & 2\sqrt{6}\sin 2\phi \\ \sqrt{2}\sin 2\phi & \sqrt{6}\sin 2\phi & \cos 2\phi \end{pmatrix} \quad . \tag{117}$$

Equating to zero the resulting coefficient of the non-diagonal element E_+ gives for the rotation angle ϕ the value

$$\tilde{\phi} = \frac{1}{2}\arctan\frac{4\tau}{\sqrt{3}\varepsilon} \quad , \tag{118}$$

and we obtain, inserting this value in (116), the diagonalized hamiltonian

$$\begin{aligned}
\tilde{H}'' &= \exp(\mathrm{ad}Z_2)(\exp(\mathrm{ad}Z_1)(\mathcal{H})) \\
&= \frac{1}{4}(\sqrt{3}\varepsilon - \sqrt{\Delta}) H_1 + \frac{1}{4}(\varepsilon - \sqrt{3\Delta}) H_2 \quad ,
\end{aligned} \tag{119}$$

with $\Delta := 3\varepsilon^2 + 16\tau^2$. The partition function can be immediately obtained from (119) as

$$\mathcal{Z} = \sum_{N_1,N_2,N_3=0,1} \exp(-\beta\,\tilde{H}'') \quad . \tag{120}$$

Predictions for physical quantities can then be obtained from $\mathcal{Z}$ once the average number of electrons n_0 is fixed through the chemical potential, according to

$$n_0 =< N_1 + N_2 >= \frac{1}{\beta\mathcal{Z}}\frac{\partial\mathcal{Z}}{\partial\mu} \,, \tag{121}$$

where as usual $< \hat{O} >$ stays for the thermodynamical average in the Gibbs ensemble of the operator $\hat{O}$.

Morever, a value for c, depending on temperature $T = (k_B\beta)^{-1}$, has to be determined either self-consistently, by

$$< A_3^\dagger A_i + A_i^\dagger A_3 >= 2\sqrt{c} < A_3^\dagger A_3 > \,, \tag{122}$$

where translational invariance of the lattice has been assumed, or variationally, minimizing the free energy:

$$\frac{\partial}{\partial c}\ln\mathcal{Z} = 0 \,. \tag{123}$$

4.2 The Falikov-Kimball model

In the present context, the hamiltonian for the FK model reads $H_{FK} = \sum_i H_{FK}^{(i)}$, with

$$H_{FK}^{(i)} = -\mu_A N_i - \mu_B D_i + U\,D_i N_i - tq\sqrt{c}\,\,(\bar{\Theta} A_i + A_i^\dagger \Theta) \quad . \tag{124}$$

The two possible eigenvalues 0 and 1 of the variable D_i label the two orthogonal projections of $H_{FK}^{(i)} = H^{(0)} \oplus H^{(1)}$. It is easy to check that the dynamical algebra $\mathcal{A}_{FK}$ of (124) coincides with the Lie algebra $u(2)$ generated by

$$\mathcal{A}_{FK} = \{N_i \pm \bar{\Theta}\Theta, \bar{\Theta} A \pm A\Theta\}\,. \tag{125}$$

A rotation similar to (116) leads to the diagonal form

$$\tilde{H}_{FK}^{(i)} = \frac{1}{2}\{\varepsilon_i(\bar{\Theta}\Theta + N_i) \pm \sqrt{\varepsilon_i^2 + 4\tau^2}(\bar{\Theta}\Theta - N_i)\} - \mu_B D_i \,, \tag{126}$$

with $\varepsilon_i = UD_i - \mu_B$. The partition function $\mathcal{Z} = \sum_{N_i,D_i,\bar{m}m=0,1} \exp(-\beta\,\tilde{H}_{FK})$, allows us to derive the chemical potentials by imposing the band filling constraints $n_A =< N_i >= \frac{1}{\beta\mathcal{Z}}\frac{\partial\mathcal{Z}}{\partial\mu_A}$, and $n_B =< D_i >= \frac{1}{\beta\mathcal{Z}}\frac{\partial\mathcal{Z}}{\partial\mu_B}$. c can once more be obtained either by implementing a self-consistency condition analogous to (122) or in variational way.

4.3 The Extended Falikov-Kimball model in the Clifford scheme

We finally study the structure of dynamical symmetry which the use of Clifford-like mean-fields defined in the previous sections leads to, in the case of the extended Falicov model [4], introduced in sect. 3. The linearized hamiltonian $\mathcal{H}_D$ is now written in the form

$$\begin{aligned}
\mathcal{H}_D &= \sum_{\alpha=1}^{2} \varepsilon_\alpha N_\alpha + V N_1 N_2 - t\left(A_1^\dagger A_2 + A_2^\dagger A_1\right) + \tilde{C} \\
&+ \sum_{\alpha=1}^{2}\left(A_\alpha^\dagger\Theta_\alpha + \bar{\Theta}_\alpha A_\alpha - \bar{\Theta}_\alpha\Theta_\alpha\right) \quad,
\end{aligned} \tag{127}$$

with $\bar{\alpha} \equiv \alpha + 1 \,(\mathrm{mod}\,2)$, and

$$
\begin{aligned}
\Theta_\alpha &= t(q-1)\vartheta_\alpha \quad, \\
\varepsilon_\alpha &= -\mu_A + Vn_0(q-1) + UD_\alpha + VD_{\bar{\alpha}} \doteq \mathcal{E} + W_\alpha \quad, \\
\tilde{C} &= (\mathcal{E} + \mu_A - \mu_B)(D_1 + D_2) + VD_1D_2 - V(q-1)n_0^2 \quad.
\end{aligned} \tag{128}
$$

Let us recall explicitly the properties of the variables $\{\Theta_\alpha, \bar{\Theta}_\alpha \,|\, \alpha = 1, 2\}$:

- generators $\{\Theta_\alpha, \bar{\Theta}_\alpha | \alpha \in \mathcal{D}\}$ anticommute with all fermion operators, *i.e.* $\{A_\beta, \Theta_\alpha\} = \{A_\beta^\dagger, \Theta_\alpha\} = \{A_\beta, \bar{\Theta}_\alpha\} = \{A_\beta^\dagger, \bar{\Theta}_\alpha\} = 0$, $\forall \alpha, \beta \in \mathcal{D}$, and anticommute with each other for $\alpha \neq \beta$: $\{\Theta_\alpha, \Theta_\beta\} = \{\bar{\Theta}_\alpha, \bar{\Theta}_\beta\} = \{\Theta_\alpha, \bar{\Theta}_\beta\} = \{\bar{\Theta}_\alpha, \Theta_\beta\} = 0$;

- locally the generators $\{\Theta_\alpha | \alpha \in \mathcal{D}\}$ are nilpotent Clifford variables: $\Theta_\alpha^2 = \bar{\Theta}_\alpha^2 = 0$, $\{\Theta_\alpha, \bar{\Theta}_\alpha\} = c_\alpha^2$, where the $\{c_\alpha\}$'s are c-numbers;

- $\bar{\Theta}_\alpha$ is the conjugate of Θ_α, $\forall \alpha \in \mathcal{D}$: in other words, the mean field *is* itself a fermionic *operator*.

Upon noticing how $\{\Theta_\alpha, \bar{\Theta}_\alpha | \alpha \in \mathcal{D}\}$ enter into play in $\mathcal{H}_\mathcal{D}$, we introduce, besides the fermionic operators A_α, $A_\alpha^\dagger$, $\alpha = 1, 2$ the auxiliary fermionic variables $A_{\lambda+2} \equiv \dfrac{1}{c_\lambda}\Theta_\lambda$, $A_{\lambda+2}^\dagger \equiv \dfrac{1}{c_\lambda}\bar{\Theta}_\lambda$; $\lambda = 1, 2$, as was already done in sect. **4.1**. The linearized hamiltonian $\mathcal{H}_\mathcal{D}$ will therefore be written in the form

$$
\mathcal{H}_\mathcal{D} = \sum_{\alpha=1}^{2} \varepsilon_\alpha N_\alpha + VN_1N_2 + \frac{1}{2}\sum_{\nu=1}^{4}\sum_{\substack{\gamma=1 \\ \gamma\neq\nu}}^{4} \tau_{\nu,\gamma}\, A_\nu^\dagger A_\gamma \quad, \tag{129}
$$

where the antisymmetric matrix **T** of elements $\mathbf{T}_{\nu,\gamma} \equiv \tau_{\nu,\gamma}$ is given by

$$
\mathbf{T} = \sigma_x \otimes [(c_1 + c_2)\mathbb{I}_2 + (c_1 - c_2)\sigma_z] - [(t + c_1c_2)\mathbb{I}_2 + (t - c_1c_2)\sigma_z] \otimes \sigma_x \,, \tag{130}
$$

σ_κ, $\kappa = x, y, z$ denoting the usual Pauli matrices. It is worth pointing out that ε_1 is in general $(V \neq U)$ different from ε_2.

The form (129) of hamiltonian (127) is particularly interesting, in that it exhibits quite manifestly the whole set of dynamical symmetries characteristic of the system in its linearized form. In order to show this, we observe first that, in the absence of the V term, we should recover (with $n = 4$) the well known $su(n)$ symmetry characteristic of a system with n fermions, generated by all number conserving bilinear forms of creation and annihilation operators [1]. The n.n. Coulomb coupling term, N_1N_2, can be expected to lead us to an extended dynamical algebra $\mathcal{A}_n$, generated by all number preserving multilinear forms of type $\{A_{\alpha_1}^\dagger \ldots A_{\alpha_n}^\dagger A_{\beta_1} \ldots A_{\beta_n}\}$, $n = 1, \ldots, 4$. One can show (by induction, after working out explicitly the cases $n = 2$ and $n = 3$) that in general such an algebra is

$$
\mathcal{A}_n = \bigoplus_{\kappa=0}^{n} u_\kappa(1) \oplus \bigoplus_{\kappa=1}^{n-1} su\left(\binom{n}{\kappa}\right) \quad. \tag{131}
$$

In present application $(n = 4)$ the dynamical algebra is therefore $\mathcal{A}_\mathcal{D} \equiv \mathcal{A}_4 \equiv 5u(1) \oplus 2 \cdot su(4) \oplus su(6)$. $\mathcal{A}_\mathcal{D}$ has 70 generators, 16 of which are Cartan, and 5 are central. These generators can naturally be obtained by commuting in all possible ways the operators entering the hamiltonian $\mathcal{H}_\mathcal{D}$.

Upon introducing the auxiliary variables $E_\beta^\alpha \doteq A_\alpha^t A_\beta$ (notice that $E_\alpha^\beta = E_\beta^{\alpha\dagger}$, $E_\alpha^\alpha \equiv N_\alpha$), the generators of $\mathcal{A}_\mathcal{D}$ turn out to be given by the following tensor operators:

$$F^{(l)\{\alpha_i\}}_{\{\beta_i\}} := \prod_{i=1}^{l} E_{\beta_i}^{\alpha_i} \quad , \quad l = 1, 2, 3, 4 \quad , \tag{132}$$

where the indices α_i, β_i all different from one another range from 1 to 4. The five central elements C_l, generating the 5 $u(1)$'s, are given by:

$$C_0 = 1 \quad , \quad C_l = \sum_{\alpha_1 < \cdots < \alpha_l} \left(\prod_{i=1}^{l} N_{\alpha_i} \right) \quad , \quad l = 1, \ldots, 4 \quad . \tag{133}$$

In order to recognize the direct product structure of the algebra, let us define the following set of tensor operators which are subsets of the $F^{(l)}$'s in (132):

$$\begin{aligned}
J_\beta^\alpha &\equiv N_\gamma N_\delta E_\beta^\alpha \quad ; \quad \hat{J}_\beta^\alpha \equiv (1 - N_\gamma)(1 - N_\delta) E_\beta^\alpha \quad ; \\
{}_\beta^{}P_\delta^\gamma &\equiv E_\beta^\alpha E_\delta^\gamma \quad ; \quad {}_\delta^\gamma M_\beta^\alpha \equiv N_\gamma (1 - N_\delta) E_\beta^\alpha \quad ;
\end{aligned} \tag{134}$$

where $\alpha \neq \beta$ and $\gamma \neq \delta$. In the first two relations of (134), we introduced the convention that whenever in the definition of some operator appear two extra indices besides those labelling the operator itself, one should think of them as assuming the complementary values in the set $\{1, 2, 3, 4\}$.

We may immediately obtain for J_β^α the commutation relations of $su(4)$:

$$\begin{aligned}
[J_\beta^\alpha, J_\delta^\gamma] &= \delta_{\beta\gamma} J_\delta^\alpha - \delta_{\alpha\delta} J_\beta^\gamma \quad , \\
[J_\beta^\alpha, J_\alpha^\beta] &= N_\gamma N_\delta (N_\alpha - N_\beta) \quad .
\end{aligned} \tag{135}$$

The first relation above holds whenever it is not simultaneously $\alpha = \delta$ and $\beta = \gamma$, in which case one should use the second.

Quite similar relations hold for and $\hat{J}_\beta^\alpha$ (upon replacing N_j with $(N_j - 1)$ in the second commutator of (135)), generating the second $su(4)$, manifestly orthogonal – due to the presence of the projection operators – to the previous one.

Analogously, the algebra generated by the set $\{{}_\delta^\gamma M_\beta^\alpha, {}_\delta^\gamma P_\beta^\alpha\}$ is:

$$\begin{aligned}
[{}_\delta^\gamma M_\beta^\alpha, {}_\sigma^\rho M_\nu^\mu] &= \delta_{\beta\mu}(\delta_{\alpha\rho}\delta_{\gamma\nu} {}_\delta^\beta M_\gamma^\alpha + \delta_{\alpha\sigma}\delta_{\gamma\rho} {}_\beta^\gamma M_\delta^\alpha) \\
&\quad - \delta_{\alpha\nu}(\delta_{\beta\rho}\delta_{\gamma\mu} {}_\delta^\alpha M_\beta^\gamma + \delta_{\beta\sigma}\delta_{\gamma\rho} {}_\alpha^\gamma M_\beta^\delta) \\
&\quad + \delta_{\alpha\sigma}\delta_{\beta\rho}\delta_{\gamma\mu} {}_\beta^\alpha P_\delta^\gamma - \delta_{\alpha\rho}\delta_{\beta\sigma}\delta_{\gamma\nu} {}_\beta^\alpha P_\gamma^\delta \quad ,
\end{aligned} \tag{136}$$

$$\begin{aligned}
[{}_\delta^\gamma P_\beta^\alpha, {}_\sigma^\rho P_\nu^\mu] &= (\delta_{\beta\mu}\delta_{\delta\rho} - \delta_{\beta\rho}\delta_{\delta\mu})(\delta_{\alpha\nu} - \delta_{\alpha\sigma}) \\
&\quad (N_\alpha N_\gamma(1 - N_\beta)(1 - N_\delta) - N_\beta N_\delta(1 - N_\alpha)(1 - N_\gamma)) \quad ,
\end{aligned} \tag{137}$$

$$\begin{aligned}
[{}_\delta^\gamma P_\beta^\alpha, {}_\sigma^\rho M_\nu^\mu] &= (\delta_{\alpha\nu}\delta_{\gamma\sigma}(\delta_{\beta\mu} - \delta_{\beta\rho}) + \delta_{\alpha\sigma}\delta_{\gamma\nu}(\delta_{\beta\mu} + \delta_{\beta\rho})) {}_\mu^\nu M_\rho^\sigma \\
&\quad + (\delta_{\alpha\nu}\delta_{\gamma\rho}(\delta_{\beta\sigma} - \delta_{\beta\mu}) - \delta_{\alpha\rho}\delta_{\gamma\nu}(\delta_{\beta\mu} + \delta_{\beta\sigma})) {}_\nu^\mu M_\sigma^\rho \quad ,
\end{aligned} \tag{138}$$

which can be easily recognized to be isomorphic with that of $su(6)$.

The operators $E_1 \doteq J_2^1$, $E_2 \doteq J_4^3$, $E_3 \doteq J_3^2$, $E_4 \doteq J_4^1$, $E_5 \doteq J_4^2$, $E_6 \doteq J_3^1$ and the analogous $\hat{E}_\kappa$, defined with the $\hat{J}$'s, together with

$$\begin{aligned}
H_1 &\doteq \frac{1}{2}(N_1 N_4 (N_2 - N_3) + N_2 N_3 (N_1 - N_4)) \quad , \\
H_2 &\doteq \frac{1}{2}(N_1 N_2 (N_3 - N_4) + N_3 N_4 (N_1 - N_2)) \quad , \tag{139} \\
H_3 &\doteq \frac{1}{2}(N_2 N_4 (N_1 - N_3) - N_1 N_3 (N_2 - N_4)) \quad , \tag{140}
\end{aligned}$$

and the analogous $\hat{H}_\kappa$ obtained from the H_κ once more replacing N_j with $(N_j - 1)$, provides us with the Cartan-Weyl realization of the two $su(4)$'s :

$$\begin{aligned}
[\mathbf{H}, E_\kappa] &= \mathbf{\Gamma}^{(\kappa)} E_\kappa \quad ; \quad [E_\kappa, E_{\kappa'}] = \mathcal{N}_{\kappa\kappa'} E_{\kappa+\kappa'} \quad ; \\
[H_i, H_j] &= 0 \quad ; \quad [E_\kappa, E_{-\kappa}] = \mathbf{\Gamma}^{(\kappa)} \cdot \mathbf{H} \quad ;
\end{aligned} \tag{141}$$

where $\mathbf{H} \equiv |H_1 H_2 H_3\rangle$, and both the vectors $\mathbf{\Gamma}^{(\kappa)}$ and the matrix $\mathcal{N}_{\kappa\kappa'}$ are straightforwardly obtained from the definitions by explicit computation.

Similarly, suitably renaming the fifteen operators M and P as:

$$\begin{aligned}
G_1 &\doteq {}_4^3 M_2^1 \quad ; \quad & G_2 &\doteq {}_3^4 M_2^1 \quad ; \quad & G_3 &\doteq {}_2^1 M_4^3 \quad ; \\
G_4 &\doteq {}_1^2 M_4^3 \quad ; \quad & G_5 &\doteq {}_4^1 M_3^2 \quad ; \quad & G_6 &\doteq {}_1^4 M_3^2 \quad ; \\
G_7 &\doteq {}_4^2 M_3^1 \quad ; \quad & G_8 &\doteq {}_2^4 M_3^1 \quad ; \quad & G_9 &\doteq {}_3^1 M_4^2 \quad ; \\
G_{10} &\doteq {}_1^3 M_4^2 \quad ; \quad & G_{11} &\doteq {}_3^2 M_4^1 \quad ; \quad & G_{12} &\doteq {}_2^3 M_4^1 \quad ; \\
G_{13} &\doteq {}_2^1 P_4^3 \quad ; \quad & G_{14} &\doteq {}_2^1 P_3^4 \quad ; \quad & G_{15} &\doteq {}_3^1 P_4^2 \quad ;
\end{aligned} \tag{142}$$

and defining moreover the five Cartan operators

$$\begin{aligned}
J_1 &\doteq \frac{1}{\sqrt{2}} (N_1 N_2 (1 - N_3 - N_4) - N_3 N_4 (1 - N_1 - N_2)) \,, \\
J_2 &\doteq \frac{1}{\sqrt{2}} (N_1 N_3 (1 - N_2 - N_4) - N_2 N_4 (1 - N_1 - N_3)) \,, \\
J_3 &\doteq \frac{1}{\sqrt{2}} (N_1 N_4 (1 - N_2 - N_3) - N_2 N_3 (1 - N_1 - N_4)) \,, \\
J_4 &\doteq \frac{1}{2\sqrt{3}} (2(N_1 - N_2)(N_3 - N_4) + (N_1 - N_4)(N_2 - N_3)) \,, \\
J_5 &\doteq \frac{1}{2} (N_1 - N_4)(N_2 - N_3) \,,
\end{aligned} \tag{143}$$

we obtain also for $su(6)$ the standard Cartan-Weyl form

$$\begin{aligned}
[\mathbf{J}, G_\kappa] &= \mathbf{\Omega}^{(\kappa)} G_\kappa \quad ; \quad [H_i, H_j] = 0 \quad ; \\
[G_\kappa, G_{\kappa'}] &= \mathcal{M}_{\kappa\kappa'} G_{\kappa+\kappa'} \quad \text{with} \quad \mathbf{\Omega}^{(\kappa)} + \mathbf{\Omega}^{(\kappa')} = \mathbf{\Omega}^{(\kappa+\kappa')} \quad ; \\
[G_\kappa, G_{-\kappa}] &= \mathbf{\Omega}^{(\kappa)} \cdot \mathbf{J} \quad ; \quad G_{-\kappa} = G_\kappa^\dagger \quad ,
\end{aligned} \tag{144}$$

where $\mathbf{J} \equiv |J_1 J_2 J_3 J_4 J_5\rangle$ together with the vectors $\mathbf{\Omega}^\kappa$ and the matrix $\mathcal{M}$ completely define the algebra structure. The vectors $\mathbf{\Gamma}^{(\kappa)}$ and $\mathbf{\Omega}^\kappa$ as well as the matrix $\mathcal{N}_{\kappa\kappa'}$ are explicitly given in ref. [4].

With these identifications, we can rewrite the Hamiltonian (129) in the form

$$\mathcal{H}_D = \mathcal{H}_{u(1)} \oplus \mathcal{H}_{su(4)} \oplus \mathcal{H}_{su(\bar{4})} \oplus \mathcal{H}_{su(6)} \quad , \tag{145}$$

where (see equation (133))

$$\mathcal{H}_{u(1)} = \bigoplus_{\ell=0}^{4} \mathcal{H}_{u(1)}^{(\ell)} \quad , \quad \mathcal{H}_{u(1)}^{(\ell)} \equiv X^{(\ell)} \mathcal{C}_\ell \quad , \tag{146}$$

with:

$$X^{(0)} = \tilde{C} \; ; \; X^{(1)} = \frac{1}{4}(\varepsilon_1 + \varepsilon_2) \; ; \; X^{(2)} = \frac{1}{6} V \; ; \; X^{(3)} = 0 = X^{(4)} \,. \tag{147}$$

One finds moreover

$$\mathcal{H}_{su(4)} = \sum_{k=1}^{3} h_k H_k + \sum_{i=1}^{2} (g_i E_i + c_i E_{i+3} + \text{h.c.}) \quad , \tag{148}$$

$$\mathcal{H}_{su(6)} = \sum_{k=1}^{5} \tilde{h}_k J_k + \sum_{i=1}^{2} \sum_{k=2i-1}^{2i} (g_i G_k + c_i G_{k+6} + \text{h.c.}) \quad , \tag{149}$$

$$\tag{150}$$

with

$$g_1 = t \,,\; g_2 = -c_1 c_2 \,,\; h_1 = \frac{1}{4}(\varepsilon_1 + \varepsilon_2 + 2V) \,,\; h_2 = h_3 = \frac{1}{4}(\varepsilon_1 - \varepsilon_2) \,,$$

$$\tilde{h}_\kappa = 2\sqrt{2}\, h_\kappa \,(V \mapsto V/2) \,,\; \kappa = 1,2,3 \,;\; \tilde{h}_4 = \frac{1}{2\sqrt{3}} V \,;\; \tilde{h}_5 = \frac{1}{2} V \quad ,$$

whereas $\mathcal{H}_{su(\bar{4})}$ is obtained from $\mathcal{H}_{su(4)}$ by replacing in it "hatted" operators and setting $V = 0$.

Last step to be performed in order to find the spectrum of $\mathcal{H}_D$ is the (independent) diagonalization of the three hamiltonians $\mathcal{H}_{su(4)}$, $\mathcal{H}_{su(\bar{4})}$, and $\mathcal{H}_{su(6)}$. This is done – as customary – by a generic inner automorphism in $\mathcal{A}_D$ (generalized Bogolubov transformation), with the procedure shown in sect. **1.2**.

Such a procedure, if one aims to obtaining only the eigenvalues of the hamiltonian, simplifies to either one of the following schemes:

1. if the fundamental faithful representation for each dynamical algebra is available, one simply writes the hamiltonian corresponding to $su(n)(n = 4, \bar{4}, 6)$ as a matrix of rank n, and the spectrum of $\mathcal{H}_{su(n)}$ is simply given by the eigenvalues $\{\omega_\kappa | \kappa = 1, \ldots, n\}$ of such matrix;

2. since the complete set of Casimir operators $\{\Gamma_\ell(\{H_i; E_{\pm\kappa}\}) | \ell = 2, 3, 4\}$ for $su(4)$ (and obvious analogous for $su(\bar{4})$, with $\hat{H}_i$ and $\hat{E}_{\pm\kappa}$ replacing H_i and $E_{\pm\kappa}$), or, alternatively, $\{\Gamma_\ell(\{J_i; G_{\pm\kappa}\}) | \ell = 2, \ldots, 6\}$ for $su(6)$, where Γ_ℓ is multilinear of order ℓ in the operators, is known [23], one can write directly the secular polynomial for $\mathcal{H}_{su(n)}$ as
$$\omega^n + \sum_{k=0}^{n-2} \gamma_{n-k} \omega^k = 0.$$
Here the coefficients γ_ℓ are equal to Γ_ℓ in which the operatorial arguments are replaced by the coefficient they have in $\mathcal{H}_{su(n)}$ [7].

Of course the complete solution of the problem would still require the determination in a self-consistent or variational way of the two mean-field parameter c_1 and c_2.

References

[1] J.L. Birman, and A.I. Solomon, *Spectrum Generating Algebras in Condensed Matter Physics*, in *Dynamical Groups and Spectrum Generating Algebras*, A. Bohm, Y. Néeman, and A.O. Barut, eds.; World Scientific, Singapore, 1988

[2] U. Brandt and C. Mielsch, *Z. Phys. B: Condensed Matter* **75**, 365 (1989); **79**, 295 (1990)

[3] A. Danani, A. Montorsi, and M. Rasetti, *Intl. J. Mod. Phys.* **B**, **6**, 3529 (1992)

[4] A. Danani and M. Rasetti, *Mod. Phys. Lett.* **B**, **6**, 1583 (1992)

[5] L.M. Falikov, J.C. Kimball, *Phys. Rev. Lett.* **22**, 997 (1969)

[6] V. Janiš, *Z. Phys. B: Condensed Matter* **83**, 227 (1991)

[7] R. Gilmore, *Lie Groups, Lie Algebras and Some of Their Applications*, J.Wiley, NY, 1974.

[8] L. Gruber, J. Wanski, J. Jedrzejiski, and P. Lemberger, *Phys. Rev.* **B41**, 2198 (1990); P. Lemberger, *J. Phys. A: Math. Gen.* **25**, 715 (1992)

[9] J.E. Hirsch, *Phys. Rev.* **B31**, 4403 (1985)

[10] *The Hubbard Model-A Reprint Volume*, edited by A. Montorsi; World Scientific, Singapore, 1992.

[11] T. Kennedy, and E.H. Lieb, *Physica* **138A**, 320 (1986)

[12] S. Kivelson, W.P. Su, J.R. Schrieffer, and A.J. Heeger, *Phys. Rev. Lett.* **58**, 1899 (1987)

[13] F.H.L. Eβler, V.E. Korepin, and K. Schoutens, *Complete Solution of the One-dimensional Hubbard Model*, Institute for Theoretical Physics, State University of New York at Stony Brook, preprint ITP-SB-91-30 (1991)

[14] F.H.L. Eβler, V.E. Korepin, and K. Schoutens, *Completeness of the SO(4)-extended Bethe Ansatz for the One-dimensional Hubbard Model*, Institute for Theoretical Physics, State University of New York at Stony Brook, preprint ITP-SB-91-51 (1991)

[15] F.H.L. Eβler, V.E. Korepin, and K. Schoutens, *New Exactly Solvable Model of Strongly Correlated Electrons Motivated by High T_c Superconductivity*, Institute for Theoretical Physics, State University of New York at Stony Brook, preprint ITP-SB-92-03 (1992)

[16] R. Livi, A. Montorsi, and M. Rasetti, *Mod. Phys. Lett.* **B6**, 151 (1992)

[17] A. Montorsi and A. Pelizzola, *J. Phys. A: Math. Gen.* **25**, 5818 (1992)

[18] A. Montorsi, M. Rasetti, and A.I. Solomon, *Phys. Rev. Lett.* **59**, 2243 (1987)

[19] A. Montorsi, and M. Rasetti, *Mod. Phys. Letters* **B4**, 613 (1990)

[20] A. Montorsi, and M. Rasetti, *Phys. Rev. Letters* **66**, 1383 (1991)

[21] A. Montorsi, and M. Rasetti, *The Cluster Fermi-linearized Hubbard Model*, in *Dynamical symmetries and chaotic behavior in physical systems*, L. Fronzoni, G. Maino, and M. Pettini, eds.; World Scientific Publ. Co., Teanek, N.J., 1990

[22] A. Pelizzola, and C. Topi, *Intl. J. Mod. Phys.* **B5**, 3073 (1991)

[23] A.M. Perelomov and V.S. Popov,*J. Nucl. Phys.* **3**, 924(1966)

[24] S. Robaszkiewicz, R. Micnas, and J. Ranninger, *Phys. Rev.* **B36**, 180 (1987)

[25] A.I. Solomon, *J. Math. Phys.* **12**, 390 (1971)

[26] A.I. Solomon, *Ann. N.Y. Acad. Sci.* **410**, 63 (1983)

[27] J.A. Verges, E. Louis, P.S. Lomdahl, F. Guinea, and A.R. Bishop, *Phys. Rev.* **B43**, 6099 (1991); K. Yonemitsu, I. Batistic, and A. R. Bishop, *Phys. Rev.* **B44**, 2652 (1991)

[28] C.N. Yang, and S.C. Zhang, *Mod. Phys. Lett.* **B4**, 759 (1990)

NUMBER AND ISOSPIN DEPENDENCE OF THE IBM3 HAMILTONIAN

J.P. Elliott, J.A. Evans and G.L. Long

School of Mathematical and Physical Sciences
University of Sussex
Brighton, Sussex, BN1 9QH, UK

INTRODUCTION

In two key papers, Arima, Otsuka, Iachello and Talmi[1] and Otsuka, Arima and Iachello[2] (OAI) gave a microscopic foundation to the interacting boson model(IBM) by associating the s-boson with a $J = 0$ pair of like nucleons in a single j-shell and the d-boson with a $J = 2$ pair. The N-boson space was then mapped onto a sub-space S in the n-nucleon shell-model, with $n = 2N$ and with the number of d-bosons related to the shell-model seniority by $2N_d = v$. The quasi-spin group $SU(2)$ provides general formulae in the shell-model for the matrix elements of an arbitrary interaction in the seniority basis, as functions of n. This allowed OAI to deduce a corresponding IBM hamiltonian which, for any n, would exactly reproduce the shell-model matrix elements in that part of the sub-space S with seniority not exceeding $v = 4$. The IBM hamiltonian contained only one and two-boson interactions but, since its parameters were functions of N, it included some many-body effects that may, for example, originate in the Pauli principle in the shell-model.

The main practical benefit from the OAI idea was that it predicted the general form of the N-dependence of the IBM hamiltonian parameters to be expected in an empirical analysis of data. The idea was also applied to the IBM2 model for neutrons and protons, with the parameters now depending on the separate numbers N_ν and N_π for neutron and proton bosons.

In a heavy nucleus, where the neutron excess ensures that valence neutrons and protons occupy different shells, isospin purity is guaranteed but, in lighter nuclei, spurious mixing of isospin is avoided either by using a basis with good isospin or by using a sufficiently complete basis that states of good isospin emerge from the diagonalisation of an isospin invariant interaction. The shell-model subspace corresponding to IBM2 through the OAI mapping does not satisfy either of these criteria and so any application of IBM2 to lighter nuclei will be liable to errors due to spurious mixing of isospin. The

Symmetries in Science VII, Edited by B. Gruber
and T. Otsuka, Plenum Press, New York, 1994

incompleteness of the IBM2 space is clear from the fact that the ν and π bosons are just two of the three components of a $T = 1$ triplet with $M_T = \pm 1$. By including a third type of boson δ with $M_T = 0$, corresponding to a neutron-proton pair, one has the isospin invariant form[3,4,5] IBM3 of the boson model. Whereas IBM2 has a direct product $U(2) \times U(6)$, where $U(2)$ refers to the charge space of ν and π and $U(6)$ refers to the sd-space, IBM3 has the product $U(3) \times U(6)$ where $U(3)$ refers to the space of ν, π and δ. Because of the overall boson symmetry, the irreducible representations of the two factors must be described by the same partition of N. Thus in IBM2 the F-spin, for $U(2)$, determines the sd-symmetry partition in $U(6)$ while in IBM3, the sd-symmetry partition determines the $U(3)$ representation. The isospin group $O(3)$ is a sub-group of $U(3)$ and hence a set of T-values is associated with each sd-symmetry partition.

In this note we summarise some work in progress to extend the OAI idea to the IBM3 model. The parameters in the boson hamiltonian now depend not only on N, but also on the isospin T.

THE SHELL-MODEL MAPPING IN IBM3

The IBM3 generalisation of the OAI mapping was given by Evans, Elliott and Szpikowski[5]. Again, the seniority basis was used for the shell-model states with the mapping equation $2N_d = v$ but, with a shell of both neutrons and protons, the seniority basis also carries the "reduced isospin" label t, which is the isospin of the unpaired nucleons. The mapping is continued by identifying t with the isospin of the d-bosons, which represent the unpaired nucleons.

Group theoretically, this seniority basis is described by the generalised quasi-spin group $O(5)$ whose ten generators are the pair creation and destruction operators and the isospin and number operators. The irreducible representations of $O(5)$ are labelled by v and t but it is conventional to use the notation (ωt) where $\omega = j + 1/2 - v/2$. The isospin structure of T-values for given v and t is not simple but a set of rules have been given by Hecht[6], based on the possible values of T_p, the isospin of the pairs. (This is a reduction from $O(5)$ to the isospin subgroup $O(3)$.) Although one may construct a set of shell-model states with definite T_p and t they are not generally orthogonal or linearly independent. The more recent use of vector coherent state theory for the group $O(5)$ by Hecht and Elliott[7] introduces a K-matrix transformation which produces an orthonormal set labelled by T_p with clear rules for identifying any overcompleteness.

In the IBM3 mapping one naturally associates T_p with the isospin of the s-bosons and in the boson framework T_p is clearly an orthonormal label. Hence the coherent state work[7] enables us to map one orthonormal set onto another using the label T_p. This completes the mapping. The table shows the mapping for states with even $N - T$ up to seniority $v = 4$. The first three columns give the shell-model $O(5)$ labels and the fourth column the IBM3 configuration. The T_p values refer to both the shell-model and IBM3. The final column lists the $U(3)$ partitions. States with definite $U(3)$ partitions are formed by summing over the relevant T_p and t values in the previous column with $U(3)$ Wigner coefficients. The suffix 0 or 2 on $[N - 2, 2]$ refers to the Vergados label which distinguishes multiplicities of T in a $U(3)$ representation. We stress that this $U(3)$ basis is constructed in both the IBM and shell-model frameworks. Thus, for $v = 4$, the sum runs over $t = 0$ and $t = 2$ which implies mixing of two $O(5)$ representations. Such mixing is found to occur in realistic cases with the lowest eigenvector being close[5,12] to the state with maximum $U(6)$ symmetry $[N]$.

Table 1. The correspondence between shell-model and IBM3 states.

Shell-model					
v	t	(ω,t)	IBM3	T_p	$U(3)$
0	0	$(j+1/2,0)$	s^N	T	$[N]$
2	1	$(j-1/2,1)$	$s^{N-1}d$	$T+1$	$[N]$
				$T-1$	$[N-1,1]$
4	0	$(j-3/2,0)$	$s^{N-2}d^2$	T	$[N]$
	2	$(j-3/2,2)$	$s^{N-2}d^2$	$T+2$	$[N-1,1]$
				T	$[N-2,2]_0$
				$T-2$	$[N-2,2]_2$
	1	$(j-3/2,1)$	$s^{N-2}d^2$	T	$[N-1,1]$

VECTOR COHERENT STATES FOR THE $O(5)$ GROUP

In principle, the existence of the generalised quasi-spin group $O(5)$ should allow us to derive formulae for the matrix elements of an arbitrary shell-model interaction in the seniority basis as functions of the two conserved quantities N and T. A few simple formulae have been known for a long time[8] and considerable work was done in this direction by Hecht[9,10]. However, the complexity of the $O(5)$ to $O(3)$ group reduction has prevented the derivation of sufficient formulae to extend the OAI idea to IBM3. We have recently[11] used the vector coherent state theory to obtain a general formula which has been evaluated for seniorities not greater than $v = 4$. This has opened the way to the construction of the IBM3 hamiltonian.

It is significant that the coherent state method makes use of a set of $T = 1$ bosons which correspond to the s-bosons of IBM3 and the formulae involve Wigner coefficients for the $U(3)$ group of the three-dimensional $T = 1$ space. In IBM3 this $U(3)$ group is part of a larger $U(3)$ group involving both s and d-bosons which plays a role analogous to the F-spin group in IBM2 which distinguishes states of different mixed symmetry. There seems to be no precise role for this wider $U(3)$ group in the group-theoretical framework of the shell-model although, as we have remarked above, it may be introduced in an approximate way, through the IBM3 mapping.

In deriving the general coherent state formula we naturally had to analyse the shell-model interaction V into components $V^{(ab)}$ belonging to definite irreducible representations $(ab) = (00), (11), (20)$, and (22) of $O(5)$. Of course, $V^{(ab)}$ is just one member of each (ab), distinguished by being invariant with respect to both the number operator and the isospin. To avoid the complications associated with the multiplicities which occur in the reduction of product representations in non-simply-reducible groups such as $O(5)$ we found it convenient to introduce the concept of the "extensions" of the interaction, defined as the other members $V^{(ab)}_{hu}$ of the representation (ab) containing $V^{(ab)}$ where u is the isospin and h is the change in particle number. In this notation $V^{(ab)} \equiv V^{(ab)}_{00}$. The formula is given in terms of reduced matrix elements of these extensions between states of full seniority. Thus for example a matrix element between states of seniority 2 and 4 involves a reduced matrix element of an extension $V^{(ab)}_{-1u}$ between states of two and four nucleons. These simple reduced matrix elements may be calculated by standard methods once the interaction V is chosen.

THE IBM3 HAMILTONIAN

With the help of these $O(5)$ formulae, the OAI idea may be extended to deduce an NT-dependent form of the IBM3 hamiltonian and the full results will be published soon[12]. Here we give only a summary.

We choose to write the general isoscalar one- and two-body IBM3 hamiltonian in the form

$$
\begin{aligned}
H = & \; H_0 + \epsilon \hat{N}_d + a_0((d^\dagger s^\dagger)_{20}.(\tilde{s}\tilde{d})_{20}) + a_1((d^\dagger s^\dagger)_{21}.(\tilde{s}\tilde{d})_{21}) - x a_0 \hat{N}_s \hat{N}_d + \\
& + \frac{1}{2} \sum_{L_2 T_2} c_{L_2 T_2}((d^\dagger d^\dagger)_{L_2 T_2}.(\tilde{d}\tilde{d})_{L_2 T_2}) + \frac{1}{2}\sum_{T_2} b_{T_2}[((s^\dagger s^\dagger)_{0 T_2}.(\tilde{d}\tilde{d})_{0 T_2}) + \text{herm.con.}] + \\
& + \sqrt{\frac{1}{2}} \sum_{T_2} d_{T_2}[((d^\dagger s^\dagger)_{2 T_2}.(\tilde{d}\tilde{d})_{2 T_2}) + \text{herm.con.}]
\end{aligned} \tag{1}
$$

where the first sum runs over $T_2 = 0$ and 2 with $L_2 = 0, 2$ and 4 and $T_2 = 1$ with $L_2 = 1$ and 3 while the last two sums have $T_2 = 0$ and 2 only. The dot denotes a scalar product in both isospin and angular momentum. Equation (1) may not appear to be the most general form since, for example, the one-body term $\hat{N}_s$ is absent but this can be written $\hat{N}_s = N - \hat{N}_d$ with N absorbed into the zero-body term H_0. In fact, four terms were eliminated in this way, a generalisation of the well-known reduction of the IBM1 hamiltonian from nine parameters to six. We took advantage of this elimination to write the sd-interaction (the terms involving a_0 and a_1) in the form given in (1) with $x = (N - T)(N + T + 1)/3N(N - 1)$ to ensure that it made no contribution in states of full symmetry, i.e. states with the partition $[N]$ for both the $U(6)$ and $U(3)$ groups, which are known from experience to dominate at low energy. This gives a simple physical interpretation to ϵ in (1). It is the excitation energy of the fully symmetric state of the configuration $s^{N-1}d$ above s^N, the same interpretation for ϵ as in IBM1. The excitation of the mixed symmetry state $[N - 1, 1]$ of $s^{N-1}d$ involves a_0 and a_1 as well as ϵ. Had we carried out the eliminations in a different way the hamiltonian would have been precisely equivalent but the interpretation of the parameters would have been less simple. For example ϵ, a_0 and a_1 would all have contributed to the excitation of the fully symmetric state of $s^{N-1}d$.

It is now a relatively straightforward task to calculate formulae for the matrix elements of the IBM3 hamiltonian (1) up to $N_d = 2$ and to equate them to the corresponding shell-model formulae from the coherent state work. This procedure provides formulae for the parameters in (1) as functions of N, T and the reduced matrix elements of the shell-model interaction. In contrast to the IBM1 work of OAI there are now more matrix elements(48) than parameters(16) but, in a numerical example, we find that a reasonably good fit is achieved. This supports the validity of the boson approximation. Because of the greater number of matrix elements than parameters, the fitting procedure needs to be discussed. Comments relating to the relative strengths of different terms are based on a numerical example in which a typical shell-model interaction was used in a $j = 21/2$ shell.

The zero-body term H_0 is found directly by equating it to the shell-model formula for $v = t = 0$. It takes the form $AN(N + B) + CT(T + 1)$ where A, B and C are constants.

The three parameters ϵ, a_0 and a_1 were found from the 2×2 shell-model matrix with $v = 2$, $t = 1$ corresponding to the boson configuration $s^{N-1}d$. In the numerical example we find that $2a_0 + 3a_1$ is small compared with a_0 and that a_0 is negative and approximately linear in N, decreasing smoothly as T increases. ϵ is roughly quadratic in N with a minimum value near the middle of the shell and increases with T.

Of the parameters off-diagonal in n_d we find that b_2 and d_2 are negligible and that b_0 is nearly proportional to a_0. The parameter d_0 is approximately linear in the first half of the shell with a zero near mid-shell. (In practice, the boson number would not exceed the mid-shell number $N = j + 1/2$, with hole bosons being used in the second half.)

The remaining parameters $c_{L_2 T_2}$ are all of the same order of magnitude and they depend strongly on both L_2 and T_2 as well as N.

In reaching these conclusions about $b, c,$ and d we have given priority to states with greatest $U(3)$ symmetry and hence greatest $U(6)$ symmetry. For example, the table shows that there are four off-diagonal matrix elements involving b_0 and b_2 and we fit the two with symmetries $[N]$ and $[N-1,1]$ in $s^{N-2}d^2$. In fact the other two matrix elements are then found to fit remarkably well also. There can be no coupling here to the states with $t = 1$ since they have odd $L = 1$ and 3 while s^N has only $L = 0$. For the two parameters d_0 and d_2 there are eight matrix elements and the inclusion of a small, simple three-body force helps to achieve an excellent fit.

CONCLUSION

Vector coherent state theory for the $O(5)$ group has been used to obtain a closed formula for the n-nucleon shell-model matrix elements in the seniority basis with isospin. This has been used with a generalised OAI mapping to obtain an IBM3 hamiltonian, depending on N and T, which reproduces the shell-model matrix elements up to seniority $v = 4$. The known N, T-dependence will serve as a guide to the choice of IBM3 parameters in empirical calculations.

REFERENCES

1. A.Arima, T.Otsuka, F.Iachello and I.Talmi, *Phys.Lett.* 66B:205 (1977)
2. T.Otsuka, A.Arima and F.Iachello, *Nucl.Phys.* 309:1 (1978)
3. J.P.Elliott and A.P.White, *Phys.Lett.* 97B:169 (1980)
4. J.P.Elliott, *Prog.Part.Nucl.Phys.* 25:325 (1990)
5. J.A.Evans, J.P.Elliott and S.Szpikowski, *Nucl.Phys.* A435:317 (1985)
6. K.T.Hecht, *Phys.Rev.* 139B:794 (1965)
7. K.T.Hecht and J.P.Elliott, *Nucl.Phys.* A438:29 (1985)
8. A.deShalit and I.Talmi, "Nuclear Shell Theory", Academic Press,New York (1963)
9. K.T.Hecht, *Nucl.Phys.* A102:11 (1967)
10. R.P.Heminger and K.T.Hecht, *Nucl.Phys.* A145:468 (1970)
11. J.P.Elliott, J.A.Evans and G.L.Long, *J.Phys.A* (1992)
12. J.A.Evans, J.P.Elliott and G.L.Long, (to be published)

SCATTERING AND DYNAMICAL SYMMETRY*

Herman Feshbach
Institute Professor Emeritus

Center for Theoretical Physics
Laboratory for Nuclear Science
and Department of Physics
Massachusetts Institute of Technology
Cambridge, Massachusetts 02139   U.S.A.

INTRODUCTION

Dynamical symmetry as formulated by Arima and Iachello (IBM) focuses on the properties of bound states. Their results for the energies of these states can be expressed as a linear combination of the Casimir invariants of a group and appropriate subgroups when the IBM Hamiltonian satisfies corresponding symmetries. In this note we consider the extension of this formalism to scattering and reactions and to the doorway states which are collective states in the positive energy continuum.

The extension to scattering is straightforward. Our attention is focused on the transition matrix. One assumes that one can expand the transition matrix into a sum of Casimir invariants including unity of the appropriate group. Symmetry breaking is accomplished when the coefficients in the expansion are functions of the Casimir invariants of the subgroups. Using the transition matrix one can calculate the cross sections for various possible reactions making it possible to determine *from experiment* the extent to which a symmetry is valid.

The Hamiltonian can also be expanded in Casimir invariants with spatially or momentum dependent coefficients. One can study the symmetry experimentally by choosing the appropriate kinematics and probes to investigate the *local symmetry* which will occur over a domain in with the symmetry breaking terms are negligible or ineffective.

Doorway states are formed from the simplest excitations (*p-h* usually) of the target nucleus. In this note one asks if there are simple algebraic operators which operating on the ground or excited states generate approximate eigenstates of the Hamiltonian. If there are, there are candidates for doorway states which will be seen in a reaction as giant resonances.

Examples of these general remarks will be described in the remainder of this note.

---

* This work is supported in part by funds provided by the U. S. Department of Energy (D.O.E.) under contract #DE-AC02-76ER03069.

AN SU(3) EXAMPLE

This section draws upon studies made by C. Dover and myself[1] on the validity of $SU(3)$ symmetry on baryon-baryon scattering. Each is a member of an $SU(3)$ octet. The combination, the direct product, $8 \otimes 8$, decomposes as follows:

$$8 \otimes 8 = 1 \oplus 8_a \oplus 8_s \oplus 10 \oplus \overline{10} \oplus 27 \tag{1}$$

where 8_a, 10 and $\overline{10}$ are antisymmetric, and 1, $8_s + 27$ are symmetric. Corresponding to each of these are a number of degenerate states under $SU(3)$ symmetry. These states are characterized by their value of the isospin, I and hypercharge Y. $SU(3)$ symmetry is broken so that the degeneracy is lifted. The states of interest are given in Table I. The partial waves listed are chosen so as to satisfy the Pauli principle. From the table, one notes that $^1S_0$ scattering will involve only the transition matrix $T(27)$ and $T(8_a)$, i.e. only two amplitudes for all possible interactions in this partial wave. Similarly the $^3S_1$ amplitude will involve three amplitudes $T(10)$, $T(\overline{10})$, and $T(8_a)$.

TABLE I

Partial waves for $8 \otimes 8$ baryon-baryon
scattering for $Y = 1, 2$ channels.

BB	(Y, I)	$^{2S+1}L_J$	D
NN	$(2,0)$	$^2S_1, {}^1P_1 \ldots$	$\overline{10}$
NN	$(2,1)$	$^1S_0, {}^3P_{0,1,2} \ldots$	27
$\Sigma N, \Lambda N$	$(1,1/2)$	$^3S_1, {}^1P_1 \ldots$	$\overline{10}_a, 8_a$
$\Sigma N, \Lambda N$	$(1,1/2)$	$^1S_0, {}^3P_{0,1,2} \ldots$	$27_s, 8_s$
ΣN	$(1,3/2)$	$^3S_1, {}^1P_1 \ldots$	10
ΣN	$(1,3/2)$	$^1S_0, {}^3P_{0,1,2} \ldots$	27

The transition amplitude may be written in terms of the quadratic and cubic Casimir operators:

$$T = a + bF^2 + cG^3 \ . \tag{2}$$

If $SU(3)$ holds, a, b and c are constants. If not, these coefficients will be functions of I and Y. Using the eigenvalues of F^2 and G^3 for each representation, one finds in the $^1S_0$ state that

$$T(27) = a_0 + 8b_0 \ , \qquad T(8_s) = a_0 + 3b_0 \ . \tag{3}$$

For the $^3S_1$ state one has

$$\begin{aligned} T(8_a) &= a_1 + 3b_1 \\ T(10) &= a_1 + 6b_1 + 18c_1 \\ T(\overline{10}) &= a_1 + 6b_1 - 18c_1 \ . \end{aligned} \tag{4}$$

TABLE II

NN and YN amplitudes

Channel	$T_0(^1S_0)$	$T_1(^3S_1)$
$np \to np$	$a_0 + 8b_0$	$a_1 + 6b_1 - 18c_1$
$\Sigma^+ p \to \Sigma^+ p$	$a_0 + 6b_0$	$a_1 + 6b_1 + 18c_1$
$\Sigma^- p \to \Sigma^- p$	$a_0 + 5b_0$	$a_1 + 5b_1$
$\Sigma^- p \to \Sigma^0 n$	$3b_0/\sqrt{2}$	$b_1/\sqrt{2} + a\sqrt{2}\,c_1$
$\Sigma^- p \to \Lambda n$	$\sqrt{3/2}\,b_0$	$-\sqrt{3/2}(b_1 - 6c_1)$
$\Lambda p \to \Lambda p$	$a_0 + (15/2)b_0$	$a_1 + (9/2)b_1 - 9c_1$

Using the $SU(3)$ Clebsch–Gordon coefficients one can now express the experimentally interesting amplitudes in terms of a_1, b_1 and c_1. These are listed in Table II.

Under $SU(3)$ symmetry, a_i and b_i are constants and therefore for the $^1S_0$ case there are only two independent amplitudes while for the $^3S_1$ partial wave are three. For the $^1S_0$ channel one has as a consequence the following relations among the cross sections

$$\sigma_0\,(np \to np) = \sigma_0\,(\Sigma^+ p \to \Sigma^+ p) \tag{5a}$$

$$\sigma_0\,(\Sigma^- p \to \Lambda n) = \frac{1}{3}\sigma_0\,(\Sigma^- p \to \Sigma^0 n) \tag{5b}$$

$$\sigma_0\,(\Lambda p \to \Lambda p) = \frac{1}{6}\left[5\sigma_0\,(\Sigma^+ p \to \Sigma^+ p) + \sigma_0\,(\Sigma^- p \to \Sigma^- p) - \frac{5}{3}\,(\Sigma^- p \to \Sigma^0 n)\right] \tag{5c}$$

For the $^3S_1$ channel we have

$$\sigma_1\,(\Lambda p \to \Lambda p) = \frac{1}{2}\left[3\sigma_1\,(\Sigma^- p \to \Sigma^- p) + 3\sigma_1\,(\Sigma^- p \to \Sigma^0 p) - \sigma_1\,(\Sigma^+ p \to \Sigma^+ p)\right] \quad . \tag{6}$$

This analysis is an extension of the analysis used to relate cross sections because of isospin symmetry as exemplified by pion-nucleon scattering.

Unfortunately, the data available for low-energy hyperon-nucleon scattering is sparse so that it is not possible to compare Eq. (5) and Eq. (6) with experiment. Instead, they have been compared with the predictions of a model[2] which has been adjusted so as to fit all available experiments including nucleon-nucleon scattering. As expected one finds that the diagonal scattering cross sections given by Eq. (5a), Eq. (5c) and Eq. (6) are not satisfied. We find that the principle symmetry breaking mechanism for Eq. (5a) is provided by the hyperon-nucleon mass difference. For Eq. (5c) and Eq. (6) the model cross-sections are 40 – 50% smaller than predicted, indicating a substantial but not enormous symmetry breaking. The interesting result is the absence of symmetry breaking for the strangeness exchange reaction given by Eq. (5b) when the kinematic dependence is factored out. This has been confirmed by a calculation[4] of this cross-section when the symmetry breaking interaction is given by the one-gluon exchange between the quarks making up the baryons, a description valid at short-range. We find in this case the possibility that the diagonal cross-section can be shifted considerably but the non-diagonal matrix elements still preserve $SU(3)$ symmetry. The one-gluon exchange is valid only for small interparticle distances. For larger interparticle distance the one-pion exchange model is used with similar results. A similar situation occurs for isobar analog states and the giant dipole state (see below). This conjecture regarding the baryon-baryon interaction is of fundamental importance and has interesting consequences for hypernuclei.

DYNAMICAL SYMMETRY AND DOORWAY STATES

We look for an operator O which upon operating on an eigenstate of the Hamiltonian generators, approximately, another eigenstate of that Hamiltonian. Formally

$$HO\psi = EO\psi \quad \text{or} \quad [H, O]\psi = (E - E_0)O\psi \quad \text{where} \quad H\psi = E_0\psi \; .$$

The requirements on O will be satisfied if

$$[H, O] = \lambda O \tag{6'}$$

where λ is a constant. We shall now give examples of O which lead to isobar analog, giant electric resonance state, spin and Gamow–Teller states. The nucleon-nucleon interaction within the nucleus is taken to be[3]

$$
V_{ij} = v_{00} + v_{01}\vec{\tau}_i \cdot \vec{\tau}_j + v_{10}\vec{\sigma}_i \cdot \vec{\sigma}_j + v_{11}(\vec{\sigma}_i \cdot \vec{\sigma}_j)(\vec{\tau}_i \cdot \vec{\tau}_j) \\
+ \left(v_0^T + v_1^T(\vec{\tau}_i \cdot \vec{\tau}_j)\right)S_{ij} + \left(v_0^{LS} + v_1^{LS}\vec{\tau}_i \cdot \vec{\tau}_j\right)(\vec{\sigma}_i + \vec{\sigma}_j) \cdot \vec{L}_{ij} \tag{7}
$$

where v_{ij} depend only on $|\vec{r}_i - \vec{r}_j|$ and are short-ranged. The last term is the spin-orbit interaction while the tensor interaction is proportional to S_{ij} taken here to be

$$S_{ij} = 3\frac{\vec{\sigma}_i \cdot \vec{r}_{ij}\, \vec{\sigma}_j \cdot \vec{r}_{ij}}{|r_{ij}|^2} - \vec{\sigma}_i \cdot \vec{\sigma}_j \; . \tag{8}$$

In the absence of the spin orbit and tensor terms, V_{ij} is $SU(4)$ asymmetric, a symmetry used by Wigner in his study of nuclei whose mass number is less than that of ^{40}Ca.

Isobar Analog States and Electric Multipole Resonances

For these cases we take O to be

$$O = \sum_\alpha \vec{a} \cdot \vec{\tau}_\alpha v_\alpha \; . \tag{9}$$

The vector $\vec{a}$ is in the three-direction for the discussion of the electric resonances and in the one- and two-direction for the isobar analog case. The quantity v_α is a function of $\vec{r}_\alpha$.

Only the commutator of O in EQ. (9) with the $\vec{\tau}_i \cdot \vec{\tau}_j$ components of V_{ij} need to be considered:

$$C \equiv \left[\sum \vec{\tau}_i \cdot \vec{\tau}_j w_{ij}, \sum \vec{a} \cdot \vec{\tau}_\alpha v_\alpha\right] \; .$$

Only $\alpha = i$ or j survive so that

$$C = \sum_{i,j} w_{ij}\left[\vec{\tau}_i \cdot \vec{\tau}_j, \vec{a} \cdot \vec{\tau}_i v_i + \vec{a} \cdot \vec{\tau}_j v_j\right] \; .$$

Evaluating the commutator yields

$$C = \sum \vec{a} \cdot (\vec{\tau}_i \times \vec{\tau}_j)\left[w_{ij}(v_i - v_j)\right] \; . \tag{10}$$

Thus if v_i is a constant or of much longer range than w_{ij} the commutator will vanish or be relatively small. As expected, these resonant states involve long-range correlations. We have not included the effects of the commutation of O in the kinetic energy and the Coulomb energy. As regards the latter, the Coulomb energy is known to be approximately diagonal between states of a given isospin. The Coulomb energy shifts the energy of state according to nuclear charge but the matrix elements of the long-range Coulomb energy between states deferring isospin are approximately zero.

The commutator of O with the kinetic energy operator depends on v_α. If v_α is constant, as is the case for isobar analog resonance, the commutator is zero. If v_α is $(\vec{r}_\alpha)$ as in the case of the electric dipole resonance, the commutator yields $\vec{p}_\alpha$. But the matrix element of $\vec{p}_\alpha$ is proportional to $\vec{r}_\alpha$ so that the commutator satisfies Eq. (6) so that again the energy of $O\psi$ differs from that of ψ, but the non-diagonal matrix elements are zero.

Spin Resonance States

The operator O in this case is

$$O = \sum_\alpha \vec{\sigma}_\alpha v_\alpha \quad . \tag{11}$$

The results as far as the $\vec{\sigma}_i \cdot \vec{\sigma}_j$ terms in v_{ij} are the same as those which were obtained for case (a). There is therefore no surprise that calculations which neglect spin orbit and tensor terms predict the presence of giant resonances.

Before going on we examine the term which is to be applied to the operator O:

$$V_{ij}^{(LS)} = (\vec{\sigma}_i + \vec{\sigma}_j) \cdot \vec{L}_{ij} v_{ij}^{(LS)} \quad .$$

This may be expressed in terms of J_{ij}, the angular momentum of the pair

$$J_{ij} = \frac{1}{2}(\vec{\sigma}_i + \vec{\sigma}_j) + \vec{L}_{ij}\vec{S}_{ij} + \vec{L}_{ij}$$

as follows

$$V_{ij}^{(LS)} = \left(J_{ij}^2 - L_{ij}^2 - S_{ij}^2\right) v_{ij}^{(LS)} \quad . \tag{12}$$

We can remove S_{ij}^2 from consideration since it involves only $\vec{\sigma}_i \cdot \vec{\sigma}_j$ which has already been discussed. For those nuclei for which the pairing interaction dominates, $J_{ij}^2 = 0$ and we are left with only the L_{ij}^2 term. Its commutation with O requires v_α to be a scalar. Thus for a weak tensor force one fins magnetic dipole resonances.

Turning to the tensor force, an O which commutes with it has the form

$$O = \sum (\vec{\sigma}_i + \vec{\sigma}_j) \cdot (\vec{r}_{ij}) f_{ij} \quad . \tag{13}$$

Its commutator with L_{ij}^2 of Eq. (12) is proportional to $\vec{r}_{ij}$ since the matrix element of the commutator is given by

$$\langle \ell' | [\vec{r}_{ij}, L_{ij}^2] | \ell \rangle = [\ell'(\ell'+1) - \ell(\ell+1)] \langle \ell' | \vec{r}_{ij} | \ell \rangle \quad . \tag{14}$$

However, Eq. (6) is not satisfied as the coefficient of the $\vec{r}_{ij}$ matrix element is ℓ-dependent. If upon averaging that state dependence disappears, Eq. (6) is satisfied. Note that the factor $[\ell'(\ell'+1) - \ell(\ell+1)]$ equals -2ℓ for $\ell' = \ell - 1$ and $2(\ell+1)$ for $\ell' = \ell+1$, the transitions allowed by the vector $\vec{r}_{ij}$. An average which is ℓ-independent thus appears possible.

In conclusion, we expect spin resonances to be possible if the spin-orbit and/or the tensor components of the nucleon-nucleon force are relatively small, and for special cases even if they are not small. Note that symmetry breaking residing in the kinetic energy have not been discussed in detail.

These results apply as well to Gamow–Teller resonances where one would now use for O a collective operator based on the $SU(4)$ generators $\sigma\tau$. Note that the nucleon-nucleon interaction, Eq. (7), is incomplete. A term is needed in order to obtain the spin-charge flip characteristic of the Gamow–Teller resonances. Such a term would permit the derivation of a Lane-type equation for these resonances. At the present time, these Gamow–Teller terms are treated by a distorted wave perturbation approximation.

CONCLUSION

We have shown how one can generate approximate eigenstates by applying an operator which reflects long-range correlations to a known eigenstate. It is expected that these eigenstates can be excited by suitable probes and will thus be observed as giant resonances if they are unbound. Generally, this can be the case because of the impact of the Colomb and/or kinetic energy. We also note that "double" and higher resonances are obtained using O^n. Their energy and widths will increase with n.

REFERENCES

1. C. B. Dover and H. Feshbach, *Ann. Phys.* **198**, 321 (1990).

2. C. B. Dover and H. Feshbach, *Ann. Phys.*, 217 (1992).

3. H. Feshbach, *Theoretical Nuclear Physics: Nuclear Reactions* (John Wiley & Sons, New York, 1992).

VIBRON MODEL DESCRIPTION OF

ATOM-MOLECULE COLLISIONS

A. Frank[1,a,b], R. Lemus[2] and R.D. Santiago[2]

[1]Departamento de Física Atómica, Molecular y Nuclear
Facultad de Física, Universidad de Sevilla
Apdo. 1065, 41080 Sevilla, España
[2]Instituto de Ciencias Nucleares, UNAM
Apdo. Postal 70-543, Circuito Exterior, C.U.
México, D.F., 04510 México

Using the vibron model and a coherent-state method we construct a three-dimensional Hamiltonian for the description of atom-diatom collisions. In the limit of one-dimensional collisions we are able to reproduce the energy transfer behavior for the interaction between an atom and a Morse oscillator.

INTRODUCTION

The vibron model was introduced by F. Iachello more than a decade ago.[1] Its conceptual framework was based upon the interacting boson model of nuclear structure,[2] where the algebraic techniques achieved impressive results. The original model was based on a $U(4)$ boson algebra which is able to accurately describe the rotation-vibration spectra of diatomic molecules.[3] The extension of the model to polyatomic molecules was carried out later on,[4] as well as its generalization to introduce the electronic excitations in diatomic molecules.[5] The algebraic techniques (based on Lie-algebra transformations) can be also appropriately combined with discrete symmetries, either to incorporate the many-centered nature of polyatomic electronic molecular spectra[6] or to provide a simple classification of stretching modes in these molecules.[7] The vibron model has thus proved to be a useful tool for the study of molecular structure.

---

[a] On sabbatical leave from Instituto de Ciencias Nucleares, UNAM.
[b] Guggenheim fellow.

Symmetries in Science VII, Edited by B. Gruber
and T. Otsuka, Plenum Press, New York, 1994

With respect to dynamical processes, such as electron-molecule scattering, the vibron model provides simple wave functions which can be coupled to scattering descriptions such as the eikonal approximation.[8]

On the other hand, atom-molecule collisions have been analyzed using both numerical and algebraic methods. Vibrational transition probabilities in one-dimensional collisions between an atom and a diatomic molecule represented by harmonic,[9] anharmonic[10] and Morse oscillators[11] have been studied using algebraic techniques in conjunction with time evolution operator methods.[9-11] For realistic three-dimensional systems, however, no systematic algebraic scheme has been implemented and one has to resort to a number of computational methods to describe atomic and molecular collisions.[12]

The purpose of this paper is to introduce a new method, based on the vibron model framework, to generalize to three-dimensional space the one-dimensional algebraic approach to atom-molecule collisions. Our basic idea is to first consider the atom-diatom system as a triatomic molecule described algebraically by means of the vibron model. We then define a classical (vector) distance from one of the atoms to the center of mass of the other two using a coherent state method. This gives rise to a Hamiltonian formally analogous to the ones used in one-dimensional schemes,[9-11] albeit a fully three-dimensional one. The derivation of this Hamiltonian is presented in the next Section. We then indicate the steps needed to solve the problem in the interaction picture,[12] including the evaluation of matrix elements and classical trajectories. Finally, in the last Section we analyze the suitability of the model Hamiltonian by considering the limit of one-dimensional collisions and comparing with known results.

THE ATOM-DIATOM INTERACTION HAMILTONIAN

The vibron model assumes that to each chemical bond between atoms in a molecule one associates a $U(4)$ algebra, realized in terms of $s(L = 0)$ and $P(L = 1)$ bosons.[3,4] For the case of a triatomic molecule, the rotation-vibration degrees of freedom are described by means of a $U^1(4) \times U^2(4)$ dynamical algebra. This means that the Hamiltonian and any other physical operators should be expressible as functions of the generators of this group. The one- and two-body form of the Hamiltonian is given by

$$\hat{\mathcal{H}}_{mol} = \hat{\mathcal{H}}_1 + \hat{\mathcal{H}}_2 + \hat{V}^{12} \quad , \tag{1}$$

where $\hat{\mathcal{H}}_i$, $i = 1, 2$, are the one- and two-body Hamiltonians for bonds 1 and 2 and $\hat{V}^{12}$ may be written in the form

$$\hat{V}^{12} = \delta_1 C_{2O^{12}(4)} + \delta_2 C_{2U^{12}(4)} + \delta_3 C_{2U^{12}(3)} + \delta_4 C_{2O^{12}(3)} \quad , \tag{2}$$

where $C_{2G(n)}$ are second-order invariant operators of the algebras $G(n)$, all of them possible subalgebras of the dynamical algebra containing the angular momentum algebra, denoted by $O^{12}(3)$ in eq. (2). The notation $G^{12}(n)$ indicates the coupling of the isomorphic algebras $G^1(n)$ and $G^2(n)$ through addition of the corresponding generators. $\hat{\mathcal{H}}_1$ and $\hat{\mathcal{H}}_2$ may in principle have the general vibron Hamiltonian form for diatomic molecules,[1,3] but in order to simplify the discussion, in this work we consider the dynamical symmetry expressions

$$\hat{\mathcal{H}}_1 = \alpha_1 C_{2O^1(4)} + \alpha_2 \hat{L}_1{}^2 \quad , \tag{3a}$$

$$\mathcal{H}_2 = \beta_1 C_{2O^2(4)} + \beta_2 \hat{L}_2^2 \ , \tag{3b}$$

which are known to correspond to Morse-like potentials.[13] This will be explicitly shown below.

We note that the interaction term $\hat{V}^{12}$ in (2) does include the $U^{12}(3)$ second order invariant, thus breaking the overall $O(4)$ dynamical symmetry of the molecule. A more convenient form for this interaction is given by a multipole expansion[14]

$$\hat{V}^{12} = \gamma_1 \hat{n}_{p_1} \cdot \hat{n}_{p_2} + \gamma_2 \hat{D}_1 \cdot \hat{D}_2 + \gamma_3 \hat{Q}_1 \cdot \hat{Q}_2 + \gamma_4 \hat{L}_1 \cdot \hat{L}_2 \ , \tag{4}$$

where $\hat{n}_{p_i}$, $\hat{D}_i$, $\hat{Q}_i$ and $\hat{L}_i$ correspond to monopole, dipole, quadrupole and orbital angular momentum operators for the bond i, respectively. The explicit form of both the Casimir invariants in (2) and (3) and the multipole operators in (4) can be found in ref. [14].

If we were interested in fitting the rovibrational spectrum of a triatomic molecule, we should then diagonalize (1) in a complete set of states and determine the appropriate value of the parameters.[4] We shall follow a different procedure, however, since our aim is to extract from eqs. (1)-(4) an interaction Hamiltonian for an atom-molecule collision process. We thus consider that one of the atoms may be located at an arbitrary distance from the other two and define a classical distance between this atom and the center of mass of the others.

A connection between the algebraic formulation (vibron model) and the usual integro-differential techniques has been established through the use of coherent states.[15,16] For triatomic molecules, this procedure is applied to both bonds and leads to the definition of radial and angular variables that completely specify the geometrical content of the vibron Hamiltonian (1).[16] In our case, however, we wish to keep the algebraic formulation for bond 1 and apply the procedure only to bond 2. In this way we shall associate a classical variable r_2 to the latter bond while the diatomic molecule is still described by the $U^1(4)$ algebra. The coherent state associated to bond 2 is given by[16]

$$|[N_2]\rho_2\theta\phi> = \frac{1}{\sqrt{N_2!(1+\rho_2^2)^{N_2}}} \left(s_2^\dagger + \rho_2 p_{2,0}^\dagger \cos\theta - \frac{\rho_2}{\sqrt{2}}[p_{2,1}^\dagger \exp(-i\phi) \right.$$

$$\left. - p_{2,-1}^\dagger \exp(i\phi)] \sin\theta \right)^{N_2} |0> \ , \tag{5}$$

where $s_2^\dagger$ and $p_{2,\mu}^\dagger$ are the $U^2(4)$ boson operators,[1,3] $|0>$ is the boson (of type 2) vacuum, ρ_2 is a radial parameter which is related to the distance r_2 between the atom and the diatom center of mass by[13]

$$\rho_2 = \sqrt{\frac{\exp(-r_2/a_0)}{2 - \exp(-r_2/a_0)}} \ , \tag{6}$$

and θ and ϕ denote the other spherical coordinates that specify the atomic position. We have also introduced in (6) the distance parameter a_0 in order to assign units to the "physical" variable r_2. We now compute the expectation value of the molecular Hamiltonian (1) with respect to (5). We need to evaluate expectation values for all opertors associated to bond 2. Two such values are given by[17]

$$< \hat{D}_{2,\mu} > = \frac{2N_2\rho_2}{(1+\rho_2^2)} \cdot \sqrt{\frac{4\pi}{3}} \ Y_{1,\mu}(\theta,\phi) \ , \tag{7a}$$

and

$$<\hat{Q}_{2,\mu}>= \frac{-N_2 \cdot \rho_2^2}{(1+\rho_2^2)} \sqrt{\frac{8\pi}{15}} \; Y_{2_\mu}(\theta,\phi) \; , \tag{7b}$$

where $<>$ indicates expectation values with respect to (5) and the $Y_{l\mu}(\theta,\phi)$ are spherical harmonics. If we take bond 1 to define the direction of the z axis, we arrive at the following expression for the "interaction" Hamiltonian:

$$\hat{\mathcal{H}}(r_2,\theta) = \hat{\mathcal{H}}_0 - 2\beta_1 N_2(N_2-1)\{e^{-2r_2/a_0} - 2e^{-r_2/a_0}\}$$

$$+ (\beta_2 N_2 + \frac{1}{2}\gamma_1 N_2 \hat{n}_{p_1})e^{-r_2/a_0}$$

$$+ \gamma_2 N_2 e^{-r_2/2a_0}\{2 - e^{-r_2/a_0}\}^{1/2} \cos\theta \, \hat{D}_{1,0} \tag{8}$$

$$- \frac{1}{4\sqrt{6}} \, \gamma_3 N_2 e^{-r_2/a_0}\{1 + 3\cos\theta\}\hat{Q}_{1,0} \; ,$$

$$\equiv \hat{\mathcal{H}}_0 + \hat{V}(r_2,\theta,\hat{n}_{p_1},\hat{D}_{1,0},\hat{Q}_{1,0})$$

where only the $\mu = 0$ components of $\hat{D}_1$ and $\hat{Q}_1$ remain due to axial symmetry around the molecular axis and where

$$\hat{\mathcal{H}}_0 = \hat{\mathcal{H}}_1 + \beta_1 N_2(4 - N_2) \; . \tag{9}$$

Note that the second term in (8), arising from the $C_{2_{0^2(2)}}$ Casimir invariant, which gives rise to a Morse potential in the r_2 variable, as mentioned earlier.

The form of (8) is analogous to the one-dimensional interaction Hamiltonians proposed in the literature,[9-11] but now in three-dimensional space. The term $\hat{\mathcal{H}}_0$ is time-independent and expressed in terms of the $U^1(4)$ generators, while the interaction potential $\hat{V}$ depends implicitly on time, through the classical trajectory $(r_2(t),\theta(t))$ of the approaching atom, as will be discussed in the next Section.

THE INTERACTION PICTURE

We now apply a semiclassical method to analyze the Hamiltonian (8). The method consists of first "freezing" the molecular coordinates by considering appropriate expectation values for the operators and then calculating a classical trajectory, specified by $r(t), \theta(t)$. Given these trajectories, the Hamiltonian takes the form

$$\hat{H}(t) = \sum_i f_i(t)\hat{X}_i \; , \tag{10}$$

where the coefficients $f_i(t)$ are obtained directly by substitution of the classical trajectory into (8) and the $\hat{X}_i$ are the $U^1(4)$ generators $\hat{n}_{p_1}$, $\hat{D}_{1,0}$ and $\hat{Q}_{1,0}$. The time-dependent Hamiltonian (10) is then solved in the sudden approximation using the interaction scheme.[9-11] We implement these steps below. Substituting in (8) the expectation values of the $U^1(4)$ operators

$$< [N]\sigma l m|\hat{n}_{p_1}|[N]\sigma l m > \equiv \bar{n}_p \; ,$$

$$< [N]\sigma, \; l+1, \; m|\hat{D}_{1,0}|[N]\sigma l m > \equiv \bar{D}_{10} \; , \tag{11}$$

$$< [N]\sigma l m|\hat{Q}_{1,0}|[N]\sigma l m > \equiv \bar{Q}_{10} \; ,$$

where $\|[N]\sigma l m >$ are the $U(4) \supset O(4)$ eigenstates associated to the molecular Hamiltonian $\hat{\mathcal{H}}_0$, we arrive at the classical Lagrangian

$$\mathcal{L} = \frac{\mu}{2}(\dot{r}^2 + r^2\dot{\theta}^2) - Ae^{-r_2/a_0} - Be^{-r_2/a_0}\cos\theta - C(e^{-2r_2/a_0} - 2e^{-r_2/a_0})$$

$$- De^{-r_2/2a_0}(2 - e^{-r_2/a_0})^{1/2}\cos\theta \quad , \tag{12}$$

where

$$A = \beta_2 N_2 + \frac{1}{2}\gamma_1 N_2 \overline{n}_{p_1} - \frac{1}{4\sqrt{6}}\overline{Q}_{1,0} \quad ,$$

$$B = -\frac{3}{4\sqrt{6}}\gamma_3 N_2 \overline{Q}_{1,0} \quad , \tag{13}$$

$$C = -2\beta_1 N_2(N_2 - 1) \quad ,$$

$$D = \gamma_2 N_2 \overline{D}_{1,0} \quad .$$

From (12) we may calculate the corresponding equations of motion, which in general require a numerical approach. In next section, however, we shall find analytic solutions for the one dimensional limit of (12), i.e., $\theta = 0$, $\overline{D}_{1,0} = 0$, $\overline{Q}_{1,0} = 0$. In the remaining part of this Section we discuss the general form of the solutions of (8) in the interaction scheme. Once the classical trajectories have been computed for a given set of parameters (13), we return to (8) which now takes the general form (10)

$$\hat{H} = \hat{\mathcal{H}}_0 + \sum_{i=1}^{5} f_i(t)\hat{X}_i \quad , \tag{14}$$

where

$$\hat{X}_1 = \hat{n}_{p_1}, \; \hat{X}_2 = \hat{Q}_{1,0}, \; \hat{X}_3 = \hat{D}_{1,0}, \; \hat{X}_4 = \hat{A}_{1,0}, \; \hat{X}_5 = 1 \tag{15}$$

and

$$f_1(t) = \frac{1}{2}\gamma_1 N_2 e^{-r_2(t)/a_0}$$

$$f_2(t) = -\frac{1}{4\sqrt{6}}\gamma_3 N_2 e^{-r_2(t)/a_0}(1 + 3\cos\theta(t))$$

$$f_3(t) = \gamma_2 N_2 e^{-r_2(t)/a_0}\cos\theta(t)(2 - e^{r_2(t)/a_0})^{1/2} \tag{16}$$

$$f_4(t) = 0$$

$$f_5(t) = \beta_2 N_2 e^{-r_2(t)/a_0} + 2\beta_1 N_2(N_2 - 1)(2e^{-r_2(t)/a_0} - e^{-2r_2(t)/a_0}) \quad .$$

The operator $\hat{X}_4 = \hat{A}_{1,0}$ is a generator of the $\overline{O}(4)$ group[14] and was included in (14) since it plays a role in the solution as we show below. In the interaction scheme, we look for the evolution operator $\hat{U}(t, t_0)$, which satisfies the equation[18]

$$i\hbar\frac{d\hat{U}(t, t_0)}{dt} = \hat{H}_I(t)\hat{U}(t, t_0) \quad , \tag{17}$$

where

$$\hat{H}_I(t) = e^{it\hat{\mathcal{H}}_0/\hbar}V_I(t)e^{-it\hat{\mathcal{H}}_0/\hbar} \quad . \tag{18}$$

The general solution of (17) is complicated by the fact that the operators in $\hat{\mathcal{H}}_0$ (see eqs. (3a) and (9)) do not close under commutation with the $\hat{X}_i$ in (15). The latter do close under commutation (and that was the reason to include $\hat{A}_{1,0}$). When $\hat{H}_I(t)$ is composed of operators that close under commutation, a general solution for $\hat{U}(t, t_0)$ is known.[18] We may, however, still find solutions of (17) in the sudden approximation, where, to first order in t

$$H_I(t) \simeq V_I(t) \ . \tag{19}$$

This corresponds to a situation where the projectile's speed is large with respect the molecule's movement. In that case we may substitute eq. (19) in (17) and the general solution is of the form[18]

$$\hat{U}(t, t_0) = e^{-i/\hbar g_1(t)\hat{n}_{P1}} e^{-i/\hbar\, g_2(t)\hat{Q}_{1,0}} e^{-i/\hbar\, g_3(t)\hat{D}_{1,0}} e^{-i/\hbar\, g_4(t)\hat{A}_{1,0}} e^{-i/\hbar\, g_5(t)} \ . \tag{20}$$

Substitution of (19) and (20) in (17) leads to the set of coupled differential equations

$$\begin{bmatrix} \dot{g}_1(t) \\ \dot{g}_2(t) \\ \dot{g}_3(t) \\ \dot{g}_4(t) \\ \dot{g}_5(t) \end{bmatrix} = \begin{bmatrix} 1 & 0 & \frac{8g_3(t)}{3\hbar}\sin(h(t)) & 0 & 0 \\ 0 & 1 & \frac{-2\sqrt{6}g_3(t)}{3\hbar}\sin(h(t)) & 0 & 0 \\ 0 & 0 & \cos(h(t)) & 0 & 0 \\ 0 & 0 & \sin(h(t)) & 0 & 0 \\ 0 & 0 & \frac{-2N_2}{\hbar}\sin(h(t)) & 0 & 1 \end{bmatrix} \begin{bmatrix} f_1(t) \\ f_2(t) \\ f_3(t) \\ f_4(t) \\ f_5(t) \end{bmatrix} . \tag{21}$$

where $h(t) = (g_1(t) - \sqrt{\frac{2}{3}}\, g_2(t))/\hbar$. Solving these equations, transition probabilities to the molecular excited states can be computed from the S-operator[18]

$$\hat{S} = \hat{U}(\infty, -\infty) \ , \tag{22}$$

as

$$P_{|N\sigma lm> \to |N\sigma'l'm'>} = |< N\sigma'l'm'|\, \hat{S}\, |N\sigma lm >|^2 \ . \tag{23}$$

In the last Section we test our model predictions by solving eqs. (21) and computing (22) and (23) for the limit of one-dimensional collision processes.

ONE-DIMENSIONAL COLLISIONS

In the limit of head-on collisions and $l = 0$ molecular states, the Lagrangian (12) takes the simpler form

$$\mathcal{L} = \frac{1}{2}\mu\dot{r}^2 - \beta e^{-2r_2/a_0} + 2\beta e^{-r_2/a_0} + \gamma e^{-r_2/a_0} \ , \tag{24}$$

where

$$\beta = -2\beta_1 N_2(N_2 - 1)$$

$$\gamma = \beta_2 N_2 + \frac{1}{2}\gamma_1 N_2 \bar{n}p \tag{25}$$

and

$$\bar{n}_p = \frac{N_1 - 1}{2} \ .$$

The corresponding equations of motion can be solved in this case, giving the general form

$$\text{Cosh}\left(\sqrt{\frac{2E}{\mu}}\,\frac{t}{a_0}+C\right) = \frac{e^{r_2/a_0} + \frac{2\beta-\gamma}{2E}}{\sqrt{\frac{\beta}{E}+\frac{(2\beta-\gamma)^2}{4E^2}}} \quad , \tag{26}$$

where C is a constant to be determined. Assuming that $t = 0$ corresponds to the classical turning point, $\dot{r}_2 = 0$, $r_2 = \delta$, the total energy is given by

$$E = \gamma e^{-\delta/a_0} + \beta(e^{-2\delta/a_0} - 2e^{-\delta/a_0}) \quad . \tag{27}$$

Substitution of (27) in (26) gives $C = 0$ and (26) can be rewritten in the form

$$e^{-r_2(t)/a_0} = \frac{\text{Sech}^2\left(\sqrt{\frac{E}{2\mu}}\,\frac{t}{a_0}\right)}{A_1 + A_2 \tanh^2\left(\sqrt{\frac{E}{2\mu}}\,\frac{t}{a_0}\right)} \quad , \tag{28}$$

where

$$A_1 = \sqrt{\frac{\beta}{E}+\frac{(2\beta-\gamma)^2}{4E^2}} + \frac{\gamma-2\beta}{2E} = e^{\delta/a_0} \quad ,$$

$$A_2 = \sqrt{\frac{\beta}{E}+\frac{(2\beta-\gamma)^2}{4E^2}} - \frac{\gamma-2\beta}{2E} = \frac{\beta}{E}\,e^{-\delta/a_0} \quad . \tag{29}$$

The solution (28) is a generalization of well known one-dimensional classical trajectories for purely exponential interactions.[9-11,19] Substitution of (28) leads to a time-dependent Hamiltonian of the form (10), with the evolution operator taking the form

$$\hat{U}(t,t_0) = e^{-i/\hbar(\alpha(t)\hat{n}_{p_1}+\omega(t))} \quad . \tag{30}$$

In the sudden approximation we arrive at equations analogous to (21) which can be solved exactly, leading to

$$\alpha(t,t_0) = \frac{1}{2}\gamma_1 N_2 a_0 \sqrt{\frac{2\mu}{A_1 A_2 E}}\left\{ \arctan\left(\sqrt{\frac{A_2}{A_1}}\,\tanh\left(\sqrt{\frac{E}{2\mu}}\,\frac{t}{a_0}\right)\right)\right.$$

$$\left. - \arctan\left(\sqrt{\frac{A_2}{A_1}}\,\tanh\left(\sqrt{\frac{E}{2\mu}}\,\frac{t_0}{a_0}\right)\right)\right\} \quad . \tag{31}$$

The transition probabilities are given by

$$P_{i\to f} = |< N v_f 00|e^{-i/\hbar\alpha(\infty,-\infty)\hat{n}_{p_1}}|N v_i 00 >|^2 \tag{32}$$

where $v_i = \frac{N_i-\sigma_i}{2}$, $v_f = \frac{N_f-\sigma_f}{2}$ are the initial and final vibrational quantum numbers associated to the molecule.[1,3] From (29) and (31) we see that the energy dependence of α is given by

$$\alpha(\infty,-\infty) = \gamma_1 N_2 a_0 \sqrt{\frac{2\mu}{\beta}}\arctan\left(\sqrt{\frac{\beta}{E}}\,e^{-\delta/a_0}\right) \quad , \tag{33}$$

where δ depends on E through (27). The matrix elements in (32) can be computed exactly,[17] but for our purposes it is enough to present a closed formula, valid for large N_1, corresponding to the evaluation in a coherent state basis:[17,19]

$$\langle N_1; v_f|\exp[(-i/\hbar)\alpha(\infty,-\infty)\hat{n}_{p_1}]|N_1; v_i\rangle$$

$$= \frac{\sqrt{v_f!(N_1-v_f)!v_i!(N_1-v_i)!}}{(1+i\rho^{1/2})^{N_1}} \times \sum_{l=0}^{v_i}\frac{(-i\rho^{1/2})^{v_i+v_f-2l}}{l!(v_i-l)!(v_f-l)!(N_1-v_i-v_f+l)!} \quad , \tag{34}$$

where $\rho^{1/2} = \tan[\alpha(\infty, -\infty)/2\hbar]$. For excitations from the ground state we obtain

$$P_{0 \to \nu_f} = \frac{N_1!}{\nu_f!(N_1 - \nu_f)!} \frac{\rho^{\nu_f}}{(1 + \rho)^{N_1}} \quad , \tag{35}$$

which is the expression derived by Levine and Wulfman[11] for the vibrational energy transfer in a one-dimensional collision between an atom and a Morse oscillator, derived using an $SU(2)$ algebra. We also reproduce the scaling laws obtained by these authors:

$$P_{\nu_i \to \nu_i + 1}/P_{0 \to 1} = (\nu_i + 1)(1 - \nu_i/N) \quad , \tag{36}$$

valid when $\rho \to 0$, and

$$P_{\nu_i \to \nu_i - 1} = \nu_i P_{1 \to 0} \left[(N_1 - \nu_i + 1)/N_1\right] \left[1 - (N_1 - \nu_i)(\nu_i - 1)(\frac{1}{2}\rho)\right]^2 \quad , \tag{37}$$

valid for $\rho \to 0$ or $\frac{1}{2}\nu_1 N_1 \rho \ll 1$. We thus find that our model gives the expected results in this limit.

We have presented in this paper an algebraic procedure to describe atom-molecule collisions bases on the vibron model for triatomic molecules. We may introduce potentials different to the Morse oscillator by considering more general Hamiltonians in (3) that fit the spectroscopic data of real diatomic molecules.[3] We are currently implementing the general procedure explained here in order to attempt a full three-dimensional description of experimental atom-diatom collisions.[19]

ACKNOWLEDGMENTS

The authors wish to acknowledge the many important suggestions of Prof. Franco Iachello, who has been a source of constant inspiration in our work. This work was supported in part by DGAPA, UNAM under project IN101889 and Ministerio de Educación y Ciencia, España.

REFERENCES

1. F. Iachello, *Chem. Phys. Lett.* 8:581 (1980).
2. F. Iachello and A. Arima, "The Interacting Boson Model", Cambridge University Press, (1987).
3. F. Iachello and R.D. Levine, *J. Chem. Phys.* 77:3046 (1982).
4. O.S. van Roosmalen, F. Iachello, R.D. Levine and A.E.L. Dieperink, *J. Chem. Phys.* 79: 2515 (1983); F. Iachello and S. Oss, *J. Mol. Spectry* 142:85 (1990); F. Iachello, S. Oss and R. Lemus, *J. Mol. Spectry* 146:56 (1991); F. Iachello, S. Oss and R. Lemus, *J. Mol. Spectry* 149:132 (1991).
5. A. Frank, F. Iachello and R. Lemus, *Chem. Phys. Lett.* 131:380 (1986); A. Frank, R. Lemus and F. Iachello, *J. Chem. Phys.* 91:29 (1989).
6. R. Lemus and A. Frank, *Phys. Rev. Lett.* 66:2863 (1991); "Symmetries in Science V", Ed. B. Gruber 122 (1991).

7. F. Iachello and S. Oss, *Phys. Rev. Lett.* 66:2976 (1991); F. Iachello and S. Oss, *Chem. Phys. Lett.* ; A. Frank and R. Lemus, *Phys. Rev. Lett.* 68:413 (1992).

8. R. Bijker, R.D. Amado and D.A. Sparrow, *Phys. Rev.* A33:871 (1986); R. Bijker and R.D. Amado, *Phys. Rev.* A34:71 (1986); R. Bijker, "Symmetries in Science V", Ed. B. Gruber 15 (1991).

9. D. Rapp, *J. Chem. Phys.* 32:735 (1960); D. Secrest and B.R. Johnson, *J. Chem. Phys.* 45:4556 (1966); J. Récamier, D. Micha and B. Gazdy, *Chem. Phys. Lett.* 119:383 (1985); F.M. Fernández, D. Micha and J. Echave, *Phys. Rev.* A40:74 (1989).

10. J. Récamier, *Chem. Phys. Lett.* 133:259 (1987); R.T. Skodje and D. Truhlar, *J. Chem. Phys.* 80:3123 (1984).

11. C.E. Wulfman and R.D. Levine, *Chem. Phys. Lett.* 97:361 (1983).

12. J.Z.H. Zhang and W.H. Miller, *Chem. Phys. Lett.* 88:4549 (1988); J.M. Launay and B. Lepetit, *Chem. Phys. Lett.* 144:364 (1988); "Dynamics of Molecular Collisions", Ed. W.H. Miller, Plenum Press (1979).

13. S. Levit and U. Smilansky, *Nucl. Phys.* A389:56 (1982).

14. R. Lemus and A. Frank, *Ann. Phys.* 206:122 (1991).

15. R. Lemus, A. Leviatan and A. Frank, *Chem. Phys. Lett.* 194:327 (1992).

16. A. Leviatan and M.W. Kirson, *Ann. Phys.* 188:142 (1988).

17. A. Frank, R. Lemus and R.D. Santiago, to be published.

18. J. Wei and E. Norman, *Proc. Am. Math. Soc.* 15:327 (1964).

19. A. Frank, R. Lemus, J. Récamier and A. Amaya, *Chem. Phys. Lett.* 193:176 (1992).

AN INTRINSIC STATE WITH DEFINITE ISOSPIN FOR IBM-3

Joseph N. Ginocchio
Theoretical Division
Los Alamos National Laboratory
Los Alamos, NM 87545

Amiram Leviatan
Racah Institute of Physics
The Hebrew University
Jerusalem 91904 Israel

*"During these moments of abstraction he seemed more intimately absolved,
in the sense of being linked anew with the universe."*

Giuseppe de Lampedusa, **"The Leopard"**.

INTRODUCTION

No doubt Francesco Iachello, like the Prince in "The Leopard", has found conso-
lation in his many contributions to physics. One of his original contributions is the
interacting boson model[1] (IBM-1). The model, based on interacting monopole and
quadrupole bosons, was initially interpreted as a quantization of the Bohr-Mottelson
quadrupole liquid drop model. Later a distinction was made between neutron and
proton bosons[2] (IBM-2). These bosons were interpreted as correlated monopole and
quadrupole pairs of neutrons and pairs of protons, a generalization of the pairing
model which was restricted to monopole correlated pairs only. This paved the way
for a microscopic interpretation of the boson model by establishing its link with the
nuclear shell model. The IBM-1 states were shown to correspond to the subset of
states of IBM-2 which are completely symmetric in the neutron and proton degrees
of freedom and belong to the highest representation of the F-spin group[3]. The latter

Symmetries in Science VII, Edited by B. Gruber
and T. Otsuka, Plenum Press, New York, 1994

is an SU(2) group which classifies the states of IBM-2 according to their symmetry properties with respect to a rotation of neutron bosons into proton bosons and vice versa. The connection of both IBM-1 and IBM-2 to the Bohr-Mottelson-like quadrupole deformations is made by introducing an intrinsic state[4-7] for the boson system. With this intrinsic state, collective motion in the IBM can then be given a geometrical visualization in terms of quadrupole deformation parameters.

IBM-2 is valid for nuclear states which have the valence protons (neutrons) filling a major shell which is closed with respect to protons (neutrons). Such states automatically have a definite isospin. This is the situation encountered in heavy nuclei where because of the neutron excess, beyond neutron number 82, the valence protons occupy orbits which are full of neutrons while the valence neutrons occupy orbits empty of protons. It follows that the subset of states which correspond to the IBM-2 states all have good isospin and so for these states the isospin label is trivial and of no interest. However, when the valence neutrons and protons are filling the same major shell such as in light nuclei, then isospin must be introduced. To take account of isospin conservation in the framework of boson models, a neutron-proton pair must be included in addition to the proton-proton and neutron-neutron pairs. The resulting IBM-3 model[8,9] with three types of monopole and quadrupole bosons will be reviewed in the next section.

Although IBM-3 was introduced for light nuclei, there are some heavy nuclei which also have neutrons and protons filling the same major shell, namely the lighter isotopes of Tellurium (Z=52, A=110-134), Xenon (Z=54, A=122-136) and Barium (Z=56, A=130-138). Furthermore, with the advent of radioactive beams, more nuclei may be studied which fall into this category. Since these heavy nuclei have collective low-lying states, it is illuminating to determine the intrinsic motion which produces collectivity in a similar manner as that was done for IBM-1 and IBM-2.[4-7] Because isobaric analog states are at high excitation energy, we want an intrinsic state for each isospin multiplet, so in the third section, we introduce an intrinsic state for IBM-3 which has a definite isospin. An interesting group structure associated with this intrinsic state is exploited in calculating intrinsic matrix elements of transition operators and in deriving an expression for the energy surface of an isospin-invariant IBM-3 Hamiltonian. The energy surface depends on quadrupole shape parameters, on the isospin, and total number of bosons, as well as on coefficients of the Hamiltonian. The different minima of the energy surface determine the possible types of nuclear shapes that the model can accommodate. Next we determine some of the parameters of the Hamiltonian and conclude with a summary and future outlook.

INTERACTING BOSON MODEL WITH ISOSPIN (IBM-3)

The building blocks of the interacting boson model with isospin (IBM-3) are three sets of monopole (s) and quadrupole (d) bosons. The three sets correspond to proton-

proton (π), neutron-neutron (ν) and proton-neutron ($\pi\nu$) pairs

$$s^\dagger_\nu \ (L=0) \quad ; \quad d^\dagger_{\nu,m} \ (L=2)$$

$$s^\dagger_{\pi\nu} \ (L=0) \quad ; \quad d^\dagger_{\pi\nu,m} \ (L=2)$$

$$s^\dagger_\pi \ (L=0) \quad ; \quad d^\dagger_{\pi,m} \ (L=2). \tag{1}$$

For each angular momentum ($L=0$ and $L=2$ projection m) the three types of bosons (ν, $\pi\nu$ and π) form an isospin triplet ($T=1, T_z=1,0,-1$)

$$\begin{pmatrix} s^\dagger_\nu \\ s^\dagger_{\pi\nu} \\ s^\dagger_\pi \end{pmatrix} \begin{pmatrix} d^\dagger_{\nu,m} \\ d^\dagger_{\pi\nu,m} \\ d^\dagger_{\pi,m} \end{pmatrix} \begin{matrix} T=1 \ \ T_z=1 \\ T=1 \ \ T_z=0 \\ T=1 \ \ T_z=-1. \end{matrix} \tag{2}$$

Spherical tensor operators both in angular momentum and isospin space can now be defined as

$$s^\dagger_\tau \quad d^\dagger_{m,\tau} \quad \tilde{s}_\tau = (-)^\tau s_\tau \quad \tilde{d}_{m,\tau} = (-)^{m+\tau} d_{-m,-\tau}, \tag{3}$$

where $m = (-2,-1,0,1,2)$ is the angular momentum projection (for $L=2$) and $\tau = (1,0,-1)$ is the isospin projection indicating bosons of type (ν, $\pi\nu,\pi$) respectively.

Boson-pair operators are constructed by coupling the bosons in (3) both in angular momentum and isospin

$$\begin{matrix} \left(s^\dagger s^\dagger\right)^{(0,T)} & L=0 & T=0,2 \\ \left(s^\dagger d^\dagger\right)^{(2,T)} & L=2 & T=0,1,2 \\ \left(d^\dagger d^\dagger\right)^{(L,T)} & L=0,2 & T=0,2 \\ \left(d^\dagger d^\dagger\right)^{(L,1)} & L=1,3 & T=1. \end{matrix} \tag{4}$$

The restrictions on the total angular momentum L and total isospin T when the boson-pair is composed of two s- or two d- bosons arise from the necessity to have a totally symmetric wave function for a system of identical bosons.

Bilinear combinations formed by one creation and one destruction operators are closed under commutation and from the generators of $U(18)$. In angular momentum and isospin coupled form they are given by

$$\left(s^\dagger \tilde{s}\right)^{(0,T)} \quad , \quad \left(s^\dagger \tilde{d} + d^\dagger \tilde{s}\right)^{(2,T)} \quad , \quad i\left(s^\dagger \tilde{d} - d^\dagger \tilde{s}\right)^{(2,T)} \quad , \quad \left(d^\dagger \tilde{d}\right)^{(L,T)} \tag{5}$$

with $L = 0,1,2,3,4$ and $T = 0,1,2$. Subsets of these operators may be identified as generators of various subgroups of $U(18)$. For example, the generators of the angular momentum group are the three isoscalar operators

$$\hat{L}_m = -\sqrt{30}\left(d^\dagger \tilde{d}\right)^{(1,0)}_{m,0}. \tag{6}$$

The following operators which change isospin but not angular momentum

$$G_\tau^{(T)} = \left(s^\dagger \tilde{s}\right)_{0,\tau}^{(0,T)} + \sqrt{5}\left(d^\dagger \tilde{d}\right)_{0,\tau}^{(0,T)}, \quad T = 1,2, \tag{7}$$

generate an SU(3) group which is a generalization of the F-spin SU(2) group encountered in IBM-2.[3] Like the F-spin group, this SU(3) group classifies the states of IBM-3 with respect to interchange of the neutron, proton, and neutron-proton bosons. The isospin SU(2) group is a subgroup of the above SU(3), and is generated by $\hat{T}_\tau = \sqrt{2}G_\tau^{(1)}$ ($\tau = 1, 0, -1$). The standard Cartan form of the isospin generators is given by

$$\hat{T}_+ = \sqrt{2}\left[s_{\pi\nu}^\dagger s_\pi + s_\nu^\dagger s_{\pi\nu} + \sum_m \left(d_{\pi\nu,m}^\dagger d_{\pi,m} + d_{\nu,m}^\dagger d_{\pi\nu,m}\right)\right]$$

$$\hat{T}_- = \left(\hat{T}_+\right)^\dagger \tag{8}$$

$$\hat{T}_z = s_\nu^\dagger s_\nu - s_\pi^\dagger s_\pi + \sum_m \left(d_{\nu,m}^\dagger d_{\nu,m} - d_{\pi,m}^\dagger d_{\pi,m}\right)$$

The Hamiltonian will be invariant under isospin but not necessarily under the larger SU(3) group given in (7). A normal-ordered form of a rotational and isospin-invariant Hamiltonian may be constructed by scalar products of the bosons in (3) and of the boson-pairs in (4). The most general such Hamiltonian with one and two-boson interactions which conserves the total number of bosons is given by $H = H_0 + V$, where

$$H_0 = \epsilon_s s^\dagger : \tilde{s} + \epsilon_d d^\dagger : \tilde{d} \tag{9}$$

and

$$V = u_0^{(0)}\left(s^\dagger s^\dagger\right)^{(0,0)} : \left(\tilde{s}\tilde{s}\right)^{(0,0)} + u_0^{(2)}\left(s^\dagger s^\dagger\right)^{(0,2)} : \left(\tilde{s}\tilde{s}\right)^{(0,2)}$$

$$+ u_2^{(0)}\left(s^\dagger d^\dagger\right)^{(2,0)} : \left(\tilde{d}\tilde{s}\right)^{(2,0)} + u_2^{(1)}\left(s^\dagger d^\dagger\right)^{(2,1)} : \left(\tilde{d}\tilde{s}\right)^{(2,1)} + u_2^{(2)}\left(s^\dagger d^\dagger\right)^{(2,2)} : \left(\tilde{d}\tilde{s}\right)^{(2,2)}$$

$$+ v_0^{(0)}\left[\left(s^\dagger s^\dagger\right)^{(0,0)} : \left(\tilde{d}\tilde{d}\right)^{(0,0)} + \text{H. c.}\right] + v_0^{(2)}\left[\left(s^\dagger s^\dagger\right)^{(0,2)} : \left(\tilde{d}\tilde{d}\right)^{(0,2)} + \text{H. c.}\right]$$

$$+ v_2^{(0)}\left[\left(s^\dagger d^\dagger\right)^{(2,0)} : \left(\tilde{d}\tilde{d}\right)^{(2,0)} + \text{H. c.}\right] + v_2^{(2)}\left[\left(s^\dagger d^\dagger\right)^{(2,2)} : \left(\tilde{d}\tilde{d}\right)^{(2,2)} + \text{H. c.}\right]$$

$$+ \sum_{L=0,2,4}\sum_{T=0,2} c_L^{(T)}\left(d^\dagger d^\dagger\right)^{(L,T)} : \left(\tilde{d}\tilde{d}\right)^{(L,T)} + \sum_{L=1,3} c_L^{(1)}\left(d^\dagger d^\dagger\right)^{(L,1)} : \left(\tilde{d}\tilde{d}\right)^{(L,1)}. \tag{10}$$

The symbol : denotes the scalar product in both angular momentum and isospin space and H. c. stands for Hermitian conjugate. The superscript (subscript) in any of the coefficients of the two-body terms in V (e.g. $u_L^{(T)}$) indicates the isospin (angular momentum) of the boson pairs.

AN INTRINSIC STATE AND ITS GROUP STRUCTURE

Quadrupole shape parameters can be introduced in the IBM-3 by means of an intrinsic state. In view of the importance of isospin, we are interested in constructing an intrinsic state with a definite isospin. The basic ingredient in this construction is a deformed (in angular momentum space) condensate boson for each isospin projection $\tau = (1, 0, -1) \leftrightarrow (\nu, \pi\nu, \pi)$

$$b_\tau^\dagger = (1 + \beta^2)^{-1/2} \left[s_\tau^\dagger + \beta \cos \gamma\, d_{0,\tau}^\dagger + \beta \sin \gamma\, \frac{1}{\sqrt{2}} \left(d_{2,\tau}^\dagger + d_{-2,\tau}^\dagger \right) \right]. \tag{11}$$

The condensate bosons in (11) have the same quadrupole deformation β, γ for all isospin projections $\tau = 1, 0, -1$. The intrinsic state is then a deformed state (in angular momentum space) of N of these bosons (representing one-half of the total number of valence nucleons) coupled to good isospin T

$$|N, T, T_Z = T, \beta, \gamma> = \eta_{N,T} (b^\dagger \cdot b^\dagger)^{(N-T)/2} \left(b_1^\dagger \right)^T |0>,$$

$$\eta_{N,T} = \sqrt{\frac{(2T + 1)!!}{(N - T)!(N + T + 1)!!\, T!}} \tag{12}$$

where the dot $\cdot$ in (12) means scalar product only in isospin space and $|0>$ is the boson vacuum representing the doubly-closed shell. The same intrinsic state would be obtained by defining a generalized condensate boson as a linear combination of the three bosons in (11), constructing a condensate wavefunction from N such bosons and projecting components with good isospin and isospin projection.

Intrinsic states with other isospin projection (T_z) can be obtained by repeated application of $\hat{T}_-$ on the state in (12)

$$|N, T, T_z, \beta, \gamma> = \sqrt{\frac{(T + T_z)!}{(T - T_z)!(2T)!}} \left(T_- \right)^{(T - T_z)} |N, T, T_z = T, \beta, \gamma> . \tag{13}$$

The intrinsic states in (12)-(13) of good (N, T, T_z) have also definite SU(3) symmetry, the highest SU(3) symmetry (N,0).

There is a very interesting group structure associated with the above intrinsic states. The intrinsic state in (12) is composed of two factors. The first factor determines its isospin character and is formed by a condensate of T deformed $b_\nu^\dagger$ bosons. The second factor is composed of a product of $(N-T)/2$ deformed boson-pairs with isospin zero, and therefore does not affect the isospin of the state. Considering the deformed boson-pair in (12), it is useful to define the following isoscalar operators

$$\hat{S}_+ = \frac{1}{2}\left(b^\dagger \cdot b^\dagger\right) = b_\pi^\dagger b_\nu^\dagger - \frac{1}{2}\left(b_{\pi\nu}^\dagger\right)^2,$$

$$\hat{S}_- = \left(\hat{S}_+\right)^\dagger$$

$$\hat{S}_0 = \frac{1}{2}\left[\left(b^\dagger \cdot \tilde{b}\right) + \frac{3}{2}\right] = \frac{1}{2}\left[b_\nu^\dagger b_\nu + b_\pi^\dagger b_\pi + b_{\pi\nu}^\dagger b_{\pi\nu} + \frac{3}{2}\right] \tag{14}$$

where $\tilde{b}_\tau = (-)^\tau b_{-\tau}$. The three operators in (14) are closed under commutation relations and are the generators of an $SU(1,1)$ quasi-spin algebra[10]

$$\left[\hat{S}_0, \hat{S}_\pm\right] = \pm\hat{S}_\pm \quad ; \quad \left[\hat{S}_+, \hat{S}_-\right] = -2\hat{S}_0. \tag{15}$$

Irreducible representations of $SU(1,1)$ are labeled by the two quasi-spin quantum numbers (S, S_0). The deformed bosons in (11) transform under $SU(1,1)$ as

$$\begin{pmatrix} b_\nu^\dagger \\ b_\pi \end{pmatrix} \quad \begin{pmatrix} b_\pi^\dagger \\ b_\nu \end{pmatrix} \quad \begin{pmatrix} b_{\pi\nu}^\dagger \\ -b_{\pi\nu} \end{pmatrix} \quad \begin{matrix} S = \frac{1}{2} \quad S_0 = \frac{1}{2} \\ S = \frac{1}{2} \quad S_0 = -\frac{1}{2}. \end{matrix} \tag{16}$$

From these transformation properties or by noting that $\hat{S}_-$ annihilates the intrinsic state (12) with $N = T$, it follows that this intrinsic state transforms under $SU(1,1)$ as

$$S = \frac{1}{2}\left(T + \frac{3}{2}\right) \quad ; \quad S_0 = \frac{1}{2}\left(N + \frac{3}{2}\right). \tag{17}$$

The same holds for the intrinsic states in (13) since the $SU(1,1)$ generators in (14) commute with the isospin generators (8). These convenient transformation properties of the intrinsic states under $SU(1,1)$ facilitates considerably the calculation of intrinsic matrix elements. Since operators can be classified under $SU(1,1)$, the Wigner Eckart theorem can be used to extract the S_0 (or equivalently the N) dependence of such matrix elements.

Each condensate boson in (11) can be complemented with five additional deformed bosons to form a complete orthonormal basis[11] for each isospin projection τ. Any operator written in terms of s- and d-bosons can be expressed in terms of members of these deformed bases. Clearly, in the resulting expression, only those terms that contain just the condensate bosons (11) will contribute to the expectation value of the operator in the intrinsic state (12). These specific terms can be further expanded as a sum of multipole operators with well defined rank S and projection S_0 with respect to $SU(1,1)$. Since according to eq. (17) the intrinsic state belongs to a given irreducible representation of $SU(1,1)$, one can now apply for each of the multipoles in the above sum, the Wigner Eckart theorem with respect to $SU(1,1)$. Thus all the N (or S_0) dependence of intrinsic matrix elements is given in terms of known $SU(1,1)$ Clebsch Gordan coefficients.[10] One is therefore left with the need to calculate only the reduced matrix elements or equivalently to calculate matrix elements in the state (12) but with $N = T$. This a much simpler task since for $N = T$ the intrinsic state (12) has a simpler form of a condensate of T deformed neutron bosons. Matrix elements in intrinsic states with a different isospin projection (13), are found by applying the Wigner Eckart theorem now in isospin space.

To illustrate the above procedure consider a quadrupole operator ($L = 2$, projection m) and isospin k ($k = 0, 1, 2$)

$$R^{(2,k)}_{m,0} = \left(d^\dagger \tilde{s} + s^\dagger \tilde{d}\right)^{(2,k)}_{m,0} + \chi_k \left(d^\dagger \tilde{d}\right)^{(2,k)}_{m,0}. \tag{18}$$

The matrix element in the intrinsic state (12) can be written as

$$< N, T, T_z = T, \beta, \gamma \mid R^{(2,k)}_{m,0} \mid N, T, T_z = T, \beta, \gamma > =$$

$$F^{(k)}_m(\beta, \gamma) \Big[\frac{1}{\sqrt{3}} < N, T, T_z = T \mid \left(b^\dagger_1 b_1 + b^\dagger_{-1} b_{-1} + b^\dagger_0 b_0\right) \mid N, T, T_z = T > \delta_{k,0}$$

$$+ \frac{1}{\sqrt{2}} < N, T, T_z = T \mid \left(b^\dagger_1 b_1 - b^\dagger_{-1} b_{-1}\right) \mid N, T, T_z = T > \delta_{k,1}$$

$$+ \frac{1}{\sqrt{6}} < N, T, T_z = T \mid \left(b^\dagger_1 b_1 + b^\dagger_{-1} b_{-1} - 2b^\dagger_0 b_0\right) \mid N, T, T_z = T > \delta_{k,2} \Big]$$

$$\tag{19}$$

where

$$F^{(k)}_m(\beta, \gamma) = -(1 + \beta^2)^{-1} \beta \Big\{ 2 \Big[\cos \gamma \, \delta_{m,0} + \sin \gamma \, \frac{1}{\sqrt{2}} \left(\delta_{m,2} + \delta_{m,-2}\right) \Big]$$

$$-\sqrt{\frac{2}{7}} \chi_k \beta \Big[\cos(2\gamma) \delta_{m,0} - \sin(2\gamma) \frac{1}{\sqrt{2}} \left(\delta_{m,2} + \delta_{m,-2}\right) \Big] \Big\}. \tag{20}$$

The first matrix element inside the square brackets in (19) involves the operator $2\hat{S}_0 - 3/2$ as can be seen from eq. (14). The second and third matrix elements involve boson operators which can be shown to transform as a scalar ($S = S_0 = 0$) and a vector ($S = 1$, $S_0 = 0$) respectively under $SU(1,1)$. Taking into account the classification (17) of $|N, T, T_z = T >$ under $SU(1,1)$, one can relate these matrix elements via the Wigner Eckart theorem to the corresponding matrix elements in the state $|N = T, T, T_z = T >$ and obtain

$$< N, T, T_z = T, \beta, \gamma | R_{m,0}^{(2,k)} | N, T, T_z = T, \beta, \gamma > =$$

$$F_m^{(k)}(\beta, \gamma) \Big[\frac{1}{\sqrt{3}} N \, \delta_{k,0} + \frac{1}{\sqrt{2}} T \, \delta_{k,1} + \frac{1}{\sqrt{6}} \Big(\frac{2N+3}{2T+3} \Big) T \, \delta_{k,2} \Big]. \tag{21}$$

To get the matrix element in an intrinsic state with a different isospin projection, one applies the Wigner Eckart theorem in isospin space

$$< N, T, T_z, \beta, \gamma | R_{m,0}^{(2,k)} | N, T, T_z, \beta, \gamma > =$$

$$\frac{(T, T_z; k, 0 | T, T_z)}{(T, T; k, 0 | T, T)} < N, T, T_z = T, \beta, \gamma | R_{m,0}^{(2,k)} | N, T, T_z = T, \beta, \gamma > \tag{22}$$

where the proportionality factor involves a ratio of the relevant isospin $SU(2)$ Clebsch Gordan coefficients.

ENERGY SURFACE

The computational procedure outlined in the previous section can also be used to calculate intrinsic matrix elements of two-body operators. Most importantly, the expectation value of the Hamiltonian in the intrinsic states (12)-(13) defines an energy surface $E_{N,T}(\beta, \gamma)$. (Since the Hamiltonian in (9)-(10) is isospin-scalar, its expectation value is diagonal in T, T_z and is independent of T_z). The resulting energy surface is

$$E_{N,T}(\beta, \gamma) = K_{N,T} + (1 + \beta^2)^{-2} \beta^2 \Big[a_{N,T} - b_{N,T} \, \beta \cos 3\gamma + c_{N,T} \, \beta^2 \Big], \tag{23}$$

where

$$K_{N,T} = \epsilon_s N + u_0^{(2)} N(N-1) + \Big(u_0^{(0)} - u_0^{(2)} \Big) A_{N,T},$$

$$a_{N,T} = \Big(\epsilon_d - \epsilon_s \Big) N + \Big(u_2^{(2)} + \frac{2}{\sqrt{5}} v_0^{(2)} - 2 u_0^{(2)} \Big) N(N-1)$$

$$+ \Big[\Big(u_2^{(0)} - u_2^{(2)} \Big) + \frac{2}{\sqrt{5}} \Big(v_0^{(0)} - v_0^{(2)} \Big) - 2 \Big(u_0^{(0)} - 2 u_0^{(2)} \Big) \Big] A_{N,T},$$

$$b_{N,T} = 2 \sqrt{\frac{2}{7}} \Big[v_2^{(2)} N(N-1) + \Big(v_2^{(0)} - v_2^{(2)} \Big) A_{N,T} \Big], \tag{24}$$

$$c_{N,T} = \left(\epsilon_d - \epsilon_s\right)N + \left(\frac{1}{5}c_0^{(2)} + \frac{2}{7}c_2^{(2)} + \frac{18}{35}c_4^{(2)} - u_0^{(2)}\right)N(N-1)$$

$$+\left[\frac{1}{5}\left(c_0^{(0)} - c_0^{(2)}\right) + \frac{2}{7}\left(c_2^{(0)} - c_2^{(2)}\right) + \frac{18}{35}\left(c_4^{(0)} - c_4^{(2)}\right) - \left(u_0^{(0)} - u_0^{(2)}\right)\right]A_{N,T}$$

and

$$A_{N,T} = \frac{(N-T)(N+T+1)}{3}. \tag{25}$$

Although the most general IBM-3 Hamiltonian with one- and two- body interactions depends on 18 parameters, the shape of the energy surface (i.e. its β-γ dependence) is determined by only 3 combinations of parameters: $a_{N,T}$, $b_{N,T}$, $c_{N,T}$. The N and T dependence in the energy surface are manifested in three functional forms: N, $N(N-1)$ and $A_{N,T}$ with the one-boson terms contributing only to the N dependence, only the isotensor interactions contributing to the $N(N-1)$ dependence, and only the difference of isoscalar and isotensor interactions contributing to the $A_{N,T}$ dependence. This means that, if the corresponding strengths of the isoscalar and isotensor interactions are equal, there is no isospin dependence in the energy surface. This (unrealistic) situation would be the case for an $SU(3)$ invariant Hamiltonian. The $A_{N,T}$ dependence is essential to reproduce the excitation energies of the isobaric analog states. The three isovector interactions in eq. (10), $u_2^{(1)}$, $c_1^{(1)}$ and $c_3^{(1)}$, do not affect the energy surface at all.

The energy surface in (23) has the same functional dependence on β, γ as the corresponding energy surface[5] for IBM-1. Hence it follows that the only types of ground state shapes allowed for an isospin invariant IBM-3 Hamiltonian with at most two-boson interactions are spherical shapes ($\beta = 0$) or deformed shapes ($\beta > 0$) which can be either gamma unstable or axially symmetric ($\gamma = 0$ prolate, $\gamma = \frac{\pi}{3}$ oblate). However, because of the explicit isospin dependence of the energy surface, the deformation parameters will depend on isospin T as well as on the total numbers of valence pairs N, and hence different types of quadrupole shapes can co-exist in the same nucleus. The conditions on the parameters $a_{N,T}, b_{N,T}, c_{N,T}$ to obtain a local/global, spherical/deformed minima are the same as those given[11] for IBM-1.

We expect the nuclei with N = T (all neutrons or all protons) to be spherical. In this subspace only the isotensor interactions contribute to the dynamics since they are the only interactions which have terms that can destroy two neutron or two proton bosons. Hence we expect the choice of parameters in the Hamiltonian to be such that the energy surface will have a global spherical minimum for $N = T$.

SEPARATION ENERGIES

The Sn isotopes (N=T) are spherical and hence have an equilibrium value of $\beta = 0$. The two-neutron separation energies for spherical nuclei ($\beta = 0$) with boson number N and isospin $T = N - 2k$ $(k = 0, 1, \ldots)$ are given by

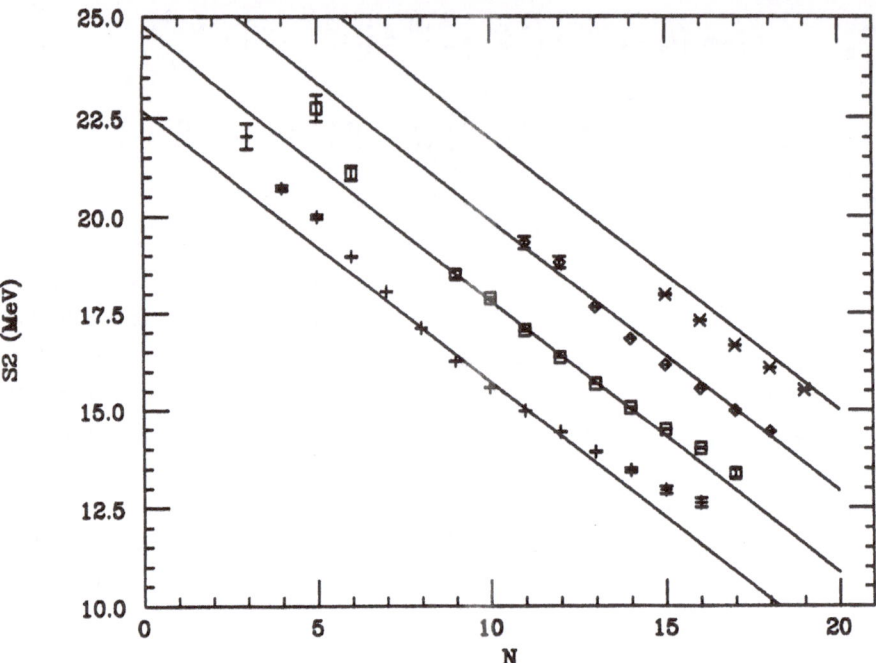

Figure 1. Two-neutron separation energies for $Sn(+)$, Te ($\square$), Xe ($\diamond$) and Ba ($\times$) isotopes as a function of the total number of bosons N. Data taken from ref. [20]. The solid curves are drawn using eq. (26) with $\epsilon_s = -22MeV$, $u_0^{(2)} = 0.348MeV$ $u_0^{(0)} = -1.213MeV$ and $k = 0, 1, 2, 3$ for Sn, Te, Xe, Ba respectively.

$$S2(N,k) = -\left[E_{N,T=N-2k}(\beta = 0, \gamma) - E_{N-1,T=N-1-2k}(\beta = 0, \gamma) \right]$$

$$= -\epsilon_s - 2(N-1)u_0^{(2)} - \frac{4}{3}\left(u_0^{(0)} - u_0^{(2)} \right) k. \qquad (26)$$

This separation energy has the same N and T dependence as that derived from generalized seniority with two-nucleon interactions only.[12] In Figure 1 we show a fit of S2 for the Sn ($k = 0$) isotopes using the expression in (26), and we extract the monopole boson energy to be $\epsilon_s = -22$ MeV. This indicates a strongly bound monopole pair which is consistent with the assumptions of the IBM. The extracted isotensor interaction between the monopole bosons is weak and repulsive and is given by $u_0^{(2)} = 0.348$ MeV. These results are consistent with those of generalized seniority.[12] The fit indicates a quadratic N dependence which may come from three-boson interactions (or three-nucleon interactions in generalized seniority).[13]

If we assume that the isotopes of Te ($k = 1$), Xe ($k = 2$), and Ba ($k = 3$) are also spherical, we see from (26) that the separation energies as a function of N will be parallel with those of the Sn isotopes, with the displacement in energy being proportional to the difference in the isotensor and isoscalar interactions. In Figure 1 we also plot the separation energies for these isotopes plotted as a function of N using the same ϵ_s and $u_0^{(2)}$ determined from the Sn isotopes and determining $u_0^{(0)}$ from a fit to the Te isotopes. As in the Sn isotopes, there is evidence for a three-body interaction among the monopole bosons for the Te isotopes (or a three-nucleon interaction in generalized seniority) contributing to the separation energy of the Te isotopes. Nevertheless, the Te isotopes are parallel to the Sn isotopes as predicted. The Xe and Ba separation energies are then predicted, and good agreement is found for the known isotopes. This agreement suggests that these isotopes are not well deformed. The strength of the isoscalar interaction extracted from the fit to the Te isotopes is $u_0^{(0)} = -1.213 \ MeV$, implying that the monopole bosons have a strong attractive interaction in the isoscalar channel.

BEYOND HALF-FILLED MAJOR SHELL

For nuclei for which both valence neutrons and protons are below the half-filled major shell, the bosons represent pairs of valence nucleons and the number of bosons N is one-half the number of valence nucleons. On the other hand, for nuclei for which both valence neutrons and protons are above the half-filled major shell, then the bosons represent pairs of valence holes and the number of bosons is one-half the number of valence holes. This prescription is a way of enforcing the Pauli principle on a boson system. In both cases, the intrinsic state (12) is well defined.

However for **either** neutrons or protons above half-filled shell, in order to respect the Pauli principle, these bosons then represent neutron or proton hole pairs. But these hole bosons do not form an isospin multiplet with the pairs below the half-filled shell. In this case Elliott and Evans[14] argue that IBM-2 should be used for these cases and isospin should be forsaken. Our application in the last section did include nuclei with valence neutrons above the half-filled shell but valence protons below. However we used zero deformation for which there is no problem, because the state of monopole bosons with definite isospin is always Pauli allowed. In practice, for small deformations, there may not be a problem either.

On the other hand there may be a more useful intrinsic state than (12). One possibility is to have the bosons which are maximally coupled to isospin, to be un-deformed and to have only the bosons coupled to isospin zero to be deformed. Such ideas need more study.

SUMMARY AND CONCLUSIONS

The field of Radioactive Nuclear Beams offers the opportunity of exploring new regions of nuclei far from the valley of stability. Before data on such exotic nuclei become available, it is clearly desirable to explore their properties to gain theoretical insight and provide guidance for experiments to come. Since there is an element of uncertainty in any extrapolation of existing models from their domain of validity, it makes sense to preserve in the process those symmetries and systematics which are known to be observed in ordinary nuclei. In the framework of boson models several extrapolation schemes have been suggested, e.g. the $N_\pi N_\nu$ scheme[15] or F-spin multiplets.[16] In this contribution we have presented an additional scheme based on combining isospin with quadrupole collectivity in a boson description of nuclei for which protons and neutrons fill the same major shell. It is suggested to use the version of the boson model with isospin (IBM-3) not just for light nuclei, but rather for heavier and possibly deformed nuclei like those which will be accessed by radioactive beams.

We have shown that quadrupole collective motion with isospin as a conserved quantum number can be given a geometric interpretation in terms of quadrupole deformation parameters by introducing an intrinsic boson state which is made up of deformed IBM-3 bosons coupled to a definite isospin. We have derived the energy surface of such a system and showed that, for interactions limited to one- and two-boson interactions, the only possible collective motion are those of a spherical vibrator, gamma unstable rotor, and axially symmetric rotor, similar to IBM-1. However the equilibrium deformation parameters will depend on isospin as well as the total number of valence pairs. Hence different collective shapes can co-exist in the same nucleus and will be distinguished by isospin. We have extracted some of the Hamiltonian parameters from separation energies of nuclei for which neutrons and protons are both filling the 50-82 major shell region of the periodic table. Although the data on heavy nuclei for which IBM-3 is applicable are at present scarce, we expect that more data may be forthcoming with the advent of radioactive beams.

The original IBM-1 version of the the interacting boson model, developed by Francesco Iachello and his colleagues, has unified different types of nuclear collective motions in one algebraic framework. The extension of IBM-1 to IBM-2 has revealed a rich variety of new proton-neutron shapes[7] and modes[17,18] in nuclei (e.g. the experimentally observed[19] scissors mode). With further inclusion of isospin as in IBM-3, there are ample reasons to expect new interesting patterns of coexisting shapes as well as novel types of excitation modes. These issues of quadrupole collectivity with isospin relevant for studying nuclei with radioactive beams, are currently under investigation.

ACKNOWLEDGMENTS

This work is supported in part by the United States Department of Energy.

REFERENCES

1. F. Iachello and A. Arima, "The Interacting Boson Model", Cambridge Univ. Press, Cambridge (1987).

2. T. Otsuka, A. Arima, F. Iachello and I. Talmi, Phys. Lett. **B76**: 139 (1978).

3. A. Arima, T. Otsuka, F. Iachello and I. Talmi, Phys. Lett. **B66**: 205 (1977).

4. J.N. Ginocchio and M.W. Kirson, Phys. Rev. Lett. **44**: 1744 (1980); A.E.L. Dieperink, O. Scholten and F. Iachello, Phys. Rev. Lett. **44**: 1747 (1980).

5. J.N. Ginocchio and M.W. Kirson, Nucl. Phys. **A350**: 31 (1980).

6. A. Leviatan and M.W. Kirson, Ann. Phys. **201**: 13 (1990).

7. J.N. Ginocchio and A. Leviatan, Ann. Phys. **216**: 152 (1992).

8. J.P. Elliott and A.P. White, Phys. Lett. **B97**: 169 (1980).

9. P. Halse J.P. Elliott and J.A. Evans, Nucl. Phys. **A417**: 301 (1984).

10. H. Ui, Ann. Phys. **49**: 69 (1968).

11. A. Leviatan, Ann. Phys. **179**: 201 (1987).

12. I. Talmi, Nucl. Phys. **A172**: 1 (1979); S. Shlomo and I. Talmi, Nucl. Phys. **A198**: 81 (1972).

13. I. Talmi, "Simple Models of Collective Motion: The Shell Model and the Interacting Boson Model", Harwood, New York (1993).

14. J.P. Elliott and J.A. Evans, Phys. Lett. **B195**: 1 (1987).

15. R.F. Casten, Phys. Lett. **B152**: 145 (1985); Phys. Rev. Lett. **54**: 1991 (1985); Nucl. Phys. **A443**: 1 (1986).

16. N.V. Zamfir, R.F. Casten, P. von Brentano and W.-T. Chou Phys. Rev. **C46**: R393 (1992).

17. F. Iachello, Phys. Rev. Lett. **53**: 1427 (1984).

18. A. Leviatan and J.N. Ginocchio, Phys. Lett. **B267**: 7 (1991).

19. D. Bohle, A. Richter, W. Steffen, A.E.L. Dieperink, N. Lo Iudice, F. Palumbo and O. Scholten, Phys. Lett. **B137**: 27 (1984).

20. A. H. Wapstra, G. Audi, and R. Hoekstra, Atomic Data and Nuclear Data Tables **39**: 281 (1988); P. Moller, Private Communication.

QUANTUM MOTION AND ALGEBRAIC
GENERATOR COORDINATE METHOD

Andrzej Góźdź

Institute of Physics
Department of Theoretical Physics
University of M. Curie–Skłodowska
pl. M. Curie–Skłodowskiej 1
20–031 Lublin, Poland

1.CLASSICAL AND QUANTUM MOTION

The symmetries, dynamical symmteries and other noninvariance groups of physical systems play very important role in both classical and quantum physics. It is worth-while idea to notice that the same group structure is in many cases responsible for description of classical system and its quantum counterpart. This seems to force an idea of *a group of motion*. The group of motion G would be responsible for 'shift' of the physical system under consideration from one to another physical state.

For classical case the G–motion (the motion generated by the group G) of a physical system can be identify with the group elements $g \in G$. A composition g of two group elements g_1 and g_2, i.e. $g = g_1 g_2$, can be interpreted as a composition of two subsequent movements determined by 'shifts' g_1 and g_2. The inverse element g^{-1} represents an inverse G–motion. A composition of g and g^{-1} gives the neutral element in G which corresponds to the motion which does not change the physical state of the system.

For quantum systems we have no unique path of the motion but the system can choose different paths with some probability amplitudes which implies that quantum systems should be described by the formal integrals over the group manifold G of the following form

$$\int_G dg u(g) g, \tag{1}$$

where $u \in L^1(G)$ describe amplitudes of the G–motion (G–amplitudes) and dg denotes the left invariant Haar measure. A classical motion corresponds to the amplitude of motion u with non–zero values strongly concentrated around a given point g_0 (for a given moment of time) to have

$$\int_G dg u(g) g \sim g_0. \tag{2}$$

Symmetries in Science VII, Edited by B. Gruber
and T. Otsuka, Plenum Press, New York, 1994

The composition of two motions (1)

$$\int_G dg u_1(g) g \int_G dg' u_2(g') g' = \int_G dg(u_1 \star u_2)(g) g, \qquad (3)$$

where

$$(u_1 \star u_2)(g) = \int_G dg' u_1(g') u_2(g'^{-1} g) \qquad (4)$$

leads to the element of the form (1) but with G–amplitude of G–motion given by the convolution (4) of the partial G–amplitudes u_1 and u_2.

An analogue of the inverse motion one can get defining involution of the element (1) as follows

$$\left(\int_G dg u(g) g \right)^{\sharp} = \int_G dg u^*(g) g^{-1} \overset{classical}{\longrightarrow} g_0^{-1}. \qquad (5)$$

In addition, because of existence of quantum interference effect one can define a sum (interference) of motions by the equation

$$\int_G dg u_1(g) g + \int_G dg u_2(g) g \equiv \int_G dg(u_1 + u_2)(g) g. \qquad (6)$$

This way we came to the algebra of quantum motions generated by the group of motion G. This algebra can be identify with the convolutive group algebra of complex functions (G–amplitudes) belonging to $L^1(G, dg)$ with involution defined by [1]

$$u^{\sharp}(g) \equiv \Delta_G(g) u^*(g^{-1}), \qquad (7)$$

where Δ_G denotes the modular function of the group G. We define the modular function by the relation

$$d(g'g) \equiv \Delta_G(g^{-1}) dg' \qquad (8)$$

2. THE METASTATE

In the first section we have derived the Banach algebra with involution

$$\mathcal{R} \equiv (L^1(G), \star, \sharp) \qquad (*)$$

which can describe quantum motion. The group of motion determines what kind of motion is under consideration. The question arise what is the physical system moving? This is determined by the metastate known from algebraic approach to quantum mechanics [2-4]. The metastate is a non–negative, appropriately normalized continous linear functional on the algebra of motions $\mathcal{R}$. In our case of $L^1(G)$ type algebra the general form of the metastate is [4,5]

$$\langle \rho; u \rangle = \int_G dg u(g) \langle \rho; g \rangle, \qquad (9)$$

where the complex function $\langle \rho; g \rangle$ called the metastate kernel (m.k.) satisfies the following conditions:
(10a) $\langle \rho; g \rangle : G \to C$
(10b) $\langle \rho; e \rangle = 1$, where e denotes the unity in G

(10c) $\langle \rho; g^{-1} \rangle = \langle \rho; g \rangle^*$

(10d) for every finite sequences of complex numbers $\alpha_1, \alpha_2, \ldots$, and points on the group manifold $g_1, g_2, \ldots$, the following relation holds

$$\sum_{i,j} \alpha_i^* \langle \rho; g_i^{-1} g_j \rangle \alpha_j \geq 0$$

In most cases the m.k. can be written in quite familiar form

$$\langle \rho; g \rangle = \mathrm{Tr}(\rho T(g)), \qquad (10)$$

where ρ is a density operator and $T(g)$ denotes an unitary representation of the group G.

Let us consider rather general case of m.k. , namely

$$\langle \Psi; g \rangle = \langle \Psi | T(g) | \Psi \rangle. \qquad (11)$$

This is rather general case because even the m.k. (10) can be expressed in the form (11) by appropriate choice of a Hilbert space. Then the metastate (9) can be rewritten as

$$\langle \rho; u \rangle = < \Psi | \left\{ \int_G dg u(g) T(g) | \Psi > \right\} \qquad (12)$$

which allows to interpret the value of the metastate as the probability amplitude that the wave packet $\int_G dg u(g) T(g) |\Psi\rangle$ (i.e. the basic state vector $|\Psi\rangle$ moved with the G-amplitude u) is in the state $|\Psi\rangle$. One can also think about $\langle \rho; u \rangle$ given by eq.(12) as about the probability amplitude that G-motion determined by the G-amplitude u does not change a basic state (like $|\Psi\rangle$ in (12)) of the quantum system.

Taking the above interpretation into account and the fact that $u^\sharp$ represent the G-amplitude of the 'inverse' motion to the motion described by u the expression $\langle \rho; u^\sharp \star v \rangle$ has a meaning of the probability amplitude that our quantum system being in the motion v is in the motion u. This implies the normalization condition for the G-amplitudes in the following form

$$\langle \rho; u^\sharp \star u \rangle = 1. \qquad (13)$$

However, one can happen that for a given $z \in \mathcal{R}$

$$\langle \rho; z^\sharp \star z \rangle = 0 \qquad (14)$$

and the normalization condition (13) cannot be fulfilled. The elements (14) create a left-ideal $\mathcal{R}_\rho$ in the algebra $\mathcal{R}$. Because of the probabilistic interpretation of the metastate the elements $z \in \mathcal{R}_\rho$ correspond to the motion with probability amplitude equal to zero and consequently the motions determined by G-amplitudes u and $u + z$ should be physically equivalent.

At this moment one can use well known GNS procedure [3,4,7] to determine the space of states generated by our group of motion G and the metastate $\langle \rho; \rangle$. The space consists of classes of equivalent in respect to left-ideal $\mathcal{R}_\rho$ G-amplitudes which can be identify with different G-motions of the quantum system. Following the GNS strategy our space of states $\mathcal{K}$ is the completed quotient space $\mathcal{R}/\mathcal{R}_\rho$ with the scalar product induced by the metastate

$$(cl_\mathcal{K}(u)|cl_\mathcal{K}(v))_\mathcal{K} \equiv \langle \rho; u^\sharp \star v \rangle, \qquad (15)$$

where the vector $cl_K(u)$ denotes the class of equivalent G-amplitudes. Following a good tradition of quantum mechanics in most case one can write u instead of $cl_K(u)$.

3. AGCM

The procedure of construction of the state space $K \leftarrow R/R_\rho$ we call the Algebraic Generator Coordinate Method (AGCM) because while using the metastates of the form (11) the obtained space K is unitary equivalent to the Generator Coordinate Method (GCM) space [8-10]. More extensive description of the AGCM approach is given in the papers [5,6]. There is also shown a few examples of applications of the method.

Following [5] we show the construction of states space generated by a compact group of motion G. In the paper [5] a specific example of s– and d–boson SU(6) group of popular Interacting Boson Model [11] is also presented. For the sake of simplicity we consider here a special, diagonal case.

Let us denote by $|\mu A a >$ the eigenbasis of a hamiltonian H; where μ denotes an invariant with respect to G set of quantum numbers, A labels the irreducible representations of G and a denotes a set of remaining quantum numbers required for unique specification of the basis for irreducible representation A of G. In addition we assume that the density operator ρ is a function of the hamiltonian H. Thus, let us denote by $\rho(\mu A a)$ and $E(\mu A a)$ the eigenvalues of the operators ρ and H, respectively. The m.k. function we choose in the form (10).

Using the overlap operator N defined by the equation

$$(Nu)(g) = \int_G dg' \langle \rho; g^{-1}g' \rangle u(g') \tag{16}$$

one can find the appropriate left–ideal R_ρ and the orthonormal basis in the state space K by solving its eigenequation

$$Nu = \Lambda u. \tag{17}$$

In our case this equation can be easily solved in the same way as in the paper [10] by making use of an expansion of the searched eigenfunctions in the matrix elements $D^A_{b'b}(g)$ of the irreducible representation of the group G. After some algebra the solutions of the eigenproblem of the overlap operator can be found in the form:

$$u \to u^A_{aa'}(g) = \sqrt{\dim_G[A]} D^{A*}_{aa'}(g) \tag{18a}$$

$$\Lambda(Aa') = \frac{\sum_\mu \rho(\mu A a')}{\dim_G[A]}, \tag{18b}$$

where $\dim_G[A]$ denotes the dimension of the irreducible representation A. The collective space is now spanned by the all eigenvectors (more precisely by their classes that are elements of the quotient space R/R_ρ) $u^A_{aa'}(g)$ for that $\Lambda(Aa') \neq 0$.

One can show directly that the vectors (for nonzero Λ)

$$e^A_{aa'} \equiv \Lambda(Aa')^{-1/2} u^A_{aa'}(g) \tag{19}$$

furnish the ortonormalized basis in K. The states (19) correspond to the so-called natural states in standard GCM approach. The hamiltonian of our system can be now obtained by the projection of the hamiltonian H onto the space K. For this purpose it is sufficient to calculate the matrix elements of H within the states (19). The result of this calculations is given below:

$$H_{(Aaa')(Aa_1a'_1)} = \frac{1}{\sqrt{\Lambda(Aa')\Lambda(Aa'_1)}} \left\langle \rho; e^\sharp_{aa'} \star H e_{a_1a'_1} \right\rangle =$$

$$= \delta_{aa_1}\delta_{a'a'_1} \frac{\sum_\mu \rho(\mu Aa')E(\mu Aa)}{\sum_\mu \rho(\mu Aa')}. \tag{20}$$

The matrix elements between the vectors belonging to different irreducible representations vanish i.e. for $A \neq A'$. It means that the projected hamiltonian is diagonal in the basis (19) and the matrix elements (20) are equal to the energies of the G–motion.

The energies (20) are labelled by three sets of quantum numbers: A that denotes the irreducible representation of the group G and two sets of additional quantum numbers a and a', where a is required to distinguish states within a given irreducible representation A. The set of extra quantum numbers a' describes some internal motions of the system in respect to the group of motion G. The very known example of such type of quantum number is the number K representing a projection of the angular momentum vector onto an internal axis in the asymmetric top.

As an example, let us consider a sytem of quadrupole bosons for which the total group of motion is the boson group SU(5) [12]. This case corresponds e.g. to the standard five dimensional harmonic oscillator and to so called vibrational limit of the IBM model [11]. Using of the group SU(5) reproduces all properties of the quadrupole bosons expressed in terms of AGCM space.

It is now interesting to restrict the motion to only SO(3)-motion and consider the SO(3) group as a subgroup of the group of motion. The eigenvalues and eigenfunctions of the overlap operator and the hamiltonian can be found acoording to the formulas (18–20) and for the rotational energy one gets the expression :

$$\mathcal{E}_{rot}(LMK) = \frac{\sum_{Nvx} E(NvxLM)\rho(NvxLK)}{\sum_{Nvx} \rho(NvxLK)}, \tag{21}$$

where N is the number of quadrupole bosons, v denotes the seniority number, x can be interpreted as a maximal number of boson triplets coupled to zero angular momentum, and L and M are the usual angular momentum quantum numbers [11,13].

The formula (21) describes, in general, rotational K–bands and M dependent energies for hamiltonians containing the third component angular momentum operator L_z. The K-bands can be not degenerated in K for the density operators dependent on L_z.

For the special, but important case of 5–D harmonic oscillator and the canonical density operator one can obtain more explicit expression for the rotational energies

$$\mathcal{E}_{rot}(LK;T) = -\hbar\omega \frac{\frac{\partial}{\partial a}\sum_N s_{NL}\exp(-aN)}{\sum_N s_{NL}\exp(-aN)} + \frac{5}{2}\hbar\omega, \tag{22}$$

where $a = \beta\hbar\omega$, $\beta = 1/T$ is an inverse boson temperature and s_{NL} denote the multiplicities of the states for given N and L listed for $L < 21$ and $N < 31$ in the paper [12].

The expression (22) describes the temperature dependent rotational spectrum projected out of the considered harmonic oscillator. This spectrum, as it could be expected, has no special regularities typical for rotators because of strong coupling between the rotational and vibrational degrees of freedom in H. However, one needs to remember that the group $G=SO(3)$ constrains the hamiltonian H and reduces degrees of freedom of the system to three angles of rotations only. After some algebra one can notice that for each temperature the spectrum corresponding to odd angular momenta is shifted by a constant value in respect to even angular momenta states, namely:

$$\mathcal{E}_{rot}(L+3) - \mathcal{E}_{rot}(L) = 3\hbar\omega. \tag{23}$$

For example, in the vibrational nuclei like ^{106}Pd the energy of the first 3^+ state is just $3\hbar\omega$ above the ground state corresponding to $L = 0$.

The obtained rotational energies are temperature dependend. In general they increase with temperature. However, usually gaps between energy levels are less sensitive on the temperature than their total energies. For our presentation of the method this dependence is irrelevant, more details is given in [5,6,12].

The $T = 0$ case is of great interest here. The spectrum for zero temperature can be obtained as the limit of the expression (22) calculated with $T \to 0^+$. Direct use of the density operator ρ for $T = 0$ to generate the rotational spectrum of the 5-D harmonic oscillator gives no effect because this state is the rotationally invariant state vector with the total angular momentum $L = 0$ and by rotations one can obtain only the ground state vector This analysis shows that using of more general metastates than those determined by the pure states, like (11), allows for generation of the rotational spectra even for undeformed nuclei. For 5-D harmonic oscillator with $T = 0$ the SO(3) spectrum is described by two simple sequences:

$$\mathcal{E}_{rot}(L, T = 0) = \begin{cases} \hbar\omega(\frac{1}{2}L + \frac{5}{2}) & ;L = 0, 2, 4, \ldots \\ \hbar\omega(\frac{1}{2}(L + 3) + \frac{5}{2}) & ;L = 3, 5, 7, \ldots \end{cases} \tag{24}$$

The states that survive after reduction of the full harmonic oscillator to only 3 rotational degrees of freedom consist of the boson configurations having the lowest number of bosons that can be coupled to required angular momenta, i.e. for an even angular momentum $L = 2N$, $N = v$ and $x = 0$ and for an odd one $L = 2N - 3$ with the same relations for the seniority number v and the maximal number of boson triplets x. The rotational spectrum for $T = 0$ is equidistant like vibrational one. For $T > 0$ the rotational spectrum varies with temperature. The levels of odd L are degenerated with levels of $L + 3$ for $T = 0$ and with $L + 1$ for $T = \infty$. For $T > 0$ each level is $(2L + 1)^2$ times degenerated, for $T = 0$ the levels with $L \geq 3$ and $L \neq 4$ have degeneracy two times higher.

We mentioned above that for $T = 0$ the levels which survive after blocking other degrees of freedom than those allowing for the rotational motion have the possible lowest energy for the given angular momentum. The question arises what is the structure of these rotational levels for $T > 0$. This problem leads to another property of the AGCM formalism that allows to represent the state vectors obtained in one collective space into another collective space which can be constructed from the algebra $\mathcal{R}$ but with $G =$SO(5), SU(5) and eventually different metastates. This feature of the formalism enables to consider the phenomena whose description, in general, requires the variable state spaces.

Returning back to our example we can construct the full space of states for 5-D harmonic oscillator and its group of motion $G =$SU(5) using equations (18) and (19). Now the states are labelled by N and the double set of quantum numbers , v, x, L and M :

$$e^N_{vxLM,v'x'L'K}(T). \tag{25}$$

The rotational states $e_{MK}^L(T)$ for the temperature T can be expanded in the basic vectors (25). After some algebra one can get that the corresponding rotational vectors within the five dimensional harmonic oscillator space which are given by the formula:

$$e_{MK}^L(T) \rightarrow e_{LMK}^{SU(5)}(T) =$$

$$= \left\{ \sum_{Nvx} \rho(NvxLK;T) \right\}^{-1/2} \sum_{Nvx} \{\rho(NvxLK;T)\}^{1/2} \, e_{vxLM,vxLK}^N(T). \qquad (26)$$

The rotational energies calculated with the states (26) are obviously given by the the formulas (21-22). On the other hand the eigenenergies of the 5-D harmonic oscillator within the space spanned by the vectors (26) are obviously independent of the temperature and are given by the usual formula $\hbar\omega(N + 5/2)$. This allowes to interpret the squared expansion coefficients in eq.(26) as the occupation probabilities of the harmonic oscillator states:

$$p_{vxLMK}^N(T) = \frac{\rho(NvxLK;T)}{\sum_{Nvx} \rho(NvxLK;T)}. \qquad (27)$$

For example, for $L = 2$ the occupation probabilities are done by the analytical formula [12]

$$p_{vx2MK}^N(T) = \frac{y^{N-1}(1 - y^3)^2}{1 + y + y^2} \quad \text{for } N > 0, \qquad (28)$$

where $y = \exp(-\hbar\omega/T)$. One can notice that for the lowest allowed shell, i.e. for $N = 1$, for T=0 the function (28) is equal to 1 and it is a decreasing function with T while other energy levels at this moment are unoccupied and their occupation probabilities vanish. The probability functions (28) for $N > 1$ have a characteristic shape with a single maximum.

4. CONCLUSIONS

The AGCM method opens new possiblities of group theoretical analysis of quantum systems. Below we summarize some of this possiblities (for more detail discussion please refer to the papers [5,6]).

The AGCM method allows for construction of the spaces of states generated by a density matrix and a given group of motions. The structure of the states space is dependent on the shape of the metastate which, in turn, can be dependent on some external parameters. This property of the formalism should, in principle allow for description of the phenomena for which the state space of a system is changing during the process.

The formalism determines uniquely the representations of the group of motion G which are involved in description of the quantum system.

This algebraical approach gives a tool for classification of the energy spectra with respect to subgroups of the group of motions giving also information about the states internal structure - in other way by reduction of degrees of freedom (constraints) one can conclude which kind of motion is responsible for the given energy level.

In some cases one can obtain, in natural way, additional quantum numbers generated by the group of motion G. The additional set of quantum numbers is responsible for an internal structure of the physical system and allows for appearance of extra energy bands, e.g. the rotational K-bands in the asymmetric top are of this nature.

Taking into account the results obtained till now one can conclude that this method collects advantages of the standard Generator Coordinate Method and the power of group theoretical approaches.

REFERENCES

1. A. A. Kirillov, "Eliemienty Tieorii Priedstavlienij", Nauka, Moskva, 1978.
2. A. Sudbery, "Quantum Mechanics and the Particles of Nature", Cambridge University Press, 1986, Ch.5.
3. G. G. Emch, "Algebraic Methods in Statistical Mechanics and Quantum Field Theory", Wiley and Sons, Inc. New York 1972.
4. O. Bratteli, D. W. Robinson," Operator Algebras and Quantum Statistical Mechanics", Springer-Verlag, New York 1979.
5. A. Bogusz, A. Góźdź, The Algebraic Generator Coordinate Method as the Constrained Quantum Mechanics, J.Phys.A., (1992) in print.
6. A. Góźdź, M. Rogatko, The Algebraic Generator Coordinate Method for Locally Compact Lie Groups, J.Phys.A., (1992) in print.
7. K. Schmuedgen," Unbounded Operator Algebras and Representation Theory", Akademie-Verlag, Berlin, 1990.
   J. M. G. Fell, R. S. Doran," Representation of *-Algebras, Locally Compact Groups, and Banach–Algebraic Bundles", Academic Press, Inc.San Diego 1988, vol I and II.
8. D. L. Hill, J. A. Wheeler, Phys.Rev. 89:112 (1953) ;
   J. J. Griffin and J. A. Wheeler, Phys.Rev. 108:311 (1957).
9. A. F. R. de Toledo Piza et al., Phys.Rev. C15:1477 (1977);
   Il Nuovo Cim. 45B:1 (1978).
10. A. Góźdź, Acta Phys. Polonica B20:235 (1989).
11. A. Arima, F. Iachello, Ann.Phys. (N.Y.) 99:253 (1976).
12. A. Bogusz, A. Góźdź, Rotational States Generated by SU(5) Dynamical Symmetry with Constraints, Annales UMCS sec. AAA vol. XLVI (1991), in print.
13. S. Szpikowski, A. Góźdź, Nucl. Phys. A340:76 (1980).

BOSON AND FERMION OPERATOR REALISATIONS
OF $su(4)$ AND ITS SEMISIMPLE SUBALGEBRAS

Bruno Gruber[1] and Michael Ramek[2]

[1]Department of Physics
Southern Illinois University at Carbondale
Carbondale, IL 62901, USA
and
Arnold Sommerfeld Institute for Mathematical Physics
Technical University Clausthal
D-3392 Clausthal, Germany
[2]Institut für Physikalische und Theoretische Chemie
Technische Universität Graz
A-8010 Graz, Austria

INTRODUCTION

The simple Lie algebra $su(4)$ is isomorphic to the simple Lie algebra $so(6)$. The various compact and noncompact real forms of $su(4) \sim so(6)$ are of considerable significance to physical applications. This is also the case for many of their semisimple subalgebras.

Among the physically relevant algebras and subalgebras of the compact and noncompact real forms of $su(4) \sim so(6)$ are found the conformal algebra, the de-Sitter algebra, the Lorentz algebra, the spin and quasispin algebras, etc. .... In addition, the non-simple Poincaré algebra is also among the physically relevant subalgebras of a noncompact real form of $su(4) \sim so(6)$.

In this article we consider the algebra $su(4) \sim so(6)$ as a complex algebra. Any desired real form is easily obtained from the complex form by standard procedures. The representation theory can then be carried out summarily for all real forms, while the question of unitarizability rests with the particular scalar product which is introduced.[1,2]

We first list all semisimple $su(4) \sim so(6)$ symmetry chains. Then we obtain a boson operator realisation, as well as a fermion operator realisation for all the symmetry chains.[3] Due to its great physical significance we include into our analysis also a non-semisimple Poincaré algebra.

The boson operator realisation is based upon a set of four boson creation and annihilation operators a_i^+, a_j, $i,j = 1,2,3,4$, satisfying the usual commutation relations. The physical interpretation of these bosons depends upon the symmetry

Symmetries in Science VII, Edited by B. Gruber
and T. Otsuka, Plenum Press, New York, 1994

chain which is chosen. The fermion operator realisation is obtained in terms of two fermion creation and annihilation operators f_i^+, f_j, $i,j = 1,2$, satisfying the usual anticommutation relations. This is the smallest set of fermions which permits a realisation of $su(4)$ and $u(4)$. Moreover, the Lie algebra $u(4)$, in this particular realisation, also closes with respect to anticommutation and represents simultaneously the 16-dimensional Clifford algebra.

The semisimple symmetry chains are defined through their embedding matrices.[4-6] Fig. 1 lists all semisimple symmetry chains and their embedding matrices.

NOTATION AND DEFINITIONS

The notation used in this article, as well as the definitions, are given below.

The Algebra $u(4)$:

Irreducible representations are labelled by the partition of an integer N,

$$[N_1, N_2, N_3, N_4], \quad N_1 \geq N_2 \geq N_3 \geq N_4 \geq 0, \quad N_i \text{ integers}$$
$$N_1 + N_2 + N_3 + N_4 = N.$$

This partition corresponds to the highest weight of the irreducible representation which it labels. The partition $[N_1, N_2, N_3, N_4]$ also characterizes the property of the states of the irreducible representation of $u(4)$ with respect to the symmetric group S_N. That is, all states of the irreducible representation $[N_1, N_2, N_3, N_4]$ of $u(4)$ have the symmetry property of the irreducible representation $[N_1, N_2, N_3, N_4]$ of the symmetric group S_N. An arbitrary weight is of the form

$$[n_1, n_2, n_3, n_4], \quad n_1 + n_2 + n_3 + n_4 = N, \quad n_i \geq 0, \text{ integers}.$$

The Algebra $su(4)$:

The weights (m_1, m_2, m_3, m_4) of $su(4)$ are related to the partition $[n_1, n_2, n_3, n_4]$ in the following manner,

$$m_i = n_i - (1/4)N, \quad \text{and}$$
$$n_i = m_i - m_4.$$

It holds for any weight

$$m_1 + m_2 + m_3 + m_4 = 0 \quad m_i = (1/4)n, \quad n \text{ integer},$$

and the irreducible representations are labelled by the highest weight

$$(M_1, M_2, M_3, M_4), \quad M_1 \geq M_2 \geq M_3 \geq M_4, \quad M_i = N_i - (1/4)N.$$

The Algebra $so(6)$:

The weights of the irreducible representations of $so(6)$ are of the form

$$(m_1', m_2', m_3'), \quad m_i' = (1/2)n, \quad n \text{ integer},$$

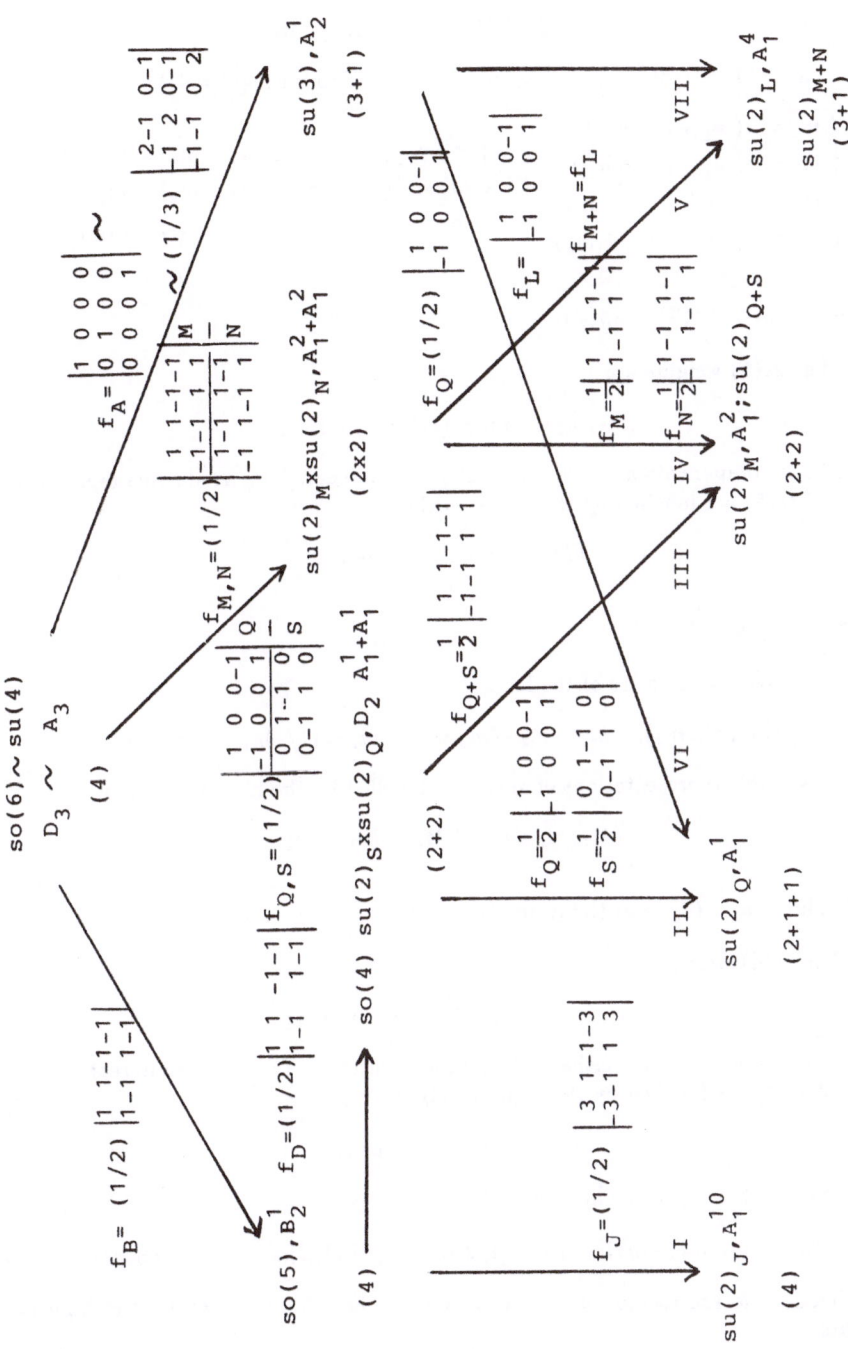

Figure 1. Semisimple symmetry chains from $so(6) \sim su(4)$. Embedding matrices and branching of 4-dimensional representation of $so(6) \sim su(4)$.

225

with all components m_i' either half-integer or integer. The irreducible representations are labelled by the highest weight

$$(M_1', M_2', M_3'), \quad M_1' \geq M_2' \geq |M_3'|.$$

The $su(4)$ weights (m_1, m_2, m_3, m_4) are related to the $so(6)$ weight (m_1', m_2', m_3'),

$$\begin{cases} m_1 = \frac{1}{2}(m_1' + m_2' + m_3') \\ m_2 = \frac{1}{2}(m_1' - m_2' - m_3') \\ m_3 = \frac{1}{2}(-m_1' + m_2' - m_3') \\ m_4 = \frac{1}{2}(-m_1' - m_2' + m_3') \end{cases} \qquad \begin{cases} m_1' = \frac{1}{2}(m_1 + m_2 - m_3 - m_4) = m_1 + m_2 \\ m_2' = \frac{1}{2}(m_1 - m_2 + m_3 - m_4) = m_1 + m_3 \\ m_3' = \frac{1}{2}(m_1 - m_2 - m_3 + m_4) = m_1 + m_4 \end{cases}$$

The Algebra $so(5) \sim sp(4)$:

The $so(5)$ weights are

$$(m_1, m_2), \quad m_i = (1/2)n, \quad n \text{ integer},$$

with both components m_i either half-integer or integer. The irreducible representations of $so(5)$ are labelled by the highest weight

$$(M_1, M_2), \quad M_1 \geq M_2 \geq 0.$$

The Algebra $su(3)$:

The $su(3)$ weights are of the form

$$(m_1, m_2, m_3), \quad m_1 + m_2 + m_3 = 0, \quad m_i = (1/3)n, \quad n \text{ integer}.$$

The irreducible representations of $su(3)$ are labelled by the highest weights

$$(M_1, M_2, M_3), \quad M_1 \geq M_2 \geq M_3.$$

The Algebra $so(4) \sim su(2) \times su(2)$:

The $so(4)$ weights are

$$(m_1, m_2), \quad m_i = (1/2)n, \quad n \text{ integer},$$

with both components m_i either half-integer or integer. The irreducible representations of $so(4)$ are labelled by the highest weight

$$(M_1, M_2), \quad M_1 \geq |M_2|.$$

The $su(2) \times su(2)$ weights are of the form

$$(m_1, -m_1; m_1', -m_1'), \quad m_1 = (1/2)n, \quad m_1' = (1/2)n', \quad n, n' \text{ integers}.$$

The irreducible representations of $su(2) \times su(2)$ are characterized by the highest weights

$$(M_1, -M_1; M_1', -M_1'), \quad M_1 \geq 0, \quad M_1' \geq 0.$$

The second component of these weights is obviously redundant. It is, however, useful for purposes of the embedding theory to keep the $su(\ell+1)$ notation identical for all ℓ.

The Algebra $su(2) \sim so(3)$:

The weights of $su(2)$ are of the form
$$(m, -m), \quad m = (1/2)n, \quad n \text{ integer.}$$
The irreducible representations of $su(2)$ are characterized by the highest weights
$$(M, -M), \quad M \geq 0.$$

THE ROOT SYSTEMS

We denote with $\sum$ the system of roots, and with $\prod$ the system of simple negative roots. The symbol W denotes the Weyl symmetry group of an algebra. We will describe its action upon a weight. $O(W)$ denotes the order of the Weyl group.

The Algebra $su(4)$:

$$\sum = \{e_i - e_j, \quad i \neq j, \quad i, j = 1, 2, 3, 4\}$$
$$\prod = \{-e_1 + e_2, -e_2 + e_3, -e_3 + e_4\}$$
$W \cdot (m_1, m_2, m_3, m_4)$:   all permutations of components m_i
$O(W) = 4!$

The Algebra $so(6)$:

$$\sum = \{e_i - e_j, e_i + e_j, \quad i \neq j, \quad i, j = 1, 2, 3, 4\}$$
$$\prod = \{-e_1 + e_2, -e_2 + e_3, -e_2 - e_3\}$$
$W \cdot (m_1, m_2, m_3)$:   all permutations of components m_i,
                         and all changes of signs in pairs

$O(W) = 2^2 \cdot 3!$

The Isomorphism $su(4) \sim so(6)$:

$su(4)$: $(-1, 1, 0, 0)$, $(0, -1, 1, 0)$, $(0, 0, -1, 1)$ ⎫
$so(6)$: $(0, -1, -1)$, $(-1, 1, 0)$, $(0, -1, 1)$ ⎬ simple neg. roots

$su(4)$: $(-1, 0, 1, 0)$, $(0, -1, 0, 1)$, $(-1, 0, 0, 1)$ ⎫
$so(6)$: $(-1, 0, -1)$, $(-1, 0, 1)$, $(-1, -1, 0)$ ⎬ , plus positive roots.

To each $su(4)$ root corresponds the $so(6)$ root listed below it.

The Algebra $so(5)$:

$$\sum = \{e_i - e_j, e_i + e_j, \quad i \neq j, \quad i, j = 1, 2; \quad \pm e_i, \quad i = 1, 2\}$$
$$\prod = \{-e_1 + e_2, -e_2\}$$
$W \cdot (m_1, m_2)$:   permutations of components m_i,
                  and all changes of sign.

$O(W) = 2^2 \cdot 2!$

The Algebra $so(4) \sim su(2)_S \times su(2)_Q$:

$$\sum = \{\pm(e_1 - e_2), \pm(e_1 + e_2)\}$$
$$\prod = \{-e_1 + e_2, -e_1 - e_2\}$$

$W \cdot (m, -m; m', -m')$:   permutations of components $m, -m$,

                                       and of components $m', -m'$

$O(W) = 4$

The Algebra $su(3)$:

$$\sum = \{e_i - e_j, \quad i \neq j, \quad i,j = 1,2,3\}$$
$$\prod = \{-e_1 + e_2, -e_2 + e_3\}$$

$W \cdot (m_1, m_2, m_3)$:   all permutations of components m_i

$O(W) = 3!$

The Algebra $su(2) \sim so(3)$:

$$\sum = \{\pm(e_1 - e_2)\}$$
$$\prod = \{-e_1 + e_2\}$$

$W \cdot (m, -m)$:   permutations of components

$O(W) = 2$

A BOSON OPERATOR REALISATION

The Algebra $u(4)$ and $su(4)$:

We introduce boson creation and boson annihilation operators

$$\{b_i^+, b_j, \quad i,j = 1,2,3,4\}$$

which satisfy the Lie products (commutator)

$$[b_i, b_j^+] = \delta_{ij}, \quad [b_i^+, b_j^+] = [b_i, b_j] = 0.$$

The set of elements

$$\{H_i = b_i^+ b_i; \quad E_{e_i - e_j} = b_i^+ b_j, \quad i < j; \quad E_{-e_i + e_j} = b_j^+ b_i;$$
$$E_{e_j - e_i} = E_{e_i - e_j}^+; \quad i < j, \quad i,j = 1,2,3,4\}$$

closes with respect to the Lie product and forms a realisation of $u(4)$ in terms of boson operators. The elements H_i form a basis for the Cartan subalgebra of $u(4)$, the $E_{e_i - e_j}$, $i < j$ form a realisation of the raising operators of $u(4)$, while the $E_{-e_i + e_j}$ form a relization of the $u(4)$ lowering operators. The e_i denote cartesian coordinate vectors in 4-dimensional euclidean space. Thus the subscripts of the E's correspond to the roots of $u(4)$. The algebra $su(4)$ is obtained by subtracting $(1/4)\sum_i H_i$ from the 4 basis elements H_i of $u(4)$. One obtains

$$H_1 \rightarrow \tfrac{1}{4}(3H_1 - H_2 - H_3 - H_4)$$
$$H_2 \rightarrow \tfrac{1}{4}(-H_1 + 3H_2 - H_3 - H_4)$$
$$H_3 \rightarrow \tfrac{1}{4}(-H_1 - H_2 + 3H_3 - H_4)$$
$$H_4 \rightarrow \tfrac{1}{4}(-H_1 - H_2 - H_3 + 3H_4)$$

The root system of $su(4)$, and the corresponding shift operators, are illustrated in Fig. 2.

Consider the linear span V_N of the set of elements

$$V_N : \left\{ X(n_1, n_2, n_3, n_4) \equiv \frac{(b_1^+)^{n_1}}{\sqrt{n_1!}} \frac{(b_2^+)^{n_2}}{\sqrt{n_2!}} \frac{(b_3^+)^{n_3}}{\sqrt{n_3!}} \frac{(b_4^+)^{n_4}}{\sqrt{n_4!}}, \quad n_1 + n_2 + n_3 + n_4 = N \right\}$$

with $X(0,0,0,0) = 1$ (the identity operator) and N some non-negative integer. If we require $b_i 1 = 0$, $i = 1, 2, 3, 4$, then each vector space V_N carries an irreducible representation of $su(4)$. (The action of $u(4)$ does not change the number N, while it exhausts all possible values n_i.) Then

$$V_1 : \{ b_1^+, b_2^+, b_3^+, b_4^+ \}$$

forms the carrier space for the 4-dimensional (defining) representation of $su(4)$, with the b_i^+ as basis vectors for the space V_1. Representing the b_i^+ by the 4-dimensional cartesian vectors e_i,

$$\begin{cases} b_1^+ \rightarrow [1000] \\ b_2^+ \rightarrow [0100] \\ b_3^+ \rightarrow [0010] \\ b_4^+ \rightarrow [0001] \end{cases} \quad \begin{cases} [H_i, b_j^+] = \delta_{ij} b_j^+, \\ [E_{-e_i + e_k}, b_j^+] = \delta_{ij} b_k^+, \end{cases}$$

one obtains, applying the three <u>simple</u> lowering operators $E_{(-1100)} = b_2^+ b_1$, $E_{(0-110)} = b_3^+ b_2$, $E_{(00-11)} = b_4^+ b_3$,

$$[1000] \xrightarrow[1]{E_{(-1100)}} [0100]$$
$$E_{(0-110)} \downarrow 1$$
$$[0010] \xrightarrow[1]{E_{(00-11)}} [0001]$$

This is an equivalent statement to the fact that the b_i^+ transform, as tensor operators, like the irreducible representation $[1000]$ of $su(4)$. The matrix representation of the E's is easily obtained as

$$E_{(-1100)} = \begin{bmatrix} 0 & 0 & 0 & 0 \\ 1 & 0 & 0 & 0 \\ 0 & 0 & 0 & 0 \\ 0 & 0 & 0 & 0 \end{bmatrix}, \quad \text{etc.}$$

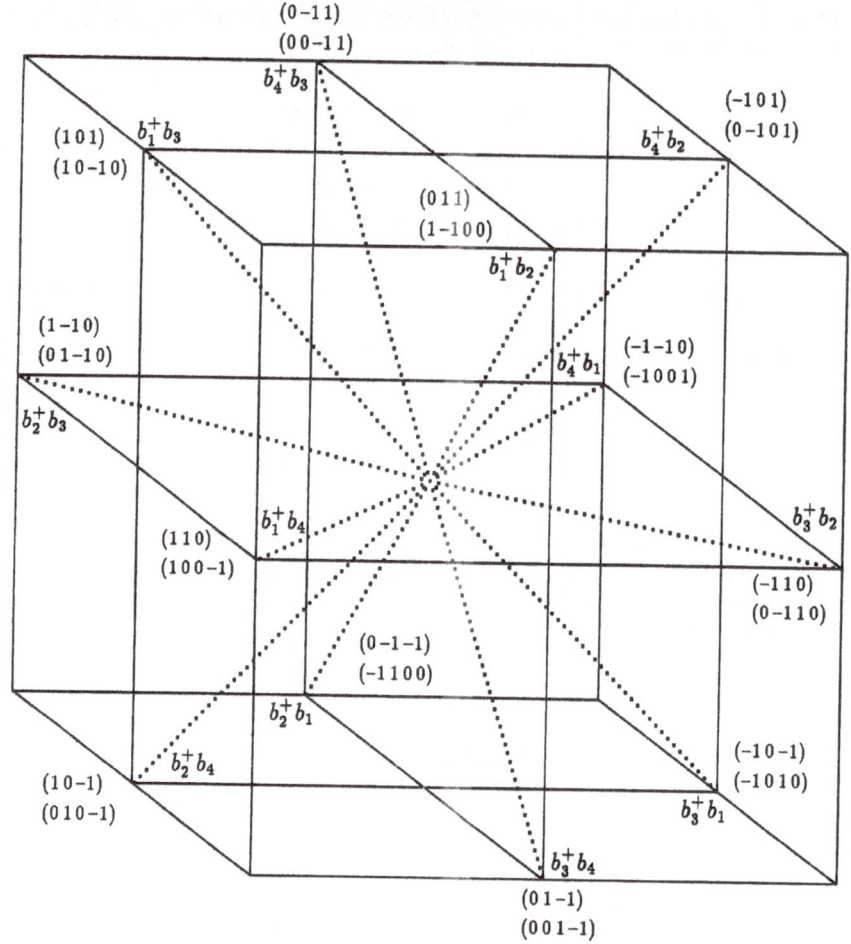

Figure 2. Root system $su(4) \sim so(6)$ boson operator realisation.

The Algebra $so(6)$:

$$\left\{ H_1' = \tfrac{1}{2}(H_1 + H_2 - H_3 - H_4), \quad H_2' = \tfrac{1}{2}(H_1 - H_2 + H_3 - H_4), \right.$$
$$H_3' = \tfrac{1}{2}(H_1 - H_2 - H_3 + H_4)$$

$$E_{(1-10)} = E_{(01-10)} = b_2^+ b_3, \quad E_{(10-1)} = E_{(010-1)} = b_2^+ b_4,$$
$$E_{(01-1)} = E_{(001-1)} = b_3^+ b_4,$$
$$E_{(110)} = E_{(100-1)} = b_1^+ b_4, \quad E_{(011)} = E_{(1-100)} = b_1^+ b_2,$$
$$E_{(101)} = E_{(10-10)} = b_1^+ b_3,$$

$$\left. \text{plus the hermitian conjugate for the negative root shift operators} \right\}$$

The H_i' are the basis elements of the Cartan subalgebra of $so(6)$, the H_i the corresponding elements of $u(4)$ (or $su(4)$). The shift operators E with three component vector subscript are those of $so(6)$, the operators E with four component vector subscripts are their "partners" in $u(4)$ (or $su(4)$). The relation for the H_i in terms of the H'_j is

$$H_1 = \tfrac{1}{2}(H'_1 + H'_2 + H'_3)$$
$$H_2 = \tfrac{1}{2}(H'_1 - H'_2 - H'_3)$$
$$H_3 = \tfrac{1}{2}(-H'_1 + H'_2 - H'_3)$$
$$H_4 = \tfrac{1}{2}(-H'_1 - H'_2 + H'_3).$$

Note that $H_1 + H_2 + H_3 + H_4 = 0$ for $su(4)$, and thus

$$H'_1 = H_1 + H_2, \quad H'_2 = H_1 + H_3, \quad H'_3 = H_1 + H_4.$$

$su(4) \rightarrow so(5)$:

The algebra of $so(5)$ in $su(4)$ is obtained as

$$\{f(H'_1) = \tfrac{1}{2}(b_1^+ b_1 + b_2^+ b_2 + b_3^+ b_3 + b_4^+ b_4),$$
$$f(H'_2) = \tfrac{1}{2}(b_1^+ b_1 - b_2^+ b_2 + b_3^+ b_3 - b_4^+ b_4),$$
$$f(E_{(-11)}) = f(E_{(1-1)})^+ = b_3^+ b_2,$$
$$f(E_{(0-1)}) = f(E_{(01)})^+ = (1/\sqrt{2})(b_2^+ b_1 + b_4^+ b_3),$$
$$f(E_{(-1-1)}) = f(E_{(11)})^+ = b_4^+ b_1,$$
$$f(E_{(-10)}) = f(E_{(10)})^+ = (1/\sqrt{2})(b_2^+ b_4 - b_1^+ b_3)\}$$

The subscripts of the shift operators E are the roots of $so(5)$. For their particular embedding in $su(4)$ see Fig. 2. The action of the two simple lowering operators $f(E_{(-11)})$ and $f(E_{(0-1)})$ on the $su(4)$ states yields

$$
\begin{array}{ccc}
su(4): & [1000] & \xrightarrow[1/\sqrt{2}]{f(E_{(0-1)})} & [0100] \\
so(5): & (\tfrac{1}{2},\tfrac{1}{2}) & & (\tfrac{1}{2},-\tfrac{1}{2})
\end{array}
$$

$$f(E_{(-11)}) \Big\downarrow 1$$

$$
\begin{array}{ccc}
[0010] & \xrightarrow[1/\sqrt{2}]{f(E_{(0-1)})} & [0001] \\
(-\tfrac{1}{2},\tfrac{1}{2}) & & (-\tfrac{1}{2},-\tfrac{1}{2})
\end{array}
$$

Beneath the $su(4)$ weights representing the $su(4)$ vectors (uniquely, since for completely symmetric representations of $su(4)$ the weight subspace dimension is 1) we have given the corresponding $so(5)$ weights. These are obtained by acting with the embedding matrix f_B, Fig. 1, upon the $su(4)$ weights. Again, the space V_1 remains irreducible. The matrix representation for the shift operators can be read off as

$$f(E_{(0-1)}) = \begin{bmatrix} 0 & 0 & 0 & 0 \\ 1/\sqrt{2} & 0 & 0 & 0 \\ 0 & 0 & 0 & 0 \\ 0 & 0 & 1/\sqrt{2} & 0 \end{bmatrix}, \quad f(E_{(-11)}) = \begin{bmatrix} 0 & 0 & 0 & 0 \\ 0 & 0 & 0 & 0 \\ 0 & 1 & 0 & 0 \\ 0 & 0 & 0 & 0 \end{bmatrix}, \text{ etc.}$$

$su(4) \rightarrow so(5) \rightarrow su(2)_J$:

$$\{f(J_0) = \tfrac{1}{2}(3b_1^+ b_1 + b_2^+ b_2 - b_3^+ b_3 - 3b_4^+ b_4),$$
$$f(J_-) = f(J_+)^+ = \sqrt{2}b_3^+ b_2 + \sqrt{\tfrac{3}{2}}(b_2^+ b_1 + b_4^+ b_3)\}$$

The roots and shift operators of $su(2)_J$ in $su(4)$ can be identified in Fig. 2.

$$su(4): \quad [1000] \xrightarrow{\frac{f(J_-)}{\sqrt{\frac{3}{2}}}} [0100] \xrightarrow{\frac{f(J_-)}{\sqrt{2}}} [0010] \xrightarrow{\frac{f(J_-)}{\sqrt{\frac{3}{2}}}} [0001]$$
$$su(2)_J: \quad (\tfrac{3}{2}, -\tfrac{3}{2}) \qquad\quad (\tfrac{1}{2}, -\tfrac{1}{2}) \qquad\quad (-\tfrac{1}{2}, \tfrac{1}{2}) \qquad\quad (-\tfrac{3}{2}, \tfrac{3}{2})$$

Again, the representation remains irreducible. Beneath the $su(4)$ weights are listed the corresponding $su(2)$ weights. The second component of these weights is redundant. For the matrix representation of $f(J_-)$ one obtains

$$f(J_-) = \begin{bmatrix} 0 & 0 & 0 & 0 \\ \sqrt{3/2} & 0 & 0 & 0 \\ 0 & \sqrt{2} & 0 & 0 \\ 0 & 0 & \sqrt{3/2} & 0 \end{bmatrix}$$

The relationship between representations and embeddings was discussed elsewhere.[7]

$$su(4) \rightarrow so(5) \rightarrow so(4) \sim su(2)_S \times su(2)_Q:$$

$$\{ f(H_1') = \tfrac{1}{2}(b_1^+ b_1 + b_2^+ b_2 + b_3^+ b_3 + b_4^+ b_4),$$
$$f(H_2') = \tfrac{1}{2}(b_1^+ b_1 - b_2^+ b_2 + b_3^+ b_3 - b_4^+ b_4),$$
$$f(E_{(-11)}) = f(E_{(1-1)})^+ = b_3^+ b_2,$$
$$f(E_{(-1-1)}) = f(E_{(11)})^+ = b_4^+ b_1 \}$$

It is easily observed that the 6 elements can be separated into two commuting subsets of three elements each,

$$su(2)_S: \quad \{ f(S_0) = \tfrac{1}{2}(b_2^+ b_2 - b_3^+ b_3),$$
$$f(S_-) = (1/\sqrt{2}) b_3^+ b_2, \quad f(S_+) = (1/\sqrt{2}) b_2^+ b_3 \}$$
$$su(2)_Q: \quad \{ f(Q_0) = \tfrac{1}{2}(b_1^+ b_1 - b_4^+ b_4),$$
$$f(Q_-) = (1/\sqrt{2}) b_4^+ b_1, \quad f(Q_+) = (1/\sqrt{2}) b_1^+ b_4 \}.$$

Again the root system of $su(2)_S \times su(2)_J$ in $su(4)$ can be identified in Fig. 2. The $su(4)$ states transform according to $so(4) \sim su(2)_S \times su(2)_Q$,

$$su(4): \quad [1000] \xrightarrow{\frac{f(Q_-)}{1/\sqrt{2}}} [0001]$$
$$so(5): \quad (\tfrac{1}{2}, \tfrac{1}{2}) \qquad\quad (-\tfrac{1}{2}, -\tfrac{1}{2})$$
$$su(2)_S \times su(2)_Q: \quad (0,0; \tfrac{1}{2}, -\tfrac{1}{2}) \qquad\quad (0,0; -\tfrac{1}{2}, \tfrac{1}{2})$$

$$[0100] \xrightarrow{\frac{f(S_-)}{1/\sqrt{2}}} [0010]$$
$$(\tfrac{1}{2}, -\tfrac{1}{2}) \qquad\quad (-\tfrac{1}{2}, \tfrac{1}{2})$$
$$(\tfrac{1}{2}, -\tfrac{1}{2}; 0,0) \qquad\quad (-\tfrac{1}{2}, \tfrac{1}{2}; 0,0)$$

The above implies that the carrier space V_1 remains no longer irreducible but decomposes into the direct sum of two 2-dimensional spaces, each of which is irreducible with respect to the action of $so(4) \sim su(2)_S \times su(2)_Q$.

$su(4) \to so(5) \to so(4) \sim su(2)_S \times su(2)_Q \to su(2)_Q$:

$$
\begin{array}{cccc}
su(4): & [1000] & \xrightarrow[1/\sqrt{2}]{f(Q_-)} & [0001] \\
so(5): & (\frac{1}{2}, \frac{1}{2}) & & (-\frac{1}{2}, -\frac{1}{2}) \\
su(2)_S \times su(2)_Q: & (0,0; \frac{1}{2}, -\frac{1}{2}) & & (0,0; -\frac{1}{2}, \frac{1}{2}) \\
su(2)_Q & (\frac{1}{2}, -\frac{1}{2}) & & (-\frac{1}{2}, \frac{1}{2})
\end{array}
$$

$$
\begin{array}{ccc}
[0100] & \; ; \; & [0010] \\
(\frac{1}{2}, -\frac{1}{2}) & & (-\frac{1}{2}, \frac{1}{2}) \\
(\frac{1}{2}, -\frac{1}{2}; 0, 0) & & (-\frac{1}{2}, \frac{1}{2}; 0, 0) \\
(0, 0) & & (0, 0)
\end{array}
$$

The 4-dimensional irreducible representation of $su(4)$ thus decomposes still further, namely into a two-dimensional and two one-dimensional irreducible representations of $su(2)_Q$.

$su(4) \to so(5) \to so(4) \sim su(2)_S \times su(2)_Q \to su(2)_{S+Q}$:

$$\{ f(S_0 + Q_0) = \tfrac{1}{2}(b_1^+ b_1 + b_2^+ b_2 - b_3^+ b_3 - b_4^+ b_4),$$
$$f(S_- + Q_-) = f(S_+ + Q_+)^+ = (1/\sqrt{2})(b_4^+ b_1 + b_3^+ b_2) \}$$

$$
\begin{array}{cccc}
su(4): & [1000] & \xrightarrow[1/\sqrt{2}]{f(S_- + Q_-)} & [0001] \\
so(5): & (\frac{1}{2}, \frac{1}{2}) & & (-\frac{1}{2}, -\frac{1}{2}) \\
su(2)_S \times su(2)_Q: & (0,0; \frac{1}{2}, -\frac{1}{2}) & & (0,0; -\frac{1}{2}, \frac{1}{2}) \\
su(2)_{S+Q}: & (\frac{1}{2}, -\frac{1}{2}) & & (-\frac{1}{2}, \frac{1}{2})
\end{array}
$$

$$
\begin{array}{cccc}
[0100] & \xrightarrow[1/\sqrt{2}]{f(S_- + Q_-)} & [0010] \\
(\frac{1}{2}, -\frac{1}{2}) & & (\frac{1}{2}, \frac{1}{2}) \\
(\frac{1}{2}, -\frac{1}{2}; 0, 0) & & (-\frac{1}{2}, \frac{1}{2}; 0, 0) \\
(\frac{1}{2}, -\frac{1}{2}) & & (-\frac{1}{2}, \frac{1}{2})
\end{array}
$$

The 4-dimensional irreducible representation of $su(4)$ thus decomposes into the direct sum of two 2-dimensional irreducible representations of $su(2)_{S+Q}$.

$su(4) \to su(2)_M \times su(2)_N$:

$$\{ f(M_0) = \tfrac{1}{2}(b_1^+ b_1 + b_2^+ b_2 - b_3^+ b_3 - b_4^+ b_4), \; f(M_-) = f(M_+)^+ = \frac{1}{\sqrt{2}}(b_3^+ b_1 + b_4^+ b_2) \},$$

$$
\begin{array}{cccc}
su(4): & [1000] & \xrightarrow[1/\sqrt{2}]{f(M_-)} & [0010] \\
su(2)_M \times su(2)_N: & (\frac{1}{2}, -\frac{1}{2}; \frac{1}{2}, -\frac{1}{2}) & & (-\frac{1}{2}, \frac{1}{2}; \frac{1}{2}, -\frac{1}{2}) \\
& \Big\downarrow f(N_-) \; 1/\sqrt{2} & & \Big\downarrow f(N_-) \; 1/\sqrt{2} \\
& [0100] & \xrightarrow[1/\sqrt{2}]{f(M_-)} & [0001] \\
& (\frac{1}{2}, -\frac{1}{2}; -\frac{1}{2}, \frac{1}{2}) & & (-\frac{1}{2}, \frac{1}{2}; -\frac{1}{2}, \frac{1}{2})
\end{array}
$$

The 4-dimensional irreducible $su(4)$ representation remains irreducible.

$su(4) \to su(2)_M \times su(2)_N \to su(2)_M$:

$$\{f(M_0) = \tfrac{1}{2}(b_1^+ b_1 + b_2^+ b_2 - b_3^+ b_3 - b_4^+ b_4),\ f(M_-) = f(M_+)^+ = (1/\sqrt{2})(b_3^+ b_1 + b_4^+ b_2)\}.$$

For the transformation properties of the states one obtains the last diagram with the arrows corresponding to $f(N_-)$ missing. The 4-dimensional representation [1000] thus decomposes into a direct sum of two 2-dimensional irreducible representations. It holds $f(M_0) = f(S_0 + Q_0)$, and the branching of the representation [1000] is the same for both $su(2)_M$ and $su(2)_{S+Q}$, namely $4 \to 2+2$ (in terms of dimensions). The two subalgebras $su(2)_M$ and $su(2)_{S+Q}$ of $su(4)$ are in fact equivalent.

The basis states of $su(2)_S \times su(2)_Q$ are related to the basis states of $su(2)_M$ by the similarity transformation

$$T = \begin{bmatrix} 1 & 0 & 0 & 0 \\ 0 & 1 & 0 & 0 \\ 0 & 0 & 0 & 1 \\ 0 & 0 & 1 & 0 \end{bmatrix}, \quad T^2 = 1.$$

The shift operator $f(M_-)$ is represented, in the 4-dimensional representation [1000] of $su(4)$ as

$$f(M_-) = \begin{bmatrix} 0 & 0 & 0 & 0 \\ 0 & 0 & 0 & 0 \\ 1/\sqrt{2} & 0 & 0 & 0 \\ 0 & 1/\sqrt{2} & 0 & 0 \end{bmatrix}.$$

Then

$$T f(M_-) T = 1/\sqrt{2} \begin{bmatrix} 0 & 0 & 0 & 0 \\ 0 & 0 & 0 & 0 \\ 0 & 1 & 0 & 0 \\ 1 & 0 & 0 & 0 \end{bmatrix} = \frac{1}{\sqrt{2}}(b_3^+ b_2 + b_4^+ b_1) = f(S_- + Q_-)$$

$$T f(M_0) T = f(M_0) = f(S_0 + Q_0).$$

$su(4) \to su(2)_M \times su(2)_N \to su(2)_{M+N}$:

$$\{f(M_0 + N_0) = b_1^+ b_1 - b_4^+ b_4,$$

$$f(M_- + N_-) = f(M_+ + N_+)^+ = \frac{1}{\sqrt{2}}(b_3^+ b_1 + b_4^+ b_2 + b_2^+ b_1 + b_4^+ b_3)\}$$

$$su(4): \quad [1000] \quad \xrightarrow[\sqrt{2}]{f(M_-+N_-)} \quad \tfrac{1}{\sqrt{2}}([0010] + [0100]) \quad \xrightarrow[\sqrt{2}]{f(M_-+N_-)} \quad [0001]$$
$$su(2)_{M+N}: \quad (1,-1) \qquad\qquad (0,0) \qquad\qquad\qquad (-1,1)$$

$$\tfrac{1}{\sqrt{2}}([0010] - [0100])$$
$$(0,0)$$

The representation [1000] decomposes into a triplet and a singlet.

234

$su(4) \rightarrow u(3)(su(3))$:

$$u(3) : \{H_i = b_i^+ b_i, \ E_{e_i - e_j} = b_i^+ b_j, i < j, \ E_{-e_i + e_j} = b_j^+ b_i, i < j; \ i, j = 1, 2, 4\}$$

Subtraction of $\frac{1}{3}\sum_j H_j$ from each of the H_i yields $su(3)$. The shift operators corresponding to the simple negative roots are $f(E_{(-110)}) = E_{(-1100)} = b_1^+ b_2$, $f(E_{(0-11)}) = E_{(0-101)} = b_4^+ b_2$.

$$
\begin{array}{cccccc}
su(4): & [1000] & \xrightarrow[1]{f(E_{(-110)})} & [0100] & \xrightarrow[1]{f(E_{(0-11)})} & [0001] \\
su(3): & [100] & & [010] & & [001]
\end{array}
$$

$$
\begin{array}{cc}
\text{and} & [0010] \\
& [000]
\end{array}
$$

Thus $4 \rightarrow 3 + 1$.

$su(4) \rightarrow su(3) \rightarrow su(2)_Q$:

$$\{f(Q_0) = \tfrac{1}{2}(b_1^+ b_1 - b_4^+ b_4), \ f(Q_-) = f(E_{(-101)}) = (1/\sqrt{2})b_4^+ b_1, \ f(Q_+) = f(Q_-)^+\}$$
$$4 \rightarrow 2 + 1 + 1.$$

$su(4) \rightarrow su(3) \rightarrow su(2)_L$:

$$\{f(L_0) = b_1^+ b_1 - b_4^+ b_4, \ f(L_-) = f(L_+)^+ = b_2^+ b_1 + b_4^+ b_2\}$$

We have

$$
\begin{array}{cccccc}
su(4): & [1000] & \xrightarrow[1]{f(L_-)} & [0100] & \xrightarrow[1]{f(L_-)} & [0001] \\
su(3): & \tfrac{1}{3}(2,-1,-1) & & \tfrac{1}{3}(-1,2,-1) & & \tfrac{1}{3}(-1,-1,2) \\
su(2)_L & (1,-1) & & (0,0) & & (-1,1)
\end{array}
$$

$$
\begin{array}{cc}
\text{and} & [0010] \\
& (0,0,0) \\
& (0,0)
\end{array}
$$

Since $f(L_0) = f(M_0 + N_0)$, and for both subalgebras $4 \rightarrow 3 + 1$, the two subalgebras $su(2)_L$ and $su(2)_{M+N}$ must be equivalent. The similarity transformation T relating the basis states of $su(2)_{M+N}$ to the basis states of $su(2)_L$ is

$$T = \begin{bmatrix} 1 & 0 & 0 & 0 \\ 0 & 1/\sqrt{2} & 1/\sqrt{2} & 0 \\ 0 & 1/\sqrt{2} & -1/\sqrt{2} & 0 \\ 0 & 0 & 0 & 1 \end{bmatrix}, \quad T^2 = 1,$$

and one obtains, with $f(L_-) = b_2^+ b_1 + b_4^+ b_2$,

$$Tf(L_-)T = T \begin{bmatrix} 0 & 0 & 0 & 0 \\ 1 & 0 & 0 & 0 \\ 0 & 0 & 0 & 0 \\ 0 & 1 & 0 & 0 \end{bmatrix} T$$

$$= \begin{bmatrix} 0 & 0 & 0 & 0 \\ 1/\sqrt{2} & 0 & 0 & 0 \\ 1/\sqrt{2} & 0 & 0 & 0 \\ 0 & 1/\sqrt{2} & 1/\sqrt{2} & 0 \end{bmatrix}$$

$$= \frac{1}{\sqrt{2}}(b_2^+ b_1 + b_3^+ b_1 + b_4^+ b_2 + b_4^+ b_3) = f(M_- + N_-)$$

$$Tf(L_0)T = f(L_0) = f(M_0 + N_0)$$

This change of basis for the chain $su(4) \to su(3) \to su(2)_L$ brings $su(2)_L$ into identical form with $su(2)_{M+N}$. However, this change of basis has consequences for the Cartan subalgebras of $su(4)$ and $su(3)$ in the chain $su(4) \to su(3) \to su(2)_L$. We obtain for the Cartan subalgebra of $u(4)$ in the new basis

$$f(H_1)' = Tf(H_1)T = b_1^+ b_1,$$
$$f(H_2)' = Tf(H_2)T = \tfrac{1}{2}(b_2^+ b_2 + b_3^+ b_3) + \tfrac{1}{2}(b_2^+ b_3 + b_3^+ b_2)$$
$$f(H_3)' = Tf(H_3)T = \tfrac{1}{2}(b_2^+ b_2 + b_3^+ b_3) - \tfrac{1}{2}(b_2^+ b_3 + b_3^+ b_2)$$
$$f(H_4)' = Tf(H_4)T = b_4^+ b_4.$$

Note the occurence of shift operators in these relations. That is, by choosing the same bases for the L and $M + N$ chains, their Cartan subalgebras in $su(4)$ are no longer identical.

Poincaré Algebra P:

This is a non semi-simple algebra. We have

$$\{f(H_1') = H_1 + H_2 = b_1^+ b_1 + b_2^+ b_2, \; f(H_2') = H_1 + H_3 = b_1^+ b_1 - b_3^+ b_3,$$
$$E_{(1-1)} = b_2^+ b_3, E_{(-11)} = b_3^+ b_2, \; E_{(11)} = b_1^+ b_4, E_{(-1-1)} = b_4^+ b_1,$$
$$P_1 = E_{(10)} = b_1^+ b_3, \; P_2 = E_{(01)} = b_1^+ b_2, \; P_3 = E_{(-10)} = b_4^+ b_2, \; P_4 = E_{(0-1)} = b_4^+ b_3\}$$

For the root system and shift operator of P in $su(4)$ see Fig. 2.

The Poincaré algebra contains $so(4)$ as a subalgebra, and the four commuting elements P_i transform like a vector with respect to the Lie algebra $so(4)$. Symbolically

$$[so(4), so(4)] \sim so(4)$$
$$[so(4), P] \sim P$$
$$[P, P] \sim 0$$

For the representation [1000] of $su(4)$ on obtains with respect to P,

[1000] $\xrightarrow{E_{(-1,-1)}}$ [0001]

$(\tfrac{1}{2}, \tfrac{1}{2})$ $(-\tfrac{1}{2}, -\tfrac{1}{2})$

$P_2 \uparrow \; \nwarrow \; P_3 \qquad P_1 \; \nearrow \; \uparrow P_4$

[0100] $\xrightarrow{E_{(-11)}}$ [0010]

$(\tfrac{1}{2}, -\tfrac{1}{2})$ $(-\tfrac{1}{2}, \tfrac{1}{2})$

that is, [1000] becomes a 4-dimensional indecomposable (i.e. reducible, but not completely reducible) representation of P.

A FERMION OPERATOR REALISATION

In the following we give a fermion operator realisation of $so(6) \sim su(4)$, and the semisimple symmetry chains originating from it. In addition we give a fermion operator realisation of the Poincaré algebra in $so(6) \sim su(4)$. Following the definition of the Fermi operators we will merely list the results.

Consider the set of operators

$$\{f_1^+, f_2^+, f_1, f_2\}$$

which satisfy the anticommutation relations $\{a, b\} = ab + ba$,

$$\{f_i, f_j^+\} = \delta_{i,j}, \quad \{f_i^+, f_j^+\} = \{f_i, f_j\} = 0.$$

These relations generate a 16-dimensional Clifford algebra for which a basis can be chosen, as the set of ordered products $(f_1^+)^i (f_2^+)^j (f_1)^k (f_2)^\ell$, $i, j, k, \ell = 0, 1$,

$$\{1, f_1^+, f_2^+, f_1, f_2, f_1^+ f_2^+, f_1^+ f_1, f_1^+ f_2, f_2^+ f_1, f_2^+ f_2, f_1 f_2,$$
$$f_1^+ f_2^+ f_1, f_1^+ f_2^+ f_2, f_1^+ f_1 f_2, f_2^+ f_1 f_2, f_1^+ f_2^+ f_1 f_2\}$$

where the identity operator 1 represents the case $i = j = k = \ell = 0$.

The linear span of the set

$$V : \{1 \equiv |0\rangle, \; f_1^+, \; f_2^+, \; f_2^+ f_1^+\},$$

forms a 4-dimensional vector space. We will require that

$$f_i 1 = 0.$$

This means that the identity operator represents the familiar physical vacuum state $|0\rangle$. In fact, the relation $f_i 1 = 0$ generates a left ideal in the space of the Clifford algebra, and V can be viewed as being equivalent to the quotient space of the Clifford algebra modulo this ideal.

Since the discussion is analogous to the boson operator case we merely state the results. For the fermion operator embeddings see Fig. 3.

The Algebra $su(4)$:

$$\{H_1 = -f_1^+ f_2^+ f_1 f_2 - \frac{1}{4}, \quad H_2 = f_1^+ f_2^+ f_1 f_2 + f_1^+ f_1 - \frac{1}{4},$$

$$H_3 = f_1^+ f_2^+ f_1 f_2 + f_2^+ f_2 - \frac{1}{4}, \quad H_4 = -f_1^+ f_2^+ f_1 f_2 - f_1^+ f_1 - f_2^+ f_2 + \frac{3}{4},$$

$$\begin{array}{llll}
E_{(-1100)} = f_2 f_1^+ f_1, & E_{(-1010)} = f_1 f_2^+ f_2, & E_{(-1001)} = f_1 f_2, & \\
& E_{(0-110)} = f_2^+ f_1, & E_{(0-101)} = f_1(1 - f_2^+ f_2), & \\
& & E_{(00-11)} = f_2(1 - f_1^+ f_1), & \\
E_{(1-100)} = f_2^+ f_1^+ f_1, & E_{(10-10)} = f_1^+ f_2^+ f_2, & E_{(100-1)} = f_1^+ f_2^+, & \\
& E_{(01-10)} = f_1^+ f_2, & E_{(010-1)} = f_1^+(1 - f_2^+ f_2), & \\
& & E_{(001-1)} = f_2^+(1 - f_1^+ f_1)\}
\end{array}$$

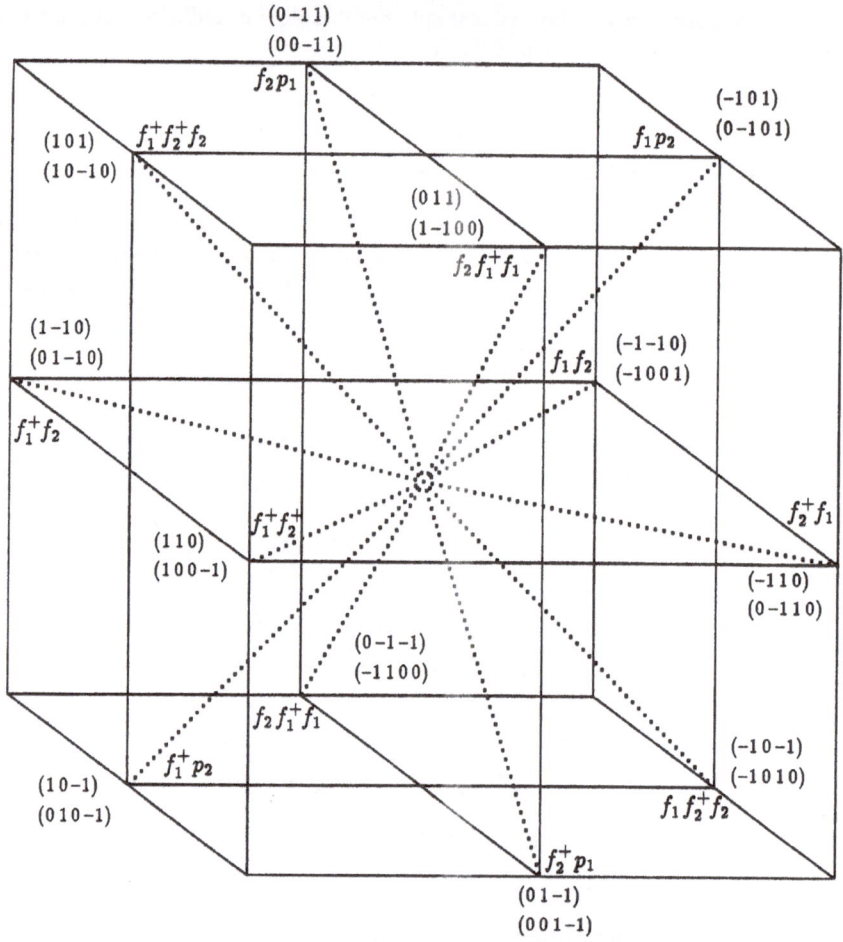

Figure 3. Root system $su(4) \sim so(6)$, fermion operator realisation. $p_i = 1 - f_i^+ f_i$.

The algebra $su(4)$ acts upon the basis elements of V in the following manner,

$$f_2^+ f_1^+ \xrightarrow[1]{E_{(-1100)}} f_1^+$$

$$E_{(0-110)} \Big\downarrow 1$$

$$f_2^+ \xrightarrow[1]{E_{(00-11)}} 1$$

Comparing with the [1000] representation of the boson operator realisation we see that the two representations are the same. We can set

$$[1000] \longleftrightarrow f_2^+ f_1^+$$
$$[0100] \longleftrightarrow f_1^+$$
$$[0010] \longleftrightarrow f_2^+$$
$$[0001] \longleftrightarrow 1.$$

The Algebra $so(6)$:

$$\{H_1' = f_1^+ f_1 - \tfrac{1}{2}, \ H_2' = f_2' f_2 - \tfrac{1}{2}, \ H_3' = 2H_1' H_2' = -2f_1^+ f_2^+ f_1 f_2 - f_1^+ f_1 - f_2^+ f_2 + \tfrac{1}{2}$$

$$E_{(1-10)} = E_{(01-10)}, \quad E_{(01-1)} = E_{(001-1)}, \quad E_{(10-1)} = E_{(010-1)},$$
$$E_{(110)} = E_{(100-1)}, \quad E_{(011)} = E_{(1-100)}, \quad E_{(101)} = E_{(10-10)},$$
$$\text{plus the operators with the negative roots}\}.$$

The E's with a 3-dimensional vector label are $so(6)$ operators, the E's with a 4-dimensional vector label are the corresponding $su(4)$ operators. For other relations see the boson operator case.

$su(4) \rightarrow so(5)$:

$$\{f(H_1') = f_1^+ f_1 - \tfrac{1}{2}, \ f(H_2') = f_2^+ f_2 - \tfrac{1}{2}, \ f(E_{(-11)}) = f_2^+ f_1, \ f(E_{(-1-1)}) = f_2 f_1,$$
$$f(E_{(-10)}) = (1/\sqrt{2})f_1, \ f(E_{(0-1)}) = \frac{1}{\sqrt{2}}f_2, \ f(E_{-\alpha}) = f(E_\alpha)^+\}$$

$$\begin{array}{cc}
f_2^+ f_1^+ & \xrightarrow[1/\sqrt{2}]{f(E_{(0-1)})} & f_1^+ \\
[1000] & & [0100] \\
(\tfrac{1}{2},\tfrac{1}{2}) & & (\tfrac{1}{2},-\tfrac{1}{2})
\end{array}$$

$$f(E_{(-11)}) \Big\downarrow 1$$

$$\begin{array}{cc}
f_2^+ & & \\
[0010] & \xrightarrow[1/\sqrt{2}]{f(E_{(0-1)})} & 1 \\
(-\tfrac{1}{2},\tfrac{1}{2}) & & [0001] \\
& & (-\tfrac{1}{2},-\tfrac{1}{2})
\end{array}$$

$su(4) \rightarrow so(5) \rightarrow su(2)_J$:

$$\{f(J_0) = 2f_1^+ f_1 + f_2^+ f_2 - \frac{3}{2}, \ f(J_-) = f(J_+)^+ = \sqrt{2}f_2^+ f_1 + \sqrt{\frac{3}{2}}f_2\}$$

$$\begin{array}{cccccccc}
f_2^+ f_1^+ & \xrightarrow[\sqrt{3/2}]{f(J_-)} & f_1^+ & \xrightarrow[\sqrt{2}]{f(J_-)} & f_2^+ & \xrightarrow[\sqrt{3/2}]{f(J_-)} & 1 \\
[1000] & & [0100] & & [0010] & & [0001] \\
(\tfrac{1}{2},\tfrac{1}{2}) & & (\tfrac{1}{2},-\tfrac{1}{2}) & & (-\tfrac{1}{2},\tfrac{1}{2}) & & (-\tfrac{1}{2},-\tfrac{1}{2}) \\
(\tfrac{3}{2},-\tfrac{3}{2}) & & (\tfrac{1}{2},-\tfrac{1}{2}) & & (-\tfrac{1}{2},\tfrac{1}{2}) & & (-\tfrac{3}{2},\tfrac{3}{2})
\end{array}$$

$su(4) \rightarrow so(5) \rightarrow so(4) \sim su(2)_S \times su(2)_Q$:

$so(4):$
$$\{f(H_1') = f_1^+ f_1 - \tfrac{1}{2}, \ f(H_2') = f_2^+ f_2 - \tfrac{1}{2}, \ f(E_{(-11)}) = f_2^+ f_1 = f(E_{(1-1)})^+,$$
$$f(E_{(-1-1)}) = f_2 f_1 = f(E_{(11)})^+\}$$

$$su(2)_S \times su(2)_Q : \quad \{f(S_0) = \tfrac{1}{2}(f_1^+ f_1 - f_2^+ f_2), \ f(S_-) = (1/\sqrt{2})f_2^+ f_1 = f(S_+)^+,$$
$$f(Q_0) = \tfrac{1}{2}(f_1^+ f_1 + f_2^+ f_2 - 1), \ f(Q_-) = (1/\sqrt{2})f_2 f_1 = f(Q_+)^+\}$$

$$
\begin{array}{ccc}
f_2^+ f_1^+ & \xrightarrow[-1/\sqrt{2}]{f(Q_-)} & 1 \\
[1000] & & [0001] \\
(\tfrac{1}{2}, \tfrac{1}{2}) & & (-\tfrac{1}{2}, -\tfrac{1}{2}) \\
(0,0; \tfrac{1}{2}, -\tfrac{1}{2}) & & (0,0; -\tfrac{1}{2}, \tfrac{1}{2})
\end{array}
$$

$$
\begin{array}{ccc}
f_1^+ & \xrightarrow[1/\sqrt{2}]{f(S_-)} & f_2^+ \\
[0100] & & [0010] \\
(\tfrac{1}{2}, -\tfrac{1}{2}) & & (-\tfrac{1}{2}, \tfrac{1}{2}) \\
(\tfrac{1}{2}, -\tfrac{1}{2}; 0,0) & & (-\tfrac{1}{2}, \tfrac{1}{2}; 0,0)
\end{array}
$$

$su(4) \to so(5) \to so(4) \sim su(2)_S \times su(2)_Q \to su(2)_S:$

$$\{f(S_0) = \tfrac{1}{2}(f_1^+ f_1 - f_2^+ f_2), \ f(S_-) = (1/\sqrt{2})f_2^+ f_1, \ f(S_+) = (1/\sqrt{2})f_1^+ f_2\}$$

$$
\begin{array}{ccc}
f_1^+ & \xrightarrow[1/\sqrt{2}]{f(S_-)} & f_2^+ \\
[0100] & & [0010] \\
(\tfrac{1}{2}, -\tfrac{1}{2}) & & (-\tfrac{1}{2}, \tfrac{1}{2}) \\
(\tfrac{1}{2}, -\tfrac{1}{2}) & & (-\tfrac{1}{2}, \tfrac{1}{2})
\end{array}
$$

$$
\begin{array}{ccc}
f_2^+ f_1^+ & , & 1 \\
[1000] & & [0001] \\
(\tfrac{1}{2}, \tfrac{1}{2}) & & (-\tfrac{1}{2}, -\tfrac{1}{2}) \\
(0,0) & & (0,0)
\end{array}
$$

$su(4) \to so(5) \to so(4) \sim su(2)_S \times su(2)_Q \to su(2)_{S+Q}:$

$$\{f(S_0 + Q_0) = f_1^+ f_1 - \tfrac{1}{2}, \ f(S_- + Q_-) = (1/\sqrt{2})(f_2^+ f_1 + f_2 f_1) = f(S_+ + Q_+)^+\}$$

$$
\begin{array}{ccc}
f_2^+ f_1^+ & \xrightarrow[-1/\sqrt{2}]{f(S_- + Q_-)} & 1 \\
[1000] & & [0001] \\
(\tfrac{1}{2}, \tfrac{1}{2}) & & (-\tfrac{1}{2}, -\tfrac{1}{2}) \\
(0,0; \tfrac{1}{2}, -\tfrac{1}{2}) & & (0,0; -\tfrac{1}{2}, \tfrac{1}{2}) \\
(\tfrac{1}{2}, -\tfrac{1}{2}) & & (-\tfrac{1}{2}, \tfrac{1}{2})
\end{array}
$$

$$f_1^+ \quad \xrightarrow[1/\sqrt{2}]{f(S_-+Q_-)} \quad f_2^+$$

$$[0100] \qquad\qquad [0010]$$

$$(\tfrac{1}{2},-\tfrac{1}{2}) \qquad\qquad (-\tfrac{1}{2},\tfrac{1}{2})$$

$$(\tfrac{1}{2},-\tfrac{1}{2};0,0) \qquad\qquad (-\tfrac{1}{2},\tfrac{1}{2};0,0)$$

$$(\tfrac{1}{2},-\tfrac{1}{2}) \qquad\qquad (-\tfrac{1}{2},\tfrac{1}{2})$$

$su(4) \to su(2)_M \times su(2)_N$:

$$\{f(M_0) = f_1^+ f_1 - \tfrac{1}{2}, \; f(M_-) = (1/\sqrt{2})f_1 = f(M_+)^+,$$
$$f(N_0) = f_2^+ f_2 - \tfrac{1}{2}, \; f(N_-) = (1/\sqrt{2})(-f_2 + 2f_2 f_1^+ f_1) = f(N_+)^+\}$$

$$f_2^+ f_1^+ \quad \xrightarrow[-1/\sqrt{2}]{f(M_-)} \quad f_2^+$$

$$[1000] \qquad\qquad [0010]$$

$$(\tfrac{1}{2},-\tfrac{1}{2};\tfrac{1}{2},-\tfrac{1}{2}) \qquad (-\tfrac{1}{2},\tfrac{1}{2};\tfrac{1}{2},-\tfrac{1}{2})$$

$$f(N_-)\Big|{-(1/\sqrt{2})} \qquad -(1/\sqrt{2})\Big| f(N_-)$$

$$f_1^+ \quad \xrightarrow[f(M_-)]{1/\sqrt{2}} \quad 1$$

$$[0100] \qquad\qquad [0001]$$

$$(\tfrac{1}{2},-\tfrac{1}{2};-\tfrac{1}{2},\tfrac{1}{2}) \qquad (-\tfrac{1}{2},\tfrac{1}{2};-\tfrac{1}{2},\tfrac{1}{2})$$

$su(4) \to su(2)_M \times su(2)_N \to su(2)_M$:

$$\{f(M_0), \; f(M_-), \; f(M_+)\}$$

It holds $f(M_0) = f(Q_0 + S_0)$. Moreover, for both cases, $4 \to 2+2$. The transformation

$$T = \begin{bmatrix} 1 & 0 & 0 & 0 \\ 0 & 1 & 0 & 0 \\ 0 & 0 & 0 & 1 \\ 0 & 0 & 1 & 0 \end{bmatrix}$$

has the property

$$T f(M_-) T = f(S_- + Q_-),$$

and it follows that the two embeddings are (mathematically) equivalent.

$su(4) \to su(2)_M \times su(2)_N \to su(2)_{M+N}$:

$$\{f(M_0 + N_0) = f_1^+ f_1 + f_2^+ f_2 - 1,$$
$$f(M_- + N_-) = (1/\sqrt{2})(f_1 - f_2 + 2f_2 f_1^+ f_1) = f(M_+ + N_+)^+\}$$

$$f_2^+ f_1^+ \quad \xrightarrow[1]{f(M_-+N_-)} \quad \tfrac{1}{\sqrt{2}}(f_1^+ - f_2^+) \quad \xrightarrow[1]{f(M_-+N_-)} \quad \mathbf{1}$$

$$[1000] \qquad\qquad \tfrac{1}{\sqrt{2}}([0100] - [0010]) \qquad\qquad [0001]$$

$$(\tfrac{1}{2}, -\tfrac{1}{2}; \tfrac{1}{2}, -\tfrac{1}{2}) \quad \tfrac{1}{\sqrt{2}}((\tfrac{1}{2}, -\tfrac{1}{2}; -\tfrac{1}{2}, \tfrac{1}{2}) - (-\tfrac{1}{2}, \tfrac{1}{2}; \tfrac{1}{2}, -\tfrac{1}{2})) \quad (-\tfrac{1}{2}, \tfrac{1}{2}; -\tfrac{1}{2}, \tfrac{1}{2})$$

$$(1, -1) \qquad\qquad\qquad (0, 0) \qquad\qquad\qquad (-1, 1)$$

$$\tfrac{1}{\sqrt{2}}(f_1^+ + f_2^+)$$

$$\tfrac{1}{\sqrt{2}}([0100] + [0010])$$

$$\tfrac{1}{\sqrt{2}}((\tfrac{1}{2}, -\tfrac{1}{2}; -\tfrac{1}{2}, \tfrac{1}{2}) + (-\tfrac{1}{2}, \tfrac{1}{2}; \tfrac{1}{2}, -\tfrac{1}{2}))$$

$$(0, 0)$$

$su(4) \to su(3)$:

$$\left\{ f(H_1') = -\frac{2}{3} f_1^+ f_2^+ f_1 f_2 + \frac{1}{3} f_2^+ f_2 - \frac{1}{3}, \ f(H_2') = \frac{4}{3} f_1^+ f_2^+ f_1 f_2 + f_1^+ f_1 + \frac{1}{3} f_2^+ f_2 - \frac{1}{3}, \right.$$

$$f(H_3') = -\frac{2}{3} f_1^+ f_2^+ f_1 f_2 - f_1^+ f_1 - \frac{2}{3} f_2^+ f_2 + \frac{2}{3},$$

$$f(E_{(-110)}) = f_2 f_1^+ f_1, \ f(E_{(0-11)}) = f_1(1 - f_2^+ f_2),$$

$$\left. f(E_{(-101)}) = f_2 f_1, \ f(E_\alpha) = f(E_{-\alpha}))^+ \right\}$$

$$f_2^+ f_1^+ \quad \xrightarrow[1]{f(E_{(-110)})} \quad f_1^+ \quad \xrightarrow[1]{f(E_{(0-11)})} \quad \mathbf{1}$$

$$[1000] \qquad\qquad [0100] \qquad\qquad [0001]$$

$$(\tfrac{1}{3})(2, -1, -1) \qquad (\tfrac{1}{3})(-1, 2, -1) \qquad (\tfrac{1}{3})(-1, -1, 2)$$

$$f_2^+$$

$$[0010]$$

$$(0, 0, 0)$$

$su(4) \to su(3) \to su(2)_Q$:

$$\left\{ f(Q_0) = \tfrac{1}{2}(f(H_1') - f(H_3')) = \tfrac{1}{2}(f_1^+ f_1 + f_2^+ f_2 - 1), \right.$$

$$f(Q_-) = \left(\frac{1}{\sqrt{2}}\right) f(E_{(-101)}) = \frac{1}{\sqrt{2}} f_1^+ f_2^+, \ f(Q_+) = \left(\frac{1}{\sqrt{2}}\right) f(E_{(10-1)}) \frac{1}{\sqrt{2}} f_2 f_1 \right\}$$

with $(f_2^+ f_1^+, 1)$ a doublet and f_1^+, f_2^+ singlets.

$su(4) \to su(3) \to su(2)_L$:

$$\left\{ f(L_0) = f_1^+ f_1 + f_2^+ f_2 - 1, \ f(L_-) = f_2 f_1^+ f_1 + f_1(1 - f_2^+ f_2), \ f(L_+) = f(L_-)^+ \right\}$$

242

$$f_2^+ f_1^+ \quad \xrightarrow{\frac{f(L_-)}{1}} \quad f_1^+ \quad \xrightarrow{\frac{f(L_-)}{1}} \quad 1$$

$$[1000] \qquad\qquad\qquad [0100] \qquad\qquad\qquad [0001]$$

$$(\tfrac{1}{3})(2,-1,-1) \qquad (\tfrac{1}{3})(-1,2,-1) \qquad (\tfrac{1}{3})(-1,-1,2)$$

$$(1,-1) \qquad\qquad\quad (0,0) \qquad\qquad\quad (-1,1)$$

$$f_2^+$$

$$[0010]$$

$$(0,0,0)$$

$$(0,0)$$

As it was the case for the boson operator realisation, the transformation T transforms $f(L)$ into $f(M+N)$.

Poincaré Algebra P:

$$\{f(H_1') = f_1^+ f_1 - \tfrac{1}{2},\ f(H_2') = f_2^+ f_2 - \tfrac{1}{2},\ E_{(-11)} = f_2^+ f_1,\ E_{(1-1)} = f_1^+ f_2,$$

$$E_{(-1-1)} = f_1 f_2,\ E_{(11)} = f_2^+ f_1^+,\ P_1 = E_{(10)} = f_1^+ f_2^+ f_2,\ P_2 = E_{(01)} = f_2^+ f_1^+ f_1,$$

$$P_3 = E_{(-10)} = f_1(1 - f_2^+ f_2),\ P_4 = E_{(0-1)} = f_2(1 - f_1^+ f_1)\}$$

REFERENCES

1. B. Gruber, R. Lenczewski, M. Lorente, On induced scalar products and unitarization, *J. Math Phys.* 31:587 (1990).

2. H.D. Doebner, B. Gruber, M. Lorente, Boson operator realizations of $su(2)$ and $su(1,1)$ and unitarization, *J. Math Phys.* 30:594 (1989).

3. B. Gruber, M. Ramek, "Symmetrization of $su(4)$ Quantum States", SIUC Internal Report (1988).

4. E. Dynkin, The maximal subgroups of the classical groups, *Am. Math. Soc. Transl. (2)* 6:111 (1965); Semisimple subalgebras of semisimple Lie algebras, *Am. Math. Soc. Transl. (2)* 6:245 (1965).

5. M. Lorente, B. Gruber, Classification of semisimple subalgebras of simple Lie algebras, *J. Math. Phys.* 13:1639 (1972).

6. B. Gruber, M.T. Samuel, Semisimple subalgebras of semisimple Lie algebras: the algebra $A_5(su(6))$ as a physically relevant example, *in*: "Group Theory and its Applications", Vol III, E.M. Loebl, ed., Academic Press, New York (1975).

7. B. Gruber, On the correlation between representations and embeddings of simple Lie algebras, *Arab J. Phys.* 2:19 (1981).

THE PROPAGATOR METHOD TO SOLVE
THE FOKKER—PLANCK EQUATION

Jian-Zhong Gu and Yin-Sheng Ling

Department of Physics, Suzhou University
Suzhou, 215006, China

I . SOLVING THE EVOLUTION EQUATION BY LIE ALGEBRA METHOD

In a lot of physical problems one should solve the evolution equation

$$\begin{cases} \dfrac{dU(t)}{dt} = A(t)U(t), & (1.a) \\ U(0) = 1 \quad . & (1.b) \end{cases}$$

If the $A(t)$ belongs to a n dimensional Lie algebra G

$$A(t) = \sum_{i=1}^{m} a_i(t) T_i, \qquad (m \leqslant n) \tag{2}$$

in which $\{T_1, T_2, ...T_n\}$ are the generators of the Lie algebra G, then, the evolution operator $U(t)$ can be expressed in the following way[1]

$$U(t) = \prod_{j=1}^{n} exp\{g_j(t)\ T_j\} \quad , \quad (g_j(0) = 0, \ j = 1-n) \tag{3}$$

Substituting (3) into (1), leads to

$$\sum_{j=1}^{n} \dot{g}_j(t) \prod_{k=1}^{j-1} exp\{g_k(t)\ ad\ T_k\}\ T_j = \sum_{i=1}^{n} a_i(t)\ T_i. \tag{4}$$

Comparing the coefficients of the two sides before the generators T_i, we can obtain a set of differential equations which is satisfied by the unknown functions $g_j(t)$.

II . THE SOLUTION OF FOKKER—PLANCK EQUATION WITH LINEAR FORCE

The Fokker—Planck equation (FPE) with the linear force

$$\frac{\partial P(x,t)}{\partial t} = \frac{\partial}{\partial x}[(ax+b)P(x,t) + \frac{\partial P(x,t)}{\partial x}] \tag{5}$$

can be solved by the Lie algebra method.

Symmetries in Science VII, Edited by B. Gruber
and T. Otsuka, Plenum Press, New York, 1994

Suppose

$$P(x, t) = U(t) P(x, 0) .$$

(6)

where the evolution operator $U(t)$ satisfies

$$\begin{cases} \dfrac{dU(t)}{dt} = \left\{ a + ax \dfrac{d}{dx} + b \dfrac{d}{dx} + \dfrac{d^2}{dx^2} \right\} U(t), \\[2mm] U(0) = 1 . \end{cases}$$

(7.a)

(7.b)

The operator $U(t)$ can be represented in the factorized form

$$U(t) = exp[\alpha(t)] exp[\beta(t) x \frac{d}{dx}] exp[\gamma(t) \frac{d}{dx}] exp[\delta(t) \frac{d^2}{dx^2}].$$

(8)

The unknown functions $\alpha(t), \beta(t), \gamma(t), \delta(t)$ obey

$$\begin{cases} \dot{\alpha} = a, \\ \dot{\beta} = a, \\ \dot{\gamma} exp(-\beta) = b, \\ \dot{\delta} exp(-2\beta) = 1, \\ \alpha(0) = \beta(0) = \gamma(0) = \delta(0) = 0 . \end{cases}$$

(9.a)

(9.b)

(9.c)

(9.d)

(9.e)

Equation (6) can be reformed as the integral form

$$P(x, t) = \int_{-\infty}^{+\infty} U(t) \delta(x - \xi) P(\xi, 0) d\xi$$

$$= \int_{-\infty}^{+\infty} K(x, t; \xi, 0) P(\xi, 0) d\xi.$$

(10)

here $K(x, t, \xi, 0)$ is a propagator:

$$K(x, t; \xi, 0) = U(t) \delta(x - \xi).$$

(11)

Solving equation (9), we can find out the explicit expressions of $\alpha(t), \beta(t), \gamma(t)$ and $\delta(t)$. The analytical form K can be written out as[2]

$$K(x, t; \xi, 0) = \sqrt{\frac{a}{2\pi[1 - exp(-2at)]}}$$

$$exp\left\{ -\frac{a[(x + \frac{b}{a}(1 - exp(-at)) - \xi exp(-at)]^2}{2[1 - exp(-2at)]} \right\}.$$

(12)

III. THE SOLUTION OF THE FPE WITH THE GENERAL POTENTIAL

In general case, the FPE

$$\frac{\partial P(x, t)}{\partial t} = \frac{\partial}{\partial x} [V'(x) + \frac{\partial}{\partial x}] P(x, t)$$

(13)

has no analytical solution.

We can write the distribution function $P(x, t)$ in the following form $(t_n = t, x_n = x, t_0 = 0)$[3]

$$P(x_n, t_n) = U(t_n - t_{n-1}) U(t_{n-1} - t_{n-2}) .... U(t_1 - t_0) P(x, t_0)$$

$$= \int_{-\infty}^{+\infty} dx_{n-1} K(x_n, t_n; x_{n-1}, t_{n-1}) \int_{-\infty}^{+\infty} dx_{n-2} K(x_{n-1}, t_{n-1}; x_{n-2}, t_{n-2})$$

$$.... \int_{-\infty}^{+\infty} dx_0 K(x_1, t_1; x_0, t_0) P(x_0, t_0) .$$

(14)

Here the propagator $K(x, t; x_0, t_0)$ obeies the same equation as (13) with an initial condition

$$\underset{t \to t_0}{Lim} K(x, t; x_0, t_0) = \delta(x - x_0) \ . \tag{15}$$

If the interval of time $[t_0, t]$ is small, the distribution of the propagator $K(x, t; x_0, t_0)$ lies in the narrow region in the phase space near the source point x_0. Then, the potential $V(x)$ can be approximated by:

$$V(x) \approx V(x_0) + V'(x_0)(x - x_0) + \frac{V''(x_0)}{2!}(x - x_0)^2$$

$$= \frac{V''(x_0)}{2!}x^2 + [V'(x_0) - x_0 V''(x_0)]x + V(x_0) - x_0 V'(x_0) + \frac{x_0^2}{2!}V''(x_0). \tag{16}$$

In this way, $K(x, t; x_0, t_0)$ can be written out by the formulae (12), we can use the simple recurrence relation

$$P(x_{j+1}, t_{j+1}) = \int_{-\infty}^{+\infty} K(x_{j+1}, t_{j+1}; x_j, t_j) \, P(x_j, t_j) dx_j \tag{17}$$

to obtain the distribution function at arbitrary time t supposing the time interval $|t_{j+1} - t_j| \ll 1$.

References

[1] Fritz wolf. J. Math. Phys. 29 (2) : 305 (1988).

[2] Yin—Sheng Ling. Physica Energiae Fortis and Physica
    Nuclearis. 8 : 743 (1992).

[3] H. Hofmann. Nucl. Phys. A 394 : 477 (1983).

SUPERDEFORMATION AND
INTERACTING BOSON MODEL

Michio Honma and Takaharu Otsuka

Department of Physics
University of Tokyo
Hongo 7-3-1, Bunkyo-ku, Tokyo, 113, Japan

INTRODUCTION

The superdeformation[1, 2] is one extremity of nuclear quadrupole deformation with the axis ratio 2:1:1 (for ^{152}Dy region), 5:3:3 (for ^{192}Hg region) *etc.* Since the first experimental discovery of the beautiful superdeformed rotational band in 1986, the structure of such significantly deformed nucleus has been one of the most important and intriguing problems in the field of nuclear structure both theoretically and experimentally. For normal quadrupole deformation, the Interacting Boson Model[3] (IBM) has been successfully established as one of the most powerful frameworks for the study of nuclear quadrupole collective motion. Thus it is quite interesting to study the superdeformation from the viewpoint of IBM.

The characteristic idea of the IBM is the correspondence between bosons and collective nucleon pairs with definite angular momentum[4, 5]. Based on this spirit, the standard Nilsson wave function is analyzed in terms of collective nucleon pairs. A suitable boson picture for the superdeformation is considered in the first half of this article. In the latter half, a simple phenomenological hamiltonian which can describe both superdeformed states and low-lying states simultaneously is searched. The angular momentum is treated exactly throughout the boson calculation. We investigate especially the spin dependence in the superdeformed states. The stabilization mechanism of superdeformed states is discussed from the viewpoint of IBM.

BOSON PICTURE FOR SUPERDEFORMED STATES

In order to investigate the relation between the superdeformation and the IBM, we solve the standard Nilsson potential with a large deformation parameter which corresponds to the superdeformation. As an example, we take ^{152}Dy (Z=66, N=86). It is suggested experimentally that this nucleus shows the superdeformation with the axis ratio 2:1:1, which is equivalent to the deformation parameter δ=0.50 . Because of this large deformation, it is insufficient to take only one active major shell by considering the correction from the core-polarization effect.

Thus we first consider the suitable model space for the description of the superdeformation. The Nilsson wave function is obtained by putting nucleons in the

Symmetries in Science VII, Edited by B. Gruber
and T. Otsuka, Plenum Press, New York, 1994

Nilsson orbits from the bottom. One Nilsson orbit can be expanded as a linear combination of many spherical harmonic oscillator orbits and the square of the expansion coefficients give the occupation probability of each spherical orbit. We expand all occupied Nilsson orbits and sum up the occupation probabilities which belong to the same spherical orbits, then we obtain the total occupation probability in terms of spherical harmonic oscillator basis. Due to the strong quadrupole field, one Nilsson orbit spreads over many spherical orbits. Thus the orbits with very high single particle energy show some finite occupation probability and the occupation of lower orbits becomes incomplete. Nevertheless by putting many nucleons, several lowest spherical orbits are occupied almost completely. We can consider these almost completely occupied orbits as the inert spherical core for the superdeformed states.

The occupation probability of each spherical harmonic oscillator orbit is shown in Table 1. The case of $\delta=0.25$ which corresponds to the normal deformation is also listed for comparison. As for proton orbits, in the case of $\delta=0.25$, the occupation is almost complete up to $Z=50$ magic number, while the occupation probability is almost vanished for the orbits above $Z=82$ magic number. This results support the validity of usual treatment with a $Z=50$ inert core and one major valence shell. On the other hand, in the case of $\delta=0.50$, the orbits are completely occupied only up to $Z=40$, while above $Z=40$ the occupation probability drops suddenly. It does not, however, become zero but remains about 10% over very many orbits up to very high energy. It is impossible to neglect the contribution from these high energy orbits. Thus it is reasonable to take a $Z=40$ spherical inert core and include quite many orbits above $Z=40$ as active valence orbits. In the same way, it turns out that the spherical inert core for neutron orbits is $N=50$.

Table 1. Occupation probabilities of spherical orbits for ^{152}Dy. The case of normal deformation ($\delta=0.25$) and of superdeformation ($\delta=0.50$) are compared.

proton orbit	occup. prob. [%]			neutron orbit	occup. prob. [%]	
	$\delta=0.50$	$\delta=0.25$			$\delta=0.50$	$\delta=0.25$
$0s_{1/2}$	100	100	2	$0s_{1/2}$	100	100
$0p_{3/2}$	100	100		$0p_{3/2}$	100	100
$0p_{1/2}$	100	100	8	$0p_{1/2}$	100	100
$0d_{5/2}$	99	100		$0d_{5/2}$	99	100
$1s_{1/2}$	99	100		$1s_{1/2}$	99	100
$0d_{3/2}$	99	100	20	$0d_{3/2}$	99	100
$0f_{7/2}$	94	100	28	$0f_{7/2}$	98	100
$1p_{3/2}$	91	98		$1p_{3/2}$	96	99
$0f_{5/2}$	90	98		$0f_{5/2}$	95	99
$1p_{1/2}$	93	98		$1p_{1/2}$	95	99
$0g_{9/2}$	60	94	50	$0g_{9/2}$	88	97
$0g_{7/2}$	30	66		$1d_{5/2}$	55	97
$1d_{5/2}$	25	56		$0g_{7/2}$	58	97
$0h_{11/2}$	24	42		$2s_{1/2}$	36	98
$1d_{3/2}$	17	16		$1d_{3/2}$	42	98
$2s_{1/2}$	25	21	82	$0h_{11/2}$	40	73
$0h_{9/2}$	11	2		$1f_{7/2}$	20	20
$0i_{13/2}$	16	3		$0h_{9/2}$	19	14
$1f_{7/2}$	11	6		$0i_{13/2}$	40	12
$1f_{5/2}$	9	1		$2p_{3/2}$	18	16
$2p_{3/2}$	6	1		$1f_{5/2}$	16	10
$0j_{15/2}$	10	3		$2p_{1/2}$	10	7
$2p_{1/2}$	7	0		$1g_{9/2}$	11	4
$0i_{11/2}$	6	2		$0j_{15/2}$	13	4
$1g_{9/2}$	6	1		$0i_{11/2}$	9	2

From the viewpoint of IBM the number of bosons is determined by the half of the number of valence nucleons. Because of the small inert core the boson number increases significantly in comparison with the usual IBM. In fact in the present case the boson number becomes (66–50)/2+(86–82)/2=10 for the usual IBM. On the other hand (66–40)/2 +(86–50)/2=31 bosons are needed for the description of the superdeformation. These bosons (superdeformed bosons) are defined within the extended valence space and carry the collectivity of many major shells. We propose in the following the extended IBM (super-IBM) containing these superdeformed bosons.

Next we study the structure of the valence wave function. The Nilsson (+number conserving BCS) wave function can be expressed as the condensation of the coherent Cooper-pairs in the Nilsson potential[6, 7]

$$|\Phi_{Nilsson}\rangle \propto (\Lambda_\pi{}^\dagger)^{N\pi} (\Lambda_\nu{}^\dagger)^{N\nu} | \text{ spherical inert core} \rangle . \tag{1}$$

In this expression $\Lambda_\pi{}^\dagger$ $(\Lambda_\nu{}^\dagger)$ denotes the creation operator of the Cooper-pair of protons (neutrons) and N_π (N_ν) means the half of the valence proton (neutron) number. These Λ-pairs can be decomposed into a linear combination of collective nucleon pairs with good angular momenta

$$\Lambda^\dagger = x_0 S^\dagger + x_2 D_0{}^\dagger + x_4 G_0{}^\dagger + \cdots\cdots, \tag{2}$$

where $S^\dagger, D_0{}^\dagger, G_0{}^\dagger, \cdots$ denote the collective nucleon pairs with spin-parity J^π=0+, 2+, 4+, $\cdots$ and x_J $(J$=0, 2, 4, $\cdots)$ are amplitudes.

The probability of each pair in the Λ-pairs is given by the square of each amplitude and listed in Table 2 for the case of δ=0.25 and δ=0.50 . It is well known that in the case of δ=0.25 the dominant components are the S-pair and the D-pair[8]. In fact the total probability of these two components is more than about 85%. In the case of δ=0.50, although the probability of G-pair increases, the total probability of the S-pair and the D-pair is about 70% and these pairs still dominate the Λ-pair. This result implies that we can take sd-boson model as a starting point, except for the description of extremely high spin states. It should be noted that the ratio of the S-pair to the other pairs is quite similar to that of s boson to the other bosons in the SU(3)-limit of IBM (see Table 2). This fact suggests the validity of taking the SU(3)-limit for the description of superdeformation. This assumption of SU(3)-limit is reasonable because this limit corresponds to the rotational limit in the usual IBM.

We investigate further the structure of collective nucleon pairs such as $S, D, G, \cdots$ in detail. These pairs are written as

$$S^\dagger = \sum_{i\le j} \alpha_{ij} A^\dagger_{ij00} , \quad D_0{}^\dagger = \sum_{i\le j} \beta_{ij} A^\dagger_{ij20} , \quad G_0{}^\dagger = \sum_{i\le j} \gamma_{ij} A^\dagger_{ij40} , \quad \cdots, \tag{3}$$

where

$$A^\dagger_{ijJM} = \frac{[c_i^\dagger c_j^\dagger]_M^{(J)}}{\sqrt{1+\delta_{ij}}} \tag{4}$$

denotes the creation operator of the pair of nucleons in the two spherical orbits i and j coupled to the total angular momentum J and magnetic quantum number M. The quantities $\alpha_{ij}, \beta_{ij}, \gamma_{ij}, \cdots$ are the amplitudes of S-pair, D-pair, G-pair, $\cdots$, respectively. The square of each amplitude means the probability of the corresponding pair. Fig. 1 shows this probability in the S-pair, D-pair and G-pair, respectively, of the ^{152}Dy proton orbit for the case of δ=0.25 and δ=0.50 .

In the case of the S-pair, there is no remarkable difference between the cases of δ=0.25 and δ=0.50 . The main constituent orbits are those within the normal valence shell ($0g_{7/2}$, $1d_{5/2}$, $1d_{3/2}$, $2s_{1/2}$, $0h_{11/2}$) and those below it (only $0g_{9/2}$ in this case). One difference can be seen in the existence of small contribution from the higher orbits (1~3$\hbar\omega$) in the case of δ=0.50 .

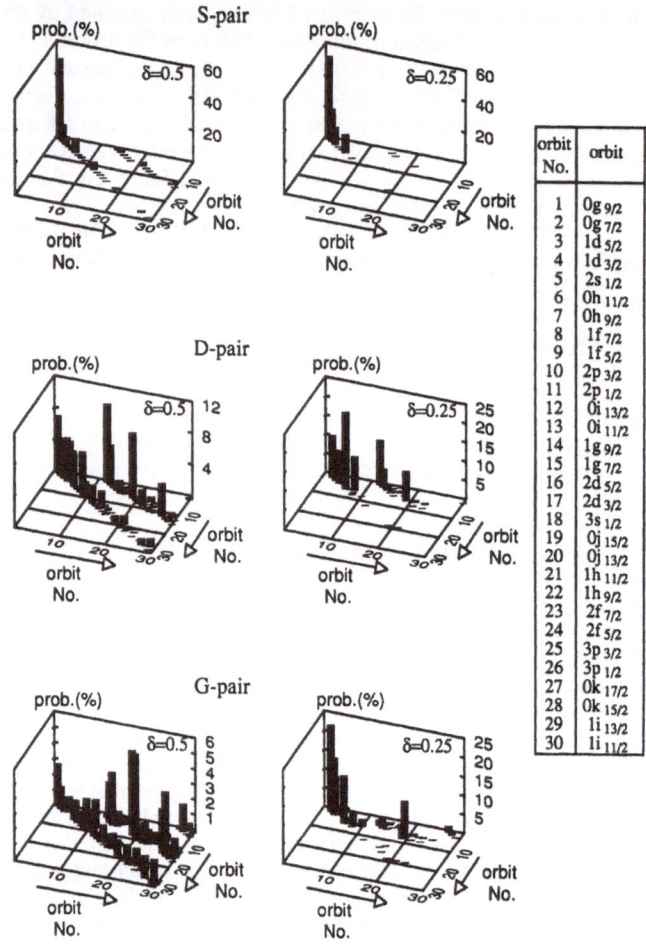

orbit No.	orbit
1	$0g_{9/2}$
2	$0g_{7/2}$
3	$1d_{5/2}$
4	$1d_{3/2}$
5	$2s_{1/2}$
6	$0h_{11/2}$
7	$0h_{9/2}$
8	$1f_{7/2}$
9	$1f_{5/2}$
10	$2p_{3/2}$
11	$2p_{1/2}$
12	$0i_{13/2}$
13	$0i_{11/2}$
14	$1g_{9/2}$
15	$1g_{7/2}$
16	$2d_{5/2}$
17	$2d_{3/2}$
18	$3s_{1/2}$
19	$0j_{15/2}$
20	$0j_{13/2}$
21	$1h_{11/2}$
22	$1h_{9/2}$
23	$2f_{7/2}$
24	$2f_{5/2}$
25	$3p_{3/2}$
26	$3p_{1/2}$
27	$0k_{17/2}$
28	$0k_{15/2}$
29	$1i_{13/2}$
30	$1i_{11/2}$

Figure 1. Probabilities of S, D and G component in the collective nucleon pair is shown (see eq. (2)). The orbit number is listed in the table.

On the other hand, in the structure of the D-pair significant differences appear between the cases of δ=0.25 and δ=0.50 . In the case of δ=0.25 the orbits within the usual valence shell mainly contribute to the construction of the collective D-pair as in the case of the S-pair. Although the other orbits contribute considerably, they still can be treated as secondary effects. However in the case of δ=0.50 the contribution from higher orbits becomes substantial and appears to be comparable to that from the orbits within the usual valence shell. Another difference can be seen in the peaks which appear in the case of δ=0.50 in the off-diagonal line in addition to the diagonal components. These off-diagonal components correspond to the nucleon pairs composed of two orbits which are $2\hbar\omega$ apart. These $2\hbar\omega$-components would not appear in usual treatment which includes only one major shell. Although such components can be seen also in the case of δ=0.25, their contribution is only less than half of the usual $0\hbar\omega$-components and can be regarded as a correction term. However in the case of δ=0.50 the contribution from the $2\hbar\omega$-components reaches up

Table 2. Probability of each angular-momentum component in the Cooper pair is listed for the cases of normal deformation (δ=0.25) and superdeformation (δ=0.50) . The probability of each boson included in the intrinsic boson of IBM in the SU(3) limit is also shown.

pair/	δ=0.25		δ=0.50		IBM-SU(3)	
boson	proton	neutron	proton	neutron	sd	sdg
S	64	71	28	29	33	20
D	27	16	40	41	67	57
G	6	7	16	14	–	23

to the same order of magnitude as that from the $0\hbar\omega$-components. Therefore these components should be treated explicitly by extending the active valence space. In the case of the G-pair the difference between δ=0.25 and δ=0.50 becomes more significant. In the case of δ=0.50 dominant constituent orbits are those above the usual valence shell. The $2\hbar\omega$-components become more dominant than the $0\hbar\omega$-components and the $4\hbar\omega$-components appear to be sizable.

The results of detailed analysis of collective pairs suggest that, in the description of the superdeformation, it is insufficient to include usual s and d bosons defined within the usual valence shell and that the boson-image of the high-j, $2\hbar\omega$-components should be included explicitly.

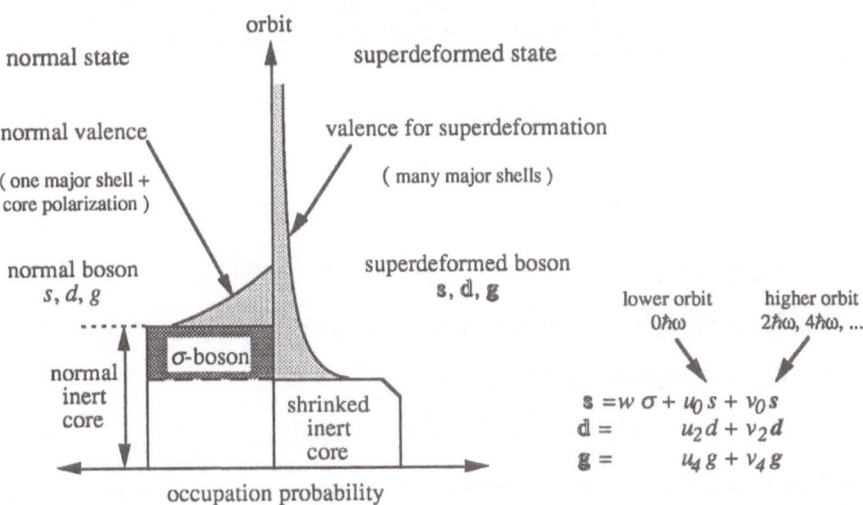

Figure 2. A boson picture for the superdeformation is shown schematically.

The boson picture for the superdeformation is shown schematically in Fig. 2. In order to describe large deformation, spherical inert core is taken to be smaller than usual, and significantly wider valence space is considered. As a result the number of superdeformed bosons becomes, in general, about three times larger than that of normal bosons. The superdeformed bosons carry the collectivity of many major shells. The superdeformed boson is the image of the collective nucleon pair which contains quite many highly excited pair components such as $2\hbar\omega$ pair and $4\hbar\omega$ pair in the sense of the spherical harmonic oscillator potential. The underlying structure is quite different between superdeformed bosons and normal bosons. The superdeformed bosons can be written as the linear combination of normal bosons (denoted s, d, g, $\cdots$) and their orthogonal components (denoted $\bar{s}$, $\bar{d}$, $\bar{g}$, $\cdots$). These orthogonal components are mainly the boson-images of highly excited nucleon pairs mentioned above. In order to treat the relation between normal states and superdeformed states, we compensate the difference of boson number between these two cases by introducing a new type of boson with $J^\pi=0^+$, denoted σ. The σ bosons describe nucleon pairs constituting the upper part of the normal inert core which become active in the case of the superdeformation. The number of σ bosons, N_0, is determined by $N_0=N_{super}-N_{normal}$, where N_{super} (N_{normal}) denotes the number of superdeformed (normal) bosons.

PHENOMENOLOGICAL HAMILTONIAN FOR SUPER-IBM

We have discussed the relation between superdeformed states and normal deformed states based on the Nilsson wave function, and obtained the simple boson picture by considering the correspondence between bosons and collective nucleon pairs. Thus we try to describe the superdeformed states in relation to the normal low-lying states based on this boson picture. As a first step of our super-IBM approach, we introduce a phenomenological hamiltonian which includes the boson single particle energy and a quadrupole-quadrupole interaction,

$$H=\varepsilon_\sigma n_\sigma+\varepsilon_s n_s+\varepsilon_d n_d+\varepsilon_g n_g+\varepsilon_{\bar{s}} n_{\bar{s}}+\varepsilon_{\bar{d}} n_{\bar{d}}+\varepsilon_{\bar{g}} n_{\bar{g}}+\kappa Q\cdot Q. \tag{5}$$

This hamiltonian can be understood as the natural extension of the usual IBM hamiltonian. In this expression $n_b=b^\dagger\cdot\bar{b}$ ($b=\sigma, s, d, g, \bar{s}, \bar{d}, \bar{g}$) denotes the boson number operator of each kind, and Q denotes the boson quadrupole operator, which includes four terms:

$$Q=e_1 T_1+e_2 T_2+e_3 T_3+e_4 T_4 \ , \tag{6}$$

where e_i ($i=1\sim4$) are boson charges and

$$T_{1\mu}=\sigma^\dagger \bar{d}_\mu+d_\mu^\dagger \sigma, \tag{7}$$

$$T_{2\mu}=s^\dagger \bar{d}_\mu+d_\mu^\dagger s+\chi_2[d^\dagger \bar{d}]^{(2)}_\mu+\lambda_2[d^\dagger \bar{g}+g^\dagger \bar{d}]^{(2)}_\mu+\omega_2[g^\dagger \bar{g}]^{(2)}_\mu \ , \tag{8}$$

$$T_{3\mu}=\bar{s}^\dagger \bar{d}_\mu+\bar{d}_\mu^\dagger \bar{s}+\chi_3[\bar{d}^\dagger \bar{\bar{d}}]^{(2)}_\mu+\lambda_3[\bar{d}^\dagger \bar{\bar{g}}+\bar{g}^\dagger \bar{\bar{d}}]^{(2)}_\mu+\omega_3[\bar{g}^\dagger \bar{\bar{g}}]^{(2)}_\mu \ , \tag{9}$$

$$T_{4\mu}=s^\dagger \bar{\bar{d}}_\mu+\bar{d}_\mu^\dagger s+\bar{s}^\dagger \bar{d}_\mu+d_\mu^\dagger \bar{s}+\chi_4[d^\dagger \bar{\bar{d}}+\bar{d}^\dagger \bar{d}]^{(2)}_\mu$$

$$+\lambda_4[d^\dagger \bar{\bar{g}}+\bar{g}^\dagger \bar{d}+\bar{d}^\dagger \bar{g}+g^\dagger \bar{\bar{d}}]^{(2)}_\mu+\omega_4[g^\dagger \bar{\bar{g}}+\bar{g}^\dagger \bar{g}]^{(2)}_\mu \ . \tag{10}$$

The term T_1 includes σ bosons. The term T_2 is composed of normal bosons only (s, d, g) and corresponds to the quadrupole operator in the usual sdg-IBM. The term T_3 is its counterparts for orthogonal components ($\bar{s}$, $\bar{d}$, $\bar{g}$). The term T_4 contains both components. In the term T_1, operators $\sigma^\dagger d_\mu$ and $d_\mu^\dagger \sigma$ are not included because σ and d are the boson-images of nucleon pairs which belong to different major shells and the matrix elements of one body operators should be vanished.

In the usual IBM, the parameters included in the hamiltonian can be determined by either completely phenomenological fit to the experimental data or more microscopic mapping techniques such as, for example, OAI mapping[9]. For the former method a large amount of systematic experimental data are necessary, and for the latter microscopic effective interaction and suitable wave function are necessary. At the starting point of super-IBM we cannot resort to either method because there are no abundant systematic data or realistic wave functions. Our aim is to investigate the possibility of unified description of both the normal states and the superdeformed states based on the viewpoint of IBM. Thus we try to make quite rough estimation according to the sprit of IBM by using the Nilsson wave function. Since this estimation may not be very realistic, we necessarily carry out the phenomenological fit to the experimental data within the extent that the consistency to the estimation is not violated strongly.

As mentioned in the previous section we can obtain collective nucleon pairs S, D, G, ··· from the Nilsson wave function. By decomposing these pairs according to the orbits of which each pair component are comprised, the fermion pair counterpart of each boson can be obtained. For example the fermion counterpart of d boson, denoted d_F, is obtained from the D-pair by taking all components made by the orbits in the normal valence shell, and the rest of the D- pair are regarded as d_F, the counterpart of d boson. The matrix elements of the quadrupole operator with respect to these single pair states are listed in Table 3 for two typical superdeformed nuclei. It can be seen that the value of $\langle 0| d_{F0}\, r^2 Y_{20}\, s_F{}^\dagger |0\rangle$ is extremely large in comparison with the other matrix elements. Thus it is reasonable to take larger value for e_3 than the other boson charges. As for the χ_i, λ_i and ω_i parameters in the quadrupole operators T_i ($i=2, 3, 4$), we use the values of the SU(3) limit in the sdg-IBM for i=3 and 4 ($\chi_i=-11\sqrt{10}/28$, $\lambda_i=9/7$, $\omega_i=-3\sqrt{55}/14$), and take O(6) limit-like values for i=2, namely $\chi_2=\omega_2=0$ and $\lambda_2=1$. This choice is reasonably understood by considering the contribution of each term to the total energy. As we see in the following, T_3 is important for the description of the superdeformed rotational band. This fact suggests that the values of the SU(3) limit is suitable for T_3. To the normal states T_2 mainly contributes and the character of the U(5) or O(6) limit of the usual IBM appears in the energy spectra around ground states of typical superdeformed nuclei. The values of parameters adopted in the following calculation are shown in Table 4. Since we need the excitation energy only, the boson single particle energies are measured from ε_s, namely we take $\varepsilon_s=0$. The core excitation effects are included in the negative value of ε_σ. The energy loss of $2\hbar\omega$ or $4\hbar\omega$ excitations are expressed by large single boson energies for s, d and g bosons.

The hamiltonian is solved by variation after projection (VAP) method. As a VAP-trial-function, we take a linear combination of many coherent states with different active boson number,

$$|\Phi; \beta_J, u_J, C_k \rangle = \sum_{k=N_{normal}}^{N_{super}} C_k\, (\sigma^\dagger)^{N-k}(s^\dagger + \beta_2 d^\dagger + \beta_4 g^\dagger)^k |0\rangle , \tag{11}$$

where

$$s' = u_0 s + v_0 s , \qquad d' = u_2 d + v_2 d , \qquad g' = u_4 g + v_4 g , \tag{12}$$

$$u_0^2 + v_0^2 = 1 , \qquad u_2^2 + v_2^2 = 1 , \qquad u_4^2 + v_4^2 = 1 . \tag{13}$$

The active boson number k runs from normal boson number N_{normal} up to superdeformed boson number N_{super}. The primed bosons s', d' and g' can become both normal bosons ($u=1$) and superdeformed bosons ($u\neq1$) by varying the values of parameters u_J ($J=0, 2, 4$). Thus this wave function can describe the change of the structure from normal deformed states to superdeformed states. The variational parameters are β_2, β_4, u_0, u_2, u_4 and C_k ($k=N_{normal} \sim N_{super}$). In order to describe the spin dependence of the structure, the angular momentum projection is carried out before each variational steps.

Table 3. Matrix elements of the quadrupole operator for these single pair states are listed. The model space for ^{194}Hg are mentioned later.

matrix element	^{152}Dy (δ=0.5)		^{194}Hg (δ=0.4)	
[efm^2]	neutron	proton	neutron	proton
$\langle 0 | d_{F0} \, r^2 Y_{20} \, \sigma_F{}^\dagger | 0 \rangle$	10.8	7.1	9.6	7.6
$\langle 0 | d_{F0} \, r^2 Y_{20} \, s_F{}^\dagger | 0 \rangle$	13.5	11.1	13.5	10.6
$\langle 0 | d_{F0} \, r^2 Y_{20} \, s_F{}^\dagger | 0 \rangle$	5.4	6.1	7.5	6.6
$\langle 0 | d_{F0} \, r^2 Y_{20} \, s_F{}^\dagger | 0 \rangle$	2.5	2.1	1.6	0.8
$\langle 0 | d_{F0} \, r^2 Y_{20} \, s_F{}^\dagger | 0 \rangle$	24.1	24.0	26.8	22.5

Table 4. Parameters used in the calculation are shown.

parameter	value	parameter	value
ε_σ [MeV]	−5.0	e_1 [efm^2]	2.5
ε_d	0.8	e_2	5.0
ε_g	2.5	e_3	15.0
ε_s	6.5	e_4	2.5
ε_d	6.5		
ε_σ	6.5	$\kappa(e_2)^2$ [MeV]	−0.015

The calculation is carried out for ^{194}Hg. In this case we take Z=50, N=82 inert core for superdeformed states, and assume Z=64, N=100 subshells for normal states. Thus the boson number is determined to be N_{super}=31 and N_{normal}=15, which leads to N_0=16. In the usual IBM this nucleus is described by 7 hole bosons. However in order to consider the relation between the normal low-lying states and the superdeformed states explicitly, the normal states should be treated also by particle bosons. For simplicity, we determine the value of β_4 by a simple function of β_2 which is examined in the preliminary calculation. The values of u_2 and u_4 are taken to be equal. The effect of highly excited pair components can be estimated even under such a constraint. Based on the result obtained in the Nilsson calculation, we take u_0 =1 which means that the s' boson is completely the same as the normal s boson. Thus the actual variational parameters become β_2, u_2 and C_k ($k=N_{normal} \sim N_{super}$).

The calculated energy is shown in Fig. 3 as a function of β_2 and u_2 for three cases with different angular momenta J=0, 20 and 40. In this case only C_k are varied for each value of β_2 and u_2 . Two separate energy minima can be seen for all three cases. One appears around $(u_2, \beta_2) \sim (1.0, 0.3)$ which corresponds to a normal state, and the other exists around $(0.4, 3.5)$ which can be understood as a superdeformed state. It should be noted that the superdeformed energy minimum becomes deeper as spin goes up, and the superdeformed state turns out to be yrast at J=40.

The structure of both minima is consistent with the analysis of the collective nucleon pair considered in the previous section. The normal states are constructed by almost completely normal bosons only, and the normal core is quite inert. In fact the expectation values of the number of orthogonal bosons s, d and g are less than 0.1 in total, and that of σ boson is 16.0 . On the other hand the superdeformed states contain both normal bosons and orthogonal bosons. The primed bosons d' and g' show the structure of "superdeformed bosons" shown in Fig. 2. The expectation value of each constituent boson is shown in Table 5 for the states in the second energy minimum. It can be seen that the orthogonal bosons d and g dominate the wave function (about 85% in total) and nearly 13 bosons are excited from the normal inert core. We can also see the stability of the second minimum. The structure is quite stable against the variation of the angular momentum, and the number of excited σ boson is almost constant.

The calculated excitation energy is plotted in Fig. 4 as a function of spin. The experimental value is also shown. In this calculation the band head energy of the superdeformed band becomes 5 MeV. Since the experimental band head energy is still not known, the same band head energy is assumed for the experimental band. It can be seen that the agreement is rather good for both low-lying states and superdeformed states. The large moment of inertia of the superdeformed states is well reproduced. The excitation energy of the superdeformed states is almost consistent with the suggestion from the experiment[10].

Two different mechanisms contribute constructively to the generation of the second (superdeformed) energy minimum. The first one is the increase of the active boson number, in other words, effect of the core excitation. The energy gain by the quadrupole-quadrupole

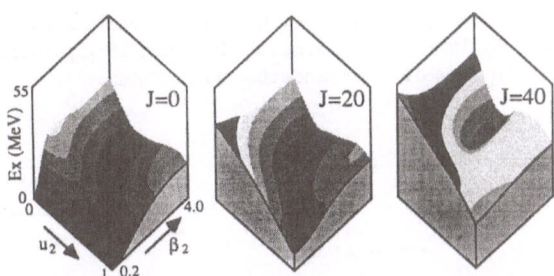

Figure 3. Excitation energy plotted as a function of u_2 and β_2. The energy minima appear at $(u_2, \beta_2)=(0.98, 0.39)$ and $(0.34, 3.13)$ for J=0, $(0.99, 0.40)$ and $(0.34, 3.14)$ for J=20, $(1.00, 0.17)$ and $(0.33, 3.15)$ for J=40. The first minimum corresponds to a normal state and the second one can be understood as a superdeformed state. The scales of axis in the middle and right figures are the same as those in the left.

Table 5. Expectation value of the number of each constituent boson for the states in the second energy minimum.

spin J	$\langle n_\sigma \rangle$	$\langle n_s \rangle$	$\langle n_s \rangle$	$\langle n_d \rangle$	$\langle n_d \rangle$	$\langle n_\varrho \rangle$	$\langle n_\varrho \rangle$
0	3.1	1.9	0	2.0	15.7	1.0	7.3
20	2.9	1.8	0	2.0	15.5	1.0	7.8
40	2.2	1.5	0	1.8	15.1	1.1	9.3

interaction is roughly proportional to the square of the active boson number. In this case, it is shown in table 5 that the active boson number is about 15 for the normal states and 28 for the superdeformed states. This difference gives rise to at least four times larger energy gain for the superdeformed states. The second reason is that the quadrupole matrix element is roughly three times larger for highly excited nucleon pairs ($2\hbar\omega$ pair, $4\hbar\omega$ pair, $\cdots$) than those in the usual valence shell ($0\hbar\omega$ pair). This leads also to larger quadrupole-quadrupole energy gain for the superdeformed states (see Table 3). Although these two processes require conside rable energy loss from single particle energies, the gained energy can exceed the loss under a certain condition of the active boson number and the mixing of highly excited pairs.

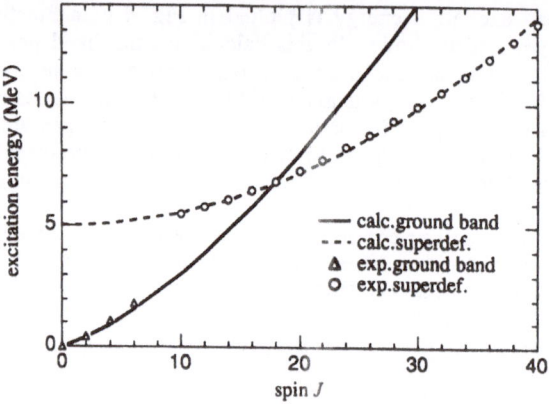

Figure 4. Calculated excitation energies of ground-state band and superdeformed band are shown by the solid and dashed lines, respectively. The experimental values are also plotted. The experimental band head energy of the superdeformed band is assumed to be the same as that of the calculation.

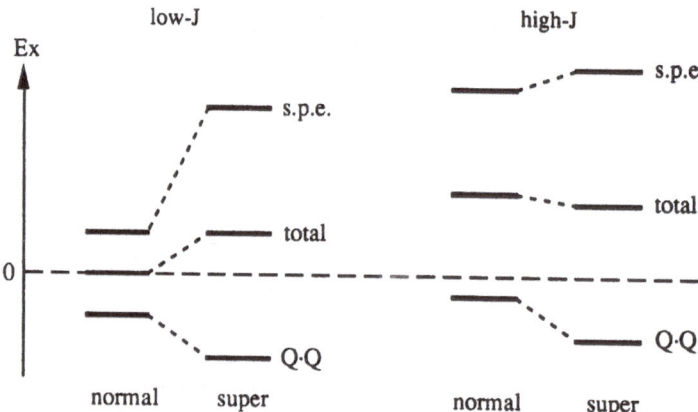

Figure 5. Mechanism of stabilization at high spin is shown schematically.

The spin dependence of the second energy minimum (stabilization at high spin) is due to the core excitation. This mechanism is schematically shown in Fig. 5 by highlighting differences between low-J and high-J states. The energy gain by the quadrupole-quadrupole interaction does not strongly depend on the angular momentum. On the other hand the boson single particle energy is significantly pushed up at high spin in the energy minimum of the normal regime. In the normal state energy minimum, as mentioned before, the core is quite inert. The angular momentum is gained only by transforming s bosons into higher (d, g) bosons with high single-particle energies. Thus the energy loss increases drastically as a function of the angular momentum within the normal deformation. In the second (superdeformed) energy minimum, the angular momentum is gained also by the core excitation (increase of the boson number). In this case many bosons contribute coherently to the construction of large angular momentum. The intrinsic structure is rather stable against the increase of the angular momentum (see Table 5), in other words, all members of the band are obtained from almost the same intrinsic state by the angular momentum projection. Since the intrinsic configuration is kept almost unchanged, the spin dependence of the single particle energy becomes rather gradual at the second energy minimum. As a result the superdeformed states can be stabilized at high spin.

SUMMARY

We summarize the present results as follows. It is needed to take the small inert core and the significantly wide valence space in order to describe the superdeformed states in terms of Interacting Bosons. As a result, the number of superdeformed bosons becomes about three times larger than that needed in the description of normal states. The collective nucleon pair obtained from the Nilsson wave function shows the S-and the D-pair dominance, and the amplitude ratios of the S-pair to the other pairs suggest the validity of SU(3)-limit of IBM as a starting point. In the S-pair the $0\hbar\omega$ pair components are dominant, while in the D-pair the contribution from the $2\hbar\omega$ pair component appear to be comparatively important This fact implies the necessity of introducing the boson-image of the $2\hbar\omega$ pair component. Based on these basic analysis, a phenomenological hamiltonian for the super-IBM is proposed which consists of the boson single particle energy and the quadrupole-quadrupole interaction. By taking into account the effects of the core excitation and the large quadrupole matrix elements for highly excited nucleon pairs, both the superdeformed rotational band and low lying states can be well reproduced simultaneously.

REFERENCES

1. B.R.Mottelson, in *Proc.Sympo.in Honor of Akito Arima: Nuclear Physics of the 1990's*, ed. by D.H.Feng, J.N.Ginocchio, T.Otsuka and D.D.Strottman, Nucl.Phys.**A522** (1991) 1c
2. P.Twin, in *Proc.Sympo.in Honor of Akito Arima: Nuclear Physics of the 1990's*, ed. by D.H.Feng, J.N.Ginocchio, T.Otsuka and D.D.Strottman, Nucl.Phys.**A522** (1991) 13c
3. F.Iachello and A.Arima, The Interacting Boson Model (Cambridge U.P., Cambridge, 1987)
4. A.Arima, T.Otsuka, F.Iachello and I.Talmi, Phys.Lett.**66B** (1977) 205
5. T.Otsuka, A.Arima, F.Iachello and I.Talmi, Phys.Lett.**76B** (1978) 139
6. T.Otsuka and N.Yoshinaga, Phys.Lett.**B168** (1986) 1
7. T.Otsuka and M.Honma, Phys.Lett.**B268**(1991) 305
8. T.Otsuka, A.Arima and N.Yoshinaga, Phys.Rev.Lett.**48** (1982) 387
9. T.Otsuka, A.Arima and F.Iachello, Nucl.Phys.**A309** (1978) 1
10. E.A.Henly, *et al.*, Lawrence Livermore National Laboratory preprint (1990)

FINITE AND INFINITE SYMMETRY IN (2+1)-DIMENSIONAL FIELD THEORY*

R. Jackiw

Center for Theoretical Physics
Laboratory for Nuclear Science
and Department of Physics
Massachusetts Institute of Technology
Cambridge, Massachusetts 02139   U.S.A.

and

So-Young Pi

Physics Department
Boston University
350 Commonwealth Avenue
Boston, Massachusetts 02215 U.S.A.

I. INTRODUCTION

These days, Franco Iachello is *the* eminent practitioner applying classical and finite groups to physics. In this he is following a Yale tradition, established by the late Feza Gursey, and he has succeeded Gursey in the Gibbs chair; Gursey in turn, had Pauli as a mentor. Iachello's striking achievement has been to find an actual realization of arcane supersymmetry within mundane adjacent even-odd nuclei. Thus far this is the only *physical* use of supersymmetry, and its fans surely must be surprised at the venue.

Here we describe the role of $SO(2,1)$ conformal symmetry in non-relativistic Chern–Simons theory: how it acts, how it controls the nature of solutions, how it expands to an infinite group on the manifold of static solutions thereby rendering the static problem completely integrable. Since Iachello has also used the $SO(2,1)$ group in various contexts, this essay is presented to him on the occasion of his fiftieth birthday.

---

* This work is supported in part by funds provided by the U. S. Department of Energy (D.O.E.) under contract #DE-AC02-76ER03069 (RJ), and #DE-AC02-89ER40509 (S-YP).

We shall discuss finite- and infinite-dimensional conformal symmetries of *field theories with non-relativistic kinematics*. Such field theories also describe the second quantization of *non-relativistic particle mechanics*. Particle mechanics, with its second order in time dynamics, has the structure of a *relativistic field theory* in one time and zero space dimensions, and a relativistic field theory in any dimension can enjoy conformal symmetry. Thus there are family relationships between the conformal symmetries of non-relativistic field theory, non-relativistic particle mechanics and relativistic field theory, and our first task is to describe these interrelations.

A conformal transformation in $(D + 1)$-dimensional relativistic field theory changes the *independent* variables, *viz.* the space-time coordinates x^μ of the fields (fields are *dependent* variables), and infinitesimally reads

$$\delta_f x^\mu = -f^\mu(x) \tag{1.1}$$

where f^μ is a *conformal Killing vector*, *i.e.* f^μ satisfies the *conformal Killing equation*.

$$\partial_\mu f_\nu + \partial_\nu f_\mu = \frac{2}{D+1} g_{\mu\nu} \partial_\alpha f^\alpha \tag{1.2}$$

Here $g_{\mu\nu}$ is the Minkowski metric tensor with signature $(1, -1, -1, \ldots)$ and D is the spatial dimensionality.

As is well-known, Eq. (1.2) has the *finite* number of $\frac{1}{2}(D + 2)(D + 3)$ solutions for $D > 1$, and conformal transformations form an $SO(2, D+1)$ group. The solutions to (1.2) comprise

$$
\begin{array}{rll}
D + 1 \text{ space-time translations} , & f^\mu(x) = a^\mu , & a^\mu \text{ constant} & (1.3\text{a}) \\
\tfrac{1}{2} D(D + 1) \text{ space-time rotations} , & f^\mu(x) = \omega^\mu{}_\nu x^\nu , & \omega_{\mu\nu} = -\omega_{\nu\mu} & (1.3\text{b}) \\
\text{a single scale transformation} , & f^\mu(x) = ax^\mu , & a \text{ constant} & (1.3\text{c}) \\
D + 1 \text{ special conformal transformations} , & f^\mu(x) = 2c \cdot x\, x^\mu - c^\mu x^2 , & c^\mu \text{ constant} &
\end{array}
$$
$$\tag{1.3d}$$

The finite versions of these are, respectively,

$$x^\mu \to x^\mu + a^\mu \tag{1.4a}$$

$$x^\mu \to \Lambda^\mu{}_\nu x^\mu , \qquad \Lambda^\mu{}_\alpha \Lambda^\nu{}_\beta g_{\mu\nu} = g_{\alpha\beta} \tag{1.4b}$$

$$x^\mu \to e^a x^\mu \tag{1.4c}$$

$$x^\mu \to \frac{x^\mu - c^\mu x^2}{1 - 2c \cdot x + c^2 x^2} \tag{1.4d}$$

The last, the finite special conformal transformation, can also be seen as an inversion, $x^\mu \to x^\mu/x^2$, followed by a translation and another inversion, *i.e.* a translation in the inverted coordinate.

At $D = 1$ there exists an *infinite* number of solutions to (1.2) corresponding to arbitrary redefinition of $x^\pm = \frac{1}{\sqrt{2}}(x^0 \pm x^1)$ and forming an infinite parameter group. Infinitesimally we have

$$\delta_f x^\pm = -f^\pm(x^\pm) , \qquad f^\pm \text{ arbitrary} \tag{1.5}$$

while the finite version reads

$$x^\pm \to X^\pm(x^\pm) , \qquad X^\pm \text{ arbitrary} \tag{1.6}$$

A linear conformal transformation on a space-time multiplet of Lorentz covariant relativistic fields φ, i.e. on the dependent variables, can be taken as

$$\delta_f \varphi = f^\alpha \partial_\alpha \varphi + \partial_\alpha f_\beta \left(\frac{\Delta}{D+1} g^{\alpha\beta} + \frac{1}{2} \Sigma^{\alpha\beta} \right) \varphi \qquad (1.7)$$

Here $\Sigma^{\alpha\beta}$ is the spin-matrix, acting on the space-time components of φ and Δ is a constant, called the scale-dimension of φ. When the Lagrange density for φ possesses a conventional relativistic kinetic term — quadratic in derivatives for Bose fields, linear for Fermi fields — the kinetic action is invariant against conformal transformations (1.7) provided

$$\Delta = \frac{D-1}{2} \qquad \text{bosons} \qquad (1.8a)$$

$$\Delta = \frac{D}{2} \qquad \text{fermions} \qquad (1.8b)$$

[These values for Δ correspond to the dimensionality of a field in units of inverse length when $\hbar$ and c are scaled to unity.] Also the Bose field monomial

$$\mathcal{L}_I = \varphi^{2\left(\frac{D+1}{D-1} \right)} \qquad (1.9)$$

leads to an invariant action $\int d^{D+1}x \mathcal{L}_I$.

At $D = 1$, Bose fields become dimensionless, see (1.8a), and the conformally invariant monomial (1.9) cannot be formed. Nevertheless, there exists a non-trivial conformally invariant theory — the completely integrable Liouville theory,

$$\mathcal{L}_{\text{Liouville}} = \frac{1}{2} \partial_\mu \varphi \partial^\mu \varphi - \frac{\mu^2}{\beta^2} e^{\beta\varphi} \qquad (1.10)$$

whose action is invariant provided the single-component scalar field φ is transformed according to an inhomogenous generalization of (1.7),

$$\delta_f \varphi = f^\alpha \partial_\alpha \varphi + \frac{1}{\beta} \partial_\alpha f^\alpha \qquad (1.11a)$$

or equivalently

$$\delta_f e^{\beta\varphi} = \partial_\alpha \left(f^\alpha e^{\beta\varphi} \right) \qquad (1.11b)$$

The hallmark of a conformally invariant theory is that an energy-momentum tensor $T^{\mu\nu}$ can be constructed, which is conserved (as a consequence of translation invariance) symmetric (as a consequence of Lorentz invariance) and traceless (as a consequence of conformal invariance). Thus conformal invariance in a relativistic field may be summarized by a relation between the energy density $\mathcal{E} = T^{00}$ and the trace of the spatial stress tensor T^{ij}.

$$\mathcal{E} = \sum_{i=1}^{D} T^{ii} \qquad (1.12)$$

The currents j_f^μ that are conserved as a consequence of conformal invariance are then constructed in the Bessel–Hagen form from a projection of $T^{\mu\nu}$ on the conformal Killing vector,

$$j_f^\mu = T^{\mu\nu} f_\nu \qquad (1.13a)$$

and the constants of motion read

$$C_f = \int d^D r \left(\mathcal{E} f^0 - \boldsymbol{\mathcal{P}} \cdot \mathbf{f} \right) \tag{1.13b}$$

where $\boldsymbol{\mathcal{P}}$ is the momentum density.

$$\mathcal{P}^i = T^{0i} \tag{1.14}$$

[Frequently it is necessary to "improve" the energy-momentum tensor obtained by Noether's theorem or by general relativistic considerations.]

The kinetic Lagrangian for non-relativistic motion of point-particles in d-dimensional space is quadratic in derivatives with respect to time, which is the single independent variable. Hence it has the structure of a $(0+1)$-dimensional relativistic "field theory," where the "field" is particle position $\mathbf{r}(t)$, now a dependent variable, while the d-spatial dimensions form an "internal" space and $\mathbf{r}$ is a vector in this space. "Conformal" transformations degenerate into reparametrization of the single independent variable, $i.e.$ time. The previous discussion can be taken at $D = 0$, but the conformal Killing equation (1.2) becomes vacuous. Nevertheless, one easily shows that (1.3) and (1.4) [with (1.3b) and (1.4b) absent] are invariances of the kinetic term provided the dependent variable $\mathbf{r}$ transforms according to

$$\delta \mathbf{r} = f \dot{\mathbf{r}} - \frac{1}{2} \dot{f} \mathbf{r} \tag{1.15}$$

when the independent variable t changes by

$$\delta t = -f(t) \tag{1.16a}$$
$$f(t) = a, at, at^2 \tag{1.16b}$$

These comprise the $D = 0$ restriction of (1.7) and (1.8a) with $\Delta = -1/2$, and form an $SO(2,1)$ group of transformations. It is seen that $\mathbf{r}$ has scale dimension of $-1/2$, $i.e.$ it scales as $\sqrt{t}$ — this is a consequence of having scaled $\hbar$ to unity and with non-relativistic kinematics one can take m to be dimensionless. Further from (1.9) at $D = 0$ one sees that the r^{-2} potential also gives an invariant action since $\dot{r}^2$ and $1/r^2$, the two terms comprising a Lagrangian or Hamiltonian, scale in the same way. When

$$H = \frac{1}{2} m \dot{r}^2 + \frac{\lambda}{r^2} \ . \tag{1.17}$$

the three constants of motion

$$C_f = Hf - \frac{m}{4} \left(\mathbf{r} \cdot \dot{\mathbf{r}} + \dot{\mathbf{r}} \cdot \mathbf{r} \right) \dot{f} + \frac{m}{4} r^2 \ddot{f} \tag{1.18}$$

also generate the transformation (1.15) when the canonical momentum $\mathbf{p}$, conjugate to $\mathbf{r}$, is taken to be $m\dot{\mathbf{r}}$, and their algebra realizes the $SO(2,1)$ Lie algebra. One can view (1.18) as the one-time, zero-space analog of the Bessel–Hagen expression (1.13).

In specific spatial dimensions d, other interactions in addition to the $1/r^2$ potential preserve conformal invariance. Examples for $d = 3$ and 2 are, respectively, interaction with a Dirac magnetic monopole and a point vortex. For these (1.17) and (1.18) retain the same form, but the relation between canonical momentum and velocity is modified by the presence of a vector potential

$$\mathbf{p} = m\dot{\mathbf{r}} + \mathbf{A} \tag{1.19}$$

where at $d = 3$, $\mathbf{A}$ is the Dirac vector potential that gives rise to a monopole magnetic field of strength g_m,

$$\mathbf{A} = \mathbf{A}_D , \qquad \mathbf{B} = \nabla \times \mathbf{A}_D = \frac{g_m \mathbf{r}}{r^3} \tag{1.20}$$

while in planar physics at $d = 2$, $\mathbf{A}$ is the vortex potential,

$$\mathbf{A} = \frac{\Phi}{2\pi} \nabla \theta , \qquad \mathbf{B} = \nabla \times \mathbf{A} = \Phi \delta^2(\mathbf{r}) \tag{1.21}$$

with $\tan \theta = y/x$ and Φ being the flux of the vortex. [In the above we use the amusing distributional formula $\nabla \times \nabla \theta = 2\pi \, \delta^2(\mathbf{r})$.]

Furthermore, at $d = 2$, the δ-function potential also scales as r^{-2}, hence it also appears to be conformally invariant. In the following, we shall consider the two-dimensional particle model, with a vortex (1.21) and δ-function interactions; *i.e.* the Hamiltonian is

$$H = \frac{1}{2} m \dot{r}^2 - g \, \delta^2(\mathbf{r}) = \frac{1}{2m} (\mathbf{p} - \mathbf{A})^2 - g \, \delta^2(\mathbf{r}) \tag{1.22}$$

Non-relativistic particle quantum mechanics may be second quantized, and in this way one is led to a non-relativistic quantum field theory, with field-theoretic symmetries that encode the above $SO(2,1)$ particle symmetries, but now realized in an action on the dependent field variable ψ, which is a function of the independent variables t, and $\mathbf{r}$, where $\mathbf{r}$ is a two-dimensional vector.

Here we shall explain these symmetries of non-relativistic field theory, at $d = 2$ — planar physics. Also we shall show how the $SO(2,1)$ group can expand to an infinite-dimensional group of conformal reparametrizations of the two-dimensional spatial plane.

Contexts, wherein recently there is encountered a non-relativistic, planar field theory, are the following two:

1) SECOND QUANTIZED, NON-RELATIVISTIC PARTICLES WITH ABELIAN OR NON-ABELIAN CHARGE, INTERACTING WITH A GAUGE FIELD WHOSE KINETIC DYNAMICS IS PROVIDED BY THE CHERN–SIMONS ACTION. The field theoretic action in the Abelian case, with gauge potentials eliminated in terms of matter variables, is

$$I = \int dt \, d^2r \left\{ i\psi^* \partial_t \psi - \frac{1}{2m} |\mathbf{D}\psi|^2 + \frac{g}{2} \rho^2 \right\} \tag{1.23}$$

where the covariant derivative $\mathbf{D}$ involves a gauge potential $\mathbf{A}$

$$\mathbf{D} = \nabla - i\mathbf{A} \tag{1.24}$$

which is determined by the matter density $\rho = |\psi|^2$

$$\mathbf{A}(t, \mathbf{r}) = \nabla \times \frac{1}{2\pi\kappa} \int d^2r' \ln |\mathbf{r} - \mathbf{r}'| \, \rho(t, \mathbf{r}') \tag{1.25}$$

so that the Chern–Simons Gauss law is satisfied.

$$\mathbf{B} = \nabla \times \mathbf{A} = -\frac{1}{\kappa} \rho \tag{1.26}$$

Here $1/\kappa$ measures the interaction strength and without loss of generality we may take it to be non-negative. [In the plane, the cross product of two vectors defines a scalar, and cross multiplication with a single vector results again in a vector; in components: $s = \epsilon^{ij} v_{(1)}^i v_{(2)}^j$; $v_{(1)}^i = \epsilon^{ij} v_{(2)}^j$.] Also there is present in (1.23) a quartic self-interaction of strength g, which is the second-quantized description of a two-body δ-function interaction. Thus (1.23) provides the second quantization of (1.22); the action may also be presented as

$$I = \int dt\, d^2r\, \{i\psi^* \partial_t \psi - \mathcal{E}\} \tag{1.27}$$

where the energy density is given by the formula[1]

$$\mathcal{E} = \frac{1}{2m} |D\psi|^2 - \frac{g}{2}\rho^2 \tag{1.28}$$

2) EFFECTIVE ACTION FOR GRAVITY OR ABELIAN VECTOR GAUGE THEORIES IN THE EIKONAL (LARGE-s, FIXED-t) LIMIT. It is found that in the eikonal regime, the conventional action [Einstein–Hilbert for gravity, Maxwell for gauge theory] can be written as a total derivative on a two-dimensional space-time plane imbedded in four-dimensional space-time. By integrating the total derivative onto a curve (parametrized by τ) forming the boundary of that two-dimensional plane, the action [without sources] becomes

$$I_{\text{eikonal}} = \frac{1}{2} \int d\tau\, d^2r \left(\partial_i \Omega^+ \partial_i \dot\Omega^- - \partial_i \Omega^- \partial_i \dot\Omega^+ \right) \tag{1.29}$$

where the overdot denotes differentiation with respect to τ, while r and ∂_i, $i = 1, 2$, refer to the remaining two spatial directions, and $\Omega^\pm$ are the surviving field (gravitational, vector) degrees of freedom.[2] Upon defining

$$\psi = \frac{1}{\sqrt{2}} (\partial_x + i\partial_y)(\Omega^+ - i\Omega^-) \tag{1.30}$$

(1.29) may be rewritten, apart from total derivative contributions, as

$$I_{\text{eikonal}} = \int d\tau\, d^2r\, i\psi^* \partial_\tau \psi \tag{1.31}$$

Precisely the same form as (1.27) is revealed, except now the energy density vanishes.

II.   SYMMETRIES

The field theoretic Lagrangian

$$L = \int d^2r\, i\, \psi^* \partial_t \psi - H \tag{2.1a}$$

$$H = \int d^2r\, \mathcal{E} \tag{2.1b}$$

represents both the non-relativistic Chern–Simons model (1.27), (1.28) and the eikonal limits of relativistic theory (1.29), (1.31), where in the latter case $\mathcal{E}$ vanishes. So we use (2.1) as the basis for our discussion of symmetries in both cases, keeping in mind that the vanishing of $\mathcal{E}$ for the latter, renders much of the analysis vacuous, but as we shall see, not without relevance to the former. Throughout we shall solely deal with the Abelian Chern–Simons theory, though as far as symmetry properties are concerned, the non-Abelian model behaves similarly. Also discussion is confined to classical symmetries of the field theory, viewed as a classical non-linear system. Anomalies in the symmetries due to quantum effects will only be mentioned in the Conclusion.

The energy density (1.28) of the Chern–Simons model

$$\mathcal{E} = \frac{1}{2m} |D\psi|^2 - \frac{g}{2}\rho^2 \tag{2.2a}$$

is identically equal to

$$\mathcal{E} = \frac{1}{2m} |D\psi|^2 - \frac{1}{2}\left(\nabla \times \mathbf{j} + \frac{1}{m}B\rho\right) - \frac{g}{2}\rho^2 \tag{2.2b}$$

where the current $\mathbf{j}$ is

$$\mathbf{j} = \frac{1}{m} \operatorname{Im} \psi^* D\psi \tag{2.3}$$

and D is the holomorphic, gauge covariant derivative.

$$D \equiv D_x - iD_y \tag{2.4}$$

The curl of $\mathbf{j}$ will not contribute to a variational derivation of the equations of motion, nor will it contribute to the integrated total energy, provided the current is sufficiently well-behaved at the edge of space [which lies at infinity]. With the assumption of requisite regularity for $\mathbf{j}$ and the use of the Chern–Simons Gauss law constraint (1.26), the energy/Hamiltonian may be presented as

$$E = H = \int d^2r \mathcal{H} \;, \qquad \mathcal{H} = \frac{1}{2m} |D\psi|^2 - \frac{1}{2}\left(g - \frac{1}{m\kappa}\right)\rho^2 \tag{2.5}$$

This is the form of the Hamiltonian density that we shall scrutinize as regards to the symmetries of the model.

The symmetries are of two kinds: a) symmetries of the action, i.e. transformations which leave the action invariant and lead to constants of motion by Noether's theorem — these are well-known and include the obvious Galileo transformations, and the $SO(2,1)$ time-reparametrization conformal symmetries specific to the planar model with which we are here concerned; and b) symmetries of the critical points of the action, i.e. transformations which leave selected equations of motion invariant, map solutions into solutions, but do not give rise to constants of motion because they do not leave invariant the action away from its critical points. These symmetries of the Chern–Simons model have not been previously studied systematically, though their occurrence in static solutions at $g = 1/m\kappa$ had been noted: they comprise conformal reparametrization symmetries of the two-dimensional plane.[3,4]

A. Finite-Dimensional Symmetry Group of the Action

As befits any respectable field theory, our model is invariant against time translation, space translation and rotation, as well as against Galileo boosts because dynamics is non-relativistic. The last invariance is perhaps unexpected in the presence of gauge fields, which conventionally are invariant against the *Lorentz* boosts of *special relativity* (indeed this led to the invention of special relativity!). What distinguishes the present situation is that the gauge dynamics are of the Chern–Simons variety, and the Chern–Simons term, being topological, is invariant against *all* space-time transformation, while the non-relativistic matter system is only Galileo invariant.

The transformation laws on the fields are familiar. For the first three,

$$
\text{time translations} \qquad
\begin{aligned}
t' &= t + a \\
\mathbf{r}' &= \mathbf{r}
\end{aligned}
\tag{2.6}
$$

$$
\text{space translation} \qquad
\begin{aligned}
t' &= t \\
\mathbf{r}' &= \mathbf{r} + \mathbf{a}
\end{aligned}
\tag{2.7}
$$

$$
\text{space rotation} \qquad
\begin{aligned}
t' &= t \\
r'^{i} &= R^{ij}(\omega)r^{j}
\end{aligned}
\tag{2.8}
$$

$[R^{ij}(\omega)$ is the rotation matrix through angle $\omega]$ the field transforms as a scalar.

$$
\psi'(t',\mathbf{r}') = \psi(t,\mathbf{r})
\tag{2.9}
$$

The last

$$
\text{Galileo boost} \qquad
\begin{aligned}
t' &= t \ , \\
\mathbf{r}' &= \mathbf{r} + \mathbf{v}t
\end{aligned}
\tag{2.10}
$$

requires a 1-cocycle in the field transformation law.

$$
\psi'(t',\mathbf{r}') = e^{im\mathbf{v}\cdot(\mathbf{r}+\frac{1}{2}\mathbf{v}t)}\psi(t,\mathbf{r})
\tag{2.11}
$$

Additionally, our system is invariant against conformal reparametrizations of time. These include three $SO(2,1)$ transformations of time: translation (2.6) and (2.9),

$$
\text{time dilation} \qquad
\begin{aligned}
t' &= at \\
\mathbf{r}' &= \sqrt{a}\,\mathbf{r}
\end{aligned}
\tag{2.12}
$$

for which the field transformation law acquires a weight factor,

$$
\psi'(t',\mathbf{r}') = \frac{1}{\sqrt{a}}\psi(t,\mathbf{r})
\tag{2.13}
$$

and translation of inverse time,

$$
\begin{aligned}
\text{conformal time} \\
\text{transformation}
\end{aligned}
\qquad
\begin{aligned}
\frac{1}{t'} &= \frac{1}{t} + a \\
\mathbf{r}' &= \frac{1}{1+at}\mathbf{r}
\end{aligned}
\tag{2.14}
$$

where the field transformation law has both a weight factor and 1-cocycle.

$$
\psi'(t',\mathbf{r}') = (1+at)\,e^{\frac{-imar^2}{2(1+at)}}\psi(t,\mathbf{r})
\tag{2.15}
$$

One can check that owing to the weight factors, which are square roots of the Jacobian, the density ρ transforms with the Jacobian, J,

$$\rho'\left(t',\mathbf{r}'\right) = J\rho(t,\mathbf{r}) \tag{2.16}$$

$$J \equiv \det\left\{\frac{\partial r^i}{\partial r'^j}\right\} \tag{2.17}$$

and the vector potential, defined by (1.25), transforms covariantly.

$$A'^i\left(t',\mathbf{r}'\right) = A^j(t,\mathbf{r})\frac{\partial r^j}{\partial r'^i} \tag{2.18}$$

The action is invariant, and the conserved generators can be obtained from Noether's theorem.

Alternatively one records the formula for the energy momentum tensor components:

energy density $\qquad T^{00} \equiv \mathcal{E} = \dfrac{1}{2m}\left|\mathbf{D}\psi\right|^2 - \dfrac{g}{2}\rho^2 \qquad$ (2.19)

momentum density $\qquad \mathcal{P} = m\mathbf{j} = \operatorname{Im}\psi^*\mathbf{D}\psi \qquad$ (2.20)

These satisfy continuity equations with energy flux $\mathbf{T}$,

energy flux $\qquad \mathbf{T} = -\dfrac{1}{2}\left(\left(D_t\psi\right)^*\mathbf{D}\psi + \left(\mathbf{D}\psi\right)^* D_t\psi\right) \qquad$ (2.21)

and momentum flux — the stress tensor T^{ij}.

momentum flux $\qquad T^{ij} = \dfrac{1}{2}\left(\left(D_i\psi\right)^*\left(D_j\psi\right) + \left(D_j\psi\right)^*\left(D_i\psi\right) - \delta^{ij}\left(D_\kappa\psi\right)^*\left(D_\kappa\psi\right)\right)$

$$+ \frac{1}{4}\left(\delta^{ij}\nabla^2 - 2\partial_i\partial_j\right)\rho + \delta^{ij}\mathcal{E} \tag{2.22}$$

Here $D_t = \partial_t + iA^0$, where A^0 solves the Chern–Simons equation that supplements (1.26).

$$A^0(t,\mathbf{r}) = -\frac{1}{2\pi\kappa}\int d^2r'\,\epsilon^{ij}\frac{\left(r^i - r'^i\right)}{\left|\mathbf{r}-\mathbf{r}'\right|^2}j^j(t,\mathbf{r}') \tag{2.23}$$

The continuity equations read

$$\partial_t\mathcal{E} + \nabla\cdot\mathbf{T} = 0 \tag{2.24}$$

$$\partial_t\mathcal{P}^i + \partial_j T^{ij} = 0 \tag{2.25}$$

Note that energy flux $\mathbf{T}$ does not equal momentum density $\mathcal{P}$, since our theory is not Lorentz invariant. But it is rotationally invariant; that is why the stress-tensor is symmetric in its spatial indices. Also T^{ij} satisfies

$$2\mathcal{E} = \sum_{i=1}^{2} T^{ii} \tag{2.26}$$

and this reflects the $SO(2,1)$ invariance, being the non-relativistic analog of (1.12).

Of course, the theory is also phase invariant; this produces one more continuity equation

$$\partial_t\rho + \nabla\cdot\mathbf{j} = 0 \tag{2.27}$$

where the proportionality of the matter flux current $\mathbf{j}$ to the momentum density (2.20) is a consequence of Galileo invariance.

The constants of motion are now constructed from moments of the energy momentum tensor and ρ. They are, respectively

$$\text{energy} \qquad E = \int d^2r\, \mathcal{E} \tag{2.28}$$

$$\text{momentum} \qquad \mathbf{P} = \int d^2r\, \mathcal{P} \tag{2.29}$$

$$\text{angular momentum} \qquad M = \int d^2r\, \mathbf{r} \times \mathcal{P} \tag{2.30}$$

$$\text{Galileo boost} \qquad \mathbf{B} = t\mathbf{P} - m \int d^2r\, \mathbf{r}\rho \tag{2.31}$$

$$\text{dilation} \qquad D = tE - \frac{1}{2} \int d^2r\, \mathbf{r} \cdot \mathcal{P} \tag{2.32}$$

$$\text{special conformal} \qquad K = -t^2 E + 2tD + \frac{m}{2} \int dr^2\, r^2 \rho \tag{2.33}$$

$$\text{matter number} \qquad N = \int d^2r\, \rho \tag{2.34}$$

It is straightforward to verify that all these are conserved, as a consequence of the continuity equations and the special properties (symmetry and trace) of the stress tensor. Note that collectively the constants may be written analogously to (1.18) and to the Bessel–Hagen expression (1.13), as

$$C_f = \int d^2r\, \mathcal{E} f_1 - \int d^2r\, \mathcal{P} f_2 + \int d^2r\, \rho f_3 \tag{2.35}$$

for suitable f_i.

The above transformations may be generalized in the following interesting manner. One may consider an *arbitrary* reparametrization of time $t \to T(t)$ [rather than the specific forms (2.6), (2.12), (2.14)]. Also one may shift $\mathbf{r}$ by a vector $\mathbf{r}_0(t)$ with *arbitrary* time dependence [rather than constant (2.7) or linear (2.10) in time]. Finally, one may rotate $\mathbf{r}$ as in (2.8), but with a *time-dependent* angle $\omega(t)$. Of course, these transformations are no longer symmetry operations of our theory, but they map our model onto another, closely related one: it is found that the above transformations introduce interactions with external fields. Specifically, after these transformations are carried out on the fields according to the rule

$$\psi'(t', \mathbf{r}') = \frac{1}{\sqrt{\dot{T}(t)}} e^{i\gamma(t,\mathbf{r})} \psi(t, \mathbf{r}) \tag{2.36}$$

where the transformed coordinates are given by

$$t' = T(t) , \qquad r'^i = \sqrt{\dot{T}(t)}\, R^{ij}(\omega(t)) (r^j + r_0^j(t)) \tag{2.37}$$

and the 1-cocycle γ takes the form

$$\gamma(t, \mathbf{r}) = \frac{m}{4} \frac{\ddot{T}(t)}{\dot{T}(t)} (\mathbf{r} + \mathbf{r}_0(t))^2 + mr^i \left(\dot{r}_0^i(t) - \dot{\omega}(t)\epsilon^{ij} r_0^j(t) \right) + \tilde{\gamma}(t)$$

$$\dot{\tilde{\gamma}}(t) = \frac{m}{2} \left(\dot{r}_0^i(t) - \dot{\omega}(t)\epsilon^{ij} r_0^j(t) \right)^2 + \frac{m}{2} r_0^2(t) \sqrt{\dot{T}(t)} \frac{d^2}{dt^2} \frac{1}{\sqrt{\dot{T}(t)}} \tag{2.38}$$

one finds that in the transformed system there arise external electric and magnetic fields, determined by the parameters of the transformations $[T(t), \omega(t), \mathbf{r}_0(t)]$, hence the fields are time-dependent but constant in space; additionally there is an external harmonic force field, with time-varying frequency.[5] [With specific time-dependence, the parameters can conspire to produce static electric and magnetic fields as well as time-independent harmonic forces; also one can suppress selectively any of the external effects.]

Note that the symmetry transformations (2.6) – (2.15) also follow the more general rules (2.36) – (2.38). One may understand the presence of a one-cocycle in (2.36) by recognizing that when the kinetic term of a *particle* Lagrangian is transformed according to (2.37), external electromagnetic and harmonic force fields are again generated, and also the Lagrangian changes by a total time derivative which is $\frac{d}{dt}\gamma(t, \mathbf{r}(t))$.

Higher symmetries are widely studied these days in field theory, but it seems that rarely do they provide specific dynamical information about a model — rather they give an elegant frame for describing solutions and other properties.

As an exception that proves the rule, we now show that the conformal symmetries allow deriving the following useful result about the highly non-linear dynamics of our Chern–Simons theory: all static solutions carry zero energy.[6] This follows immediately from (2.32) and/or (2.33): the left sides are time independent, and so are the last terms in the right sides for time independent ρ and $\mathcal{P} = m\mathbf{j}$, *i.e.* for static solutions. Thus $H = E$, and also but less importantly D, must vanish. Similarly, from (2.31) one sees that $\mathbf{P}$ must vanish, but this is not surprising — we expect static solutions to carry no momentum. Note however: angular momentum need not vanish for static configurations; it can be constructed from the current $m\mathbf{j} = \mathcal{P}$, which in the static case must be divergence-free, according to (2.27).

Since E can be given by (2.5) (provided there is sufficient regularity so that the integral $\int d^2r\, \nabla \times \mathbf{j}$ vanishes) we see that static solutions can exist only for $g \geq 1/m\kappa$. Especially interesting is the limiting case $g = 1/m\kappa$, where the integrand is non-negative and therefore must vanish on static solutions. In this way the $SO(2,1)$ conformal symmetry demands that *all* static solutions (at $g = 1/m\kappa$) satisfy

$$D\psi = 0 \qquad (2.39)$$

Together with the Chern–Simons constraint (1.26) this implies that ρ satisfies the Liouville equation,

$$\nabla^2 \ln \rho = -\frac{2}{\kappa}\rho \qquad (2.40)$$

which can be integrated explicitly in terms of two arbitrary functions, which are further specified by the physical requirements that one may wish to impose on static solutions.[1,3]

[In the non-Abelian case, the analogous equations, with ψ in the same adjoint representation as the gauge fields, realize a two-dimensional reduction of four-dimensional self-dual gauge field equations in a space with signature $(+ + --)$ and lead to many integrable systems, principally the Toda system.[4]]

As is well-known and was remarked in the Introduction, the Liouville equation is invariant against conformal redefinition of the two-dimensional plane. In Euclidean space this involves the complex variable $x + iy = z$ transforming into an arbitrary function of z, but not of z^*: Our next task will be to understand the properties of the action (1.23) (at $g = 1/m\kappa$) that are responsible for this infinite symmetry. In fact its stationary points are conformally invariant.

But before turning to this topic, we point out that the above described transformations can be used to generate interesting new solutions from the explicitly determined static solutions. First by Galileo [(2.10), (2.11)] or conformal [(2.14), (2.15)] boosting of static solutions, one obtains time-dependent solutions to the Chern–Simons model. Moreover, by performing transformations with time-dependent parameters [(2.36), (2.37)], one finds time-dependent solutions to the Chern–Simons model with external, appropriately constructed electric and magnetic fields as well as an external harmonic force field.[7]

B. Infinite-Dimensional Symmetry Group of Stationary Points of the Action

The dilation transformation (2.12) and (2.13) rescales the spatial coordinate **r**. Here we inquire about the response of the action (2.1a), (2.5) to a conformal redefinition of spatial coordinates,

$$\mathbf{r}' = \mathbf{r}'(\mathbf{r}) \tag{2.41a}$$

where

$$x' + iy' \equiv z' = z'(z) \tag{2.41b}$$

and time is unchanged, $t' = t$.

Generalizing (2.13) and (2.36), we posit a field transformation law with a weight,

$$\psi'(\mathbf{r}') = \frac{\partial z^*}{\partial z'^*}\psi(\mathbf{r}) \tag{2.42}$$

which apart from a phase is the square root of the Jacobian, as in (2.13) and (2.15), while the choice of phase is dictated by the Hamiltonian (2.5). [Since time is not transformed, we suppress the time argument.] This has the consequence that the density transforms with the Jacobian as in (2.16).

$$\rho'(\mathbf{r}') = J\rho(\mathbf{r}) \tag{2.43}$$

$$J = \det\left\{\frac{\partial r^i}{\partial r'^j}\right\} = \left|\frac{\partial z}{\partial z'}\right|^2 \tag{2.44}$$

For infinitesimal $\delta z = -f(z)$, this transformation law coincides with (1.11b), taken in Euclidean space and $e^{\beta\varphi}$ identified with ρ. It further follows that the gauge potential transforms covariantly.

$$A'^i(\mathbf{r}') = A^j(\mathbf{r})\frac{\partial r^j}{\partial r'^i} \tag{2.45}$$

This is most easily proven by first noting that **A**, when given by Eq. (1.25), is transverse and satisfies $\nabla_{\mathbf{r}} \times \mathbf{A}(\mathbf{r}) = -\frac{1}{\kappa}\rho(\mathbf{r})$, and then verifying that $\mathbf{A}'(\mathbf{r}')$ in (2.45) also is transverse and satisfies $\nabla_{\mathbf{r}'} \times \mathbf{A}'(\mathbf{r}') = -\frac{1}{\kappa}J\rho(\mathbf{r}) = -\frac{1}{\kappa}\rho'(\mathbf{r}')$. [In carrying out the differentiations it is useful to pass to complex variables.] It follows that $D\psi \equiv (\partial_x - i\partial_y - iA^x + A^y)\psi$ transforms with the Jacobian.

$$D_{\mathbf{r}'}\psi'(\mathbf{r}') = JD_{\mathbf{r}}\psi(\mathbf{r}) \tag{2.46}$$

So finally we can state the transformation law for the Lagrange density.

$$\begin{aligned}
\mathcal{L} &= i\psi^*\partial_t\psi - \frac{1}{2m}|D\psi|^2 + \frac{1}{2}\left(g - \frac{1}{m\kappa}\right)\rho^2 \\
&= i\psi^*\partial_t\psi - \mathcal{H}
\end{aligned} \tag{2.47}$$

Evidently it is true that

$$\mathcal{L}'(\mathbf{r}') = Ji\psi^*(\mathbf{r})\partial_t\psi(\mathbf{r}) - J^2\mathcal{H}(\mathbf{r}) \qquad (2.48)$$

so that the Lagrangian transforms as

$$L' = \int d^2r'\,\mathcal{L}'(\mathbf{r}') = \int d^2r\,i\psi^*(\mathbf{r})\partial_t\psi(\mathbf{r}) - \int d^2\mathbf{r}\,J\mathcal{H}(\mathbf{r}) \qquad (2.49)$$

One factor of the Jacobian has disappeared when changing spatial variables in the integration, and the symplectic form $\int d^2r\,i\psi^*\partial_t\psi$ is invariant. But the Hamiltonian density $\mathcal{H}$ remains with one factor J, hence the total Lagrangian is not in general invariant, and neither is the action, the time integral of L — because t is not changed in the present transformation rules [in contrast to (2.12) and (2.14)]. [It does not appear possible to find a transformation of time that would restore invariance.]

However, for static solutions we know that $E = \int d^2r\,\mathcal{H}$ vanishes. If this vanishing is due to the local vanishing of $\mathcal{H}$, as is true at $g = 1/m\kappa$, then the static critical points of the action are invariant. This then shows that static solutions with zero $\mathcal{H}$ will be mapped into each other by spatial conformal transformations — the dilation (2.12) expands to an infinite symmetry group on the solutions, but there are no new constants of motion.

[Since in the non-Abelian generalization, with matter in the adjoint representation, the corresponding static Chern–Simons equations are dimensional reductions of self-dual Yang–Mills equations in four dimensions,[4] the *finite-dimensional* conformal invariance of the latter[8] is seen to survive the dimensional reduction, and in two dimensions expands to the *infinite-dimensional* conformal group.]

On the other hand, in the effective field theories for the eikonal regime (1.29), (1.31), where there is no Hamiltonian to begin with, the transformations (2.41), (2.42) *are* symmetries of the action, and also τ may be arbitrarily reparametrized. Note that owing to the derivative relation (1.30) between $\Omega^\pm$ and ψ: $\psi = \sqrt{2}\frac{\partial}{\partial z^*}(\Omega^+ - i\Omega^-)$, the transformation law for $\Omega^\pm$ is without the weight factor,

$$\Omega'^\pm(\mathbf{r}') = \Omega^\pm(\mathbf{r}) \qquad (2.50)$$

which arises for ψ, as in (2.42), when the derivative is taken.

III. CONCLUSION

The rigid scale invariance of the action for non-relativistic $(2+1)$-dimensional field theory with quartic self-interaction and coupling to a Chern–Simons gauge field, expands at the static critical points of the action to the infinite conformal group on the plane. The scale symmetry allows establishing the important result that static solutions carry zero energy, and the infinite conformal symmetry "explains" why the static system is completely integrable. The kinetic action of effective eikonal field theories also possesses the infinite symmetry.

The Chern–Simons model at $g = 1/m\kappa$ is the bosonic partner of an $N = 2$ supersymmetric theory with fermions and the invariance of the extended action against the supersymmetric generalization of the bosonic symmetries (2.6) – (2.15) has been established.[9] While the invariances of the static critical points in the supersymmetric action have not been explicitly checked, they too presumably enjoy an infinite conformal symmetry, because the supersymmetric static equations retain the form of the bosonic equations.

In our considerations, the possibility of quantum symmetry breaking anomalies has been ignored. It is known that the quartic self-interaction, which as we have seen is formally scale invariant, suffers from quantum scale anomalies.[10] This is particularly clear in the first quantized framework, where the two-dimensional δ-function potential, while scaling classically as r^{-2}, does not give rise to energy-independent phase shifts, as is required by scale invariance and is explicitly realized by the scale invariant $1/r^2$ potential. There is a quantum scale anomaly — the simplest example of the anomaly phenomenon.[11] On the other hand, anomalies in the theory with *both* quartic self-coupling and Chern–Simons interaction have thus far not been assessed; in fact there is some indication of anomaly cancellation, even without supersymmetry.[12] Further research on this question would be interesting.[13]

REFERENCES

1. For a discussion of non-relativistic Chern–Simons theory and its relation through second quantization to the particle mechanics of (1.22), see *e.g.* R. Jackiw and S.-Y. Pi, *Phys. Rev. D* **42**, 3500 (1990).

2. For gravity: H. Verlinde and E. Verlinde, *Nucl. Phys.* **B371**, 246 (1992); for Maxwell theory: R. Jackiw, D. Kabat and M. Ortiz, *Phys. Lett. B* **277**, 148 (1992).

3. For Abelian Chern–Simons interactions: R. Jackiw and S.-Y. Pi, *Phys. Rev. Lett.* **64**, 2969 (1990); (C) **66**, 2682 (1992) and Ref. [1].

4. For non-Abelian Chern–Simons interactions: B. Grossman, *Phys. Rev. Lett.* **65**, 3230 (1990); G. Dunne, R. Jackiw, S.-Y. Pi and C. Trugenberger, *Phys. Rev. D* **43**, 1332 (1991); G. Dunne, *Commun. Math. Phys.* (in press).

5. S. Takagi, *Prog. Theor. Phys.* **84**, 1019 (1990), **85**, 463, 723 (1991), **86**, 783 (1991).

6. D. Freedman and A. Newell (unpublished).

7. Z. Ezawa, M. Hotta and Z. Iwazaki, *Phys. Rev. Lett.* **67**, 441 (1991); *Phys. Rev. D* **44**, 452 (1991); R. Jackiw and S.-Y. Pi, *Phys. Rev. Lett.* **67**, 415 (1991) and *Phys. Rev. D* **44**, 2524 (1991).

8. R. Jackiw and C. Rebbi, *Phys. Rev. D* **14**, 517 (1977).

9. M. Leblanc, G. Lozano and H. Min, *Ann. Phys.* (NY) (in press).

10. O. Bergman, MIT preprint CTP#2045 (1991).

11. R. Jackiw, in *M. A. B. Bég Memorial Volume*, A. Ali and P. Hoodbhoy, eds. (World Scientific, Singapore, 1991).

12. G. Lozano, *Phys. Lett. B* (in press).

13. O. Bergman, in preparation.

PRE-FREEZING PHENOMENON IN SUPERIONIC MATERIALS

M. Kobayashi[1] and F. Shimojo[2]

[1]Department of Physics
[2]Graduate School of Science and Technology
Niigata University, Niigata 950–21, Japan

INTRODUCTION

Superionic conductors (SIC) are crystalline materials which exhibit extremely high values of ionic conductivity comparable to those of liquid electrolytes at relatively low temperatures. Figure 1 shows schematically a structure of α-AgI which is a typical SIC. SIC are composed of two types of ions. One is a lattice ion. The other is a mobile ion. I ions form the bcc (body-centered cubic) lattice and Ag ions can diffuse through the lattice. These materials have high ionic conductivity in reflection with fast motion of mobile ions. The temperature dependence of ionic conductivity of α-AgI is shown in Fig.2. In the low temperature β phase, AgI takes the hexagonal wurtzite structure and has small conductivity. The $\beta \to \alpha$ phase transition occurs at $T_c = 420$K. In the α phase I ions form the bcc structure, as is shown in Fig.1 and the ionic conductivity increases by four orders of magnitude. The ionic conductivity in the α phase is comparable to that of the molten phase even though T_c is a factor of 2 below the melting temperature $T_m = 828$K. Making use of these high conductivity, these materials have given impetus to a new technology, which contains devices based upon the motion of ions in solids, for example fuel cells, solid state batteries, gas sensors, lithography etc.

Recently, we have investigated the wavenumber-dependent static dielectric function $\varepsilon(k)$ of AgI in its superionic and molten phases. These studies give us information about some ordering of ions or *symmetry* of ion configuration in the superionic conductor AgI.

We also have investigated the fractal behavior in diffusion trajectories of ions and Brownian functions. We expect the forerunning phenomenon to the crystallization or *symmetry change* of ion configuration also in a fractal behavior.

Symmetries in Science VII, Edited by B. Gruber
and T. Otsuka, Plenum Press, New York, 1994

CAUSALITY AND STABILITY

We begin by considering the time-dependent response of a system to an external charge density $\delta\rho_{ex}$. The Poisson equations read

$$\text{div } \mathbf{D} = 4\pi\delta\rho_{ex}, \quad \text{div } \mathbf{E} = 4\pi\delta\rho_t \quad . \tag{1}$$

Here $\delta\rho_t$ is the total charge density which includes the $\delta\rho_{ex}$ and the induced charge density $\delta\rho_i$. $\mathbf{D}$ and $\mathbf{E}$ are the displacement field and electric field vectors, respectively. As both $\mathbf{D}$ and $\mathbf{E}$ are purely longitudinal fields, we can define a scalar longitudinal dielectric function with its Fourier transform by means of the relation

$$\mathbf{D}(\mathbf{k},\omega) = \varepsilon(\mathbf{k},\omega)\mathbf{E}(\mathbf{k},\omega) \quad . \tag{2}$$

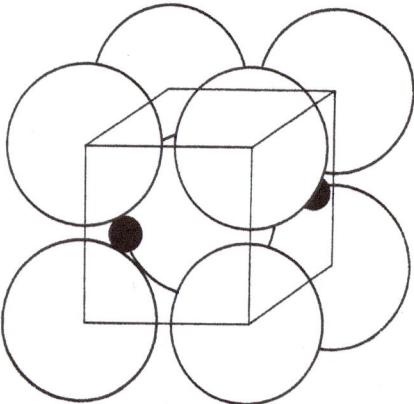

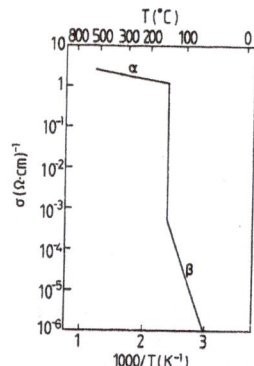

Figure 1. Structure of α-AgI. Large circles and small black circles represent I ions and Ag ions, respectively.

Figure 2. Temperature dependence of ionic conductivity σ of α-AgI.

From these equations, we have

$$\delta\rho_t(\mathbf{k},\omega) = \varepsilon(\mathbf{k},\omega)^{-1}\delta\rho_e(\mathbf{k},\omega) \quad . \tag{3}$$

We see that the inverse dielectric function is the response function of the external charge density to the total charge density. The response of the system must be causal. Then the response function satisfies the following Kramers-Kronig relations which are a direct consequence of the physical principle of causality:

$$\varepsilon(\mathbf{k},\omega)^{-1} = 1 + \frac{2}{\pi}\int_0^\infty d\omega' \frac{\omega' \text{ Im } \varepsilon(\mathbf{k},\omega')^{-1}}{\omega'^2 - \omega^2 - i\delta} \quad . \tag{4}$$

In the static limit, we have

$$\varepsilon(\mathbf{k}, 0)^{-1} \; = \; 1 \; + \; \frac{2}{\pi} \int_0^\infty \frac{d\omega'}{\omega'} \; \text{Im } \varepsilon(\mathbf{k}, \omega')^{-1} \; . \tag{5}$$

The fluctuation-dissipation theorem shows that the imaginary part of the dielectric function connects with the structure factor $S(\mathbf{k}, \omega)$. Using the structure factor, we can rewrite Eq.(4) as

$$\varepsilon(\mathbf{k}, \omega)^{-1} \; = \; 1 \; - \; \frac{8\pi e^2}{k^2} \int_0^\infty d\omega' S(\mathbf{k}, \omega') \frac{\omega'}{\omega'^2 - \omega^2} \; . \tag{6}$$

If we take into account of $S(\mathbf{q}, \omega) > 0$, we have

$$\text{Im } \varepsilon(\mathbf{k}, \omega)^{-1} \; \leq \; 0 \; . \tag{7}$$

Putting it into Eq.(5), we get

$$\varepsilon(\mathbf{k}, 0) \; \geq \; 1, \quad \varepsilon(\mathbf{k}, 0) \; < \; 0 \; . \tag{8}$$

Thus the causality conditions corresponding to the action of an external charge on the system do not preclude negative values for a static dielectric function. Only the values between 0 and 1 turn out to be forbidden.[1]

Let's ask what will happen to the system when the Kramers-Kronig relations are violated. A violation of these relations will lead to the appearance of a zero for static dielectric function. This shows that there exists a finite internal electric field E even if an external displacement field $\mathbf{D} = 0$. This leads to the appearance in the system of some kind of ordering such as charge density waves or *symmetry change* of ion configuration.

STATIC DIELECTRIC FUNCTION

The static dielectric function can be written in terms of the charge–density correlation function $S_{CC}(\mathbf{k})$ as follows:

$$\varepsilon(\mathbf{k})^{-1} \; = \; 1 \; - \; \frac{4\pi \rho_0 e^2}{k_B T k^2} S_{CC}(\mathbf{k}) \; , \tag{9}$$

where $\rho_0 = N/V$ is the number density. The charge–density correlation function $S_{CC}(\mathbf{k})$ is given by the partial structure factors as follows:

$$S_{CC}(\mathbf{k}) = \sum_\alpha \sum_\beta z_\alpha z_\beta (C_\alpha C_\beta)^{1/2} S_{\alpha\beta}(\mathbf{k}) \; , \tag{10}$$

where $C_\alpha = N_\alpha / \sum N_\beta$ and z_α are the concentration and the effective valence of α-type ions, respectively. Last, the partial static structure factors are defined by

$$S_{\alpha\beta}(\mathbf{k}) \; = \; \frac{1}{(N_\alpha N_\beta)^{1/2}} < \sum_{i\,(\alpha)} \sum_{j\,(\beta)} e^{i\mathbf{k}\cdot(\mathbf{r}_j - \mathbf{r}_i)} > \; . \tag{11}$$

Aniya *et al.*[2] have calculated the charge–density correlation function $S_{CC}(\mathbf{k})$ using the result of hypernetted-chain calculations performed by Stafford & Silbert.[3] The obtained dielectric function is shown in Fig.3 for the molten phase. Their result is shown by solid curve. The dashed curve shows the static dielectric function of molten NaCl and the dotted curve shows that of one-component plasma. Those were calculated by

Fasolino et al.[4] $\Gamma = 160$ is a plasma parameter. The plasma parameter Γ is defined by the ratio of the Coulomb potential energy to the kinetic energy and is given by

$$\Gamma = (Ze)^2/ak_BT .$$

Here

$$a = (3V/4\pi N)^{1/3}$$

is the radius of a sphere with the characteristic volume V/N and is usually referred to as the ion-sphere radius. According to calculations by Slattery et al.,[5] the fluid–solid transition in a classical one component plasma occurs at $\Gamma = 168$. Then the dotted curve is almost on its transition point.

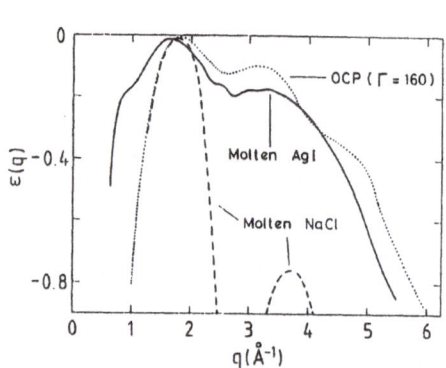

Figure 3. Static dielectric function $\varepsilon(k)$ of molten AgI(ref.2), molten NaCl(ref.4) and one-component plasma(ref.4).

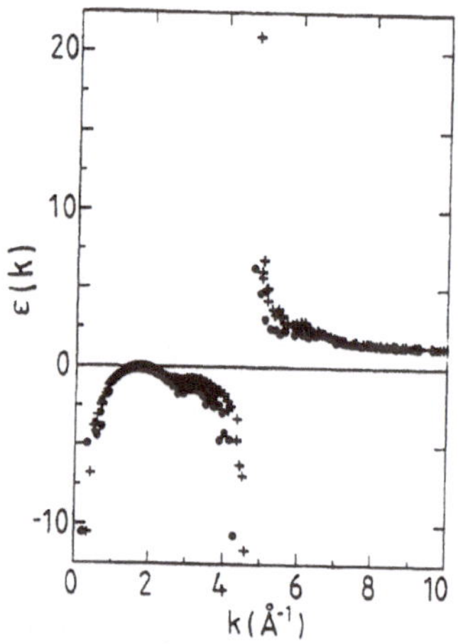

Figure 4. $\varepsilon(k)$ for two different temperatures 903K (marking with +) and 1036K (marking with black circle) in the molten phase(ref.6).

The solid curve $\varepsilon(k)$ touches the horizontal axis at about $k = 1.7\text{Å}^{-1}$. Namely $\varepsilon(k)=0$ has been realized. This shows the forerunning phenomenon to the crystallization which corresponds to the *symmetry reduction* of configuration of ions. We think that ions in the molten phase are ready to change to the solid state phase, *i.e.* α-AgI. The solid curve is similar to that of one-component plasma near the fluid–solid transition point than that of the molten NaCl. We think that this behavior reflects the covalent nature of AgI and the tendency of the ions in the molten phase to take the structure of α-AgI.

Next We have performed the computer simulation of a molecular dynamics (MD) method to investigate microscopically the characteristic behavior of the static dielectric function.[6] The computer simulation offers exact calculations. Then it may be one of the most suitable method for the study of complicated materials such as SIC in which mobile ions migrate through lattice ions.

The computer simulations have been carried out for 500 particle system (250 Ag + 250 I). The system size is about 26 Å. An integration time step is $\Delta t = 5.0 \times 10^{-15}$ s.

We used the effective pair potentials which have the same form given by Parrinello et al.[7] Their forms are given by

$$V_{ij}(r) = \frac{H_{ij}}{r^{n_{ij}}} + \frac{z_i z_j}{r} - \frac{P_{ij}}{r^4} - \frac{W_{ij}}{r^6} \quad , \tag{12}$$

with

$$H_{ij} = A(\sigma_i + \sigma_j)^{n_{ij}} \quad \text{and} \quad P_{ij} = \frac{1}{2}(\alpha_i z_j^2 + \alpha_j z_i^2) \quad ,$$

where σ_i, z_i and α_i are the particle radius, the effective charge and the electric polarizability, of the i-th ion, respectively. The potentials are composed of four interactions: repulsive interaction, Coulomb interaction, charge–dipole interaction and van der Waals interaction.

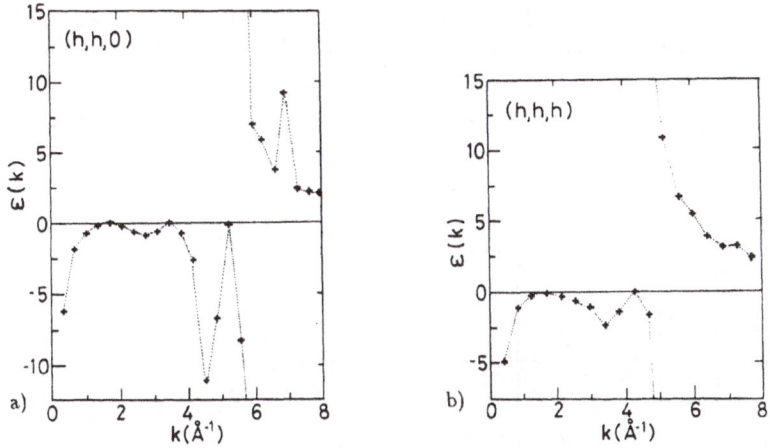

Figure 5. $\varepsilon(k)$ for a) (h, h, 0) and b) (h, h, h) directions in the α-phase (683K).

Let us discuss the dielectric behavior of the molten phase. It is shown in Fig.4. From this figure we see that $\varepsilon(k)$ is negative for $k \leq 5\text{Å}^{-1}$. A very similar behavior was found in a classical one–component plasma by Fasolino et al.,[4] in a hydrogen plasma by Hansen and McDonald[8] and in a degenerate electron liquid by Iyetomi et al.[9] Fasolino et al. calculated the static dielectric function using the structural data based upon the computer simulation. They found that the inverse dielectric function rises smoothly with increasing k from zero to unity at small values of the plasma parameter Γ, but begins to show an increasingly deep negative region for k around the main peak of the structure factor with increasing Γ, as oscillations develop in the radial distribution $g(r)$. This behavior is very close to the transition of plasma to a Wigner crystal. Figure 4 is plotted for two different temperatures 903K (marking with +) and 1036K (marking with black points). At fixed k, for k in the region where $\varepsilon(k) < 0$, $\varepsilon(k)$ increases as T decreases. This is similar to that occurring in a classical one component plasma by changing Γ, if we note that a decreasing T in the molten phase corresponds to an increasing Γ. The curve $\mathcal{E}(k)$ touches the abscissa axis at $k = 1.7\text{Å}^{-1}$. This shows the forerunning phenomenon to the crystallization, in which ions in the molten phase are ready to change to the solid state phase, $i.e.$ α-AgI.

Figure 5 shows the wave-number-dependence of $\mathcal{E}(k)$ along the (h, h, 0) and (h, h, h) in the α-phase at 683K. The values of $\mathcal{E}(k)$ become zero at $k = 1.74$, 3.49 and 5.23 Å^{-1} in Fig.5a with reflection of the Debye peaks which lead to the divergence of $S_{CC}(\mathbf{k})$. Figures 3-5 show the existence of a negative sign for the static dielectric function in the both molten phase and α-phase of AgI. These also teach us that a some kind of ordering of ions in a regular lattice system and a sign of the forerunning phenomena to an ordering from a liquid state are found in the wavenumber-dependence of $\mathcal{E}(k)$.

FRACTAL BEHAVIOR OF DIFFUSION TRAJECTORIES

1. α-Phase

We also have investigated the fractal behavior in diffusion trajectories of ions and Brownian functions. We expect the forerunning phenomenon to the crystallization also in a fractal behavior.[10]

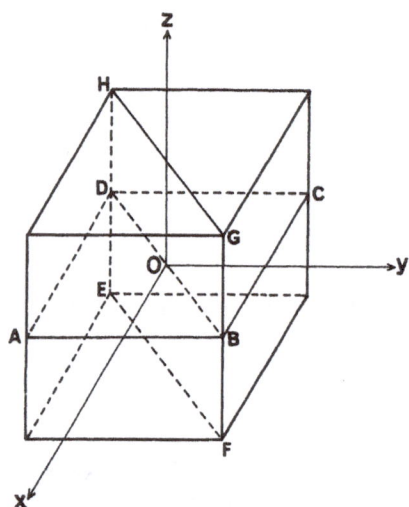

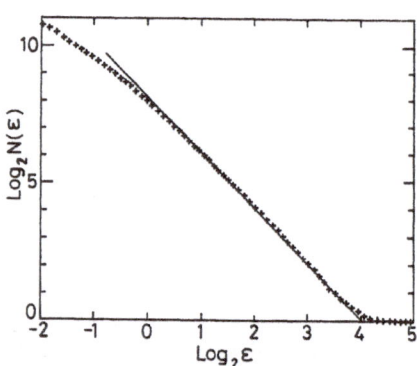

Figure 6. Division number $N(\varepsilon)$ against the length ε for the diffusion trajectory of Ag ions in α-AgI. The solid line represents a straight line with slope -2(ref.10).

Figure 7. Nine different planes for the calculation of Brownian functions.

Let us divide the diffusion trajectory of an ion with a divider of length ε and denote the division number as $N(\varepsilon)$. If $N(\varepsilon)$ is proportional to ε^{-D}, $i.e.$

$$N(\varepsilon) \propto \varepsilon^{-D} , \tag{13}$$

D is called the fractal dimension. The fractal dimension shows a measure of self-similarity in a geometric pattern. The fractal behavior of the diffusion trajectory of ions is similar to that of coast lines.[11] Figure 6 shows the variation of the division number $N(\varepsilon)$ against ε for 250 Ag ions at $T=670$ K in the α-phase. We get $D=2$ from this figure. For Brownian motion in ordinary fluids the value of D is expected to be 2.[11] The study of the fractal behavior of Ag ions in α-AgI suggests their liquid like motion in accordance with experiments on their structure.

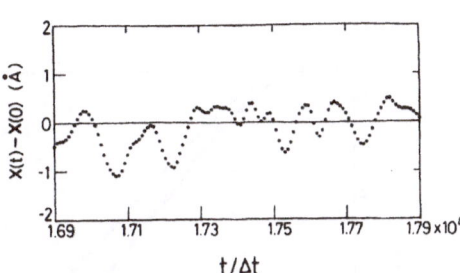

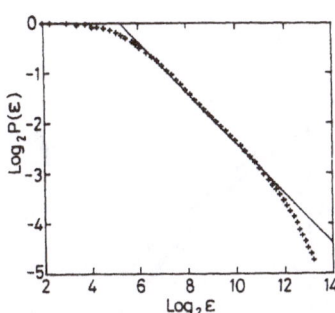

Figure 8. The Brownian function $x(t) - x(0)$ over a narrow range of Δt, plotted at interval of $12\Delta t$. A gap G is a time interval for which $x(t) - x(0) = 0$.

Figure 9. Probability function $P(g)$ of gaps in zerosets of Ag ions at 670K in the α-phase. The solid line represents a straight line with slope -0.5(ref.10).

Next we investigate the fractal behavior associated with the sets of constancy, or isosets, of coordinate functions $x(t), y(t)$ and $z(t)$. The zeroset is defined as those instants $t=\tau$ for which $x(t)=0$. We take the origin 0 of coordinates at the center of system as shown in Fig.7. The calculations are performed for nine different planes; which are ABCD plane and two other topologically equivalent planes and EFGH plane and 5 other topologically equivalent planes. Figure 8 shows the time variation of the coordinate function $x(t)$ of a Ag ion measured relative to its value at $t=0$ in the molten phase at $T=900$ K. The function $x(t) - x(0)$ is plotted at intervals of $12\Delta t$. We call the function $x(t) - x(0)$ the Brownian function. A gap G is a time interval in which the coordinate function has finite values. As mentioned above the Brownian function becomes zero at both ends of a time interval. Gaps between successive values of τ are characterized by a probability function, $Pr(G > g)$, for finding a gap of duration G greater than a given value g as follows:

$$P(g) \quad \equiv \quad Pr(G > g) \quad . \tag{14}$$

If there is self-similarity in zerosets, the probability function is expressed by

$$P(g) \quad \propto \quad g^{-D} \quad , \tag{15}$$

where $\bar{D}$ is the fractal dimension of the zeroset. In Fig.9, the probability function $P(g)$ is plotted against a gap g. We get $\bar{D}=0.5$ from Fig.9. Then we have the relation

$$D \quad = \quad (1 - \bar{D})^{-1} \quad . \tag{16}$$

Application of this relation to SIC system was confirmed by Vashishta *et al.*[12] for Ag_2S. It is known that the probability function $Pr(G > g)$ of a Brownian zeroset has the fractal dimension $\bar{D} = 0.5$.[11]

2. Molten Phase

Next we calculated $N(\varepsilon)$ and $P(g)$ for 250 Ag ions and 250 I ions at T=900 K in the molten phase. Results are plotted in Figs.10 and 11. From these figures, we get

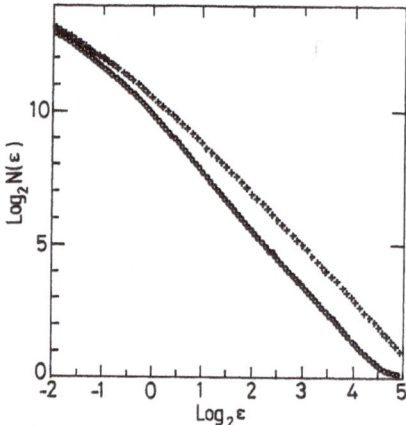

Figure 10. $N(\varepsilon)$ for the diffusion trajectory of Ag ions and I ions at 900K in the molten phase(ref.10).

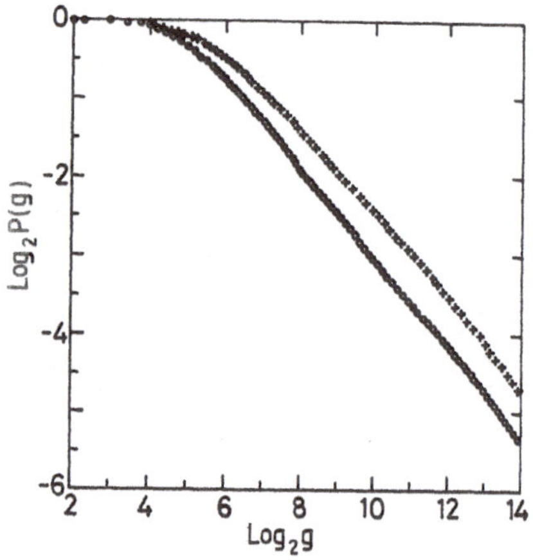

Figure 11. $P(g)$ of Ag ions and I ions at 900K in the molten phase(ref.10).

$D=2$ and $\bar{D}=0.5$ for Ag ions. These results are the same as those in the α-phase. But we get $D=2.17$ and $\bar{D}=0.58$ for I ions. These values of D and $\bar{D}$ for I ions did not change for the calculations of 0.06 million time steps and 0.12 million time steps. Then the number of time steps will be enough for the calculation of the fractal dimension of I ions in the molten phase. On the other hand the mean-square displacement (MSD) was also calculated for Ag and I ions and is shown in Fig.12. The MSD of Ag ions in the α-phase increases linearly with time, whereas that of I ions remains nearly constant. These are reflected with a liquidlike diffusion-motion of Ag ions and a lattice vibration of I ions. In the molten phase, the MSD's of Ag ions and I ions increase linearly with time, which means a liquidlike diffusion-motion of both Ag and I ions. When I ions conduct a liquidlike motion, the fractal dimension D of their trails is expected to be 2. Why has the fractional value of $D=2.17$ been realized for I ions at 900 K in the molten phase? As the time change in the MSD shows, I ions also diffuse in the system. However a local structure of I ions at 900 K in the molten phase may have a similar structure to that in the solid phase, because our temperature, 900 K, is just over the melting point, 828 K.

Now let's show another evidence that a local structure of I ions at 900K has a similar structure to that in the solid phase. The calculation of the partial pair distribution functions support this suggestion.

Figure 13 shows the partial pair distribution function of I-I in the molten phase (900 K). This shows the nearest neighbor distance between I-I and the coordination number are 4.4 Åand 14, respectively, which are the same as those in the α-phase. The coordination number means a number of I ions which are coordinated around a centered I ion.

If we perform MD simulations to the system at higher temperature than 900 K, we may get $D=2$ for I ions in the molten phase for lack of the local structure of I ions. But as it was difficult for economic reasons to perform supplementary MD simulations to our system, we were obliged to do them for the AgI system of small size with 108 particles (54Ag + 54I) using the same potential as Eq(12).

Results of supplementary MD simulations in the molten phase are as follows: We get $D=2$ for Ag ions and $D=2.17$ for I ions at 1000 K, which are the same results as those for 500 particles system. As the temperature is raised at 2000 K, we get $D=2$ for both Ag and I ions. These are the results as we expected and support our considerations.

At an extremely high temperature over the melting temperature, the system is in completely liquid state, which leads to a value of $D = 2$ for Ag and I ions. At the temperature just over the melting temperature, a local structure is similar to that in the solid state phase. This shows the forerunning phenomenon to the crystallization, in which ions in the molten phase are ready to change to the solid state phase. We got a similar phenomenon found in the wavenumber-dependent static dielectric function $\varepsilon(\mathbf{k})$.

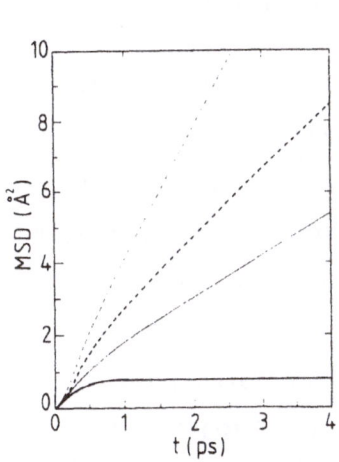

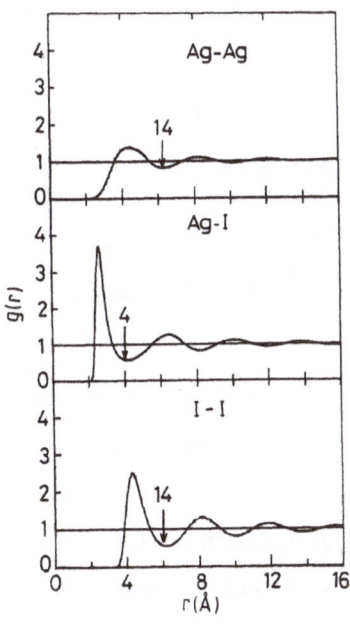

Figure 12. Mean-square displacements of Ag ions (dotted line) and I ions (solid line) at (1) $T=670K$ (bold line) and (2) $T= 900K$ (thin line)(ref.10).

Figure 13. Partial pair distribution functions for Ag-Ag, Ag-I and I-I at 903K in the molten phase. The arrows show coordination numbers.

CONCLUSION

We have studied the pre-freezing phenomenon in the static dielectric function $\varepsilon(k)$ of AgI and fractal behavior in diffusion trajectories of ions. The results are summarized as follows:

- The $\varepsilon(k)$ touched the abscissa axis at some k as the melting temperature was approached.

- This shows the tendency of ions in the molten phase to order similarly to the solid phase (*i.e.* α-AgI), which leads to the *symmetry reduction* of configuration of ions.

- In the solid phase, the values of $\varepsilon(k)$ became zero at each Debye peak.

- The study of $\varepsilon(k)$ gave us information about ordering of ions in a regular lattice system and a sign of the forerunning phenomena to an ordering from a liquid state.

- A similar phenomenon was found in the fractal behavior of diffusion trajectories of ions.

ACKNOWLEDGEMENT

This work was supported in part by a Grant-in-Aid for Scientific Research from the Ministry of Education, Science and Culture.

REFERENCES

1. O.V. Dolgov, D.A. Kirzhnits and E.G. Maksimov: Rev. Mod. Phys. **53** (1981) 81.

2. M. Aniya, H. Okazaki and M. Kobayashi: Phys. Rev. Lett. **65**(1990)1474.

3. A. J. Stafford and M. Silbert: Z. Phys. B **67**(1987)31.

4. A. Fasolino, M. Parrinello and M. P. Tosi: Phys. Lett. **66**A (1978)119.

5. W. L. Slattery, G. D. Doolen and H. E. DeWitt: Phys. Rev. A**21**(1980)2087.

6. F. Shimojo and M. Kobayashi: J. Phys. Soc. Jpn. **60** (1991) 3725.

7. M. Parrinello, A. Rahman and P. Vashishta: Phys. Rev. Lett. **50**(1983)1073.

8. J. P. Hansen and I. R. McDonald: Phys. Rev. Lett. **41**(1978)1379.

9. H. Iyetomi, K. Utsumi and S. Ichimaru: Phys. Rev. B**24**(1981)3226.

10. M. Kobayashi and F. Shimojo: J. Phys. Soc. Jpn. **60**(1991)4078.

11. B. B. Mandelbrot: *The Fractal Geometry of Nature* (Freeman, New York, 1977).

12. P. Vashishta, I. Ebbsjö, R. K. Kalia and S. W. de Leeuw: Solid State Commun. **59**(1986)873.

CLASSICAL SOLUTIONS OF SCHROEDINGER EQUATIONS AND NONSTANDARD ANALYSIS

Tsunehiro Kobayashi

Institute of Physics
University of Tsukuba
Ibaraki 305, Japan

INTRODUCTION

It is believed that classical mechanics is obtained from quantum mechanics by taking the limit of $\hbar \to 0$. Such a consideration is easily understood in the following discussions: The Schroedinger equation is generally written down as

$$i\hbar \frac{\partial \Psi(\mathbf{r}, t)}{\partial t} = H\Psi(\mathbf{r}, t), \tag{1}$$

where Hamiltonian is given by

$$H = (1/2m)[-(\hbar\nabla)^2 + V(\mathbf{r})].$$

The solution of (1) (the wave function) can be described as

$$\Psi(\mathbf{r}, t) = e^{iS(\mathbf{r},t)/\hbar}, \tag{2}$$

where $S(\mathbf{r}, t)$ is a function having the dimension of action. For the simplicity we shall discuss one-dimensinal problems here. The extention to arbitrary dimentions is straightforward. Solutions for stationary states are derived by putting $S(x, t) = W(x) - Et$, where E denotes the energy. Now the Schroedinger equation is rewritten as the equation for $W(x)$ in the following form,

$$i\frac{\hbar}{2m}\frac{d^2W}{dx^2} = \frac{1}{2m}\left(\frac{dW}{dx}\right)^2 + V(x) - E. \tag{3}$$

Taking $\hbar = 0$ in (3), we have the equation

$$\frac{1}{2m}\left(\frac{dW}{dx}\right)^2 + V(x) - E = 0. \tag{4}$$

Symmetries in Science VII, Edited by B. Gruber
and T. Otsuka, Plenum Press, New York, 1994

This equation coincides with the Hamilton-Jacobi equation of classical mechanics which apparently derives the solution for the path expected from classical mecanics, i.e.

$$W_{cl}(x) = \int^x \sqrt{2m(E - V(x'))}dx'. \tag{5}$$

It is, therefore, bilieved that classical mechanics is realized in the $\hbar \to 0$ limit of quantum mechanics.

A question arises, that is, "Is the above argument precise enough to understand solutions of the Schroedinger equation in the $\hbar \to 0$ limit?". It is important that even if $\hbar$ is small enough to be negligible as compared with other classical quantities, $\hbar$ is not exactly zero but should be taken as an infinitesimal ($\hbar \approx 0$). The difference between zero and infinitesimal for $\hbar$ is clearly seen in the wave function given by (2). In the case of $\hbar = 0$ the solutions satisfying (4) are only meaningful. Expanding $S(x,t)$ in the power of $\hbar$ as $S(x,t) = S_0(x,t) + \hbar S_1(x,t) + ..$, we see that the lowest order term $S_0(x,t)$ is only meaningful if $\hbar = 0$ is taken. However, the wave function having the diverging phase $S(x,t)/\hbar$ loses the meaning. In order to keep the meaning of the wave function we have to keep the fact that $\hbar$ is an infinitesimal but not exactly zero. In such a treatment we also have to take account of the term $\hbar S_1(x,t)$ to get the correct expression for the wave function having the phase $S(x,t)/\hbar$.

The derivation of the exact classical limit also seems to be important in the quantum theory of measurement, where the transition of pure states to mixed states (so-called wave-function collapse) proposed by von Neuman[1] must be derived. It is known that classical states are described in terms of mixed states (no interference among different classical paths), while quantum states are always represented by pure states. The derivation of the transition from pure states to mixed states will be closely connected with the problem for deriving the exact classical limit from quantum mechanics.

An interesting mathematical theory to investigate the limit is nonstandard analysis proposed by Robinson[2], where the set of real numbers R is extended to $*R$ which contains R, infinitesimals and infinities. The nonstandard analisis provides a suitable framework for the analysis by means of infinitesimals and infinities. Then in the theory $\hbar$ can be taken as an infinitesimal ($\hbar \approx 0$). We shall investigate the Schroedinger eq. in terms of the nonstandard analysis where $\hbar$ is taken as an infinitesimal and the limit of $\hbar \to 0$ is exactly defined by the operation of taking standard parts with respect to $\hbar$.[4] Hereafter we write the standard-part operation for a function g as $st_\hbar(g)$.

ULTRA EIGENVECTORS IN EXTENDED HILBERT SPACE

Following the transfer theorem by Robinson in the nonstandard analysis, usual Hilbert spaces are extended to extended Hilbert spaces ($*H$). (In details of the discussions in this section, see Ref.3.) On $*H$ we have a different equivalence relation arising from infinitesimal differences between eigenvectors.[3] That is, since infinitesimal differences represented by infinitesimal vectors cannot be distinguishable, eigenvectors f and $f + d$ should represent the same state if $d, f \in *H$, $\|f\| = 1 (\|f\| = \sqrt{<f,f>}$: the norm of f) and $\|d\| \approx 0$ (d : an infinitesimal vector). In general eigenvectors in the extended Hilbert space ($*H$) are extended to ultra eigenvectors (f) which are defined by

$$\|Af - \lambda f\|/\|f\| \approx 0$$

where A is a self-adjoint operator and $\lambda \in *R$. We should notice the difference between the ultra eigenvectors above defined and usual eigenvectors satsfying $\|Af -$

$\lambda f\| = 0$. Especially it is shown that eigenvectors for continuous spectrum are represented by ultra eigenvectors. Now we can write the equivalence relation on $*H$ as

f and g are equivalent on $*H$ ($f \longleftrightarrow g$) if and only if $f, g \in *H$ and
$\|f\| = \|g\| = 1$ and there exists a $\phi \in *R$ such that $\|e^{i\phi}f - g\| \approx 0$.

The set of physical normed states $S(*H)$ is defined as the quotient set $S(*H) = U(*H)/ \longleftrightarrow$ by means of the equivalence relation, where $U(*H)$ is the set of unit vectors in $*H$. Following the discussions of Farrukh[3], we have the relation between probabilities for equivalent states ($f \longleftrightarrow g$) on $*H$

$$st < f, E_\lambda(A)f >= st < g, E_\lambda(A)g >, \tag{6}$$

where $st <>$ denotes the operation for taking the standard part in the nonstandard analysis and $E_\lambda(A)$ is the spectral measure associated with a standard self-adjoint operator A for $\lambda \in R$. It is important that in the extened Hilbert space $*H$ the theory allows different expressions for a physical state of $S(*H)$ arising from not only the phase factor $e^{i\phi}$ for $\phi \in *R$ but also infinitesimal vectors. Hereafter we shall represent the set of the physical unit vectors under the condition $\hbar \approx 0$ by $S_\hbar(*H)$. Note that ultra eigenvectors on $*H$ satisfy the equation

$$H\phi_c^E \approx E\phi_c^E$$

instead of the exact equality, where the meaning of $\approx$ is defined by

$$\|H\phi_c^E - E\phi_c^E\| \approx 0 \tag{7}$$

for vectors satisfying $\|\phi_c^E\| = 1$.

CLASSICAL SOLUTIONS OF SCHROEDINGER EQ

In the stationary state we have the following relation for the momentum operator $p = -i\hbar\partial/\partial x$;

$$p\phi(x) = (dW/dx)\phi(x)$$

where $\phi(x) = exp(iW(x)/\hbar)$. We define the momentum function $P(x)$ as

$$P(x) \equiv dW/dx = u(x) + iv(x), \tag{8}$$

where $u(x)$ and $v(x)$, respectively, stand for the real and the imaginary parts of $P(x)$. In the eigenstate of the momentum, of course, $v(x) = 0$ must be satisfied. Considering that the momentum is an observable quantity in classical mechanics, the same relation $v(x) = 0$ has to be derived in the operation of $st_\hbar(v(x))$, whereas $st_\hbar(u(x)) \neq 0$ and ∞ should be satisfied except special cases for rest particles. Note that in $S_\hbar(*H)$ all the observed numbers must be not infinitesimals but real numbers. We expect the following orders for u and v;

$$u(x) \sim 0(1),$$

$$v(x) \sim O(\hbar),$$

where $0(1)$ and $O(\hbar)$ are, respectively, defined by

$$st_\hbar(0(1)) \neq 0 and \infty,$$

$$st_\hbar(O(\hbar)) = 0.$$

The equation for $W(x)$ is reduced to the following two equations for u and v,

$$\hbar(du/dx) \approx 2uv, \qquad (A)$$

$$\hbar(dv/dx) \approx 2m(E - V(x)) - (u^2 - v^2), \qquad (B)$$

where (A) and (B), respectively, correspond to the imaginary and the real parts of (3). From the order estimation for u and v we see that (A) is satisfied only in the following two cases except solutions for $u = 0$:

$$du/dx \sim u \sim du/dx \sim v \sim O(\hbar), \qquad (I)$$

or

$$du/dx \sim u \sim O(1),$$

$$dv/dx \sim v \sim O(\hbar). \qquad (II)$$

Taking account of (B), we see that in case (I) the relation

$$E - V(x) \sim O(\hbar^2)$$

is expected, while

$$E - V(x) \sim O(1) \qquad (9)$$

but

$$2m(E - V(x)) - u^2 \sim O(\hbar^2) \qquad (10)$$

are expected in case (II). It is apparent that the solutions (I) represent eigenfunctions for the usual quantum mechanics. We know that the very solutions (II) are what we are interested in as the solutions connecting to classical states, because even in the $\hbar \to 0$ limit (the $st_\hbar$ operations) the momentum u and the energy E still have classically observable values, that is, they are not infinitesimals but real numbers in the limit. Let us study this type of the solutions in the following discussions.

From (A) we easily derive the relation

$$v \approx (\hbar/2)(1/u)(du/dx) \approx (\hbar/2)(d\ln(u)/dx). \qquad (11)$$

It is interesting that v always has the order of $O(\hbar)$, of which property is independent from the order of u except the case for $u = $constants. We should, however, notice that even if $v(x)$ vanishes in the $\hbar \to 0$ limit the imaginary part of $W(x)$ given by the integration of $v(x)$ always has a contribution to the wave function in the limit, which is evaluated as

$$e^{-ImW(x)/\hbar} \approx (u(x))^{-1/2} \qquad (12)$$

from (11). We can easily estimate the zeroth order of $u(x)$ from (B) for the condition $E - V(x) \sim O(1)$ as

$$u_{cl}(x) = \pm\sqrt{2m(E - V(x))}. \qquad (13)$$

This is nothing but the classical momentum of the particle moving in the potential $V(x)$. We can calculate the order $O(\hbar)$ of $u(x)$ by putting the form $u(x) = u_{cl}(x) + \hbar u_1(x)$ and $v \sim O(\hbar)$ into (B). Since the left hand side of (B) is of the order of $O(\hbar^2)$ and only one term $-2u_{cl}(x)\hbar u_1(x)$ in the righthand side has the order $O(\hbar)$, we derive

$$u_1(x) = 0. \qquad (14)$$

As was previously noted, the higher order $(O(\hbar^n), n \geq 2)$ terms which have no contribution in the operation (6) for real-number eigenvalues $(\lambda \in R)$ are buried in the ambiguity on $*H$ represented by the equivalence relation. Then in the wave function we may put

$$u(x) \approx u_{cl}(x), v_c(x) \approx (\hbar/2u_{cl})(du_{cl}/dx)$$

for the region $E - V(x) \sim O(1) > 0$. The eigenfunction is given by

$$\phi_c^E(x) \approx (N/\sqrt{u_{cl}(x)})exp(i\int_c^x u_{cl}(x\prime)dx\prime/\hbar), \tag{15}$$

for $0 < E - V(x) \in R$, where N is a normalization constant.

When the potential have the points (x_o) satisfying $E - V(x_o) = 0$, we do not have any solutions representing $u \sim O(1)$ in the neighbourhood of x_o. We can easily obtain $W(x)$ directly from (3) in the region $0 < E - V(x) < O(\hbar^2)$, where the equation is reduced to

$$i\hbar\frac{d^2W}{dx^2} \approx (\frac{dW}{dx})^2. \tag{16}$$

We find that (16) has the solution

$$W(x) \approx -i\hbar ln|x_o - x|. \tag{17}$$

From the relation $exp(-ImW(x)/\hbar) = |x_o - x|$ it is shown that $\phi_c^E(x)$ behaves as $|x_o - x|$ in the neighbourhood of x_o. We may have different expressions for $\phi_c^E(x)$ satisfying the conditions (15) and (17), which are allowed on $*H$ because of the ambiguity in terms of infinitesimals. For instance, we may put

$$\phi_c^E(x) = \frac{1}{\sqrt{u_c^E(x)}}exp(iW_{cl}(x)/\hbar), \tag{18}$$

where $W_{cl}(x)$ is given in (5) and $u_c^E(x)$ is a function satisfying the conditions

$$u_c^E(x) \approx u_{cl}(x) + O(\hbar^2)$$

for $0 < E - V(x) \sim O(1)$

$$u_c^E(x) \propto |x_o - x|^{-2}$$

for $0 < E - V(x) < O(\hbar^2)$ such as

$$u_c^E(x) = \frac{u_{cl}(x)^2 + \hbar^2 A(x_o - x)^{-2}}{u_{cl}(x) + \hbar^2 B}$$

with the constants $A, B \in *R$ but being finite. This is a quite different feature of the nonstandard analysis from the usual classical limit in quantum mechanics, which has no such ambiguity. The important point is the fact that

$$\phi_c^E(x = x_o) = 0.$$

This fact tells us that we can connect the eigenfunction given in the region $E - V(x) > 0$ with the function

$$\phi_c^E(x) = 0$$

in the region satisfying $E - V(x) < 0$ at $x = x_o$. That is to say, our solution is written as

$$\phi_c^E(x) = (N^E/\sqrt{u_c^E(x)})exp(iW_{cl}(x)/\hbar)$$

for $E - V(x) \geq 0$

$$= 0 \tag{19}$$

for $E - V(x) < 0$. The property $\phi_c^E(x) = 0$ for $E - V(x) < 0$ is a quite nice feature of the solutions representing classical states, because we know that no penetration in the region $E - V(x) < 0$ must be realized in classical mechanics . Actually $\phi_c^E(x)$ looks like a quantum solution for the potential expressed by $V(x) = V(x)$ for $E - V(x) > 0$ and $V(x) = \infty$ for $E - V(x) < 0$. Note that though $cos(W_{cl}/\hbar)$ or $sin(W_{cl}/\hbar)$ may be used instead of $exp(iW_{cl}/\hbar)$ in (19) for stationary states, we shall use (19) because no essential difference arises in the following discussions.

AN EXAMPLE FOR HARMONIC OSCILLATOR

Let us compare the classical solutions above obtained with the usual quantum mechanical solution for harmonic oscillator potential. For the simplicity we compare the absolute value of $\phi_c^E(x)$ having $E = (1/2)m\omega^2 x_o^2$ with the wave function of the ground state $\phi_o(x)$ having $E = (1/2)\hbar\omega$ for the harmonic oscillator potential $V(x) = (1/2)m\omega^2 x^2$. They are given as

$$|\phi_c^E(x)| \approx \sqrt{\frac{1}{\pi(x_o^2 - x^2)^{1/2}}},$$

for $0 < E - V(x) \sim O(1) \in R$

$$\propto |x_o - x|,$$

for $0 < E - V(x) < O(\hbar^2)$ and

$$= 0$$

for $0 > E - V(x)$, and

$$\phi_o(x) = (\frac{m\omega}{\pi\hbar})^{1/4}exp(-\frac{m\omega}{\hbar}x^2).$$

for $\forall x \in {}^*R$. The difference is quite obvious, that is, the quantum mechanical solution has nonzero value only in the region of $x^2 \sim O(\hbar)$, while ϕ_c^E has nonzero values even in the region $x \sim O(1)$. The difference between the behaviours at the origin $(x = 0)$ in the $\hbar \to 0$ limit is crucial, that is, $st_\hbar(\phi_c^E(0)) = \sqrt{1/\pi x_o} \in R$ but $st_\hbar(\phi_o(0)) = \infty$ is not the element of R. In the classical solution the operation $st_\hbar(|\phi_c^E(x)|)$ only sweep off the two neighbourhood regions $x_o^2 - x^2 < O(\hbar^2)$ where $\phi_c^E(x)$ differs from the solutions expected from classical mechanics.

DECOHERENCE AMONG CLASSICAL SOLUTIONS

Let us examine the properties of $\phi_c^E(x)$.
(1) The distribution function is given by

$$\rho_c = |\phi_c^E(x)|^2 = |N^E|^2/u_c^E(x) \tag{20}$$

in the region $E - V(x) > 0$ and

$$\rho_c = 0$$

for $E - V(x) < 0$. We see that

$$st_\hbar(\rho_c) = |N^E|^2 / \sqrt{2m(E - V(x))} \qquad (21)$$

for $E - V(x) > 0$. This distribution coincides with that expected from classical mechanics for the particle moving in the potential $V(x)$ with the energy E, where, of course, $E - V(x) \in R$ have to be satisfied.
(2) The phase of ϕ_c^E is given by

$$(1/\hbar)ReW^E(x) = (1/\hbar)W_{cl}(x) = (1/\hbar) \int^x \sqrt{2m(E - V(x'))}dx'. \qquad (22)$$

It is important that $W_{cl}(x)$ always has the order of $O(1)$ for $\forall x \in R$ in the region $0 < E - V(x) \in R$ because $W_{cl}(x)$ represents the classical path.

Let us consider the superposition of two different classical states with two different energies E and E' satisfying the condition $0 \neq E - E' \in R$, i.e. $E - E' \sim O(1)$, as

$$\psi(x) = c_E \phi_c^E(x) + c_{E'} \phi_c^{E'}(x), \qquad (23)$$

where c_E and $c_{E'}$ are constants normalized with

$$|c_E|^2 + |c_{E'}|^2 = 1. \qquad (24)$$

The density matrix of ψ is derived as

$$\rho_\psi(x) = |c_E|^2 \rho_c^E(x) + |c_{E'}|^2 \rho_c^{E'}(x) + (c_E c_{E'}^* \phi_c^E(x)\phi_c^{E'}(x)^* + c.c.). \qquad (25)$$

The off-diagonal term has the phase factor

$$exp[i(1/\hbar)(W_{cl}^E(x) - W_{cl}^{E'}(x))]. \qquad (26)$$

It is shown that

$$st_\hbar((1/\hbar)(W_{cl}^E(x) - W_{cl}^{E'}(x))) = 0$$

for $E = E'$

$$= \infty \qquad (27)$$

for $0 \neq E - E' \in R$, because the difference between W_{cl}^E and $W_{cl}^{E'}$ corresponding to two different classical paths must be an observable real number. This fact tells us that expectation values for all classical observables $A \sim O(1)$ related to the $st_\hbar$ operations of (6) do not have any contributions from the off-diagonal terms, that is to say, we may write

$$st_\hbar < \psi, A\psi > = st_\hbar(|c_E|^2 < \phi_c^E, A\phi_c^E >) + st_\hbar(|c_{E'}|^2 < \phi_c^{E'}, A\phi_c^{E'} >). \qquad (28)$$

The contributions of the off-diagonal terms disappear in the $st_\hbar$ operation appearing in the expectation values because of the divergent property of the phases. We may conclude that all the pure states represented by the superposition in terms of the classical solutions $\phi_c^E(x)$ are reduced to the mixed states on the $st_\hbar$ operation in $S_\hbar(^*H)$. In other words classical states can be described in terms of wave functions of the extended quantum mechanics by means of nonstandard analysis with $\hbar \approx 0$, but we cannot observe any quantum effects arising from interferences among different states. There is no difference between pure states and mixed states on $S_\hbar(^*H)$. This fact is the most important result obtained on $S_\hbar(^*H)$.

ON THE SCHROEDINGER'S CAT

Now we can give an answer to the problem so-called the Schroedinger's cat in the quantum theory of measurement. Two states representing a living cat and a dead cat clearly have differences observed in classical mechanics and expressed by classical numbers contained in R. Then the pure state superposed by the two states is always reduced to the mixed state in the evaluations of expectation values for all physical quantities observed in classical mechanics. We may conclude that there is no physically observable difference between the pure state and the mixed state for the Schroedinger's cat in the classical limit.

The decoherence property of classical solutions may also help us to understand the transition from pure states to mixed states (so-called wave-function collapse) in the quantum theory of measurement, if it is shown that detectors in measurement processes are described in terms of the classical solutions on $S_\hbar(^*H)$.

REMARKS

I would like to comment that the nonstandard analysis can be applicable to many physical processes containing very small quantities or very large quantities as compared with other physical values .e.g. the light velocity c in nonrelativistic phenomena, Avogadro's number in thermal phenomena, size and mass of universe, gravitational coupling constant, Einstein's constant in general relativity and etc.

I would like also stress that the complete theory containing quantum and classical mechanics, and maybe gravitations also, should include detectors for every measurement processes in the theory itself. It seems to be possible that the role of detctors in physical processes can be represented by the free ultra filter in the nonstandard analysis.

REFERENCES

1. J. von Neuman,"Die mathematische Grundlagen der Quanten-mechanik" Springer, Berlin, (1932).
2. A. Robinson, "Non Standard Analysis", North-Holland, Amsterdam, (1970).
3. M.O. Farrukh, Application of nonstandard analysis to quantum mechanics, *Jour. Math. Phys.* 16:177(1975).
4. T. Kobayashi, An answer to the Schroedunger's cat, preprint of University of Tsukuba, Ibaraki 305, Japan (1992).

ATOMIC SUPERSYMMETRY, OSCILLATORS, AND THE PENNING TRAP

V. Alan Kostelecký

Physics Department
Indiana University
Bloomington, IN 47405 U.S.A.

INTRODUCTION

It is a pleasure to participate in celebrating Franco Iachello's 50th birthday. I wish him many happy returns.

At least two physical supersymmetries are known to exist in nature. One, discovered by Franco and his coworkers, makes connections between different nuclei [1]. The other, atomic supersymmetry, is the subject of part of this paper. It can be viewed as a symmetry-based approach to the construction of an effective central-potential model describing the behavior of the valence electron in atoms and ions.

This paper begins with some background information and a summary of results in atomic supersymmetry. The connection between the supersymmetric Coulomb and oscillator problems in arbitrary dimensions is outlined. Next, I treat the issue of finding a description of supersymmetry-based quantum-defect theory in terms of oscillators. A model with an anharmonic term that yields analytical eigenfunctions is introduced to solve this problem in arbitrary dimensions. Finally, I show that geonium atoms (particles contained in a Penning trap) offer a realization of a multidimensional harmonic oscillator in an idealized limit. The anharmonic theory presented here provides a means of modeling the realistic case.

SUPERSYMMETRIC QUANTUM MECHANICS

This section provides some background in supersymmetric quantum mechanics [2] and establishes notation.

A quantum-mechanical hamiltonian H_S is said to be supersymmetric if it commutes with N supersymmetry operators Q_j and if it is generated by anticommutators according to

$$\{Q_j, Q_k\} = \delta_{jk} H_S \ . \tag{1}$$

The operators H_S and Q_j form the generators of a superalgebra denoted by sqm(N).

Symmetries in Science VII, Edited by B. Gruber
and T. Otsuka, Plenum Press, New York, 1994

For the purposes of this paper it suffices to consider the special case $N = 2$, with superalgebra sqm(2). Define the linear combinations

$$Q = \sqrt{\tfrac{1}{2}}\,(Q_1 + iQ_2) \ , \qquad Q^\dagger = \sqrt{\tfrac{1}{2}}\,(Q_1 - iQ_2) \ . \tag{2}$$

Then, the supersymmetric hamiltonian can be written

$$H_S = \{Q, Q^\dagger\} \ . \tag{3}$$

For one-dimensional quantum systems, the superalgebra sqm(2) admits a two-dimensional representation. Write

$$Q = \begin{pmatrix} 0 & 0 \\ A & 0 \end{pmatrix} \ , \qquad H_S = \begin{pmatrix} h_+ & 0 \\ 0 & h_- \end{pmatrix} \ , \tag{4}$$

with

$$A = -i\partial_x - iU'/2 \ . \tag{5}$$

Here, U' denotes dU/dx for some function $U = U(x)$. In this representation, the supersymmetric hamiltonian H_S contains two components h_+ and h_-, referred to as the bosonic and fermionic hamiltonians, respectively. These satisfy the equations

$$h_\pm \Psi_{\pm n} \equiv \Big[-\frac{d^2}{dx^2} + V_\pm(x)\Big]\Psi_{\pm n} = \epsilon_n \Psi_{\pm n} , \tag{6}$$

with

$$V_\pm(x) = (\tfrac{1}{2}U')^2 \mp \tfrac{1}{2}U'' \ . \tag{7}$$

Using the above relations, some general properties of the supersymmetric system can be obtained. First, for unbroken supersymmetry the ground-state energy is zero. Second, except for the ground state (which appears in the spectrum of h_+) the bosonic and fermionic spectra are degenerate. Finally, the supersymmetry generators $Q, Q^\dagger$ map degenerate states from the two sectors into each other.

ATOMIC SUPERSYMMETRY: EXACT LIMIT

Consider the Schrödinger equation for the hydrogen atom. In spherical polar coordinates, the equation separates into angular and radial parts. The angular part gives the spherical harmonics, while the radial part can be expressed as

$$\Big[-\frac{d^2}{dy^2} - \frac{1}{y} + \frac{l(l+1)}{y^2} - \frac{1}{2}E_n\Big]\chi_{nl}(y) = 0 \ . \tag{8}$$

Here, atomic units are used, and

$$y = 2r \ , \qquad E_n = -fr12n^2 \ , \qquad \chi_{nl}(2r) = rR_{nl}(r) \ . \tag{9}$$

The radial wave functions $R_{nl}(r)$ are given by

$$R_{nl}(r) = \frac{2}{n^2}\Big[\frac{\Gamma(n-l)}{\Gamma(n+l+1)}\Big]^{\frac{1}{2}}\Big(\frac{2r}{n}\Big)^l \exp\Big(-\frac{r}{n}\Big) L_{n-l-1}^{(2l+1)}\Big(\frac{2r}{n}\Big) \ . \tag{10}$$

In this expression, $L_n^{(\alpha)}(x)$ are the Sonine-Laguerre polynomials, rather than the more restricted Laguerre polynomials (for which α must be integer). This distinction is important in subsequent sections. The Sonine-Laguerre polynomials are defined by

$$L_n^{(\alpha)}(x) = \sum_{p=0}^{n} (-x)^p \frac{\Gamma(n+\alpha+1)}{p!\,(n-p)!\,\Gamma(p+\alpha+1)} \quad . \tag{11}$$

With l fixed, the terms in brackets in Eq. (8) can be reinterpreted as $h_+ - \epsilon_n$, where h_+ is the hamiltonian of Eq. (6). The requirement that the ground-state energy be zero permits the separation of h_+ and ϵ_n, so that the supersymmetric partner hamiltonian h_- and the supersymmetry generator Q can be found [3]. These can be given by specifying the function U of Eq. (5), which is

$$U(y) = \frac{y}{l+1} - 2(l+1)\ln y \quad . \tag{12}$$

Explicitly, the hamiltonian h_- looks like h_+ but with the constant $l(l+1)$ replaced with $(l+1)(l+2)$, i.e.,

$$h_- - h_+ = \frac{2(l+1)}{y^2} \quad . \tag{13}$$

This shift implies that for each l the eigenfunctions of h_- are $R_{n,l+1}$, where $n \geq 2$. Together with the continuum states, they form a complete and orthonormal set.

This formalism can be given a useful physical interpretation as follows. Consider the case $l = 0$. The bosonic sector then describes the s orbitals of hydrogen. Since the spectrum of h_- is degenerate with that of h_+ except for the ground state, and since the supersymmetry generator Q acts on the radial part of the hydrogen wavefunctions but leaves the spherical harmonics untouched, h_- describes a physical system that appears hydrogenic but that has the 1s orbitals inaccessible. One way of realizing this in practice is to fill the 1s orbitals with electrons, thereby excluding the valence electron by the Pauli principle. The element with filled 1s orbitals and one valence electron is lithium. This suggests that h_- should be interpreted as an effective one-body hamiltonian describing the valence electron of lithium when it occupies the s orbitals. At this level, the description cannot be exact because most of the electron-electron interactions are disregarded. However, these can be introduced as supersymmetry-breaking terms. One procedure for this is outlined in a later section of this paper. Even in the absence of such terms, some experimental support for this atomic supersymmetry can be adduced; see ref. [3].

By redefining the energy of the 2s orbital in lithium to be zero, the hamiltonian h_- becomes a suitable choice for a bosonic hamiltonian of a second supersymmetric quantum mechanics. The fermionic partner can be constructed, and an analogous interpretation to the one above can be made. This suggests the s orbitals of lithium and sodium should also be viewed as supersymmetric partners. The process can be repeated for s orbitals and can also be applied for other values of the angular quantum number l, leading to supersymmetric connections among atoms and ions across the periodic table. In the exact-symmetry limit, these connections all involve integer shifts in l and are linked to the Pauli principle. See ref. [3] for more details.

OSCILLATOR REFORMULATION: EXACT CASE

Before describing a method for incorporating supersymmetry-breaking effects, it is appropriate to discuss an alternative formulation of the exact-symmetry case using harmonic oscillators. This section outlines the connections that exist between the radial equations for atomic supersymmetry generalized to arbitrary dimensions and those for the supersymmetric harmonic oscillator [4].

Consider first the d-dimensional Coulomb problem. Upon separation into an angular and a radial part, the radial equation appears:

$$\left[-\frac{d^2}{dy^2} - \frac{1}{y} + \frac{(l+\gamma)(l+\gamma+1)}{y^2} - \frac{1}{2}E_{dn}\right]v_{dnl}(y) = 0 \ . \tag{14}$$

Here, atomic units are used, and

$$y = 2r \ , \quad E_{dn} = -\frac{1}{2(n+\gamma)^2} \ , \quad \gamma = \tfrac{1}{2}(d-3) \ . \tag{15}$$

The radial wave functions are given by

$$v_{dnl}(y) = c_{dnl}\, y^{l+\gamma+1} \exp\left(-y/2(n+\gamma)\right) L_{n-l-1}^{(2l+2\gamma+1)}\left(y/(n+\gamma)\right) \ , \tag{16}$$

where c_{dnl} is a normalization constant.

As before, this one-variable equation can play the role of the bosonic hamiltonian in a supersymmetric quantum mechanics. Appropriately redefining the energy zero so that the ground state has vanishing eigenvalue permits the identification of h_+ and hence of U, Q and h_-. For example,

$$U(y) = \frac{y}{l+\gamma+1} - 2(l+\gamma+1)\ln y \ . \tag{17}$$

Just as for the $d = 3$ case, the fermionic and bosonic hamiltonians differ only by the replacement of l by $l+1$, so that

$$h_- - h_+ = \frac{2(l+\gamma+1)}{y^2} \ . \tag{18}$$

As expected, all the results of the previous section are recovered if $d = 3$, i.e., $\gamma = 0$.

Consider next the supersymmetric harmonic oscillator. For convenience, analogous variables to the Coulomb-problem quantities y, d, n, l, $h_\pm$ are now denoted by upper case symbols Y, D, N, L, $H_\pm$. Thus, upon separation of the angular and radial parts of the Schrödinger equation for the D-dimensional harmonic oscillator, the radial equation is obtained as:

$$\left[-\frac{d^2}{dY^2} + Y^2 + \frac{(L+\Gamma)(L+\Gamma+1)}{Y^2} - 2E_{DN}\right]V_{DNL}(Y) = 0 \ . \tag{19}$$

Here, atomic units have again been used for simplicity, and the oscillator is assumed to have unit frequency. The radial variable is now Y, and

$$E_{DN} = \tfrac{1}{2}(2N + 2\Gamma + 3) \ , \quad \Gamma = (D-3)/2 \ . \tag{20}$$

The radial wave functions are given by

$$V_{DNL}(Y) = C_{DNL} \, Y^{L+\Gamma+1} \exp(-Y^2/2) L_{N/2-L/2}^{(L+\Gamma+1/2)}(Y^2) \ , \tag{21}$$

with C_{DNL} a normalization constant. Note that the usual expressions for the harmonic oscillator in three dimensions are recovered when $\Gamma = 0$.

If Eq. (19) is used to define the hamiltonian H_+ of a supersymmetric quantum mechanics (a redefinition of the energy zero is again needed), then the supersymmetry is specified by a function U given by

$$U(Y) = Y^2 - 2(L + \Gamma + 1) \ln Y \ . \tag{22}$$

It then follows that H_- differs from H_+ by the replacement of L with $L+1$. Therefore,

$$H_- - H_+ = \frac{2(L + \Gamma + 1)}{Y^2} \ . \tag{23}$$

So far, four eigenspectra associated with the supersymmetric Coulomb and oscillator problems have been introduced, defined by the four hamiltonians h_+, h_-, H_+, and H_-. The hamiltonians h_+ and h_- are related by the map $l \rightarrow l + 1$, and the hamiltonians H_+ and H_- are related by $L \rightarrow L + 1$. The next step is to relate the d-dimensional Coulomb problem to the D-dimensional oscillator, i.e., connect h_+ to H_+.

It can be shown [4] that an eigenfunction of h_+ can be transformed by a one-parameter mapping into an eigenfunction of H_+. Explicitly, the functions v_{dnl} and V_{DNL} are connected by the equation

$$v_{dnl}\big((n + \gamma)Y^2\big) = K_{DNL} Y^{1/2} V_{DNL}(Y) \ , \tag{24}$$

where K_{DNL} is a proportionality constant and

$$D = 2d - 2 - 2\lambda \ , \quad N = 2n - 2 + \lambda \ , \quad L = 2l + \lambda \ . \tag{25}$$

The integer λ is the mapping parameter. Notice in particular that only oscillators in even dimensions appear.

The existence of the one-parameter map between h_+ and H_+ combined with the supersymmetry maps evidently establishes connections between any two of the four hamiltonians h_+, h_-, H_+, and H_-. More details can be found in ref. [4].

BROKEN SUPERSYMMETRY AND QUANTUM-DEFECT THEORY

As noted above, atomic supersymmetry in the exact limit is not physically realized because the valence electron interacts with the core electrons by more than the Pauli principle. This section discusses the incorporation of supersymmetry-breaking effects in the context of alkali-metal atoms.

One important effect of the interactions between the valence electron and the core is the change in energy eigenvalues relative to the hydrogenic case. In alkali-metal atoms, the Rydberg series [5] provides a simple formula for the measured energies, given by

$$E_{n^*} = -\frac{1}{2n^{*2}} \ . \tag{26}$$

In this expression,

$$n^* = n - \delta(n, l) \ , \tag{27}$$

where $\delta(n, l)$ is called the quantum defect. For a fixed value of l and increasing n, it turns out that the quantum defects rapidly attain asymptotic values: $\delta(n, l) \simeq \delta(l)$.

The changes in the energy eigenvalues imply that the exact atomic supersymmetry is broken. The breaking can be viewed as an additional contribution H_B to the supersymmetric hamiltonian H_S of Eq. (4). For example, if h_+ arises from the radial equation for hydrogen and h_- is interpreted as the radial equation for the valence electron of lithium in the exact-supersymmetry limit, then the hamiltonian H describing the two systems *including* supersymmetry-breaking effects can be taken as

$$H = H_S + H_B \ , \tag{28}$$

where H_B has the form

$$H_B = \begin{pmatrix} 0 & 0 \\ 0 & V_B(y) \end{pmatrix} \tag{29}$$

and $V_B(y)$ is such as to generate the observed energy eigenspectrum of lithium.

The determination of a suitable V_B is not straightforward. However, it turns out that a functional form for V_B can be found that yields analytical eigenfunctions as solutions to the Schrödinger equation [6]. It is

$$V_B(y) = \frac{l^*(l^* + 1) - l(l + 1)}{y^2} + \frac{n^2 - n^{*2}}{4n^2 n^{*2}} \ . \tag{30}$$

Here, l^* is a modified angular quantum number given by

$$l^* = l + i(l) - \delta(l) \ , \tag{31}$$

where $i(l)$ is an integer parameter shifting the angular quantum number in a manner characteristic of supersymmetry. (If desired, $\delta(l)$ could be replaced by $\delta(n, l)$.) This model effectively replaces the hydrogenic radial equation with one of similar form but involving n^* and l^* rather than n and l.

By construction, the energy eigenvalues are those of the physical atom. The resulting eigenfunctions $R^*_{n^* l^*}(r)$ are analytical and are given by

$$R^*_{n^* l^*}(r) = \frac{2}{n^{*2}} \left[\frac{\Gamma(n^* - l^*)}{\Gamma(n^* + l^* + 1)} \right]^{\frac{1}{2}} \left(\frac{2r}{n^*} \right)^{l^*} \exp\left(-\frac{r}{n^*} \right) L^{(2l^*+1)}_{n-l-i-1}\left(\frac{2r}{n^*} \right) \ . \tag{32}$$

The Sonine-Laguerre polynomials enter again because

$$n^* - l^* - 1 = n - l - i(l) - 1 \tag{33}$$

remains integer. For asymptotic quantum defects $\delta(l)$ and including the continuum states, these eigenfunctions form an orthonormal and complete set.

More details about this construction can be found in ref. [6]. (A connection to parastatistics is elucidated in ref. [7].) There is a reasonable body of evidence to support the notion that the analytical eigenfunctions provide a good model for the valence electron, especially in alkali-metal atoms. For instance, transition probabilities calculated with the analytical eigenfunctions agree with experiment and with accepted values [8]. (Some recursion formulae for matrix elements are given in ref. [9].) Transition probabilities for other elements, notably alkaline-earth ions, have also been obtained in this way [10]. Moreover, these analytical eigenfunctions have been used as trial wavefunctions in detailed atomic calculations [11]. Stark maps for the alkali-metal atoms can also be calculated using the model [12]. The resulting clear anticrossings and small-field quadratic Stark effects for the s and p orbitals are in agreement with experiment. For example, the model yields Stark maps for the $n = 15$ lines of lithium and sodium that are indistinguishable from the numerical and experimental results of ref. [13]. (Ref. [12] also studied other possible quantum-mechanical supersymmetries involving hydrogen. In particular, a double sqm(2) appears when the separation is carried out in parabolic coordinates.) The model is expected to break down at short distances from the nucleus, of order of the core size, but despite this some dominant features of the fine structure in alkali-metal atoms are correctly reproduced and the Landé semiempirical formula naturally appears [14].

OSCILLATOR REFORMULATION: BROKEN CASE

This section presents a reformulation of the analytical quantum-defect model in terms of oscillators with radial equation modified by an anharmonic term. For generality, the connection between the two is treated in arbitrary dimensions. The oscillator models that appear have analytical solutions. The link between the two theories is via a three-parameter map. In the limit of vanishing quantum defect, this map provides a generalization of the one of ref. [4] discussed above. For example, it can be used to connect a Coulomb problem to an anharmonic oscillator with *odd* dimensionality.

The first step is to construct the d-dimensional extension of the quantum-defect model of ref. [6]. This is done by adding to the hamiltonian arising from Eq. (14) an extra term $V_B^d(y)$ generalizing V_B in Eq. (30), with

$$V_B^d(y) = \frac{(l^* + \gamma)(l^* + \gamma + 1) - (l + \gamma)(l + \gamma + 1)}{y^2} + \frac{(n + \gamma)^2 - (n^* + \gamma)^2}{4(n + \gamma)^2(n^* + \gamma)^2} \ . \tag{34}$$

Here, n^* and l^* are modified quantum numbers given by

$$n^* = n - \delta(d, l) \ , \qquad l^* = l + i(d, l) - \delta(d, l) \ , \tag{35}$$

in analogy with Eqs. (27) and (31), with

$$\gamma = \tfrac{1}{2}(d - 3) \tag{36}$$

as before. The ensuing radial equation in atomic units is

$$\left[-\frac{d^2}{dy^2} - \frac{1}{y} + \frac{(l^* + \gamma)(l^* + \gamma + 1)}{y^2} - \frac{1}{2}E_{dn^*} \right] v_{dn^* l^*}^*(y) = 0 \ , \tag{37}$$

where

$$E_{dn^*} = -\frac{1}{2(n^* + \gamma)^2} \ . \tag{38}$$

The radial wave functions solving Eq. (37) are given by

$$v_{dn^* l^*}^*(y) = c_{dn^* l^*}^* \, y^{l^* + \gamma + 1} \exp\left[-y/2(n^* + \gamma)\right] L_{n-l-i-1}^{(2l^* + 2\gamma + 1)} \, y/(n^* + \gamma) \ , \tag{39}$$

where $c_{dn^* l^*}^*$ is a normalization constant.

To identify a mapping between the quantum-defect theory and an oscillator-type model, a (supersymmetry-breaking) term $V_B^D(Y)$ modifying the hamiltonian coming from the harmonic-oscillator radial equation (19) is needed. A suitable choice is

$$V_B^D(Y) = \frac{(L^* + \Gamma)(L^* + \Gamma + 1) - (L + \Gamma)(L + \Gamma + 1)}{Y^2} + 2(N - N^*) \ . \qquad (40)$$

In this expression,

$$\Gamma = \tfrac{1}{2}(D - 3) \qquad (41)$$

as before, and the modified quantum numbers N^* and L^* are given by

$$N^* = N - 2\Delta(D, N, L) \ , \qquad L^* = L + 2I(D, L) - 2\Delta(D, N, L) \ , \qquad (42)$$

where the integer $2I(D, L)$ is a supersymmetry-type shift and 2Δ represents a quantum anharmonicity (which can be viewed as an oscillator 'defect'). The factors of two are introduced for notational simplicity in what follows. The extra term (40) introduces an anharmonic piece into the oscillator potential, which in turn changes the energy eigenspectrum. In atomic units with a unit-frequency oscillator, the anharmonic radial equation becomes

$$\left[-\frac{d^2}{dY^2} + Y^2 + \frac{(L^* + \Gamma)(L^* + \Gamma + 1)}{Y^2} - 2E_{DN^*} \right] V_{DN^* L^*}^*(Y) = 0 \ , \qquad (43)$$

where the energy eigenvalues are shifted according to

$$E_{DN^*} = \tfrac{1}{2}(2N^* + 2\Gamma + 3) \ . \qquad (44)$$

The eigensolutions for this anharmonic oscillator are

$$V_{DN^* L^*}^*(Y) = C_{DN^* L^*}^* \ Y^{L^* + \Gamma + 1} \exp(-Y^2/2) L_{N/2 - L/2 - I}^{(L^* + \Gamma + 1/2)}(Y^2) \ , \qquad (45)$$

where $C_{DN^* L^*}^*$ is a normalization constant.

The map between the radial equations for the d-dimensional quantum-defect theory and the D-dimensional anharmonic oscillator connects the eigensolutions $v_{dn^* l^*}^*$ and $V_{DN^* L^*}^*$. It is given by

$$v_{dn^* l^*}^* \big((n^* + \gamma)Y^2 \big) = K_{DN^* L^*}^* Y^{1\,'2} V_{DN^* L^*}^*(Y) \ , \qquad (46)$$

where $K_{DN^* L^*}^*$ is a proportionality constant and

$$D = 2d - 2 - 2\lambda \ , \quad N = 2n + 2(\Delta - \delta) - 2 + \lambda \ , \quad L = 2l + 2(\Delta - \delta) - 2(I - i) + \lambda \ . \qquad (47)$$

For fixed d, n, l, δ, and i there are three quantities that effectively act as mapping parameters: λ, Δ, and I. For the eigenfunctions (45) to exist I must be integer, so the supersymmetry-type shift $2I$ in L is an even integer. Since D must also be an integer, λ is integer or half-integer. The quantum numbers N and L are also integer, which implies that $2(\Delta - \delta) + \lambda$ must be integer. Note that this generalizes the exact-symmetry case: the requirement that λ be a whole integer is no longer needed because half-integral values can be absorbed in the difference $2(\Delta - \delta)$. This means, for example, that when the Coulomb problem is treated in the exact limit ($\delta = 0$) it is now possible to map it into a (modified) oscillator in an *odd* number of dimensions, provided Δ is quarter-integer valued.

SUPERSYMMETRIC OSCILLATORS AND THE PENNING TRAP

In this section, I demonstrate that geonium atoms provide a physical realization of a $D > 1$ supersymmetric harmonic oscillator. For simplicity, the specific case $D = 2$ is considered, although under suitable conditions an oscillator might appear with $D = 3$ or larger. In practical situations the supersymmetry is broken for reasons to be described. The analytical anharmonic oscillator model introduced in the previous section should provide a good approximation to the exact wavefunctions for this case. Space limitations prevent more than a sketch of the relevant physics being given here; details will appear elsewhere.

Geonium atoms are formed by a set of charged particles bound in a Penning trap [15], which is a suitable combination of a homogeneous magnetic field and an electrostatic quadrupole potential. The simplest geonium atom has just one trapped particle of charge e and mass m [16]. Successively adding further particles in the trap generates elements of the geonium periodic table.

For simplicity, consider the idealized Penning trap with electromagnetic fields specified in cylindrical coordinates (ρ, θ, z) by

$$\mathbf{B} = B\hat{z} , \qquad \phi = \tfrac{1}{2}\frac{V}{d^2}(z^2 - \tfrac{1}{2}\rho^2) . \tag{48}$$

The quantity d is a measure of the trap dimension and is to be specified in terms of the configuration of the quadrupole electrodes. For (stable) trapping, $eV > 0$. The quantum-mechanical motion of the particle in the field $\mathbf{B}$ is that of a harmonic oscillator with binding frequency equal to the cyclotron frequency, given in SI units by

$$\omega_c = \frac{|eB|}{m} . \tag{49}$$

(In fact, there are *two* oscillators involved in this motion, but only one enters the quantum hamiltonian.) Similarly, the electrostatic field generates an axial harmonic motion independent of the cyclotron motion, with axial frequency

$$\omega_z = \sqrt{\frac{eV}{md^2}} . \tag{50}$$

These are the motions of primary interest here.

The simultaneous presence of electric and magnetic fields also generates another (unbound) circular motion, called the magnetron motion, with frequency ω_m. For simplicity, this is largely disregarded here. The eigenvalue spectrum of the system is split by all these interactions and also (for particles with spin $\mathbf{S}$) by the spin interaction $-\mathbf{S} \cdot \mathbf{B}$. The latter splitting implies the existence of a supersymmetry of the type discussed in refs. [17]. This supersymmetry is not directly relevant to the discussion here, and the spin degree of freedom is neglected in what follows.

The combination of the cyclotron and axial motions forms a system of two one-dimensional oscillators. This becomes a physical realization of a $D = 2$ harmonic oscillator when the applied electromagnetic fields are chosen such that $\omega_c = 3\omega_z/2$, i.e.,

$$V = \frac{4}{9} \frac{|e| B^2 d^2}{m} \quad .$$

(51)

The quantum problem can then be separated in polar coordinates and the radial equation has the general form of Eq. (19) with $\Gamma = -\frac{1}{2}$, and with suitable constant factors inserted to allow for non-unit binding frequency and for SI units.

For fixed L, this oscillator can be used as the bosonic partner H_+ in a supersymmetric quantum mechanics. The partner hamiltonian H_- is specified by Eq. (23). It represents a system having an eigenspectrum degenerate with the bosonic sector but with the ground state missing. As in atomic supersymmetry, one practical realization of this is to fill the ground state with particles and invoke the Pauli principle. If $L = 0$, for example, H_+ describes the S orbitals of the simplest geonium atom with one trapped particle. Then, H_- can be interpreted as an effective theory describing the behavior of the 'valence' particle in the S orbitals of a more complex geonium atom in which the lowest S orbital is filled. This interpretation invokes the approximation in which particle interactions other than those implied by the Pauli principle are disregarded. All the atomic supersymmetries of ref. [3] have analogues in this system. For example, there are connections between pairs of geonium atoms throughout the geonium periodic table.

There exists a mapping between the $d = 3$ Coulomb problem and the $D = 2$ harmonic oscillator, as discussed above. The connection between the eigenfunctions given in Eq. (24) therefore establishes a correspondence between eigenfunctions of elements in the usual periodic table and the geonium periodic table. The map (24) is fixed here by setting $\lambda = 1$, so that

$$N = 2n - 1 \; , \quad L = 2l + 1 \; .$$

(52)

Moreover, in the exact supersymmetry limit this map induces other maps involving the supersymmetric partners so that all four hamiltonians are linked.

The supersymmetries are broken by the interaction of the valence particle with the 'core' of the geonium atom. These interactions will shift the eigenenergy of the valence particle from $E_N = \hbar\omega_z(N+1)$ (in the level with quantum numbers N, L) to some other energy

$$E_{N^*} = \hbar\omega_z(N^* + 1) \; ,$$

(53)

where by definition

$$N^* = N - 2\Delta(N, L) \; .$$

(54)

A model incorporating these exact new eigenenergies and yielding analytical solutions has been introduced in the previous section. In the present case, it is obtained by adding an anharmonic term to the oscillator hamiltonian, giving the radial equation (43) with $\Gamma = -\frac{1}{2}$ and with suitable dimension-correcting factors inserted. The analytical solutions are given by a corresponding modification of Eq. (45).

It is physically plausible to conjecture that the quantum anharmonicity rapidly approaches an asymptotic value as N becomes large, i.e.,

$$\Delta(N, L) \simeq \Delta(L) \; .$$

(55)

In this case, the eigensolutions form a complete and orthogonal set. The previous section also provides a mapping between this theory and the analytical quantum-defect theory for ordinary atoms. Combined with the magnetron and spin splittings, the anharmonic model is likely to provide a simple method for calculations of physical properties of the valence particle in geonium atoms.

ACKNOWLEDGMENTS

I thank Robert Bluhm, Mike Nieto, Bob Pollock, Stuart Samuel, and Rod Truax for discussion. Part of this work was performed at the Aspen Center for Physics. This research was supported in part by the United States Department of Energy under contracts DE-AC02-84ER40125 and DE-FG02-91ER40661.

REFERENCES

1. F. Iachello, Phys. Rev. Lett. **44**, 772 (1980); A. Balantekin, I. Bars and F. Iachello, Nucl. Phys. **A370**, 284 (1981).
2. H. Nicolai, J. Phys. A **9**, 1497 (1976); E. Witten, Nucl. Phys. **B188**, 513. (1981).
3. V.A. Kostelecký and M.M. Nieto, Phys. Rev. Lett. **53**, 2285 (1984); Phys. Rev. A **32**, 1293 (1985).
4. V. A. Kostelecký, M. M. Nieto, and D. R. Truax, Phys. Rev. D **32**, 2627 (1985).
5. J. R. Rydberg, Kongl. Sven. vetensk.-akad. hand. **23**, no. 11 (1890); Philos. Mag. **29**, 331 (1890).
6. V. A. Kostelecký and M. M. Nieto, Phys. Rev. A **32**, 3243 (1985).
7. J. Beckers and N. Debergh, "On a Parastatistical Hydrogen Atom and its Supersymmetric Properties," Liège preprint, 1992.
8. W.L. Wiese, M.W. Smith, and B.M. Glennon, *Atomic Transition Probabilities, Vols. 1 and 2,* Natl. Bur. Stand. (U.S.) Natl. Stand. Ref. Data Ser. Nos. 4 and 22 (U.S. GPO, Washington, D.C., 1966 and 1969).
9. G.-W. Wen, "Matrix Elements of the Radial Operators and their Recursion Relations in Analytical Quantum Defect Theory," Hunan Normal preprint, 1992.
10. M.T. Djerad, J. Phys. II **1**, 1 (1991).
11. R.E.H. Clark and A.L. Merts, J. Quant. Spectrosc. Radiat. Transfer **38**, 287 (1987).
12. R. Bluhm and V.A. Kostelecký, Phys. Rev. A, in press.
13. M.L. Zimmerman, M.G. Littman, M.M. Kash, and D. Kleppner, Phys. Rev. A **20**, 2251 (1979).
14. V. A. Kostelecký, M. M. Nieto, and D. R. Truax, Phys. Rev. A **38**, 4413 (1988).
15. F.M. Penning, Physica (Utrecht) **3**, 873 (1936); H. Dehmelt, Rev. Mod. Phys. **62**, 525 (1990); W. Paul, *ibid.,* 531; N.F. Ramsey, *ibid.,* 541.
16. For a review of the case of a single trapped particle, see, for example, L.S. Brown and G. Gabrielse, Rev. Mod. Phys. 58, 233 (1986).
17. R. Jackiw, Phys. Rev. D **29**, 2375 (1984); R. Hughes, V.A. Kostelecký and M.M. Nieto, Phys. Lett. B 171, 226 (1986); Phys. Rev. D 34, 1100 (1986). See also B.W. Fatyga, V.A. Kostelecký, M.M. Nieto and D.R. Truax, Phys. Rev. D **43**, 1403 (1991).

DYNAMICAL SYMMETRIES IN THE
sdg INTERACTING BOSON MODEL

V.K.B. Kota and Y.D. Devi

Physical Research Laboratory
Ahmedabad 380 009, India

1. INTRODUCTION

The interacting boson model (IBM) proposed by Arima and Iachello (1975, 1978a) in 1975 provides an algebraic description of the quadrupole collective properties of low-lying states in nuclei in terms of a system of interacting bosons. In this model the bosons are assumed to be made up of correlated pairs of valence nucleons (as Balantekin *et al* (1981) put it *"the correlations are so large that the bosons effectively loose the memory of being fermion pairs − − −−"*) and they carry angular momentum $\ell = 0$ (s-boson representing pairing degree of freedom) or $\ell = 2$ (d-boson representing quadrupole degree of freedom). This model is remarkably successful in explaining experimental data (Iachello and Arima 1987; Casten and Warner 1988). In the words of Feshbach (Iachello 1981) *"IBM has yielded new insights into the behaviour of low-lying levels and indeed has generated a renaissance of the field of nuclear spectroscopy"*. The extended sdg interacting boson model (sdgIBM or simply gIBM) where one includes in addition the hexadecupole ($\ell = 4$) g-bosons is ideally suited for analyzing $E4$ data. In the last ten years different types of $E4$ data (mainly in rare - earths) are obtained and they include : (i) β_4(hexadecupole deformation parameter) data for rare-earths and several actinide nuclei; (ii) isoscalar mass transition density $B(\mathrm{IS}4; 0^+_{GS} \rightarrow 4^+_\gamma)$, which gives information on Y_{42} deformation; (iii) $E4$ transition matrix elements involving 4^+_i ($i \leq 6$) states in some of the Cd, Pd, Er, Yb, Os and Pt isotopes; (iv) $E4$ strength distributions in ^{112}Cd, ^{150}Nd and ^{156}Gd; (v) hexadecupole transition densities for exciting some of the 4^+ levels in Cd, Pd, Os and Pt isotopes; (vi) $K^\pi = 0^+_3, 3^+_1, 2^+_2$ and 4^+ bands which can be interpreted as bands built on hexadecupole vibrations. The microscopic (shell model based) theories of IBM provide another strong basis for the inclusion of g-bosons (i.e. for the importance of $G\left(L^\pi = 4^+\right)$ pairs); the conclusions from these studies is as stated by Yoshinaga *et al* (1984) *"the inclusion of hexadecupole degree of freedom in addition to the monopole and quadrupole degrees of freedom is important and sufficient to reproduce physical quantities"*. Finally it should be added that there are a large number of other signatures indicating that g-bosons should be included in the IBM (Casten and Warner 1988; Devi and Kota 1990a). As early as in 1980 Feshbach (see Iachello, 1979,1981) stressed the need to explore the sdgIBM and in particular the corresponding dynamical symmetries. The progress made in exploring, understanding and applying the dynamical symmetries of sdgIBM are briefly described in this article

Symmetries in Science VII, Edited by B. Gruber
and T. Otsuka, Plenum Press, New York, 1994

with special emphasis on the $SU_{sdg}(3)$ limit. We dedicate this artcle to Franco Iachello on his 50th birthday.

2. sdgIBM DYNAMICAL SYMMETRIES

2.1 Classification

In the sdgIBM, the states of a N - boson system belong to the totally symmetric irrep $\{N\}$ of $U(15)$ group. The general $1 + 2$ body hamiltonian (H_{gIBM}) which preserves angular momentum and conserves the boson number in the sdg space, contains 35 free parameters; three single boson energies ϵ_s, ϵ_d and ϵ_g and thirty two two-body matrix elements. The H_{gIBM} is solvable when it becomes a linear combination of the Casimir operators of the various groups in a group sub-group chain $U(15) \supset G \supset G' \supset - - - -$ $- \supset O(3)$, where $O(3)$ is the group corresponding to angular momentum in sdg space. Then H_{gIBM} is said to possess a dynamical symmetry, which is denoted by the first sub-group G in the given chain. A complete classification of the dynamical symmetries and their algebra exhaust all the analytical (limiting) solutions of gIBM. Kota (1984) and Meyer et al (1986) showed (the former using physical arguments and the latter using representation theory) that gIBM possesses seven dynamical symmetries and they correspond to four strong coupling limits $SU_{sdg}(3)$, $SU_{sdg}(5)$, $SU_{sdg}(6)$ and $O_{sdg}(15)$ and three weak coupling limits $U_{sd}(6) \oplus U_g(9)$, $U_{dg}(14)$ and $U_d(5) \oplus U_{sg}(10)$ respectively; initial studies on the $SU_{sdg}(3)$, $SU_{sdg}(5)$ and $U_{dg}(14)$ limits are due to Ratnaraju (1981; see Kota 1982 and Wu 1982), Sun et al (1983), and Ling (1983) respectively. The complete group chains and the corresponding group labels or irreducible representations (irreps) are given in figure 1. The group generators, irreps, quadratic Casimir operators $(C_2(G)$; G denoting a group) and their eigenvalues are worked out

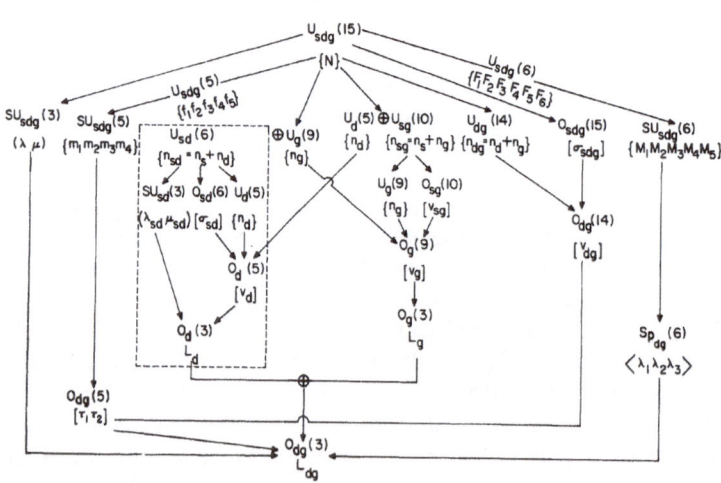

FIGURE 1. Dynamical symmetry group chains in sdgIBM. The dashed box gives the group chains in sdIBM. The irrep labels corresponding to each group in the various symmetry chains are also shown. The subscript labels (sdg, sd, d, − − −) define the space in which a given group is realized.

308

in detail by Kota *et al* (1987). It should be pointed out that the $SU_{sdg}(3)$group chain follows from the classic work of Elliott (1958a, 1958b) as the $\ell = 0, 2$ and 4 values corresponding to the s, d and g bosons can be viewed as arising out of the oscillator shell with principle quantum number $\mathcal{N} = 4$, the $SU_{sdg}(5)$ group chain by recognizing that a particle carrying angular momentum $\ell = 0, 2$ and 4 can be thought of as composed of two pseudo-bosons each carrying angular momentum $\tilde{\ell} = 2$, similarly the $SU_{sdg}(6)$ chain with two pseudo-fermions each carrying angular momentum $\tilde{j} = 5/2$ and finally the $O_{sdg}(15)$ chain corresponds to generalized seniority in sdg space. The remaining three chains $U_{sd}(6) \oplus U_g(9)$, $U_{dg}(14)$ and $U_d(5) \oplus U_{sg}(10)$ can be realized by demanding that $\{n_{sd} = n_s + n_d, n_g\}, \{n_{dg} = n_d + n_g, n_s\}$ and $\{n_{sg} = n_s + n_g, n_d\}$ respectively to be good quantum numbers.

The $SU_{sdg}(3)$ group is generated by the eight operators $\{Q_\mu^2(s), \mathbf{L}\}$ where $Q_\mu^2(s)$ is the quadrupole generator and $\mathbf{L}$ is the angular-momentum generator,

$$Q_\mu^2(s) = 4\sqrt{\frac{7}{15}}\left(s^\dagger \tilde{d} + d^\dagger \tilde{s}\right)_\mu^2 - 11\sqrt{\frac{2}{21}}(d^\dagger \tilde{d})_\mu^2 + \frac{36}{\sqrt{105}}\left(d^\dagger \tilde{g} + g^\dagger \tilde{d}\right)_\mu^2 - 2\sqrt{\frac{33}{7}}(g^\dagger \tilde{g})_\mu^2$$

$$L_\mu^1 = \sqrt{10}\{(d^\dagger \tilde{d})_\mu^1 + \sqrt{6}(g^\dagger \tilde{g})_\mu^1\} \tag{1}$$

The ground state (GS) $SU_{sdg}(3)$ irrep is $(4N, 0)$. In the $SU_{sdg}(5)$ limit the generators of $SU_{sdg}(5)$ are

$$G_\mu^{L_0=1-4} = \sum_{\ell_1,\ell_2=0,2,4} \sqrt{(2\ell_1 + 1)(2\ell_2 + 1)}(-1)^{L_0}\begin{Bmatrix} \ell_1 & \ell_2 & L_0 \\ 2 & 2 & 2 \end{Bmatrix}(b_{\ell_1}^\dagger \tilde{b}_{\ell_2})_\mu^{L_0} \tag{2}$$

The GS $SU_{sdg}(5)$ irrep is $\{2N, 0, 0, 0\}$. In the $SU_{sdg}(6)$ limit the generators of $SU_{sdg}(6)$ are

$$h_\mu^{L_0=1-4} = \sum_{\ell_1,\ell_2=0,2,4} \sqrt{(2\ell_1 + 1)(2\ell_2 + 1)}(-1)^{L_0}\begin{Bmatrix} \ell_1 & \ell_2 & L_0 \\ 5/2 & 5/2 & 5/2 \end{Bmatrix}(b_{\ell_1}^\dagger \tilde{b}_{\ell_2})_\mu^{L_0} \tag{3}$$

The GS $SU_{sdg}(6)$ irrep is $\{N, N, 0, 0, 0\}$. In the the $O_{sdg}(15)$ limit the GS belongs to the $O(15)$ irrep $[N]$. In order to understand the physical relevance of the various dynamical symmetries the Wigner-Racah algebra of the group chains given in figure 1 are to be worked out. The algebra is available for the $SU_{sdg}(3)$ limit (Elliott 1958a, 1958b; Hecht 1964,1965; Arima and Iachello 1978b) and $U_{sd}(6) \oplus U_g(9)$ limit with $n_g = 0$ (Arima and Iachello 1976,1978b,1979) and $n_g = 1$ (Devi and Kota 1991a). In addition some preliminary attempts are made for $U(5) \supset O(5) \supset O(3)$ chain relevent for $SU_{sdg}(5)$ limit (Vanthournout *et al* 1988,1989) and $U(9) \supset O(9) \supset O(3)$ chain relevant for $U_{sd}(6) \oplus U_g(9)$ limit (Yu *et al* 1986; Sen and Yu 1987). An alternative to Wigner - Racah algebra is to use the coherent state (CS) formalism. This gives information on geometric shapes and also large$-N$ limit expressions for various observables.

2.2 Geometric Shapes

The geometric shapes corresponding to sdgIBM dynamical symmetries is investigated (Devi and Kota 1990b) using the so called projective CS $\mid N; \beta_2, \beta_4, \gamma >$ where β_2 and β_4 are quadrupole and hexadecupole deformation parameters and γ is the asymmetry angle,

$$\mid N; \beta_2; \beta_4, \gamma) = \left[N! (1 + \beta_2^2 + \beta_4^2)^N\right]^{-1/2}\left\{s_0^\dagger + \beta_2\left[\cos\gamma \, d_0^\dagger + \right.\right.$$
$$\sqrt{\tfrac{1}{2}}\mathrm{Sin}\,\gamma\left(d_2^\dagger + d_{-2}^\dagger\right)\right] + \tfrac{1}{6}\beta_4\left[(5\cos^2\gamma + 1)\, g_0^\dagger\right.$$
$$\left.\left. + \sqrt{\tfrac{15}{2}}\mathrm{Sin}\,2\gamma\left(g_2^\dagger + g_{-2}^\dagger\right) + \sqrt{\tfrac{35}{2}}\mathrm{Sin}^2\gamma\left(g_4^\dagger + g_{-4}^\dagger\right)\right]\right\}^N \mid 0) \tag{4}$$

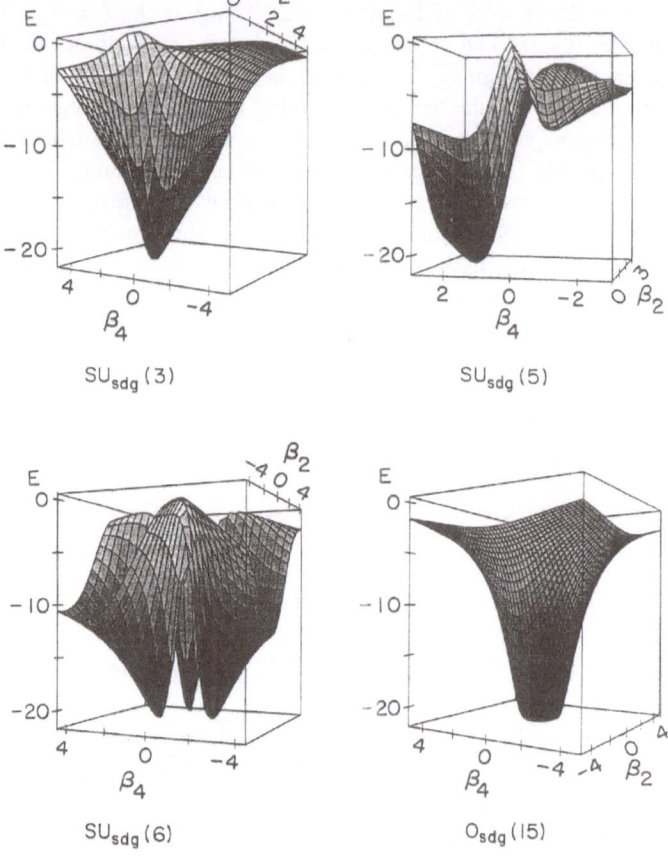

FIGURE 2. Potential energy surfaces E (the E corresponds to $E(N; \beta_2, \beta_4, \gamma)$ defined in sect. 2.2) vs β_2, β_4 in the four strong coupling limits $SU_{sdg}(3)$ with $\gamma = 0°, SU_{sdg}(5)$ with $\gamma = 60°, SU_{sdg}(6)$ with $\gamma = 0°$ and $O_{sdg}(15)$ which is independent of γ.

where $\beta_2 \geq 0, -\infty \leq \beta_4 \leq +\infty, 0° \leq \gamma \leq 60°$. The equilibrium shape parameters $(\beta_2^0, \beta_4^0, \gamma^0)$ are obtained by minimising the energy functional $E(N; \beta_2, \beta_4, \gamma)$ where $E(N; \beta_2, \beta_4, \gamma) = < N; \beta_2, \beta_4, \gamma |\mathrm{H}| N; \beta_2, \beta_4, \gamma >$. For the four strong coupling limits $SU_{sdg}(3), SU_{sdg}(5), SU_{sdg}(6)$ and $O_{sdg}(15)$ the equilibrium parameters $(\beta_2^0, \beta_4^0, \gamma^0)$ are $(\sqrt{20/7}, \sqrt{8/7}, 0°)$, $(\sqrt{10/7}, \sqrt{18/7}, 60°)$, $[(\sqrt{25/14}, \sqrt{3/14}, 0°)$, $(\sqrt{8/7}, \sqrt{6/7}, 60°)$, $(\sqrt{1/14}, -\sqrt{27/14}, 60°)]$ and $((\beta_2^0)^2 + (\beta_4^0)^2 = 1, \gamma^0$ undefined) respectively. From these parameters and the potential energy surfaces shown in figure 2 one clearly sees that $SU_{sdg}(3)$ has one minimum (thus producing a deformed surface), $SU_{sdg}(5)$ limit has two minima that are displaced in energy, $SU_{sdg}(6)$ limit has three degenerate minima while the $O_{sdg}(15)$ limit is completely γ - unstable. See (Devi and Kota 1990b) for details regarding weak-coupling limits.

From the CS analysis it is found that: (1) The weak coupling limit $U_{dg}(14)$ and the strong coupling limits $\{SU_{sdg}(3), SU_{sdg}(5)\}$ and $\{SU_{sdg}(6), O_{sdg}(15)\}$ are similar to the sdIBM $U_d(5)$, $SU_{sd}(3)$ and $O_{sd}(6)$ limits respectively, although the $SU_{sdg}(5)$ limit is not strictly stable deformed nor the $SU_{sdg}(6)$ limit completely unstable. It should be added that the $U_{sd}(6) \oplus U_g(9)$ limit is essentially (in the ground state domain) same as the sdIBM. The results clearly indicate that vibrational nuclei can be described by interpolating $[U_{sd}(6) \supset U_d(5)] \oplus U_g(9)$ and $U_{dg}(14)$ limits, deformed nuclei by the $SU_{sdg}(3)$, $SU_{sdg}(5)$ and $[U_{sd}(6) \supset SU_{sd}(3)] \oplus U_g(9)$ limits and the γ-unstable nuclei by the $O_{sdg}(15)$, $SU_{sdg}(6)$ and $[U_{sd}(6) \supset O_{sd}(6)] \oplus U_g(9)$ limits respectively; (2) One important feature of the solutions for $E(N; \beta_2, \beta_4, \gamma)$ is that we never have non-axial shapes (γ different from $0°$ or $60°$) in the dynamical symmetry limits although $E(N; \beta_2, \beta_4, \gamma)$ contain $\cos^2 3\gamma$ terms. For a general H_{gIBM} non-axial shapes are possible in sdgIBM unlike in sdIBM and one such example is worked out by Kuyucak and Morrison (1991); (3) From figure 2 it follows that $SU_{sdg}(5)$ limit admits two minima and hence in sdg model it is in principle possible, with an additional variable, to have shape phase transition (i.e the system can tunnel from one minimum to another by varying the variable). Secondly $SU_{sdg}(6)$ limit shows that it is possible to have shape coexistence in sdgIBM. It is seen that the three degenerate minima in $SU_{sdg}(6)$ limit correspond to quite different shapes (in fact one is prolate and the other two oblate). Thus unlike sdIBM, the sdgIBM admits shape phase transition and shape coexistence. Kuyucak *et al* (1991a) showed, by using angular momentum projection from single intrinsic states, that these phase transitions can be driven by angular momentum; (4) In order to visualize the geometric shapes corresponding to the CS (4), we should be able to express the shape variables $(\beta_2^p, \beta_4^p, \gamma^p)$ that define the nuclear surface in terms of the CS variables $(\beta_2, \beta_4, \gamma)$. To this end one equates the $E2$ and $E4$ matrix elements in CS and geometric model descriptions. This correspondence is used recently (Devi and Kota 1992a) in studying β_4 - systematics in rare - earth and actinide nuclei.

2.3 Two Nucleon Transfer

Two nucleon transfer (TNT) strengths provide information on relevance of sdgIBM symmetries. Ignoring the cut-off factors the TNT operators for $\ell = 0, 2, 4$ transfer in sdgIBM and the corresponding strengths $S_\ell^{(\pm)}$ for $N \to N \pm 1$ are,

$$P_+^\ell = \eta_{+\ell} b_\ell^\dagger, \quad P_-^\ell = \eta_{-\ell} \tilde{b}_\ell, \quad b_0^\dagger = s^\dagger, \quad b_2^\dagger = d^\dagger, \quad b_4^\dagger = g^\dagger, \tag{5}$$

$$S_\ell^{(\pm)} \left(N; 0_{GS}^+ \to N + 1; L_f^+ \right) = (\eta_{\pm\ell})^2 \left| \left\langle N \pm 1; L_f^+ \left\| P_{(\pm)}^\ell \right\| N; 0_{GS}^+ \right\rangle \right|^2 \delta_{\ell L_f} \tag{6}$$

where η's are free parameters. In (6), (+) is for particle addition and (-) is for particle removal. The ratio $R_\pm$ for $\ell = 0$ transfer which is the ratio of the summed cross-sections going to all excited states to that of the GS removes the uncertainties in the normalization factors in both experimental and theoretical results and can be treated as the ratio of summed strengths. Using the N-boson CS $\mid N; \beta_2 \beta_4 \gamma >$, defined in (4), large$-N$ limit expressions for $R_\pm$ are derived (Devi and Kota 1991b); $R_+ = [(\beta_2^0)^2 + (\beta_4^0)^2]/N$, $R_- = 0$. In the $SU(3)$, $SU(5)$, $SU(6)$ and $O(15)$ limits R_+ takes values $4/N$, $4/N$, $2/N$ and $1/N$ respectively. Data for $R_\pm$ all across rare-earth region is available. In the case of the deformed Gd and Er isotopes, the (t,p) data give the value of $R_+ \sim 0.3 - 0.4$ which is close to $SU_{sdg}(3)$ value $4/N$; $SU_{sd}(3)$ gives $2/N$. Thus the (t,p) data for deformed nuclei is better explained with g-bosons. It should be noted that the results for $SU_{sdg}(3)$ and $SU_{sdg}(5)$ are similar and therefore the TNT data cannot distinguish between these two limits. The (p,t) data for the γ-unstable Pt, Os isotopes give the value of R_+ (because of hole bosons) to be ~ 0.08 which is close to the $O_{sd}(6)$ or $O_{sdg}(15)$ value $1/N$. The $SU_{sdg}(3)$ description in this region, as expected, is poor and that of $SU_{sdg}(6)$ is not very much different from the $O_{sd}(6)$ and $O_{sdg}(15)$ descriptions. The peaks in the TNT data are described by detailed sdgIBM calculation (Devi and Kota 1992b).

2.4 Interpolation of dynamical symmetries and the hamiltonian H_{SYM}

The sdgIBM symmetries, except $SU_{sdg}(3)$ and $U_{sd}(6) \otimes U_g(9)$, have limited applicability. In order to proceed further it is important to obtain sdg hamiltonians and then use them in numerical calculations that are required for a majority of (transitional) nuclei. Broadly speaking, two approaches to this rather complicated problem are available: (i) phenomenological; (ii) microscopic (based on shell model and its relatives). The symmetry defined hamiltonian H_{SYM} of Devi et al (1989), Boson surface delta interaction of Chen et al (1986), hamiltonian based on the commutator method given by Kuyucak et al (1991b, 1991c) belong to the first class while the OAI mapped and IBM-2 to IBM-1 projected hamiltonian of Devi and Kota (1992c), Dyson-boson mapped and IBM-2 to IBM-1 projected hamiltonain of Navratil and Dobes (1991) and single $-j$ shell seniority mapped hamiltonian of Yoshinaga (1989) belong to the second class.

Interpolating the hamiltonians defined by the Casimir operators (linear and quadratic) of the various symmetry groups of gIBM, a schematic interaction H_{SYM} with a few parameters ($2 + 6$ instead of $2 + 32$; the s - boson energy ϵ_s can be removed for calculating excitation energies and transition strengths) can be written down,

$$
\begin{aligned}
H_{SYM} &= \epsilon_d \hat{n}_d + \epsilon_g \hat{n}_g + \alpha_1 \left(H(SU_{sdg}(3)) \right) + \alpha_2 \left(H(SU_{sdg}(5)) \right) + \\
&\quad \alpha_3 \left(H(SU_{sdg}(6)) \right) + \alpha_4 \left(H(O_{sdg}(15)) \right) + \\
&\quad \alpha_5 \left(H(O_{sd}(6)) \right) + \alpha_6 \mathbf{L} \cdot \mathbf{L} \; ;
\end{aligned}
$$

$$
H(SU_{sdg}(3)) = -C_2(SU_{sdg}(3)) + \frac{3}{4}(\mathbf{L} \cdot \mathbf{L}) = -\frac{3}{4}Q^2(s) \cdot Q^2(s) \; ;
$$

$$
H(SU_{sdg}(5)) = -C_2(SU_{sdg}(5)) + \frac{1}{2}C_2(O_{dg}(5)) = -4[G^2 \cdot G^2 + G^4 \cdot G^4] \; ;
$$

$$
H(SU_{sdg}(6)) = -C_2(SU_{sdg}(6)) + \frac{1}{2}C_2(Sp_{dg}(6)) = -4[h^2 \cdot h^2 + h^4 \cdot h^4] \; ;
$$

$$
H(O_{sdg}(15)) = C_2(O_{sdg}(15)) - C_2(O_{dg}(14)) = [I^2 \cdot I^2 + I^4 \cdot I^4] \; ;
$$

$$
H(O_{sd}(6)) = I^2 \cdot I^2 ; \; I^2 = (s^\dagger \tilde{d} + d^\dagger \tilde{s})^2 , \; I^4 = (s^\dagger \tilde{g} + g^\dagger \tilde{s})^4 \tag{7}
$$

The $Q^2(s)$, G's, h's and $\mathbf{L}$ are defined in (1,2,3). One improtant feature of H_{SYM} is, as can be seen from (7), that it is indeed a mixture of quadrupole - quadrupole and hexadecupole - hexadecupole forces. Based on the results of Sects. 2.2, 2.3 it is easily seen that the free parameters in H_{SYM} (7) are further reduced depending upon the nucleus under study. For example, for vibrational nuclei $(Pd - Cd$ region) it is appropriate to choose $\alpha_2 = \alpha_3 = \alpha_5 = 0$, for rotational nuclei (for example, ^{156}Sm, ^{168}Er) $\alpha_3 = \alpha_5 = 0$ and for γ - unstable nuclei $(Pt - Os$ region) $\alpha_2 = 0$. The H_{SYM} is employed with success to describe the properties (spectra, $E2$, $E4$, $---$ matrix elements) of $\{^{20}Ne, {}^{24}Mg, {}^{32}S, {}^{36}Ar\}$ (Devi et al 1989; Devi 1991), $\{Sm$ isotopes$\}$ (Devi and Kota 1992b) and $\{^{192}Os, {}^{194,196,198}Pt\}$ (Devi and Kota 1991c, 1992d; Devi 1991). In the later two studies mixing calculations in the $(sd)^N \oplus (sd)^{N-1}(g)^1 \oplus (sd)^{N-2}(g)^2$ space with good angular momentum are carried out (this is equivalent to using $U(1) \oplus [U(5) \supset O(5) \supset O(3)] \oplus [U(9) \supset O(9) \supset O(3)]$ group chain). As an example the predicted $E4$ strength distributions in $^{152,154}Sm$ are shown in figure 3; a one parameter $E4$ operator is used in the calculations. Similarly in ^{192}Os, ^{198}Pt, ^{196}Pt and ^{194}Pt the energies of the first $(3,3,4,3)$ 4^+ levels are $(0.58, 0.91, 1.07)$, $(0.99, 1.29, 1.79)$, $(0.88, 1.29, 1.54, 1.89)$ and $(0.81, 1.23, 1.91)$ MeV respectively and the experimental and calculated $E4$ matrix elements $(B(E4; 0^+_{GS} \rightarrow 4^+_i)$(in units of $10^{-2}e^2b^4$) are $[(4.0, 1.39, 1.21), (4.71, 1.02, 2.25)]$, $[(2.07 \pm 0.09, 1.54 \pm 0.15, 0.88 \pm 0.13), (2.16, 1.54, 0.72)]$, $[(3.06 \pm 0.25, 2.47 \pm 0.28, 0.45 \pm 0.08, 4 \pm 0.2), (3.5, 3.1, 0.92, 2.0)]$ and $[(3.8 \pm 0.27, 1.32 \pm 0.16, 5.24 \pm 0.32), (4.22, 1.79, 4.42)]$ respectively; a simple 2-parameter $E4$ operator is used in the calculations.

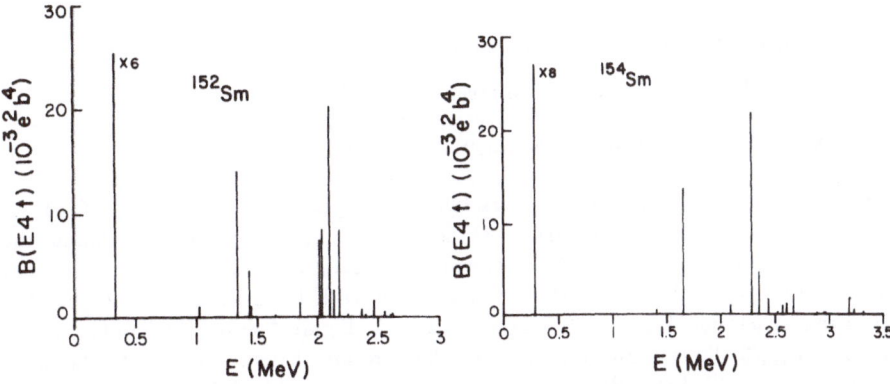

FIGURE 3. $E4$ strength distributions in ^{152}Sm and ^{154}Sm: The $B(E4) = B(E4$; $0^+_{GS} \rightarrow 4^+_i)$ values are shown in the figure as a function of the energy of 4^+_i levels. For the 4^+_1 level $B(E4$) is to be multiplied by a factor six for ^{152}Sm and factor eight for ^{154}Sm. The $E4$ transition operator and the corresponding effective charges are given in (Devi and Kota 1992b).

3. $SU_{sdg}(3)$ LIMIT

3.1 SU(3) irreps

The $SU_{sdg}(3)$ limit is special as it generates band structures that are seen in many nuclei which are not otherwise possible in sdIBM. Using the basic association $\{1\}_{U(15)} \rightarrow (40)_{SU(3)} \rightarrow L = 0, 2, 4$, the symmetric $U(15)$ irrep $\{N\}$, appropriate for N bosons, can be reduced to $SU(3)$ irreps (λ, μ); this reduction is denoted by $\{n\} \otimes \{f\}$ where $\otimes$ stands for what is known as plethysm in group theory (Wybourne 1970); $n = 4$ and

$\{f\} = \{N\}$ for sdgIBM. A complete reduction is obtained by using the result

$$\{n\} \otimes \{f\} = [r!]^{-1} \sum_k h_k C_k^{\{f\}} (\{n\} \otimes S_1)^\alpha (\{n\} \otimes S_2)^\beta \dots \tag{8}$$

In (8) k is the class of the symmetric group of order $r!$ corresponding to the partition $[f]$ of integer r specified by cyclic structure $S_1^\alpha S_2^\beta \dots$, h_k is the order of the class k, and $C_k^{\{f\}}$ is the character of the class k corresponding to the partition $[f]$. Eq. (8) is used (Kota *et al* 1987) to obtain $N \to (\lambda, \mu)$ by using the fact that $C_k^{\{N\}} = 1$ independent of k and the following result (Kota 1977),

$$\begin{aligned}
\{n\} \otimes S_r = \sum_{a,b} [&\{nr - ar, ar - br, br\} - \{nr - ar, ar - br - 1, br + 1\} \\
&+ \{nr - ar - 1, ar - br - 1, br + 2\} - \{nr - ar - 1, ar - br + 1, br\} \\
&+ \{nr - ar - 2, ar - br + 1, br + 1\} \\
&- \{nr - ar - 2, ar - br, br + 2\}]
\end{aligned} \tag{9}$$

In (9) the summation is over all positive integers a and b with the constraint that the non standard $U(3)$ irreps $\{f_1, f_2, f_3\}$ are to be ignored; note that $\lambda = f_1 - f_2$ and $\mu = f_2 - f_3$. Tables for $\{N\} \to (\lambda, \mu)$ for $N \le 10$ are given in (Kota *et al* 1987) and upto 15 are given by Kota (1986a).

3.2 SU(3) bands, energy formula and B(E2)'s

The low lying $SU(3)$ irreps for a given N(with $N \ge 4$) are

$$\{N\} \to (\lambda, \mu) = (4N, 0) \oplus (4N - 4, 2) \oplus (4N - 6, 3) \oplus (4N - 8, 4)^2 \oplus -- \tag{10}$$

Note that the $(4N - 8, 4)$ irrep occurs twice (multiplicity denoted by $\alpha = 0, 1$). It is to be mentioned that Akiyama (1985) gave a novel prescription for obtaining the multiplicities 'α' for the class of $SU(3)$ irreps that satisfy the condition $4N = \lambda + 2\mu$. The $SU(3)$ irreps (λ, μ) generate rotation bands with band head quantum number $K = \min(\lambda, \mu), \min(\lambda, \mu) - 2, -------, 0$ or 1. Figure 4 shows the band structure in the $SU_{sdg}(3)$ limit. Two important results here are that, in addition to the usual ground state (GS), beta (β), gamma (γ) bands, one has odd$-K$ bands ($K^\pi = 3^+, 1^+$) that are absent in sdIBM and two $(4N - 8, 4)$ irreps against one $(2N - 8, 4)$ irrep in sdIBM. From the structure of the $SU_{sdg}(3)$ intrinsic states (Hecht 1965) shown in the inset to figure 4, it is clear that one of the $(4N - 8, 4)$ irreps labelled $(4N - 8, 4)_{\alpha=0}$ is two phonon in character and the other $(4N - 8, 4)_{\alpha=1}$ is hexadecupole vibrational in character and they can be distinguished by TNT as discussed ahead in Sect. 3.4. Instead of the α label one can use (as it is not a group label) Akiyama's (1985) W label and the $\mathbf{S}$ operator $(\mathbf{S} = [B^\dagger(04)\tilde{B}(40)]^{(00)}; B^\dagger(04) = [b^\dagger_{(40)} \tilde{b}_{(40)})]^{(04)}$ where $b^\dagger$ is single boson creation operator) which is diagonal in the W-basis for the (λ, μ) irreps with $\lambda + 2\mu = 4N$. It should be mentioned that $W = 0$ for the irreps with no multiplicities. A simple hamiltonian and the corresponding energy formula in the $SU_{sdg}(3)$ limit are

$$\begin{aligned}
H &= \alpha C_2(SU_{sdg}(3)) + \beta S + \gamma \mathbf{L} \cdot \mathbf{L} \\
E(\lambda, \mu, W, L) &= \alpha(\lambda^2 + \mu^2 + \lambda\mu + 3(\lambda + \mu)) + \frac{\beta}{\sqrt{375}} W(2N - W + 3) + \\
&\quad \gamma L(L + 1)
\end{aligned} \tag{11}$$

Similarly the analytical formulas for the $E2$ transition strengths and the quadrupole moments of the GS band are (with $T^{E2} = aQ^2_\mu(s)$),

$$B(E2; (4N,0)L \rightarrow (4N,0)L-2) = a^2 \left[\frac{2L(L-1)(4N-L+2)(4N+L+1)}{(2L-1)(2L+1)} \right]$$

$$Q_2((4N,0)L) = -a\frac{16\pi}{15} \left[\frac{L(8N+3)}{(2L+3)} \right] \tag{12}$$

It should be clear from (12) that in the GS band the $E2$ strengths grow upto $L = 2N$ (against $L = N$ in sdIBM) and falls to zero at $L = 4N + 2$ and the band cut - off extends from $L = 2N$ of sdIBM to $L = 4N$.

3.3 Large N limit expressions for γ-band B(E4)'s

In the $SU_{sdg}(3)$ limit the GS - band is generated by $(4N,0)$ irrep and the γ - band comes from $(4N - 4,2)$ irrep. Exact expression for GS - band to γ - band transitions which are given by $< (4N-4,2) K_f = 2 L_f = \lambda ||T^{E\lambda}||(4N,0) 0^+_{GS} >$ are derived (Devi and Kota 1992e) and using the large $-N$ limit expression for the $SU_{sdg}(3)$ recoupling coefficients (Bijker and Kota 1988) and Isoscalar Factors (ISF) (Wu et al 1986; Devi and Kota 1991b) analytical expressions for $B_\gamma(IS\lambda) = B(IS4; 0^+_{GS} \rightarrow 4^+_\gamma)$ are derived,

$$T^{E\lambda} = \sum_{\ell,\ell'=0,2,4} e^{(\lambda)}_{\ell,\ell'}(b^\dagger_\ell \tilde{b}_{\ell'})^\lambda$$

$$B\left(IS\lambda; 0^+_{GS} \rightarrow K_f = 2\ L_f = \lambda\right) = 2N \left\{ \sum_{\ell,\ell'} a^0_{\ell'} a^2_\ell \langle \ell 2\ell' 0 | \lambda 2\rangle \right\}^2 \ ;$$

$$a^0_0 = \sqrt{\tfrac{7}{35}} \ ; \ a^0_2 = \sqrt{\tfrac{20}{35}} \ ; \ a^0_4 = \sqrt{\tfrac{8}{35}} \ ; a^2_0 = 0 \ ; \ a^2_2 = \sqrt{\tfrac{1}{7}} \ ; \ a^2_4 = \sqrt{\tfrac{6}{7}} \tag{13}$$

The structure factors a^k_ℓ define the intrinsic states corresponding to $(4N,0)$ and $(4N - 4,2)$ irreps; see the inset to figure 4. The expression (13) for $\lambda = 2$ is same as (31.c) of Yoshinaga (1986). Expression (13) for $\lambda = 4$ is employed (Devi and Kota 1992e) in analyzing the $B_\gamma(IS4)$ data produced recently by Ichihara et al (1987) in $^{152,154}Sm$, ^{160}Gd, ^{164}Dy, $^{166,168}Er$, ^{176}Yb, $^{182,184}W$, and ^{192}OS using 65 MeV polarized protons. The effective charges $e^{(4)}_{\ell,\ell'}$ are obtained using the results due to seniority transformed multi $-j$ shell Dyson boson mapping theory given by Navratil and Dobes (1991). The $SU_{sdg}(3)$ limit is found to describe the above data reasonably well.

3.4 TNT in $SU_{sdg}(3)$ Limit

Using (5,6), Akiyama et al (1986) derived numerically TNT strengths for the particular case of $^{166}Er(t,p)^{168}Er$ ($N = 15 \rightarrow N = 16$) in the $SU_{sdg}(3)$ limit. On the other hand, Devi and Kota (1991b) derived analytical expressions using (5,6) and the $U(15) \supset SU(3) \supset O(3)$ Wigner - Racah algebra. The results that are accurate for large N are given in figure 4. The $^{166}Er(t,p)^{168}Er$ data exhibits selection rules and certain other interesting features. In sdIBM $K^\pi = 0^+_3, 0^+_4, 2^+_2$ etc cannot be excited because $(2N,0) \otimes (20) \rightarrow (2N+2,0)K^\pi = 0^+_1 \oplus (2N-2,2)K^\pi = 0^+_2, 2^+_1$. The forbidden levels are observed experimentally. The strength to 0^+_1, 0^+_2(1.217 MeV), 0^+_3(1.422 MeV) and 0^+_4(1.833 MeV) are (100, 15, 10, 2.4) and (100, 8.4, 15.2, 0) in experiment and $SU_{sdg}(3)$ limit (from figure 4) respectively. As can be seen from figure 4, the transfer to $LK^\pi = 22^+_2$ belonging to $(4N - 8,4)_{\alpha=1}$ is six times the transfer to $LK^\pi = 22^+_1$ and it is remarkably close to data value of 5.0 (Akiyama et al 1986). It is seen in data that the strength to $LK^\pi = 44^+$ state at 2.06 MeV which belongs to $(4N-4,4)_{\alpha=0}$ is rather weak. From figure 4 one can see that the $(4N-4,4)_{\alpha=0,1}$ intrinsic states for $(N+1)$ boson system are two-phonon and 1g-boson type respectively. This leads to the selection rule, as can be seen from figure 4, that the transfer to $(4N-4,4)_{\alpha=0}$ is forbidden. Thus, the nature of the observed $(4N-4,4)$ bands can be inferred from TNT. However, with Akiyama's (Akiyama 1985) multiplicity label W the selection rule (except in large-N limit) is not exact. Another feature that provides a signature of g-bosons is that the $4_5^+(1.737 \text{ MeV})$ state (Davidson $et~al$ 1981) is strongly populated although the theoretical and experimental values do not match very well (ratio of the strength to 4_5^+ and 4_1^+ are 40 :100 and 37 : 12 respectively). The excitation to 4_5^+ cannot be one step in sdIBM while in gIBM it can be as it may belong to $LK^\pi = 43_1^+$.

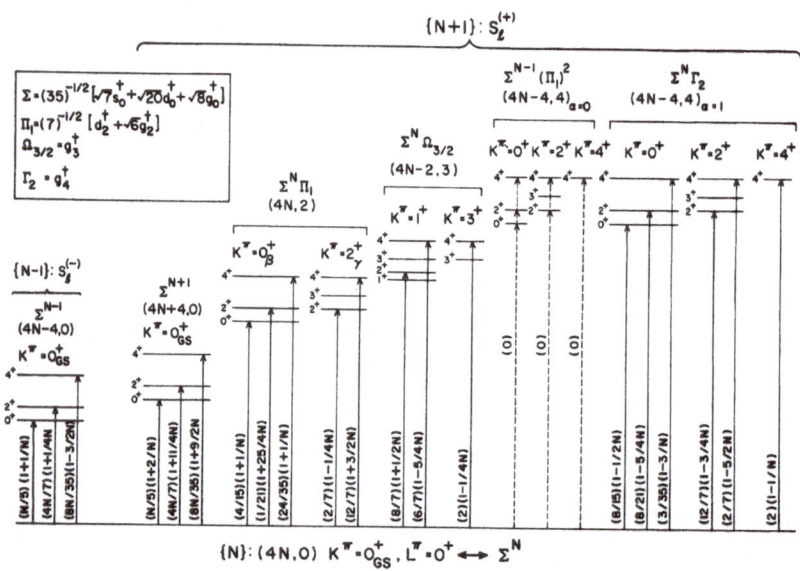

FIGURE 4. Low - lying spectrum in the $SU_{sdg}(3)$ limit for $N+1$ boson system. Also shown are the GS of N - boson system and the GS band of $N-1$ boson system. The intrinsic structure of the bands ($SU_{sdg}(3)$ irreps) are given and they are defined in terms of the single boson states Σ, Π_1, $\Omega_{3/2}$ and Γ_2 defined in the inset to the figure. Analytical expressions (Devi and Kota 1991b) that are accurate for large-$N(N \geq 8)$ for $N \to N+1$ and $N \to N-1$ TNT strengths $S_\ell^{(+)}$ and $S_\ell^{(-)}$ respectively are given in the figure. The dashed arrows indicate the TNT selection rule in the $SU_{sdg}(3)$ limit.

Thus the 4_5^+ can be interpreted as $LK^\pi = 43_1^+$ and it is in confirmity with detailed spectroscopic calculations (Yoshinaga $et~al$ 1986); see below.

3.5 Simple $SU_{sdg}(3)$ analysis

Assuming $SU_{sdg}(3)$ to be a good symmetry and using the formulas for energies, $B(E2)'s$ and quadrupole moments and $B(E4)'s$ for GS band (11-13) several well deformed nuclei are analyzed: (i) Wu (1982) analyzed the spectrum of ^{176}Hf (it has 1^+ level at 1.672 MeV) by adding $L^2(L+1)^2$ term to (11) with $\beta = 0$; (ii) Suguna $et~al.$

(1981) analyzed ^{184}W (it has 3^+ band starting from ~ 1.5 MeV) and ^{234}U (it has 3^+ band starting from ~ 1.3 MeV) with $\beta = 0$ in (11); (iii) Wu and Zhou (1984) analyzed ^{168}Er (it has 3^+ band starting from ~ 1.8 MeV) by adding perturbation corrections to (11) with $\beta = 0$; (iv) Akiyama *et al* (1987) analyzed ^{178}Hf (it has $K^\pi = 3^+$ band starting from ~ 1.6 MeV but no $K^\pi = 1^+$ band/level, it also has excited $K^\pi = 0^+, 2^+$ and 4^+ bands occuring twice and hence interpreted in terms of $(4N-8,4)_{\alpha(W)=0,1}$ and ^{234}U with non-zero β in (11); (v) Ratnaraju (1986) proposed that the elementary permissible diagrams (EPD) of Bargmann and Moshinsky (1961) can be used to distinguish multiplicity labels α of $(\lambda\mu)$ and they are used to write energy formulas for levels belonging to each $(\lambda\mu)$. The 0^+ levels in ^{170}Yb are studied using this method; (vi) With the extension of band cut - off (12) in sdgIBM, the $B(E2; L+2 \rightarrow L)/B(E2; 2^+ \rightarrow 0^+)$ grow and come closer to rotational model values. Wu (1982) analyzed the $B(E2)$ data for ^{232}Th, ^{168}Er, ^{170}Hf, ^{164}Yb while Akiyama *et al* (1987) analyzed ^{238}U using (12) and the data is in much closer agreement with the sdg model; (vii) Suguna *et al* (1981) employing (12) and its extension, fitted with a single effective charge the $B(E2; 2^+ \rightarrow 0^+)$ values taking $\eta = 2\,(sd$ bosons) for $^{120}Xe, \eta = 4\,(sdg)$ for ^{198}Pt, ^{154}Gd, ^{186}W, ^{158}Gd, ^{184}Pt, ^{156}Gd, ^{158}Dy and ^{160}Gd and $\eta = 6\,(sdgi)$ for ^{218}Ra, $^{228-232}Th$ and $^{230-236}U$ and obtained good agreement. Note that given a value of η the ground state band corresponds to the $SU(3)$ irrep $(\eta N, 0)$; (vii) occupation numbers $< n_\ell >$ take a simple form in the $SU(3)$ limit and they are studied by Wu and Zhou (1984) and Kota (1987). Occupation numbers $< \hat{n}_\ell >^{(4N,0)L=0}$ are $N(4N+1)/5(4N-3), 16N(N-1)(4N+1)/7(4N-1)(4N-3)$ and $64N(N-1)(2N-3)/35(4N-1)(4N-3)$ for $\ell = 0(s), 2(d)$ and $4(g)$ respectively. From these expressions it is seen that in the $SU_{sdg}(3)$ limit for the low - lying levels the g - boson occupancy is $\sim 20\%$. These simple calculations do not describe many important features observed in rotational nuclei, for example: (i) anharmonicity associated with the $K^\pi = 4^+$ bands $(E(K^\pi = 4_1^+ L = 4^+)/E(K^\pi = 2_1^+ L = 2^+) \simeq 2.5$ but not 2 in ^{168}Er); (ii) large splitting between $(4N-6,3)K^\pi = 3^+, 1^+$ bands; (iii) splitting between the $(4N-8,4)$ with $\alpha = 0$ and 1 irreps etc. This clearly calls for mixing calculations in $SU_{sdg}(3)$ basis.

3.6 Detailed SU$_{sdg}$(3) basis studies

Yoshinaga *et al* (1986,1988) carried out careful and successful analysis of ^{168}Er (Davidson *et al* (1981) using neutron capture γ - ray spectroscopy ((n,γ) reaction) produced precise and almost complete information about all the bands below 2 MeV excitation in ^{168}Er), employing $SU_{sdg}(3)$ basis with all $SU(3)$ irreps satisfying $\lambda + 2\mu = 4N - 3r, r = 0, 1, 2$ and $\mu \leq 8, 7, 6$ respectively (54 $SU(3)$ irreps, 20 distinct ones). The hamiltonian in (11) was extended to include: (i) two pieces (U and Z) that do not change the positions of $W(\alpha) = 0$ states; (ii) a piece (H_1) which does not connect the states $(4N,0)K = 0L = 4$ and $(4N-4,2)K = 0L = 4$ and thus H_1 can produce the anharmonicity associated with the $K^\pi = 4_1^+$ band; (iii) a piece H_2 which does not connect $(4N-4,2)K = 0L = 4$ and $(4N-8,4)K = 0W = 1L = 4$ so that mainly it lowers $0_2^+, 0_3^+$ bands; (iv) a piece P that eliminates connection between $(4N-6,3)K = 1L = 4$ with other low - lying 4^+ levels so that $K = 1$ band can be moved up easily. See Yoshinaga *et al* (1988) for the definition of U, Z, H_1, H_2 and P and also for details regarding the actual calculations in the $SU_{sdg}(3)$ basis. The spectrum obtained is in excellent agreement with data. The sdgIBM cures the problem of anharmonicity associated with the 4_1^+ band and it also predicts correctly the position of $K^\pi = 3_1^+$ band. In addition, the structure of $K^\pi = 0_3^+$ and 4_1^+ bands at 1.422 MeV and 2.06 MeV are predicted to be one g - boson ($\alpha(W) = 1$) and two - phonon ($\alpha(W) = 0$)

type respectively. The $^{166}Er(t,p)^{168}Er$ data supports the two - phonon structure of the $K^\pi = 4_1^+$ band; see sect 3.4. The two - phonon structure is confirmed recently by Borner et al (1991) who measured the ratio $R(4^+) = B(E2; K^\pi = 4_1^+ L = 4^+ \to 2_\gamma^+)/B(E2; 2_\gamma^+ \to 0_{GS}^+)$. The experimental value is $0.52 < R(4^+) < 1.61$ and sdgIBM prediction is $R(4^+) = 1.4$.

4. $[U_{sd}(6) \supset G] \oplus U_g(9)$ LIMIT WITH $n_g = 1$

The $[U_{sd}(6) \supset G] \oplus U_g(9)$ limit with $n_g = 1$ provides a natural frame-work to understand and predict about hexadecupole ($E4$) vibrational states (or the collective bands built upon them). Clear band structures emerge out when the core is described by one of the three limiting symmetries $U_d(5)$, $SU_{sd}(3)$ or $O_{sd}(6)$ of IBM. In all the three limits, classification of the band structures and the corresponding analytical formulas for energies and $E2$ aand $E4$ matrix elements are given in detail by Devi and Kota (1991a).

4.1 $SU_{sd}(3)\times 1g$ Limit

The $SU_{sd}(3)\times 1g$ limit generates rotational bands built on hexadecupole vibrations. The hamiltonian in the $SU_{sd}(3) \times 1g$ limit is

$$H_{SU_{sd}\times 1g} = -\kappa Q_{sd}^2 \cdot Q_{sd}^2 + \epsilon_g \hat{n}_g - \Gamma Q_{sd}^2 \cdot Q_g^2$$
$$-\Lambda : \left[\left(d^\dagger \tilde{g}\right)^4 \cdot \left(g^\dagger \tilde{d}\right)^4\right] : + k'\mathbf{L} \cdot \mathbf{L} \tag{14}$$

where $Q_{sd}^2 = \sqrt{8/3}\{(s^\dagger \tilde{d} + d^\dagger \tilde{s})^2 - \sqrt{7}/2(d^\dagger \tilde{d})^2\}$ is the $SU_{sd}(3)$ quadrupole operator, $Q_g^2 = \left(g^\dagger \tilde{g}\right)^2$, :: denotes normal ordering and $\mathbf{L}(1)$ is the angular momentum generator. It can be shown that (Arima and Iachello 1978b; Bijker and Kota 1988) as $N \to \infty$ and $\Gamma, \Lambda \gg \kappa$, the $\Psi(K, L)$ states given below become the eigenstates of the hamiltonian in (14),

$$\Psi(K, L) = |N; N_{sd} = (N-1), (2N-2, 0); g; KL\rangle =$$

$$\sum_{R=even} (-1)^L \sqrt{\frac{2R+1}{2L+1}} \sqrt{\frac{2}{(1+\delta_{K0})}} \langle R\,0\,4K|L\,K\rangle \times$$

$$|N_{sd} = (N-1), (2N-2, 0)R; g; KL\rangle ;$$

$$L = 0, 2, 4, ---\text{ for } K = 0 \text{ and } L = K, K+1, ---\text{for } K \neq 0. \tag{15}$$

In (15), $(2N-2, 0)$ is the leading $SU_{sd}(3)$ irrep for $(N-1)$ sd - boson system. Using the results given by Bijker and Kota (1988) and treating $\kappa Q_{sd}^2 \cdot Q_{sd}^2$ term perturbatively (Arima and Iachello 1978b), the energies of the $\Psi(K, L)$ levels can be written down,

$$E(N; (2N-2, 0); g; KL) = -\kappa\{(2N^2 - N - 1) - \tfrac{3}{8}(L(L+1) + 20 - 2K^2)\}$$

$$-\tfrac{\Gamma(N-1)}{3\sqrt{154}}(3K^2 - 20)\left\{1 + \tfrac{\Lambda(3K^2-20)}{5\sqrt{154}\Gamma}\right\} + \epsilon_g + k'L(L+1) \tag{16}$$

Thus one can have rotational bands built on hexadecupole vibrations. Their structure is given by (15) and their energies by (16). In addition to the $1g$-boson bands, there are the usual $K^\pi = 0^+$ GS band described by the $SU_{sd}(3)$ irrep $(2N, 0)$ and the β, γ bands generated by the irrep $(2N-4, 2)$ with $K^\pi = 0^+$ and 2^+ respectively. With the variation in Λ, relative positions of $1g$-boson bands can be altered, in the absence of the Λ term ($\Lambda = 0$) the bands will be arranged as $K^\pi = 4^+, 3^+, 2^+, 1^+$ and 0^+ or

$K^\pi = 0^+, 1^+, 2^+, 3^+$ and 4^+ depending upon the sign of Γ. For example with $\kappa = 15$ keV, $k' = 7.7$ keV, $\Gamma = 75$ keV and with $\Lambda \leq -0.15$ MeV one can bring down the $K^\pi = 3^+$ band and the ordering of the bands for $\Lambda = -0.25$ MeV is $K^\pi = 3^+, 2^+, 4^+, 1^+$ and 0^+. Thus it is seen that the exchange term can explain some of the low-lying $K^\pi = 3^+$

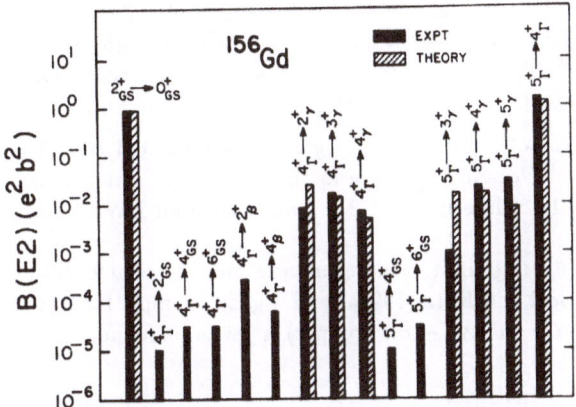

FIGURE 5. Comparison between experiment (thick bars) and theory (hatched bars) for the $B(E2; L_i^+ \to L_f^+)$ values in ^{156}Gd for the $E2$ transitions from the hexadecupole vibrational $K^\pi = 4^+(\Gamma)$ band with $L^\pi = 4^+, 5^+$ to $K^\pi = 4^+$, GS, β and γ band members. Shown also is the $E2$ strength from $2^+_{GS} \to 0^+_{GS}$. The experimental data is from Van Isacker et al (1982) and the theoretical calculations are in the $SU_{sd}(3) \times 1g$ limit (Devi and Kota 1991a). $B(E2)'s < 10^{-6}e^2b^2$ in the calculations are not shown in the figure.

bands observed in some rare-earth nuclei. In the $SU_{sdg}(3)$ limit of gIBM (Akiyama et al 1987) 3^+ and 1^+ bands come very close whereas with $\Lambda = -0.25$ MeV they are separated by about 1MeV.

The benchmark examples for bands built on one - phonon hexadecupole vibrations (described by $SU_{sd}(3) \times 1g$ limit of gIBM) come from ^{156}Gd and ^{178}Hf (Devi and Kota 1991a); ^{156}Gd is discussed below. The number of bosons is $N = 12$ for ^{156}Gd, and in this nucleus a $K^\pi = 4^+$ band (referred to as Γ-band) is observed at 1.51 MeV, in addition to the β-band at 1.05 MeV and the γ-band at 1.15 MeV. The structure of the Γ-band is $| N; (2N - 2, 0); g; K = 4 L >$. From the analytical expressions derived (Devi and Kota 1991a) for the $E2$ decay of the Γ-band numbers it is seen that the intraband transitions ($\Gamma \to \Gamma$) are strong and go as N^2 while the transitions to γ-band are independent of N. Moreover the decay to β and GS bands is forbidden. Thus $\Gamma \to \Gamma \gg \Gamma \to \gamma \gg \Gamma \to \beta$, GS; the later two being zero. This trend is clearly seen in the data given by Van Isacker et al (1982) and these authors have to perform elaborate numerical (1g-boson) mixing calculations to demonstrate that the sdg model predicts the same, while the analytical results given by Devi and Kota (1991a) bring out the above feature instantaneously. Experimental $B(E2)$ values (in units of $10^{-2}e^2b^2$) for $4^+_\Gamma \to L^+_\gamma \sim 1, 4^+_\Gamma \to$ GS, $\beta \sim 0, 5^+_\Gamma \to L^+_\gamma \sim 2, 5^+_\Gamma \to$ GS ~ 0 and $5^+_\Gamma \to 4^+_\Gamma \sim 150$. The data is well explained as shown in figure 5.

4.2 U(5) × 1g and O(6) × 1g limits

In the $U_d(5) \times 1g$ limit, with suitably chosen hamiltonian, one has N, N' bands with $1g$ boson excitations in addition to the ground Y, X, Z — bands (Arima and Iachello 1976); Y - band : $| N; N_{sd} = N; n_d, v_d = n_d, L_d = 2n_d >$. The N - band is defined by $| N; N_{sd} = N - 1; n_d, v_d = n_d, L_d = 2n_d; g; L = 2n_d + 4 >$ where $n_d = 0, 1, 2, 3, ----, N-1$ and N' - band is simailarly defined with $L = 2n_d + 3$. Cd isotopes form good examples for $U_d(5) \times 1g$ limit. The 4^+ states at 2.22 MeV, 2.38 MeV and 2.33 MeV in ^{110}Cd, ^{114}Cd and ^{116}Cd are hexadecupole vibration in nature and they correspond to the $n_d = 0$ member of the N - band in $U_d(5) \times 1g$ limit. The boson number N for ^{110}Cd, ^{114}Cd and ^{116}Cd is 7, 9 and 8 respectively. The $B(E4)$ values for the decay of these 4^+ states to the 0^+_{GS} are given by the formula $B(E4) = N(e_{04}^{(4)})^2$ with the $E4$ operator being $e_{04}^{(4)}(s^\dagger \tilde{g} + g^\dagger \tilde{s})$. The predicted $B(E4)$ values (parameter free) for ^{110}Cd, ^{114}Cd and ^{116}Cd are in the ratio 0.88 : 1.1 : 1 and they should be compared with the values 0.85 : 1.2 : 1 deduced from DWBA analysis (Koike *et al* 1969).

In the $O_{sd}(6) \times 1g$ limit, band structures and the related $E2$ and $E4$ properties are studied by Devi and Kota (1991a) and a good example for this limit (expected to come from $Pt - Os$ and $Xe - Ba$ isotopes) is yet to be found.

5. $SU_{sdg}(3)$ LIMIT OF p − n sdgIBM AND p − n sdgIBFM

The inclusion of $\pi - \nu$ degrees of freedom gives rise to F - spin where $F_{max} = (N_\pi + N_\nu)/2$ and $F_Z = (N_\pi - N_\nu)/2$; for low - lying states $F = F_{max}$, while for scissors states $F = F_{max} - 1$. The resulting $p - n$ sdgIBM admits a large variety of dynamical symmetries. Till now only the $SU_{sdg}(3)$ limit is explored and the only problem addressed here is the study of scissors state and $M1$ distributions in even-even nuclei. It should be stressed that complete study (even the classifications) of dynamical symmetries of $p - n$ sdgIBM is yet to be carried out. The growing data on several 1^+ states, the resulting $M1$ distributions (Richter 1991), and the failure of sdIBM to explain the observed fragmentation, resulted in several studies in sdgIBM framework since the $p - n$ sdgIBM accomodates much larger variety of 1^+ states. In the $SU_{sdg}(3)$ limit of $p - n$ sdgIBM the group chain and the basis states are

$$\left| \begin{array}{cccc} U_{sdg}(30) & \supset & U_{sdg}(15) & \otimes & SU(2) & \supset \\ \{N\} & & \{f\} = \{f_1, f_2\} & & F = (f_1 - f_2)/2 & \end{array} \right.$$

$$\left. \begin{array}{cccc} SU_{sdg}(3) & \otimes & SU(2) & \supset & O_{dg}(3) & \otimes & SU(2) \\ (\lambda_B, \mu_B) & & F & \cdot & KL & & \end{array} \right\rangle \tag{17}$$

In this limit the GS is denoted by $| \{N\} F = N/2 (4N, 0) K = 0 L = 0 >$. The $M1$ operator in pn - sdgIBM is

$$T^{M1} = \sum_{\rho=\pi,\nu} \sqrt{10} \{g_{d,\rho}(d_\rho^\dagger \tilde{d}_\rho)^1 + \sqrt{6}g_{g,\rho}(g_\rho^\dagger \tilde{g}_\rho)^1\} , \tag{18}$$

which has four g-factors. The $M1$ excited states from the GS are labelled by $| \{f_1, f_2\} (\lambda_f, \mu_f) K_f L_f; F', F'_Z)$. Note that the $U_{sdg}^B(15)$ irrep $\{f_1, f_2\}$ can be symmetric $\{N\}$ with $F' = N/2$ or mixed symmetric $\{N-1, 1\}$ with $F' = N/2 - 1$. The $M1$ excited core irreps (λ_f, μ_f) are $(4N-6, 3)$ with $F' = N/2$(symmetric (s)) and $(4N-2, 1)$, $(4N-6, 3)^2$ with $F' = N/2 - 1$(mixed symmetric (ms)) ; note that $(4N - 6, 3)$ irrep appears twice for $F' = N/2 - 1$ and they are denoted by $\alpha = 0$ and 1 respectively. Using the

$SU(3)$ tensorial decomposition of the $M1$ operator (18) expressions for $B(M1; F = N/2\,(4N,0)K = 0\,L = 0 \rightarrow F'\,(\lambda_f, \mu_f)K_f = 1\,L_f = 1)$ are derived (Devi and Kota 1992, Wu et al 1987) in terms of $SU(3)$ and F - spin ($SU(2)$) recoupling coefficients and $SU(3)$ ISF's. Substituting the large$-N$ limit values for the recoupling coefficients and ISF's compact analytic expressions are obtained for the above $B(M1)'s$,

$$(\lambda_f \mu_f) = (4N - 6, 3), F' = N/2 \;:\; (3/4\pi)(96/49)N\,(\,h_{R+}\,)^2$$

$$(\lambda_f \mu_f) = (4N - 2, 1), F' = N/2 - 1 \;:\; (3/4\pi)(8N_\pi N_\nu/N)\,(\,g_{R-}\,)^2$$

$$(\lambda_f \mu_f) = (4N - 6, 3)_{\alpha=0}, F' = N/2 - 1 \;:\; (3/4\pi)(96 N_\pi N_\nu/49N)\,(\,h_{R-}\,)^2$$

$$(\lambda_f \mu_f) = (4N - 6, 3)_{\alpha=1}, F' = N/2 - 1 \;:\; 0$$

$$g_{R+} = (N_\pi g_{R\pi} + N_\nu g_{R\nu})/N, \; h_{R+} = (N_\pi h_\pi + N_\nu h_\nu)/N, \; g_{R-} = (g_{R\pi} - g_{R\nu}),$$

$$h_{R-} = (h_\pi - h_\nu), \; g_{R\rho} = 7^{-1}\,(1 + 2x_0)\,g_{d\rho} + (6 - 2x_0)\,g_{g\rho}$$

$$h_\rho = (\,g_{d\rho} - g_{g\rho}\,) = -\sqrt{7/60}\,A_{3\rho}\,;\; \rho = \pi, \nu, \; x_0 = (4N + 4)/(4N - 1) \tag{19}$$

While going from sd to sdg case there is an enhancement by a factor 2 in $B(M1)'s$ as the corresponding strength given by (19) is of the form $(3/4\pi)\,(2(\eta N_\pi)(\eta N_\nu))\,/\,(\eta N)$ $(g_{R-})^2$; $\eta = 2$ for sd and $\eta = 4$ for sdg. The $B(M1)$ strength to scissors $(4N - 2, 1)$ irrep derives from both (11) and (33) tensor parts while the strengths to $(4N - 6, 3)F'$

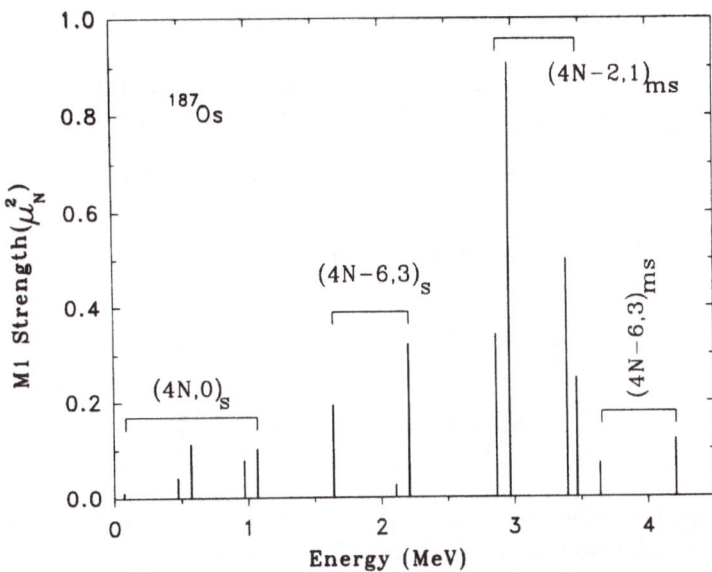

FIGURE 6. $M1$ strength distribution in ^{187}Os. For the final state core $(\lambda', \mu')\,SU^B(3)$ irreps $(4N, 0)_s$, $(4N - 6, 3)_s$, $(4N - 2, 1)_{ms}$ and $(4N - 6, 3)_{ms}$ the strengths corresponding to $(\lambda_f, \mu_f)\,SU^{BF}(3)$ irreps $\{(4N - 2, 3), (4N - 4, 4), (4N - 4, 4), (4N, 2), (4N, 2)\}$, $\{(4N - 7, 4), (4N - 6, 2), (4N - 6, 2)\}$, $\{(4N - 4, 4), (4N - 4, 4), (4N - 3, 2), (4N - 3, 2)\}$ and $\{(4N - 7, 4), (4N - 6, 2)\}$ respectively are shown in the figure; $N = 10, N_\pi = 3$ and $N_\nu = 7$. Note that strengths less than $0.01\mu_N^2$ are not shown. (Devi and Kota 1992f) gives details.

irreps arise due to (33) tensor part alone. There will be more fragmentation of $M1$ strength (compared to sdIBM) with the inclusion of g - bosons as some strength can go to $(4N-6,3)1^+$ states and this fraction of the strength depends on the g - factors h_ρ defined in (19); $h'_\rho s$ are determined from the g - factor data of γ - band members. The $h'_\rho s$ and $g'_\rho s$ are determined for ^{156}Gd by Wu $et\ al$ (1987), Sm isotopes by Mizusaki $et\ al$ (1991), ^{168}Er by Kuyucak $et\ al$ (1988) and ^{187}Os by Devi and Kota (1992f).

In order to study odd - mass nuclei in sdgIBM one has to couple the odd fermion to the even - even core described by sdgIBM. The resulting sdg interacting boson fermion model (sdgIBFM) has been studied so far by exploiting its $SU(3) \otimes U(2)$ limit. This limit is used in studying $M1$ - distributions (in particular scissors states) for even - odd nuclei (where the odd neutron occupies all the natural parity orbits) in $N = 82 - 126$ shell; till to date no experimental data is available for scissors states in odd-A nuclei. The g-bosons are included for the first time in IBFM by Devi and Kota (1992g) to study the effects due to hexadecupole degree of freedom on scissors $M1$ states in odd-A nuclei. The question of fragmentation of $M1$ strength is more severe in odd - A nuclei and hence the relevance of g - bosons. For $N = 82 - 126$ shell nuclei the odd - fermion occupies single particle natural parity orbits $j = 1/2, 3/2, 5/2, 7/2, 9/2$ which can be split into pseudo-orbital part $\tilde\ell = 0, 2, 4$ and pseudo-spin part $\tilde{s} = 1/2$ (with this the states of the odd-particle transform as the (40) irrep of $SU(3)$) and the core nucleus is described by the $SU(3)$ limit of sdgIBM, two coupling schemes giving rise to $SU(3) \otimes U(2)$ limit of sdgIBFM are possible,

$$
\left|
\begin{array}{cccc}
U^B_{sdg}(30) \otimes U^F(30) & \supset & U^B_{sdg}(15) \otimes & SU^B(2) & \otimes\ U^F(15) \otimes U^F(2) \\
\{N\} \qquad \{1\} & & \{f\} = \{f_1, f_2\} & F = \frac{1}{2}(f_1 - f_2) & \{1\}
\end{array}
\right.
$$

$$
\supset
\begin{array}{c}
I \\ \\ II
\end{array}
\begin{array}{ccc}
SU^B_{sdg}(3) \otimes & SU^B(2) \otimes & SU^F(3) \otimes\ U^F(2) \\
(\lambda_B, \mu_B) & F & (40) \\ \\
U^{BF}_{sdg}(15) & \otimes & SU^B(2) \otimes\ U^F(2) \\
\{f'\} = \{f'_1, f'_2, f'_3\} & & F
\end{array}
\begin{array}{c}
I \\ \\ II
\end{array}
\supset
$$

$$
\supset
\begin{array}{ccc}
SU^{BF}(3) & \otimes & SU^B(2) \otimes\ U^F(2) \\
(\lambda_{BF}, \mu_{BF}) & & F
\end{array}
$$

$$
\supset \left[
\begin{array}{ccc}
O^{BF}(3) & \otimes & SU^F(2) \supset \ \ Spin^{BF}(3) \\
L & & \tilde{s} = 1/2 \qquad J
\end{array}
\right] \otimes
$$

$$
\left.
\left[
\begin{array}{cc}
SU^B(2) \supset & U^B(1) \\
F & F_Z = (N_\pi - N_\nu)/2
\end{array}
\right]
\right\rangle
$$

(20)

Details regarding chain - II are given in (Kota 1986b, Devi and Kota 1992g). In chain - I the GS $SU^{BF}(3)$ irrep (λ_{BF}, μ_{BF}) which is often denoted by (λ_{GS}, μ_{GS}) is determined by IBFM vs pseudo-Nilsson correspondence given by Bijker and Kota (1988). The GS in chain - I is denoted by $|\ \{N\}\ (4N, 0) \otimes (40);\ (\lambda_{GS}, \mu_{GS})\,K\,L;\ 1/2;\ J\ :\ F = N/2,\ F_Z = (N_\pi - N_\nu)/2 >$. The $M1$ excited states from the GS are labelled by $|\ \{f'\}\,(\lambda', \mu') \otimes (40);\ (\lambda_f, \mu_f)\,K_f\,L_f;\ 1/2;\ J_f\ :\ F',\ F'_Z = (N_\pi - N_\nu)/2 >$ and the excited core irreps (λ', μ') are $(4N, 0)$, $(4N-4, 2)$, $(4N-6, 3)$ with $F' = N/2$ and $(4N-2, 1)$, $(4N-4, 2)$, $(4N-6, 3)^2$ with $F' = N/2 - 1$. The $M1$ states (λ_f, μ_f) arising out of the irrep $(4N-2, 1)$ are the scissors states. It is easy to work out the $M1$ states (λ_f, μ_f) corresponding to the core irreps $\{f'\}(\lambda', \mu')$. With the $M1$ operator given in (18) expressions for $M1$ matrix elements are derived in terms of $SU(3)$ recoupling coefficients (Devi and Kota 1992f,g). Using these results the location and $M1$ decay strength of scissors state in particular and all the $M1$ states in general are studied for nuclei with odd neutron in $N = 82 - 126$ shell. There are four classes of even-odd nuclei with the odd neutron in 82 - 126 shell that respect pseudo - spin which is a requirement for the goodness of $SU(3) \otimes U(2)$ limit of IBFM. The stable nuclei in these classes which can be probed by conventional $(e, e'), (p, p')$ and (γ, γ') experiments for $M1$ states are $\{^{187,189}Os\}$, $\{^{173}Yb, ^{177}Hf\}$ and $\{^{171}Yb\}$ and their GS belong to the pseudo-Nilsson configurations $[\widetilde{411}]$, $[\widetilde{413}]$ and $[\widetilde{420}]$ with the corresponding $SU^{BF}(3)$ irreps being $(4N - 2, 3)$, $(4N - 2, 3)$ and $(4N, 2)$ respectively. Among the nuclei listed above, ^{187}Os is the best candidate for the study of $M1$ distributions as it was recently demonstrated by Warner and Van Isacker (1990) that the $E2$ transition strengths in this nucleus are well described by the $SU(3) \otimes U(2)$ limit. Therefore, the predictions for $M1$ distributions in ^{187}Os are shown in figure 6. The results in figure 6 show that large fraction of the $M1$ strength goes to the scissors states (arising out of $(4N-2,1)_{ms}$ irrep) and these states are well separated in energy from rest of the states. This feature makes it plausible that the $M1$ states arising out of $(4N - 2, 1)$ irrep can be detected in experiments. Keeping this in mind complete results for scissors states are derived by Devi and Kota (1992g,h).

6. Future Outlook

Applications of dynamical symmetries in sdgIBM are described in this article. Many other aspects of sdgIBM are investigated by several research groups and the whole work is reveiwed in a recent article by Devi and Kota (1992i).

The $p - n$ sdg IBF^{n}M (n denoting number of quasi particles) which includes hexadecupole vibrations, two-phonon excitations, scissors configurations and quasiparticle structures appears to be the ultimate model for complete understanding of band structures and the associated $M1, E2$ and $E4$ properties in heavy nuclei and $SU(3)$ limit of this model is particularly suitable for deformed nuclei. This extension of sdgIBM will enable one to study in addition high spin states, superdeformed bands, odd-odd nuclei etc. (Iachello 1991). For even-even nuclei with $n = 0 \oplus 2 \oplus 4$ and odd-A nuclei with $n = 1 \oplus 3$ quasiparticles and using $SU(3)$ limit of $p - n$ sdgIBF^{n}M (see chain (20); here the particle $SU^F(3)$ irreps for any n and pseudo-spin value can be obtained using (8,9) and tabulation for the same are available for $\eta = 2, 3, 4, 5$ oscillator shells), some aspects of high spin states are being studied.

Recently Balantekin et al (1991) showed that, in IBM framework when $SU_{sd}(3)$ limit is employed for deformed nuclei, the sub-barrier fusion cross - sections involve group elements of $SU_{sd}(3)$. The expressions for the later are recently given by Ginocchio (1991). However with the important role played by hexadecupole degree of freedom (Fernandez et al 1991) in sub-barrier fusion, it appears that $SU_{sdg}(3)$ limit is more relevant. The relevant group elements of $SU_{sdg}(3)$ are given by the matrix elements of the operator $e^{i\alpha Q_0^2(s)}$ and they can be calculated by using the following integral representation for the GS band,

$$| N, (4N, 0)LM = 0 > =$$

$$\frac{(2L+1)}{8\pi^2 \Lambda(L)} \int d\Omega D_{00}^L(\Omega) R_\Omega(s^\dagger + \sqrt{20/7}d_0^\dagger + \sqrt{8/7}g_0^\dagger)^N | 0 > ;$$

$$\Lambda(L) = [5^N (2L + 1)N!(4N)!/(4N - L)!!(4N + L + 1)!!]^{1/2} \qquad (21)$$

In (21) R_Ω is the rotation operator. It is plausible that eq(8.7-8.9) of Ginnochio (1991) can be transcribed to the sdg case with the replacement $2N \to 4N$. This topic is being investigated.

7. Acknowledgements

The authors thank F. Iachello for his interest in the work reported in this article and also for many valuable discussions. One of the authors (VKBK) thanks B. Gruber for inviting him to write this contribution.

8. References

Akiyama Y 1985 *Nucl. Phys.* **A433** 369

Akiyama Y , Heyde K , Arima A and Yoshinaga N 1986 *Phys. Lett.* **173B** 1

Akiyama Y , von Brentano P and Gelberg A 1987 *Z. Phys.* **A326** 517

Arima A and Iachello F 1975 *Phys. Rev. Lett.* **35** 1069

Arima A and Iachello F 1976 *Ann. Phys. (NY)* **99** 253

Arima A and Iachello F 1978a *Phys. Rev. Lett.* **40** 385

Arima A and Iachello F 1978b *Ann. Phys. (NY)* **111** 201

Arima A and Iachello F 1979 *Ann. Phys. (NY)* **123** 468

Balantekin A B , Bars I and Iachello F 1981 *Nucl. Phys.* **A379** 284

Balantekin A B , Bennett J R and Takigawa N 1991 *Phys. Rev.* **C44** 145

Bargmann V and Moshinsky M 1961 *Nucl. Phys.* **23** 177

Bijker R and Kota V K B 1988 *Ann. Phys. (NY)* **187** 148

Borner H G , Jolie J , Robinson S J , Krusche B , Piepenbring R , Casten R F , Aprahamian A and Draayer J P 1991 *Phys. Rev. Lett.* **66** 691

Casten R F and Warner D D 1988 *Rev. Mod. Phys.* **60** 389

Chen H T , Kiang L L , Yang C C , Chen L M , Chen T L and Jiang C W 1986 *J. Phys.* **G12** L217

Davidson W F , Warner D D , Casten R F , Schreckenbach K , Borner H G , Simic J , Stojanovic M , Bogdanvic M , Koicki S , Gelletly W , Orr G B and Stells M L 1981 *J. Phys.* **G7** 455

Devi Y D 1991 *Some studies in sdg Interacting Boson Model of Atomic nuclei* Ph. D. Thesis (Gujarat University, India)

Devi Y D , Kota V K B and Sheikh J A 1989 *Phys. Rev.* **C39** 2057

Devi Y D and Kota V K B 1990a Physical Research Laboratory Report **PRL -TN-90-68**

Devi Y D and Kota V K B 1990b *Z. Phys.* **A337** 15

Devi Y D and Kota V K B 1991a *J. Phys. G. Nucl. Part. Phys.* **17** 465

Devi Y D and Kota V K B 1991b *J. Phys. G. Nucl. Part. Phys.* **17** L185

Devi Y D and Kota V K B 1991c *Proc. Symp. Nucl. Phys.* (Bombay, India) **34B** 47

Devi Y D and Kota V K B 1992a , *Phys. Rev.* **C46** 1370

Devi Y D and Kota V K B 1992b , *Phys. Rev.* **C45** 2238

Devi Y D and Kota V K B 1992c Physical Research Laboratory report **PRL-TH-92/13**

Devi Y D and Kota V K B 1992d Physical Research Laboratory report **PRL-TH-92/14**

Devi Y D and Kota V K B 1992e Physical Research Laboratory report **PRL-TH-92/15**

Devi Y D and Kota V K B 1992f Phys. Lett. **B287** 9

Devi Y D and Kota V K B 1992g *Nucl. Phys.* **A541** 173

Devi Y D and Kota V K B 1992h in *Group Theory and Special Symmetries in Nuclear Physics* (eds) J.P. Draayer and J.Janecke (Singapore: World Scientific) pp. 122

Devi Y D and Kota V K B 1992i Pramana - *J. Phys.* (in press)

Elliott J P 1958a *Proc. Roy. Soc. (London)* **A245** 128

Elliott J P 1958b *Proc. Roy. Soc. (London)* **A245** 562

Fernandez Niello J O , di Tada M , Macchiavelly A O , Pacheco A J , Abriola D , Elgu M , Etchegoyen A, Etchegoyen M S , Gil S and Tessoni J E 1991 *Phys. Rev.* **C43** 2303

Ginicchio J N 1991 *in Symmetries in science* V (eds) B. Gruber,L.C. Biedenharn and H.D. Doebner (New York: Plenum)

Hecht K T 1964 *in Selected Topics in Nuclear Spectroscopy* (Compiler) B J Verhaar (Amsterdam: North Holland) pp. 51

Hecht K T 1965 *Nucl. Phys.* **62** 1

Iachello F (editor) 1979 *Interacting bosons in nuclear physics* (New York: Plenum)

Iachello F (editor) 1981 *Interacting Bose-Fermi systems in nuclei* (New York: Plenum)

Iachello F and Arima A 1987 *The interacting boson model* (Cambridge: Cambridge Univ. Press)

Iachello F 1991 *Nucl. Phys.* **A522** 83c

Ichihara T , Sakaguchi H , Nakamura M , Yosoi M , Yeiri M , Tekeuchi Y , Togawa H Tsutsumi T and Kobayashi S 1987 *Phys. Rev.* **C36** 1754

Koike M , Nonaka I, Kokame J , Kamitsubo H , Awaya Y , Wada T and Nakamura H 1969 *Nucl Phys.* **A125** 161

Kota V K B 1977 *J. Phys.* **A10** L39

Kota V K B 1982 SU(3) *bands with* g - *bosons in IBA* (University of Rochester Report **UR – 811**)

Kota V K B 1984 *in Interacting Boson-Boson and Boson-Fermion systems* (ed) O Scholten (Singapore: World Scientific) pp. 83

Kota V K B 1986a Physical Research Laboratory Report **PRL–TN – 86 – 54**

Kota V K B 1986b *Phys. Rev.* **C33** 2218

Kota V K B 1987 *in Recent trends in Theoretical Nuclear Physics* (ed) K Srinivasa Rao (Delhi: Macmillan India Ltd) pp. 70, pp. 102

Kota V K B , De Meyer H , Vander Jeugt J and Vanden Berghe G 1987 *J. Math. Phys.* **28** 1644

Kuyucak S , Lac V S and Morrison I 1991a *Phys. Lett.* **263B** 146

Kuyucak S , Lac V S , Morrison I and Barrett B R 1991b *Phys. Lett.* **263B** 347

Kuyucak S and Morrison I 1991 *Phys. Lett.* **255B** 305

Kuyucak S , Morrison I, Faessler A and Gelberg A 1988 *Phys. Lett.* **215B** 10

Kuyucak S , Morrison I and Sebe T 1991c *Phys. Rev.* **C43** 1187

Ling Y 1983 *Chin. Phys. (printed by APS)* **3** 87

Meyer H De, Van der Jeugt J , Vanden Berghe G and Kota V K B 1986 *J. Phys.* **A19** L565

Mizusaki T , Otsuka T and Sugita M 1991 *Phys. Rev.* **C44** R1277

Navratil P and Dobes J 1991 *Nucl. Phys.* **A532** 223

Ratna Raju R D 1981 *Phys. Rev.* **C23** 518

Ratna Raju R D 1986 *J. Phys.* **G12** L279

Richter A 1991 *Nucl. Phys.* **A522** 139c

Sen X Y and Yu Z R 1987 *J. Math. Phys.* **28** 2192

Suguna M , Ratna Raju R D and Kota V K B 1981 *Pramana.* **17** 381

Sun H Z , Moshinsky M , Frank A and Van Isacker P 1983 *Kinam* **5** 135

Van Isacker P , Heyde K , Waroquier M and Wenes G 1982 *Nucl. Phys.* **A380** 383

Vanthournout J , De Meyer H and Vanden Berghe G 1988 *J. Math. Phys.* **29** 1958

Vanthournout J , De Meyer H and Vanden Berghe G 1989 *J. Math. Phys.* **30** 943

Warner D D and Van Isacker P 1990 *Phys. Lett.* **247B** 1

Wu H C 1982 *Phys.Lett.* **110B** 1.

Wu H C , Dieperink A E L and Pittel S 1986 *Phys. Rev.* **C34** 703.

Wu H C , Dieperink A E L and Scholton O 1987 *Phys. Lett.* **187B** 205

Wu H C and Zhou X Q 1984 *Nucl. Phys.* **A417** 67

Wybourn B G 1970 *Symmetry principles in atomic spectroscopy* (New York: Wiley Interscience)

Yoshinaga N 1986 *Nucl. Phys.* **A456** 21

Yoshinaga N 1989 *Nucl. Phys.* **A493** 323

Yoshinaga N , Akiyama Y and Arima A 1986 *Phys. Rev. Lett.* **56** 1116

Yoshinaga N , Akiyama Y and Arima A 1988 *Phys. Rev.* **C38** 419

Yoshinaga N , Arima A and Otsuka T 1984 *Phys. Lett.* **143B** 5

Yu Z R , Scholten O and Sun H Z 1986 *J. Math. Phys.* **27** 442

NON-ADIABATICITY, CHAOS AND TOPOLOGICAL EFFECTS ON NUCLEAR COLLECTIVE DYNAMICS

D. Kusnezov

Center for Theoretical Physics, Sloane Physics Laboratory
Yale University, New Haven, CT 06511-8167, USA

INTRODUCTION

I review the role played by level crossings for collective motion in the presence of intrinsic degrees of freedom. Level crossings induce monopole gauge potentials, which act as sources of chaos for the collective dynamics. As a result, collective dynamics can develop distinct components, including trapped, quasi-trapped and quasi-free motion. From the point of view of large amplitude shape diffusion, level crossings result in memory effects, which on certain time scales, results in fractional Brownian motion, a behavior not contained in the standard Fokker-Plank approach to shape diffusion. Rather, the appropriate class of diffusion processes for collective dynamics seems to be Lévy diffusion. I outline a series of open problems related to the microscopic understanding of diffusion and dissipation.

It is a pleasure to present this paper on the occasion of Francesco Iachello's fiftieth birthday. He has been a source of enthusiasm and constant support on numerous problems over the years. While this article is not on spectrum generating algebras per se, nor on the group theoretical aspects of collective motion, it does touch on some more unusual aspects of representation theory. Specifically, the quantum dynamics of the intrinsic levels when described in terms of a density matrix, can be recast as motion on a classical Lie algebra manifold, for which the Casimir operators and representations are quantum mechanically real valued. In the spirit of these proceedings, I will review some of the group theoretical aspects of this problem. The problem I would like to address is that of large amplitude nuclear shape fluctuations, such as those encountered in fission, heavy ion collisions, giant resonances and so forth. The characteristic feature in these phenomena is that the nuclear collective degrees of freedom undergo large amplitude motion in the presence of a high density of intrinsic states. I would like to argue that collective motion in the presence of intrinsic degrees of freedom can be rendered chaotic by the presence of level crossings, and that this can modify completely the expected results one would obtain from driven of adiabatic/diabatic treatments, as well as the diffusive nature of the motion.

The standard treatment of chaos in collective motion is usually related to the spectral statistics of the collective Hamiltonian[1]. However, there is a different source of chaos that can arise in collective motion. The source of the chaos can be attributed to level crossings (although this must be carefully qualified) and its source can be interpreted in terms of a Lagrangian with topological terms. The physical picture I have in mind is the interplay between collective/slow and intrinsic/fast degrees one might encounter in collective motion of strongly interacting complex systems. Such situations are encountered in atomic clusters, deformable cavities, nuclear collective motion and in general mixed classical/quantum systems. In the case of nuclear collective motion, one can consider what happens when one deforms the surface of a nucleus. In this situation, the Fermi surface will correspondingly deform, forcing a rearrangement of single particle levels. For the ground state, particles must then be rearranged between levels to generate a spherical Fermi surface, generating level crossings. A plot of the total energy of the nucleus for various nucleon configurations (in the n levels considered) versus the collective variable will then reveal a series of level crossings, which can be real or avoided[2,3]. The questions we are interested in addressing are related to how the level crossings (either real or avoided) interact with the collective variables, and the modifications to the conventional wisdom. Because this interaction seems to be generally chaotic, this has important consequences on the treatment of large amplitude collective motion and associated phenomena. We also suspect that chaos plays an important role in the microscopic description of diffusion as well as dissipation of the collective variables[4].

COLLECTIVE AND INTRINSIC DYNAMICS

For simplicity we will take the collective degrees of freedom (e.g. quadrupole degrees of freedom β, γ), denoted Q, P, to be classical. The interaction of these coordinates with n intrinsic states can be written as an $n \times n$ matrix

$$H_{ij} = H_{coll}(Q, P)\delta_{ij} + V_{ij}(Q, P), \qquad H_{coll} = \frac{1}{N}\mathrm{Tr}H \qquad (1)$$

The starting point of this article is this form of the collective/intrinsic Hamiltonian. I do not want to justify the origin of these dynamics here, since the identification of collective coordinates is not a simple matter, and the set óne chooses is certainly not unique or complete. However, in the simplest scheme, one can generally reduce the interaction of collective and intrinsic states to this type of coupled channel problem. H_{coll} is the piece of the collective Hamiltonian which is diagonal in the intrinsic basis, and V_{ij} is the interaction of intrinsic and collective degrees of freedom. With no loss of generality, we consider the interaction V_{ij} to be independent of the momenta P. The Hamiltonian, and hence V_{ij}, must be Hermitian in order to have real eigenvalues. Since the $SU(n)$ Gell-Mann matrices λ_i form a basis for the most general traceless Hermitian $n \times n$ matrix, it is convenient to expand the Hamiltonian in this basis. We can always choose $\mathrm{Tr}V = 0$ by absorbing the trace of V into H_{coll} and redefining V. Then, the interaction V_{ij} can be written as

$$V_{ij}(Q) = a(Q)_k\,(\lambda_k)_{ij}\,, \qquad a(Q)_k = \frac{1}{2}\mathrm{Tr}(V\lambda_k), \qquad (2)$$

where the $a(Q)_k$ are the matrix elements of V with respect to the basis $\{\lambda_k\}$, and $(\lambda_k)_{ij}$ is the ij component of the matrix λ_k. The general form of the Hamiltonian describing the interaction of intrinsic and collective degrees of freedom is then

$$H = H_{coll}(Q, P) \cdot 1_n + a(Q) \cdot \lambda\,, \qquad (3)$$

with 1_n the identity matrix.

The dynamics associated with the intrinsic space can be investigated in terms of the density matrix. The density matrix offers a convenient method to understand the dynamics, and allows for both pure and mixed descriptions. For the $n-$level system, the density matrix can be parameterized as

$$\rho = \frac{1}{n}(1_n + \lambda \cdot r) = \frac{1}{n}(1_n + \tilde{\rho}), \qquad \tilde{\rho} = \lambda \cdot r. \tag{4}$$

Here r is an $n^2 - 1$ dimensional vector, which contains the information of the density matrix, and $\tilde{\rho}$ will be called the *reduced density matrix*. Trace relations for the Gell-Man matrices allow the interchangeable use of $\tilde{\rho}$ and r:

$$r_i = \frac{n}{2}\text{Tr}(\rho\lambda_i) = \frac{1}{2}\text{Tr}(\tilde{\rho}\lambda_i) \tag{5}$$

The quantum equations of motion for the intrinsic degrees of freedom can then be studied in terms of the matrix equation for ρ, or in terms of a vector equation for r:

$$i\dot{\rho} = [H, \rho], \qquad \longleftrightarrow \qquad \dot{r}_i = (2f_{ijk})a_j(Q)r_k = (a(Q) \times r)_i \tag{6}$$

where f_{ijk} is the completely anti-symmetric tensor of $SU(n)$. Since H is always a linear expansion in the λ matrices, there will be no ordering ambiguities, and we can represent the quantum evolution of the system exactly as a classical Hamiltonian flow, with equations of motion given by the Lie-Poisson brackets

$$\{F(r), G(r)\} = J_{ij}(r)\frac{\partial F}{\partial r_i}\frac{\partial G}{\partial r_j}, \qquad J_{ij}(r) = 2f_{ijk}r_k, \tag{7}$$

(Notice that these provide the classical analog of the commutators, which is evident if we take $F = r_i$ and $G = r_j$, since then $\{r_i, r_j\} = 2f_{ijk}r_k$, which can be compared to $[\lambda_i, \lambda_j] = 2if_{ijk}\lambda_k$.) The (fully quantum) equations of motion can be written in terms of these Poisson brackets

$$\dot{r}_i = \{H, r_i\} = (\frac{\partial V}{\partial r} \times r)_i, \qquad V(Q, r) \equiv a(Q) \cdot r \tag{8}$$

A pure intrinsic state can be distinguished from a mixed state by the necessary (but not sufficient) condition on the magnitude of the vector r which appears in ρ:

$$r = \frac{1}{2}n(n-1) \quad [\text{pure}, \rho^2 = \rho], \qquad r < \frac{1}{2}n(n-1) \quad [\text{mixed}, \rho^2 \neq \rho]. \tag{9}$$

One of the peculiar properties of this density matrix realization is the related to the Casimir operators. It is well known that $n-1$ Casimir operators (polynomials of order $m = 2, 3, ..., n$ in λ), denoted $C_m(\lambda)$, can be constructed and commute with all the matrices λ_k. The existence of these operators is due to the nature of the structure constants, and must also be present in the Lie-Poisson realization. If we define

$$C_m(r) = \text{Tr}[(\lambda \cdot r)^m] = \text{Tr}[(\tilde{\rho})^m], \qquad m = 2, ..., n \tag{10}$$

then it is easy to check by virtue of the structure constants that

$$[C_m(\lambda), \lambda_i] = 0, \qquad \leftrightarrow \qquad \{C_m(r), r_i\} = 0. \tag{11}$$

The importance of these relations is that the $C_m(r)$ are *topological integrals of motion*; they are in involution with any Hamiltonian. These terms are present since the phase space of r is compact. Their presence will induce topological gauge potentials in the collective/intrinsic Lagrangian.

2-Level Intrinsic Dynamics. As an example, consider the purely intrinsic 2-level Hamiltonian $H = \mathbf{B} \cdot \sigma$. We can write this as $H(r) = \mathrm{Tr}H = \mathbf{B} \cdot \mathbf{r}$, where $\mathbf{r} = (x, y, z)$. The 2 state density matrix is $\rho = (1 + \sigma \cdot \mathbf{r})/2$. The quantum equations of motion and the Lagrangian from which they are derived are

$$\dot{\mathbf{r}} = \mathbf{B} \times \mathbf{r}, \qquad \mathcal{L} = \frac{1}{2} \frac{z(x\dot{y} - y\dot{x})}{r^2 - z^2} - \mathbf{B} \cdot \mathbf{r} = \frac{r}{2}\mathcal{A}_D \cdot \dot{\mathbf{r}} - \mathbf{B} \cdot \mathbf{r} \qquad (12)$$

(Note: the cross product is defined with a factor of 2 in Eq. 6). The first term in $\mathcal{L}$, $\mathcal{A}_D$, is the gauge potential of a Dirac monopole located at $r = 0$. It is the equivalent of the '$p\dot{q}$' term, but appears as a monopole since in must enforce the $r = const$ constraint in the dynamics. The origin of the monopole, $r = 0$, is the 'level crossing' of the reduced density matrix $\tilde{\rho} = \sigma \cdot \mathbf{r}$, which has eigenvalues $E_{\pm} = \pm r$. The topological integral of motion is r^2 ($\{r^2, r_i\} = 0$), can be identified as the equivalent of the total angular momentum J^2 ($[J^2, J_i] = 0$). For conventional quantum systems, $J = n/2$, ($n = 0, 1, 2, ...$), while for this quantum dynamics, a pure systems has integer 'quantum number' $r = 1$ while a mixed system has a real valued representation $0 < r < 1$. So in general, the allowed $SU(2)$ Cartan representations $[2J]$ are real valued with $[0 \leq 2J \leq 1]$. This gauge potential enforces the fact that r is an integral of motion. As we can see, the dynamics always remains a fixed distance from the monopole (r=const). This is important in drawing the distinction between Berry's gauge potential and $\mathcal{A}$. They are distinct, but the gauge potential $\mathcal{A}$ can be mapped into Berry's gauge potential when we perform a Born-Oppenheimer approximation. ∎

If the collective degrees of freedom are fixed, so that a_i is a constant vector, the intrinsic dynamics (i.e. r−dynamics) is linear and hence non-chaotic. This is expected since quantum mechanics is a linear theory. Thus there are on the order of n^2 integrals of motion for this system, roughly corresponding to the dimension of the intrinsic space (the precise number is $n - 1 + n(n - 1)/2 = (n + 2)(n - 1)/2$). Some of these integrals of motion will depend on the form of the Hamiltonian ($n(n - 1)/2$ of them), while others define the dynamical manifold and are independent of the choice of the Hamiltonian ($n - 1$ of them). It is this latter class of topological invariants which are of interest, and lead to many intriguing aspects of collective/intrinsic motion, from fractionalization of quantum numbers, to the origin of Berry's phase in the Born-Oppenheimer approximation, to anomalous diffusion and so on[5].

So we can understand the density matrix phase space in the following manner. r has $n^2 - 1$ components, but evolves on a manifold defined by $n - 1$ constraints (the Casimirs). This results in a dynamical manifold of dimension $n^2 - 1 - (n - 1) = n(n - 1)$. In terms of phase space, this corresponds to $n(n - 1)/2$ pairs of coordinates and momenta. Since the manifold is compact, we cannot define $q_i(r)$ and $p_i(r)$ globally, but we can define them locally at any point in phase space.

If we now couple n−intrinsic states with its density matrix to the collective motion, we have the equations of motion for the Hamiltonian (1):

$$\dot{Q}_i = \frac{\partial H_{coll}}{\partial P_i}, \qquad \dot{P}_i = -\frac{\partial H_{coll}}{\partial Q_i} - \frac{2}{n}\frac{\partial a_k}{\partial Q_i}r_k, \qquad \dot{r} = a(Q) \times r. \qquad (13)$$

The interesting properties of these equations are the feedback from the intrinsic motion to the collective dynamics. The $a \times r$ term is due to the monopole at $r = 0$, which causes this effective magnetic interaction. Another important consideration is whether the collective Hamiltonian is chaotic or regular. As we mention below, diffusion rates and correlations will depend strongly on this. For this general dynamics, it is also possible to demonstrate that this Hamiltonian is derivable from a Lagrangian, which has $n-1$ topological gauge potentials. Details of this construction can be found in Ref. 5.

2-Level Collective/Intrinsic Dynamics. The most general Hamiltonian of this type can be written as $H = H_c(Q, P) + a(Q) \cdot \sigma$, which has the dynamics

$$\dot{Q}_i = \frac{\partial H_{coll}}{\partial P_i}, \quad \dot{P}_i = -\frac{\partial H_{coll}}{\partial Q_i} - \frac{\partial a_k}{\partial Q_i} r_k, \quad \dot{r} = a(Q) \times r. \tag{14}$$

and is derivable from

$$\mathcal{L} = P \cdot \dot{Q} + \frac{z(x\dot{y} - y\dot{x})}{2(r^2 - z^2)} - (H_c(Q, P) + a(Q) \cdot r) \tag{15}$$

This Hamiltonian has level crossings (diabolical points) at $|a(Q)| = 0$, and eigenvalues $E_\pm = H_c \pm |a|$. The monopole gauge potential is not located at the level crossings of H, but at those of $\tilde{\rho}$, namely $r = 0$. The should be compared to the Born-Oppenheimer approximation, which generates monopole gauge potentials at the diabolical points of H. In contrast to Berry's phase which cannot exist in one collective dimension, we can have a topological term in the Lagrangian influencing the dynamics since it is a monopole located at $r = 0$, and not in (Q, P) space. Integrating out the intrinsic degrees of freedom will cause this monopole to vanish.∎

DENSITY MATRICES AND GROUP CHAINS

In the spirit of these proceedings, one can develop a notion of group chains and dynamic symmetries for the density matrix, although it is more of a curiosity than a useful identification. By taking advantage of (i) H being linear in the $SU(n)$ generators and (ii) group structure of the interaction, we can determine what type of observables might become fractionalized by the level crossings and their gauge potentials. Since the vector r can have arbitrary magnitude between $0 < r \leq n(n-1)/2$, the representation labels of $SU(n)$, which define the values of the Casimirs $C(m)$ are thus also real valued for this quantum dynamics. Hence integrals of motion for the combined system which have contributions from both collective and intrinsic spaces will in general become real valued due to level crossings.

Since the Hamiltonian is a linear expansion in the λ_i, it cannot have an expansion in terms of Casimir operators, which are at least quadratic. Dynamical symmetries are useful since they provide quantum numbers and energies of the system. In the context of density matrices, however, we are interested in the integrals of motion. For this reason, only the canonical group chains can provide any meaningful definition for dynamic symmetries, in which case the Hamiltonian is expressed in terms of generators of the Cartan subalgebra of $SU(n)$. This leads, however, to trivial equations of motion. Non-canonical decompositions $SU(n) \supset \mathcal{G} \supset \mathcal{G}' \supset \cdots$, will contain elements of the subalgebras, but these will not be integrals of motion.

4–Level System: $SU(4) \supset SU(3) \supset SU(2)$ If the interaction $a(Q)_i$ contains non-zero components for $i = 3, 8, 15$, these corresponds to the generators of the Cartan subalgebra of $SU(4)$. Since we can identify λ_3 with $SU(2)$, $\lambda_{3,8}$ with $SU(3)$ and $\lambda_{3,8,15}$ with $SU(4)$, he expansion $a \cdot \lambda = a_3 \lambda_3 + a_8 \lambda_8 + a_{15} \lambda_{15}$ can be defined as the $SU(4) \supset SU(3) \supset$

$SU(2)$ dynamical symmetry limit of the density matrix. This is, however, a special situation, and is only valid for the canonical embeddings. Non-canonical embeddings usually have Cartan generators constructed out of the raising and lowering operators. For $SU(3) \supset O(3)$, the Cartan generators for $SU(3)$ are $\lambda_{3,8}$, while that for $O(3)$ is λ_2. In this case, λ_2 is not invariant, so that dynamic symmetry is not applicable.∎

So while for certain situations one might have a 'dynamical symmetry', in which case the dynamics is particularly trivial, the more general case of a subalgebra is the situation where $a(Q)\cdot\lambda$ is written only in terms of generators of a particular subalgebra of $SU(n)$, and not a group chain. Since the Hamiltonian is linear, the identification of further subalgebras does not provide any more information, so that the reduction is limited to this largest subalgebra. In general (see Ref. 5 for additional details):

- Different (dynamical) symmetries correspond to particular classes of topological Lagrangians;

- these Lagrangians that generates the collective/intrinsic equations of motion have induced monopole gauge potentials expressed in terms of the Casimir invariants of the group chain as well as the generators of the subalgebras;

- these monopoles are located at the level crossings of the reduced density matrix $\tilde{\rho}$, and not at the diabolical points of the Hamiltonian (1). The dynamics never approaches or moves away from the monopoles – the distance of the trajectory from the various monopoles can be expressed in terms of topological invariants;

- the Casimir operators (and the representation labels) are real valued, in spite of their quantum nature;

- a consequence of the real valuedness of the Casimir operators is that quantum numbers corresponding to observables which are expressible in terms of both collective and intrinsic coordinates can become correspondingly real valued (i.e. fractionalization of quantum numbers due to level crossings.)

4–Level System: $U(4) \supset O(4)$ Consider the following model which exemplifies some of these points. Assume that there are only four relevant intrinsic states $(n = 4)$ which interacting with 3 collective coordinates $Q = (Q_x, Q_y, Q_z)$ via the Hamiltonian:

$$H = \frac{P^2}{2} + V(Q) + a(Q)\cdot\lambda. \tag{16}$$

The λ are generators of $SU(4)$. Now lets examine what happens if this interaction does not span the entire $SU(4)$ space, but those of a subalgebra of $SU(4)$. Lets consider the $SU(4) \supset O(4)$ group chain. Such a situation arises if only $\lambda_{2,5,7,9,11,13}$ are involved, since $\mathbf{L} \to (\lambda_2, -\lambda_5, \lambda_7)$ and $\mathbf{A} \to (\lambda_9, \lambda_{11}, \lambda_{13})$ are generators of $O(4)$, corresponding the 'angular momentum' and the 'Runge-Lenz vector'. For instance is we take the interaction matrix as

$$a(Q)\cdot\lambda = \kappa\mathbf{Q}\cdot\mathbf{L} + \kappa'\mathbf{Q}\cdot\mathbf{A} = \begin{pmatrix} 0 & \kappa'Q_x & \kappa'Q_y & \kappa'Q_z \\ \kappa'Q_x & 0 & -i\kappa Q_z & i\kappa Q_y \\ \kappa'Q_y & i\kappa Q_z & 0 & -i\kappa Q_x \\ \kappa'Q_z & -i\kappa Q_y & i\kappa Q_x & 0 \end{pmatrix} \tag{17}$$

There are two topological integrals of motion, whish can be identified with the Casimir invariants of $O(4)$,

$$C_2 = \mathbf{J}_1^2 + \mathbf{J}_2^2 = \frac{1}{2}(\mathbf{L}^2 + \mathbf{A}^2), \quad C_2' = \mathbf{J}_1^2 - \mathbf{J}_2^2 = \mathbf{L}\cdot\mathbf{A} \tag{18}$$

where $\{C_2, H\} = \{C_2', H\} = 0$, and four dynamical integrals of motion:

$$J = Q \times P + L, \qquad E = \frac{P^2}{2} + V(Q) + \kappa Q \cdot L + \kappa' Q \cdot A. \qquad (19)$$

where $L = (r_7, -r_5, r_2)$, $A = (r_9, r_{11}, r_{13})$ and we have used $J_1 = (L + A)/2$, $J_2 = (L - A)/2$. The conserved angular momenta picks up a contribution from the density matrix monopole. Since the magnitudes of L and A are in general real, the J_i are real valued as well. So if we were to requantize the model, L would become integral, but in general J_i would be real.

The three $SU(4)$ Casimirs can be written in terms of C_2 and C_2' as $C_{2,SU(4)} = 4C_2$, $C_{3,SU(4)} = 0$, $C_{4,SU(4)} = C_2^2 - 4C_2'^2$, from which one could compute the representation $[N_1, N_2, N_3]$. The $O(4)$ representation here is determined from the magnitudes of the vectors J_1, denoted j_i, which gives $(p, q) = (j_1 + j_2, j_1 - j_2)$. The representations are real valued in general $0 \le j_i \le 1$. The Lagrangian which generates this Hamiltonian is

$$\mathcal{L} = P \cdot \dot{Q} + \frac{j_1}{2} A_D(J_1) \cdot \dot{J}_1 + \frac{j_2}{2} A_D(J_2) \cdot \dot{J}_2 - H, \qquad A_D(J) \equiv \frac{(-J_y, J_x, 0)}{J(J^2 - J_z^2)} \qquad (20)$$

where the Dirac monopoles are located at $j_i = 0$. This is the Lagrangian for any $SU(4) \supset O(4)$ Hamiltonian H. For the full $SU(4)$ case, there are three gauge potentials whose general expression is quite complicated.[5,6] ∎

Non-Abelian Gauge Potentials: The $(S \cdot B)^2$ Interaction. A type of Hamiltonian encountered often in the discussion of non-Abelian is the NMR interaction $(S \cdot B)^2$. One can derive in a similar fashion non-Abelian gauge potentials in the intrinsic space. These are generally present when the density matrix representations are fully symmetric, in which case the dynamical manifold has peculiar properties due to the degeneracies. Details can be found in Ref. 5.∎

MULTIPLE LEVEL CROSSINGS: TOPOLOGICAL PHASES AND CHAOS

What is the relation of the density matrix monopoles to Berry's phase? This can be understood if we examine the situation of a multiply repeated avoided level crossing. Consider the simple illustration for two levels, with three collective coordinates $Q = (Q_x, Q_y, Q_z)$, $P = (P_x, P_y, P_z)$ with the Hamiltonian

$$H = \left[\frac{P^2}{2} + V(Q)\right] + A(Q) \cdot \sigma, \qquad A(Q) = (\sin Q_x, \sin Q_y, \sin Q_z) \qquad (21)$$

which has the energy eigenvalues

$$E_\pm(Q, P) = \left[\frac{P^2}{2} + V(Q)\right] \pm \sqrt{\sin^2 Q_x + \sin^2 Q_y + \sin^2 Q_z}. \qquad (22)$$

The term in the square brackets if H_{coll}, and the potential $V(Q)$ is arbitrary. The equations of motion are

$$\dot{Q} = P, \qquad \dot{P} = -\frac{\partial V}{\partial Q} - (x \cos Q_x + y \cos Q_y + z \cos Q_z), \qquad \dot{r} = a \times r. \qquad (23)$$

which can also be derived from the Lagrangian

$$\mathcal{L} = P\dot{Q} + \frac{z}{2r(r^2 - z^2)}(x\dot{y} - y\dot{x}) - \left(\frac{P^2}{2} + V(Q) + A(Q) \cdot r\right). \qquad (24)$$

The level crossings of H are at $Q_x, Q_y, Q_z = 0, \pm\pi, \pm 2\pi, ...$, while the single monopole gauge potential has singularity at $r = 0$. The Born-Oppenheimer approximation corresponds to setting $\dot\rho \sim 0$ which implies $\dot r = 0$. This defines a mapping, denoted f_{BO} which integrates out the intrinsic space, mapping the gauge potentials to the diabolical points of the Hamiltonian. For the two level system here, the mapping is

$$f_{BO} : r \to A(Q), \qquad r_i = \frac{A_i(Q)}{|A(Q)|}|r|. \qquad (25)$$

If we perform the BO approximation according to f_{BO}, we obtain the Lagrangian

$$\tilde{L} = P\dot{Q} + \frac{\sin Q_z (\sin Q_x \cos Q_y dQ_y - \sin Q_y \cos Q_x dQ_x)}{\sqrt{\sin^2 Q_x + \sin^2 Q_y + \sin^2 Q_z (\sin Q_x^2 + \sin Q_y^2)}} - \left(\frac{P^2}{2} + V \pm A\right). \qquad (26)$$

Near the diabolical points of Eq. 21, the gauge potential is precisely that for a Dirac monopole, but globally it is periodic, since the $r = 0$ monopole has been mapped to an infinite number of diabolical points. This is the origin of Berry's phase. It comes from topological terms which are always present in the full Lagrangian, and only becomes Berry's phase upon integrating out the intrinsic degrees of freedom. It should be pointed out that while the gauge potential for the original system $A(r)$ is related to the Dirac monopole $A_D(r)$ by $A(r) = rA_D(r)$, the gauge potential obtained after applying f_{BO} to $A(r)$ is precisely the Dirac monopole: $f_{BO} : A(r) \to A_D(A(Q))$. The equations of motion for $\tilde{L}$ are

$$\dot{Q} = P \qquad (27)$$

$$\dot{P} = -\left[\frac{\partial V}{\partial Q} + \frac{A(Q)}{|A(Q)|}\right] + P \times \frac{A(Q)}{|A(Q)|^3}. \qquad (28)$$

The term $B = A(Q)/|A(Q)|^3$ is the magnetic field strength of the Dirac monopole located at the points $A(Q) = 0$, which are the degeneracies of H. The term which appears in the equation for $\dot{P}$ is the magnetic force $v \times B$ from the diabolical point. Near a diabolical point, we have $\sin Q_i \sim Q_i$ and $\cos Q_i \sim 1$, which then simplifies B to the Dirac monopole $B_i = Q_i/Q^3$. This is the magnetic force which arises from Berry's non-integrable phase.

ANOMALOUS DIFFUSION: LEVY FLIGHTS AND FRACTIONAL BROWNIAN MOTION

The starting point for the analysis of stationary stochastic processes is generally the chain equation for the probability to go from one value of the collective variable q to q' in a time t, denoted $P(q'|q,t)$, which can be written as

$$P(q' - q, t) = \int P(q' - q'', t_1) P(q'' - q, t - t_1) dq'' \qquad (29)$$

or in terms of its Fourier transform

$$p(k, t) = p(k, t - t_1) p(k, t_1), \qquad p(k, t) = \int dq P(q, t) \exp(ikq) dq. \qquad (30)$$

The standard solution which one investigates is based on the Gaussian distribution, which satisfies the chain condition:

$$p(k, t) = e^{-Dk^2 t}, \qquad P(q, t) = \frac{1}{\sqrt{4\pi Dt}} e^{-q^2/4Dt}. \qquad (31)$$

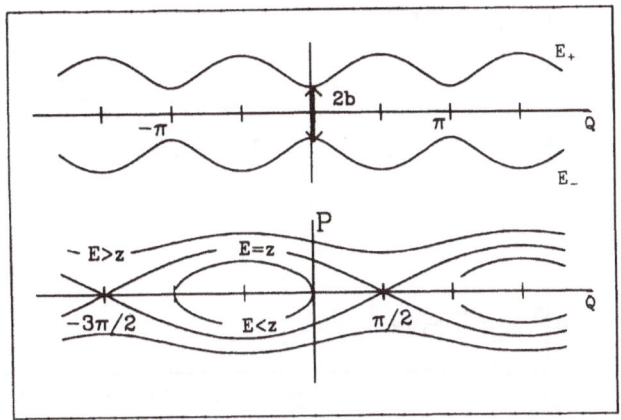

Figure 1. Top: Structure of the infinitely repeated avoided level crossings of the Hamiltonian (32). Bottom: Separatrix for $z = \rho_{11} - \rho_{22} > 0$ for collective coordinates. The difference in periodicity with the level crossings shows that only alternate level crossings are attractive/repulsive. This staggering changes as the intrinsic population shifts, and the separatrix disappears and reappears shifted by π. Consequently, collective trajectories that were trapped, become free.

Lévy showed that there is actually an infinite number of solutions of the chain equation of the form[7,8]:

$$p(k,t) = e^{-D|k|^\alpha t} \tag{32}$$

which leads to non-negative normalizable distributions $P(q,t)$ for $0 < \alpha \le 2$. This more general class of Lévy distributions has been largely overlooked in physics since the distributions have infinite moments. (Recall, it is the finiteness of the moments in the case of the Gaussian which leads to the Fokker-Plank equation and the Ornstein-Uhlebeck processes). However, if one assigns finite velocities to the random steps, one can control the moments and obtain a general class of diffusive processes with $< Q^2 > = Dt^\kappa$, referred to as fractional Brownian motion. (The fact that $\kappa \ne 1$ can be understood in terms of memory effects.) I would like to argue that Lévy diffusion is a form of diffusion which must be present in the description of large amplitude collective motion in the presence of level crossings.

The effects of level crossings on diffusive motion can be illustrated in the simple example. Consider one collective coordinate Q and an infinite number of avoided level crossings, given by

$$H = \frac{P^2}{2} + \kappa(a_x \sigma_x + a_y \sigma_y + \sigma_z \sin Q) \tag{33}$$

with $a_x = a_y$ constant. The energies are

$$E_\pm = \pm \kappa \sqrt{b^2 + \sin^2 Q} \tag{34}$$

which has avoided crossings at $Q = n\pi, n = 0, \pm 1, \pm 2, ...$, as shown in the top of Fig. 1. This model was considered in Ref. 9, and was shown to be integrable in the limits $b = \sqrt{a_x^2 + a_y^2} = 0$ and $\kappa = 0$, but otherwise chaotic. In this picture, the collective coordinate undergoes motion according to

$$\ddot{Q} + \kappa(\rho_{11} - \rho_{22}) \cos Q = 0 \tag{35}$$

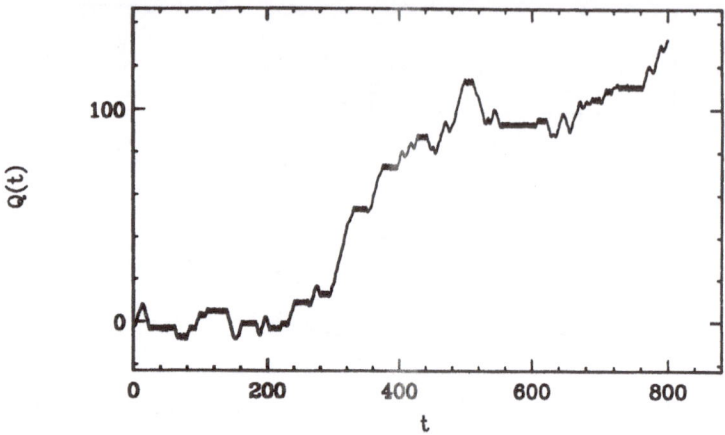

Figure 2. A quasi-trapped trajectory, indicating Lévy type diffusion. Diffusion through many level crossings results in energy loss and eventual trapping by the seperatrix (Fig. 1). Until favorable conditions exist for an intrinsic population shift, the dynamics remains localized in the well. Energy exchange allows it to eventually escapes with sufficient energy to traverse many level crossings before becoming trapped once again.

which is the pendulum equation with a dynamic frequency proportional to the intrinsic population difference, with $-r \leq \rho_{11}-\rho_{22} \leq r$. The pendulum separatrix is shown in the bottom of Fig. 1 for $\rho_{11} > \rho_{22}$. The difference in periodicity, π for the level crossings and 2π for the separatrix, indicates that alternate level crossings are repulsive or attractive. For low collective energies, some collective trajectories can be forever trapped in the pendulum potential; for higher energies, a population shift is energetically possible, and the separatrix which once trapped the trajectory $Q(t)$ vanishes, reappearing shifted by π, freeing the collective coordinate for a while until it becomes trapped again. I refer to these as quasi-trapped trajectories. For the highest energies, a spin flip is not sufficient to trap the collective coordinate, and it is free to escape. The quasi-trapped motion is characteristic of Lévy type diffusion, although this is a completely dynamical analog. A typical trajectory $Q(t)$ of this type is shown in Fig. 2. (This behavior is also characteristic of persistent fractional Brownian motion.[10]) There are clearly two distinct types of competing diffusive processes: diffusion within the well and diffusion through level crossings. If one considers a thermal ensemble of collective momenta (corresponding to a crude 'hot nucleus'), one can evolve the nucleus through the level crossings and examine the nature of the diffusion. The diffusion induced by the level crossings is shown in Fig. 3, and behaves for long times roughly as $< Q^2 > \sim t^{7/4}$. For short times, the motion is within a well, and the motion is free streaming $< Q^2 > \sim t^2$. On time scales of $o(1)$, the memory effects of level crossings become important here and persist for long times.

From the point of view of stochastic processes, one might also consider a fractional Fokker-Plank equation[11] which gives rise to Lévy type diffusion. In modeling the fission process, one can try to find the optimal diffusion consistent with both photon and neutron emmision data. It is also curious to see what the modifications of Kramer's tunneling amplitudes for the fission process would be in the presence of fractional diffusion.

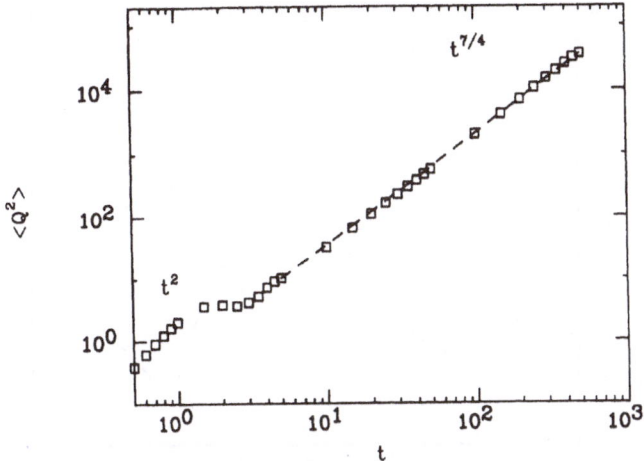

Figure 3. Fractional collective diffusion exhibited by a thermal ensemble of initial conditions. For very short time scale, the motion is Hamiltonian and the diffusion is free streaming. On longer time scale, the combination of trapped, quasi-trapped and quasi-free trajectories in the ensemble result in $t^{7/4}$ behavior, characteristic of persistent diffusion.

The essential mathematical difference is that one must replace ordinary time derivatives with fractional time derivatives in Fokker-Plank equation. Heuristically, persistent diffusion will result in diffusion constants smaller than those found by implementing the Fokker-Plank equation. This is qualitatively in the right direction, since microscopic TDHF calculation yield diffusion constants which are much smaller than those obtained by ordinary diffusion equations. Whether colored noise, multi-dimensional diffusion or level crossings are responsible is a very interesting question.

SEMICLASSICAL AND QUANTUM PROBLEMS

A few selected problems one can address concerning the semiclassical and fully quantum descriptions are discussed here.

Matrix Models One interesting approach to the fully quantum problem is to consider the mixing of collective and intrinsic states from the random matrix point of view. We can construct matrix models in the direct product space $\mid Collective\rangle \otimes \mid Intrinsic\rangle$, using a Hamiltonian of the type $H_{ia,jb}$, with ij and ab as indices in the corresponding spaces. This has the general form

$$H_{ia,bj} = H_{ij}^{coll} \otimes 1_{ab} + 1_{ij} \otimes H_{ab}^{i} + V_{ia,jb} \qquad (36)$$

One can now define random matrix correlations for components ij and ab, as well as mixed correlations between them. These will correspond the the classically chaotic nature of the collective Hamiltonian (ij), the intrinsic states (ab) and the interaction ib, aj. If one takes these as Poisson, GOE or GUE, one can examine the transition probability for traversing a certain distance in collective coordinate. The transition/diffusion will depend strongly on the statistical nature of the interactions. One might hope to extract dissipation rates for the different combinations of statistics. To see that there

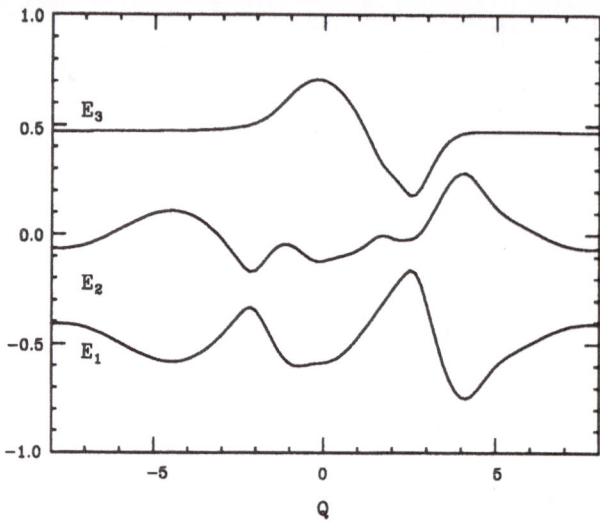

Figure 4. A collection of avoided crossings for a three level system.

is a physical difference, consider the following illustration. Take the three levels shown in Fig. 4, and consider starting the system on the left at $Q = -5$ with initial collective velocity $P \sim 1$. The survival probabilities for (a) a driven collective coordinate, $\dot{Q} = t$, (b) a regular collective Hamiltonian and (c) a chaotic collective Hamiltonian are shown in Fig. 5, using similar initial conditions. The transition and duration clearly depends on the chaotic nature of the system. If we now fully quantize the system, we can go to the random matrix models to examine exactly solvable situations as described above. It should be understood, that while case (a) always progresses from left to right, small variations in the initial conditions will cause (b) and (c) to scatter either to the left or the right. In general, the scattering of collective coordinates from the level crossings is fractal[5]. In Fig. 6, the collective phase space trajectories corresponding to the three situations in Fig. 5 are displayed. The chaotic Hamiltonian can be seen to much more strongly mix the intrinsic states that a simpler adiabatic/diabatic picture would.

Path Integral Construction The path integral derived by Pechukas[12] for the scattering of molecules provides an interesting starting point for the interaction of collective and intrinsic degrees of freedom. It was originally derived to take into account molecular excitations (intrinsic states) which occur when two molecules scatter (collective coordinates correspond to intermolecular separation). The resulting equations of motion are surprisingly similar to those discussed here, and provide an method to compute the semi-classical corrections to the purely classical collective coordinates used in the Hamiltonian (1).

Calogero-Moser Dynamics The generalized Calogero-Moser equations are molecular dynamics type equations for which the coordinates are the intrinsic energy levels. Energy level dynamics of Hamiltonians of the form $H = H_0 + tV$ can be studied easily this way.[13] If one now includes the collective interactions, one must define a suitable (vector) realization of these dynamics[5]. The resulting systems can be examined and compared to driven systems, comparing dissipation rates and diffusive behavior. One can then understand how badly one violates the adiabatic hypothesis, and how strongly one must

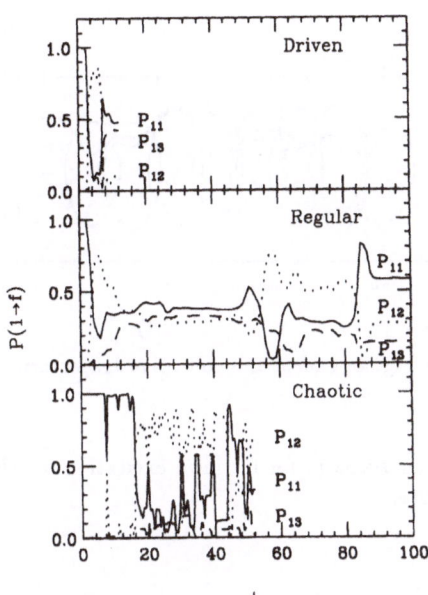

Figure 5. Transition probabilities for a collective coordinate evolving through the levels of Fig. 4, with initial position $Q_o = -5$, $P_o = 1.05$, and in the lowest eigenstate Ψ_1. We plot the transition probabilities at $Q = 5$, defined as ($\Psi_o = \Psi_1(Q = -5, t = 0)$) $P_{11} = |\langle \Psi_1(t)|\Psi_o\rangle|^2$ (solid), $P_{12} = |\langle \Psi_2(t)|\Psi_o\rangle|^2$ (dots), and $P_{13} = |\langle \Psi_3(t)|\Psi_o\rangle|^2$ (dashes) for the cases of (a) *Top:* a driven system, $Q(t) = t$, (b) *Middle:* a regular collective Hamiltonian $H_c(Q, P) = P^2/2$, and (c) *Bottom:* a chaotic collective Hamiltonian.

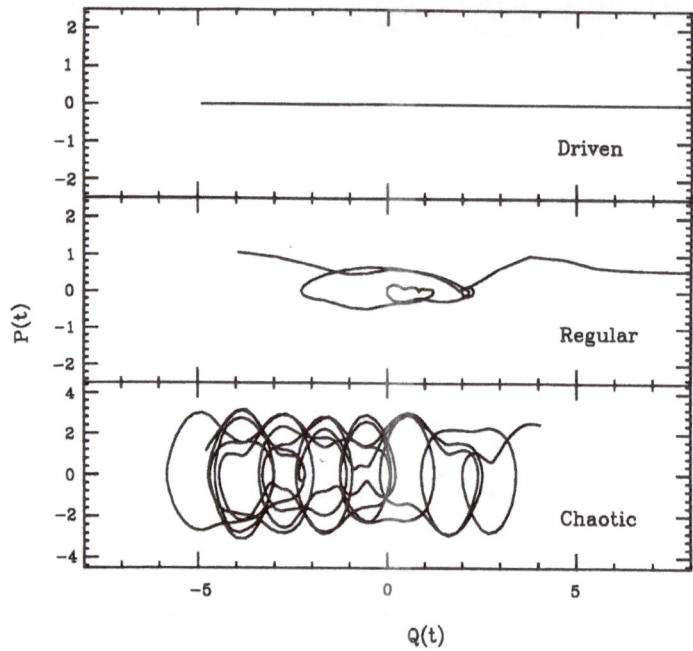

Figure 6. Same as Fig. 5, except the collective trajectories are plotted for the three situation.

modify the Landau-Zenner scenario for intrinsic excitation which leads to dissipation in the collective coordinate.

CONCLUSIONS

I have addressed a collection of problems, rewriting them in a topologically convenient manner and investigated some of the manifestations. What seems clear is that the importance of level crossings, however presented, is still an area of untapped physical wealth. There seem to be many effects, from the microscopic (topology, fractional quantum numbers, chaotic collective behavior,...), to the macroscopic (matrix models, exotic diffusion, transmission rates,...). While experimental observations are presently limited to estimates of diffusion constants, there is still much to be understood about the nature of collective diffusion and the origin of dissipation. These types of results should be developed with not only nuclei in mind, but any complex many body system, from cavities to atomic clusters. We are still left with many quite basic questions to address. For instance: To what extent is the Landau-Zenner scenario violated due to the chaotic interaction, and how useful is that concept when the level density is so high? (Since the 'true' collective dynamics is free rather than driven at constant velocity, the Landau-Zenner picture becomes strongly modified.); What is the origin of dissipation, and to what extent do level crossings play a role? Do level crossings result in ordinary diffusion, or do they induce more exotic processes such as Lévy flights and fractional diffusion? If so, one must use a modified realization of the Fokker-Plank equation which includes fractional time derivatives. (Incidentally, the notion of fractional derivatives, $d^{\alpha}f(x)/dx^{\alpha}$ where α is real valued, was introduced by Liouville in 1832[14,11].) The topological aspects are presently more of a curiosity, but allow a better understanding of the measurable quantities such as survival probabilities and transition rates. Finally, relating the level statistics to the chaotic nature of the interaction and dynamics, understanding the richness of the simple pictures of shape diffusion, collective motion, and dissipation will provide a better understanding of the underlying theory of collective/intrinsic interaction.

ACKNOWLEDGEMENTS

This work was supported under DOE grant DE-FG02-91ER40608.

REFERENCES

[1] See for example Y. ALHASSID, N. WHELAN and A. NOVOSELSKY, Mod. Phys. Lett. **A7**, 2453 (1992).

[2] F. BARRANCO, G. BERTSCH, R. A. BROGLIA and E. VIGEZZI, Nucl. Phys. **A512**, 253 (1990).

[3] W. SWIATECKI, Nuc. Phys. **A488c**, 375 (1988).

[4] A. BULGAC and D. KUSNEZOV, Nuc. Phys. **A545**, 549 (1992).

[5] D. KUSNEZOV, '*Influence of Level Crossings on Nuclear Collective Motion*', Yale University preprint (1993) submitted.

[6] A.BULGAC and D.KUSNEZOV, Ann. Phys. **199**, 187 (1990).

[7] P. LÉVY, "Calcul des Probabilities", Guthier-Villars, Paris (1925).

[8] E.W.MONTROLL and B.J.WEST, in: "Studies in Statistical Mechanics", Vol. VII, North-Holland, Amsterdam (1979).

[9] D. KUSNEZOV, '*Anomalous Diffusion and Levy Flights and Collective Motion*', Yale University preprint (1993) submitted.

[10] J. FEDER, "Fractals", Plenum, New York (1988).

[11] G.JUMARIE, J. Math. Phys. **33**, 3536 (1992).

[12] P.PECHUKAS, Phys. Rev. **181**, 174 (1969).

[13] J.BURGDORFER, X.YANG and E.ESCHENAZI, Quantum Chaology and Spectra, in: "Irregular Atomic Systems and Quantum Chaos", J.C.GAY, Ed., Gordon-Breach, Philadelphia (1992).

[14] J.LIOUVILLE, J. de l'Ecole Polytechnique **21**, 1, 71, (1832).

MICROSCOPIC APPROACH TO IBFM EXCHANGE TERM BY USING ENFORCED SYMMETRY CONDITIONS

G. Kyrchev[1] and V. Paar[2]

[1]Institute of Theoretical Physics and Astrophysics
University of Cape Town, Republic of South Africa
[2]Department of Physics, Faculty of Science
University of Zagreb, 41000 Zagreb, Croatia

INTRODUCTION

In the last two decades a significant role in nuclear physics has been played by the Interacting boson model (IBM), introduced by Arima and Iachello[1], and Interacting boson-fermion model (IBFM), introduced by Iachello and Scholten[2]. In the framework of IBFM for odd-mass nuclei it was recognized that the exchange term is important[2]. The exchange interaction has been also investigated in the context of various collective models[3-8]. The exchange term in IBFM has the form[1,9]:

$$W_{exch} = \sum_{j_1 j_2 j_3} W_{j_1 j_2}^{j_3} : \left[(d_2^+ \times \tilde{a}_{j_1})_{j_3} \times (a_{j_2}^+ \times \tilde{d}_2)_{j_3} \right]_{(0)} : \qquad (1)$$

Here, $W_{j_1 j_2}^{j_3}$ are free parameters; d_2^+ and a_j^+ are the creation operators of d-boson and of odd fermion with angular moment j, respectively; and

$$\tilde{d}_{2\mu} = (-1)^{\mu} d_{2-\mu} \quad , \quad \tilde{a}_{jm} = (-1)^{j-m} \; a_{j-m}. \qquad (2)$$

Several methods have been employed in the microscopic approach to the exchange term in the investigations by Talmi[10], Gelberg[10], Broglia et al.[11], Kaup[12], Otsuka et al.[13] and de Kock and Geyer[14]. Otsuka et al.[13] have investigated contributions of different kinds of effective forces to the exchange term in the single-j shell case by

Symmetries in Science VII, Edited by B. Gruber
and T. Otsuka, Plenum Press, New York, 1994

a method of mapping matrix elements. De Kock and Geyer[14] have investigated the origin of the exchange term in the single-j shell case employing Dyson boson mapping and a subsequent similarity transformation which reexpresses the Dyson boson images in terms of seniority bosons.

In this paper we present an approach to the exchange term by using operator method in the multi-j shell case. The advantage of our method, in comparison with the method by Otsuka et al.[13], is that it can be effected explicitly even in situations when it is not straightforward to guess the ensuing boson-fermion equivalent of a given fermion operator.

We investigate the microscopic origin of the exchange term in the framework of Extended RPA with SU(6) algebraic constraints for boson-fermion systems (to be referred to as SU(6)-BF ERPA). This is a nonstraightforward extension of the microscopic approach to the enforced SU(6) symmetry in RPA, which was previously introduced for boson system[15] (referred to as SU(6)-ERPA). The key role in this approach is played by the constrained RPA quadrupole phonon operators subjected to the constraints due to enforcement of the SU(6) symetry[15]. As a side remark we note that in addition to give the microscopic expressions for the parameters $W_{j_1 j_2}^{j_3}$ of the exchange term, the SU(6)-BF ERPA provides the microscopic foundation of IBFM and SU(6/M) super-phenomenology along the lines outlined in refs.[16,17].

We have pointed out[16] that the quadrupole-quadrupole effective interaction yields the exchange term if the kinematic effects (Pauli principle) are properly accounted for. This leads to modification of the quadrupole phonon operators $\{\overline{Q_{2\mu}^+}, \overline{Q_{2\mu}}; \mu = 0, \pm 1, \pm 2\}$ for boson-fermion system in comparison to the quadrupole phonon operators $\{Q_{2\mu}^+, Q_{2\mu}\}$ for boson system. (The notation $\overline{Q}_\mu^+$ is used to denote quadrupole phonon operators for boson-fermion system, to be distinguished from the standard notation Q_μ^+ for the quadrupole phonon operators for boson system.) Full account of SU(6) - BF ERPA will be presented in ref.[17], while here we focus on the microscopic derivation of the exchange term.

PAIRING PLUS QUADRUPOLE MODEL HAMILTONIAN IN QUASIPARTICLE REPRESENTATION

We employ the pairing plus quadrupole model Hamiltonian in the quasiparticle-phonon representation[18]

$$H = \sum_{jm} \epsilon_j \overline{\alpha_{jm}^+ \alpha_{jm}} - \frac{1}{2} \kappa_0^{(2)} \sum_\mu : \overline{M_{2\mu}^+ M_{2\mu}} : \tag{3}$$

The first term represents the single quasiparticle Hamiltonian for odd system. The second term expresses the quadrupole-quadrupole interaction. It involves the normal product of modified quadrupole phonon operators

$$\overline{M}_{2\mu}^{+} = \frac{(-1)^{\mu}}{\sqrt{5}} \sum_{jj'} f_{jj'}^{(2)} \left[u_{jj'}^{(+)} \frac{1}{2} (\Psi_{jj'} + \Phi_{jj'})(\overline{Q}_{2\mu}^{+} + (-1)^{\mu}\overline{Q}_{2-\mu}) + v_{jj'}^{(-)} \overline{B}(jj';2\mu) \right]. \quad (4)$$

Here, $f_{jj'}^{(2)}$ denote the reduced single-particle matrix elements of the quadrupole operator $r^2 Y_{2\mu}$; $u_{jj'}^{(+)}$ and $v_{jj'}^{(-)}$ are definite combinations of the Bogoliubov-Valatin u, v − factors, namely:

$$u_{jj'}^{(+)} = u_j v_{j'} + u_{j'} v_j \quad , \quad v_{jj'}^{(-)} = u_j u_{j'} - v_j v_{j'}$$

and

$$\overline{B}(jj';2\mu) = \sum_{mm'} <jm \; j'm'|2\mu> \overline{\alpha_{jm}^{+} \tilde{\alpha}_{j'm'}} \; , \quad (5)$$

where $\overline{\alpha_{jm}^{+} \tilde{\alpha}_{jm'}}$ denote the modified density quasiparticle operators, referring to the odd-mass system (cf.eq. (8c) below). The operators $\overline{Q}_{2\mu}^{+}$ ($\tilde{\overline{Q}}_{2\mu} = (-1)^{\mu}\overline{Q}_{2-\mu}$) denote the modified quadrupole phonon creation (annihilation) operators corresponding to the first collective solution of the quasiparticle RPA secular equations for odd-mass nuclei[19]. These operators are defined in a straightforward way:

$$\overline{Q}_{2\mu}^{+} = \frac{1}{2} \sum_{jj'mm'} \left[\Psi_{jj'} \langle jmj'm'|2\mu\rangle \overline{(\alpha_{jm}^{+}\alpha_{j'm'}^{+})} - \Phi_{jj'} \langle jmj'm'|2\mu\rangle \overline{(\tilde{\alpha}_{j'm'}\tilde{\alpha}_{jm})} \right] . \quad (6)$$

Generally the r.h.s. of eq.(4) includes summation over all RPA solutions. We take into account only the lowest RPA solution. It is a standard approximation, inherent to all purely collective models (in which the coupling to noncollective modes is neglected). This approximation is not arbitrary, but is symmetry dictated[15]. This, of course, restricts the scope of the microscopically derived collective model.

MODIFIED QUADRUPOLE PHONON OPERATORS IN TAMM-DANKOFF APPROXIMATION

Our main objective is to construct the modified quadrupole phonon operators $\{\overline{Q}_{2\mu}^{+}, \tilde{\overline{Q}}_{2\mu}\}$. For the sake of simplicity we first construct these operators in Tamm-Dankoff approximation, i.e. with $\Phi_{jj'} = 0$. In this case we have

$$\left(\overline{Q}_{2\mu}^{+}\right)^{TD} = \frac{1}{2} \sum_{jj'mm'} \left[\Psi_{jj'} \langle jmj'm'|2\mu\rangle \overline{(\alpha_{jm}^{+}\alpha_{j'm'}^{+})} \right] , \quad (7.a)$$

$$\left(\tilde{\overline{Q}}_{2\mu}\right)^{TD} = \frac{1}{2} \sum_{jj'mm'} \left[\Psi_{jj'} \langle jmj'm|2\mu\rangle \overline{(\tilde{\alpha}_{j'm'}\tilde{\alpha}_{jm})} \right] . \quad (7.b)$$

In this case the construction of operators is greatly facilitated by the fact that direct and inverse transformations relating the IBM quadrupole bosons $\{d_{2\mu}^+, \tilde{d}_{2\mu}\}$ and the ideal bosons $b_{jmj'm'}^+, \tilde{b}_{jmj'm'}$ are readily obtainable.

For the odd-fermion system, we have to add to the set of bifermion operators $\{\overline{\alpha^+\alpha^+}, \overline{\alpha^+\alpha}, \overline{\alpha\alpha}\}$ a set of single-fermion and unity operators $\{\overline{\alpha^+}, \overline{\alpha}, I\}$. The ensuing set of operators

$$\{I, \overline{\alpha^+}, \overline{\alpha}, \overline{\alpha^+\alpha^+}, \overline{\alpha^+\alpha}, \overline{\alpha\alpha}\}$$

forms the Lie algebra $so(2n + 1)$ of the special orthogonal group, where n is the number of relevant single particle levels.

According to Okubo[20] and Marshalek[21], the corresponding boson-fermion realization for the pair-quasiparticle and density-quasiparticle operators reads

$$\overline{(\alpha_{j'm'}\alpha_{jm})} \rightarrow b_{jmj'm'} \tag{8.a}$$

$$\overline{(\alpha_{jm}^+\alpha_{j'm'}^+)} \rightarrow b_{jmj'm'}^+ - \sum_{j_1m_1j_1'm_1'} b_{jmj_1m_1}^+ b_{j_1'm_1'j'm'}^+ b_{j_1'm_1'j_1m_1} +$$

$$+ \sum_{j_1m_1} b_{j_1m_1jm}^+ \beta_{j'm'}^+ \beta_{j_1m_1} - \sum_{j_1'm_1'} b_{j_1'm_1'j'm'}^+ \beta_{jm}^+ \beta_{j_1'm_1'} \tag{8.b}$$

$$\overline{(\alpha_{jm}^+\alpha_{j'm'})} \rightarrow \sum_{j_1m_1} b_{jmj_1m_1}^+ b_{j'm'j_1m_1} + \beta_{jm}^+ \beta_{j'm'} \ . \tag{8.c}$$

Here, the ideal boson operators $b_{jmj'm'}^+$, $b_{jmj'm'}$ satisfy the following antisymmetry and commutation relations:

$$b_{j'm'jm}^+ = -b_{jmj'm'}^+ \ , \tag{9.a}$$

$$\left[b_{jmj'm'}, b_{j_1m_1j_1'm_1'}^+ \right] = \delta_{jj_1}\delta_{mm_1}\delta_{j'j_1'}\delta_{m'm_1'} - \delta_{jj_1'}\delta_{mm_1'}\delta_{j'j_1}\delta_{m'm_1} \tag{9.b}$$

The ideal odd fermion operators β_{jm}^+ and β_{jm} commute with ideal boson operators and exhibit some characteristic features[20-23].

Replacing the operator $\overline{(\tilde{\alpha}_{j'm'}\tilde{\alpha}_{jm})}$ in eq. (7b) by its boson-fermion equivalent according to eq. (8a) we obtain

$$\left(\tilde{Q}_{2\mu}\right)^{TD} = \frac{1}{2}\sum_{jj'} \Psi_{jj'}\langle jmj'm'|2\mu\rangle \tilde{b}_{jmj'm'} \tag{10}$$

The r.h.s of this relation is, in fact, the IBM d-boson operator $\tilde{d}_{2\mu}$. Indeed, if we define

$$\tilde{d}_{2\mu} \equiv \frac{1}{2} \sum_{jj'} \Psi_{jj'} \langle jmj'm'|2\mu \rangle \tilde{b}_{jmj'm'} \tag{11.a}$$

$$d_{2\mu}^+ \equiv \frac{1}{2} \sum_{jj'} \Psi_{jj'} \langle jmj'm'|2\mu \rangle b_{jmj'm'}^+ \tag{11.b}$$

it is apparent that $\tilde{d}_{2\mu}$ and $d_{2\mu}^+$ are second-rank tensors, as the IBM d-bosons should be. On the other hand, using (11a,b), antisymmetry and commutation relations (9a,b) and the ortonormalization condition

$$\sum_{jj'} \Psi_{jj'}^2 = 2$$

we readily show that the operators $\tilde{d}_{2\mu}$ and $d_{2\mu}^+$ satisfy the ensuing commutation relations:

$$\left[\tilde{d}_{2\mu}, \tilde{d}_{2\nu}\right] = \left[d_{2\mu}^+, d_{2\nu}^+\right] = 0, \quad \left[\tilde{d}_{2\mu}, d_{2\nu}^+\right] = (-1)^\mu \delta_{\mu-\nu} \tag{12}$$

Thus, using eqs. (10) and (11a) we have

$$\left(\tilde{Q}_{2\mu}\right)^{TD} = \tilde{d}_{2\mu} \tag{13}$$

Utilizing eqs. (8b), (11a), (11b) and the inverse transformations of (11a) and (11b), and carrying out the requisite Wigner-Racah calculus on the r.h.s. of relation (7a) we obtain

$$\left(\overline{Q}_{2\mu}^+\right)^{TD} = d_{2\mu}^+ - \sum_{\nu\lambda\rho} C_{\nu\mu\lambda\rho}^{TD} d_{2\nu}^+ d_{2\lambda}^+ d_{2\rho} + \sum_{\nu jmj'm'} \Gamma_{TDj'm'}^{\mu\nu jm} d_{2\nu}^+ \beta_{jm}^+ \tilde{\beta}_{j'm'} \tag{14}$$

with

$$C_{\nu\mu\lambda\rho}^{TD} \equiv \frac{25}{2} \sum_{K\kappa} \langle 2\nu2\lambda|K\kappa \rangle \langle 2\rho2\mu|K\kappa \rangle C_K^{TD}$$

$$C_K^{TD} \equiv \sum_{j_1 j_2 j_3 j_4} (-1)^{j_1-j_3} \Psi_{j_1 j_2} \Psi_{j_3 j_1} \Psi_{j_2 j_4} \Psi_{j_3 j_4} \begin{Bmatrix} j_4 & j_3 & 2 \\ j_2 & j_1 & 2 \\ 2 & 2 & K \end{Bmatrix} \tag{15}$$

$$\Gamma^{\mu\nu jm}_{TDj'm'} =$$

$$- \sum_{LMj''} \sqrt{5}(2L+1)^{1/2}\langle jmj'm'|LM\rangle(-1)^L\langle 2\nu LM|2\mu\rangle(-1)^{j+j''}\begin{Bmatrix} 2 & 2 & L \\ j & j' & j'' \end{Bmatrix}\Psi_{jj''}\Psi_{j'j''}$$

$$(16)$$

In this way, we have provided the microscopic foundation for the ansatz of Ring and Schuck[24] (modified Dyson mapping), which is being used in the approach to the microscopic formulation of IBFM (cf.eqs. (32a)-(32c) in the paper by Gambhir et al.[25]).

As seen from eqs. (13) and (14), for odd system the form of operators $\left(\tilde{Q}_{2\mu}\right)^{TD}$ remains unchanged, while the operators $\left(\overline{Q}^+_{2\mu}\right)^{TD}$ split into two parts, the one connected with even-even core, and the kinematical part which reflects the effects of antisymmetrization. The former is, in fact, the microscopic Dyson realization of $\{Q^{+TD}_{2\mu}, \tilde{Q}^{TD}_{2\mu}\}$ (cf.ref.[26] and eqs. (45),(46) in ref.[15]). The fact that $\left(\tilde{Q}_{2\mu}\right)^{TD}$ and $\left(\overline{Q}^+_{2\mu}\right)^{TD}$, given by eqs. (13) and (14), do not transform into each other under hermitian conjugation is a consequence of Dyson nature of the boson-fermion realization of the bifermion operators $\overline{\alpha_{j'm'}\alpha_{jm}}$ and $\overline{\alpha^+_{jm}\alpha^+_{j'm'}}$ (cf.eqs.(8a) and (8b)).

MODIFIED PHONON OPERATORS FOR RANDOM PHASE APPROXIMATION

The generalization of relations (13) and (14) to the random phase approximation is not trivial. Generally, the operators $\overline{Q}^+_{2\mu}$ and $\tilde{Q}_{2\mu}$ are elements of the original $so(2n+1)$ algebra. Now, if we enforce $\{Q^+_{2\mu}, \tilde{Q}_{2\mu}\}$ to close approximately the su(6) algebra, we can use the sophisticated method elaborated in ref.[15], in order to extend the construction of Tamm-Dankoff phonon operators to the RPA phonon operators for boson-fermion system. Clearly, the form (13)-(14) of $\{(\overline{Q}^+_{2\mu})^{TD}, (\tilde{Q}_{2\mu})^{TD}\}$ is maintained for $\{\overline{Q}^+_{2\mu}, \tilde{Q}_{2\mu}\}$ as well, only the structure constants $C^{TD}_{\mu\nu\lambda\rho}$ and quantities $\Gamma^{\mu\nu jm}_{TDj'm'}$ have to be modified in a relevant way, which is described below. In addition, a set of constraints embodied in a set of non-linear conditions on $\Psi_{jj'}$ and $\Phi_{jj'}$, emerge as result of the enforcement of SU(6) symmetry (cf. eqs. (30)-(33) of ref.[15]). Following the procedure presented in ref.[15], we require that the set of operators $\{Q^+_{2\mu}, \tilde{Q}_{2\mu}, [\tilde{Q}_{2\mu}, Q^+_{2\nu}]\}$ (i.e. the part associated with the even-even system) forms an su(6) algebra.

In order to obtain collective SU(6)-symmetry we impose tha SU(6) enforcing conditions:

$$C_K = C \ , \quad K = 0, 2, 4 \tag{17}$$

$$D_K = 0 \ , \quad K = 0,1,2,3,4 \tag{18}$$

If we do so, then we derive the generalized Dyson (GD) boson-ideal fermion representation for the modified RPA phonon opreators:

$$\overline{Q}_{2\mu}^{+GD} = d_{2\mu}^+ - C d_{2\mu}^+ \sum_\lambda d_{2\lambda}^+ d_{2\lambda} + \sum_{\lambda j m j' m'} \Gamma_{j'm'}^{\mu\lambda jm} d_{2\lambda}^+ \beta_{jm}^+ \tilde{\beta}_{j'm'} \tag{19}$$

$$\tilde{\overline{Q}}_{2\mu}^{GD} = \tilde{d}_{2\mu} \tag{20}$$

The quantities C_K are the RPA-extension of the quantities C_K^{TD} which are given by eq. (15). The explicit expressions for C_K and D_K are given by eqs. (24) and (25) of ref.[15] . As to the RPA extension of $\Gamma_{TDj'm'}^{\mu\nu jm}$, it can be derived as follows. Using eqs. (19) and (20) we calculate the commutator

$$\left[\tilde{\overline{Q}}_{2\mu}^{GD}, \overline{Q}_{2\nu}^{+GD}\right] = \delta_{\mu\nu} - \delta_{\mu\nu} C \sum_\lambda d_{2\lambda}^+ d_{2\lambda} - C d_{2\nu}^+ d_{2\mu} + \sum_{jmj'm'} \Gamma_{j'm'}^{\mu\nu jm} \beta_{jm}^+ \tilde{\beta}_{j'm'} \tag{21}$$

This should be compared to the expression for the corresponding commutator in the conventional fermion representation (cf.eqs.(A1),(A4) and (17a) in ref.[15]). If we replace the operator $\overline{(\alpha^+\alpha)}$ in the r.h.s. of eq. (A1) in ref.[15] by its boson-fermion realization according to (8c), we separate the terms with operator structure $\beta^+\tilde{\beta}$. By direct comparison with eq. (21) we obtain the expression for the quantities $\Gamma_{j'm'}^{\mu\nu jm}$ which presents the RPA extension of $\Gamma_{TDj'm'}^{\mu\nu jm}$

$$\Gamma_{j'm'}^{\mu\nu jm} = -\sum_{LM} \sqrt{5}(2L+1)^{1/2}\langle jmj'm'|LM\rangle\langle 2\nu LM|2\mu\rangle S^{(-)}(jj';L) \tag{22}$$

where

$$S^{(-)}(jj';L) = \sum_{j''}(-1)^{j+j''}\begin{Bmatrix} 2 & 2 & L \\ j & j' & j'' \end{Bmatrix}\left[(-1)^L \Psi_{jj''}\Psi_{j'j''} - \Phi_{jj''}\Phi_{j'j''}\right]$$

If we impose the SU(6) enforcing conditions with respect to Tamm-Dankoff phonons (eq.(14))

$$C_K^{TD} = C^{TD} , \quad K = 0,2,4$$

then the eq.(14) acquires a form similar to that of eq. (19).

For our future purposes (providing microscopic foundation of IBFM) it is convenient to hermitize partially the GD boson-ideal fermion realization, given by eqs. (19) and (20). We would like to transcribe $\{\overline{Q}_{2\mu}^{+GD}, \tilde{\overline{Q}}_{2\mu}^{GD}\}$ (cf.eqs.(19) and (20)) in such a form, that the purely boson parts of $\{\overline{Q}_{2\mu}^{+GD}, \tilde{\overline{Q}}_{2\mu}^{GD}\}$ (i.e. $\{Q_{2\mu}^+, \tilde{Q}_{2\mu}\}$, related to the even-even core) transform into the Schwinger realization of $\{Q_{2\mu}^+, \tilde{Q}_{2\mu}\}$ (ref.[15].) The ensuing similarity transformation is, in fact, known (cf.,e.g.ref.[27]):

$$\hat{S}(\hat{N}_d) = \left[(N - \hat{N}_d)!/N!\right]^{1/2} ,$$

where

$$\hat{N}_d \equiv \sum_\lambda d_{2\lambda}^+ d_{2\lambda} , \quad N = Int \left|\frac{1}{C}\right| .$$

Utilizing the identities

$$d_{2\mu} f(\hat{N}_d) = f(\hat{N}_d + 1) d_{2\mu} , \quad d_{2\mu}^+ f(\hat{N}_d) = f(\hat{N}_d - 1) d_{2\mu}^+ ,$$

valid for arbitrary function $f(\hat{N}_d)$, we readily obtain

$$\hat{S}(\hat{N}_d) d_2 \hat{S}^{-1}(\hat{N}_d) = (\hat{N} - \hat{N}_d)^{1/2} d_2 = s^+ d_2$$

$$S(N_d) d_2^+ S^{-1}(N_d) = d_2^+ (N - \hat{N}_d)^{1/2} (N - \hat{N}_d)^{-1} = d_{2\mu}^+ s(s^+ s)^{-1}$$

We have introduced above the s-bosons according to the natural prescription

$$s^+ s \to N - \hat{N}_d , \quad d_{2\mu}^+ (N - \hat{N}_d)^{1/2} \to d_{2\mu}^+ s , \quad (N - \hat{N}_d)^{1/2} d_{2\mu} \to s^+ d_{2\mu}$$

With these preparations we are now ready to effect the similarity transformation upon $\{\overline{Q}_{2\mu}^{+GD}, \tilde{Q}_{2\mu}^{GD}\}$. Performing the transformation we arrive at the following "Generalized Schwinger" (GS) boson-ideal fermion realization for the RPA odd-mass collective quadrupole phonon operators:

$$\overline{Q}_{2\mu}^{+GD} \equiv \hat{S}(\hat{N}_d) \overline{Q}_2^{+GD} \hat{S}^{-1}(\hat{N}_d) = N^{-1} d_{2\mu}^+ s + \sum_{\lambda j m j' m'} \Gamma_{j'm'}^{\mu \lambda j m} d_{2\lambda}^+ s(s^+ s)^{-1} \beta_{jm}^+ \tilde{\beta}_{j'm'} \quad (23)$$

$$\tilde{Q}_{2\mu}^{GS} \equiv \hat{S}(\hat{N}_d) \tilde{Q}_{2\mu}^{GD} \hat{S}^{-1}(\hat{N}_d) = s^+ \tilde{d}_{2\mu} \quad (24)$$

Converting $\{\overline{Q}_{2\mu}^{+GD}, \tilde{Q}_{2\mu}^{GD}\}$ into a form, in which the operators $\{Q_{2\mu}^+, \tilde{Q}_{2\mu}\}$ occur in the standard Schwinger realization, turns out to be a crucial step in providing microscopic justification of IBFM[17]. To our knowledge the closed partially hermitian form of $\{\overline{Q}_{2\mu}^{+GS}, \tilde{Q}_{2\mu}^{GS}\}$, expressed by eqs. (23) and (24), is not available in the literature.

It follows from eqs. (4),(23) and (24) that the phonon component of our quadrupole operator $\overline{M}_{2\mu}^+$, given by eq. (4), contains as a particular case the quadrupole operator Q, given by eq. (10) of ref.[13].

This completes the construction of the RPA phonon operators $\{\overline{Q}_{2\mu}^{+GS}, \tilde{Q}_{2\mu}^{GS}\}$ for odd system.

It should be pointed out that this construction has been carried out in the spirit of recent advances in microscopic foundations of collective models[28].

DERIVATION OF EXPRESSION FOR STRENGTH OF THE EXCHANGE TERM

Let us now concentrate on the microscopic derivation of strength parameters of the exchange term.

Once we have derived eqs. (23) and (24), together with explicit forms for the quantities C and Γ, the inference of the exchange term is simply a matter of technicalities. From eqs. (4),(23) and (24) it is evident that the effective residual interaction in eq. (3), $-\frac{1}{2}\kappa_0^{(2)}\sum_\mu : \overline{M}_{2\mu}^+ \overline{M}_{2\mu} :$, produces among other terms, the exchange term (1). This term orginates from the combination of $\overline{Q}_{2\mu}^{+GS} \tilde{\overline{Q}}_{2\mu}^{GS}$ in which only the kinematic contribution of $\overline{Q}_{2\mu}^{+GS}$ is considered.

Such combinations have the ensuing operator structure. In fact:

$$(d_2^+ s(s^+s)^{-1}\beta^+ \tilde{\beta})s^+ \tilde{d}_2 =: d_2^+ \tilde{\beta}\beta^+ \tilde{d}_2 :$$

since

$$s(s^+s)^{-1}s^+ = ss^+(s^+s+1)^{-1} = (s^+s+1)(s^+s+1)^{-1} = I \quad .$$

Selecting this type of structure from the second term in the Hamiltonian (3) and performing the requisite recouplings, we obtain the geometrical part precisely in the form (1). The final result reads:

$$W_{EXCH} =$$

$$\kappa_0^{(2)}M^2 \sum_{j_1 j_2 j_3 L} (2L+1)\begin{Bmatrix} 2 & 2 & L \\ j_1 & j_2 & j_3 \end{Bmatrix} S^{(-)}(j_1 j_2; L) : \left(\left(d_2^+ \times \beta_{j_2}\right)_{(j_3)} \times \left(\beta_{j_1}^+ \times \tilde{d}_2\right)_{(j_3)}\right)_{(00)} :$$

with

$$M = \frac{1}{2}\sum_{jj'} f_{jj'}^{(2)}u_{jj'}^{(+)}(\Psi_{jj'} + \Phi_{jj'}) \quad .$$

In this way we have expressed tha parameters $W_{j_1 j_2}^{j_3}$ of the exchange interaction in terms of the microscopic quantities $f_{j_1 j_2}^{(2)}, u_{j_1 j_2}^{(+)}$ and amplitudes of the constrained RPA phonon operators $\{\Psi_{j_1 j_2}, \Phi_{j_1 j_2}\}$. The latter can be obtained by solving numerically nonlinear algebraic system of equations which have been derived[17] from variational principle accounting for both the dynamics (given by Hamiltonian (3)) and the kinematical constraints dictated both by the enforcement of the SU (6) symmetry (cf.eqs.(17), (18) and by Pauli corrections induced by the presence of an odd particle (cf.eq.(22)). This system of nonlinear equations is an extension for odd-mass nuclei of the system of equations derived in ref.[29] for even-even nuclei. The main modifications in the former system of equations are due to the exchange term (25).

CONCLUSIONS

We have presented the microscopic derivation of IBFM exchange term in the framework of SU(6)-BF ERPA. Using as a starting point the boson-fermion realization of $so(2n + 1)$ algebra due to Marshalek and Okubo, we have constructed the constrained TDA/RPA quadrupole phonon operators for odd mass nuclei both in Generalized Dyson (cf.eqs.(19) and (20)) and in Generalized Schwinger boson-ideal fermion realizations (cf.eqs. (23) and (24)). The latter is an apt representation which was missing in the published literature.

Microscopic expressions for the parameters of the IBFM exchange term have thus been inferred. In addition, a justification of the ansatz of Ring and Schuck (the modified Dyson mapping)[24] has been provided. We note that our operator method could be employed in other situations as well.

As pointed out in ref.[16], the exchange term plays a distinguished role in models based on dynamical supersymmetries. Namely, it measures deviation from the usual dynamical symmetry $SU(6) \times SU(M)$, i.e. in the present framework the exchange term generates the supersymmetry. This problem is in the course of investigations[17], and is relevant for microscopic foundation of SU(6/M) superphenomenology[30].

ACKNOWLEDGEMENTS

We are indebted to F. Iachello for his participation at some stages of this work. We thank to R.Bijker, R.F.Casten, D.H.Feng, R.Gilmore, M.W.Guidry, R.V.Jolos, A.Leviatan, V.M.Mikhailov, S.Pittel, P.Ring, R.Russev, I.Talmi, V.V.Voronov and C.L.Wu for useful discussions.

REFERENCES

1. A.Arima and F.Iachello, Phys.Rev.Lett. 35: 1069 (1975).

2. F.Iachello and O.Scholten, Phys.Rev.Lett. 43: 679(1979).

3. B.R.Mottelson in: Proc.Intern.Conf. on Nuclear Structure, Phys.Soc. Japan, Tokyo (1968).

4. P.A.Simard and M.Banvill, Nucl.Phys. A158: 51 (1970).

5. A.Kuriyama, T.Marumori and K.Matsuyanagi: Progr.Theor.Phys.Jap. 47, 498 (1972);

   A.Kuriyama, T.Marumori and K.Matsuyanagi: Progr. Theor.Phys.Suppl. 58:1 (1975).

6. F.Donau and D.Janssen, Nucl.Phys. A209:109 (1973);

   F.Donau and U.Hagemann, Nucl.Phys. A256: 27(1976).

7. O.Civitarese, R.A.Broglia and D.R.Bes, Phys.Lett. B72: 45 (1977).

8. K. Allaart, P.Hofstra and V.Paar, Nucl.Phys. A366:384 (1981).

9. "Interacting Bose-Fermi Systems in Nuclei", F.Iachello, ed., Plenum, New York (1981).

10. I.Talmi, in: "Interacting Bose-Fermi Systems in Nuclei ", F.Iachello, ed., Plenum, New York (1981);

    A.Gelberg, Z.Phys. A310:117 (1983).

11. R.A.Broglia, E.Maglione and A.Vitturi, Nucl.Phys. A376:45 (1982).

12. U.Kaup, in: Progress in Particle and Nuclear Physics, Vol.9;

    D.Wilkinson,ed., Pergamon, Oxford (1983).

13. T.Otsuka, N.Yoshida, P.Van Isacker, A.Arima nad O.Scholten, Phys.Rev. C35:358 (1987).

14. E.A. de Kock and H.B.Geyer, Phys. Rev. C38:2887 (1988).

15. G.Kyrchev and V.Paar, Phys.Rev. C37:838 (1988).

16. G.Kyrchev and V.Paar, Phys.Lett. B195: 107 (1987).

17. V.Paar and G.Kyrchev (unpublished).

18. V.G.Soloviev, "Theory of Complex Nuclei", Pergamon, Oxford (1976).

19. A.I.Vdovin, V.V.Voronov, V.G.Soloviev and Ch.Stoyanov, Particles and Nuclei 16:246 (1985).

20. S.Okubo, Phys.Rev. C10:2048 (1974).

21. E.R.Marshalek, Nucl.Phys. A224: 221, 245 (1974).

22. E.R.Marshalek, Nucl.Phys. A347: 253 (1980)

23. A.Klein and E.R.Marshalek, Z.Phys. A329: 441 (1988).

24. P.Ring and P.Schuck, "The nuclear many body problem", Springer, Berlin (1980).

25. Y.K.Gambhir, P.Ring and P.Schuck, Nucl.Phys. A423: 35 (1984).

26. R.V.Jolos and D.Janssen, Particles and Nuclei 8: 330 (1977).

27. J.Dobaczewski, Nucl.Phys A 380: 1 (1982).

28. A.Klein, E.R.Marshalek, J.Math.Phys. 30: 219 (1989).

29. D.Karadjov, G.Kyrchev and V.V.Voronov, in: "Shell Model and Nuclear Structure: Where do we stand?",A.Covello, ed., World Scientific, Singapore (1988).

30. D.Warner, in "Nuclear Structure, Reactions and Symmetries" R.A.Meyer and V.Paar, eds., World Scientific, Singapore (1986) and references therein;

    J.Vervier, Rivista del Nuovo Cimento 10: No.9 (1987) and references therein.

ALGEBRAIC APPROACH TO VIBRATIONAL EXCITATIONS IN AN ANHARMONIC LINEAR CHAIN

R. Lemus[1] and A. Frank[2, a, b]

[1]Instituto de Ciencias Nucleares, UNAM
 Apdo. Postal 70-543, Circuito Exterior, C.U.
 México, D.F., 04510 México
[2]Departamento de Física Atómica, Molecular y Nuclear
 Facultad de Física, Universidad de Sevilla
 Apdo. 1065, 41080 Sevilla, España

A description of vibrational excitations in one-dimensional monoatomic and di-atomic Bravais lattices is presented by means of an algebraic model of n-coupled anharmonic oscillators. The anharmonic interaction corresponds to a Morse-like potential. The energy expression $E(k)$ for one-phonon excitations is obtained analytically in both cases. In addition, we present numerical results for two-phonons.

INTRODUCTION

Algebraic techniques have been applied to the study of numerous physical and chemical systems. They prove to be very useful in the analysis of very complex situations, where their versatility brings about important advantages over the usual integro-differential techniques. In the case of molecular physics, many new developments are due to F. Iachello who, together with others, proposed algebraic models to describe the spectroscopic properties of molecules. In 1981 Iachello introduced the vibron model,[1] which gives a simple, yet accurate description of the rotational and vibrational structure of diatomic molecules. The model was later extended to include linear polyatomic[2] and non-linear triatomic molecules.[3] In recent years the electronic degrees of freedom of diatomic molecules have been also incorporated into the vibron model framework.[4] One of the underlying reasons why the vibron model has

---

[a] On sabbatical leave from Instituto de Ciencias Nucleares, UNAM.
[b] Guggenheim Fellow.

succeeded in describing the rotation-vibration spectra of molecules is the existence in the model of an $O(4)$ dynamical symmetry, which is associated with a Morse potential interaction.[5] It is well known that in the Born-Oppenheimer approximation the interactions between nuclei are well described by this kind of potential. Although this is a three-dimensional result, analogous relations hold in one and two dimensions.

In a one dimensional system the realization of $SU(2)$ on the sphere can be associated with a Morse potential, which means that the eigenstates of a one dimensional Morse-like Hamiltonian may be put in a one to one correspondence with the states associated with the chain $U(2) \supset O(2)$.[6] This result has been exploited to study complex situations, such as atom-molecule collision processes.[7]

The $U(2)$ algebraic model may be considered as the one dimensional limit of the $U(4)$ three-dimensional vibron model. In the same way as the vibron model was extended to describe polyatomic molecules, the $U(2)$ model can be generalized and applied to molecules with several bonds. O.S. van Roosmalen *et al.* analyzed the case of two bonds by considering the vibrational states of specific molecules like H_2O, SO_2, O_3 and others.[8] Because of the one dimensionality of the $U(2)$ model, however, this method can only describe stretching modes. The extension of the $U(2)$ model to polyatomic molecules was carried out recently by Iachello and Oss,[9] who applied it to benzene and to several octahedral molecules. A remarkable feature of this procedure is that the diagonalization automatically provides states carrying irreducible representations of the appropriate point group. This is achieved by expanding the Hamiltonian as a suitable linear combination of Casimir operators.[10]

We show in this work that in addition to molecular systems the $U(2)$ approach to n-coupled oscillators can be used to describe linear anharmonic chains. We present here a study of the vibrational degrees of freedom in one dimensional monoatomic and diatomic Bravais lattices.

ALGEBRAIC MODEL OF COUPLED OSCILLATORS

The model is based on the isomorphism between the Lie algebra $U(2)$ and the one dimensional Morse oscillator[6]

$$\hat{\mathcal{H}} = -\frac{\hbar^2}{2\mu}\frac{d^2}{dx^2} + D(e^{-2x/d} - 2e^{-x/d}) \ . \tag{II.1}$$

The eigenstates of (II.1) may be put into a one to one correspondence with the $U(2) \supset O(2)$ states, characterized by the quantum numbers $|N, m>$, as long as the value of m is restricted to non-negative values. The Morse Hamiltonian (II.1) has the algebraic realization[6]

$$\hat{\mathcal{H}} = A[\hat{C}_{O(2)}^2 - N^2] \tag{II.2}$$

where A is related to the Morse parameters[6] and $\hat{C}_{O(2)}$ is the invariant operator of $O(2)$, with eigenvalues $m = N, N-2, \ldots, 1$ or 0. The term $-AN^2$ in (II.2) was introduced in order to place the ground state at null energy.

We now consider a system of atoms where η chemical bonds are involved. In the algebraic model we associate to each bond i a $U^i(2)$ algebra. Therefore the product $U^1(2) \times U^2(2) \times \ldots \times U^\eta(2)$ establishes the dynamical group of the system. The Hamiltonian (and every other operator) may be expanded in terms of generators of the $U^i(2)$ groups.

The contributions to the Hamiltonian are then taken to be the Casimir operators of the groups involved in the different reductions of the dynamical algebra into its subalgebras. A possible decomposition involves the reduction

$$U^1(2) \times U^2(2) \times \ldots \times U^\eta(2) \supset O^1(2) \times \ldots \times O^\eta(2) \supset O(2) \tag{II.3}$$

where the coupling to the final $O(2)$ group is carried out through the different intermediate couplings $O^{ij}(2)$. A second chain arises from all the possible couplings of the $U^i(2)$ groups to obtain a total $U(2)$ which in turn contains the full $O(2)$ group. The expansion of the Hamiltonian, up to two body interactions and conserving the quantum number of the total $O(2)$, can be given in terms of η contributions $h_i = A_i[\hat{C}^2_{O(2)} - N_i^2]$, representing the η one-dimensional independent Morse oscillators, plus two types of bond-bond interactions: $[\hat{C}^2_{O^{ij}(2)} - (N_i + N_j)^2]$ and $\hat{M}_{ij}$, which correspond to the Casimir operators $\hat{C}_{O^{ij}(2)}$ of the $O^{ij}(2)$ groups and the Majorana operators, respectively. The latter are related to the $U^{ij}(2)$ Casimir operators $\hat{C}_{2U^{ij}(2)}$ by the relation

$$\hat{M}_{ij} = -\frac{1}{2}[\hat{C}_{2U^{ij}(2)} - \hat{C}_{2U^i(2)} - \hat{C}_{2U^j(2)} - 2N_i N_j] \quad, \tag{II.4}$$

where N_k corresponds to the number of bosons associated to the $U^k(2)$ group. The Hamiltonian has thus the general form[9]

$$\hat{\mathcal{H}} = h_0 + \sum_{i=1}^{\eta} A_i[\hat{C}^2_{O^i(2)} - N_i^2] + \sum_{i>j}^{\eta} B_{ij}[\hat{C}^2_{O^{ij}(2)} - (N_i + N_j)^2] + \sum_{i>j}^{\eta} \lambda_{ij}\hat{M}_{ij} \quad. \tag{II.5}$$

The simplest basis to diagonalize the Hamiltonian (II.5) is the one associated to the local modes chain (II.3)

$$|[N], m_1, m_2, \ldots, m_\eta; M > \quad, \tag{II.6}$$

where the quantum numbers m_i and M correspond to the eigenvalues of the Casimir operators $\hat{C}_{O^i(2)}$ and $\hat{C}_{O(2)}$, respectively. In (II.6) we considered that all the bonds are equivalent, therefore $N_i = N$, for all i. It is convenient to introduce in (II.6) the equivalent quantum numbers

$$v_i = \frac{N_i - m_i}{2}, \quad V = \frac{N\eta - M}{2} \quad,$$

since they correspond[9] to the number of quanta in each oscillator

$$v_i = 0, 1, 2, \ldots, \frac{N}{2} \quad \text{or} \quad \frac{N-1}{2} \quad,$$

and the total number of quanta

$$V = v_1 + v_2 + \ldots + v_\eta \quad.$$

The wave functions are then labeled as

$$|[N], v_1 v_2, \ldots, v_\eta; V > \quad. \tag{II.7}$$

The operators involved in the two first sums of the Hamiltonian (II.5) are diagonal in the basis (II.7)

$$< [N]v_1 v_2, \ldots, v_\eta; V |(\hat{C}^2_{O^i(2)} - N^2)|[N]v_1 v_2, \ldots, v_\eta; V > = 4(v_i^2 - N v_i) \quad, \tag{II.8a}$$

$$< [N]v_1 v_2, \ldots, v_\eta; V |(C^2_{O^{ij}(2)} - (N + N)^2)|[N]v_1 v_2, \ldots, v_\eta; V >$$

$$= -8N(v_i + v_j) + 4(v_i + v_j)^2 \quad, \tag{II.8b}$$

while the Majorana operator $\hat{M}_{ij}$ has both diagonal and nondiagonal matrix elements[8]

$$< [N]v_1v_2,\ldots,v_\eta; V|\hat{M}_{ij}|[N]v_1v_2,\ldots,v_\eta; V >= N(v_i + v_j) - 2v_iv_j \qquad (II.8c)$$

$$< [N]\ldots,v'_i,\ldots,v'_j,\ldots; V|\hat{M}_{ij}|[N]\ldots,v_i,\ldots,v_j,\ldots; V >=$$

$$- \sqrt{(N - v_i)(v_i + 1)(N - v_j + 1)v_j}\; \delta_{v'_i,\, v_i+1}\, \delta_{v'_j,\, v_j-1}$$

$$- \sqrt{(N - v_i + 1)v_i(N - v_j)(v_j + 1)}\; \delta_{v'_i,\, v_i-1}\, \delta_{v'_j,\, v_j+1} \quad . \qquad (II.8d)$$

These simple results for the matrix elements allow us to carry out its diagonalization in a straightforward way. The Hamiltonian (II.5), however, is completely general and does not satisfy the symmetry requirements for a linear chain with periodic boundary conditions. In the next sections we present the particular form of the Hamiltonian (II.5) for monoatomic and diatomic Bravais lattices, as well as the results of its diagonalization.

MONOATOMIC BRAVAIS LATTICE

A monoatomic lattice with periodic boundary conditions consists of η equivalent atoms distributed along a line separated by the same distance a. Numbering the bonds of the chain, we have

$$
\begin{array}{cccccccc}
1 & 2 & & \eta/2 & \eta/2+1 & & \eta-1 & \eta \\
o---o---o--- & \ldots & o---o--- & \ldots & ---o----o \\
\end{array}
\qquad (III.1)
$$

$$/-a-/ \qquad\qquad \mathcal{O}$$

where, due to the boundary conditions, the atom located at the end of the chain corresponds to the first one. The space group $\mathcal{G}$ associated with this system has two kinds of elements, namely, traslations and the inversion. The set of traslations $\mathcal{J}$ forms an invariant subgroup of $\mathcal{G}$. The space group can thus be resolved into cosets with respect to $\mathcal{J}$:

$$\mathcal{G} = \mathcal{J} + \{I|0\}\mathcal{J} \quad , \qquad (III.2)$$

where

$$\mathcal{J} = \{ \{E|t_n\},\ n = 0, 1, 2,\ldots,\eta - 1\} \quad .$$

We have used the Seitz notation[11] $\{R|t_n\}$, where the first symbol refers to the point group operation and t_n stands for a traslation vector $t_n = na$. By conjugating $\{E|t_n\}$ and $\{I|t_n\}$ with all the elements of $\mathcal{G}$, we obtain its classification in classes.

According to Bloch's theorem, the wave functions that provide a basis for the irred. reps. of $\mathcal{G}$ have the form (in one dimension)

$$\Psi_k = e^{-ikx}u_k(x) \quad , \qquad (III.3)$$

where $u_k(x) = u_k(x - a)$. The subindex k labels the irred. reps. and is a wave vector belonging to the reciprocal lattice

$$k = \left(\frac{2\pi}{a}\right)k_x, \quad k_x = -\frac{1}{2} + \frac{1}{\eta}, \quad -\frac{1}{2} + \frac{2}{\eta},\ldots,\frac{1}{2} \quad . \qquad (III.4)$$

The interval $-\pi/a < k \leq \pi/a$ defines the first Brillouin zone. It is possible to obtain explicit irred. reps. for $\mathcal{G}$ by applying its elements on the wave functions Ψ_k.[11] In Table I we present the irreducible representations of the group (III.2).

We now derive the Hamiltonian of the system III.1 by imposing the invariance of (II.5) under the elements of the space group $\mathcal{G}$. Following ref. [10] we start establishing an isomorphism between $\mathcal{G}$ and a subgroup of the symmetric group S_η:

$$\{E|t_1\} \rightarrow (123,\ldots,\eta)$$

$$\{I|0\} \rightarrow (1\eta)(2,\eta-1),\ldots,(\frac{\eta}{2},\frac{\eta}{2}+1) \qquad (III.5)$$

where $\{E|t_1\}$ and $\{I|0\}$ are generators of $\mathcal{G}$. Note that we have taken the inversion at point $\mathcal{O}$ in figure (III.1). The isomorphism allows to deduce the effect of the elements of $\mathcal{G}$ on the interactions in (II.5).

The procedure to obtain the symmetry adapted Hamiltonian consists of applying the elements of $\mathcal{G}$ on a minimum set of independent interactions in (II.5). This action generates a first contribution of $\hat{\mathcal{H}}$. A second contribution is obtained by taking a second set of independent interactions and aplying to them the elements of $\mathcal{G}$. The process is repeated until there are no more independent interactions. The method is explained in ref. [10] and discussed in detail in ref. [12]. The final expression of the Hamiltonian is

$$\hat{\mathcal{H}} = E_0 + A_1 \sum_{i=1}^{\eta} \hat{C}'_{2O^i(2)} + \sum_{\omega=0}^{\eta/2-1} B_{1,2+\omega} \sum_{i=1;j}^{\eta} \hat{C}'_{2O^{ij}(2)} + \sum_{\omega=0}^{\eta/2-1} \lambda_{1,2+\omega} \sum_{i=1;j}^{\eta} \hat{M}_{ij} \quad ,$$

$$j = mod(i+\omega,\eta)+1 \quad j = mod(i+\omega,\eta)+1 \qquad (III.6)$$

where we have defined

$$C'_{2O^i(2)} = C^2_{O^i(2)} - N^2_i \quad ,$$

$$C'_{2O^{ij}(2)} = C^2_{O^{ij}(2)} - (N_i + N_j)^2 \quad .$$

The sum over ω in (III.6) leads to the interactions corresponding to first neighbors ($\omega = 0$), second neighbors ($\omega = 1$),..., and so on. The function mod (a,b) has the usual meaning (remainder a/b).

The Hamiltonian (III.6) commutes with $\mathcal{G}$. Consequently, its eigenfunctions carry the irred. reps. of the space group. As mentioned before, a simple basis to perform the diagonalization of (III.6) is the local mode basis (II.7), where the only non-diagonal contributions correspond to the Majorana operators. Since these operators commute with the Casimir operator of the total $O(2)$ group, the Hamiltonian conserves the total number of quanta V. The matrix of $\mathcal{H}$ in the basis (II.7) has thus a blocked form characterized by V. The dimension ND of each submatrix is given by $ND = (\eta + V - 1)!/[(\eta - 1)!V!]$.

From the numerical point of view, we would have to carry out the diagonalization of matrices of dimension $ND(\eta, V)$ for every multiplet characterized by the total number of quanta V. It is possible, however, to simplify the procedure by carrying out a symmetry projection before diagonalization. In this way each matrix of dimension ND reduces to $\eta/2 + 1$ submatrices. The case $V = 1$ deserves special attention, since

Table I. Irreducible representations of the one dimensional Bravais lattice.[11]

Class \\ Irrep. \\ k Parity	$\{E\|0\}$	$\{E\|t_{\frac{1}{2}\eta}\}$	$\{E\|t_{2n}\}$, $\{E\|t_{n-2n}\}$. $n=1,2,$ $..,\frac{n}{2}-1$	$\{E\|t_{2n-1}\}$, $\{E\|t_{n-(2n-1)}\}$. $n=1,2$ $..,n/2$	$\{I\|t_{2n}\}$ $n=0,1$ $..,\frac{n}{2}-1$	$\{I\|t_{2n-1}\}$. $n=1,2$ $..,n/2$
$0\ g$	1	1	1	1	1	1
$0\ u$	1	1	1	1	-1	-1
$1/2\ g$	1	$(-)^{n/2}$	1	-1	1	-1
$1/2\ u$	1	$(-)^{n/2}$	1	-1	-1	1
gen k	$\begin{pmatrix} 1 & 0 \\ 0 & 1 \end{pmatrix}$	$\begin{pmatrix} \lambda_k^{\frac{1}{2}n} & 0 \\ 0 & \lambda_{-k}^{\frac{1}{2}n} \end{pmatrix}$	$\begin{pmatrix} \lambda_k^{2n} & 0 \\ 0 & \lambda_{-k}^{2n} \end{pmatrix}$	$\begin{pmatrix} \lambda_k^{2n-1} & 0 \\ 0 & \lambda_{-k}^{2n-1} \end{pmatrix}$	$\begin{pmatrix} 0 & \lambda_k^{2n} \\ \lambda_{-k}^{2n} & 0 \end{pmatrix}$	$\begin{pmatrix} 0 & \lambda_k^{2n-1} \\ \lambda_{-k}^{2n-1} & 0 \end{pmatrix}$
Number of classes	1	1	$\frac{1}{2}\eta - 1$ (Two	elements per class.)	1	1

$$\lambda_k^n = exp(-2\pi i k_z n)$$

after projection the submatrices are of dimension one, i.e., in this case the Hamiltonian is already diagonal in the projected basis.

For one quantum the basis (II.7) has the simple form

$$|[N]1000\ldots0;1> \equiv |1>, \quad |[N]0100\ldots0;1> \equiv |2>, \quad \ldots \quad, |[N]00\ldots001;1> \equiv |\eta > \quad . \quad (III.7)$$

Since the set $|\alpha >, \alpha = 1,2,\ldots,\eta$ is equivalent under the action of $\mathcal{G}$, we can choose the first function $|1>$ as the one to be projected. The projection operator for a given wave vector k is[11]

$$P_\mu^k = \frac{n_k}{g} \sum_{k \subset \mathcal{G}} D_{\mu\mu}^{k*}(k) \hat{O}_R \quad , \quad\quad\quad\quad (III.8)$$

where μ corresponds to k or $-k$ for a general k, and g is the order of $\mathcal{G}$. The irr. reps. $D^k(R)$ are given in Table I. The action of the operator (III.8) over the function $|1>$ produces the following projected function for a general k:

$$\Psi_\mu^k = \frac{1}{\sqrt{\eta}} \sum_{\alpha=1}^{\eta} e^{2\pi i k_x(\alpha-1)} |\alpha > \quad , \quad\quad\quad\quad (III.9)$$

where a normalization factor was included. The same procedure for the special wave vectors gives rise to

$$\Psi^{0,g} = \frac{1}{\sqrt{\eta}} \sum_{\alpha=1}^{\eta} |\alpha > \quad , \quad \Psi^{\frac{1}{2},u} = \frac{1}{\sqrt{\eta}} \{ \sum_{\alpha\,odd}^{\eta-1} |\alpha > - \sum_{\alpha\,even}^{\eta} |\alpha > \} \quad . \quad\quad (III.10)$$

The projections over $k = \frac{1}{2}, g$ and $k = 0, u$ are null for the one-phonon case.

We now compute the expectation value of $\hat{\mathcal{H}}$ with respect to the functions (III.9 - III.10), to obtain the expression for the energy $E(k)$. The procedure is straightforward and can be found in detail in ref. [12]. Here we only present the results. Taking into account the expressions (II.8) for the matrix elements of the operators involved in the Hamiltonian, we obtain for a general k

$$E(k) = < \Psi_\mu^k | \hat{\mathcal{H}} | \Psi_\mu^k > = E_0 + 4A_1(1 - N) + 8(1 - 2N) \sum_{\omega=0}^{\eta/2-1} B_{1,2+\omega}$$

$$+ 2N \sum_{\omega=0}^{\eta/2-1} \lambda_{1,2+\omega}[1 - \cos(2\pi k_x(\omega + 1))] \quad , \quad\quad\quad (III.11)$$

while for the special values of k

$$E(0,g) = E_0 + 4A_1(1 - N) + 8(1 - 2N) \sum_{\omega=0}^{\eta/2-1} B_{1,2+\omega} \quad , \quad\quad\quad (III.12)$$

$$E(\frac{1}{2},u) = E(0,g) + 2N \sum_{\omega=0}^{\eta/2-1} (\lambda_{1,2+\omega} + (-)^\omega \lambda_{1,2+\omega}) \quad . \quad\quad\quad (III.13)$$

It is worthwhile noting that both $E(0,g)$ and $E(\frac{1}{2},u)$ are contained in $E(k)$. Substitution of $k_x = 0$ and $k_x = \frac{1}{2}$ in (III.11) gives the energies (III.12) and (III.13) for special k wave vectors. This was to be expected, since for the limit of infinite number of atoms in the chain the wave vector k becomes a continuous variable. The energies (III.11 - III.13) should then belong to a continuous curve.

We wish to point out the differences between the harmonic oscillator chain and our results. In the harmonic approximation for first-neighbor interactions, the energy

$E(k)$ for one phonon excitations is proportional to the frequency $\omega(k)$ of the particular normal mode k. For a monoatomic lattice[13]

$$\omega(k) = 2\sqrt{\frac{\gamma}{m}}|\ sen\ \pi k_x|\ , \qquad\qquad (III.14)$$

where m is the atomic mass and γ corresponds to the potential energy parameter. Comparing (III.14) with (III.11) for $\omega = 0$, we note that the one phonon energy effectively corresponds to the square of the harmonic energy. Consequently, a dramatic difference arises for $k = 0$, where the group velocity ($v_g = \hbar\ dE(k)/dk$) is null in the Morse case, while for the harmonic oscillator it is given by $a\sqrt{\gamma/m}$.[13] This result is in

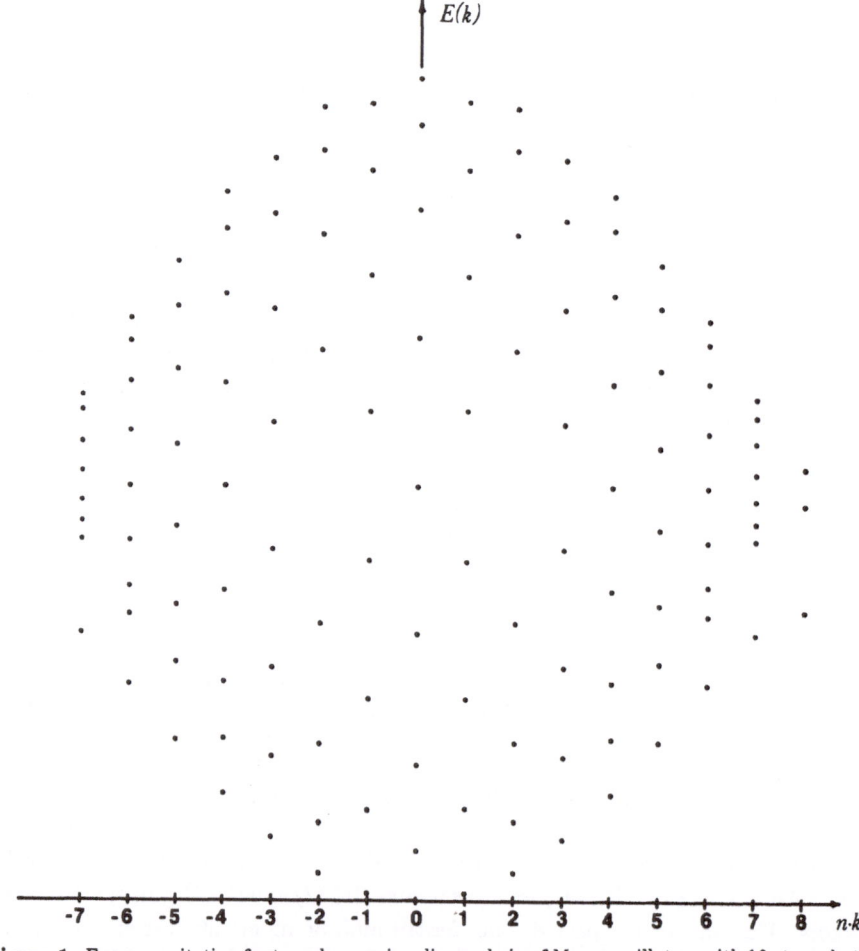

Figure 1. Energy excitation for two phonons in a linear chain of Morse oscillators with 16 atoms in the monoatomic Bravais lattice. The parameters used are $A_1 = -5$, $B_{12} = -2$, $\lambda_{12} = 1$. All parameters in cm^{-1}. Boson number $N = 50$. Maximum energy value 2961 cm^{-1}. Minimum value 2546 cm^{-1}. One phonon energy $E(0, g)$ is taken to be zero.

accordance with the analysis of local modes in anharmonic solids presented by Bruinsma et al.[14] In order to build up the correlations needed for normal mode excitations, we need to introduce n-neighbor interactions.[12]

Two phonon excitations cannot be treated analytically, since the Hamiltonian contains terms involving phonon interactions (Majorana operators). We thus perform a numerical diagonalization of (III.6). In figure 1 we show the spectrum for two-phonon excitations, taking into account only first-neighbor interactions. The addition of higher order interactions does modify, however, the results in a significant way.[12] In the harmonic two-phonon case (III.14) does not correspond directly to the spectrum. In order to obtain it we need to take into account that the product of representations

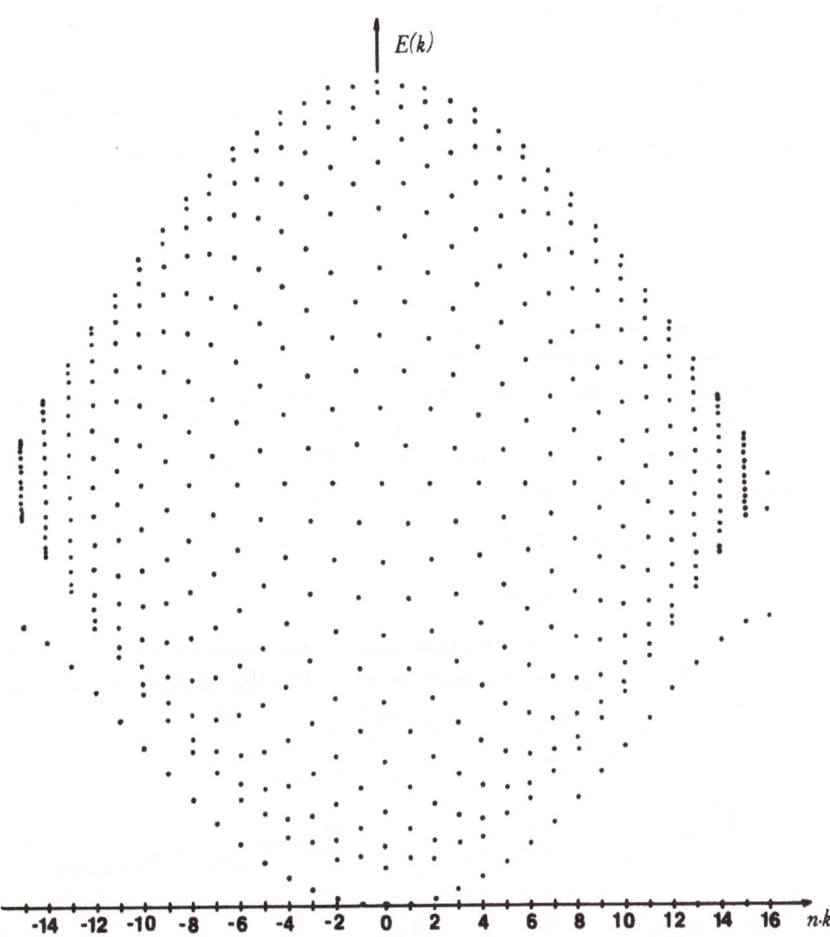

Figure 2. Two phonon energy excitation for Morse oscillators including 32 atoms in the monoatomic Bravais lattice. The parameters used are $A_1 = -5$, $B_{12} = -2$, $\lambda_{12} = 1$. All parameters in cm^{-1}. Boson number $N = 50$. Maximum energy value 2963 cm^{-1}. Minimum value 2546 cm^{-1}. One phonon energy $E(0, g)$ is taken to be zero.

$k_1 \otimes k_2$ reduces according to the following rule:

$$k_1 \otimes k_2 \rightarrow \sum q \; ; \qquad q = k_1 + k_2 + g \; , \qquad (III.15)$$

where g is either a null vector or a reciprocal lattice vector

$$g = \frac{2\pi}{a} m \; , \qquad m \; \text{integer},$$

for which the wave vector q remains in the first Brillouin zone.[13] This expression is equivalent to momentum conservation in crystals.

In order to study the effect of the number of atoms on the spectrum, in Figure 2 we show the result of doubling the number from 16 to 32. It is evident that the energy distribution is essentially the same, although with an increase in the number of states. In the 32-atom case new points appear between the ones already present for $\eta = 16$.

DIATOMIC BRAVAIS LATTICE

A diatomic Bravais lattice is also defined by the vectors $t_n = na$ introduced in the previous section, but now the basis consists of an A atom (white points) located at $P_A = 0$ and a B atom (black points) at $P_B = \frac{1}{2}a$, as shown in the figure below

where again the last A atom coincides with the first one due to the cyclic boundary conditions. We shall consider again an even number η of cells.

The space groups for the monoatomic and diatomic lattices are the same, but the bases are different since we are dealing with different systems. The fact that we are describing a different system is reflected in the isomorphism between $\mathcal{G}$ and the subgroup of the symmetric group $S_{2\eta}$ (instead of S_η)

$$\{E|t_n\} \rightarrow (123,\ldots,2\eta)^{2n} \qquad n = 0,1,\ldots,\eta-1$$

$$\{I|0\} \rightarrow (1,2\eta)(2,2\eta-1)\ldots(\eta,\eta+1) \; . \qquad (IV.1)$$

The isomorphism determines the Hamiltonian for this system, which is established in accordance with ref. [10], as in the previous case. The Hamiltonian takes the form

$$\hat{\mathcal{H}} = E_0 + A_1 \sum_{i=1}^{2\eta} \hat{C}'_{20^i(2)} + \sum_{\omega=0}^{\eta-1} B_{1,2+\omega} \sum_{i=1,3,5,\ldots;j=mod(i+\omega,2\eta)+1}^{2\eta-1} \hat{C}'_{20^{ij}(2)}$$

$$+ \sum_{\omega=0}^{\eta-1} B_{2,3+\omega} \sum_{i=2,4,6,\ldots;j=mod(i+\omega,2\eta)+1}^{2\eta} \hat{C}'_{20^{ij}(2)} + \sum_{\omega=0}^{\eta-1} \lambda_{1,2+\omega} \sum_{i=1,3,5,\ldots;j=mod(i+\omega,2\eta)+1}^{2\eta-1} \hat{\mathcal{M}}_{ij}$$

$$+ \sum_{\omega=0}^{\eta-1} \lambda_{2,3+\omega} \sum_{i=2,4,6,\ldots;j=mod(i+\omega,2\eta)+1}^{2\eta} \hat{\mathcal{M}}_{ij} \; , \qquad (IV.2)$$

with the following conditions in the interaction parameters

$$\begin{aligned} B_{1,2+\omega} &= B_{2,3+\omega} \\ \lambda_{1,2+\omega} &= \lambda_{2,3+\omega} \qquad \omega = 1,3,5,\ldots,\eta-1 \end{aligned} \; .$$

The sums over ω have the same meaning as in (III.6). We note, however, that in this case the bond-bond interactions split into two separate contributions: One that involves a sum over even values of i and the other over the odd values. This is a consequence of the $2n$ power in the isomorphism (IV.1).

The diagonalization of (IV.2) is carried out again in the basis (II.7), although it now takes the form

$$|[n]v_1 \; v_2 \ldots v_{2n}; V > \; , \qquad (IV.3)$$

since the number of degrees of freedom corresponds to twice the number of cells η. The dimension of each matrix block ND, characterized by the total number of quanta, is now given by $ND = (2\eta + V - 1)!/[(2\eta - 1)!V!]$, leading to a higher dimensionality for the matrices to be diagonalized. It is again possible, however, to project the bases for the irred. reps. of $\mathcal{G}$, which for $V = 1$ still leads to analytic expressions for $E(k)$. Using the notation (III.7) we have for (IV.3)

$$|\alpha > \; , \quad \alpha = 1, 2, \ldots, 2\eta \; .$$

We now proceed to project the basis state $|1 >$ over the irred. reps. of $\mathcal{G}$. For a general wave vector, the action of (III.8) over $|1 >$ gives rise to the wave function

$$\Psi_\mu^k = \frac{1}{\sqrt{\eta}} \sum_{\beta=1,3,5}^{2\eta-1} e^{2\pi i k_x \left(\frac{\beta+1}{2} - 1\right)} |\beta > \; . \qquad (IV.4)$$

Note that the sum runs only over β odd. The reason for this is that the sets of functions $\{|\beta >, \; \beta \; even\}$ and $\{|\beta >, \; \beta \; odd\}$ form invariant subspaces under $R \subset J$, which in this case constitutes the relevant group in (III.8), since $\mathbf{D}_{\mu\mu}^k(R) = 0$ for $R \subset \{I|0\}J$. We then project over $|2 >$ in order to obtain the independent set of functions

$$\Phi_\mu^k = \frac{1}{\sqrt{\eta}} \sum_{\alpha=2,4,6\ldots}^{2\eta} e^{2\pi i k_x \left(\frac{\alpha}{2} - 1\right)} |\alpha > \; . \qquad (IV.5)$$

We now turn our attention to the special wave vectors. The projection over the state $|1 >$ gives four wave functions, corresponding to the one dimensional representations of Table I,

$$\Psi^{0,g} = \frac{1}{\sqrt{2\eta}} \sum_{\alpha=1}^{2\eta} |\alpha > \; , \qquad (IV.6a)$$

$$\Psi^{0,u} = \frac{1}{\sqrt{2\eta}} \left\{ \sum_{\alpha=1,3,5\ldots}^{2\eta-1} |\alpha > - \sum_{\alpha=2,4,6,\ldots}^{2\eta} |\alpha > \right\} \; , \qquad (IV.6b)$$

$$\Psi^{\frac{1}{2},g} = \frac{1}{\sqrt{2\eta}} \sum_{\alpha=0}^{\eta/2-1} \left\{ |1 + 4\alpha > + |2\eta - 4\alpha > - |4\alpha + 3 > - |2\eta - 4\alpha - 2 > \right\} \; , \qquad (IV.6c)$$

$$\Psi^{\frac{1}{2},u} = \frac{1}{\sqrt{2\eta}} \sum_{\alpha=0}^{\eta/2-1} \left\{ |1 + 4\alpha > - |2\eta - 4\alpha > - |4\alpha + 3 > + |2\eta - 4\alpha - 2 > \right\} \; . \qquad (IV.6d)$$

The projection shows that for special vectors the Hamiltonian is already diagonal, while for general k vectors a diagonalization of 2×2 matrices is necessary, the energies are thus given by the roots of the determinant

$$\begin{vmatrix} \mathcal{H}_{11} - E & \mathcal{H}_{12} \\ \mathcal{H}_{21} & \mathcal{H}_{22} - E \end{vmatrix} = 0 \; , \qquad (IV.7)$$

where

$$\mathcal{H}_{11} = \; < \Psi_\mu^k |\hat{\mathcal{H}}| \Psi_\mu^k > \quad ,$$

$$\mathcal{H}_{22} = \; < \Phi_\mu^k |\hat{\mathcal{H}}| \Phi_\mu^k > \quad ,$$

$$\mathcal{H}_{12} = \mathcal{H}_{21}^* = \; < \Psi_\mu^k |\hat{\mathcal{H}}| \Phi_\mu^k > \quad . \tag{IV.8}$$

The solution of (IV.7) leads to

$$E_\pm(k) = \frac{H_{11} + H_{22}}{2} \pm \frac{1}{2} \sqrt{(H_{11} - H_{22})^2 + 4H_{12}H_{12}^*} \quad , \tag{IV.9}$$

where the explicit form of the matrix elements (IV.8) is given by

$$H_{11} = \Delta + 8(1 - 2N) \sum_{\omega=1,3,5\ldots}^{\eta-1} B_{1,2+\omega} + 2N \sum_{\omega=1,3,5,\ldots}^{\eta-1} \lambda_{1,2+\omega\ldots}$$

$$- 2N \sum_{\omega=1,3,5,\ldots}^{\eta-1} \lambda_{1,2+\omega} \cos[2\pi k_x(\omega + 1)/2] \quad ,$$

$$H_{22} = \Delta + 8(1 - 2N) \sum_{\omega=1,3,5,\ldots}^{\eta-1} B_{2,3+\omega} + 2N \sum_{\omega=1,3,5,\ldots}^{\eta-1} \lambda_{2,3+\omega}$$

$$- 2N \sum_{\omega=1,3,5}^{\eta-1} \lambda_{2,3+\omega} \cos[2\pi k_x(\omega + 1)/2]$$

$$H_{12} = -N \sum_{\omega=0,2,4,\ldots}^{\eta-2} \left\{ \lambda_{1,2+\omega} e^{2\pi i k_x \frac{\omega}{2}} + \lambda_{2,3+\omega} e^{-2\pi i k_x \left(\frac{\omega+2}{2}\right)} \right\} \quad ,$$

where

$$\Delta = E_0 + 4 \cdot A_1 \cdot (1 - N) + 4(1 - 2N) \sum_{\omega=0,2,4,\ldots}^{\eta-2} (B_{1,2+\omega} + B_{2,3+\omega})$$

$$+ N \sum_{\omega=0,2,4,\ldots}^{\eta-2} (\lambda_{1,2+\omega} + \lambda_{2,3+\omega}) \quad .$$

For special k vectors, the matrix elements of $\hat{\mathcal{H}}$ with respect to (IV.6) lead to the following expressions for the energy

$$E(0, g) = \Gamma \quad ,$$

$$E(0, u) = \Gamma + 2N \sum_{\omega=0,2,4,\ldots}^{\eta-1} (\lambda_{1,2+\omega} + \lambda_{2,3+\omega}) \quad ,$$

$$E\left(\frac{1}{2}, g\right) = \Gamma + 2N \left\{ \sum_{\substack{\omega=0,4,8,12,\ldots \\ 1,5,9,13,\ldots}}^{\eta-1} \lambda_{1,2+\omega} + \sum_{\substack{\omega=1,5,9,\ldots \\ 2,6,10,\ldots}}^{\eta-1} \lambda_{2,3+\omega} \right\} \quad ,$$

$$E\left(\frac{1}{2}, u\right) = \Gamma + 2N \left\{ \sum_{\substack{\omega=1,5,9,\ldots \\ 2,6,10,\ldots}} \lambda_{1,2+\omega} + \sum_{\substack{\omega=0,4,8,\ldots \\ 1,5,9,13,\ldots}} \lambda_{2,3+\omega} \right\} \quad ,$$

where

$$\Gamma = E_0 + 4A_1(1 - N) + 4(1 - 2N) \sum_{\omega=0}^{\eta-1} (B_{1,2+\omega} + B_{2,3+\omega}) \quad .$$

Like in the monoatomic Bravais lattice, the substitution of $k_x = 0$ and $k_x = \frac{1}{2}$ in (IV.9) reproduces the energies presented above for the special k vectors.

For first-neighbors interactions, the energy expression (IV.9) acquires the simple form

$$E_{\pm}(k) = E_0 + 4(1-N)A_1 + 4(1-2N)(B_{12} + B_{23}) + N(\lambda_{12} + \lambda_{23})$$

$$\pm N\sqrt{\lambda_{12}^2 + \lambda_{23}^2 + 2\lambda_{12}\lambda_{23}\cos 2\pi k_x} \quad . \tag{IV.10}$$

This result is similar to the expression obtained for the harmonic oscillator chain. In that case, when two atoms of masses m_1 and m_2 are involved per primitive cell, two branches are present for the frequency $\omega(k)$ of the normal modes:[13]

$$\omega_s(k) = \left\{ 2\gamma \left[\frac{m_1 + m_2}{m_1 m_2} + (-)^s \frac{1}{m_1 m_2} \sqrt{m_1^2 + m_2^2 + 2m_1 m_2 \cos 2\pi k_x} \right] \right\}^{1/2} , \tag{IV.11}$$

where s denotes the branch; when $s = 1$ the branch is called acoustic, while for $s = 2$ it is called optical. Note that, as in the monoatomic chain, the energy of the Morse-

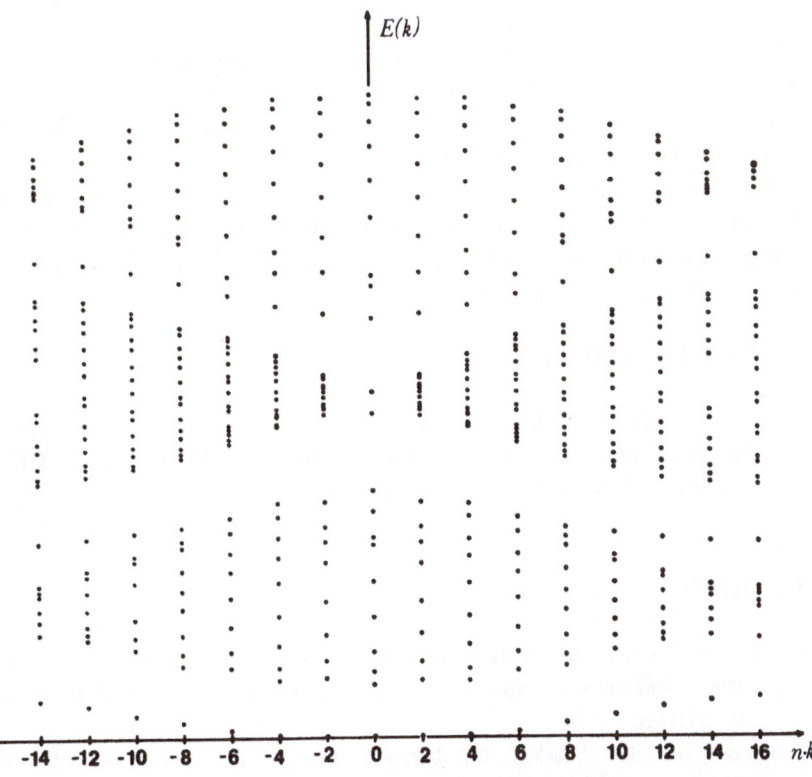

Figure 3. Two phonon excitation energies for a Morse diatomic Bravais lattice, using 32 atoms. The parameters used are $A_1 = -5$, $B_{12} = -2$, $B_{23} = -2.5$, $\lambda_{12} = 0.5$, $\lambda_{23} = 1.0$. All parameters in cm^{-1}. Boson number $N = 50$. Maximum energy value 3088 cm^{-1}. Minimum value 2735 cm^{-1}. One phonon energy $E(0, g)$ is taken to be zero.

potential chain essentially corresponds to the square of the harmonic oscillator case, where the masses play an equivalent role to that of the interaction parameters.

As for the monoatomic system, it is not possible to obtain analytically the energy for excitations of two or more phonons. In this case a numerical calculation is again needed. In figure 3 we present the energy distribution $E(k)$ for two phonon excitations in the first Brillouin zone.

CONCLUSIONS

We have presented an algebraic approach based in the $SU(2)$ group to describe the vibrational modes of a linear Bravais lattice. The method is completely general and may be applied to any linear chain. In this paper we have presented results for monoatomic and diatomic Bravais lattices. Generalization to chains containing more than two types of atoms is strightforward.

A remarkable result of the model is that for one phonon excitations the energy expression $E(k)$ associated to the Morse potential corresponds to the square of the harmonic oscillator energy. This fact implies not only quantitative but important qualitative differences between these systems, particularly at $k = 0$, where the derivative (associated with the group velocity) $\frac{dE}{dk}$ is null. This effect has been predicted by Bruinsma et al. using a very different approach.[14]

In the case of the monoatomic linear chain we studied the effect of the number of atoms for two phonon excitations. It has been shown that increasing the number of atoms does not affect the structure of the spectra. In other words, the density of states remains essentially unaffected.

The algebraic model presented here is capable of describing the stretching modes of anharmonic linear systems. It is possible, however, to generalize the model to include bending modes in the plane by using $U(3)$ groups, which describe Morse potentials in the plane. This work is in progress.

ACKNOWLEDGEMENTS

We are grateful to Prof. F. Iachello for his continuous interest and many suggestions. This work was supported in part by DGAPA, UNAM under project IN101889 and Ministerio de Educación y Ciencia, España.

REFERENCES

1. F. Iachello, *Chem. Phys. Lett.* 78:581 (1981); F. Iachello and R.D. Levine, *J. Chem. Phys.* 77:576 (1982); O.S. van Roosmalen, A.E.L. Dieperink and F. Iachello, *Chem. Phys. Lett.* 85:32 (1982).

2. O.S. van Roosmalen, F. Iachello, R.D. Levine and A.E.L. Dieperink, *J. Chem. Phys.* 79:2515 (1983); J. Hornos and F. Iachello, *J. Clhem. Phys.* 90:5284 (1989); F. Iachello, S. Oss and R. Lemus, *J. Mol. Spectrosc.* 146:l56 (1991); ibidem, 149:132 (1991).

3. F. Iachello and S. Oss, *J. Mol. Spectrosc.* 142:85 (1990).

4. A. Frank, F. Iachello and R. Lemus, *Chem. Phys. Lett.* 131:380 (1986); A. Frank, R. Lemus and F. Iachello, *J. Chem. Phys.* 91:29 (1989); R. Lemus and A. Frank, *Ann. of Phys.* 206:122 (1991); A. Frank, R. Lemus and F. Iachello, "Algebraic Model for Molecular Electronic Spectra", Symmetries in Science V, Edited by B. Gruber et al. Plenum Press, N.Y. 1991; R. Lemus, A. Leviatan and A. Frank, *Chem. Phys. Lett.* 194:327 (1992); R. Lemus and A. Frank, *J. Chem. Phys.* "Constrained Calculations in the Electron-Vibron Model and the Born-Oppenheimer Approximation", *J. Chem. Phys.*, in press.

5. S. Levit and V. Smilansky, *Nuclear Physics* A389:56 (1982).

6. M. Berrondo and A. Palma, *J. Phys. A: Math. Gen.* 13:773 (1980); Y. Alhassid, F. Gürsey and F. Iachello, *Ann. of Phys.* 148:346 (1983).

7. C.E. Wulfman and R.D. Levine, *Chem. Phys. Lett.* 97:361 (1983).

8. O.S. van Roosmalen, I. Benjamin and R.D. Levine, *J. Chem. Phys.* 81:5986 (1984).

9. F. Iachello and S. Oss, *Phys. Rev. Lett.* 66:2976 (1991);
   F. Iachello and S. Oss, *Chem. Phys. Lett.* 187:500 (1991).

10. A. Frank and R. Lemus, *Phys. Rev. Lett.* 68:413 (1992).

11. S.L. Altmann, "Induced Representations in Crystals and Molecules Point, Space and Nonrigid Molecule Groups", Academic Press, 1977.

12. R. Lemus and A. Frank, "Algebraic Description of a Linear Chain of n-coupled Anharmonic Oscillators", to be published.

13. A.S. Davidov, Solid State Theory. Mir. Moscow. 1981;
    Neil W. Ashcroft and N. David Mermin, Solid State Physics. Holt-Saunders International Editions 1976.

14. R. Bruinsma, K. Maki and J. Wheatley, *Phys. Rev. Lett.* 57:1773 (1986).

QUADRATIC ALGEBRAS IN QUANTUM MECHANICS

Pascal Létourneau and Luc Vinet

Laboratoire de Physique Nucléaire
Université de Montréal
Montréal, Canada H3C 3J7

INTRODUCTION

Dynamical symmetries prove extremely useful and have applications in many physical situations[1]. Customarily, these symmetries are described mathematically in terms of Lie groups and algebras. It has been appreciated howewer that more general structures can sometimes be required. A celebrated example is that of Lie superalgebras; these arise when supersymmetries are encountered. Another type of algebras, that of quantum groups, is also being recognized as potentially relevant to describe dynamical symmetries, especially in the context of integrable models. These quantum groups belong to the class of quadratic algebras which are defined by subjecting the generators to quadratic relations. The purpose of this paper is to provide a simple example where the symmetries are most naturally discussed using such a quadratic algebra.

We shall consider a two-dimensional anisotropic harmonic oscillator with a 2 to 1 frequency ratio. This problem is known to admit accidental degeneracies that can be explained with the help of Lie algebraic techniques [2,3]. It is first shown that the associated Halmiltonian possesses two constants of motion that are quadratic in the momenta. A realization of the su(2) algebra is then constructed using these conserved quantities. Since the dimensions of the unitary representations of su(2) coincide with the degree of the degeneracies, these are thus accounted for. This resolution is not completely satisfactory howewer. The reason is that under commutation the constants of motion do not actually form a Lie algebra; one needs in fact, to divide these quantities by the square root of certain operators to obtain closure. As we shall explain, such difficulties do not show up if a quadratic algebra framework is used instead of the Lie algebraic one. It therefore seems that quadratic (and polynomial) algebras could be useful tools to deal with dynamical symmetries. Such a claim has been made also in Refs. [4-6], where other interesting examples are discussed.

We shall also make another interesting observation from this analysis of the anisotropic harmonic oscillator. Recently, quasi-exactly solvable problems have attracted a lot of attention. (For a review see Ref. [7].) These quantum systems are characterized by the fact that part of their spectra can be obtained algebraically. This is made possible by the existence of a hidden symmetry algebra generated by first

Symmetries in Science VII, Edited by B. Gruber
and T. Otsuka, Plenum Press, New York, 1994

order differential operators. Hamiltonians that can be expressed as bilinear combinations of these operators will leave invariant any finite-dimensional representation space of the symmetry algebra and as a result, the corresponding part of their spectra can be computed through a finite-matrix eigenvalue problem. In one dimension, more specifically, one uses the su(2) realization

$$T^0 = -j + z\frac{\partial}{\partial z}, \qquad T^+ = 2jz - z^2\frac{\partial}{\partial z}, \qquad T^- = \frac{\partial}{\partial z},$$
$$[T^0, T^\pm] = \pm T^\pm, \qquad [T^+, T^-] = 2T^0. \tag{1}$$

which for j as emi-integer, entails a $(2j+1)$-dimensional representation over the space spanned by the monomials $1, z, \ldots, z^{2j}$. One then takes for Hamiltonians

$$\tilde{H} = \sum_{a,b=0,\pm} C_{ab}T^aT^b + \sum_{a=0,\pm} C_aT^a. \tag{2}$$

Through a similarity transformation and possibly, a change of variable $x = x(z)$, these Hamiltonians can be cast in the form $-\frac{1}{2}\frac{d^2}{dx^2} + V(x)$ and a list of quasi-exactly solvable Schrödinger operators is thus obtained. We may remark that exactly solvable systems result if the parameter j does not appear in the Hamiltonian. This happens for instance in the case of the Morse potential $V(x) = A(e^{-2ax} - 2e^{-x})$ which plays an important rôle in molecular physics. The simplest example of a quasi-exactly solvable problem in one dimension is provided by the anharmonic oscillator with potential $V(x) = \frac{1}{2}\omega^2x^6 - 2\beta\omega^2x^4 + [2\beta^2\omega^2 - \omega(4j + \frac{3}{2})]x^2$. This system, as it turns out, is related to the two-dimensional anisotropic oscillator with Hamiltonian $H = -\frac{1}{2}(p_1^2 + p_2^2) + \frac{1}{2}\omega^2[x_1^2 + 4(x_2 - \beta)^2]$, $p_i = -i\frac{\partial}{\partial x_i}$ $i = 1, 2$. Indeed, owing to the higher symmetry of the Hamiltonian, the Schrödinger equation $H\psi(x_1, x_2) = E\psi(x_1, x_2)$ separates in parabolic coordinates as well in Cartesian coordinates[3]. We shall see that when parabolic coordinates are used, the separated equations coincide exactly with the one-dimensional equation associated to the anharmonic oscillator with potential $V(x)$.

We take great pleasure in dedicating this paper to Franco Iachello who has so brilliantly applied algebraic methods to domains as diverse as nuclear, high energy, atomic, and molecular physics. We hope that he will find interesting the occurrence of quadratic algebras in quantum mechanics. Let us also recall that Professor Iachello himself studied (among others) the dynamical symmetries of the Morse Hamiltonian[8]. He has shown that these can be traced back to the su(2) symmetry of the two-dimensional isotropic harmonic oscillator Schrödinger equation. In this case, one separates the variables in polar coordinates and after a change of variables, identifies the radial part as the Schrödinger equation for the particle in the one-dimensional Morse potential. It is in close analogy with this approach that the relation between the quasi-exactly anharmonic oscillator and the two-dimensional anisotropic oscillator with a 2 to 1 frequency ratio is established here.

The remainder of the paper is organized as follows. We first set our notation and identify the accidental degenaracies of a generalized 2-d anisotropic oscillator. We then present the quadratic dynamical algebra and show how the dynamics is resolved by constructing the representations of this algebra. We indicate in particular how the degeneracies are explained in this framework. Finally, we make the connection between the 2-d anisotropic oscillator and the quasi-exactly solvable anharmonic oscillator. Conclusions follow.

THE GENERALIZED 2-D ANISOTROPIC OSCILLATOR

We shall be considering a two-dimensional quantum system with Hamiltonian

$$H = H_1 + H_2 \,, \tag{3}$$

where

$$H_1 = \frac{1}{2}(p_1^2 + \omega^2 x_1^2) + \frac{1}{8}(\alpha^2 - 1)\frac{1}{x_1^2} \,, \tag{4a}$$

$$H_2 = \frac{1}{2}\left[(p_2^2 + 4\omega^2(x_2 - \beta)^2\right] \,, \tag{4b}$$

$p_i = -i\frac{\partial}{\partial x_i}$, $i = 1, 2$, and $\alpha, \beta, \omega \in \mathbb{R}$. In the special case $\alpha = 1$, H governs the dynamics of an anisotropic oscillator with a 2 to 1 frequency ratio.

It will be convenient to introduce the annihilation and creation operators

$$a_1 = \frac{1}{\sqrt{2\omega}}(\omega x_1 + ip_1) \,, \qquad a_1^\dagger = \frac{1}{\sqrt{2\omega}}(\omega x_1 - ip_1) \,, \tag{5a}$$

$$a_2 = \sqrt{2\omega}\left[(x_2 - \beta) + \frac{i}{2\omega}p_2)\right] \,, \qquad a_2^\dagger = \sqrt{2\omega}\left[(x_2 - \beta) - \frac{i}{2\omega}p_2)\right] \,, \tag{5b}$$

that satisfy the commutation relations

$$[a_k, a_l^\dagger] = k\delta_{kl} \qquad [a_k, a_l] = [a_k^\dagger, a_l^\dagger] = 0 \quad k, l = 1, 2 \,. \tag{6}$$

In terms of these operators,

$$\frac{H_1}{\omega} = a_1^\dagger a_1 + \frac{1}{4}(\alpha^2 - 1)(a_1 + a_1^\dagger)^{-2} + \frac{1}{2} \,, \tag{7a}$$

$$\frac{H_2}{\omega} = a_2^\dagger a_2 + 1 \,. \tag{7b}$$

The spectrum of $H = H_1 + H_2$ is easily obtained by separating the Schrödinger equation in Cartesian coordinates. With $\psi(x_1, x_2) = \phi(x_1)\chi(x_2)$, the equation $H\psi = E\psi$ amounts to $H_1\phi(x_1) = E_1\phi(x_1)$ and $H_2\chi(x_2) = E_2\chi(x_2)$ with $E = E_1 + E_2$. We recognized in H_2, the Hamiltonian of a 1-d harmonic oscillator; its eigenfunctions are of the form $< x_2|n_2 > \propto < x_2|(a_2^\dagger)^{n_2}|0 >$ and have energy $E_2 = 2n_2 + 1$. The eigenvalues of H_1 can be determined by exploiting the conformal symmetry of this operator. Indeed, define the operators

$$B_1^+ = (a_1^\dagger)^2 - \frac{1}{4}(\alpha^2 - 1)(a_1 + a_1^\dagger)^{-2}$$

$$= a_1^\dagger(a_1 + a_1^\dagger) - \frac{H_1}{\omega} + \frac{1}{2} \tag{8a}$$

$$B_1^- = (B_1^+)^\dagger \,. \tag{8b}$$

Clearly,

$$[H_1, B_1^\pm] = \pm 2\omega B_1^\pm \,. \tag{9}$$

It is also straightforward to check that

$$B_1^\pm B_1^\mp = \left[(\frac{H_1}{\omega} \mp 1) + \frac{|\alpha|}{2}\right]\left[(H_1 \mp 1) - \frac{\alpha}{2}\right] \,, \tag{10}$$

from where it follows that

$$[B_1^+, B_1^-] = -4\frac{H_1}{\omega}. \tag{11}$$

One immediatly identifies (9) and (11) as the commutation relations of $so(2,1)$. The relevant representations of this algebra are easily constructed. We known that H_1 is bounded below. There must therefore be states annihilated by B_1^- from which the eigenstates of H_1 are to be obtained by repeated action with B_1^+; correspondingly, the energy eigenvalue will raise above the minimal values in steps of 2ω. One thus sets

$$H_1|E_1> = E_{E_1}|E_1>, \tag{12a}$$

$$B_1^+|E_1> = s_{E_1+2\omega}|E_1 + 2\omega>, \tag{12b}$$

$$B_1^-|E_1> = s_{E_1}|E_1 - 2\omega>. \tag{12c}$$

From (12), one has $B_1^+ B_1^-|E_1> = s_{E_1}^2|E_1>$ and with the help of (10) one obtains

$$s_{E_1}^2 = \left(\frac{E_1}{\omega} - 1 + \frac{|\alpha|}{2}\right)\left(\frac{E_1}{\omega} - 1 - \frac{|\alpha|}{2}\right). \tag{13}$$

The lowest energies are the zeros of $s_{E_1}^2$ and following our discussion the spectrum of H_1 is found to be

$$E_1 = \left(2n_1 + 1 \pm \frac{|\alpha|}{2}\right)\omega \qquad n_1 = 0, 1, \dots \tag{14}$$

Note that the set $(2n_1 + 1 - |\alpha|/2)\omega$ is admissible only if $0 \le |\alpha| < 2$, otherwise there would be states with non-positive norms. When $\alpha = 1$, the two series in (14) combine to give the standard oscillator spectrum $E_1 = (n_1 + \frac{1}{2})\omega$, $n_1 = 0, 1, \dots$ The total energies are thus

$$E = E_1 + E_2 = \left[2(n_1 + n_2 + 1) \pm \frac{|\alpha|}{2}\right]\omega$$
$$= (2N \pm \frac{|\alpha|}{2})\omega \qquad n_1, n_2 = 0, 1, \dots, \quad N = 1, 2, \dots \tag{15}$$

Since the integer N is made out of the sum of the two integer-valued quantum numbers n_1 and n_2, accidental degeneracies occur and their multiplicity is in fact given by N. We shall now show that a dynamical quadratic algebra can be used to explain these degeneracies.

THE DYNAMICAL QUADRATIC ALGEBRA

Consider the operators

$$D = H_1 - H_2, \tag{16a}$$

$$C_+ = \omega B_1^+ a_2, \tag{16b}$$

$$C_- = C_+^\dagger = \omega B_1^- a_2^\dagger. \tag{16c}$$

It is easy to see that these quantities are conserved. Together with the Hamiltonian H and the identity operator I, these operators (16) generate the symmetry algebra of the anisotropic oscillator. The stucture relations are straightfowardly determined to be:

$$[H, D] = [H, C_+] = [H, C_-] = [H, I] = 0 \,, \tag{17a}$$

$$[D, I] = [C_+, I] = [C_-, I] = 0 \,, \tag{17b}$$

$$[D, C_\pm] = \pm 4\omega C_\pm \,, \tag{17c}$$

$$[C_+, C_-] = \frac{3}{2} D^2 + DH - \frac{1}{2} H^2 + \left(2 - \frac{\alpha^2}{2} \right) \omega^2 I \,. \tag{17d}$$

We note in (17d) that the commutator of C_+ and C_- is given as a quadratic combination of D and H. The operators H, D, C_+, C_- and I do not therefore define a Lie algebra. In a representation on an energy eigenstate, H is constant; it may then not be taken as a generator inequivalent to I. Since we are dealing with differential operators, the Jacobi identity is satisfied. It is also not difficult to see that this algebra possesses a Casimir operator K commuting with all the generators. Explicitly,

$$K = \frac{\omega}{2} \{C_+, C_-\} + \frac{1}{8} D^3 + \frac{1}{8} HD^2 - \left[\frac{1}{8} H^2 + (\frac{\alpha^2}{8} - \frac{3}{2})\omega^2 \right] D \,. \tag{18}$$

Of course since H is central, we might equivalently take as Casimir operator

$$\tilde{K} = K - \frac{1}{8} H[H^2 - (4 - \alpha^2)\omega^2] \,. \tag{19}$$

It is checked that $\tilde{K} = 0$ in our specific realization. Clearly the operators $C_\pm$ transform the degenerate eigenstates of H among themselves. In fact, the representations of the quadratic algebra (17) that we need, can be easily worked out; they will be seen to determine the spectrum of H and to completly account for its degeneracies.

Define the basis vectors $|E, d>$ as the simultaneous eigenstates of H and D:

$$H|E, d> = E|E, d>, \qquad D|E, d> = d|E, d> \,. \tag{20}$$

As far as the Schrödinger equation is concerned, separation of variables in Cartesian coordinates is associated to this choice: using still $\psi(x_1, x_2) = \phi(x_1)\chi(x_2)$, $D\psi(x_1, x_2) = d\psi(x_1, x_2)$ with $d = E_1 - E_2$.

In view of (17c) and the fact that $C_+ = C_-^\dagger$, we must have

$$C_+|E, d> = t_{d+4\omega}|E, d + 4\omega> \,, \tag{21a}$$

$$C_-|E, d> = t_d|E, d - 4\omega> \,. \tag{21b}$$

It follows of course that $C_+ C_-|E, d> = t_d^2|E, d>$. Now with the help of (17d), (18) and (19), we find that

$$C_+ C_- = \frac{1}{\omega} \tilde{K} + \frac{1}{8\omega}(H + D - 2\omega + \omega|\alpha|)(H + D - 2\omega - \omega|\alpha|)(H - D + 2\omega) \,. \tag{22}$$

We have already pointed out that in our realization $\tilde{K} = 0$, we thus get here

$$t_d^2 = \frac{1}{8\omega}(d - d_+)(d - d_-)(\bar{d} - d) \,, \tag{23}$$

where

$$d_{\pm} = -E + \omega(2 \pm |\alpha|), \qquad (24a)$$

$$\tilde{d} = E + 2\omega. \qquad (24b)$$

(Naturally, this result could have been obtained by working directly with the expressions (16) and using (7b) and (10).) At this point, two requirements on the representations of the symmetry algebra allow to determine the spectrum of D and H. First, these representations should be unitary and hence we should have $t_d^2 \geq 0$. Second, they should be finite-dimensional as there are only finitely many degenerate energy eigenstates. This last condition implies that there should be D-eigenvalues d_{min} and d_{max}, such that $C_-|E, d_{min}\rangle = 0$, $C_+|E, d_{max}\rangle = 0$ or equivalently such that $t_{d_{min}} = 0$, $t_{d_{max}+4\omega} = 0$. The eigenvalues of D would thus run from d_{min} to d_{max} in steps of 4ω. A first series meeting the above two critera has

$$d = d_+, d_+ + 4\omega, \dots, \tilde{d} - 4\omega. \qquad (25)$$

If we denote by N $(= 1, 2, \dots)$ the number of degenerate H-eigenstates and hence, the dimension of the representations, we have

$$(\tilde{d} - 4\omega) - d_+ = 4(N-1)\omega. \qquad (26)$$

Using the definitions (24), we see that (26) entails the energy spectrum formula:

$$E = (2N + \frac{|\alpha|}{2})\omega \qquad N = 1, 2, \dots \qquad (27)$$

If $0 \leq |\alpha| < 2$, another series is present:

$$d = d_-, d_- + 4\omega, \dots, \tilde{d} - 4\omega. \qquad (28)$$

Note that t_d^2 is also non-negative here because the gap $-2|\alpha|\omega$ between d_- and d_+ does not dominate the increment 4ω. As before, if there are N eigenstates of D with the eigenvalues (28), they are seen to have energy

$$E = \left(2N - \frac{|\alpha|}{2}\right)\omega \qquad N = 1, 2, \dots \qquad (29)$$

This is how the quadratic symmetry algebra (17) enables one to obtain the spectrum of H - compare (15) with (27) and (29).

SEPARATION OF VARIABLES AND THE QUASI-EXACTLY SOLVABLE ANHARMONIC OSCILLATOR

For most known systems with accidental degeneracies and hidden symmetries, the Schrödinger equation admits seperation in more than one coordinate systems. The present anisotropic oscillator is no exception. In two dimensions, there is a correspondance between coordinate systems in which variables separate in the Schrödinger equation and the constants of motions that are quadratic in momenta[3]. So far with our problem, we have only exploited the fact that $H\psi = E\psi$ separates in Cartesian coordinates. We have indicated that this coordinate system is singled out when the constant of motion $D = \frac{1}{2}(p_1^2 - p_2^2) + \frac{1}{2}\omega^2 x_1^2 + \frac{1}{8}(\alpha^2 - 1)\frac{1}{x_1^2} - 2\omega^2(x_2 - \beta)^2$ is diagonalized. Consider now the following hermitian combination of the symmetry generators of H:

$$R = \sqrt{2\omega}(C_+ + C_-) - 2\beta(H + D). \tag{30}$$

This conserved quantity is also quadratic in the momenta; in fact, in Cartesian coordinates, it takes the form

$$R = p_1 L + L p_1 + 2\omega^2 x_1(x_2 - 2\beta) + \frac{1}{2}(1 - \alpha^2)\frac{x_2}{x_1^2}, \tag{31}$$

where $L = x_1 p_2 - x_2 p_1$. Diagonalizing R instead of D leads to separation in the parabolic coordinates (ξ_+, ξ_-) that are related to the Cartesian ones as follows

$$x_1 = \xi_+ \xi_- \qquad x_2 = \frac{1}{2}(\xi_+^2 - \xi_-^2). \tag{32}$$

From (30) we see that R is tridiagonal in the bases $|E, d_n^\pm>$, $d_n^\pm = d_\pm + 4n\omega$, $n = 0, 1, \ldots, N - 1$:

$$R|E, d_n^\pm> = \sqrt{2\omega}(t_{d_{n+1}^\pm}|E, d_{n+1}^\pm> + t_{d_n^\pm}|E, d_{n-1}^\pm>) - 2\beta(E + d_n^\pm)|E, d_n^\pm>. \tag{33}$$

Its eigenvalues $\omega\Lambda$ are given by the zeroes of the characteristic polynomial of Nth order associated to this matrix. Let $|E, \Lambda>$ be the corresponding eigenstates

$$R|E, \Lambda> = \omega\Lambda|E, \Lambda>. \tag{34}$$

A three-term recurrence relation for the overlap function $< E, \Lambda|E, d_n^\pm>$ is simply obtained in our algebraic framework. Take

$$< E, \Lambda|E, d_n^\pm> = < E, \Lambda|E, d^\pm> P_n^\pm(\Lambda). \tag{35}$$

Clearly $P_0^\pm(\Lambda) = 1$. From (33) and (34), the coefficients $P_n^\pm(\Lambda)$ are seen to satisfy

$$\omega\Lambda P_n^\pm(\Lambda) = \sqrt{2\omega}(t_{d_{n+1}^\pm} P_{n+1}^\pm(\Lambda) + (t_{d_n^\pm} P_{n-1}^\pm(\Lambda)) - 2\beta(E + d)P_n^\pm(\Lambda). \tag{36}$$

If we now set

$$P_n^\pm(\Lambda) = (2\omega)^{n/2}(\prod_{i=1}^{n} t_{d_i^\pm})^{-1} Q_n^\pm(\Lambda) \qquad n = 1, \ldots, N - 1 \tag{37}$$

and use the definitions (24) as well as $E = (2N \pm \frac{|\alpha|}{2})\omega$, we find that the factors $Q_n^\pm(\Lambda)$ obey the three-term recurrence relation

$$Q_{n+1}^\pm(\Lambda) = [\frac{\Lambda}{2} + \beta(4n + 2 \pm \alpha)]Q_n^\pm(\Lambda) - 4\omega n(n \pm \frac{|\alpha|}{2})(N - n)Q_{n-1}^\pm(\Lambda) \tag{38}$$

$$n = 0, 1, \ldots, N - 1; \qquad Q_0^\pm(\Lambda) = 1.$$

These functions of Λ, thus define orthogonal polynomial sets. (This is true for the $Q_n^-(\Lambda)$ as long as $|\alpha| < 2$.) Let us now make explicit our claim that a quasi-exactly solvable anharmonic oscillator descents from the 2-d anisotropic oscillator. In the parabolic coordinates (32), H and R take the form

$$H = \frac{1}{2}(\xi_+^2 + \xi_-^2)^{-1} \times [\pi_+^2 + \pi_-^2 + \omega^2(\xi_+^6 + \xi_-^6) - 4\beta\omega^2(\xi_+^4 - \xi_-^4)$$

$$+ 4\beta^2\omega^2(\xi_+^2 + \xi_-^2) + \frac{1}{4}(\alpha^2 - 1)(\xi_+^{-2} + \xi_-^{-2})], \tag{39a}$$

$$R = (\xi_+^{-2} + \xi_-^{-2})^{-1}$$

$$\times [\xi_+^{-2}\pi_+^2 - \xi_-^{-2}\pi_-^2 + \omega^2(\xi_+^4 - \xi_-^4) - 4\beta\omega^2(\xi_+^2 + \xi_-^2) + \frac{1}{4}(\alpha^2 - 1)(\xi_+^{-4} - \xi_-^{-4})], \tag{39b}$$

where $\pi_\pm = -i\frac{\partial}{\partial\xi_\pm}$. If we set

$$< \xi_+, \xi_- | E, \Lambda > = \psi_{E,\Lambda}(\xi_+, \xi_-) = u_\Lambda^{(+)}(\xi_+)u_\Lambda^{(-)}(\xi_-), \tag{40}$$

we readily see that the eigenvalue equations $H\psi_{E,\Lambda} = E\psi_{E,\Lambda}$ and $R\psi_{E,\Lambda} = \omega\Lambda\psi_{E,\Lambda}$ amount to the following two separated equations for the functions $u_\Lambda^{(\pm)}(\xi_\pm)$:

$$\left[\frac{1}{2}\pi_\pm^2 + V_E(\xi_\pm)\right]u_\Lambda^{(\pm)} = \pm\omega\Lambda u_\Lambda^{(\pm)}, \tag{41}$$

where

$$V_E(\xi_\pm) = \frac{1}{2}\omega^2\xi_\pm^6 + \frac{1}{8}(\alpha^2 - 1)\frac{1}{\xi_\pm^2} - 2\beta\omega^2\xi_\pm^4 + (2\beta^2\omega^2 - E)\xi_\pm^2. \tag{42}$$

We recognize that the equations (41) have the form of one-dimensional Schrödinger equations in a (generalized) oscillator potential with anharmonicity terms in ξ^4 and ξ^6. (Note that $E = (2N \pm \frac{|\alpha|}{2})\omega$ enters as a parameter in the potential.) Remarkably these equations coincide with the Schrödinger equation associated to the simplest example of a quasi-exactly solvable system. To see this, return to the realization given in (1) of the $su(2)$ algebra. If we make the change of variable $z = \xi^2$ and take

$$H = e^{-a}\tilde{H}e^a, \tag{43}$$

with

$$\tilde{H} = -2T^0T^- - (N + 1 \pm |\alpha|)T^- - 4\beta\omega T^0 - 2\omega T^+ \tag{44a}$$

and

$$a = \omega\frac{\xi^4}{4} - \beta\omega\xi^2 - \frac{1}{2}(1 \pm |\alpha|)\ln\xi, \tag{44b}$$

we find after a short computation that

$$H = \frac{1}{2}\pi^2 + V_E(\xi) + \beta\omega\left(E \pm \frac{|\alpha|}{2}\right) \tag{45}$$

for $\pi = -i\frac{\partial}{\partial\xi}$. Up to a constant this agrees with (41) – (42). So, if one is considering the one-dimensional Schrödinger equation

$$[\frac{1}{2}\pi^2 + V_E(\xi)]u(\xi) = \epsilon u(\xi) \tag{46}$$

with $E = (2N \pm \frac{|\alpha|}{2})\omega$, the eigenvalues and eigenfunctions of the N lowest levels can be obtained exactly. This is done as follows using the connection between (46) and the 2-d anisotropic oscillator. On the one hand, the first N energies ϵ are given by the eigenvalues of R. On the other hand, the two-dimensional wave functions $\psi_{E,d_n^\pm}(x_1, x_2) = < x_1, x_2 | E, d_n^\pm >$ are very easily determined in Cartesian coordinates. In terms of the overlap coefficients $< E, \Lambda | E, d_n^\pm >$ for which we have a recursion relation, the eigenfunctions of R are given by

$$< x_1, x_2 | E, \Lambda > = \sum_{n=0}^{N-1} < E, \Lambda | E, d_n^\pm > < x_1, x_2 | E, d_n^\pm > . \qquad (47)$$

One then passes to the parabolic coordinate system (32), in which we known that separation occurs and write $< x_1, x_2 | E, \Lambda > = u^{(+)}(\xi_+) u^{(-)}(\xi_-)$ to identify $u^{(+)}(\xi)$ as the wave function associated to the energy level $\epsilon = \omega\Lambda$. For an exposition of how these same eigenvalues and eigenfunctions are found by exploiting the realization $(43) - (44)$ of H see Ref. [7].

CONCLUSIONS

Summarizing, we have provided here a novel algebraic analysis of the much studied[2,3,9] 2-dimensional anisotropic harmonic oscillator(3)-(4). We have illustrated with this very simple example that quadratic algebras do arise naturally in the description of systems with accidental degeneracies. We hope to have shown that the prejudice according to which symmetry algebras should be Lie algebras or superalgebras is not well founded. We have also explained how quasi-exactly solvable systems could be obtained through dimensional reduction by exploiting the fact that (solvable) systems with higher symmetries usually separate in more than one coordinate system.

For two dimensions, there exists a classification of potentials with accidental degeneracies for which the Schrödinger equation separates in two coordinate systems[3]. In addition to the one that we considered in detail, the list consists of

1. $V(x_1, x_2) = \alpha(x_1^2 + x_2^2) + \dfrac{\beta}{x_1^2} + \dfrac{\gamma}{x_2^2}$

2. $V(\xi_+, \xi_-) = \dfrac{\alpha + \beta/\xi_+^2 + \gamma/\xi_-^2}{\xi_+^2 + \xi_-^2}$

3. $V(\xi_+, \xi_-) = \dfrac{\alpha + \beta\xi_+ + \gamma\xi_-}{\xi_+^2 + \xi_-^2}$

In all of these cases, we have shown that the symmetries responsible for the accidental degeneracies are again naturally described by quadratic algebras. These results will be reported in full in a forthcoming publication[10].

ACKNOWLEDGMENTS

We thank Jean LeTourneux for discussions. One of us (P. L.) gratefully acknowledges receiving a NSERC University Undergraduate Student Research Award. This work is supported in part through funds provided by the Natural Sciences and Engineering Council (NSERC) of Canada and the Fonds FCAR of Québec.

REFERENCES

1. *Dynamical Groups and Spectrum generating Algebras*, vol. 1 and 2, Bohm, A., Barut, A. and Ne'eman, Y., eds. (World Scientific, Singapore, 1988).

2. Demkov, Yu., The definition of the symmetry group of a quantum system. The anisotropic oscillator, Soviet Phys. JETP **17**, 1349-1351 (1963).

3. Winternitz, P., Smorodinskii.Ya. A., Uhlir, M. and Fris, I., Symmetry groups in classical and quantum mechanics, Soviet J. Nucl. Phys. **4**, 444-450 (1967).

4. Gal'bert, O. F., Granovskii, Ya. I. and Zhedanov, A. S., Dynamical symmetry of anisotropic singular oscillator, Phys. Lett. **A 153**, 177-180 (1991); Granovskii, Ya. I., Zhedanov A. S. and Lutzenko, I. M., Quadratic algebra as a "hidden" symmetry of the Hartmann potential, J. Phys. **A 24**, 3887-3894 (1991).

5. Higgs, P. W., Dynamical symmetries in a spherical geometry I, J. Phys. **A 12**, 309-323 (1979); Zhedanov, A. S., The "Higgs algebra" as a quantum deformation of su(2), Mod. Phys. Lett **A 7**, 507-512 (1992).

6. Granovskii, Ya. I., Lutzenko, I. M., and Zhedanov A. S., Mutual integrability, quadratic algebras and dynamical symmetry, Ann. Phys. **217**, 1-20 (1992).

7. Shifman, M. A., New findings in quantum mechanics (partial algebraization of the spectral problem), Int. J. Mod. Phys. A **4**, 2897-2952 (1989).

8. Alhassid, Y., Gürsey, F. and Iachello F., Group theory approach to scattering, Ann. Phys. **148**, 346-380 (1983).

9. Duimo, F. and Zambotti, G., Dynamical group of the anisotropic harmonic oscillator, Nuovo Cimento A **43**, 1203-1207 (1966); Cisneros, A. and McIntosh, H. V., Search for universal symmetry group in two dimensions, J. Math. Phys. **11**, 870-895; Moshinsky, M., Patera, J. and Winternitz, P., Canonical transformation and accidental degeneracy. III A unified approach to the problem, J. Math. Phys. **16**, 82-92 (1975).

10. Létourneau, P. and Vinet, L., Quadratic algebras in two-dimensional quantum mechanics, in preparation.

PARTIAL DYNAMICAL SYMMETRIES

Amiram Leviatan

Racah Institute of Physics, The Hebrew University
Jerusalem 91904, Israel

INTRODUCTION

The pioneering work of F. Iachello and his colleagues on the interacting boson model (IBM) of nuclei[1] has given a great impetus to the use of algebraic techniques in diverse areas of physics. Notable examples are the use of spectrum generating algebras to construct models for describing spectra and decays in nuclei,[1-2] molecules[3] and hadrons[4]. The extensive use of group theory makes such models an efficient computational tool for tractable yet detailed calculations of observables. In addition, the algebraic approach has the capacity to emphasize and illuminate the role played by underlying symmetries in determining properties of complex many-body systems. In the present contribution we show that the combined computational simplicity and symmetry structure embedded in algebraic models such as the IBM, make them an ideal testing ground for further theoretical developments of new concepts of symmetry.

We begin by introducing a novel concept of symmetry which we call "partial dynamical symmetry" and motivate the reasons for exploring it. As the name suggests, we are interested in dynamical systems for which a particular symmetry is obeyed by only a subset of states in the Hilbert space. We clarify the concept via an example, using a specific family of IBM Hamiltonians exhibiting partial dynamical SU(3) symmetry. This example will then serve as a guideline for proposing a general algorithm how to construct Hamiltonians with partial dynamical symmetry for any semi-simple group. We note that, though we are considering a generalization of partial dynamical symmetry, the construction presented below can also be used to generalize the concept of symmetry to that of partial symmetry. The work[5] reported here was done in collaboration with Y. Alhassid (Yale University).

EXACT, DYNAMICAL AND PARTIAL SYMMETRIES

Symmetry plays an important role in solving for the eigenstates of the Hamiltonian. For an exact symmetry to occur, the Hamiltonian is required to commute with all the generators of the symmetry group. Consequently, the Hamiltonian admits a block structure in a basis labeled by the irreducible representations (irreps) of the symmetry group, so that inequivalent irreps do not mix. Furthermore, eigenstates which belong to the same irrep of the symmetry group are degenerate.

Another symmetry concept which has been used extensively in physics is that of a dynamical symmetry[1,6]. It is a situation in which the Hamiltonian is written in terms of the Casimir operators of a chain of subgroups

$$G_0 \supset \ldots \supset G \supset \ldots \supset G' \quad . \tag{1}$$

When a dynamical symmetry occurs, the following features are observed: (i) The labels of irreps of the groups in the chain serve as quantum numbers to classify the eigenstates of the Hamiltonian. This group-theoretical classification allows the use of the Wigner Eckart theorem to obtain selection rules and to facilitate evaluation of matrix elements. (ii) The wave functions, eigenvalues and other observables (e.g. transition rates) are known analytically. (iii) The structure of the wave-functions is dictated by symmetry and is therefore independent of the Hamiltonian's parameters (i.e. independent of the coefficients of the various Casimir operators forming the Hamiltonian). Only the last group, G', in the chain (1) is a symmetry group of the Hamiltonian. An intermediate group in the chain, say G, does not leave the Hamiltonian invariant, so that eigenstates which belong to an irrep of G are usually not degenerate. However, in a dynamical symmetry the Hamiltonian commutes with the Casimir invariants of the groups in the chain, so that states which belong to different irreps are not mixed. The Hamiltonian still has a block structure in a basis characterized by the irreps of the groups.

In applications of group theoretical methods to realistic systems, one often finds that the assumed symmetry is only approximate and is fulfilled by only some of the states but not by others. For example, certain degeneracies implied by the symmetry are not always realized in nature. To conform with the experimental data, one is therefore compelled to break the symmetry. Under such circumstances one usually expects that all eigenstates will now be mixed and that none of the virtues of a dynamical symmetry would be retained. The question we wish to investigate is whether it is possible to break the symmetry but in a special way so that some of the eigenstates of the Hamiltonian (but not all) would still exhibit the previously mentioned properties (i) to (iii) of a dynamical symmetry. We refer to such a symmetry structure as "partial dynamical symmetry". An Hamiltonian with the above property is not invariant under the group G, nor does it commute with the Casimir invariants of G, so that various irreps are in general mixed in its eigenstates. However, there is a subset of eigenstates for which no such mixing occurs. These states can still be labeled by irreps of G, and their eigenvalues and wave functions are known analytically. It is by no means obvious that such Hamiltonians exist nor how to construct them. In the next section we will use the interacting boson model to gain insights on these questions.

PARTIAL DYNAMICAL SYMMETRY IN THE IBM

The interacting boson model (IBM) has been used extensively to describe properties of quadrupole collective states in a wide range of nuclei[7]. The building blocks of the model are one monopole boson ($s^\dagger$) and five quadrupole bosons ($d_\mu^\dagger$). The model has $U(6)$ as a spectrum generating algebra for which the Hamiltonian is rotational invariant and conserves the total number of bosons N. One of the possible dynamical symmetries

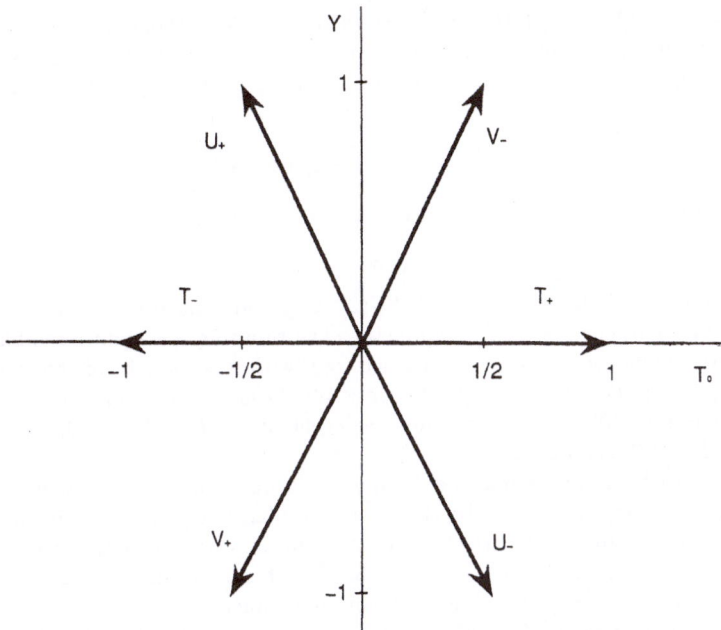

Figure 1. The Cartan-Weyl basis of generators of $SU(3)$. Their expressions in terms of the multipoles of eq. (3) are: $Y = -(2\sqrt{2}/3)Q_0^{(2)}$, $T_0 = -(1/2)L_0^{(1)}$, $T_- = -(2/\sqrt{3})Q_2^{(2)}$, $U_- = -\sqrt{2/3}\,Q_{-1}^{(2)} + (1/2)L_{-1}^{(1)}$, $V_+ = -\sqrt{2/3}\,Q_1^{(2)} - (1/2)L_1^{(1)}$, $T_+ = (T_-)^\dagger$, $U_+ = (U_-)^\dagger$, $V_- = (V_+)^\dagger$.

of the model corresponds to the chain[1]

$$
\begin{array}{ccccc}
U(6) & \supset & SU(3) & \supset & O(3) \\
{[N]} & & (\lambda, \mu)\ (K) & & L
\end{array}
\tag{2}
$$

which describes rotations and vibrations in axially deformed nuclei. Thus the groups G and G' of eq. (1) are $SU(3)$ and $O(3)$, respectively. The labels of irreps are indicated below each group in the chain. (The $O(2)$ subgroup with its associated label M of the projection of the angular momentum L, does not play a role for an $O(3)$ scalar Hamiltonian. K is an additional label needed for a complete classification of states. In a geometric description of nuclei it corresponds to the projection of the angular momentum on the symmetry axis).

The generators of the $SU(3)$ group are the quadrupole $Q^{(2)}$ and angular momentum $L^{(1)}$ operators given by

$$
Q_\mu^{(2)} = d_\mu^\dagger s + s^\dagger \tilde{d}_\mu - \frac{1}{2}\sqrt{7}\,(d^\dagger \tilde{d})_\mu^{(2)} \quad ; \quad L_\mu^{(1)} = \sqrt{10}(d^\dagger \tilde{d})_\mu^{(1)} \quad .
\tag{3}
$$

Here $\tilde{d}_\mu = (-)^\mu d_{-\mu}$ and standard notation of angular momentum coupling is used. An alternative Cartan-Weyl basis for the $SU(3)$ generators is displayed in Fig. 1.

In case of an exact $SU(3)$ symmetry, the Hamiltonian can be written as

$$
H(SU(3)) = h_0\left[-\hat{C} + 2\hat{N}(2\hat{N} + 3)\right]
\tag{4}
$$

where $\hat{C} = 2Q^{(2)} \cdot Q^{(2)} + (3/4)L^{(1)} \cdot L^{(1)}$ is the quadratic Casimir operator of $SU(3)$ and $\hat{N}$ is the total boson number operator which has a constant value N for any N- boson state. The eigenvalues of $H(SU(3))$ are given in terms of the (λ, μ) labels of irreps of $SU(3)$

$$E_{SU(3)} = h_0\left[-(\lambda^2 + \mu^2 + \lambda\mu + 3\lambda + 3\mu) + 2N(2N+3)\right] \quad . \tag{5}$$

The allowed values of $\lambda \geq 0$, $\mu \geq 0$ are $(2N - 4k - 6m, 2k)$ with m, k non-negative integers.

Adding the Casimir operator of $O(3)$ ($\vec{L}^2$) to the Hamiltonian in (4), produces an $L(L+1)$ splitting of the states in each $SU(3)$ irrep and transforms the exact $SU(3)$ symmetry into a dynamical $SU(3)$ symmetry. The resulting spectrum resembles that of an axially-symmetric quadrupole-deformed rotor with rotational bands built on each $SU(3)$ irrep (λ, μ). Thus the irrep $(2N, 0)$ constitute the ground state band ($K = 0$; $L = 0, 2, 4\ldots$) and the $(2N - 4, 2)$ irrep comprise degenerate β ($K = 0$; $L = 0, 2, 4\ldots$) and γ ($K = 2$; $L = 2, 3, 4\ldots$) bands.

Since in actual deformed nuclei the β and γ bands are not degenerate[7], one is obliged to break the $SU(3)$ symmetry. To carry out the plan outlined in the introduction to examine the possibility of partial symmetries, we may wish to retain good $SU(3)$ character for members of the ground state band, but not for all members of excited bands. This may be accomplished in the following manner.

The ground state band of an $SU(3)$ nucleus is represented in the IBM by the following condensate of N bosons[8]

$$|c; N > = (N!)^{-1/2}[(s^\dagger + \sqrt{2}d_0^\dagger)/\sqrt{3}]^N|0 > \quad . \tag{6}$$

where $|0 >$ is the bare vacuum (no bosons). The above deformed intrinsic state is known[8,9] to transform as the lowest weight state in the $(2N, 0)$ irrep of $SU(3)$ (i.e. it is annihilated by the operators T_-, U_- and V_+ in Fig. 1). Members of the ground state band with good angular momentum L span the $(2N, 0)$ irrep and can be obtained from $|c; N >$ by standard angular momentum projection. If we now find an IBM $O(3)$ scalar Hamiltonian, which is not an $SU(3)$ scalar, but still has $|c; N >$ as an eigenstate, this would fulfill the requirements of a partial $SU(3)$ symmetry. Fortunately such Hamiltonians exist and were already encountered in a study of the intrinsic structure of the interacting boson model of nuclei[10]. They have the generic form

$$H = h_0 P_0^\dagger P_0 + h_2 \sum_\mu P_{2,\mu}^\dagger P_{2,\mu} , \tag{7}$$

where h_0, h_2 are arbitrary constants. The boson pair operators of angular momentum $L = 0$ and 2 transform as $(0, 2)$ under $SU(3)$ and are defined as

$$P_0^\dagger = d^\dagger \cdot d^\dagger - 2(s^\dagger)^2 ,$$
$$P_{2,\mu}^\dagger = \sqrt{2}\, s^\dagger d_\mu^\dagger + \sqrt{\frac{7}{2}}(d^\dagger d^\dagger)_\mu^{(2)} \quad . \tag{8}$$

It is possible to show that for $h_2 = 2h_0$, The Hamiltonian in (7) reduces to that in (4), $H = H(SU(3))$, and thus becomes an $SU(3)$ scalar. For $h_2 = -2h_0/5$, H in (7) is a $(2, 2)$ $SU(3)$ tensor component. For arbitrary h_0, h_2 coefficients H is therefore not an $SU(3)$ scalar, nevertheless it always has $|c; N >$ (which has good $SU(3)$ $(2N, 0)$ species) as an exact zero-energy eigenstate. This property is a direct outcome of the following relations observed by the bosons pair operators

$$P_{L,\mu}|c; N > = 0 \qquad L = 0, 2 . \tag{9}$$

Since H in (7) is an $O(3)$ scalar, it follows that states of good L projected from $|c; N >$ are also zero energy eigenstates which span the $(2N, 0)$ representation. Thus when the coefficients h_0, h_2 are positive, H (7) becomes positive definite by construction, and even though it is not an $SU(3)$ scalar, it still has an exactly degenerate zero-energy ground state band whose rotational members possess good $SU(3)$ symmetry. Even more surprising is the presence (to be shown below) of additional eigenstates in excited bands of H (7) with good $SU(3)$ character. Further examination shows that the boson pair operators (8) satisfy also the following relations for single commutators acting on the condensate

$$\left[P_{L,\mu}, P_{2,2}^\dagger\right]|c; N > \; = \; \delta_{L,2}\delta_{\mu,2}(6N + 9)|c; N > \tag{10}$$

and the following operator- identities for double commutators

$$\left[\left[P_{L,\mu}, P_{2,2}^\dagger\right], P_{2,2}^\dagger\right] \; = \; \delta_{L,2}\delta_{\mu,2}\, 12 P_{2,2}^\dagger \tag{11}$$

By combining relations (9)-(11) we immediately find that the sequence of states

$$|k > \propto \left(P_{2,2}^\dagger\right)^k |c; N > \quad , \tag{12}$$

are eigenstates of H (7) with eigenvalues

$$E_k \; = \; 3h_2\left(2N + 1 + 2k\right)k \tag{13}$$

Relations (9)-(11) hold for any number of bosons in the condensate. If we want to consider states with a fixed total boson number N we can take in (12) a condensate with $N - 2k$ bosons. The resulting states would still be eigenstates of H (7) but now with eigenvalues $E_k = 3h_2(2N + 1 - 2k)k$. By comparing the latter eigen energies with the $SU(3)$ eigenvalues (5), and recalling that $H(h_2 = 2h_0) = H(SU(3))$, it is easy to verify that these $|k >$ states are in the $SU(3)$ irreps $(2N - 4k, 2k)$ with $2k \leq N$. It can be further shown that they are lowest weight states in these representations.

The indicated $|k >$ states are deformed and in the nuclear physics terminology[9] are referred to as "intrinsic states" associated with rotational bands of an axially deformed nucleus. They have well defined angular momentum projection $(K = 2k)$ along the symmetry axis and represent γ^k bands. In particular, $|k = 0 >$ represents the ground-state band $(K = 0)$ and $|k = 1 >$ is the γ-band $(K = 2)$. The intrinsic states break the $O(3)$ symmetry but since the Hamiltonian in (7) is an $O(3)$ scalar, the projected states $|(2N - 4k, 2k) K = 2k; L, M >$, with good $L \geq K$ have good $SU(3)$ symmetry and are also eigenstates of H. It should be noted that for $k \neq 0$ the states with good $SU(3)$ and good L projected from $|k >$ span only part of the corresponding $SU(3)$ irreps. There are other states originally in these irreps which do not preserve the $SU(3)$ symmetry and therefore get mixed. This situation corresponds precisely to that of partial $SU(3)$ symmetry. An Hamiltonian H which is not an $SU(3)$ scalar has a subset of eigenstates which continue to have good $SU(3)$ symmetry. All of the above discussion is applicable also to the case when we add to the Hamiltonian (7) the Casimir operator of $O(3)$, and by doing so converting the partial $SU(3)$ symmetry into partial dynamical $SU(3)$ symmetry. The additional $\vec{L}^2$ term contributes just an $L(L + 1)$ splitting but does not affect the wave functions.

An example of this $SU(3)$ partial dynamical symmetry is displayed in Fig. 2 for $N = 7$ bosons. Typical spectra of the Hamiltonian (7) are shown for two cases: (a) $h_2 = 2h_0$ and (b) $h_2 = 1.25h_0$. As noted above, case (a) corresponds to a full $SU(3)$ symmetry

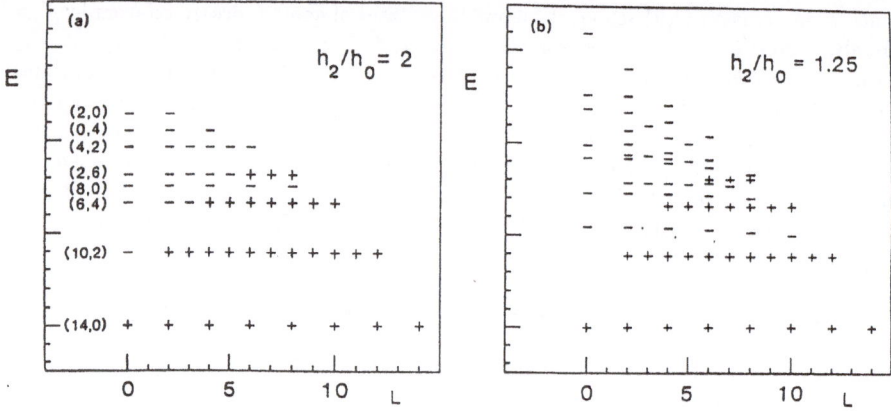

Figure 2. Spectra (energy E vs. angular momentum L) of the Hamiltonian (7): (a) for $h_2/h_0 = 2$. $SU(3)$ symmetry labels (λ, μ) are shown on the left. Some levels with $L \neq 0$ exhibit multiplicity which is not shown. (b) For $h_2/h_0 = 1.25$. Levels which continue to exhibit $SU(3)$ symmetry are marked by a $(+)$ symbol. The energy scale is arbitrary and the same value of h_2 was used in both cases.

and all states are arranged in degenerate $SU(3)$ multiplets. For case (b) the $SU(3)$ symmetry is broken, so that most of the eigenstates have a mixture of $SU(3)$ irreps and the above $SU(3)$ degeneracy is lifted. However, some of the eigenstates (marked by a $+$ in Fig. 2) continue to carry good $SU(3) \supset O(3)$ representation labels and are arranged in multiplets. Within each such multiplet the L degeneracy can be lifted by adding the Casimir invariant of $O(3)$ ($\vec{L}^2$) to the Hamiltonian (7), contributing an $L(L+1)$ splitting. The $SU(3)$ symmetry of case (a) then becomes an $SU(3)$ dynamical symmetry, with rotational bands built on each of the $SU(3)$ irreps. In case (b) we obtain an $SU(3)$ partial dynamical symmetry. The ground state $K = 0$ band continues to carry good $SU(3)$ labels $(14, 0)$. The $\beta(K = 0)$ and $\gamma(K = 2)$ bands, which originally were degenerate in the $SU(3)$ dynamical symmetry limit (and belonging to the $(10, 2)$ representation), split. Only the members of the γ-band continue to carry good $SU(3)$ labels $(10, 2)$, while the β-band has a mixture of several $SU(3)$ irreps. Similarly, the $\gamma^2(K = 4)$ band and the $\gamma^3(K = 6)$ band preserve their $SU(3)$ character. All other states exhibit mixing.

To further visualize the phenomena of partial dynamical symmetry, we show in Fig. 3 the $SU(3)$ content of three eigenstates of case (b). These are the seventh and eighth $L = 6$ states and the fourth $L = 0$ state, all of which have a dominant $(2, 6)$ component. The state $L = 6_7$ is one of the solvable states in the $\gamma^3(K = 6)$ band and is 100% in the $(2, 6)$ irrep. The state $L = 6_8$ which in case (a) was a pure $(2, 6)$ state, degenerate with $L = 6_7$, is now a mixture of several $SU(3)$ representations and only 73% $(2, 6)$. Other states are even more strongly mixed; the fifth $L = 0$ state in case (a) which belonged to the irrep $(2, 6)$, had crossed the fourth $L = 0$ level to become the $L = 0_4$ level of case (b). It has only 51% in the irrep $(2, 6)$ and exhibits a significant spread over six $SU(3)$ irreps.

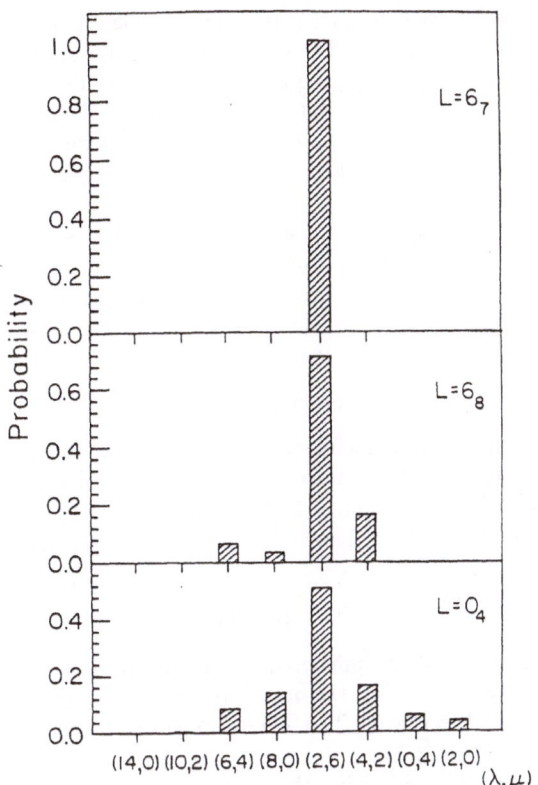

Figure 3. $SU(3)$ decomposition (probability vs. $SU(3)$ irrep (λ, μ)) for selected states in the spectrum of Fig. 2 (b).

AN ALGORITHM FOR CONSTRUCTING PARTIAL SYMMETRIES

The previous example of $SU(3)$ partial dynamical symmetry in the IBM, provides the necessary clues how to formulate a systematic procedure for constructing Hamiltonians with partial dynamical symmetry G, for any semi-simple group G, which still have G' ($G \supset G'$) as their symmetry group. For simplicity we take $G' \equiv O(3)$.

A semi-simple algebra G (of rank ℓ) can be described[11] in terms of its Cartan-Weyl basis composed of maximally commuting subset of generators H_i ($i = 1, \ldots, \ell$) and ladder operators E_α. The vector $\alpha = (\alpha_1, \ldots, \alpha_\ell)$ is called a root and is defined by $\left[H_i, E_\alpha \right] = \alpha_i E_\alpha$. An irrep of G can be described in a basis of common eigenstates $|\lambda >$ of the H_i's such that $H_i |\lambda > = \lambda_i |\lambda >$. The vector $\lambda = (\lambda_1, \ldots, \lambda_\ell)$ is called a weight. A weight is positive or negative according to the sign of its first non-zero component. The highest weight Λ in an irrep is the weight for which $\Lambda - \lambda$ is positive for all other weights λ. It is used to characterize and label the corresponding irrep. The representation which is conjugate to $[\Lambda]$ is denoted by $[\Lambda]^*$. Its weights are obtained by reversing the sign of the weights of $[\Lambda]$, so that $[\Lambda]^*$ can be characterized by its lowest weight $-\Lambda$. In this context, $SU(3)$ is a semi-simple group of rank $\ell = 2$. Its Cartan-Weyl basis is shown in Fig. 1. The two commuting H_i's are $H_1 = (\sqrt{3}/2)Y$ and $H_2 = T_0$, Y being the hypercharge, T_0 the isospin projection. There are three positive roots: $\alpha = (0,1)$, $\beta = (\sqrt{3}/2, -1/2)$, $\alpha + \beta = (\sqrt{3}/2, 1/2)$ and the corresponding ladder operators are $E_\alpha = T_+/\sqrt{2}$, $E_\beta = U_+/\sqrt{2}$, $E_{\alpha+\beta} = V_-/\sqrt{2}$ expressed in terms of T-, U-,V- spin components as defined in Fig. 1. The maximal weight of an irrep (λ, μ) of $SU(3)$ is $\Lambda = \big((\lambda + 2\mu)/2\sqrt{3}, \lambda/2\big)$.

The key elements in the IBM example of partial dynamical $SU(3)$ symmetry, were the vacuum state (6) and the relations (9)-(11). Accordingly, in order to generalize the formalism to an arbitrary semi-simple group G, we choose the building blocks of our construction to be: (i) A vacuum state $|vac>$, which is assumed to be a lowest weight vector $-\Lambda_0$ in the irrep $[\Lambda_0]^*$ of G and therefore satisfies

$$E_{-\alpha}|vac> = 0 \qquad \text{for all positive roots } \alpha \quad . \tag{14}$$

(ii) The components $T_{[\Lambda]\lambda}$ of an irreducible tensor operator under the group G belonging to the representation $[\Lambda]$. Henceforth, to keep the notation simple, we shall denote $T_{[\Lambda]\lambda}$ by T_λ. The highest weight component of T is then T_Λ. Note that $T_\Lambda^\dagger$ transform according to the irrep $[\Lambda]^*$, and that $T_\Lambda^\dagger$ is a lowest weight component in that representation satisfying

$$\left[E_{-\alpha}, T_\Lambda^\dagger\right] = 0 \qquad \text{for all } \alpha > 0 \quad . \tag{15}$$

In the IBM $SU(3)$ example of the previous section, the vacuum state $|vac>$ is the $SU(3)$ boson condensate $|c; N>$ (6), the operators P_0 and $P_{2,\mu}$ (8) transform like the irrep $(2,0)$ and correspond to the T_λ. The highest weight component is $T_\Lambda = P_{2,2}$ with the maximal weight $\Lambda = (1/\sqrt{3}, 1)$.

Our algorithm is based on requiring the vacuum $|vac>$, and the highest weight component T_Λ, to satisfy the following conditions

$$T_\Lambda|vac> = 0 \quad ; \quad \left[T_\Lambda, T_\Lambda^\dagger\right]|vac> = a|vac> \quad ;$$
$$\left[\left[T_\Lambda, T_\Lambda^\dagger\right], T_\Lambda^\dagger\right] = bT_\Lambda^\dagger \quad , \tag{16}$$

where a, b are constants. In the IBM $SU(3)$ example such conditions are satisfied with $T_\Lambda^\dagger = P_{2,2}^\dagger$ (which is the lowest weight component in the irrep $(0,2)$) and $a = 6N + 9$, $b = 12$.

Using the properties of the algebra and of the tensor operator it is possible to show that the following relations result from (16)

$$\left[E_{-\alpha}, T_\Lambda\right]|vac> = 0 \quad ; \quad \left[\left[E_{-\alpha}, T_\Lambda\right], T_\Lambda^\dagger\right]|vac> = 0 \quad ;$$
$$\left[\left[\left[E_{-\alpha}, T_\Lambda\right], T_\Lambda^\dagger\right], T_\Lambda^\dagger\right] = 0 , \tag{17}$$

for any positive root α. To get the first relation we use (14), while the other two relations are obtained by using the Jacobi identity once and twice respectively, together with properties (14) and (15). Eqs. (17) are similarly satisfied when $\left[E_{-\alpha}, T_\Lambda\right]$ is replaced by $\left[E_{-\gamma}, \ldots, \left[E_{-\beta}, \left[E_{-\alpha}, T_\Lambda\right]\right] \ldots\right]$ for any positive roots $\alpha, \beta, \ldots, \gamma$. Since any weight vector can be obtained from the highest one by a linear combination of such repetitive applications, it follows that

$$T_\lambda|vac> = 0 \quad ; \quad \left[T_\lambda, T_\Lambda^\dagger\right]|vac> = 0 \quad ;$$
$$\left[\left[T_\lambda, T_\Lambda^\dagger\right], T_\Lambda^\dagger\right] = 0 . \tag{18}$$

for all $\lambda \neq \Lambda$.

Eqs. (16) and (18) are the counterpart of the IBM $SU(3)$ eqs. (9)-(11). They can be viewed as a generalization of the conditions satisfied by harmonic oscillators operators. Note, however, that the first two relations in (16) and (18) are not an operator identity and that no further assumption is made on the vacuum state nor on the operator T_Λ, which may be composite objects (as is indeed the case for the condensate (6) and the boson pair operators (8) in the IBM $SU(3)$ example discussed before).

In analogy to eq. (12) we consider the sequence of states

$$|k> \propto \left(T_\Lambda^\dagger\right)^k |vac> \ . \tag{19}$$

Due to (16) these are eigenstates of $T_\Lambda^\dagger T_\Lambda$, and because of (18) they have the property that $T_\lambda |k>= 0$ for any $\lambda \neq \Lambda$. We conclude that $|k>$ are exact eigenstates of any Hamiltonian of the form

$$H = h_{\Lambda\Lambda} T_\Lambda^\dagger T_\Lambda + \sum_{\lambda,\sigma\neq\Lambda} h_{\lambda\sigma} T_\lambda^\dagger T_\sigma \ , \tag{20}$$

with eigenvalues

$$E_k = h_{\Lambda\Lambda}\left[ka + \frac{1}{2}k(k-1)b\right] \ , \tag{21}$$

that are independent of the parameters $h_{\lambda\sigma}$. Eq. (21) is the counterpart of the IBM $SU(3)$ expression (13). It is straightforward to show that $|k>$ is the lowest weight vector in the representation $[\Lambda_k]^* \equiv [\Lambda_0 + k\Lambda]^*$ of G. Since for a general set of coefficients $h_{\lambda\sigma}$, the Hamiltonian (20) is not a G scalar, all of its eigenstates, except $|k>$, will be a mixture of irreps of G. We have thus accomplished our initial goal, except that the Hamiltonian (20) is generally not rotational (or G') invariant. To remedy that, we decompose the tensor operator T_Λ into rotational tensor operators T_{LM} of good angular momentum L and projection M. We can then construct rotational scalar Hamiltonians

$$H = \sum_{LM} h_L T_{LM}^\dagger T_{LM} \ , \tag{22}$$

where we have have omitted possible multiplicity indices of the representations. If in addition we assume that the highest weight vector Λ has a well defined, multiplicity free, $\bar{L}$-value in the decomposition to $O(3)$, then (22) is of the form (20) with $h_{\Lambda\Lambda} = h_L$. $|k>$ are therefore eigenstates of the scalar Hamiltonian (22). When $\bar{L} \neq 0$ and/or $|vac>$ does not have zero angular momentum, the states $|k>$ do not have good L. The $O(3)$ symmetry is thus spontaneously broken in the sense that the eigenstates $|k>$ do not have the same symmetry as the Hamiltonian. States of good angular momentum can be obtained by projection from $|k>$. Since H is a scalar, the projection operator commutes with it, so that the projected states $|[\Lambda_k] L, M>$ are also eigenstates of (22). A special case for (22) may be a Casimir invariant of G for which a dynamical symmetry occurs. However, with arbitrary coefficients h_L, a generic Hamiltonian in (22) is not a G scalar, yet a subset of its eigenstates $|[\Lambda_k] L, M>$ are characterized by good quantum numbers of $G \supset O(3)$. It should be noted that these special states belong to particular irreps of G (of the form $[\Lambda_0 + k\Lambda]^*$), and in general they span only part of these irreps i.e. other states, originally in these special irreps, no longer have good G symmetry. We also note that the above discussion holds for any compact group G' (replacing $O(3)$), since appropriate projection operators can be constructed in such a case.

SUMMARY AND CONCLUSIONS

In this contribution we have used the interacting boson model of Iachello and coworkers to explore a new symmetry construction of partial dynamical symmetry. Such type of symmetry occurs when an Hamiltonian is not invariant under a symmetry group but a subset of its eigenstates do have good symmetry. Next a general algorithm was presented for constructing Hamiltonians which possess this property. The algorithm is phrased in terms of algebraic relations and as such is applicable to any semi-simple group and does not rely on any particular realizations of the algebra (e.g. bosons, fermions etc.). It is remarkable that this novel type of symmetry is observed in Hamiltonians which have been used to describe band-structure and spectra in nuclei[10,12]. Thus an empirical phenomenon, which occurs naturally in nuclear physics has been raised in this work to the level of an exact, general concept.

The partial dynamical symmetry discussed in this work is exact. In that sense it differs from approximate partial dynamical symmetries that were noticed elsewhere[13]. As explicitly shown, one of the striking features of Hamiltonians having exact partial dynamical symmetry, is the possibility that the special, exact eigenstates that retain the good symmetry, span only part of the corresponding irreps. This is not the case in the quasi-exactly solvable Hamiltonians constructed recently[14], in which the solvable states form complete representations. The coexistence of solvable and unsolvable states, together with the availability of an algorithm, distinguish the notion of partial dynamical symmetry from the notion of accidental degeneracy[15], where all levels are arranged in degenerate multiplets.

The concept of partial dynamical symmetry may play an important role in exploring systems which exhibit both regular and irregular (chaotic) behavior, and in constructing models of complex systems in which some of the elementary excitations are described by solvable states. The present contribution is a vivid demonstration of the ability of algebraic models such as the IBM to provide a proper environment for nurturing new concepts of symmetry.

ACKNOWLEDGEMENTS

It is a pleasure and a privilege for me to dedicate this contribution to F. Iachello in the occasion of his 50th birthday. His approach to physics, advocating a persistent search for guiding symmetry principles has provided a decisive impact and inspiration for the ideas discussed here.

REFERENCES

[1] F. Iachello and A. Arima, "The Interacting Boson Model" (Cambridge Univ. Press, Cambridge, 1987).

[2] F. Iachello and P. Van Isacker, "The Interacting Boson-Fermion Model" (Cambridge Univ. Press, Cambridge, 1991).

[3] F. Iachello, Int. J. Quantum Chem. **41**, 77 (1992).

[4] F. Iachello, Nucl. Phys. **A518**, 173 (1990).

[5] Y. Alhassid and A. Leviatan, J. Phys. A., in press.

[6] For a general review see "Dynamical Groups and Spectrum Generating Algebras", Eds. A. Bohm, Y. Néeman and A.O. Barut, World Scientific 1988.

[7] R. F. Casten and D. D. Warner, Rev. Mod. Phys. **60**, 389 (1988).

[8] J.N. Ginocchio and M.W. Kirson, Nucl. Phys. **A350**, 31 (1980).

[9] H.T. Chen and A. Arima, Phys. Rev. Lett. **51**, 447 (1983).

[10] A. Leviatan, Z. Phys. **A321**, 467 (1985);
M.W. Kirson and A. Leviatan, Phys. Rev. Lett. **55**, 2846 (1985);
A. Leviatan, Ann. Phys. **179**, 201 (1987).

[11] R. Gilmore, "Lie Groups, Lie Algebras and some of their Applications", Wiley, New York, 1974.

[12] A. Leviatan, J.N. Ginocchio and M.W. Kirson, Phys. Rev. Lett. **65**, 2853 (1990).

[13] Y. Alhassid, E. Hinds and D. Meschede, Phys. Rev. Lett. **59**, 1545 (1987).

[14] A.V. Turbiner, Sov. Phys. JETP **67**, 230 (1988);
M.A. Schifman, Int. J. Mod. Phys. **4**, 2897 (1989).

[15] M. Moshinsky and C. Quesne, Ann. Phys. **148**, 462 (1983);
H. Ui and G. Takeda, Prog. Theor. Phys. **72**, 266 (1984);
M. Moshinsky, C. Quesne and G. Loyola, Ann. Phys. **198**, 103 (1990).

GROUP O(4) IN THE NUCLEAR IBA MODEL

P. O. Lipas

Institute for Nuclear Theory, University of Washington
Seattle, WA 98195, USA
and
Department of Physics, University of Jyväskylä
SF-40351 Jyväskylä, Finland

ABSTRACT

The group O(4) is established as a link in the group chains of the interacting-boson model (IBA) of nuclear structure. It is obtained as a subgroup of O(5) by excluding the $m = 0$ component of the d boson. Since only the third component of the angular momentum is then conserved, O(4) symmetry is limited to intrinsic states. No O(3) can be sandwiched between this O(4) and the physical O(2). Branching rules are derived for groups U(4), O(4) and O(2). The dynamic-symmetry contributions of O(4) and U(4) to the energy are given. The structure $O(4) \approx O(3) \times O(3) \supset O(3)$ is established.

1. INTRODUCTION

The interacting-boson approximation/model (IBA/IBM) of nuclear structure, initiated by Iachello and Arima in 1974 [1] and thoroughly described in their 1987 book [2], owes its appeal in large measure to the exploitation of group theory. The basic model, IBA-1, builds nuclear states from identical bosons which have six states of good angular momentum: s $(J = 0)$ and d_M $(J = 2, M = \pm 2, \pm 1, 0)$, collectively called b_i $(i = 1, 2, \ldots, 6)$. The Hamiltonian and other physical operators of the model are constructed from the bilinear creation and annihilation operators $b_i^\dagger b_j$.

The IBA-1 model has U(6) as its spectrum-generating algebra, which is spanned by the 36 boson-number-conserving operators $b_i^\dagger b_j$. The physically relevant subgroup chains are

$$U(6) \supset U(5) \supset O(5) \supset O(3) \supset O(2), \tag{1}$$

Symmetries in Science VII, Edited by B. Gruber
and T. Otsuka, Plenum Press, New York, 1994

$$U(6) \supset SU(3) \supset O(3) \supset O(2),\tag{2}$$

$$U(6) \supset O(6) \supset O(5) \supset O(3) \supset O(2),\tag{3}$$

where $O(3) \supset O(2)$ refers to the physical angular momentum and its third component. The special cases of the model where the Hamiltonian can be expressed as a linear combination of the Casimir operators belonging to any one of the chains (1)–(3) constitute the dynamic symmetries of the model. They are commonly named after the first link. Thus the IBA Hamiltonian with dynamic symmetry U(5) has the form

$$H = \epsilon C_1(U5) + \alpha C_2(U5) + \beta C_2(O5) + \gamma C_2(O3),\tag{4}$$

where the C_n are linear and quadratic Casimir operators; $C_{n>2}$ do not occur because only two-boson interactions are assumed. Moreover, U(6) would only contribute an additive constant for a given nucleus characterized by a fixed N, and $C_2(O2)$ would only occur in an anisotropic environment. The eigenstates of the Hamiltonian (4) are labeled by the appropriate quantum numbers as $|[N]n_d \tau \nu_\Delta J M\rangle$, where the Young-pattern notation $[N]$ means that the states are fully symmetric under exchange of bosons. The eigenvalues are

$$E = \epsilon n_d + \alpha n_d(n_d + 4) + 2\beta\tau(\tau + 3) + 2\gamma J(J + 1).\tag{5}$$

For details see e.g. refs. [2, 3]. The other two dynamic symmetries SU(3) and O(6) of the model yield analogous simple analytic solutions. The three dynamic symmetries are identified as representing three basic types of nuclear collective motion: U(5) ↔ spherical vibrator, SU(3) ↔ deformed rotor–vibrator, O(6) ↔ γ-unstable nucleus.

One may note that the group O(4) does not appear in the chains (1) and (3), although one would expect to have it as a subgroup of O(5). This is due to the requirement that the O(3) representing *physical angular momentum* be included in the chains. The possible role of O(4) in the *nuclear* IBA context has not been considered beyond recognizing its incompatibility with conserved angular momentum [4]. However, in Iachello's group-theoretic *molecular* model, the so-called vibron model, the group O(4) plays a prominent role [5]. It arises specifically from the U(4) spectrum-generating algebra realized in terms of bosons designated as σ ($J = 0$) and π_μ ($J = 1$, $\mu = \pm 1, 0$). For completeness, it is appropriate to remark that, quite apart from any boson models, O(4) is the symmetry of the bound states of the $1/r$ potential, as is extensively discussed by Gilmore [6].

The present study addresses the possible role of O(4) in the nuclear IBA model based on U(6). It turns out that O(4) *can* play a role when the angular-momentum requirement is relaxed so that only *axial* symmetry rather than full spherical symmetry prevails. This situation is relevant to the physical model in its alternative formulation in terms of an *intrinsic system* [2], from which states of good angular momentum can be obtained by projection.

In sect. 2 below, I examine the O(5) algebra of the IBA and define from it a physically reasonable O(4) algebra. The O(4) Casimir operator and the associated dynamic symmetries are discussed in sect. 3. In sect. 4 the present O(4) is cast into the isomorphic form O(3) × O(3), and sect. 5 is a conclusion.

2. THE IBA LIE ALGEBRAS OF O(5) AND O(4)

The generic, uncoupled boson operators commute according to

$$[b_i, b_j^\dagger] = \delta_{ij} , \quad [b_i, b_j] = 0 = [b_i^\dagger, b_j^\dagger] . \tag{6}$$

The generators $b_i^\dagger b_j$ close under commutation according to

$$[b_i^\dagger b_j, b_k^\dagger b_l] = \delta_{jk} b_i^\dagger b_l - \delta_{il} b_k^\dagger b_j . \tag{7}$$

When $i = 1, 2, \ldots, n$, this is the Lie algebra of U(n). Any linearly independent n^2 linear combinations of the $b_i^\dagger b_j$ generate the same algebra. Incidentally, the same Lie algebra (7) would also be defined by analogous *fermion* operators [3].

The IBA Lie algebra is normally expressed in terms of boson generators which are angular-momentum tensors. The generic tensorial boson operators are $b_{lm}^\dagger$ and $\tilde{b}_{lm} \equiv (-)^{l-m} b_{l,-m}$ with commutation relations

$$[\tilde{b}_{lm}, b_{\lambda\mu}^\dagger] = (-)^{l-m} \delta_{l\lambda} \delta_{m,-\mu} , \quad [\tilde{b}_{lm}, \tilde{b}_{\lambda\mu}] = 0 = [b_{lm}^\dagger, b_{\lambda\mu}^\dagger] . \tag{8}$$

The tensorial generators are then

$$[b_l^\dagger \tilde{b}_\lambda]_{LM} \equiv \sum_{m\mu} (lm\lambda\mu|LM) \, b_{lm}^\dagger \tilde{b}_{\lambda\mu} , \tag{9}$$

where $(lm\lambda\mu|LM)$ is an O(3) Clebsch–Gordan coefficient. The number of coupled generators is the same as that of the uncoupled ones; thus, if the number of the states b_{lm} is n, the number of generators is n^2 and the algebra is U(n). The Lie algebra, though equivalent to that of eq. (7), has a more complicated appearance [3]:

$$\begin{aligned}
[[b_{\lambda_1}^\dagger \tilde{b}_{\lambda_2}]_{l_1 m_1}, [b_{\lambda_3}^\dagger \tilde{b}_{\lambda_4}]_{l_2 m_2}] = \sum_{lm} & \sqrt{(2l_1 + 1)(2l_2 + 1)} \, (l_1 m_1 l_2 m_2 | lm) \\
\times \Big[& \delta_{\lambda_2\lambda_3} (-)^{\lambda_1 - \lambda_4 + l} \begin{Bmatrix} l_1 & l_2 & l \\ \lambda_4 & \lambda_1 & \lambda_2 \end{Bmatrix} [b_{\lambda_1}^\dagger \tilde{b}_{\lambda_4}]_{lm} \\
& - \delta_{\lambda_1\lambda_4} (-)^{\lambda_2 - \lambda_3 + l_1 + l_2} \begin{Bmatrix} l_1 & l_2 & l \\ \lambda_3 & \lambda_2 & \lambda_1 \end{Bmatrix} [b_{\lambda_3}^\dagger \tilde{b}_{\lambda_2}]_{lm} \Big] ,
\end{aligned} \tag{10}$$

where the coefficient $\{\cdots\}$ is a $6j$ symbol of angular-momentum recoupling. The equation is valid not only for bosons b_λ of any multipolarities but care has been taken with the phases so that it holds also for fermion operators. In standard IBA only $\lambda = 0$ (s) and $\lambda = 2$ (d) occur, and the corresponding creation operators are called $s^\dagger$ and $d^\dagger$.

The generators of U(5), as it appears in eq. (1), consist of d bosons only. With the abbreviation $T_{lm} \equiv [d^\dagger \tilde{d}]_{lm}$, the 25 coupled generators of U(5) are the T_{lm} with $l = 4, 3, 2, 1, 0$. Of these, the ten T_{3m} and T_{1m} by themselves close under commutation and form the Lie algebra of O(5). The commutation relations from eq. (10) are

$$[T_{3m}, T_{3\mu}] = \sqrt{\tfrac{3}{5}} \, (3m3\mu|3, m+\mu) \, T_{3,m+\mu} - \sqrt{\tfrac{14}{5}} \, (3m3\mu|1, m+\mu) \, T_{1,m+\mu} , \tag{11}$$

$$[T_{3m}, T_{1\mu}] = -\sqrt{\tfrac{6}{5}} \, (3m1\mu|3, m+\mu) \, T_{3,m+\mu} , \tag{12}$$

$$[T_{11}, T_{1,-1}] = -\sqrt{\tfrac{1}{10}}\, T_{10}\,, \quad [T_{10}, T_{1,\pm1}] = \pm\sqrt{\tfrac{1}{10}}\, T_{1,\pm1}\,. \tag{13}$$

The last commutators are essentially those of angular momentum, $J_m = \sqrt{10}\, T_{1m}$, and the angular-momentum O(3) is indeed a subgroup of the present O(5).

Now one would like to search for an O(4) subalgebra which contained the angular-momentum O(3). This would require six closing generators which included the three T_{1m}. Possible candidates for the other three would be e.g. $T_{3,\pm1}$, T_{30}, but they do not yield a closing set since, by eq. (12), $[T_{3,\pm1}, T_{1,\pm1}] = \pm 2^{-1/2} T_{3,\pm2}$. Altogether eqs. (11)–(13) show that it is impossible to find an O(4) which contains the O(3) of angular momentum [4].

Mathematically it is obvious that O(5) has O(4) as a subgroup, although rotational invariance excludes it from standard IBA. The next best thing is to seek an O(4) which still yields *axial* symmetry, i.e. contains the physical O(2) as a subgroup.

As a first attempt at this goal, consider the uncoupled generators $d_m^\dagger d_\mu$ of U(5) and define the ten antisymmetric generators of O(5) by $A_{m\mu} \equiv d_m^\dagger d_\mu - d_\mu^\dagger d_m = -A_{\mu m}$. These generators are essentially Cartesian, since they could be relabeled with indices 1–5. The $A_{m\mu}$ lead in a clearcut fashion to a subgroup O(4) of O(5) when one of the five values of m is dropped [7]. There are five possible sets of six generators of O(4), one for each omitted value of m. Their algebras have the same appearance. When another m value is dropped, the three remaining $A_{m\mu}$ generate one of four possible O(3) subgroups.

Their straightforward properties notwithstanding, the $A_{m\mu}$ are not well suited for the physics at hand because they not only fail to be physical O(3) tensors but they do not even have a good third component of the angular momentum. Specifically this is seen from the relation

$$A_{\mu\nu} = (-)^\nu \sum_{lm} (2\mu 2, -\nu|lm)[T_{lm} - (-)^m T_{l,-m}]\,, \tag{14}$$

derived through well-known properties of the Clebsch–Gordan coefficients. No single T_{lm} can be expressed in terms of the $A_{\mu\nu}$; an inverse transformation is possible only for the $\pm m$ combination:

$$T_{lm} - (-)^m T_{l,-m} = \sum_{\mu\nu} (-)^\nu (2\mu 2, -\nu|lm)\, A_{\mu\nu}\,. \tag{15}$$

Note that for $m = 0$ this gives $0 = 0$.

For the physically vital connection to angular momentum, it is necessary to abandon the $A_{\mu\nu}$ and proceed in terms of an alternative set of generators, where the annihilation operators appear in the tensorial form $\tilde{d}_m$. So, with all five m values initially present, I define modified antisymmetric generators

$$O_{\mu\nu} \equiv \sqrt{\tfrac{1}{2}}\,(d_\mu^\dagger \tilde{d}_\nu - d_\nu^\dagger \tilde{d}_\mu)\,, \tag{16}$$

with the factor $\sqrt{1/2}$ inserted for normalization. Since $O_{\nu\mu} = -O_{\mu\nu}$, I choose to list these operators below as $O_{\mu>\nu}$. The $O_{\mu\nu}$ are related to the tensorial generators T_{lm} according to

$$O_{\mu\nu} = \sqrt{2} \sum_{\substack{l=1,3 \\ m}} (2\mu 2\nu|lm)\, T_{lm}\,, \tag{17}$$

with the inverse relation

$$T_{lm} = \sqrt{2} \sum_{\mu>\nu} (2\mu 2\nu|lm)\, O_{\mu\nu}\,, \quad l = 1, 3\,. \tag{18}$$

Contrary to the situation for the Cartesian generators $A_{\mu\nu}$ as seen in eqs. (14) and (15), the T_{lm} with odd l can be expressed in terms of the new generators $O_{\mu\nu}$. It is important to note that the $A_{\mu\nu}$ and the $O_{\mu\nu}$ have fundamentally different symmetries so that there is no transformation between them.

Equation (17) yields the ten O(5) generators

$$O_{21} = T_{33}, \qquad O_{20} = T_{32},$$

$$O_{2,-1} = \sqrt{\tfrac{3}{5}}\,T_{31} + \sqrt{\tfrac{2}{5}}\,T_{11}, \qquad O_{2,-2} = \sqrt{\tfrac{1}{5}}\,T_{30} + \sqrt{\tfrac{4}{5}}\,T_{10},$$

$$O_{10} = \sqrt{\tfrac{2}{5}}\,T_{31} - \sqrt{\tfrac{3}{5}}\,T_{11}, \qquad O_{1,-1} = \sqrt{\tfrac{4}{5}}\,T_{30} - \sqrt{\tfrac{1}{5}}\,T_{10},$$

$$O_{1,-2} = \sqrt{\tfrac{3}{5}}\,T_{3,-1} + \sqrt{\tfrac{2}{5}}\,T_{1,-1}, \qquad O_{0,-1} = \sqrt{\tfrac{2}{5}}\,T_{3,-1} - \sqrt{\tfrac{3}{5}}\,T_{1,-1},$$

$$O_{0,-2} = T_{3,-2}, \qquad O_{-1,-2} = T_{3,-3}. \qquad (19)$$

The nontrivial inverse relations are

$$T_{31} = \sqrt{\tfrac{3}{5}}\,O_{2,-1} + \sqrt{\tfrac{2}{5}}\,O_{10}, \qquad T_{30} = \sqrt{\tfrac{1}{5}}\,O_{2,-2} + \sqrt{\tfrac{4}{5}}\,O_{1,-1},$$

$$T_{3,-1} = \sqrt{\tfrac{3}{5}}\,O_{1,-2} + \sqrt{\tfrac{2}{5}}\,O_{0,-1}, \qquad T_{11} = \sqrt{\tfrac{2}{5}}\,O_{2,-1} - \sqrt{\tfrac{3}{5}}\,O_{10},$$

$$T_{10} = \sqrt{\tfrac{4}{5}}\,O_{2,-2} - \sqrt{\tfrac{1}{5}}\,O_{1,-1}, \qquad T_{1,-1} = \sqrt{\tfrac{2}{5}}\,O_{1,-2} - \sqrt{\tfrac{3}{5}}\,O_{0,-1}. \qquad (20)$$

The transformation between the $O_{\mu\nu}$ and the T_{lm} is unitary. Hence the generators $O_{\mu\nu}$ are unitarily equivalent to the T_{3m}, T_{1m}, so that they generate the standard O(5) of the IBA. In particular I note that the three angular-momentum components $\propto T_{1m}$ can be expressed as linear combinations of the $O_{\mu\nu}$. Thus the alternative generators $O_{\mu\nu}$, unlike the $A_{\mu\nu}$, are compatible with good angular momentum.

To proceed from O(5) to O(4) as outlined above [7], the only $\pm m$-symmetric possibility is to omit $m = 0$, so that the four remaining boson states are $b_{\pm 2, \pm 1}$. Then the six O(4) generators of type (16) are

$$O_{21}, \ O_{2,-1}, \ O_{2,-2}, \ O_{1,-1}, \ O_{1,-2}, \ O_{-1,-2}. \qquad (21)$$

As seen from eq. (19), they can be expressed in terms of eight T_{lm} operators: $T_{3,\pm 3}$, $T_{3,\pm 1}$, T_{30}, $T_{1,\pm 1}$ and T_{10}. There is now no complete inverse transformation; in particular $T_{1,\pm 1}$ cannot be expressed in terms of the remaining $O_{\mu\nu}$, as can be seen explicitly from eq. (20). However, T_{10} can be expressed as a linear combination of the generators (21), so that this O(4) contains a good third component of the angular momentum. With O(4) included, the subgroup chains (1) and (3), respectively, are then modified to

$$U(6) \supset U(5) \supset O(5) \supset O(4) \supset O(2), \qquad (22)$$

$$U(6) \supset O(6) \supset O(5) \supset O(4) \supset O(2), \qquad (23)$$

where the O(2) refers expressly to physical angular momentum.

3. THE O(4) CASIMIR OPERATOR

The operators $O_{\mu\nu}$ satisfy the commutation relation

$$[O_{\mu\nu}, O_{\omega\sigma}] = (-)^\omega \delta_{\nu,-\omega}\sqrt{\tfrac{1}{2}}O_{\mu\sigma} - (-)^\sigma \delta_{\mu,-\sigma}\sqrt{\tfrac{1}{2}}O_{\omega\nu}$$
$$- (-)^\sigma \delta_{\nu,-\sigma}\sqrt{\tfrac{1}{2}}O_{\mu\omega} + (-)^\omega \delta_{\mu,-\omega}\sqrt{\tfrac{1}{2}}O_{\sigma\nu}. \tag{24}$$

For the six O(4) generators defined above this yields the following 15 commutation relations:

$$[O_{21}, O_{2,-1}] = 0, \qquad [O_{21}, O_{2,-2}] = -\sqrt{\tfrac{1}{2}}O_{21}, \qquad [O_{21}, O_{1,-1}] = \sqrt{\tfrac{1}{2}}O_{21},$$

$$[O_{21}, O_{1,-2}] = 0, \qquad [O_{21}, O_{-1,-2}] = -\sqrt{\tfrac{1}{2}}O_{2,-2} + \sqrt{\tfrac{1}{2}}O_{1,-1},$$

$$[O_{2,-1}, O_{2,-2}] = -\sqrt{\tfrac{1}{2}}O_{2,-1}, \qquad [O_{2,-1}, O_{1,-1}] = -\sqrt{\tfrac{1}{2}}O_{2,-1},$$

$$[O_{2,-1}, O_{1,-2}] = -\sqrt{\tfrac{1}{2}}O_{2,-2} - \sqrt{\tfrac{1}{2}}O_{1,-1}, \qquad [O_{2,-1}, O_{-1,-2}] = 0,$$

$$[O_{2,-2}, O_{1,-1}] = 0, \qquad [O_{2,-2}, O_{1,-2}] = -\sqrt{\tfrac{1}{2}}O_{1,-2},$$

$$[O_{2,-2}, O_{-1,-2}] = -\sqrt{\tfrac{1}{2}}O_{-1,-2}, \qquad [O_{1,-1}, O_{1,-2}] = -\sqrt{\tfrac{1}{2}}O_{1,-2},$$

$$[O_{1,-1}, O_{-1,-2}] = \sqrt{\tfrac{1}{2}}O_{-1,-2}, \qquad [O_{1,-2}, O_{-1,-2}] = 0. \tag{25}$$

With rank 2, the group O(4) has two commuting generators for labeling its basis states within an irreducible representation (irrep). Altogether five of the commutators (25) vanish, but there are no three (or more) $O_{\mu\nu}$ which commute with one another. A natural choice for *the* two commuting generators are $O_{2,-2}$ and $O_{1,-1}$. As seen from eq. (19), they are two orthogonal linear combinations of T_{10} and T_{30}, which commute according to eq. (12). Thus the choice may be viewed as an extension of the standard choice of T_{10} for state labeling in physical O(3).

In addition to two state labels within an irrep, the rank 2 of O(4) implies the existence of two Casimir operators, one quadratic and one cubic, and two quantum numbers for labeling an irrep. With generic orthogonal-group generators $A_{ij} = b_i^\dagger b_j - b_j^\dagger b_i$, the quadratic Casimir operator can be written as $\sum_{kl} A_{kl} A_{lk}$; it satisfies its defining property of commuting with all the A_{ij}. An analogous construction from the $O_{\mu\nu}$ generators (16) yields for the cases of present relevance

$$C_2(On) = 4\sum_{\mu > \nu}(-)^{\mu+\nu}O_{\mu\nu}O_{-\nu,-\mu} = 4\sum_{\mu > \nu}O_{\mu\nu}O_{\mu\nu}^\dagger, \qquad n = 5, 4. \tag{26}$$

For $n = 5$ all values of μ, ν are included in the sum, while for $n = 4$ the value 0 is excluded. The arbitrary factor 4 in front has been chosen so that eq. (26) yields precisely the conventional IBA result [2, 3] $C_2(O5) = 4(T_3^2 + T_1^2)$.

Equation (26) does not work for the physical O(3) because, as seen from eq. (20), the three T_{1m} cannot be expressed in terms of three $O_{\mu\nu}$. Likewise it does not work for the physical O(2) because its one generator T_{10} is a linear combination of two $O_{\mu\nu}$. The cubic Casimir operator, of generic form $\sum_{ijk} A_{ij} A_{jk} A_{ki}$, is not needed for the IBA with two-boson interactions.

The next task is to find the eigenvalues of

$$C_2(O4) = 4(-O_{21}O_{-1,-2} - O_{2,-1}O_{1,-2} + O_{2,-2}^2$$
$$+ O_{1,-1}^2 - O_{1,-2}O_{2,-1} - O_{-1,-2}O_{21}). \qquad (27)$$

From the general result for orthogonal groups [2] I can write immediately the eigenvalue

$$\langle C_2(O4) \rangle = 2[\omega_1(\omega_1 + 2) + \omega_2^2]. \qquad (28)$$

Here ω_1 and ω_2 are the two quantum numbers labeling an irrep of O(4). In the context of IBA-1 only *fully symmetric* representations are allowed. Then only $\omega_1 \equiv \omega$ appears and the eigenvalue is

$$\langle C_2(O4) \rangle = 2\omega(\omega + 2). \qquad (29)$$

The branching rule for O(5) $\supset$ O(4) is

$$\omega = \tau, \tau - 1, \ldots, 1, 0, \qquad (30)$$

in complete analogy to the O(6) $\supset$ O(5) branching rule familiar in the IBA [2, 3].

The results (29) and (30) can be derived directly by means of boson quasispins analogously to the O(5) case [3, 8]. In the present O(4) case the quasispin operators are

$$S_+ \equiv \tfrac{1}{2} \sum_{m \neq 0} (-)^m d_m^\dagger d_{-m}^\dagger, \quad S_- \equiv S_+^\dagger,$$

$$S_0 \equiv \tfrac{1}{4} \sum_{m \neq 0} (-)^m (d_m^\dagger \tilde{d}_{-m} + \tilde{d}_{-m} d_m^\dagger) = S_0^\dagger. \qquad (31)$$

The component S_0 can also be expressed in terms of the appropriate number operator as

$$S_0 = \tfrac{1}{2}\hat{n}_4 + 1, \quad \hat{n}_4 \equiv \sum_{m \neq 0} d_m^\dagger d_m. \qquad (32)$$

The quasispin operators are seen to satisfy the commutation relations

$$[S_+, S_-] = -2S_0, \quad [S_\pm, S_0] = \mp S_\pm. \qquad (33)$$

This is the algebra of SU(1, 1) $\approx$ O(2, 1). Its Casimir operator is

$$C \equiv -S_+ S_- - S_0 + S_0^2, \qquad (34)$$

with eigenvalue equations

$$C|S\mu\rangle = S(S - 1)|S\mu\rangle, \quad S_0|S\mu\rangle = \mu|S\mu\rangle, \qquad (35)$$

and the raising/lowering relation is

$$S_\pm|S\mu\rangle = \sqrt{\mu(\mu \pm 1) - S(S - 1)}\,|S, \mu \pm 1\rangle. \qquad (36)$$

Equations (32) and (35) give $\mu = \tfrac{1}{2}n_4 + 1$, which has an absolute minimum value of 1. For a given S, μ has a minimum value μ_{min} such that eq. (36) yields $S_-|S\mu_{min}\rangle = 0$, which is satisfied by $\mu_{min} = S \geq 1$. Then μ takes on the values $\mu = S, S + 1, \ldots$ To a given $\mu_{min} = S$ corresponds a certain $(n_4)_{min} \equiv \omega$, so that

$$S = \tfrac{1}{2}\omega + 1. \qquad (37)$$

It follows that

$$n_4 = 2\mu - 2 = 2S - 2, 2S, 2S + 2, \ldots = \omega, \omega + 2, \ldots . \tag{38}$$

The quantum numbers ω and n_4 are seen to provide state labels equivalent to S and μ so that $|S\mu\rangle \equiv |\omega n_4\rangle$.

Equation (38) can be inverted to give the values of ω that belong to a fixed n_4:

$$\omega = n_4, n_4 - 2, n_4 - 4, \ldots, 0 \text{ or } 1. \tag{39}$$

This is the branching rule for $U(4) \supset O(4)$, and it is analogous to the $U(5) \supset O(5)$ branching rule $\tau = n_d, n_d - 2, \ldots$. Since $n_4 = n_d, n_d - 1, \ldots, 0$, the branching rule (30) follows.

The branching rule for $O(4) \supset O(2)$ is easily deduced by examining the chain $U(4) \supset O(4) \supset O(2)$. The situation is analogous to the standard IBA branching $U(5) \supset O(5) \supset O(3)$ [2, 3], only simpler. The $O(2)$ quantum number is the angular-momentum component M, which is simply the sum of the m values of the d_m present in the state. For $n_4 = 0$, M and ω are trivially 0. For $n_4 = 1$, the possible M values are $\pm 2, \pm 1$ and ω equals 1. The first nontrivial case is $n_4 = 2$: $|M| = 4, 3, 2, 1, 0^2$; these states have $\omega = 2$, except that one of the two $M = 0$ states has $\omega = 0$. For $n_4 \geq 3$, another label, analogous to the label ν_Δ of standard IBA, becomes necessary.

The eigenvalues of the $O(4)$ Casimir operator (27) also follow from the preceding quasispin study. Straightforward algebraic manipulation of eq. (27) gives the somewhat surprising result that

$$C_2(O4) = 8C, \tag{40}$$

where C is precisely the quasispin Casimir operator (34). Thus, by eqs. (35) and (37), the auxiliary $SU(1,1)$ algebra immediately furnishes the eigenvalues of the physical operator $C_2(O4)$ as

$$\langle C_2(O4)\rangle = 8\langle C\rangle = 8S(S - 1) = 2\omega(\omega + 2), \tag{41}$$

in agreement with eq. (29). For $O(5)$ and $O(6)$, the same technique applies, but not quite in the clean form of eq. (40). Thus for $O(5)$ the result is $C_2(O5) = 8C - \frac{5}{2}$.

What role would $O(4)$ play in the dynamic symmetries (22) and (23)? The d-boson interaction term in an axially symmetric IBA Hamiltonian could be written in the multipole form

$$H_{\text{int}} = \sum_{l\lambda m} c_{l\lambda m} T_{lm} T_{\lambda - m} . \tag{42}$$

Equation (17) shows that the Casimir operator $C_2(O4)$ of eq. (27) is of this form and thus is a special case of eq. (42). If the complete IBA Hamiltonian can be expressed as a linear combination of the Casimir operators of chain (22) or (23), the $O(4)$ contribution to the energy is proportional to $\omega(\omega + 2)$.

As is evident from the above construction, yet another group chain emerges, namely

$$U(6) \supset U(5) \supset U(4) \supset O(4) \supset O(2). \tag{43}$$

The $U(4) \supset O(4)$ branching rule is given by eq. (39). By the general rule [2, 3] the linear and quadratic Casimir operators of $U(4)$ are $\hat{n}_4$ and $\hat{n}_4(\hat{n}_4 + 3)$, respectively, with eigenvalues n_4 and $n_4(n_4 + 3)$.

4. ISOMORPHISM O(4) ≈ O(3) × O(3)

It is of interest to establish the well-known isomorphism $O(4) \approx O(3) \times O(3)$ in the present context. This isomorphism has been extensively exploited by Van Isacker [9] in developing the Racah algebra of $O(4)$. In fact, ref. [9] has provided important guidelines for the present examination.

Since the $O(4)$ generated by the operators (27) does not have good physical angular momentum, but only its third component, the groups $O(3)$ that now appear cannot restore angular momentum. The first idea to get at a mathematical $O(3) \subset O(4)$ would be to drop one more m from those still present, i.e. $m = \pm 2, \pm 1$. In the Cartesian basis with generators $A_{\mu\nu}$ (sect. 2), there are four such $O(3)$ subalgebras, one for each dropped m. With the generators $O_{\mu\nu}$, however, there are none. One can see this by inspection of the $O(4)$ commutators (25). Dropping $m = 2$, for example, leaves only the last three commutators, and one of them is zero, so they do not generate an $O(3)$. The symmetry of the $O_{\mu\nu}$ is so constrained that they are not equivalent to the $A_{\mu\nu}$, as was discussed in sect. 2.

To elucidate the constrained nature of the $O_{\mu\nu}$, I digress to consider the subgroup relation $O(5) \supset O(3)$. Inspection of the $O(5)$ commutation relations (24), 45 in all, shows that $O(5)$ has only *two* $O(3)$ subgroups (not angular momentum) generated by sets of three $O_{\mu\nu}$. They are $O_{10}, O_{0,-1}, O_{1,-1}$ and $\pm O_{20}, \mp O_{0,-2}, O_{2,-2}$. When, on the other hand, $O(5)$ is generated by the $A_{\mu\nu}$, it has *ten* $O(3)$ subgroups (also unphysical) generated by sets of three $A_{\mu\nu}$.

To establish the structure $O(4) \approx O_1(3) \times O_2(3)$, one needs two commuting sets of $O(3)$ generators. The zeros present in eq. (25) suggest that O_{21} and $O_{-1,-2}$ belong to one set, and $O_{2,-1}$ and $O_{1,-2}$ to another. Call these sets 1 and 2, respectively. The third generators for sets 1 and 2, respectively, are found to be

$$O_{\mathrm{I}} \equiv \sqrt{\tfrac{1}{2}}\,(O_{2,-2} - O_{1,-1}), \qquad O_{\mathrm{II}} \equiv \sqrt{\tfrac{1}{2}}\,(O_{2,-2} + O_{1,-1}). \tag{44}$$

The generators $O_{21}, O_{-1,-2}, O_{\mathrm{I}}$ of set 1 and $O_{2,-1}, O_{1,-2}, O_{\mathrm{II}}$ of set 2, respectively, obey the canonical $O(3)$ commutation relations of the angular-momentum components J_1, J_{-1} and J_0, cf. eq. (13). This correspondence is of course only mathematical. Expressed in terms of the T_{lm}, by eq. (19), the "zero" components are

$$O_{\mathrm{I}} = \sqrt{\tfrac{1}{10}}\,(-T_{30} + 3T_{10}), \qquad O_{\mathrm{II}} = \sqrt{\tfrac{1}{10}}\,(3T_{30} + T_{10}). \tag{45}$$

Although they are axially symmetric, $m = 0$, they do not represent the physical $O(2)$ because they contain not only T_{10} but also T_{30}.

One might think that physical $O(2)$ could be recovered by forming linear combinations $3G_1 + G_2$ of the generators of sets 1 and 2. The zero component would then be $3O_{\mathrm{I}} + O_{\mathrm{II}} = J_0$, i.e. precisely the angular-momentum component. However, such new generators do not produce an $O(3)$ algebra. An $O(3)$ *can* be produced by the direct-sum algebra, $G_1 + G_2$. The generators are then

$$K_1 \equiv O_{21} + O_{2,-1} = T_{33} + \sqrt{\tfrac{3}{5}}\,T_{31} + \sqrt{\tfrac{2}{5}}\,T_{11},$$

$$K_{-1} \equiv O_{-1,-2} + O_{1,-2} = T_{3,-3} + \sqrt{\tfrac{3}{5}}\,T_{3,-1} + \sqrt{\tfrac{2}{5}}\,T_{1,-1},$$

$$K_0 \equiv O_{\mathrm{I}} + O_{\mathrm{II}} = \sqrt{2}\,O_{2,-2} = \sqrt{\tfrac{2}{5}}\,(T_{30} + 2T_{10}). \tag{46}$$

The K_m generate the group $O_{1+2}(3)$ such that

$$O(4) \approx O_1(3) \times O_2(3) \supset O_{1+2}(3) \supset O_{1+2}(2). \qquad (47)$$

The $O(2)$ once again is not the physical angular-momentum component because it contains T_{30}. I conclude that the isomorphism discussed here corresponds to the required mathematical structure but does not have direct physical content.

In fact, it is impossible to find an $O(3)$ subgroup of the present $O(4)$ which would contain the physical $O(2)$. This conclusion is reached by setting $K_0 \equiv T_{10}$ and letting $K_{\pm 1}$ be general linear combinations of the $O(4)$ generators (21). Such K_m cannot satisfy the $O(3)$ commutation relations which have the form (13).

5. CONCLUSION

The group $O(4)$ is absent from the well-known group chains of the IBA. This is due to the requirement that the $O(3)$ of physical angular momentum be contained in them, as is necessary in an isotropic environment. However, this requirement does not apply when the IBA is formulated in terms of an intrinsic system. Nevertheless, even then it is desirable to maintain the requirement of *axial* symmetry, i.e. conservation of the third component of the angular momentum.

The main result of the present study is that it is indeed possible to sandwich an $O(4)$ between the $O(5)$ of the standard IBA and the physical $O(2)$. This gives rise to the three group chains (22), (23) and (43), applicable to an intrinsic system with axial symmetry. Rewritten with the quantum numbers included, these chains are

$$
\begin{array}{cccccc}
U(6) & \supset U(5) & \supset O(5) & \supset O(4) & \supset O(2) \\
{[N]} & n_d & \tau & \omega & \nu_4 & M
\end{array}
\qquad (48)
$$

$$
\begin{array}{cccccc}
U(6) & \supset O(6) & \supset O(5) & \supset O(4) & \supset O(2) \\
{[N]} & \sigma & \tau & \omega & \nu_4 & M
\end{array}
\qquad (49)
$$

$$
\begin{array}{cccccc}
U(6) & \supset U(5) & \supset U(4) & \supset O(4) & \supset O(2) \\
{[N]} & n_d & n_4 & \omega & \nu_4 & M
\end{array}
\qquad (50)
$$

The reduction rules of the quantum numbers are contained in eqs. (30) and (39) and in the adjoining text. The additional quantum number needed for a complete reduction $O(4) \supset O(2)$ is called ν_4.

If the IBA intrinsic Hamiltonian has one of the dynamic symmetries associated with the group chains (48)–(50), the $O(4)$ link of the chain contributes an energy term proportional to $\omega(\omega + 2)$, and the $O(2)$ contribution is $\propto M^2$. If the dynamic symmetry (50) applies, the $U(4)$ contributions are $\propto n_4$ and $\propto n_4(n_4 + 3)$.

When good angular-momentum component is required, it is impossible to sandwich an $O(3)$ between $O(4)$ and $O(2)$, in complete analogy with standard IBA where $O(4)$ cannot be sandwiched between $O(5)$ and $O(3)$. The present $O(4)$ does have $O(3)$ subgroups, but they do not contain the physical $O(2)$. In particular, the standard isomorphic structure $O(4) \approx O(3) \times O(3) \supset O(3)$ is possessed by the d-boson groups studied here.

ACKNOWLEDGMENTS

I wish to thank Piet Van Isacker for his decisive suggestion of using an uncoupled basis as a starting point in the present study. I am most grateful to George Rosensteel for highly valuable advice and innovative suggestions.

I thank the Institute for Nuclear Theory at the University of Washington for its hospitality and the Department of Energy for partial support during the completion of this work.

REFERENCES

1. F. Iachello and A. Arima, *Phys. Lett.* **53B**, 309 (1974).

2. F. Iachello and A. Arima, *The Interacting Boson Model*, Cambridge University Press, Cambridge (1987).

3. R. F. Casten, J. P. Draayer, K. Heyde, P. O. Lipas, T. Otsuka and D. D. Warner, *Algebraic Approach to Nuclear Structure: Interacting Boson and Fermion Models*, Gordon and Breach, New York (1992), Ch. 2.

4. F. Iachello, private communication (1983).

5. F. Iachello, *Symmetries in Science V*, eds. B. Gruber, L. C. Biedenharn and H. D. Doebner, Plenum Press, New York (1991), p. 305.

6. R. Gilmore, *Lie Groups, Lie Algebras, and Some of Their Applications*, Wiley-Interscience, New York (1974).

7. P. Van Isacker, private communication (1991).

8. A. Arima and F. Iachello, *Ann. Phys., NY* **99**, 253 (1976).

9. P. Van Isacker, *Symmetries in Science V*, eds. B. Gruber, L. C. Biedenharn and H. D. Doebner, Plenum Press, New York (1991), p. 323.

SYMMETRY AND SYMMETRY BREAKING LATENT IN THE UNCERTAINTY PRINCIPLE

Koichiro Matsuno

Department of BioEngineering
Nagaoka University of Technology
Nagaoka 940-21, Japan

INTRODUCTION

Symmetry and symmetry breaking are both sides of the same coin when viewed from the perspective of quantum mechanics. If the quantum state constituting the initial and boundary conditions is given by whatever means, the quantum mechanical development specified by the quantum mechanical equation of motion will preserve the symmetry property latent in the boundary conditions (Ne'eman, 1990). The present scheme of preserving the symmetry property with regard to the quantum state or the quantum mechanical wavefunction would, however, lose its underpinning if the initial preparation of the quantum state constituting the boundary conditions for its subsequent development is not completed for whatever reasons. Although it keeps identifying how the quantum state develops once the initial state is given along with its boundary conditions, quantum mechanics, though legitimate in itself, is intrinsically incompetent in preparing and reasoning the boundary conditions (Kuppers, 1992). A molecule in the atmosphere certainly obeys quantum mechanics, but the boundary conditions applying to the molecule have to be imposed. Quantum mechanics specifies how the quantum state develops under given boundary conditions rather than how the boundary conditions to be imposed upon the quantum state would develop. This unavailability of definite boundary conditions within the framework of quantum mechanics will provide the process of measurement with its unique role (Matsuno, 1990).

Molecules in the atmosphere can interact with each other regardless of whether theoretically imposed boundary conditions would be available. Even if the quantum state of the whole system of interacting molecules is not identified for definite, these molecules can interact with each other. Since molecular interaction refers to the process of how each molecule detects others and reacts upon them accordingly, measurement on the part of interacting molecules is found to be ubiquitous even if the quantum state of the whole system is not identifiable as a matter of principle. Molecular measurement proceeding internally in the absence of the identifiable quantum state is intrinsically irreversible in the sense that there is no means of predetermining what will be measured beforehand. Of course, if the total quantum state is available on the other hand, the time development of any observable can be identified and predetermined with the use of the quantum mechanical dynamic law that guarantees a unique development of the quantum state. There would be no irreversibility distinguishing between before and after actual measurement, because what will be measured can be identified well in advance in terms of the quantum state whose development can uniquely be predetermined.

Internal measurement proceeding in the absence of the quantum state to be identified on a global scale is intrinsically irreversible and breaks the translational symmetry in time (Matsuno, 1985), whereas time development of the quantum state, once imposed, maintains its temporally translational symmetry. This difference between symmetry-breaking and symmetry-preservation in time rest solely upon whether or not the insistence on the availability of the quantum state on a global scale could be legitimate. It is one thing to theoretically impose the quantum state, but quite another to actually prepare it in a manner such that the theoretical imposition could be justified.

Insistence on the quantum state on a global scale requires at the least the preparation of the initial quantum state that implements the initial and boundary conditions for the dynamic development. Once the initial and boundary conditions are fixed, the subsequent dynamic development uniquely determines how the quantum state evolves. However, there is an insurmountable difficulty in preparing the initial and boundary conditions. Preparation of the initial condition on a global scale implies that every interacting element constituting the initial quantum state has to be coordinated in a global manner instantaneously at the time of preparing the initial condition. Theoretical imposition of the initial condition requires global coordination and identification in an instantaneous manner, in spite of the fact that there is no means of coordination nor identification proceeding beyond light velocity in the realm of material processes. Although there have been countless experiments showing that theoretical imposition of the quantum state on a global scale is really successful, these previous successes alone could not dismiss the imposed theoretical nature of the quantum state. Instead, it is required to admit that theoretical imposition of the quantum state is no more than a matter of approximation. In order to explicate what is approximated by such a theoretical imposition, one would be asked to retreat to the point where molecular interaction proceeding internally is regarded as being intrinsically irreversible.

Internal measurement that recognizes the asymmetry between before and after the actual measurement can reduce to the typical time development of the quantum state if theoretical impostion of the quantum state is taken to be approximately valid. Otherwise, internal measurement should not be dismissed even though its theoretical implications still remain to be seen. In particular, one of the advantages for paying attention to internal measurement is its capacity of breaking temporal symmetry. Evolution of matter leading to biological organizations is an indication of increase of complexity and order in spatial organization. Such an increase in the complexity and order in time is certainly a case of breaking temporal symmetry (Matsuno, 1984). Internal measurement thus provides a vehicle for enhancing material complexity and organization. Further elucidation on what internal measurement looks like requires a quantum mechanical scrutiny of itself, though without imposing the quantum state arbitrarily. One of the attempts in this direction is to scrutinize the role of the uncertainty principle, because it refers to the process of measurement without specifying the nature of the quantum state available or conceivable.

THE UNCERTAINTY PRINCIPLE IN INTERNAL MEASUREMENT

In order to examine how interacting molecules participate in internal measurement, let us consider a case of weakly interacting non-degenerate gas having N degrees of freedom contained in a box with its volume V. The weakly interacting non-degenerate gas is characterized by the energy allotted to each degree of freedom in motion, e_i with $i=1,2,...N$. The energy e_i of the i-th degree of freedom is measured by the whole gas except the very degree of freedom through its interaction with all of the other degrees of freedom.

The process of measuring e_i internally is certainly subject to the uncertainty principle

$$\Delta e_i \Delta t \simeq \hbar \qquad (i=1,2,...,N),$$

where Δe_i is the amount of uncertainty in the energy e_i to be measured internally over the time interval Δt which is to be determined, and $\hbar$ is Planck's constant divided by 2π. What is specific to the uncertainty Δe_i in the energy e_i to be measured is that both the object of and the means of measurement consist of the weakly interacting non-degenerate gas. The uncertainty originates both in the measuring and in the measured.

The uncertainty Δe_i associated with the internal measurement in fact consists of two factors; one is due to the fluctuation in the measured energy $\Delta_m e_i$ and the other is due to the deviation of the reference point of the measurement apparatus formed internally, $\Delta_r e_i$. The equality

$$\Delta e_i = \Delta_m e_i + \Delta_r e_i$$

tells the fact that the uncertainty Δe_i in the energy of the i-th degree of freedom is the resultant of both fluctuations in the energy to be measured and in the pointer of the measurement apparatus, though not legible to the anthropocentric bystander.

At the same time, the total energy E of the weakly interacting non-degenerate gas is also measurable by connecting it to a measurement apparatus provided externally. The uncertainty ΔE in measuring the total energy follows the uncertainty principle

$$\Delta E \Delta t \simeq \hbar$$

409

with

$$E = \sum_{i=1}^{N} e_i \ .$$

Since the total uncertainty is the sum total of the fluctuations in the energy of each degree of freedom in motion, the relationship

$$\Delta E = \sum_{i=1}^{N} \Delta_m e_i$$

follows due to the fact that the contribution to the total uncertainty from each degree of freedom is simply additive.

Noting that each degree of freedom in the gas is allotted an almost equal amount of energy depending upon its temperature, one can have approximate equalities

$$\Delta_m e_i \simeq \Delta_m e$$

$$\Delta_r e_i \simeq \Delta_r e \qquad\qquad (i=1,2,...,N)$$

where $\Delta_m e$ is the measurable average fluctuation or uncertainty in the energy of each degree of freedom in motion and similarly $\Delta_r e$ is the average fluctuation in the reference point of the apparatus formed internally for measuring each degree of freedom. We thus have the uncertainty principle expressed in the form

$$\Delta_m e \Delta t \simeq \frac{\hbar}{N}$$

$$\Delta_r e \Delta t \simeq \frac{N-1}{N} \ \hbar \ .$$

If one considers the limit $N \to \infty$ as with the case of taking the thermodynamic limit, only the fluctuation in the reference point of internal measurement would survive. The thermodynamic limit does not leave internal measurement any irreversible changes. Internal measurement at the thermodynamic limit preserves the symmetry property latent in the system. On the other hand, however, if the total number N of degrees of freedom in motion is literally taken to be finite, internal measurement leaves irreversible changes on the internally measured object through the uncertainty principle. Internal measurement in the system with a finite number of degrees of freedom constantly breaks the symmetry property latent in the system with time. Further details on symmetry-preservation or symmetry-breaking during internal measurement will be worked out when the time scale of internal measurement (Salthe, 1989) is made explicit.

TIME SCALE OF INTERNAL MEASUREMENT

The uncertainty principle suggests that when the time scale of measurement Δt is given, the process of measurement gives rise to

410

fluctuations in the energy to be measured by the amount of $\hbar/\Delta t$. In particular, if the weakly interacting non-degenerate gas with N degrees of freedom in motion at temperature T is the case, the energy quantum associated with internal measurement that is of order of $\hbar/\Delta t$ would have to be found within the energy excitations available to the gas at temperature T. Quantum fluctuations unique to internal measurement are thus embedded within those fluctuations intrinsic to the gas at temperature T. This fact indicates that the time scale of internal measurement, Δt cannot be much less than the inverse of thermal frequency $\hbar/\kappa T$, where κ is Boltzmann's constant. Otherwise, quantum fluctuations unique to internal measurement would have to have the energy greater than that of a thermal quantum. At the same time, the time scale of internal measurement cannot be much greater than the inverse of thermal frequency, because thermal fluctuations would allow only those fluctuations corresponding to excitation or absorption of thermal quanta. Consequently, the time scale of internal measurement turns out to be of the same order with the inverse thermal frequency as expressed in

$$\Delta t \sim \hbar/\kappa T .$$

The standard interpretation of the uncertainty principle expressed as $\Delta e_i \Delta t \sim \hbar$ is that the energy fluctuation Δe_i is counted in units of and is of order of $\hbar/\kappa T$. In fact, that thermal quanta of energy $\sim \kappa T$ are statistically independent with each other is consistent with the choice of $\hbar/\kappa T$ as the time scale of successive internal measurement. If the time scale of successive internal measurement were set to be either greater or lesser than the inverse thermal frequency $\sim \hbar/\kappa T$, the internal identification of the thermal quantum having energy $\sim \hbar/\kappa T$ would lose its material underpinning. For the quantum having the energy differing significantly from κT would come to serve as a counting unit there.

In contrast, the uncertainty principle of internal measurement in the form of $\Delta_m e \Delta t \sim \hbar/N$ for each degree of freedom suggests a renewed interpretation of the very principle itself. Since the time scale of successive internal measurement is set to be the inverse thermal frequency $\Delta t \sim \hbar/\kappa T$ and since the fluctuation $\Delta_m e$ is counted in units of the thermal quantum having the energy $\sim \kappa T$, expression $\Delta_m e \Delta t \sim \hbar/N$ turns out to imply that one thermal quantum of energy $\sim \kappa T$ would be either emitted or absorbed over the period of $N\Delta t$ $\sim N\hbar/\kappa T$ for each degree of freedom. The sequence of emitting and absorbing thermal quanta undoubtedly breaks the translational symmetry in time in the sense that there is no means for foretelling when next thermal quantum will be emitted or absorbed, though those events in the past are definite in the record.

The temporal symmetry breaking is also expressible as that the event that one thermal quantum would be either emitted or absorbed over the period of $N\hbar/2\kappa T$ with respect to each degree of freedom in motion could come to occur with probability 1/2. The present probabilistic interpretation of the uncertainty principle now gives internal measurement a quantitative figure measuring the extent of breaking the temporal symmetry in terms of information bits. In essence, each degree of freedom in motion is to generate information or to transform the prior indefiniteness into the posterior definiteness by the amount of 1 bit on average over the period $N\hbar/2\kappa T$. Internal measurement constrained by the uncertainty principle thus comes to generate information at the rate

$$i_R \sim \frac{2\kappa T}{N\hbar} \qquad\qquad \text{(bits/s/degree-of-freedom)}.$$

Information generation rate is certainly a quantitative figure measuring the extent of temporal symmetry breaking, because it distinguishes between the indefiniteness yet to be specified and the definiteness already identified.

EMPIRICAL SIGNIFICANCE OF TEMPORAL SYMMETRY BREAKING

Generation of information has long been discarded in the standard practicing of quantum mechanics as in the case of taking the thermodynamic limit. This is because practical significance of internal measurement has not yet fully been worked out. Internal measurement not in the thermodynamic limit breaks temporal symmetry and should be examined in the light of biological material organization, because biological evolution is an indisputable empirical testimony to temporal symmetry breaking. For this purpose, we try to estimate the rate of information generation per degree of freedom of the radiation field because biological material organizations evolving on the earth have been immersed within the radiation field.

Estimating the rate of information generation requires the total number N of the available degrees of freedom of the radiation field. The total number N for the radiation field at temperature T is related to the volume V of the box which the field occupies through the relationship (cf., Landau and Lifshitz, 1969)

$$N \sim V(\frac{\kappa T}{2\pi \hbar c})^3$$

where c is light velocity. The volume of the region in which the radiation field of biological significance could be found on the earth is that of the region in which the radiation field of temperature of roughly 300K could be situated. This region would be roughly 1km width of the layer surrounding the surface of the earth, or $V \sim 10^{24}$ cm^3. The total number of the available degrees of freedom of the radiation field thus yields $N \sim 10^{31}$. These figures result in the rate of information generation

$$i_R \sim 10^{-17} \qquad\qquad \text{bits/s/degree-of-freedom}.$$

The rate of information generation per degree of freedom can be thought to be independent of the particularity which each degree of freedom represents so long as all of the available degrees of freedom interact with each other. The radiation field lets its each degree of freedom interact with others through atoms and molecules serving as intermediaries of interaction. In particular, if an arbitrary pair of two degrees of freedom having different rates of information generation comes to interact, the difference of the rates would finally disappear in due course of time because of the de facto strong correlation between the two.

One of the empirical facts that provide the rate of information generation is available from nucleotide substitutions in DNA molecules. The genes encoding cytochrome c recorded roughly 5 nucleotide substitutions per 100 codons per 100 million years (Goodman, 1981). Since a nucleotide substitution occurs out of four alternatives of A, T, G and C and since each nucleotide in a DNA molecule represents one degree of freedom whose value is taken out of only the four alternatives, the resulting rate of information generation gives (Matsuno, 1990)

$$i_{R:cytochrome\ c} \sim 10^{-17} \qquad \text{bits/s/degree-of-freedom.}$$

Although there have been many point mutations that could not survive in the population in the long run, their net contribution to information generation vanishes after all because of the mutual cancellation between the positive and negative contribution.

We thus observe that there is a rough coincidence of the rates of information generation between the radiation field on the surface of the earth and DNA molecules found in biological organizations that have evolved and functioned there. This coincidence is more than just being a mere accident. The radiation field on the earth that preceded the emergence of the earth-bound biological organizations had already been equipped with the mechanism of generating information whose functioning could also be confirmed later in the biological realm that has evolved there.

SYMMETRY AND SYMMETRY BREAKING IN INTERNAL MEASUREMENT

Internal measurement can be global as well as being local. Measurement of each degree of freedom in motion by all the others is local, whereas measuring the total body is global. What is unique to local measurement is that it can have only the limited access to the complete situation of the whole system at any moment (Gunji, 1991; Konno, 1992). In contrast, global measurement permitting the full access to the complete situation has to follow what each local measurement has accomplished. There is no means of coordinating both local and global measurement instantaneously in a manner that the uncertainty principle of each of them may be fulfilled at the same time. Each local measurement subjected to the uncertainty principle proceeds so as to let global measurement fulfill the similar principle subsequently. The present interplay between local and global measurement gives the uncertainty principle a specific connotation such that it is not the uncertainty principle which upholds internal measurement, but internal measurement itself which upholds the uncertainty principle.

Needless to say, quantum mechanics supplemented by the definite initial and boundary quantum state determines its observables in a manner that the uncertainty principle is observed and that their symmetry property is preserved. Both fulfillment of the uncertainty

413

principle and preservation of symmetry property are a consequence of the availability of the quantum state constituting the initial and boundary conditions. It thus turns out that quantum mechanics addresses itself to two independent problems at the same time. One is how to prepare the quantum state initially, and the other is how the once prepared quantum state develops with time. Dynamics of the quantum mechanical state takes the preparation of the definite initial state for granted and concentrates on the second problem of its development. The present assertion for no dynamics for preparing its initial state , however, leads to a queer conclusion such that the process of preparing the initial state is not actually a process in time at all. The problem of measurement in quantum mechanics in fact centers around whether the initial setup of the measurement apparatus which experimenters manage to control could really be separated from the subsequent development of the quantum state to be examined. Mere proclamation of the initial state does not guarantee its separation from the subsequent dynamic development.

Whether the process of preparing the initial state could proceed independently of its later development has to be examined without employing the scheme of state dynamics, because in the latter the complete preparation of the initial state is required as being indispenable from the very beginning. Quantum mechanics certainly provides itself with the dynamic scheme that does dispense with definite quantum states. Internal measurement supplemented by the uncertainty principle does neither separate nor distinguish between the initial preparation of what is to observed and its subsequent development. Quantum mechanics of internal measurement, instead of that of the definite quantum state, is thus capable of examining to what extent it could reduce to a form of state dynamics as allowing for the preparation of its initial state.

In essence, symmetry breaking dynamics of internal measurement underlies symmetry preserving dynamics of the quantum mechanical state. This nested structure of symmetry breaking and preservation provides a unified view on how to reconcile these apparently conflicting operations. It is the process of symmetry breaking which is responsible for generating those states that preserve a symmetry property over a limited time interval. Both symmetry breaking and preservation are latent in the uncertainty principle in quantum mechanics.

REFERENCES

Goodman, M., 1981. Decoding the pattern of protein evolution. Progr. Biophys. Molec. Biol. 37, 105-167.

Gunji, Y. P., 1991. The form of life. 1. It is possible but not necessary. Appl. Math. Comp. 47, 267-288.

Konno, N., 1992. A formalism of multidimensional living systems based on nonstandard analysis. Appl. Math. Comp. 49, 231-238.

Kuppers, B. -O., 1992. Understanding complexity. In: Beckermann, A., Flohr, H., and Kim. J. (Eds.) Emergence or Reduction: Essays on the Prospects of Nonreductive Physicalism (Walter de Gruyter, Berlin) pp. 241-256.

Laundau, L. P., and Lifshitz, E. M., 1969. Statistical Physics, 2nd edn. (Pergamon Press, Oxford).

Matsuno, K., 1984. Is matter inanimate? : protobiological information from within. Origins Life 14, 489-496.

Matsuno, K., 1985. How can quantum mechanics of material evolution be possible?: symmetry and symmetry breaking in protobiological evolution. BioSystems 17, 179-192.

Matsuno, K., 1990. Nonlocality and symmetry in quantum mechanics versus localizability and symmetry breaking in protobiology. In: Gruber, B., and Yopp, J. H. (Eds.) Symmetries in Science IV (Plenum Press, New York) pp. 125-146.

Ne'eman, Y. 1990. The interplay of symmetry, order and informa tion in physics and the impact of gauge symmetry on algebraic topology. Symmetry: Culture & Science 1, 229-255.

Salthe, S. N., 1989. Self-organization of/in hierarchically structured systems. Syst. Res. 6, 199-208.

ELECTRON-MOLECULE COLLISION PROCESS:
REVIEW AND PERSPECTIVE OF ALGEBRAIC APPROACHES

Alberto Mengoni[1]

Department of Physics
Japan Atomic Energy Research Institute
Tokai-mura, Ibaraki 319-11
Japan

INTRODUCTION

The calculation of the scattering cross section for the electron molecule collision process can be a formidable and not rarely an untreatable task using conventional methods based on integro-differential techniques. Recently, an alternative procedure has been proposed (Bijker et al., 1986) which makes use of an algebraic description of the molecular structure, the Vibron Model (Iachello, 1981), in conjunction with the Glauber diffractive theory for the dynamic of the collision process. This approach, named algebraic-eikonal, has been applied to the calculation of scattering cross sections for several dipolar diatomic targets. The long-range dipole interaction, certainly dominant for the vibrationally elastic process, has been included in the formalism and the relative scattering amplitudes derived.

More recently, improvements have been considered to this first basic theory. Higher order types of interactions like quadrupole and polarization, have been considered. The vibrational excitation process has been included into the approach. The treatment of the scattering process within the Glauber approximation has been extended to the case of triatomic target.

In addition to the Glauber approximation, a different description of the dynamic of the collision process has also been proposed which combines the eikonal approximation together with coupled-channel calculations.

In the following paragraph we will review the *basic* algebraic-eikonal approach and subsequently describe the most recent developments of the theory. We will not go into many details (these can be found in the references). Instead, here our intention is to point out the most important steps that lead to the application of the algebraic methods in the field of the electron-molecule collision processes.

It is a pleasure to dedicate this contribution to Prof. F. Iachello on the occasion of his 50th birthday.

---

[1]Permanent address: ENEA, INN.SVIL, Computational Division, V.le G. B. Ercolani 8, 40138 Bologna, Italy.

THE ALGEBRAIC–EIKONAL APPROACH

The algebraic–eikonal approach has been introduced by Bijker et al. (1986) and described in detail in several subsequent papers by the same group (Bijker and Amado, 1986, 1988).

Let us consider the collision process $e^- + AB(vJ) \rightarrow AB(v'J') + e^-$ for a diatomic molecule AB where the molecular initial and final rotational–vibrational states have been indicated respectively by vJ and v'J'. The scattering process is generally treated in the adiabatic–nuclei approximation, i.e. the collision time $t_{coll} \ll t_v$, where t_v is the characteristic time for vibrational motion (the rotational motion is much slower than the vibrational one). If, in addition, the electron wavelength is small compared with the range of variation of the interaction (eikonal approximation), the scattering amplitude can be written as

$$F_{f,i}(q) = \frac{ik}{2\pi} \int d^2b \, e^{iq \cdot b} <f|\, 1 - e^{i\chi(b,s)}\,|i> \tag{1}$$

where f and i indicate respectively the final and the initial molecular state. **k** and **k'** are the incident and final electron wave vectors, $q = k - k'$, and **b** is the impact parameter.

The eikonal phase is given by

$$\chi(b,\Omega_s) = - \frac{m_e}{\hbar^2 k} \int_{-\infty}^{\infty} V(r,\Omega_s)dz \tag{2}$$

with the geometry defined in such a way that the z–axis is on the direction of **k** and the direction of the molecular axis is defined by Ω_s. m_e is the electron mass.

The interaction potential can be given (see for example Itikawa, 1971) in term of a multipole expansion of type

$$V(r,\Omega_s) = \sum_{\lambda=0}^{\infty} \sum_{\mu=-\lambda}^{\lambda} v_\lambda(r)(-)^\lambda Y_{\lambda,\mu}(\Omega_e) Y_{\lambda,-\mu}(\Omega_s) \tag{3}$$

where $Y_{\lambda\mu}$ are spherical harmonics and $v_\lambda(r)$ appropriate functions of the distance between the incoming electron and the center of mass of the molecule. Ω_e defines the direction of the incoming electron in the same frame.

There are essentially two steps to be performed in order to introduce the algebraic methods in the simple theory so far described: 1) replace the molecular wave functions with the algebraic molecular states, 2) define appropriate algebraic operators for the interaction.

The Vibron Model

The algebraic description of the target is provided by the Vibron Model (Iachello, 1981). Its basic assumption is that the vibrational and rotational degrees of freedom of a single molecular bond can be described by the U(4) algebra.

The two dynamical symmetries of this model are

$$U(4) \begin{array}{c} \nearrow O(4) \subset O(3) \\ \searrow U(3) \subset O(3) \end{array} \tag{4}$$

The Hamiltonian representing realistic molecular species contains only generators of O(4) and O(3). Therefore, it is diagonal in the basis provided by the chain decomposition given above. Its spectrum is given by

$$E(v,J) = h_0 - 4A(N+2)(v+\tfrac{1}{2}) + 4A(v+\tfrac{1}{2})^2 + BJ(J+1) \tag{5}$$

where h_0, A and B, together with the total number of vibrons N, are to be determined from the molecular constants. The energy expression contains the vibrational harmonic as well as the anharmonic terms, in addition to the rotational energy term. The classical limit of the Hamiltonian, as well as the energy spectrum of equation 5, correspond to that of a one-dimensional Morse oscillator (van Roosmalen, 1982). A complete description of the Vibron Model for diatomic and triatomic molecules has been given by Iachello and Levine (1982) and van Roosmalen et al. (1983).

The literature concerning the Vibron Model has recently grown as a consequence of the fact that the model has been successfully applied to the description of rotational and vibrational degrees of freedom in polyatomic molecules. A list of the papers recently published is given in table 1 together with a list of the molecules treated in the frame of the algebraic methods therein described.

Table 1. Molecular species analyzed with the Vibron Model

Molecule	Reference
HCN, CO_2, H_3^+	van Roosmalen et al. (1983)
H_2O^{16}, H_2O^{18}, D_2O^{16}, H_2S^{32}, $S^{32}O_2^{16}$	Iachello and Oss (1990)
HCN, OCP, OCS	Cooper and Levine (1991)
NO_2, $C^{12}O_2$, $C^{13}O_2$, HCN, OCS	Iachello et al. (1991a)
C_2H_2	Hornos and Iachello (1989)
C_2H_2, C_2D_2, C_2HD	Iachello et al. (1991b)
HCCF	Viola (1991), Iachello et al. (1992a)
HCNO	Iachello et al. (1992b)
SF_6, WF_6, UF_6	Iachello and Oss (1991)
C_6H_6, $C_6H_{6-n}D_n$ $_{(n=1,...6)}$	Iachello and Oss (1992a)

The dipole interaction

In the algebraic-eikonal approach, the spherical harmonics of equation 3 are replaced by suitable algebraic operators. If the molecular structure is described by the Vibron Model in its O(4) dynamical symmetry, the appropriate operator for dipole transitions is of the type

$$\hat{D}_\mu = [\pi^+ \times \tilde{\pi}]_\mu^{(1)} \tag{6}$$

where π_μ^+ and π_μ, $\mu=0,\pm1$, are vibron creation and annihilation operators.

This operator is an element of the O(4) algebra and its matrix elements, as well as the matrix elements of the eikonal, with phase

$$\chi(\boldsymbol{b}) = \eta_1(b)\,\hat{b}\cdot\hat{D} \ , \tag{7}$$

can be calculated analytically. Their expressions have been worked out first by Bijker et al. (1986). In equation 7, $\eta_1(b)$ is an appropriate function of the impact parameter b.

Calculations using this theory have been performed for some dipolar diatomic molecules (Bijker et al. 1986, Bijker and Amado, 1986, 1992, Mengoni and Shirai, 1988, 1991, Alhassid and Shao, 1992). As an example, we show in figure 1 the algebraic–eikonal calculation of the differential cross section for electron scattering on HF molecule at incident energies of 2 eV and 3 eV.

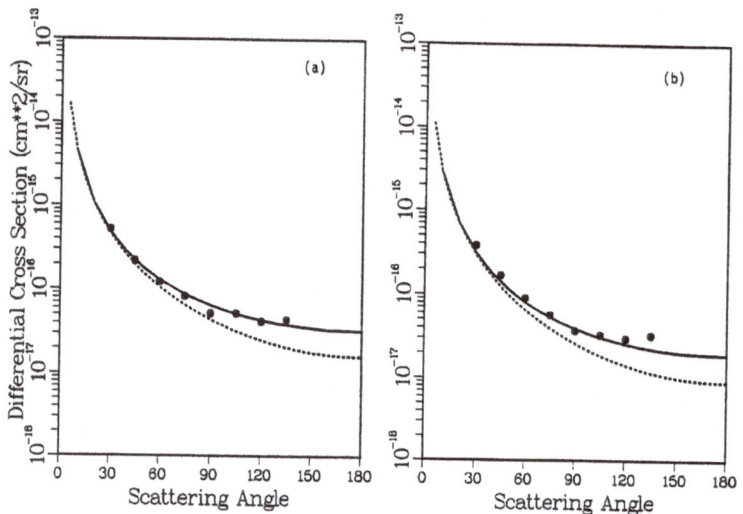

Figure 1. Differential cross section for electron scattering on HF molecule at 2 eV (a) and 3 eV (b). The experimental points are from Rädle et al. (1989). See the text for explanations.

The parameters used in this calculation are those given in the reference (Mengoni and Shirai, 1991). In the figure, the dotted lines represent the calculation done with the algebraic–eikonal formalism using dipole interaction only. The solid lines are calculated including the quadrupole interaction.

The calculated values compare favorably with the measurements and the inclusion of quadrupole interaction is needed for the reproduction of the large scattering angles part (see below).

For dipole interaction only, the algebraic–eikonal approach has been extended to the treatment of the scattering by triatomic molecules (Bijker and Amado, 1988). The calculation of the scattering cross section has been made only for the vibrational elastic scattering process on e⁻ + HCN.

The quadrupole interaction

This interaction has been included into the algebraic–eikonal formalism in order to improve the description of the collision process (Mengoni and Shirai, 1991), particularly at large scattering angles.

420

The quadrupole operator in the Vibron Model is

$$\hat{Q}_\mu = [\pi^+ \times \tilde{\pi}]_\mu^{(2)} \tag{8}$$

and the eikonal phase is given by

$$\chi(b) = \eta_2(b) \, (\hat{Q}_2 e^{-2i\phi_*} + \hat{Q}_{-2} e^{+2i\phi_*}) \tag{9}$$

where $\eta_2(b)$ is a function only of the impact parameter b.

Several methods have been developed to calculate the eikonal matrix elements for the quadrupole interaction containing the operator given above (Mengoni and Shirai, 1991, Alhassid and Shao, 1992). The effect of the quadrupole interaction on the vibrational elastic differential cross section can be seen in figure 1. Similar results have been obtained by Alhassid and Shao (1992).

The polarization interaction

This type of interaction has been included in the algebraic–eikonal approach using a distorted–wave approximation (Bijker and Amado, 1988, Alhassid and Shao, 1992). The polarization potential contains a tensorial as well as a scalar component. Its effect is important at large angles because the interaction potential is effective at shorter distances compared to the dipole component. However a strict algebraic treatment of this type of interaction has not being worked out, so far.

The vibrational excitation process

The inclusion of the vibrational excitation process requires special attention. In fact, in addition to the long–range dipole interaction, short–range interactions play a fundamental role for this process. For the dipole interaction, the operator of equation 6 is not appropriate to describe vibrational excitation. In fact, it has been shown that a good description of the electromagnetic transition probabilities between molecular rotational and vibrational states can obtained using the operator

$$\hat{T}_\mu = d_0 \hat{D}_\mu + d_1 \frac{1}{2} [e^{\lambda \hat{n}_*} \cdot \hat{D} + \hat{D} e^{\lambda \hat{n}_*}]_\mu \tag{10}$$

where $\hat{n}_\pi = \pi^+ \cdot \tilde{\pi}$ is the π–vibron number operator and d_0, d_1 and λ are constants (Iachello et al., 1991c). The inclusion of the second term in equation 10 allows for vibrational transitions and represents a first extension of the algebraic–eikonal approach. Also here, different techniques have been applied for the calculation of the eikonal matrix elements (Mengoni and Shirai, 1991, Bijker and Amado, 1992, Alhassid and Shao, 1992).

However, in order to describe this excitation process, a more realistic description of the short–range interaction is needed together with a better representation of the dynamic of the collision process. Both these requests can be satisfied using the hybrid approach.

THE HYBRID APPROACH

This approach has been proposed by Bijker et al. (1990) in order to better represent the collision dynamic. It essentially consists in adopting a full coupled–channel calculation for *low* partial-waves and the algebraic–eikonal approach for *high* partial-waves, in a partial–wave expansion of the scattering amplitudes.

In this way, the coupled–channel calculation effort is concentrated in a relatively small number of terms and can be handled by traditional techniques. On the other hand, the algebraic–eikonal approach is used for the high partial–wave terms which are dominated by the long–range dipole interaction.

This approach has been so far developed and tested against full coupled–channel calculations only for a diatomic molecular target (Bijker et al. 1990). Its capabilities will be most valuable in the treatment of electron collision with polyatomic molecules.

A different approach has been proposed very recently (Alhassid and Shao, 1992). It makes use of the rotating–frame approximation which transforms a full coupled–channel into a single–channel problem for a fluctuating potential. Its capabilities are being studied at the present.

CONCLUDING REMARKS

We would like to add here a few comments on the present developments of the theories reviewed above. As far as the algebraic–eikonal approach is concerned, the most important development needed is the inclusion of higher multipolarities in the interaction potential. The treatment of short–range interactions, like exchange and correlation-polarization, could in principle be handled using a multipole expansion similar of that of equation 3. In this case many more terms, in addition to the dipole and quadrupole, would be needed. However, within this frame, questions on the validity of the Glauber approximation at large angles arise. In this respect, the hybrid approach and the more recent rotating–frame approximation, are the most suitable for treatment of realistic situations. These last two methods are the appropriate basis for the treatment of electron collision processes with polyatomic molecules, where short–range interactions play the key role.

ACKNOWLEDGMENTS

This work has been supported by the Commission of the European Communities under the EC–STF programme in Japan. I would like to thank Dr. I. Shimamura of RIKEN and Prof. Y. Alhassid of Yale University for many valuable discussions on the subject. The continuous support of Dr. T. Shirai and Dr. Y. Kikuchi of JAERI during the several stages spent at JAERI/NDC is especially appreciated.

REFERENCES

Alhassid, Y., and Shao, B., 1992, *Phys. Rev. A*, in press.

Bijker, R., and Amado, R. D., 1986, *Phys. Rev. A* 34, pag. 71.

Bijker, R., and Amado, R. D., 1988, *Phys. Rev. A* 37, pag. 1425.

Bijker, R., and Amado, R. D., 1992, *Phys. Rev. A* 46, pag. 1388.

Bijker, R., Amado, R. D., and Sparrow, D. A., 1986, *Phys. Rev. A* 33, pag. 871.

Bijker, R., Amado, R. D., and Collins, L. A., 1990, *Phys. Rev. A* 42, pag. 6414.

Cooper, I. L., and Levine, R. D., 1991, *J. Mol. Spectr.* 148, pag. 391.

Hornos, J., and Iachello, F., 1989, *J. Chem. Phys.* 90, pag. 1989.

Iachello, F., 1981, *Chem. Phys. Lett.* 78, pag. 581.

Iachello, F., and Levine, R. D., 1982, *J. Chem. Phys.* 77, pag. 3046.

Iachello, F., and Oss, S., 1990, *J. Mol. Spectr.* 142, pag. 85.

Iachello, F., and Oss, S., 1991, *Phys. Rev. Lett.* 66, pag. 2976.

Iachello, F., and Oss, S., 1992a, *J. Mol. Spectr.* 153, pag. 225.

Iachello, F., Oss, S., and Lemus, R., 1991a, *J. Mol. Spectr.* 146, pag. 56.

Iachello, F., Oss, S., and Lemus, R., 1991b, *J. Mol. Spectr.* 149, pag. 132.

Iachello, F., Leviatan, A., and Mengoni, A., 1991c, *J. Chem. Phys.* 95, pag. 1449.

Iachello, F., Oss, S. and Viola, L., 1992a, *Mol. Phys.* in press.

Iachello, F., Manini N., and Oss, S., 1992b, *J. Mol. Spectr.* in press.

Itikawa, Y., 1971, *J. Phys. Soc. J.* 30, pag. 835.

Mengoni, A., and Shirai, T., 1988, *J. Phys. B: At. Mol. Opt. Phys.* 21, L576.

Mengoni, A., and Shirai, T., 1991, *Phys. Rev. A* 44, pag. 7258.

Rädle, M., Knoth G., Jung K., and Ehrhardt H., 1989, *J. Phys. B: At. Mol. Opt. Phys.* 22, pag. 1455.

van Roosmalen, O., Iachello, F., Levine, R. D., and Dieperink, A. E. L., 1983, *J. Chem. Phys.* 79, pag. 2515.

van Roosmalen, O., 1982, *Algebraic description of nuclear and molecular rotation–vibration spectra*, PhD thesis, Rijksuniversiteit Groningen, The Netherlands, unpublished.

Viola, L., 1991, *Spettroscopia rotovibrazionale di molecole quadriatomiche lineari nel modello a vibroni: monofluoroacetilene*, PhD thesis, Trento University, Italy, unpublished.

DYNAMIC SYMMETRY IN HADRON PHYSICS:

A MESONIC PUZZLE FOR THE $U(4) \supset SO(4)$ CHAIN

Nimai C. Mukhopadhyay and L. Zhang

Department of Physics, Rensselaer Polytechnic Institute
Troy, New York 12180-3590

ABSTRACT

We show that the dynamical $U(4)$ symmetry of the mesonic color string has interesting spectroscopic implications, independent of other details of color dynamics. We classify the mesonic spectral interval according to the subalgebras in the $U(4)$ chains into three types: O, U and N. The O type, in which the spectral interval decreases as a function of excitation energies, is shown to be more numerous in mesonic spectra than the other ones. The most frequent vibrational spectroscopic interval in the dynamical symmetry $U(4) \supset SO(4)$ should be N type in the limit of infinite number of mesonic states. The relative infrequent occurence of the N-type interval in the mesonic vibrational spectroscopy is a puzzle.

INTRODUCTION

It is with great pleasure that we join our colleagures from around the world to honor our friend, collaborator and esteemed colleague, Professor Francesco Iachello on his fiftieth birthday. We wish Franco many happy years of good health and exciting excursions into the beauties of theoretical physics. We also say "Grazie!" to him for selflessly sharing with us his enthusiasm and knowledge of physics. What follows below is to greet him with flowers of our common garden that we have jointly been cultivating for some time[1], along with others. We hope he would approve.

One of Iachello's major contributions has been to bring to bear upon problems of molecular, atomic, nuclear and particle physics, the power of group theory[2] through the consideration of symmetry of the appropriate system. Even though such symmetries are not perfect in many cases, the ideas of dynamic symmetry (DS) and spectrum generating algebras are helpful to write mass formulas and to compute transition rates in the appropriate quantum systems[3]. This has led us[1] to consider similiar methods for hadron spectroscopy. Though such methods were tried in the early years[4], these fell into disuse, after the gauge theory of quantum chromodynamics (QCD) became a part of the standard model framework as a universally accepted theory of strong interaction. The reason for our revival of venerable group-theoretic techniques in hadron spectroscopy is not to challenge QCD, but to see if we can gain *new insights* into the dynamics of quarks and gluons, hidden in the QCD. At a minimum, we hope to learn about properties of hadrons reasonably quickly through the elegant simplicity of the group-theoretical methods. This is also a way to check and extend results of the successful quark model approach[5,6].

What follows below is a discussion of the various mesonic vibrational intervals expected in the model with the $U(4)$ group as a starting symmetry for the mesonic color string. We shall show that one expects three spectral classes in the dynamic symmetry chains $U(4) \supset SO(4)$ and $U(4) \supset U(3)$: we call them O-type, the U-type and the N-type, depending on the nature of the vibrational interval. In the limit of infinite number of vibrational states of the color string, we should expect the N-type spectrum in the vibrational context for the chain $U(4) \supset SO(4)$. We shall examine the mesonic vibrational intervals for the validity of this conclusion and point out a puzzle: there is precious little evidence for this anticipation. We believe the understanding of this to be of basic importance in the further development of the dynamic symmetry approach for hadrons, particularly for the $U(4)$ chains.

MESON SPECTRAL CLASSES FOR THE DYNAMIC SYMMETRY

OF THE $U(4)$ ALGEBRAS OF THE COLOR STRING

Treatments motivated by QCD, from bags to lattices, suggest[7,8] a string-like structure of mesons. A perusal of this aspect of the $q\bar{q}$ color bond has led us[1] to consider the consequences of the DS of the $U(4)$ group, which is the compactified approximate realization of the relevant non-compact group $U(3,1)$. It must be stated that this symmetry assignment is not unique. There could be other symmetry groups that might also serve as a good starting point. The experience of Iachello and collaborators in molecular[9], and nuclear physics[10] is a very helpful guide here. Now if we insist on the orbital angular momentum L of the $q\bar{q}$ system as a good quantum number, the spectrum generating algebra of $U(4)$ admits only two subalgebra chains:

$$\begin{aligned} I \quad & U(4) \supset SO(4) \supset SO(3) \supset SO(2), \\ II \quad & U(4) \supset U(3) \supset SO(3) \supset SO(2). \end{aligned} \tag{1}$$

Here we shall explore some characteristic spectroscopic implications of the DS via either of the two alternatives (1), *without introducing any other dynamical considerations of the quark-gluon dynamics*. We shall show that this immediately leads to certain characteristic spectral interval rules that lend themselves to direct experimental scrutiny. We shall then examine their relevance to known mesons.

The only difference between the chains I and II comes at the first subalgebra, which are respectively $SO(4)$ and $U(3)$. Thus, the DS mass formula for mesons, without lifting the degeneracies in $SO(3)$ and $SO(2)$, are[9]:

$$M^2(N, \, \omega) = \alpha_a + \alpha_b \omega(\omega + 2) \,, \tag{2}$$

$$M^2(N, \, n) = \beta_0 + \beta_1 n + \beta_2 n(n + 3) \,, \tag{3}$$

where α_i, β_i are group constants, N, ω and n denote the quantum numbers for $U(4)$, $SO(4)$ and $U(3)$ respectively. Using the vibrational quantum number ν for the mesonic string,

$$\nu = \frac{(N - \omega)}{2} \,, \tag{4}$$

Eq.(2) can be written as

$$M^2(N, \, \nu) = \alpha_0 + \alpha_1[(N + 1)\nu - \nu^2] \,, \tag{2'}$$

where we have absorbed terms involving N only in α_0, and $\alpha_1 = -4\alpha_b$, another group constant.

In a confined $q\bar{q}$ system, the total number of meson states must be infinite. Thus, in our description of the hadron spectra, we must take the quantum number N in Eqs. (2), (3) to be infinitely large. In practice, we can take N to be a large number to accomondate states accessible readily to experiments. We shall take N to be a fixed quantity for mesons, the effective value of which would be large(≈ 100).

We can immediately find two kinds of characteristic meson intervals, entirely fixed by the DS property of the $U(4)$ subalgebra chain. We shall call these the O type and the U type, standing for $SO(4)$ and $U(3)$ respectively, key point of difference in the two subalgebra chains in (1). From (2'), we have

$$M^2(\nu + 1) - M^2(\nu) = \alpha_1(N - 2\nu) \,, \tag{5}$$

suppressing the label N on the left-hand side hereafter. Clearly (5) must be positive. Since $N > 2\nu$, *this interval monotonically deceases as ν increases.* We shall call this the *O-type interval.* From (3), we get

$$M^2(n + 2) - M^2(n) = \beta_1 + \beta_2(4n + 10) \,. \tag{6}$$

Here $n = N$, $N - 1$, ...1, 0. Positivity of this interval requires

$$\beta_2(4n + 10) > -\beta_1 \,.$$

Thus, *this interval monotonically increases as n increases,* a feature quite distinct from the O-type interval. We shall call this the *U-type interval.* A third special case is possible when *the interval, defined either in (5) or (6), reaches a constant value, independent of the quantum number ν or n.* We shall term this the *N-type interval.* This happens in (5), when

$$N \gg \nu, \tag{7a}$$

yielding

$$M^2(\nu + 1) - M^2(\nu) \approx \alpha_1 N = const. \tag{7b}$$

It occurs in (4), when

$$\beta_2 \to 0, \tag{7c}$$

giving

$$M^2(n+2) - M^2(n) \approx \beta_1 = const. .$$ (7d)

Note that any allowed value of ν can be combined with that of orbital angular momentum quantum number L, but in latter case n can only be L, $L+2$ etc. From the spectral interval alone, we cannot distinguish between the two special cases (7a), (7c).

Note that the requirement that the $q\bar{q}$ system has an infinite number of bound states, forces N to be large, so as to meet the condition (7a) easily. Thus, *if $U(4) \supset SO(4)$ is the DS of the $q\bar{q}$ meson system, we should expect the mesonic spectral interval (5) to be the N-type for low values of ν.* As we shall see below, this is hardly the case. Quite a few examples of the O-type behavior are seen in the meson families. This is our puzzle.

EXAMPLES FROM MESON FAMILIES:

WHERE ARE THE N-TYPE VIBRATIONAL INTERVALS?

The interesting point about *the interval test* for the DS characteristic of the meson string, discussed above, is that *it can be used directly in conjuction with the spectroscopic data without invoking any other theoretical consideration, if appropriate meson states are selected.* A point to keep in mind here is that any chosen interval should not involve physics of lifting degeneracies in the $SO(3)$ and $SO(2)$ subalgebras. Thus, it should not involve different L, S and J values. Most complete spectroscopic informations[11] on such states that we can utilize involve the $J^{PC} = 0^{-+}$ and 1^{--} states: the former are in the π and K families, the latter in the heavy meson families as well. For the former, let us consider the states involving the spectroscopic designations 1^1S_0, 2^1S_0 and 3^1S_0, using the standard notation $n^{2S+1}L_J$ of the quark model. Two examples available are from the π and K families. In the π family, $\pi^{\pm}$, $\pi(1300)$ and $\pi(1700)$ can be safely assigned 1^1S_0, 2^1S_0 and 3^1S_0 respectively. Within experimental errors, these are consistent with the *O-type behavior*:

$$M^2[\pi(1770)] - M^2[\pi(1300)] < M^2[\pi(1300)] - M^2(\pi^{\pm}).$$ (8)

Since the left and right sides of the inequality (8) are $1.44 \pm 0.37 GeV^2$ and $1.67 \pm 0.26 GeV^2$, *we cannot rule out here the N-type behavior.* In the K-family, we can assign 1^1S_0, 2^1S_0 and 3^1S_0 to $K^{\pm}$, $K(1460)$ and $K(1830)$ respectively. These again seem to display an O-type behavior:

$$M^2[K(1830)] - M^2[K(1460)] < M^2[K(1460)] - M^2(K^{\pm}),$$ (9)

the left-hand side of this inequality giving $1.2 GeV^2$, while the right-hand side yielding $1.9 GeV^2$. Since the PDG do not assign any errors for the masses of $K(1830)$ and $K(1460)$, we cannot yet estimate the errors on the intervals involved in (7), but the difference of mass squared looks clear-cut here.

For examples of O-type behavior in the 1^{--} mesons, we now go to the charmonium and bottomonium families. In the former, we can take the states $J/\psi(3097)$, $\psi(3685)$ and $\psi(4040)$. Quark model assignments of the first two states are unmistakable: 1^3S_1 and 2^3S_1 respectively. Due to substantial threshold effects, there are difficulties in the quantum number assignment for the state $\psi(4040)$. We shall use the likely assignment[12]

3^3S_1, which needs confirmation. We then read off the *O-type* interval rule:

$$M^2[\psi(4040)] - M^2[\psi(2S)] < M^2[\psi(2S)] - M^2[J/\psi(1S)], \tag{10}$$

the left-side of this inequality being $2.7 GeV^2$ and the right-hand side being $4.0 GeV^2$. This rule is even more firm in the Υ-family, if we consider the 1^{--} states $\Upsilon(9460)$, $\Upsilon(10023)$ and $\Upsilon(10365)$, which have confirmed spectroscopic assignments 1^3S_1, 2^3S_1 and 3^3S_1 respectively. These follow the *O-type* interval rule:

$$M^2[\Upsilon(3S)] - M^2[\Upsilon(2S)] < M^2[\Upsilon(2S)] - M^2[\Upsilon(1S)], \tag{11}$$

where the left-hand side of the inequality is $6.8 GeV^2$, and the right-hand is $11 GeV^2$. This progression continues with the PDG assignment of $\Upsilon(10580)$ as $\Upsilon(4S)$. Thus, we find an extension of the *O-type* interval rule:

$$M^2[\Upsilon(4S)] - M^2[\Upsilon(3S)] < M^2[\Upsilon(3S)] - M^2[\Upsilon(2S)], \tag{12}$$

the left-hand side of the inequality (10) being $4.7 GeV^2$. Thus, (11) and (12) together provide a three-interval inequality of the *O-type*, the most elaborate of our sequences so far:

$$\Delta(4,3) < \Delta(3,2) < \Delta(2,1), \tag{12'}$$

where $\Delta(n_1, n_2) = M^2(n_1) - M^2(n_2)$, n_i is the principal quantum number of the Υ state i.

We shall now examine the probable evidence for the *U-type* interval in the meson spectrum. Let us consider the π-family first and take the sequence of states $\rho(770)$, $\rho(1450)$ and $\rho(2150)$. Taking them as 1^3S_1, 2^3S_1 and 3^3S_1 respectively, we find here an *inversion from the O-type rule*:

$$M^2[\rho(2150)] - M^2[\rho(1450)] > M^2[\rho(1450)] - M^2[\rho(770)], \tag{13}$$

where the left-hand side is $2.5 GeV^2$ and the right-hand side is $1.5 GeV^2$. This is a candidate for the *U-type* sequence. One uncertainty from the spectroscopy here: no experimental errors for the mass uncertainty of $\rho(2150)$ is given in PDG, so it is possible that $\rho(2150)$ is a mixture of 3^3S_1 and 2^3D_1. In the charmonium family, consider the states $\psi(4415)$, $\psi(4040)$ and $\psi(3685)$. Taking our previous assignments of the last two of these as 3^3S_1 and 2^3S_1 respectively, a *U-type* sequence emerges if $\psi(4415)$ turns out to be 4^3S_1, as is suspected[1]; this is so, since

$$M^2[\psi(4415)] - M^2[\psi(4040)] > M^2[\psi(4040)] - M^2[\psi(3685)], \tag{14}$$

the left-hand side being $3.2 GeV^2$, and the right-hand side $2.7 GeV^2$. Again, this is not without some spectroscopic ambiguity. In a detailed DS approach that we have considered elsewhere[1], we have assigned $\psi(4040)$, $\psi(4160)$ and $\psi(4415)$ the $O(4)$ quantum numbers of $\nu = 2, 1, 3$ respectively, corresponding to $L = 0, 2, 0$, in that order. These, in the $U(3)$ model, should then have $n = 4, 4, 6$ also. In general, QCD arguments[13] would suggest that annihilation of the flavor diagonal $q\bar{q}$ systems should be *inhibited* in states of higher L, compared with $L = 0$. However, experimentally[14,11] $\Gamma(e^+e^-)$, the width of decay of these mesons into e^+e^- pair, in keV, is known to be 0.75 ± 0.15, 0.77 ± 0.23 and 0.47 ± 0.10 respectively, the first two numbers showing practically no difference at all within experimental errors. A much more precise experimental determination of these widths would be crucial to settle this point.

In the $b\bar{b}$ family[15,11], two groups have reported $5S$ and $6S$ excitation states at 10.85GeV and 11.02GeV respectively. The state at 10.85GeV breaks the O-type interval rule of the sequence (12'), as

$$M^2[\Upsilon(108500)] - M^2[\Upsilon(4S)] > M^2[\Upsilon(4S)] - M^2[\Upsilon(3S)], \qquad (15)$$

wherein the left-hand side is 5.73GeV^2 versus 4.7GeV^2 of the right-hand side of the inequality (15). This is an evidence of the U-type rule. However the $6S$ state at 11.02GeV resumes the O-type interval sequence established in (12'):

$$M^2[\Upsilon(6S)] - M^2[\Upsilon(108500)] > M^2[\Upsilon(108500)] - M^2[\Upsilon(4S)]. \qquad (16)$$

Here is yet another spectrocopic question to be settled. If the assignment of $\Upsilon(108500)$ is confirmed to be $\Upsilon(5S)$, (13), (14) and (15) would establish a pattern: excited states at very high energy might correspond to an U-type sequence rather than an O-type. This would definitely *disagree* with the expectations of best variant of the quark shell model calculations for mesons: the calculation by Godfrey and Isgur[16] predicts an O-type sequence for the (nS) states of $c\bar{c}$ and $b\bar{b}$ mesons in a monotonic fashion. This is an important physics point to be settled by future precision spectroscopy.

CONCLUDING REMARKS

Our works on meson spectroscopy[1] using the dynamic symmetry $U(4) \supset SO(4)$ suggests, with the possible exception of heavy flavored mesons, that the best overall fit for all meson families is obtained when the quantum number N is chosen to have a large value. Given the fact that $U(4)$ is taken to be a compactification of $U(3,1)$, large N is indeed the limit needed[3] to justify the $U(4)$ choice in the context of mesons as color strings. However, our previous analysis indicates that we cannot as yet establish the numerically large value of N from the vibrational spectrum, except possibly for one N-type interval in (8), from the known mesonic spectra. Thus, our phenomenological need for large N in mass fits could be an indication of an inversion of the interval from the O-type to the U-type in the region of high excitation, the N-type fit being the best numerical compromise between these two. Further experimental and theoretical works are needed to explore this intriguing prospect of possible inversion of mesonic spectral interval connected with the $U(4)$ dynamical symmetry of the meson color bond. Most important of all, we also need to explore the underlying dynamical mechanisms that select one type of interval over another.

We should point out that our interval classification scheme depends on the orbital angular momentum L being a good quantum number. Tensor color hyperfine interaction[17], of course, destroys that assumption. Quark shell model investigations[16] for mesons show relatively small mixing of L values in most cases. This should help preserve the goodness of our scheme.

In summary, we have demonstrated that the dynamical $U(4)$ symmetry of the mesonic color string has interesting spectrocopic implications which are relatively insensitive to the other aspects of color dynamics. This allows us to classify mesonic spectral interval into three types. From experimental informations on mesons[11], we have demonstrated the definite presence of the O-type intervals, and shown possible examples of the other two. A crucial question that we are unable to answer, in the context of the dynamic symmetry $U(4) \supset SO(4)$, as to why the mesonic vibrational interval are not all N-type, the natural type in the limit $N \to \infty$, which, by the way, is not expected from the quark model either. It calls for further exploration of mesonic spectroscopic properties both experimentally and theoretically. This has implications for baryons as well, which promise to be exciting.

ACKNOWLEDGEMENTS

We thank Professor F. Iachello for many valuable discussions. This research is supported by the U.S. Department of Energy Grant #$DE - FG02 - 88ER40448.A004$.

REFERENCES

[1] F. Iachello, N. C. Mukhopadhyay and L. Zhang, Contributed Paper to the International Conference on Particles and Nuclei(PANIC-XII), Cambridge, Massachusetts(1990); *Phys.Lett.* $\underline{B256}$: 295(1991); *Phys.Rev.* $\underline{D44}$: 898(1991). See also F. Iachello and D. Kusnezov, *Phys.Lett.* $\underline{B255}$: 493(1991) and *Phys.Rev.* $\underline{D45}$: 4156(1992); F. Iachello and T.-S. H. Lee, preprint(1992). For a discussion of mesonic vibrational states in the quark model, see N. C. Mukhopadhyay and L. Zhang, *Phys.Rev.* $\underline{D44}$: 2085(1991).

[2] F. Dyson, "Symmetry Groups in Nuclear and Particle Physics", W.A. Benjamin, New York(1966).

[3] F. Iachello, *Nucl.Phys.* $\underline{A497}$: 23c(1989).

[4] F. Gürsey and L. A. Radicati, *Phys.Rev.Lett.* $\underline{13}$: 173(1964); Y. Dothan, M. Gell-Mann and Y. Ne'eman, *Phys.Lett.* $\underline{17}$: 148(1965).

[5] R. P. Feynman, M. Kislinger and F. Ravndal, *Phys.Rev.* $\underline{D3}$: 2706(1971).

[6] N. Isgur and G. Karl, *Phys.Rev.* $\underline{D18}$: 4187(1979).

[7] K. Johnson and C. B. Thorn, *Phys.Rev.* $\underline{D13}$: 1934(1976).

[8] I. Bars and A. J. Hanson, *Phys.Rev.* $\underline{D13}$: 1744(1974).

[9] F. Iachello and R. D. Levine, *J.Chem.Phys.* $\underline{77}$: 3046(1982); O. S. van Roosmalen, F. Iachello, R. D. Levine and A. E. L. Dieperink, *J.Chem.Phys.* $\underline{79}$: 2515(1983).

[10] F. Iachello and A. Arima, "Interacting Boson Model", Cambridge Univ. Press, Cambridge(1987), and references therein.

[11] K. Hikasa *et al.* (Particle Data Group[PDG]), *Phys.Rev.* $\underline{D45}$: I.1(1992).

[12] K. Gottfried and V. F. Weisskopf, "Concepts of Particle Physics", Vol II, Oxford, New York(1986), p.328.

[13] F. Close, "An Introduction to Quarks and Partons", Academic, New York and London (1979), p.405.

[14] R. Brandelik *et al.* , *Phys.Lett.* $\underline{76B}$: 361(1978); J. Siegrist *et al.* *Phys.Rev.Lett.* $\underline{36}$: 700(1976).

[15] D. M. J. Loveloch *et al.* *Phys.Rev.Lett.* $\underline{54}$: 377(1985); see also D. Besson *et al.*, *ibid* $\underline{54}$: 381(1985).

[16] S. Godfrey and N. Isgur, *Phys.Rev.* $\underline{D32}$: 189(1985).

[17] A. De Rújula, H. Georgi and S. L. Glashow, *Phys.Rev.* $\underline{D12}$: 147(1975).

QCD FOUNDATIONS FOR HADRON REGGE TRAJECTORIES

AND FOR THE ARIMA-IACHELLO SYMMETRIES OF NUCLEI

Yuval Ne'eman[1]*$ and Djordje Sijacki[2]

[1]Raymond and Beverley Sackler Faculty of Exact
Sciences Tel-Aviv University, Tel-Aviv, Israel 69978

*Wolfson Chair Extraordinary in Theoretical Physics
Tel-Aviv University

$Supported in part by the US DOE Grant
DE-FG05-85ER40200; also on leave from Center for
Particle Physics, University of Texas, Austin

[2]Institute of Physics, P.O.Box 57, Belgrade, Yugoslavia

This work is dedicated to Franco Iachello, for whose important contributions to the study of nuclear structure, through the phenomenological and kinematical identification of the symmetries of nuclei, we are trying to provide a dynamical interpretation.

ABSTRACT

We review our Pseudo-Gravity hypothesis which points to the two (or more) gluon exchange in QCD as the origin of Regge excitations and a variety of other hadronic features resembling gravity. We present a detailed dynamical study. One effect in nuclei is the emergence of the Arima-Iachello model, with its 2^+, 0^+ ground state. We explain the relevant dynamics.

1. APPROXIMATING QUANTUM CHROMODYNAMICS

Quantum Chromodynamics (QCD) is the component of the "Standard Model" (SM) describing the Strong Interactions. It had originally been suggeste [1] as a force explaining the zero-SU(3) triality through a dynamical saturation mechanism, but still with the additional role of preserving integer-valued electric charges, so that the quarks' fractional values would represent averages. The latter attempt was not vindicated by experimental results relating to sum-rules, in which the two sets of charges could be distinguished. However, success in the renormalization of the Yang-Mills interaction [2] led to the proof of asymptotic freedom

Symmetries in Science VII, Edited by B. Gruber
and T. Otsuka, Plenum Press, New York, 1994

[3-5], providing direct evidence for an SU(3)$_{colour}$ gauge interaction at short distances (SD) as compared to hadron size. This explains, in principle, the successes of the Non-Relativistic Quark Model (NRQM) and the observation of scaling in deep-inelastic electron-nucleon scattering experiments; these successes were deemed sufficient to assume that UV Asymptotic Freedom implied, at the other end of the spectrum, Infrared Confinement - or the justification of the extrapolation to long distances (LD), on the scale of hadronic sizes. QCD was repostulated [6,7] with fractional charges; yet one more element which went into this anzatz is the seggregation of the flavour-SU(3) breaking component of what used to be considered as the Strong Interactions, a postulate I had suggested in 1964 [8]. It was now simply left as a theoretical input for the SM, in the form of a set of quark masses - with a similar set of lepton masses. Note that the idea that flavour-SU(3) breaking quark masses and lepton masses have a single origin is also contained in my original postulate [8].

The non-linear nature of the Yang-Mills force - and the strength of the effective couplings involved - have hampered until now any direct comparison between experiments and the theory. Qualitatively, the observation of jets fits well with what the theory says about the gluons; as to confinement, there is at this stage no complete proof that Yang-Mills theory enforces it, though it has been proven on a discrete lattice (replacing spacetime) or for SU(n), n → ∞.

The present application of QCD, anywhere but in the (perturbative) UV region, thus relies exclusively on approximations. "Horizontally" *, this has recently involved {flavour-SU(3) x heavy-quark-SU(3)}, an extension of our 1961 SU(3) (which is "light-quark SU(3)"), i.e. of the NRQM. Vertically*, the structure is much more complicated and there is a need for a good approximation. The "Bag" model did provide at the time a qualitative picture. Studies based on the Skyrme model in an improved version have provided for sample non-perturbative solutions, mostly qualitatively. More recently, three methods have been utilized to provide quantitative information about the vertical structure. Following Lovelace and Lipatov-Kirschner, McGuigan and Thorn [9] have used the n → ∞ approximation; Capstick and Isgur [10] have developed a method of relativistic corrections, based on QCD, to NRQM predictions.

The present paper reviews a vertical approximation, which we put forward in 1990 [11], mostly to understand hadron structure. This approximation is based on <u>a precise identification of a gravity-like component, active in the IR region of QCD and dominating the LD zone.</u> We then applied the method to nuclei [12], showing that it provides the dynamical underpinnings for the emergence of the Arima-Iachello "IBM" symmetry [13-14] and other nuclear schemes. In further work, we showed how these dynamics explain the J ~ M² pattern of Regge trajectories [15]. Here, we shall analyze the static potential fitting this picture. We also studied the extension of these symmetries by C, P, CP [16].

2. PRECURSORS: GRAVITY-LIKE THEORIES FOR THE STRONG INTERACTIONS

Pseudo-Gravity - as we named our method - also explains the partial success of pre-QCD treatments, in which some form of Gravity was invoked in explaining Strong Interactions. In the current-algebra (1962-66) phase, the coupling of the f⁰ (2⁺,1235) meson, a hadron, to other hadron matter, was shown to be approximately universal [17-18] due to its 2⁺ spin, using a dispersion relations analysis. Salam [19-22] and collaborators later chose to replace this phenomenological approach by a direct postulate of "Strong Gravity", a geometrical doubling of the General Theory of Rela-

---

* horizontally - relating to the flavour structure; vertically - to the angular momentum excitation structure.

434

tivity, generating both ordinary Gravity and the Strong Interactions. A mixing between the two "gravitons" results in diagonal states, namely the existing zero-mass Einstein graviton, plus the f^0. In a later version of this theory, the derivation exploits the merger of colour-SU(3) with the Lorentz group, as in static (1964) Gursey-Radicati-Sakita SU(6). This is then extended to an Einstein-like symmetry of the local frames, with SL(6,C) replacing SL(2,C). Part of the justification in assigning a special role to the f^0 in Strong Interactions derived from its position on the Pomeranchuk trajectory, i.e. to a feature dominating high energy scattering and total cross-sections. This in itself points to a possible link between Gravity and the Strong Interactions - namely that both are dominated (though in apparently different ways) by a mechanism with the internal quantum numbers of the vacuum. Other phenomenological theories explored this feature, providing highly successful fits with experiments; they relied very much on the "universality" displayed by the "charges" carried by "high energy currents" [23], i.e. Regge vertices for poles or cuts with $J^P = 1^-$, 2^+. They also involved the Pomeron [24] and the universality of its couplings [25].

Salam's Strong Gravity depended on an extra input, either the doubling of the Einstein geometry or the SL(6,C) merger. In a different approach, it was assumed [26] that Einsteinian Gravity, with its unrenormalizability, malizability, represents just the low-energy residue of a larger structure. The Strong Interactions might then originate in some part of that hidden quantum structure. In the Poincaré Gauge Theory [27], a candidate theory for this "fuller" Gravity, we found [28] that the theory indeed provides for a confining potential, with components going ~ r or ~ r^2, aside from the Newtonian ~ 1/r. An Affine model [29] appeared to yield additional characteristic features, fitting this marriage between Strong Interactions and Gravity. Stelle [30,31] indeed showed that even when staying Riemannian and with Einstein's Lagrangian, quadratic terms emerge anyhow, from quantum renormalization effects, making the theory renormalizable, though not unitary.

Another partially successful QCD precursor with Gravity-like features is the String [32,33]. Here was a theory of the Strong Interactions yielding an excitation spectrum which did look like a stylization of the observations, with infinite linear arrays, etc.. This was not the first such description. Back in 1965, we had applied the techniques of non-compact groups [34] to the description of Regge trajectories. The unitary irreducible representations of SL(3,R), and more specifically the multiplicity-free subset, provided a perfect setting for the systematics of boson excitations (we did not know about the representations of the double-covering at that time) in hadrons and nuclei (sequences of levels with spins 0,2,4,.. and 1,3,5,..). In deformed nuclei, inertial effects were involved, triggered by the quadrupole moments, leading to the SU(3) symmetry of Elliott [35], now supplemented by SL(3,R) [36] and incorporated in the generalized systematics of Sp(3,R) [37,38]. Thus, for the hadron Regge sequences, we conjectured a similar effect due to deformative pulsations, i.e. a concrete gravitational effect producing what one would have considered as a highly typical result of Strong Interactions [34].

We now return to the String. Aside from the spectrum pattern, here again, as in Salam's theories, a 2^+ particle appeared to play a special role (for closed strings; open strings would point to a 1^- Yang-Mills-like state). Between 1974 (the Yoneya/Schwarz-Scherk suggestion that the string be used as a theory of Quantum Gravity) and 1984 (proof of the vanishing of anomalies) the Superstring indeed moved into its present role as a candidate Theory of Everything - in fact a theory of Quantum Super-Gravity with Unification. The smoothness of the transition in this change of roles testifies to the apparent overlap and resemblances between the two interactions, when one disregards the difference in coupling strengths.

In the pursuit of the above mentioned Affine Gravity program, aiming at a renormalizable theory of Gravity - a result we attained in the sequence represented by ref. [39-42], though unitarity is not yet achieved to date - we were impressed by the good experimental fit with a classification based on the manifields structure [43] - fields whose structure is given by the tensorial and spinorial infinite representations of SL(4,R) and of its Double-Covering $\overline{SL}$(4,R) [44], just as ordinary tensor fields represent finite representations of SL(4,R). The Hilbert space hadron spectrum is given by the unitary representations of $\underline{SA}$(4,R), induced over the little group $\underline{SL}$(3,R) with its infinite irreducible representations. This completed a full circle from [34], but in between, we had discovered the double-coverings and its unirreps.

As a result we assumed a phenomenological approach, in which we adopted the group representations structure of $\underline{SL}$(4,R) [45] without a proven dynamical motivation, except for a general linkage between colour confinement, a geometric feature (the conservation of a volume) with a geometric gauge theory [45].

3. PSEUDO-GRAVITY: A COMPONENT OF QCD

Pseudo-Gravity involves nothing but QCD. It isolates, in the IR region - both in the confinement zone and in the longer range inter-hadron and nuclear zones - a subset of Feynman diagrams generating a Gravity-like force. Once this is recognized, most of the clues which appeared to point to a link between the Strong Interactions and Gravity can now be understood as the effects of this component of QCD.

We can describe our approach as an application of the theorem $1 + 1 = 2$ (this was also the way in which the J =2 state appeared in the closed string). The key to our approach is the fact that in the interaction between two (colourless) hadrons, there is a contribution represented by the exchange of 2 gluons (or more), thus states with spins 2,1,0.

The local colour-SU(3) variation of the $B^a{}_\mu(x)$ gluon in QCD is given by

$$\delta_\epsilon B^a{}_\mu = \partial_\mu \epsilon^a + B^b{}_\mu (\lambda_b)^a{}_c \epsilon^c \tag{3.1}$$

We now separate the constant component $N^a{}_\mu$ with vanishing field strength,

$$B^a{}_\mu = N^a{}_\mu + A^a{}_\mu \ , \ \partial_\mu N^a{}_\nu - \partial_\nu N^a{}_\mu = if^a{}_{bc} N^b{}_\mu N^c{}_\nu \tag{3.2}$$

Between hadrons, i.e. between zero-colour systems, the simplest gluonic exchange is thus given by the two-gluon combination,

$$G_{\mu\nu}(x) = B^a{}_\mu B^b{}_\nu \eta_{ab} \tag{3.3}$$

whose colour-SU(3) local gauge variation is then,

$$\delta_\epsilon G_{\mu\nu} = \delta_\epsilon \{\eta_{ab}(N^a{}_\mu + A^a{}_\mu)(N^b{}_\nu + A^b{}_\nu)\} =$$

$$= \eta_{ab}(\partial_\mu \epsilon^a N^b{}_\nu + N^a{}_\mu \partial_\nu \epsilon^b + \partial_\mu \epsilon^a A^b{}_\nu + A^a{}_\mu \partial_\nu \epsilon^b)$$

$$+ i\eta_{ab} \{f^a{}_{cd} B^c{}_\mu \epsilon^d B^b{}_\nu + f^b{}_{cd} B^a{}_\mu B^c{}_\nu \epsilon^d\} \tag{3.4}$$

The last bracket represents the homogeneous SU(3) transformation

$$if_{bcd} (B^b{}_\mu B^c{}_\nu + B^c{}_\mu B^b{}_\nu) \epsilon^d \tag{3.4a}$$

which vanishes for a scalar expression - or more technically, in this case, by total antisymmetry of the (compact) SU(3) structure constants. The terms involving the constant $N^a{}_\mu, N^b{}_\nu$ can be rewritten as effective "pseudo-diffeomorphisms", defined by

$$\xi_\mu \cong \eta_{ab} \, \epsilon^a \, N^b{}_\mu, \tag{3.5}$$

For terms in $A^a{}_\mu$, $A^b{}_\nu$, integration by parts yields $\eta_{ab}(\epsilon^a \partial_\mu A^b{}_\nu + \partial_\nu A^a{}_\mu \epsilon^b)$. Taking Fourier transforms, i.e. matrix elements for these gluonic fluctuations, we note that these terms vanish in the IR region. As a result, $G_{\mu\nu}(x)$ in (3.4) transforms in fact as a "world tensor" under our pseudo-diffeomorphisms (3.5),

$$\delta_\epsilon G_{\mu\nu} = \partial_\mu \xi_\nu + \partial_\nu \xi_\mu \tag{3.6}$$

It was shown long ago [47] that any massless spin-2 field will behave and couple like a graviton. Here it will be massless as a combined result of Poincaré invariance (and as its corollary, conservation of the energy-momentum tensor) together with its own role in "gauging" "Pseudo-Covariance". Note that the definition (3.3) ensures that since

$$G_{\mu\nu} = \eta_{ab} \, N^a{}_\mu \, N^b{}_\nu + .. \tag{3.6a}$$

it is <u>invertible</u>, a necessary condition for it to act as a metric.

In pondering about a gravity-like force resulting from quantum coherent "gluing" (of QCD gluons), it is encouraging to remember that there now exist models in which <u>gravity itself is an effective force</u>, induced through a similar mechanism in some other interaction [48-50]. Returning to Pseudo-Gravity, it is important to note that we are not describing a (massive) "gluonium" two-gluon bound state. Such a system does appear in hadron systematics, a "glue-ball" whose mass [51] is generally estimated (from lattice calculations) to lie around 1.7 GeV. Its exchange would represent yet another Yukawa force, competing with the exchanges of various quark-antiquark combinations, with a somewhat shorter range. This has nothing to do with the action of $G_{\mu\nu}(x)$, an effective potential whose action resembles that of gravity and is long-ranged. Moreover, as a result of Lorentz invariance,

$$D_\sigma G_{\mu\nu} = 0 \tag{3.7}$$

and we note that $G_{\mu\nu}$ is a Riemannian pseudo-metric, further stressing its gravitational-like action. All of this is also in contradistinction to the f^0 "strong graviton" of [19]: the f^0 is a massive particle (1,250 MeV) and is now thought to represent yet another quark-antiquark exchange, i.e. a Yukawa short range force. This is very different from the long range coherent effects of gravity, emulated in "pseudo-gravity" by the massless $G_{\mu\nu}$. As to the f^0, since it does possess the Lorentz quantum numbers of $G_{\mu\nu}(x)$, it will at best contribute a "nearby" pole to the dispersion relations for the interaction with the "pseudo-graviton" [17,18].

4. THE EFFECTIVE ACTION

The effective action for this IR (zero-colour) hadron sector of QCD, written as a pseudo-gravitational theory, with matter in $\underline{SL}(4,R)$ mani-fields, was derived in ref.[15] by requiring Riemannian structure, with the pseudo-metric $G_{\mu\nu}$, in the presence of the $\underline{SL}(4,R)$ hadronic currents. The action reads,

$$I = \int d^4x \, \sqrt{-G}\{ -a \, R_{\mu\nu} \, R^{\mu\nu} + b \, R^2 - c l_G^{-2} R + l_S^{-2} \, \Sigma_{\alpha\beta}{}^\gamma \, \Sigma^{\alpha\beta}{}_\gamma +$$

$$+ l_Q^{-2} \, \Delta_{\alpha\beta}{}^\gamma \, \Delta^{\alpha\beta}{}_\gamma + \mathcal{L}_M \} \tag{4.1}$$

The first three terms constitute the Lagrangian which Stelle [30,31] showed to be renormalizable, though not unitary. R is the scalar curvature and $R_{\mu\nu}$ the Ricci tensor. The fourth and fifth terms are spin-spin and shear-shear contact interaction terms (with tangent-plane or anholonomic indices); (a,b,c) are dimensionless constants; l_G, l_S and l_Q have the dimensions of lengths; from our knowledge of the hadrons we estimate them to be of hadron size ~ 1 GeV. $\mathcal{L}_M$ is the matter Lagrangian. The pseudo-metric field equations are given by the expressions (; stands for a covariant derivative, with non-propagating shear and torsion, i.e. with just the Christoffel connection):

$$- a\, R^{\mu\nu}{}_{;\sigma}{}^{;\sigma} + (a-2b)\, R^{;\mu;\nu} - (a/2 - 2b)\, G^{\mu\nu}\, R^{;\eta}{}_{;\eta} + 2a\, R^{\mu\sigma\nu\tau}\, R_{\sigma\tau} -$$

$$- 1/2\, G^{\mu\nu}\, (a\, R^{\eta\lambda}\, R_{\eta\lambda} - b\, R^2) - 2b\, R\, R^{\mu\nu} + c\, l_G^{-2}\, (R^{\mu\nu} - 1/2\, R\, G^{\mu\nu}) -$$

$$- 1/2\, l_S^{-2}\, \Sigma_{\alpha\beta}{}^{\gamma}\, \Sigma^{\alpha\beta}{}_{\gamma} - 1/2\, l_Q^{-2}\, \Delta_{\alpha\beta}{}^{\gamma}\, \Delta^{\alpha\beta}{}_{\gamma} = 1/2\, \theta^{\mu\nu} \tag{4.2}$$

It is obvious from this expression that $R_{\mu\nu}\, R^{\mu\nu}$ and R^2 terms in the action yield the fourth-order derivative equation terms, while R yields the usual second-order derivative terms. The latter terms, with their ~ 1/r static solution, would have been relevant at short distances, if it were not for the fact that this would then relate to a region lying outside of the zone of applicability of pseudo-gravity. At large distances, in the IR region itself, they can be neglegted owing to the rather soft r-dependence. For practical purposes we can thus set c = 0 and neglect the R term (Riemannian, with only { } as connection).

We linearize our theory in terms of

$$H_{\mu\nu}(x) = G_{\mu\nu}(x) - \eta_{\mu\nu}, \tag{4.3}$$

where $\eta_{\mu\nu}$ is the Minkowski metric. The appropriate action has the form,

$$I = \int d^4x\, \{(a/4)\, H_{\rho\sigma}\, \Box^2\, P_{(2)}{}^{\rho\sigma\mu\nu}\, H_{\mu\nu} + (a - 3b)\, H_{\rho\sigma}\, \Box^2\, P_{(0)}{}^{\rho\sigma\mu\nu}\, H_{\mu\nu} -$$

$$- (1/2)\, l_S^{-2}\, H^{\mu\nu}\, \Sigma_{\alpha\beta}{}^{\gamma}\, \Sigma^{\alpha\beta}{}_{\gamma}\, H_{\mu\nu} - (1/2)\, l_Q^{-2}\, H^{\mu\nu}\, \Delta_{\alpha\beta}{}^{\gamma}\, \Delta^{\alpha\beta}{}_{\gamma}\, H_{\mu\nu} -$$

$$- (1/2)\, \theta^{\mu\nu}\, H_{\mu\nu}\, \} \tag{4.4}$$

The two completely transverse projectors for J = 2 and J = 0 respectively, $P_{(2)}{}^{\rho\sigma\mu\nu}$ and $P_{(0)}{}^{\rho\sigma\mu\nu}$, are defined in terms of $\theta^{\mu\nu} = \eta^{\mu\nu} - (\partial^\mu\, \partial^\nu / \Box)$,

$$P_{(2)}{}^{\rho\sigma\mu\nu} = 1/2\, (\theta^{\rho\mu}\, \theta^{\sigma\nu} + \theta^{\rho\nu}\, \theta^{\sigma\mu}) - P_{(0)}{}^{\rho\sigma\mu\nu}$$

$$P_{(0)}{}^{\rho\sigma\mu\nu} = 1/3\, \theta^{\rho\sigma}\, \theta^{\mu\nu} \tag{4.5}$$

Raising and lowering indices is done with the Minkowski metric $\eta_{\mu\nu}$. Note that the Abelian gauge invariance of the linearized action involves the absence of differential operators containing $(\partial^\mu\, \partial^\nu / \Box)$. The direct consequence - an important point - is the absence of the J = 1 part of $H_{\mu\nu}$ in the action. For simplicity, we choose b = a/4. With the absence of the J = 1 mode we get for the action,

$$I = \int d^4x\, \{H_{\rho\sigma}\, [(a/4)\Box^2 - 1/2\, (l_S^{-2}\, \Sigma_{\alpha\beta}{}^{\gamma}\, \Sigma^{\alpha\beta}{}_{\gamma} +$$

$$+ l_Q^{-2}\, \Delta_{\alpha\beta}{}^{\gamma}\, \Delta^{\alpha\beta}{}_{\gamma})\,]P_{(2+0)}{}^{\rho\sigma\mu\nu}\, H_{\mu\nu} - (1/2)\, \theta^{\mu\nu}\, H_{\mu\nu}\}, \tag{4.6}$$

The projector $P_{(2+0)}{}^{\rho\sigma\mu\nu} = P_{(2)}{}^{\rho\sigma\mu\nu} + P_{(0)}{}^{\rho\sigma\mu\nu}$ insures that the only physical, propagating modes are $J^P = 2^+$ and $J^P = 0^+$. Restricting to the homogeneous part, as required for the evaluation of the propagator, we get for the $H_{\mu\nu}$ field the equation of motion,

$$[(a/4)\, \Box^2 - (1/2)\, l_S^{-2}\, \Sigma_{\alpha\beta}{}^{\gamma}\, \Sigma^{\alpha\beta}{}_{\gamma} - l_Q^{-2}\, \Delta_{\alpha\beta}{}^{\gamma}\, \Delta^{\alpha\beta}{}_{\gamma}]\{P_{(2+0)}H\}_{\mu\nu}(x) = 0 \tag{4.7}$$

438

which becomes in momentum space

$$[(a/4)(p^2)^2 - (1/2) \, l_S^{-2} \, f_S \, M_\alpha{}^\beta \, M^\alpha{}_\beta -$$

$$- (1/2) \, l_Q^{-2} \, f_Q \, T_\alpha{}^\beta \, T^\alpha{}_\beta]\{P_{(2+0)}H\}_{\mu\nu}(p) = 0 \qquad (4.8)$$

In these expressions, we have factored out the SL(4,R) group factors (the bilinear forms in the algebra's generators). The factors f_S and f_Q represent the residual parts of the configuration space integrals, while $\bar{P}_{(2+0)}{}^{\rho\sigma\mu\nu} = 1/2 \, (\theta^{\rho\mu} \, \theta^{\sigma\nu} + \theta^{\rho\nu} \, \theta^{\sigma\mu})$, where $\theta^{\mu\nu} = \eta^{\mu\nu} - p^\mu p^\nu / p^2$. In the following, we shall use $H_{\mu\nu}{}'$ for $\{P_{(2+0)} H\}_{\mu\nu}$.

For Pseudo-Gravity, we may regard (4.7) and (4.8) as the dynamical equations above the theory's "vacuum", as represented by hadron matter itself. The equations represent the excitations produced over that ground state by the Pseudo-Gravity potential; i.e. they are like equations for the $H_{\mu\nu}{}'$ field in an external field of hadron matter.

5. REGGE TRAJECTORIES

We have recently shown [15] that in a rest frame, these equations yield Regge trajectories in the Chew-Frautschi plot, one for each unitary irreducible representation of SL(3,R), the "little group" for this set of unitary (induced) representations of SA(4,R). Evaluating [52] the (single) Casimir invariant of SA(4,R), we have demonstrated that for these representations (type IIA) the Casimir invariant vanishes and there are thus no kinematical constraints on the mass/spin relation. Denoting by $C^2{}_{sl(3,R)}$ the quadratic Casimir invariant of SL(3,R), and by α' the (asymptotic) trajectory slope, we found,

$$(J + 1/2)^2 = (\alpha' m^2)^2 + \alpha_0^2, \qquad (5.1)$$

$$(\alpha')^2 = [(2/a)(l_S^{-2} \, f_S + l_Q^{-2} \, f_Q)]^{-1} \qquad (5.2)$$

$$\alpha_0^2 = 1/4 + (l_Q^{-2} \, f_Q)(l_S^{-2} \, f_S + l_Q^{-2} \, f_Q)^{-1} \, C^2{}_{sl(3,R)} \qquad (5.3)$$

Neglecting a slight bending at small m^2, i.e. the α_0^2 term, yields the Regge trajectory

$$J = \alpha' m^2 - 1/2 \qquad (5.4)$$

Note that the $J \sim m^2$ behaviour results directly from the quartic momenta appearing in quadratic gravity - i.e. $(m^2)^2$ - as juxtaposed in the action with (torsion)^2, the latter becoming (matter spin)^2, through an algebraic equation of motion.

6. STATIC POTENTIALS

We now consider the static solutions of the Pseudo-Gravity equation. I) We first focus on the IR zone within hadronic matter, $r \le .8$ fm $< r_h$. Pseudo-Gravity dominates, far from the UV zone. The momentum-space Green function is given by

$$G^{\rho\sigma\mu\nu}(p) = (4/a) \, \{\bar{P}_{(2+0)}{}^{\rho\sigma\mu\nu} /(p^2)^2\} \qquad (6.1)$$

while the Pseudo-Gravity potential, in the static limit, takes the form, for a point source with typical hadronic strength

$$H := H^\mu{}_\mu{}', \quad (a/4)(\nabla^2)^2 \, H(r) = (1/\alpha') \, \delta^3(r) \qquad (6.2)$$

The solution is the canonical linear confining potential

$$H(r) = (1/2\pi a a') \, r \, , \qquad r_{UV} \ll r \lesssim .8 \text{ fm} < r_h \tag{6.3}$$

II) We now turn to distances larger than the typical hadronic size r_h. In this region, the shear and spin currents make a non-trivial contribution. The momentum space Green function is given by

$$G^{\rho\sigma\mu\nu}(p) = (4/a) \, \tilde{P}_{(2+0)}{}^{\rho\sigma\mu\nu} / \{(p^2)^2 - 2 l_S^{-2} \, f_S \, M.M - 2 l_Q^{-2} \, f_Q \, T.T\} \tag{6.4}$$

which can be rewritten as

$$G^{\rho\sigma\mu\nu}(p) = (2/a\mu^2)[\{1/(p^2 - \mu^2)\} - \{1/(p^2 + \mu^2)\}] \, \tilde{P}_{(2+0)}{}^{\rho\sigma\mu\nu} \tag{6.4a}$$

$$(\mu^2)^2 \cong 2\{l_S^{-2} \, f_S \, M.M + l_Q^{-2} \, f_Q \, T.T\} = [1/(a')^2] \, [\{J + (1/2)\}^2 - (\alpha_0)^2] \tag{6.4b}$$

where we have used (5.1-4). For $r \geq r_h$, the static potential induced by a point source $(1/a') \, \delta^3(r - r_h)$ reads,

$$H(r - r_h) = \{1/(2\pi a a' \mu^2)\}[- [\{e^{-\mu(r - r_h)}\}/(r - r_h)] +$$

$$+ [\{\{e^{i\mu(r - r_h)}\}/(r - r_h)\} + \{\{e^{-i\mu(r - r_h)}\}/(r - r_h)\}]] \tag{6.5}$$

We required that the potential does not rise at infinity. The first term is a Yukawa potential (glueballs) overlapping with all other meson exchanges (outside Pseudo-Gravity). The other two terms represent a new long-range (proper Pseudo-Gravity) potential generating multi-hadron systems at distances $r > r_h$,

$$H_{1r}(r - r_h) = \{1/(\pi a a' \mu^2)\} \, [\cos\mu(r - r_h)]/(r - r_h) \, , \, r \geq r_h \tag{6.6}$$

This is the pairing force for nucleons, in heavy nuclei preferably. It represents, in that range, the main contribution of $H(r)$, $J^P = 2^+$, 0^+ quanta. We can now collect our results and add, for a more complete picture, the two main non-Pseudo-Gravity contributions of QCD:

$$H(r) = \begin{cases} - (\alpha_s/r) \, \lambda.\lambda \; + A \, r & (r < r_h) \\[2mm] - (A/\mu^2)\{e^{-\mu(r - r_h)}\}/(r - r_h) + (2A/\mu^2)\{\cos\mu(r - r_h)\}/(r - r_h) \\ \qquad\qquad (r \geq r_h) \end{cases} \tag{6.7}$$

where $A = 1/(2\pi a a')$. We have included the 1-gluon QCD potential, dominating the UV region, in the shorter distance expression. In the long distance contribution, the Yukawa term overlaps with the higher order QCD contributions providing for quark-antiquark exchanges; it stands for the glueball exchanges [51]. Note that the plus sign between the longer range contributions provides for a smooth transition from the $\sim r$ behaviour to the $\sim \cos(\mu \, \Delta r)/\Delta r$ zone.

The H(r) potential can thus be described by the graph in Fig. 1.

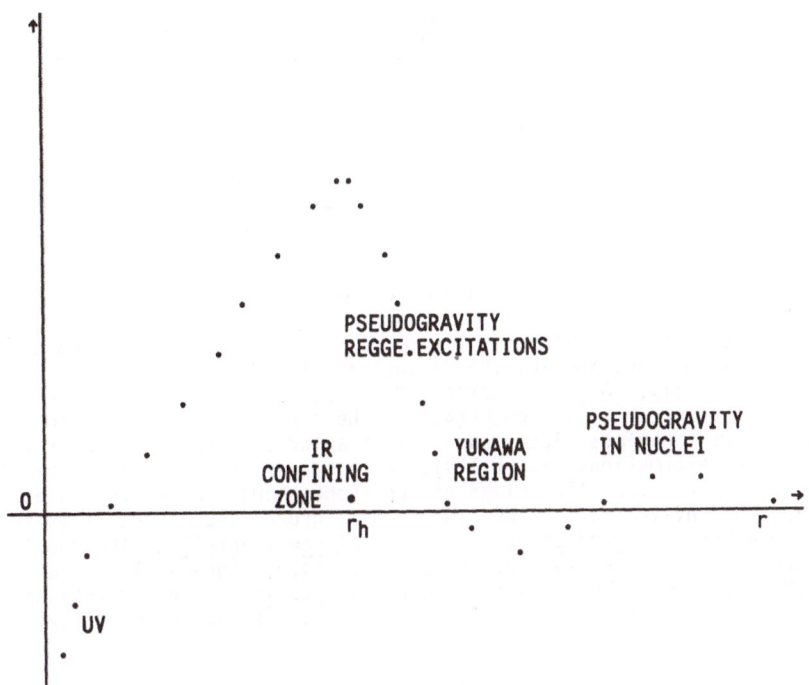

Fig. 1. The Pseudo-Gravity Static Potential

7. NUCLEI AND THE IBM GROUND STATE

The Interacting Boson Model [13,14] has been very successful as a dynamical symmetry, "correlating as well as providing an understanding of a large amount of data which manifest the collective behavior of nuclei" [53]. The model's point of departure is the observation that the two lowest levels in the great majority of even-even nuclei are the 0^+ and 2^+ levels, with relatively close excitation energies, realized by proton or neutron pairs. The model postulates a phenomenological U(6) symmetry between the six states in 0,2. The 0^+, 2^+ excitations in nuclei already appear in the presence of a single nucleon pair above a closed shell. Such "collective" excitations have to originate in QCD, with long-range coherent contributions. We claim that (6.7) represents such a component.

The conventional quasi-long-range binding mechanism due to QCD, i.e. the exchange of quark-antiquark pairs (mesons, mosly with spins 0, 1) does not generate quadrupole excitations. Skipping the 1^- dipole is generally intimately connected with tensor (gravitational) forces [54].

In the Pseudo-Gravity action (4.6), the nucleon-pseudo-graviton vertex is provided by the spin-spin and shear-shear terms. Matter enters through the nucleon manifields $\Psi(x)$, i.e. a tower containing the nucleon with all its excitations (an infinite set). The spin and shear currents are bilinears $\Psi^+\{J\}\Psi$, $\Psi^+\{T\}\Psi$; the squared bilinears are multiplied by H_{pg}, the pseudo-graviton. One simple vertex is thus ($\Psi^+T\Psi$ H $\underline{\Psi^+T\Psi}$), with contraction of the underlined bilinear, a closed loop radiative correction. The formation of the IBM ground state results from diagrams as in Fig.2.

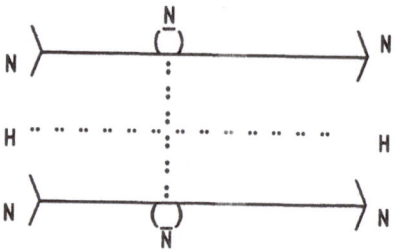

Figure 2. pseudo-graviton H in NN bound state.

The two nucleons pair kinematically to J = 0, and with the dynamical pseudo-graviton coupling in, the 2^+ and 0^+ pairs form.

Algebraically, $G_{\mu\nu}(x)$ carries the 10-dimensional (non-unitary) irreducible representation of GL(4,R). The non-relativistic subgroup of SL(4,R) (the traceless piece, whose algebra includes shears, aside from Lorentz transformations) is SL(3,R). We also saw that this is indeed the little group for the IIA class of representations that we regard as fitting the physical situation. Under this group, the 0^+ and 2^+ states span together one irreducible 6-dimensional representation. The couplings to Pseudo-Gravity are given by the SL(4,R) group; they will thus be SL(3,R) invariant. There is therefore full justification, in this picture, for the IBM postulate of a U(6) symmetry between the basic excitations.

8. REFERENCES

1. Y. Nambu, in "Preludes in Theoretical Physics", A. de-Shalit, H. Feshbach and L. Van Hove editors, North Holland, Amsterdam (1966), pp. 133-142; M.Y. Han and Y. Nambu, Phys. Rev. 139 (1965) B1006.
2. G. 't Hooft, Nucl. Phys. B33 (1971) 173 and B35 (1971) 167.
3. D.J. Gross and F. Wilczek, Phys. Rev. Lett. 30 (1973) 1343.
4. H. Politzer, Phys. Rev. Lett. 30 (1973) 1346.
5. G. 't Hooft, Aix Conf., 1971, unpub.
6. S. Weinberg, Phys. Rev. Lett. 31 (1973) 494.
7. H. Fritzsch, M. Gell-Mann and Leutwyler, Phys. Lett. B47 (1973) 365.
8. Y. Ne'eman, Phys. Rev. 134 (1964) B1355.
9. M. McGuigan and C.B. Thorn, Phys. Rev. Lett. 69 (1992) 1312.
10. S. Godfrey and N. Isgur, Phys. Rev. D32 (1985) 189; S. Capstick and N. Isgur, Phys. Rev. D34 (1986) 2809.
11. Dj. Šijački and Y. Ne'eman, Phys. Lett. B247 (1990) 571.
12. Dj. Šijački and Y. Ne'eman, Phys. Lett. B250 (1990) 1.
13. A. Arima and F. Iachello, Phys. Rev. Lett. 35, (1975) 1069.
14. same authors, Ann. Phys. 99, 253 (1976); 111, 201 (1978); 123, 468 (1979).
15. Y. Ne'eman and Dj. Šijački, Phys. Lett. B276 (1992) 173.
16. Y. Ne'eman and Dj. Šijački, "Hadrons II", to be pub.
17. P.G.O. Freund, Phys. Lett. 2 (1962) 136.
18. M. Gell-Mann, Phys. Rev. 125 (1962) 1067, footn.38.
19. C.J. Isham, A. Salam and J. Strathdee, Phys. Rev. D3 (1971) 867.
20. same authors, Phys. Rev. D8 (1973) 2600 and D9 (1974) 1702.
21. same authors, Lett. Nuo. Cim. Ser. 3, 1 (1972) 969.
22. C. Sivaram and K. Sinha, Phys. Rep. 51 (1979) 111.
23. N. Cabibbo, L.P. Horwitz and Y. Ne'eman, Phys. Lett. 22 (1966) 336.
24. N. Cabibbo, L.P. Horwitz, J.J.J. Kokkedee, Nuo. Cim. 45 (1966) 275.
25. L.P. Horwitz and Y. Ne'eman, Phys. Lett. B26 (1967) 88.
26. Y. Ne'eman, in To Fulfill a Vision, Y. Ne'eman, ed., Addison-Wesley Pub., Reading, Mass. (1981), pp. 99-114.

27. F.W. Hehl, in Spin, Torsion, Rotation and Supersymmetry, P.G.
    Bergmann and V. de Sabbata eds., Plenum Press, NY (1980), p. 5.
28. F.W. Hehl, Y. Ne'eman, J. Nitsch and P. Von Der Heyde, Physics Lett.
    B78 (1978) 102.
29. Y. Ne'eman and Dj. Šijački, Ann. Phys.(NY) 120 (1979) 292.
30. K.S. Stelle, Phys. Rev. D16, 953 (1977).
31. K.S. Stelle, Gen. Rel. Grav., 9 (1978) 353.
32. M. Jacob, ed., Dual Theory, North Holland Pub. Co., Amsterdam (1974).
33. J.H. Schwarz, ed., Superstrings: The First 15 Years, World Scientific
    Pub. Co., Singapore (1985), 2 volumes.
34. Y. Dothan, M. Gell-Mann and Y. Ne'eman, Phys. Lett. 17 (1965) 148.
35. J.P. Elliott, Proc. Roy. Soc. A245 (1958) 128, 562.
36. L. Weaver and L.C. Biedenharn, Phys. Lett, 32B (1970) 326.
37. S. Goshen and H.J. Lipkin, Ann. Phys. 6 (1959) 301.
38. D.J. Rowe, in Dynamical Groups and Spectrum Generating Algebras,
    A. Bohm, Y. Ne'eman and A.O. Barut editors, World Scientific,
    Singapore (1989), p. 287; G. Rosensteel and D.J. Rowe, Phys. Rev.
    Lett. 47 (1981) 223; J.P. Draayer and K.J. Weeks, Phys. Rev.
    Lett. 51, 1422 (1983).
39. Y. Ne'eman and Dj. Šijački, Phys. Lett. B157 (1985) 267.
40. Y. Ne'eman and Dj. Sijacki, Phys. Lett. B200 (1988) 489.
41. C.Y. Lee and Y. Ne'eman, Phys. Lett. B233 (1989) 286.
42. C.Y. Lee and Y. Ne'eman, Phys. Lett. B242 (1990) 59.
43. Dj. Šijački and Y. Ne'eman, J. Math. Phys. 16 (1985) 2457.
44. Y. Ne'eman and Dj. Šijački, Int. J. Mod. Phys., A2 (1987) 1655.
45. Y. Ne'eman and Dj. Šijački, Phys. Lett. B157 (1985) 275.
46. Y. Ne'eman and Dj. Šijački, Phys. Rev. D37 (1988) 3267.
47. S. Weinberg, in Particles and Field Theory (Brandeis lectures 1964)
    S. Deser and K. Ford eds., Prentice-Hall, Englewood, NJ (1965),
    p. 405.
48. S.L. Adler, Rev. Mod. Phys. 54 (1982) 729.
49. A. Zee, Phys. Rev. D23 (1981) 858.
50. Yu.V. Novozhilov and D.V. Vassilevich, Lett. Math. Phys. 21 (1991)
    253.
51. P. de Forcrand and K.F. Liu, Phys. Rev. Lett. 69 (1992) 245.
52. J. Lemke, Y. Ne'eman and J. Pecina-Cruz, J. Math. Phys., 33 (1992)
    2656.
53. D.H. Feng and R. Gilmore, in Dynamical Groups and Spectrum
    Generating Algebras, A. Bohm, Y. Ne'eman and A.O. Barut editors,
    World Scientific Pub., Singapore (1989), p. 209.
54. W.E. Couch and E.T. Newman, J. Math. Phys. 13 (1972) 929.

BOSON-FERMION OPERATORS FOR DESCRIPTION
OF YOUNG DIAGRAMS AND RELATED TOPICS

Masao Nomura

Institute of Physics
College of Arts and Sciences
University of Tokyo
Komaba, Tokyo, 153, Japan

INTRODUCTION

We discuss Young diagrams and symmetric group characters[1,2] (SGC) from a new physical point of view.[3-7] We investigate also close relationship among Young diagrams, symmetry-fixed mean traces of a many-body system,[8-11] Yang-Baxter (Y-B) relations[12-14] and Jones-Ocneanu traces in the knot theory.[15-17]

Discussion begins with a quantum mechanical model[3,4] for Young diagrams. We emphasize the following points:

(a) A solvable Hamiltonian of interacting bosons realizes Young diagrams. Various group-theoretical terms correspond in novel ways to physical terms. SGC acts as the transformation coefficient between the basis function of non-interacting bosons and that of interacting bosons.

(b) A unitary operator[4] exists which transforms the boson operators into fermion operators. The basis operator of non-interacting fermions is transcribed into a basis operator of interacting bosons. Alternatively, a single boson operator is transcribed into linear combination of fermion-pair operators.

(c) Quantized Young diagrams obey an extended Wick theorem.[7]

Next discussion is on reducing many-body mean trace with symmetry $[\lambda]$ being fixed.[8-11] The reduction factor, called propagation coefficient Z, is described in terms of SGC.[5] We present a reduction formula[3,5] for SGC, and apply it to a class of SCG which Heisenberg[8] discussed in theories of ferromagnetism.

A surprise is that the propagation coefficient Z is a solution to Y-B relation for vertex models.[12-14] The rapidity parameter here runs over discrete values, expressed in terms of the numbers of particles.

In the last part, we point out that Ocneanu's trace $X_L(q, \dot{\lambda})$ in the knot theory,[15-17] an extension of Jones' polynomial,[15] is described as the symmetry-fixed mean trace. Our formalism provides Young-diagrammatic implication to the parameters q and $\dot{\lambda}$.

Symmetries in Science VII, Edited by B. Gruber
and T. Otsuka, Plenum Press, New York, 1994

A DYNAMICAL MODEL OF YOUNG DIAGRAM

A Brief Review on Schur Functions

Schur- (S-) function is the primitive character of irreducible representation (IR) of $GL(n)$. It is a generating function of SGC. Littlewood[1] generalized concept of S-function in the following way. We expand a given function $F(x)$ $(F(0)=1)$ as

$$F(x) = 1 + \sum_{j=1}^{\infty} h_j x^j = exp(\sum_{j=1}^{\infty} s_j x^j/j). \tag{1}$$

A well-known example of $F(x)$ is

$$F(x) = \{\prod_{i=1}(1 - \epsilon_i x)\}^{-1}. \tag{2}$$

In physics, we understand $F(x)$ as the partition function. The factor $h_j j!$ is the j-th moment and $s_j(j-1)!$ implies the j-th cumulant. Let us define S-function $\{\lambda\}$ by

$$\{\lambda\} \equiv \{\lambda_1 \lambda_2 \ldots \lambda_n\} \tag{3}$$
$$= |h_{\lambda_i - i + j}| \tag{4}$$
$$= \sum_{\ell} g_\ell \chi_{(\ell)}^{[\lambda]} s_1^{\ell_1} s_2^{\ell_2} \ldots s_n^{\ell_n} / n!. \tag{5}$$

where $|h(i,j)|$, seen in (4), stands for the determinant with the (i,j) element $h(i,j)$. The symbol (ℓ) denotes the class $(1^{\ell_1} 2^{\ell_2} \ldots n^{\ell_n})$ of S_n, and g_ℓ is the number of elements in (ℓ):

$$g_\ell = n!/(\prod_{j=1}^{n}(\ell_j! j^{\ell_j})). \tag{6}$$

The S-function is expressed in terms of h_j and s_j, respectively, in (4) and (5). The symbol $\chi_{(\ell)}^{[\lambda]}$ denotes SGC, called also the element of a transition matrix.[2] SGC is free from $F(x)$ with non-vanishing s_j and h_j. We cite in Table 1 the SGC of S_4:[1]

Outer product of S-functions is reduced as[1]

$$\{\lambda\}\{\mu\} = \sum_{\nu} \Gamma_{\lambda\mu\nu} \{\nu\}, \tag{7}$$

which means in terms of SGC,

$$\chi_{(\ell_1 \ell_2)}^{[\lambda]} = \sum_{\mu\nu} \Gamma_{\mu\nu\lambda} \chi_{(\ell_1)}^{[\mu]} \chi_{(\ell_2)}^{[\nu]}. \tag{8}$$

Table 1. Character table of S_4.

Class	(1^4)	$(1^2 2)$	(13)	(2^2)	(4)
g_ℓ	1	6	8	3	6
[4]	1	1	1	1	1
[31]	3	1	0	-1	-1
[2^2]	2	0	-1	2	0
[21^2]	3	-1	0	-1	1
[1^4]	1	-1	1	1	-1

Outer division (called also the skew diagram) means

$$\{\nu\}/\{\lambda\} = \sum_\mu \Gamma_{\lambda\mu\nu}\{\mu\} \qquad (9)$$

$$= \lambda(\frac{\partial}{\partial s_1}, 2\frac{\partial}{\partial s_2}, 3\frac{\partial}{\partial s_3}, \ldots)\,\nu(s_1, s_2, s_3, \ldots). \qquad (10)$$

The second equality[18] is of importance in studies of classical solitons.[19,20,6] Operations other than outer product and outer division are out of concern. The Heisenberg algebra, seen in (10), is transcribed into the algebra of boson operators such as

$$\{\lambda\}\{1\}/\{1\} - (\{\lambda\}/\{1\})/\{1\} = \{\lambda\} \qquad \leftrightarrow \qquad bb^+ - b^+b = 1. \qquad (11)$$

Frobenius Notation and Fermion Operator Description

As alternative to (3) it is suggestive to use Frobenius notation,[1]

$$\{\lambda\} = \begin{pmatrix} \alpha_1 & \alpha_2 & \cdots & \alpha_r \\ \beta_1 & \beta_2 & \cdots & \beta_r \end{pmatrix}. \qquad (12)$$

Here, $\alpha_j = \lambda_j - j$ and $\beta_j = \bar{\lambda}_j - j$: $\bar{\lambda}$ is the partition conjugate to λ. The suffix r means the number of boxes on the leading diagonal (NW to SE) of the Young diagram. For example,

$$\{431\} = \begin{pmatrix} 3 & 1 \\ 2 & 0 \end{pmatrix}. \qquad (13)$$

Frobenius notation is depicted by Maya diagram.[19] We show in Figure 1 both Young and Maya diagrams for the above example.

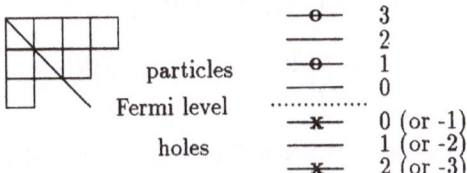

Figure 1. Young diagram vs. Maya diagram. The symbols o and x stand for a particle and a hole, respectively.

Maya diagram stands for the state of spinless fermions occupying one at most for each level. The ground state $\{0\}$ defines Fermi level. Excitation energy of a state is equal to the sum of λ_i's, i.e. the number of boxes. The Young diagram with r hooks is represented as the state of r particles and r holes (rp-rh). Coleman[21] presented a device similar to Maya diagram. In his device, Fermi level is set below the lowest vacant level, i.e. the diagonal line passing on the SW corner of the diagram.

The rp-rh basis function of fermions is describable (p.112, 114 of ref.1) as the determinant with each element being a single hook i.e. a 1p-1h basis function:

$$\{\lambda\} = \left| \begin{pmatrix} \alpha_i \\ \beta_j \end{pmatrix} \right|. \qquad (14)$$

Boson Hamiltonian of a Hidden-Symmetry Type

It is shown[1] that Young diagram is realized by boson Hamiltonian,

$$H = \sum_{k=1}^{\infty} k b_k^+ b_k + G \sum_{i,j=1}^{\infty} \{ij(i+j)\}^{1/2}(b_i^+ b_j^+ b_{i+j} + h.c.).\tag{15}$$

The interaction term means a resonant condition of three waves:

and

Current (energy) is conserved at each vertex. The Hamiltonian (15) with special initial condition can be a model of coagulation (or cascade process).

Let us solve (15), using the normalized states of non-interacting bosons,

$$|\ell >_n \equiv B_n^+(\ell)|0 >,\tag{16}$$

in which

$$B_n^+(\ell) = \prod_{j=1}^{n} (b_j^+)^{\ell_j}/\sqrt{\ell_j!}.\tag{17}$$

We find that the normalized eigenstates $|\lambda >_n$ of (15) are, irrespective of G,

$$|\lambda >_n = A_n^+(\lambda)|0 >.\tag{18}$$

Here, A^+ is a cluster of boson operators,

$$A_n^+(\lambda) = \sum_{\ell} \sqrt{g_\ell/n!}\, \chi_{(\ell)}^{[\lambda]} B_n^+(\ell)\tag{19}$$

$$= \sum_{\ell} \frac{g_\ell}{n!} \chi_{(\ell)}^{[\lambda]} b_1^{+\ell_1} b_2^{+\ell_2} \ldots b_n^{+\ell_n}.\tag{20}$$

We notice that A^+ is quantum version of S-functions. Eigenvalue of $|\lambda >$ is described with the expectation value of Casimir operator of $U(n)$,

$$E(\lambda) = n + G\sum_{i} \lambda_i(\lambda_i - 2i + 1); \quad n \equiv \sum_{i} \lambda_i.\tag{21}$$

Using orthogonalities of SGC, we invert (19) such that

$$B_n^+(\ell) = \sum_{\lambda} \sqrt{g_\ell/n!}\, \chi_{(\ell)}^{[\lambda]} A_n^+(\lambda).\tag{22}$$

With $(\ell) = (n)$ it gives

$$b_n^+ = \sum_{k=0}^{n-1} (-1)^k A_n^+(\lambda = n-k, 1^k).\tag{23}$$

In Figure 2 are depicted (20) for $|\lambda >$ with $n = 4$.

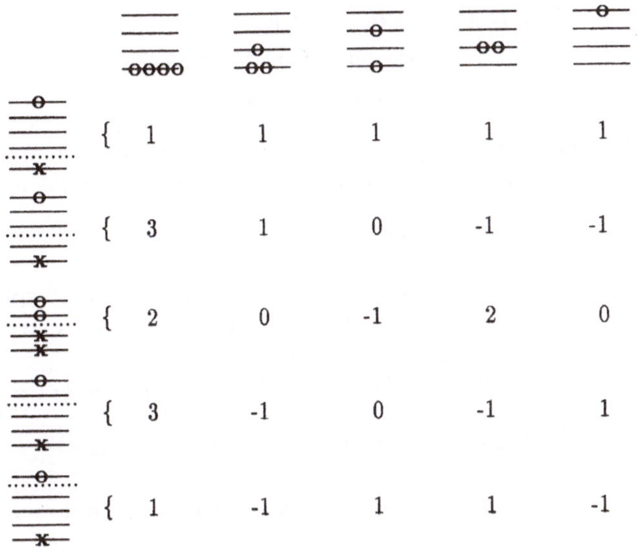

$$B^+(\ell)|0> \quad \{ \ |1^4> \quad \sqrt{6}|(21)> \quad \sqrt{8}|(31)> \quad \sqrt{3}|(2^2)> \quad |(4)> \ \}/\sqrt{4!}.$$

	$|1^4>$	$\sqrt{6}|(21)>$	$\sqrt{8}|(31)>$	$\sqrt{3}|(2^2)>$	$|(4)>$	
$A^+(4)|0>$ {	1	1	1	1	1	}$/\sqrt{4!}$
$A^+(31)|0>$ {	3	1	0	-1	-1	}$/\sqrt{4!}$
$A^+(22)|0>$ {	2	0	-1	2	0	}$/\sqrt{4!}$
$A^+(21^2)|0>$ {	3	-1	0	-1	1	}$/\sqrt{4!}$
$A^+(1^4)|0>$ {	1	-1	1	1	-1	}$/\sqrt{4!}$

Figure 2. Eigenstates of the Hamiltonian (15).
The cases of $n = 4$ are shown.

We emphasize in Figure 2 that SGC appears as the transformation coefficient. The first and the second orthogonalities of SGC appear as the orthogonalities of the states.

As seen in (23), a single boson operator corresponds to a collective states of 1p-1h fermions: An example is seen in the last column of Figure 2;

$$b_4^+ = (A^+(4) - A^+(31) + A^+(21^2) - A^+(1^4))/\sqrt{4!}. \qquad (24)$$

Boson-Fermion Transformation

Boson-Fermion correspondence so far discussed is realized by unitary operator,[4]

$$U(\epsilon_1\epsilon_2) = exp[\sum_{j=1}^{\infty} \frac{\pi}{2} (\rho_j b_j^+ - \rho_j^+ b_j)/\sqrt{j}]. \qquad (25)$$

Here, ρ^+ is a combination of fermion-pair operators,

$$\rho_j^+ = \sum_{n=-\infty}^{\infty} C_{n+j}^+ C_n \quad and \quad \rho_j = \sum_{n=-\infty}^{\infty} C_n^+ C_{n+j}, \qquad (26)$$

in which C^+ and C are creation and annihilation operators of fermions, respectively. The operator U was discussed first by Mattis and Lieb.[22] It is shown that

$$[\rho_j, \rho_k^+] = k\delta(j,k), \quad [\rho_j^+, C_n^+] = C_{n+j}^+, \quad [\rho_j, C_n^+] = C_{n-j}^+. \qquad (27)$$

The cluster operators A^+, (20), is transformed into fermion operators:

$$U^{-1} A_n^+(\lambda) U = \sum_{\ell} \frac{g_\ell}{n!} \chi_{(\ell)}^{[\lambda]} \rho_1^{+\ell_1} \rho_2^{+\ell_2} \dots \rho_n^{+\ell_n}. \qquad (28)$$

This acting on the vacuum $|0>$ is transformed as

$$U^{-1}A_n^+(\lambda)U|0>= U^{-1}A_n^+(\lambda)|0> \tag{29}$$
$$= \text{a monomialof fermions}|0> \tag{30}$$
$$= \text{Maya diagram.} \tag{31}$$

Boson-fermion correspondence here is very different from that of Kyoto School.[23] The latter is based on *a non-unitary operator* called time evolution operator.

Algebras of the Cluster Operators $\{A^+, A\}$

We can manipulate cluster operators A^+, A without referring explicitly to single-particle operators b^+, b. This statement is more than what outer products and divisions imply, as we exemplify later in (36).

The rule for outer product, (7), is preserved in the algebras of A^+ and A:[7]

$$A_n^+(\mu)A_m^+(\lambda) = \sum_\nu < m+n\,\nu|A_n^+(\mu)|m\lambda > A_{m+n}^+(\nu). \tag{32}$$

Here, the matrix element (called spectroscopic amplitude) is just multiplicity, Γ;

$$< m+n\,\nu|A_n^+(\mu)|m\lambda > = \Gamma_{\lambda\mu\nu}. \tag{33}$$

A surprise comes out when we commute A with A^+. All the results of ref. 7, which concerns Wick theorem for cluster operators, are applicable to A^+ and A:

$$A_m(\lambda)A_n^+(\mu) = \sum_{\lambda'\mu'} < m\lambda|A_{m'}^+(\lambda')A_{n'}(\mu')|n\mu > A_{n'}^+(\mu')A_{m'}(\lambda'), \tag{34}$$

in which $m-m' = n-n'$. The matrix element M appearing in (34) is rewritten as

$$M = \sum_\nu <m\lambda|A_{m'}^+(\lambda')|n-n'\,\nu><n-n'\,\nu|A_{n'}(\mu')|n\mu>= \sum_\nu \Gamma_{\nu\lambda'\lambda}\,\Gamma_{\nu\mu'\mu}. \tag{35}$$

In the following is an example of (34):

$$A(21)A^+(31) = A^+(31)A(21)+A^+(1)+(A^+(11)+2A^+(2))A(1)$$
$$+(A^+(21)+A^+(3))A(2), \tag{36}$$

which is to be compared with the outer division,

$$\{31\}/\{21\} = \{1\}. \tag{37}$$

Applying $|0>$ to both sides of (36) yields quantum version of (37).

Remarks

We recall a simplified Tomonaga-Luttinger Hamiltonian of fermions,[4]

$$H = \sum_{k=1}^\infty k\rho_k^+\rho_k + \sum_{k=1}^\infty g_k\rho_k^+\rho_k. \tag{38}$$

Using the unitary operator (25), we can transcribe H into a non-interacting boson Hamiltonian, the reverse direction to (15): The eigenstate $\rho_1^{+\ell_1}\rho_2^{+\ell_2}\ldots\rho_n^{+\ell_n}|0>$ of (38) is transcribed into the non-interacting boson state $B^+(\ell)|0>$.

Maya diagram is not suitable for depicting outer product, (7). We understand that the diagram describes a quantum state rather than the operator itself: e.g. $\{431\} \leftrightarrow C_{-1}C_{-3}C_3^+ C_1^+ |0>$, but not $\{431\} \leftrightarrow C_{-1}C_{-3}C_3^+ C_1^+$.

In recent years the boson model for Young diagram, (15), has been discussed by Stone[24] in relation with edge excitation of quantum Hall effects and by Jevicki[25] in terms of matrix models.

TRACES IN A MANY-BODY SYSTEM AND SYMMETRIC GROUP CHARACTERS

Heisenberg[8] investigated global distribution of spectra in a many-body spin-1/2 system. He evaluated a class of SGC's to deduce the first and the second central moments of spectra in the n-body system with the total spin s: From these moments he determined spectral distribution with the definite s, assuming Gaussian distribution. We cite here a result of ref. 8 on evaluation of a class of SGC:

$$
\chi_{(22)}^{[\lambda_1 \lambda_2]} \equiv \chi_{\bar{n}-s, \bar{n}+s}^{(12)(34)} = \frac{4(2\bar{n}-4)!}{(\bar{n}-s)!(\bar{n}+s+1)!}\{2s^5 + 5s^4 + 4s^2(\bar{n}^2 - 5\bar{n}+4)
$$
$$
+ s^2(6\bar{n}^2 - 30\bar{n}+19) + 2s(\bar{n}^4 - 6\bar{n}^3 + 15\bar{n}^2 - 14\bar{n}+3)
$$
$$
+ \bar{n}^4 - 6\bar{n}^3 + 14\bar{n}^2 - 9\bar{n}\}, \tag{39}
$$

where $\bar{n} = n/2$, $n = \lambda_1 + \lambda_2$ and $s = (\lambda_1 - \lambda_2)/2$.

In the last twenty years, this Heisenberg method has been renewed as so-called spectral distribution method (SDM): See refs. 5, 9-11 and works cited therein. The first step of SDM is to express a powers of a given Hamiltonian (which need not be a spin Hamiltonian.) as a sum of many-body operators.[10,11] Subsequently, we evaluate mean trace of a general k-body operator V_k over the anti-symmetrized n-body states with the quantum number $U(N)$ being fixed:[5]

$$
< V_k >^{n\lambda} \equiv \sum_{\alpha\beta} < n\lambda\alpha\beta |V_k| n\lambda\alpha\beta > / \{d_N(n\lambda)m_N(\lambda)\}. \tag{40}
$$

Here, β specifies the row of the IR of dimensionality $m_N(\lambda)$, and α is the additional quantum number with dimensionality $d_N(n\lambda)$. Of importance is the case of $n \gg k$. The key to evaluate (40) lies in the reduction relation,

$$
< V_k >^{n\lambda} = \sum_{\mu \vdash k} Z(n\lambda, k\mu) < V_k >^{k\mu} . \tag{41}
$$

The factor Z is called propagation coefficient. It is a polynomial of degree k in n, and satisfies the discrete-analog Chapman-Kolmogorov equation,[5]

$$
\sum_{\nu \vdash n'} \{Z(n\lambda, n'\nu)/\binom{n}{n'}\}\{Z(n'\nu, k\mu)/\binom{n'}{k}\} = Z(n\lambda, k\mu)/\binom{n}{k}, \tag{42}
$$

and the condition

$$
Z(n\lambda, n\lambda') = \delta(\lambda, \lambda'), \quad \sum_\mu Z(n\lambda, k\mu) = \binom{n}{k}. \tag{43}
$$

We solve[5] (42) with (43) to get

$$
Z(n\lambda, k\mu) = \binom{n}{k}\frac{f(\lambda, \mu)f(\mu, 0)}{f(\lambda, 0)} = \frac{\chi_{(1^k)}^{[\mu]}}{k!}\sum_\rho \frac{g_\rho \chi_\rho^{[\mu]} \chi_\rho^{[\lambda]}}{\chi_{(1^n)}^{[\lambda]}}. \tag{44}
$$

We have defined f such that

$$f(\lambda,\mu) = (n-k)! \left| \frac{1}{(\lambda_i - i - \mu_j + j)!} \right| ; \quad \lambda \vdash n, \ \mu \vdash k. \tag{45}$$

in which $m! = 0$ if $m < 0$. In the second step of (44) we have used the relation,[3,5]

$$\chi^{[\lambda]}_{(1^n-k\ell)} = \sum_{\mu \vdash k} f(\lambda,\mu) \chi^{[\mu]}_{(\ell)}, \tag{46}$$

where (ℓ) indicates a class of S_k. We understand $f(\lambda,0) = \chi^{[\lambda]}_{(1^n)}$. In the case of $[\lambda] = [\lambda_1 \lambda_2]$ the relation (46) gives

$$\chi^{[\lambda_1\lambda_2]}_{(1^n-k\ell)} = (n-k)! \sum_{\mu_1\mu_2\vdash k} \left(\frac{1}{(\lambda_1-\mu_1)!(\lambda_2-\mu_2)!} - \frac{1}{(\lambda_1-\mu_2+1)!(\lambda_2-\mu_1-1)!} \right) \chi^{[\mu_1\mu_2]}_{(\ell)}. \tag{47}$$

Combining (42) and (44), we get

$$\sum_\mu f(\lambda,\mu)f(\mu,\nu) = f(\lambda,\nu); \quad \lambda \vdash n, \ \nu \vdash k. \tag{48}$$

It is shown that

$$f(\bar{\lambda},\bar{\mu}) = f(\lambda,\mu), \quad \{1\}^{n-k}\{\mu\} = \sum_\lambda f(\lambda,\mu)\{\lambda\}, \quad f(\lambda,\mu) = \sum_\nu f(\nu,0)\Gamma_{\mu\nu\lambda}. \tag{49}$$

We substitute the first equality of (44) into (41) and obtain

$$f(\lambda,0) < V_k >^{n\lambda} = \binom{n}{k} \sum_{\mu\vdash k} f(\lambda,\mu)f(\mu,0) < V_k >^{k\mu}, \tag{50}$$

a generalization of (46). The binomial coefficient in n and k in (50) owes to the fact that V_k is symmetric in particle indices. In the case when the operator V_k is not symmetrized, the binomial coefficient should be excluded. This is applied to (51) and (60)-(75). We can deduce (46) from (50), using

$$\ll P(\ell) \gg^{n\lambda} = \chi^{[\lambda]}_{(\ell)}, \quad \text{e.g.} \quad \ll P_{12}P_{34} \gg^{n\lambda} = \chi^{[\lambda]}_{(22)}, \quad \ll P_{12}P_{13} \gg^{n\lambda} = \chi^{[\lambda]}_{(3)}, \tag{51}$$

in which $\ll \gg^{n\lambda} \equiv f(\lambda,0) < >$. SGC here acts a very different role from SGC in the quantized Young diagram. In particular, (1^{n-k}) of the class $(1^{n-k}\ell)$ stands for $(n-k)$ *non-interacting particles*, as we observe in (46) and (50).

In applying (46) and (50) to evaluation of the second moment of spectra discussed in ref. 8, we express V^2 in terms of two-, three- and four-body operators:

$$< (\textstyle\sum V_{ij})^2 >^n = < (n-1)(V_{12})^2 + 2(n-2)V_{12}V_{23} + (n-3)(n-2)V_{12}V_{34} >^n. \tag{52}$$

We subsequently substitute $V_{ij} = 2P_{ij} - 1$ of ref. 8 into (52), and use (51) to get

$$\ll \textstyle\sum (V_{ij})^2 \gg^{n\lambda}$$
$$= (n+3)(n-1)\chi^{[\lambda]}_{(1^n)} - 4(n-1)^2\chi^{[\lambda]}_{(2)} + 8(n-2)\chi^{[\lambda]}_{(3)} + 4(n-3)(n-2)\chi^{[\lambda]}_{(22)}. \tag{53}$$

Next, we evaluate SGC's on the right-hand side of (53), using (47). We then get (39). A mystery is that the polynomials appearing in (39) is factorized with $2s + 1$ and remaining factors are polynomials in $s(s + 1)$:

$$\{ \ \} \text{ in } (39) = (2s+1)\{16s^2(s+1)^2 + 8s(s+1)(n^2-10n+12)$$
$$+ n(n-2)(n^2-10n+36)\}/16. \tag{54}$$

NOVEL SOLUTIONS TO YANG-BAXTER EQUATION

We point out here that the propagation coefficient Z, appearing in (41) is a solution to Y-B equation. Let us consider the case of $su(2)$, i.e. $[\lambda] = [\lambda_1 \lambda_2]$:

$$< V_k >^{nSS_z S'_z} \equiv \sum_\alpha < nSS_z \alpha |V_k| nSS'_z \alpha > / d_{N=2}(nS). \tag{55}$$

Here, V_k is a k-body operator non-scalar in isospin space. It is shown, by using device of ref. 11, that

$$\sum_\alpha A(nSS_z \alpha) A^+(nSS'_z \alpha)$$

$$= \sum_{kss_z} (-1)^k Z(nSS_z S'_z; kss_z s'_z) \frac{d(nSS)}{d(kss)} \sum_\gamma A^+(kss'_z \gamma) A(kss_z \gamma). \tag{56}$$

The propagation coefficient Z here is explicitly written as

$$Z(nSS_z S'_z; kss_z s'_z)$$

$$= \sum_{s''} (ss''s_z s''_z |SS_z)(ss''s'_z s''_z |SS'_z) \binom{n}{k} \frac{m_{N=2}(k,s) m_{N=2}(n-k,s'')}{m_{N=2}(n,S)}, \tag{57}$$

where $s''_z = S_z - s_z = s'_z - s'_z$. The factor $m_{N=2}$, appearing in (57), is dimensionality of $(1/2)^n S$:

$$m_{N=2}(n,S) = (2S+1)\binom{n+1}{n/2-S} / (n+1). \tag{58}$$

In a heuristic way, let us compare Z, (57), with the R matrix element, which satisfies Y-B relation, of the XXX model investigated by by Babujian and Tsvelick :[26] To this end we use the identity,

$$m_{N=2}(2\Lambda-2, S) = \frac{2S+1}{2\Lambda-1}\binom{2\Lambda-1}{\Lambda-a+b} \prod_{p=a-b}^{S} \frac{\Lambda-p}{\Lambda+p}, \tag{59}$$

in which a and b are taken arbitrary ($\Lambda \neq a-b$), and Λ implies rapidity[26]. We find

$$(R^{Ss})^{S'_z s'_z}_{S_z s_z} = Z(nSSzS'_z; k,s,-s_z,-s'_z)/G(S,s,n,k), \tag{60}$$

in which

$$G(S,s,n,k) = (2S+1)\binom{n}{k} \frac{g(k,s)g(n-k,s-S)}{(2s-2S+1)g(n,S)}, \tag{61}$$

and

$$\text{the rapidity parameter of ref. 26} = (n-k)/2 + 1. \tag{62}$$

It shows that R and Z agree with each other apart from the normalization factor G.

JONES-OCNEANU TRACE AS A SYMMETRY-FIXED TRACE

Ocneanu's trace[16] $X_L(q, \dot\lambda)$, which is a generalization of Jones' polynomial invariant $V_L(q)$ in the knot theory, is expanded[15] as a sum of weighted normal traces indexed by Young diagrams:

$$X_L(q, \dot\lambda)/C = tr(x) = \sum_{\mu \vdash k} W_\mu(q, \dot\lambda)\, tr_\mu(x). \qquad (63)$$

Here, $x \in H(q, k) \subset \overset{\infty}{\underset{}{U}} H(q, n)$, where $H(q, n)$ is Hecke algebra[15,16] with generators g_1, $g_2, \ldots, g_{n-1}$: With $q = 1$, the element g_i is the transposition $(i, i+1)$ of S_n. The scale factor C is to fulfill symmetry between $tr(g_i)$ and $tr(g_i^{-1})$, which is out of concern, here. The weight W_μ is an S-function:[15]

$$W_{\bar\mu} = (\frac{1-q}{1-\dot\lambda q})^k \frac{|q^{(i-1)(\mu_j+p-j)}|}{|q^{(i-1)(p-j)}|} = (\frac{1-q}{1-\dot\lambda q})^k \{\mu\}. \qquad (64)$$

Here, $\bar\mu$ is conjugate partition to μ. We have defined $\{\mu\}$ with respect to the function,

$$F(x) \equiv \{\prod_{i=1}^{p}(1 - q^{i-1}x)\}^{-1} = \sum_{i=0}^{\infty} h_i x^i, \qquad (65)$$

where

$$h_i = \frac{(1 - \dot\lambda q^i)(1 - \dot\lambda q^{i-1})\ldots(1 - \dot\lambda q)}{(1-q)(1-q^2)\ldots(1-q^i)}. \qquad (66)$$

The objective of this section is to express the expansion (63) as the reduction relation for symmetry-fixed trace, (41), and to grasp q and $\dot\lambda$ in diagrammatic ways. Let us consider the specific partition $[\Lambda]$ of n,

$$[\Lambda] = \begin{cases} [\epsilon q^{i-1}] \text{ i.e. } [\epsilon, \epsilon q, \epsilon q^2, \ldots, \epsilon q^{p-1}]; & \text{if } q \leq 1, \\ [\epsilon q^{1-i}] \text{ i.e. } [\epsilon, \epsilon q^{-1}, \epsilon q^{-2}, \ldots, \epsilon q^{-(p-1)}]; & \text{if } q > 1. \end{cases} \qquad (67)$$

Here, p is expressed in terms of q and $\dot\lambda$ by

$$\dot\lambda = \begin{cases} q^{p-1}; & \text{if } \dot\lambda \leq q \leq 1, \text{ or } 1 < q \leq \dot\lambda, & \cdots \text{ case (a)}, \\ q^{-p-1}; & \text{if } q < 1 < \dot\lambda q^2 \text{ or } \dot\lambda q^2 < 1 < q, & \cdots \text{ case (b)}. \end{cases} \qquad (68)$$

That p is restricted to $1, 2, 3, \ldots$ yields the condition between q and $\dot\lambda$, which agrees with the result of ref. 17. The parameter ϵ is given by

$$\epsilon = \frac{n(1-q)}{1 - \dot\lambda q}, \qquad (69)$$

so as to hold $\Sigma\Lambda_i = n$. We evaluate $f(\Lambda, \mu)$, (45), to get

$$\frac{f(\Lambda, \mu)}{f(\Lambda, 0)} = \frac{(n-k)!}{n!} \frac{\Pi_{j=1}^{p} \mu'_j!}{|(\alpha_i)^{p-j}|} \left|\binom{\alpha_i}{\beta_j}\right|, \qquad (70)$$

where $\alpha_i \equiv \dot{\lambda} - i$ and $\beta_i \equiv \mu - i$ (the convention, (12)). The $n \to \infty$ limit of (70) gives

$$\lim_{n \to \infty} \frac{f(\Lambda, \mu)}{f(\Lambda, 0)} = W_{\hat{\mu}}, \tag{71}$$

in which

$$\hat{\mu} = \bar{\mu} \text{ and } \mu, \text{ respectively, for the cases of (a) and (b).} \tag{72}$$

The relation (71) indicates that W_μ, is a propagation coefficient:

$$W_\nu(q, \dot{\lambda}) = \sum_{\mu \vdash n} f(\mu, \nu) W_\mu(q, \dot{\lambda}), \quad W_\lambda(q, \dot{\lambda}) W_\mu(q, \dot{\lambda}) = \sum_\nu \Gamma_{\lambda\mu\nu} W_\nu(q, \dot{\lambda}). \tag{73}$$

The second relation is due to W_μ as a S-function. Let us put

$$tr_{\hat{\mu}}(x) \equiv f(\mu, 0) < x >^{k\mu} f(\mu, 0) = \ll x \gg^{k\mu}, \tag{74}$$

a relation corresponding to (51). With $q = 1$, $tr_\mu(g_1 g_3)$, for example, is reduced to SGC, $\chi^\mu_{(22)}$: The following discussion does not concern further specification of tr_μ such as $tr_\nu(x_1 x_2) = \Sigma tr_\lambda(x_1) tr_\mu(x_2) \Gamma_{\lambda\mu\nu}$ to ensure $tr(x_1) tr(x_2) = tr_{(x_1 x_2)}$. Combining (71) with (50), we obtain

$$tr(x) = \lim_{n \to \infty} < x >^\Lambda : \tag{75}$$

We can describe Jones trace in terms of the symmetry-fixed trace.

From (50) and the fact that $x \vdash k$ implies also $x \vdash n$ with $n > k$, we get

$$tr_\nu(x) = \sum_{\mu \vdash k} f(\nu, \mu) tr_\mu(x). \tag{76}$$

It is shown that W_μ satisfy the relation,

$$\sum_{\mu \vdash k} (-1)^c W_\mu(q, \dot{\lambda}) = \frac{(1 - q)^k (1 - (\dot{\lambda} q)^k)}{(1 - q^k)(1 - \dot{\lambda} q)^k}, \tag{77}$$

where μ is the partition of a single hook, $[k - c, 1^c]$. For proof, we make use of the fact that the expectation value of the k-body Casimir operator of $U(n)$ in the $n \to \infty$ limit is transformed as

$$\lim_{n \to \infty} k! \binom{n}{k} \chi^{[\Lambda]}_{(k)} / f(\Lambda, 0) = \sum_{i=1} \lambda_i^k = \epsilon \sum_{i=1}^p q^{(i-1)k}. \tag{78}$$

We have used (46) and $\chi^{[k-c,1^c]}_{(k)} = (-1)^c$.

The functions W_μ with $\mu \vdash k$ is a polynomial of degree k in $\dot{\lambda}$. We can express them as a sum of single-hook diagrams: For example,

$$W_{[33]} = \frac{q^2(1-q)}{(1+q^2)(1-q^5)} \{(1+q^2) W_{[51]} + (1+q+q^2) W_{[41^2]} + q W_{[31^3]} \}. \tag{79}$$

We have emphasized new correspondence among (22), inverse of (5), (50) and (63). Further investigation is in progress.

Acknowledgments

This paper is dedicated to Professor F. Iachello in commemoration of his fiftieth anniversary. The author wishes to express appreciation to Professor L.C. Biedenharn for his recommendation to attend the Symposium and to Professor B. Gruber for the hospitality at the Symposium. Last but not least the author would like to thank Professor A. Arima for his guidance to symmetric group at seminars in 1963.

References

1. D.E. Littlewood, The Theory of Group Characters, 2nd ed.(1950).
2. I. Macdonald, "Symmetric Functions and Hall-Littlewood Polynomials," Oxford, Clarendon, (1979).
3. M. Nomura, Phys. Lett. 117A:289(1986).
4. M. Nomura, J. Phys. Soc. Japan. 56:3821(1987).
5. M. Nomura, Phys. Lett. 126A:1(1987).
6. M. Nomura, J. Phys. Soc. Japan. 56:4265(1987).
7. M. Nomura, J. Math. Phys. 26:732(1985).
8. W. Heisenberg, "Zur Theorie des Ferromagnetismus," Z. Phys. 49:619(1928).
9. J.B. French, in "Nuclear Structure," A.Hossain, Harum-ar-Raschid and M. islam, ed., North-Holland, Amsterdam,1969.
10. M. Nomura, Prog. Theor. Phys. 51:489(1974).
11. M. Nomura, J. Math. Phys. 27:536(1986).
12. C.N. Yang, Phys. Rev. Lett. 19:1312(1967).
13. R.J. Baxter, "Exactly Solved Models in Statistical Mechanics," Academic, London, 1982.
14. M. Nomura, J. Phys. Soc. Japan. 58:2694(1989).
15. V.F.R. Jones, Bull. Amer. Math. Soc. 12:103(1985). "Hecke Algebra Representations of Braid Groups and Link Polynomials," Hand-written manuscript.
16. P. Freyd, D. Yetter, J. Hoste, W.B.R. Lickorish, K. Millett and A. Ocneanu, Bull. Amer. Math. Soc. 12:239(1985).
17. H. Wenzl, Invent. Math. 92:349(1989).
18. H.Foulkes, J. London Math. Soc. 24:136(1949).
19. M. Sato and M. Noumi, "Soliton Equations and the Universal Grassmann Manifold," Sugaku Kokyuroku, Sophia Univ, No.18, 1984 (In Japanese).
20. R. Hirota, J. Phys. Soc. Japan. 55:2137(1986).
21. A.J. Coleman, Advances in Quantum Chemistry, P.O. Löwdin ed., Academic, New York, 4:83 (1968).
22. D.C. Mattis and E.H. Lieb, J. Math. Phys. 6:304(1965).
23. E. Date, M. kashiwara, M. Jimbo and T. Miwa in "Non-linear Integrable Systems," M. Jimbo and T. Miwa, ed., World Scientific, 1983, p. 943.
24. Stone: Preprint "Edge Waves in the Quantum Hall Effect," IL-TH-8 (1990).
25. Jevicki: in "Differential Geometric Methods in Theoretical Physics," S.Catto and A. Rocha ed., World, Singapore(1992), Vol.1, p.1021.
26. H.M. Babujian and A.M. Tsvelick, Nucl. Phys. B265:24(1986).

HYPERINFLATION AND SELF-SIMILARITY IN QUASIPERIODIC ONE-DIMENSIONAL LATTICES

T. Odagaki

Department of Liberal Arts and Sciences
Kyoto Institute of Technology
Matsugasaki, Kyoto 606, Japan

INTRODUCTION

According to the existence or lack of translational symmetry, the structure of condensed matter is commonly classified into two classes, regular crystals and disordered systems. Translational symmetry in regular crystals plays a pivotal role in determining their properties. For example, one-electron states in crystals described by a hamiltonian with a periodic potential energy must be Bloch states, i.e. an eigenfunction of the hamiltonian must be a product of a plane wave and a periodic function of the lattice, which is extended in the entire crystal. This makes the energy band periodic in the reciprocal lattice space and, in turn, the density of states shows the van Hove singularities.[1] On the other hand, randomness in the hamiltonian yields properties qualitatively different from those of crystals. In fact, it is known that eigenfunctions may be localized when the randomness exceeds a critical strength.[2] It is also known that the van Hove singularities are smeared out by randomness.[3]

Recently, a new structure of the condensed state has been introduced,[4] which does not belong to either of crystals or random systems. This structure called a quasiperiodic lattice or a quasicrystal is not a crystal because it does not have any translational symmetry. Nor it is a random system because the structure is determined by a definite algorithm. In order to see this point more clearly, let us consider a one-dimensional lattice produced by a projection method as shown in Fig. 1. Projection of a strip with slope α in the first quadrant of a square lattice onto a line with slope ρ yields, after adjusting the length scale, a one-dimensional lattice

$$x_n = n + \rho[\alpha n] \qquad (n = 0, 1, ...), \qquad (1)$$

where $[x]$ denotes the integer part of x. It is apparent that there are two lattice spacings, 1 and $1 + \rho$, which are placed in a certain order determined by α. When α is rational, then the sequence becomes periodic and when α is irrational, the sequence cannot be periodic and becomes quasiperiodic since it is determined uniquely by α. It is instructive to see that the difference among crystals, quasicrystals and random arrays

Symmetries in Science VII, Edited by B. Gruber
and T. Otsuka, Plenum Press, New York, 1994

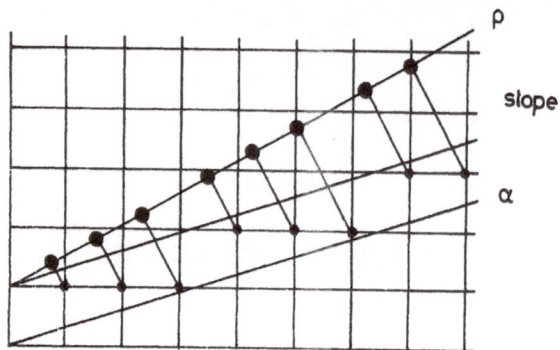

Fig. 1. The projection method to produce a quasiperiodic lattice. A strip with slope α in the first quadrant of a square lattice is projected onto a line with slope ρ, leaving lattice points with quasiperiodic order when α is irrational.

exists in the number of possible configurations for a segment with a given number of units. For crystals, the number of different configurations for N units is always determined by the size of the unit cell and must be $\sim O(1)$, independent of N. For a random array of two elements the number of configuration of N units is $2^N \sim O(e^N)$. On the other hand, for quasicrystals, it is straightforward to see that the corresponding number is $\sim O(N)$.

In this paper, I consider a quasiperiodic sequence of 0 and 1 given by

$$F(\alpha) = \{F_n(\alpha)\} \qquad (n \geq 1) \tag{2}$$

with

$$F_n(\alpha) = [(n+1)\alpha] - [n\alpha], \tag{3}$$

where $n = 1, 2, 3, \cdots$ and α is a real parameter in $(0, 1)$ characterizing the sequence. It should be remarked that this sequence of 0 and 1 is isomorphic to the one-dimensional lattice (1) and the results obtained in the following hold for any system isomorphic to $F(\alpha)$. The sequence defined by Eq. (2) is periodic when α is a rational number a/b ($a < b$ have no common deviser), whose periodic unit denoted by $\Pi(a/b)$ consists of a 1's and $b - a$ 0's.

In the following, I study the structure of one-dimensional quasiperiodic sequences $F(\alpha)$ and show that the quasiperiodic sequences can be classified into two classes according to the self-similarity. In particular, I show that the self-similarity exists in and only in quasiperiodic sequences determined by algebraic numbers of degree two. I also give a description of the structure of arbitrary quasiperiodic sequences based on the hyperinflation symmetry. An application of the hyperinflation is also explained to derive the density of states of a tight-binding hamiltonian.

INFLATION AND SELF-SIMILARITY

First, I introduce an inflation for the sequence (2). Suppose $F(\alpha)$ consists of two segments $S(0, 1)$ and $T(0, 1)$ which are units of k_s 0's, ℓ_s 1's and k_t 0's, ℓ_t 1's, respectively. Simultaneous transformations for the two segments

$$S(0,1) \to S'(0,1)$$
$$T(0,1) \to T'(0,1)$$

(4)

are called an inflation of the sequence, where the numbers of 0 and 1 in $S'(0,1)$, k'_s, ℓ'_s, and those in $T'(0,1)$, k'_t, ℓ'_t, are assumed to satisfy $k'_s + \ell'_s \geq k_s + \ell_s$ and $k'_t + \ell'_t \geq k_t + \ell_t$.

When a sequence $F(\alpha)$ is invariant under the inflation (4), then it is called *self-similar* with respect to the inflation (4). [When the inflation (4) changes the sequence $F(\alpha)$ to some other sequence $F(\alpha')$, it is called a hyperinflation.[5]] When $F(\alpha)$ is self-similar, the ratio of the frequencies of 0 and 1 in the sequence should not change when the inflation (4) is applied. Therefore, α must satisfy a quadratic equation

$$[(k_s + \ell_s)(k'_t + \ell'_t) - (k_t + \ell_t)(k'_s + \ell'_s)]\alpha^2 +$$

$$[(k_t + \ell_t)\ell'_s - (k'_t + \ell'_t)\ell_s + (k'_s + \ell'_s)\ell_t - (k_s + \ell_s)\ell'_t]\alpha + \ell_s\ell'_t - \ell_t\ell'_s = 0, \quad (5)$$

and hence α must be an algebraic number of degree two or a quadratic number. From the contraposition of this statement, the sequence $F(\alpha)$ for α other than quadratic numbers can not be self-similar with respect to the inflation (4). As I discuss below, the converse is also shown to hold, namely the sequence $F(\alpha)$ for any quadratic number has self-similarity.

As an example, let us consider $\alpha = 2 - \sqrt{3}$, for which $F(\alpha)$ becomes

$$F(2 - \sqrt{3}) = 0\ 0\ 1\ 0\ 0\ 0\ 1\ 0\ 0\ 0\ 1\ 0\ 0\ 1\ 0\ 0\ 0\ 1\ 0\ 0\ 0\ 1\ 0\ 0\ 0\ 1\ 0\ 0\ 1\ 0\ 0\ 1\ 0 \cdots.$$

This sequence can be viewed as consisting of two segments, 0 and 10, and it is self-similar with respect to inflation

$$0 \to 0010,$$
$$10 \to 0010010.$$

(6)

SELF-SIMILARITY IN QUADRATIC QUASIPERIODIC SEQUENCES

The existence of the self-similarity in $F(\alpha)$ for arbitrary quadratic numbers can be proved by explicitly writing down the algorithm to find it.[6] To this end, I exploit the continued-fraction expansion of α, which is known to be periodic beyond a certain level.[7] Let us write the continued-fraction expansion of a quadratic number α as

$$\alpha = \frac{1}{k_1 +} \frac{1}{k_2 +} \cdots \frac{1}{k_{n-1} + \theta}, \quad (7)$$

$$\theta = \frac{1}{h_1 +} \frac{1}{h_2 +} \cdots \frac{1}{h_{m-1} + \theta}. \quad (8)$$

The periodic part θ is known to be the inverse of a reduced quadratic number, namely $\theta \in (0,1)$ and its conjugate $\bar{\theta} < -1$. When the continued fraction is truncated at a certain level, it gives an approximant to the irrational number. I denote the i-th approximant of α by p_i/q_i

$$\frac{p_i}{q_i} = \frac{1}{k_1 +} \frac{1}{k_2 +} \cdots \frac{1}{k_{i-1}}, \quad i = 2, \cdots, n. \quad (9)$$

Since θ is the stable fixed point of a modular transformation like

$$\gamma(\theta) = \frac{p\theta + q}{r\theta + s}, \tag{10}$$

with $ps - qr = 1$, α is the stable fixed point of a modular transformation

$$\delta(\alpha) = \frac{p_{n-1}\gamma[\theta(\alpha)] + p_n}{q_{n-1}\gamma[\theta(\alpha)] + q_n} \equiv \frac{A\alpha + B}{C\alpha + D}. \tag{11}$$

Here,

$$\theta(\alpha) = \frac{p_n - \alpha q_n}{\alpha q_{n-1} - p_{n-1}} \tag{12}$$

and $AB - CD = ps - qr = 1$ is shown to hold.

I consider a series of transformation

$$\frac{p_n}{q_n} \rightarrow \delta\left(\frac{p_n}{q_n}\right) \rightarrow \delta^2\left(\frac{p_n}{q_n}\right) \rightarrow \delta^3\left(\frac{p_n}{q_n}\right) \rightarrow \cdots \tag{13}$$

which converges to α monotonically. Since

$$\delta\left(\frac{p'_{n-2}}{q'_{n-2}}\right) = \frac{p_n(s-r) + p_{n-1}(q-p)}{q_n(s-r) + q_{n-1}(q-p)}, \tag{14}$$

$$\delta\left(\frac{p_n}{q_n}\right) = \frac{p_n s + p_{n-1} q}{q_n s + q_{n-1} q}, \tag{15}$$

$$\delta\left(\frac{p_{n-1}}{q_{n-1}}\right) = \frac{p_n r + p_{n-1} p}{q_n r + q_{n-1} p}, \tag{16}$$

where $p'_{n-2}/q'_{n-2} \equiv (p_n - p_{n-1})/(q_n - q_{n-1})$, it can be proved[6] that for $i = 1, 2, 3, \cdots$, $\Pi[\delta^{(i)}(p_n/q_n)]$ is obtained from $\Pi[\delta^{(i-1)}(p_n/q_n)]$ by inflation

$$\begin{aligned}\Pi(p_{n-1}/q_{n-1}) &\rightarrow \Pi[\delta(p_{n-1}/q_{n-1})] \\ \Pi(p'_{n-2}/q'_{n-2}) &\rightarrow \Pi[\delta(p'_{n-2}/q'_{n-2})].\end{aligned} \tag{17}$$

When $p'_{n-2}/q'_{n-2} = 1$ for even n or $p_{n-1}/q_{n-1} = 1$ for odd n, the transformation for $\Pi(1)$ in Eq. (17) must be understood as

$$\Pi(1) \rightarrow \bar{\Pi}[\delta(1)], \tag{18}$$

where $\bar{\Pi}(\cdots)$ denotes the same sequence as $\Pi(\cdots)$ except for 10 at the right end changed to 01. It should be noted that the same modification must apply in the following. As the inflation (17) holds at the fixed point of the modular transformation $\delta(\alpha)$, it is the *self-similar transformation* of $F(\alpha)$ for the quadratic number α satisfying $\alpha = \delta(\alpha)$. Therefore, $F(\alpha)$ is self-similar when α is a quadratic number.

As an example, let us consider $\alpha = 2 - \sqrt{3}$ again. It is easy to see

$$\alpha = \frac{1}{3 + \theta} \tag{19}$$

$$\theta = \frac{1}{1+} \frac{1}{2 + \theta}, \tag{20}$$

and thus

$$\gamma(\theta) = \frac{\theta + 2}{\theta + 3} \tag{21}$$

$$\delta(\alpha) = \frac{1}{4 - \alpha}. \tag{22}$$

Therefore, $p_n/q_n = 1/3$, $p_{n-1}/q_{n-1} = 0/1$ and $p'_{n-2}/q'_{n-2} = 1/2$ with $n = 2$, and the inflation rule (17) becomes

$$\begin{aligned}
\Pi(0) = 0 &\rightarrow \Pi[\delta(0)] = \Pi[1/4] = 0010 \\
\Pi(1/2) = 10 &\rightarrow \Pi[\delta(1/2)] = \Pi[2/7] = 0010010.
\end{aligned} \tag{23}$$

The inflation rule (17) can be reduced to simpler forms in certain cases. Suppose p''_{n-2} and q''_{n-2} defined by

$$\frac{p'_{n-2}}{q'_{n-2}} = \frac{(k_{n-1} - 1)p_{n-1} + p_{n-2}}{(k_{n-1} - 1)q_{n-1} + q_{n-2}} = \frac{p_{n-1} + p''_{n-2}}{q_{n-1} + q''_{n-2}}, \tag{24}$$

is in the converging region Γ_α of the fixed point iteration $\alpha = \delta(\alpha)$. Then, the inflation (17) can be reducible to

$$\begin{aligned}
\Pi(p_{n-1}/q_{n-1}) &\rightarrow \Pi[\delta(p_{n-1}/q_{n-1})] \\
\Pi(p''_{n-2}/q''_{n-2}) &\rightarrow \Pi[\delta(p''_{n-2}/q''_{n-2})].
\end{aligned} \tag{25}$$

This process can be repeated to reach

$$\begin{aligned}
\Pi(p_{n-1}/q_{n-1}) &\rightarrow \Pi[\delta(p_{n-1}/q_{n-1})] \\
\Pi(p_{n-2}/q_{n-2}) &\rightarrow \Pi[\delta(p_{n-2}/q_{n-2})],
\end{aligned} \tag{26}$$

as long as p_{n-2}/q_{n-2} is in Γ_α. Furthermore, noting that

$$\frac{p_{n-1}}{q_{n-1}} = \frac{k_{n-2}p_{n-2} + p_{n-3}}{k_{n-2}q_{n-2} + q_{n-3}}, \tag{27}$$

one can reduce the inflation rule when p_{n-3}/q_{n-3} is in Γ_α, and so on. Since Γ_α for quasi-reduced quadratic numbers (a quadratic irrational in $(0, 1)$ whose conjugate is not in $(0, 1)$) contains $(0, 1)$, the inflation rule can always be reducible to[8]

$$\begin{aligned}
\Pi(0) = 0 &\rightarrow \Pi[\delta(0)] \\
\Pi(1) = 1 &\rightarrow \bar{\Pi}[\delta(1)].
\end{aligned} \tag{28}$$

For example, it is easy to verify that the inflation (23) is reducible to

$$\begin{aligned}
\Pi(0) = 0 &\rightarrow \Pi[\delta(0)] = \Pi[1/4] = 0010 \\
\Pi(1) = 1 &\rightarrow \bar{\Pi}[\delta(1)] = \bar{\Pi}[1/3] = 001.
\end{aligned} \tag{29}$$

For non-quasi-reduced quadratic numbers, the reduction to the rule (28) is not possible because the converging region Γ_α does not contain $(0, 1)$. For example the simplest inflation for $F[(16 - 2\sqrt{2})/31]$ is

$$\begin{aligned}
\Pi(0) &= 0 \rightarrow \Pi(8/19) = 01010010101001010 \\
\Pi(1/2) &= 10 \rightarrow \Pi(3/7) = 0101010.
\end{aligned} \tag{30}$$

GENERALIZED HYPERINFLATION

For an arbitrary irrational number α in $(0, 1)$ other than quadratic numbers, there is no self-similar transformation in the form (4). However, one can have a hyperinflation which generates a sequence of rationals converging to the corresponding quasiperiodic sequence. Let us write the continued fraction expansion of α as

$$\alpha = \frac{1}{k_1+} \frac{1}{k_2+} \cdots \frac{1}{k_{n-1}+} \cdots \tag{31}$$

and the n-th approximant to α as $(p_1/q_1 = 0/1)$

$$\frac{p_n}{q_n} = \frac{1}{k_1+} \frac{1}{k_2+} \cdots \frac{1}{k_{n-1}}. \tag{32}$$

For these approximants, the periodic unit $\Pi(\frac{p_{n+1}}{q_{n+1}})$ for $\frac{p_{n+1}}{q_{n+1}}$ is given by the following generalized hyperinflation rule: When n is even

$$\Pi\left(\frac{p_{n+1}}{q_{n+1}}\right) = \begin{cases} \underbrace{\Pi(\frac{p_{n-1}}{q_{n-1}}) \Pi(\frac{p_n}{q_n}) \cdots \Pi(\frac{p_n}{q_n})}_{k_n} & p_n/q_n \neq 1 \\ \underbrace{1 \cdots 1}_{k_n} \Pi(\frac{p_{n-1}}{q_{n-1}}) & p_n/q_n = 1 \end{cases}, \tag{33}$$

and when n is odd

$$\Pi\left(\frac{p_{n+1}}{q_{n+1}}\right) = \begin{cases} \underbrace{\Pi(\frac{p_n}{q_n}) \cdots \Pi(\frac{p_n}{q_n})}_{k_n} \Pi(\frac{p_{n-1}}{q_{n-1}}) & p_{n-1}/q_{n-1} \neq 1 \\ \underbrace{\Pi(\frac{p_n}{q_n}) \cdots \Pi(\frac{p_n}{q_n})}_{k_{n-1}} 1 \Pi(\frac{p_n}{q_n}) & p_{n-1}/q_{n-1} = 1 \end{cases}. \tag{34}$$

As an example, I take $\alpha = (e^2 - 1)/(e^2 + 1) = \frac{1}{1+} \frac{1}{3+} \frac{1}{5+} \frac{1}{7+} \cdots$, a transcendental number. As the order of approximant becomes higher, the periodic unit is transformed as

$$0 \to 1 \to 1\ 1\ 1\ 0 \to \underline{1\ 1\ 1\ 0}\ \underline{1\ 1\ 1\ 0}\ \underline{1\ 1\ 1\ 0}\ \underline{1\ 1\ 1\ 0}\ 1\ \underline{1\ 1\ 1\ 0} \to \cdots. \tag{35}$$

The hyperinflation can be used to find the transformation of various physical quantities along the sequence of approximants, which include the trace mapping of unimodular matrices appearing in many one-dimensional systems[6] and the partition function for the Ising spin system.[9]

DENSITY OF STATES OF APPROXIMANT CRYSTALS

As an example of applications of the inflation, I consider the density of states of the tight-binding hamiltonian

$$H(\alpha) = \sum_i \{|i > \epsilon_i < i| + |i > t_i < i+1| + |i+1 > t_i < i|\}, \tag{36}$$

where the transfer energy is given by $t_i \equiv t_{F_i(\alpha)}$ with $F_i(\alpha)$ in Eq. (3). The site energy ϵ_i is set to zero for a given system to be analyzed. I investigate approximant quasiperiodic sequences generated by Eq. (22) with initial condition $\alpha = 0$. Equation (22) yields $\alpha^{(n)} = G_n/G_{n+1}$ for the n-th entry. Here, $\{G_n\}$ $(G_n = [(2+\sqrt{3})^{n-1} - (2-\sqrt{3})^{n-1}]/2\sqrt{3})$ is the series generated by the recurrence relation $G_{n+2} = 4G_{n+1} - G_n$ with $G_1 = 0$ and $G_2 = 1$. Noting that there are three types of site energies and two types of transfer

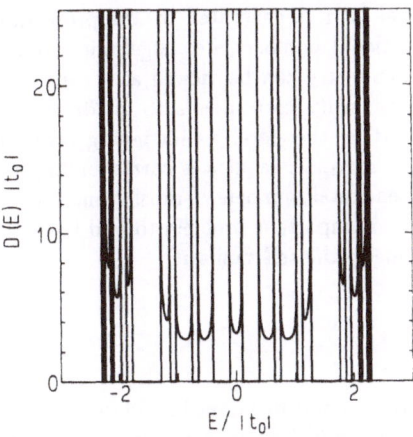

Fig. 2. The density of states $D(E)$ of hamiltonian (36) for $\alpha^{(3)} = 4/15$, normalized as $\int_{-\infty}^{\infty} D(E)dE =$ the number of sites in a unit cell. $t_1/t_0 = 1.5$.

energies, it is straightforward to derive a decimation transformation from $H(\alpha^{(n)})$ to $H(\alpha^{(n-1)})$. After n transformations, the system becomes a regular chain $H(\alpha^{(1)})$ with site energy $\bar{\epsilon}$ and transfer energy $\bar{t}$ which are functions of energy E. The density of states $D(E)$ for $H(\alpha^{(n)})$ is given by

$$D(E) = \frac{\ell_n}{2\pi} \frac{d}{dE} \int_{E(k) \le E} dk = \frac{1}{\pi} \frac{|df(E)/dE|}{\sqrt{1 - |f(E)|^2}}, \tag{37}$$

where

$$f(E) \equiv \frac{E - \bar{\epsilon}}{2\bar{t}} = \cos(k\ell_n), \tag{38}$$

defines $E(k)$ and $\ell_n \equiv G_{n+1}a$, a being the lattice constant for the chain, is the length of a unit cell. I show in Fig. 3 the density of states $D(E)$ for $\alpha^{(3)} = 4/15$. As the order of the approximant gets higher, the the density of states becomes more spiky, and eventually point-like at $n = \infty$.

CONCLUDING REMARKS

In this article, I showed that the quasiperiodic lattice generated by the projection method can be classified into two classes, self-similar ones and non-self-similar ones, and proved that the self-similar quasiperiodic sequences can be related to quadratic irrationals or algebraic numbers of degree two. There have been numerous studies on the Fibonacci sequence $F(\frac{\sqrt{5}-1}{2})$.[4] The Fibonacci sequence belongs to the quadratic quasicrystals and could be a typical example for it. However it can hardly be a typical system for arbitrary quasiperiodic chains.

The self-similarity is a kind of symmetry operation existing in the quadratic quasi-periodic sequence. It can be used in the evaluation of various physical quantities. Generalized hyperinflation generates a series of approximants to a quasiperiodic sequence for arbitrary irrationals and is also useful in elucidating properties of the quasiperiodic system.

In the one-dimensional lattices generated by the projection method (1), regular crystals correspond to rational numbers, i.e. algebraic numbers of degree one, and hence the translational symmetry can be viewed as a consequence of rationality. As I showed here, the self-similarity corresponds to quadratic numbers, i.e. algebraic numbers of degree two. It is, therefore, an intriguing problem to study what kind of symmetry exists in quasiperiodic sequences corresponding to algebraic numbers of higher degree and transcendental numbers. Translational symmetry yields the Bloch theorem in crystals. It is an important problem to find if any basic theorem exists in quadratic quasicrystals due to the self-similarity.

ACKNOWLEDGEMENT

This work was supported in part by Grant-in-Aid for Scientific Research on Priority Areas from the Ministry of Education, and Science and Culture. The text processing of this paper was carried out using the Data Processing System for Perceptive Informations at Kyoto Institute of Technology.

REFERENCES

1. N. W. Ashcroft and N. David Mermin, "Solid State Physics", Holt, Reinhart and Winston, Philadelphia, (1976).

2. P. W. Anderson, Phys. Rev. **109**, 1492–1505 (1958).

3. See, for example, A. Gonis, "Green Function For Ordered And Disordered Systems", North Holland, Amsterdam (1992).

4. D. Shechtman, I. Blech, D. Gratias, and J. W. Cahn, Phys. Rev. Lett. **53**, 1951–1954 (1984). See also "The Physics of Quasicrystals", edited by P. J. Steinhardt and S. Ostlund, World-Scientific, Singapore (1987); "Introduction to Quasicrystals", edited by M. V. Jaric, Academic, SanDiego (1988); "Quasicrystals", edited by T. Fujiwara and T. Ogawa, Springer, Berlin (1990); "Quasicrystals", edited by K. H. Kuo and T. Ninomiya, World Scientific, Singapore (1991).

5. T. Odagaki and H. Aoyama, Phys. Rev. Letters, **61**, 775–778 (1988); Phys. Rev., **B39**, 475-487 (1989).

6. T. Odagaki and M. Kaneko, preprint.

7. A. Baker, "A Concise Introduction To The Theory Of Numbers", Cambridge University Press, London (1986).

8. M. Kaneko and T. Odagaki, preprint.

9. T. Odagaki, J. Math. Phys. **31**, 421–422 (1990).

ALGEBRAIC MODELS:
NEW DEVELOPMENTS IN MOLECULAR SPECTROSCOPY

Stefano Oss

Department of Physics
University of Trento
and INFN
38050 POVO (TN) Italy

INTRODUCTION

In this paper, dedicated to Franco Iachello on occasion of his 50th birthday, an overview of dynamical symmetries in molecular spectroscopy and its applications will be given. Since the original formulation of the algebraic model for simple molecules[1-4], Iachello and co-workers started to improve and generalize this model to reproduce spectroscopical and dynamical properties of more and more complex systems. The present status of art is such that algebraic techniques can provide a non trivial contribution to usual approaches in this well consolidated (but still open) field of physics. As it will be pointed out in a subsequent section of this paper, simplified one-dimensional dynamical symmetries can even attack and solve problems beyond the possibilities of traditional models.

The present paper is structured as follows. After some general remarks on the algebraic approach to the molecular spectroscopical problem, two algebraic models will be explained in distinct sections. A section devoted to the most recent applications of these models will follow. The style of the present paper is thought to give a not too technical overview on this subject. More detailed papers already published are listed in the references. Another key point of this paper is that the theoretical formulations here expressed should be constantly referred to the specific framework of molecular spectroscopy. This area of research is extremely active and on its theoretical side a well-settled formalism and language must be respected. For this reason it could happen sometimes that the rigorous group-theoretical language is slightly softened in favour of an important reciprocal dialogue with "traditional" molecular physicists.

Symmetries in Science VII, Edited by B. Gruber
and T. Otsuka, Plenum Press, New York, 1994

DYNAMICAL SYMMETRIES AND MOLECULAR SPECTROSCOPY

Algebraic theory is used in many areas of modern physics as a general framework to attack different problems. One of the most striking examples of such strategy is given by the interacting boson model used since many years in nuclear physics[5]. More recently an algebraic model for the molecular spectroscopical problem has been introduced[1]. It must be stressed here that the development of new techniques in molecular spectroscopy is still worth being done. As it is well known, traditional approaches to this problem come across serious difficulties when dealing with more than three interacting atoms in a molecule. More specifically, those methods based on the numerical solution of the Schrödinger equation can result in slow or missing convergence of the solution itself. Moreover, troubles concerning a proper definition of both potential and kinetic couplings often arise. A different approach to the same class of problems is based on the use of semiempirical expansions. Here, like in the Dunham series[6], the spectroscopical term values are obtained as a power series in the rotovibrational quantum numbers. In this case a certain number of parameters must be fixed by using a fitting procedure over an experimental data base of levels. As it usually happens for medium-size and large molecules, few experimental data are available. This makes simply impossible to determine the large number of the expansion parameters. As a second point, these semiempirical expansions do not provide the eigenfunctions of the physical system.

Algebraic approaches to the molecular rotovibrational problem seem to offer an alternative and competitive solution to the above mentioned problems. An algebraic Hamiltonian operator (thus providing both energies and wavefunctions) is constructed based on a relatively small set of arbitrary parameters. The diagonalization procedure is usually of reasonable dimensionality as it will be discussed later. Other quantities of interest, like IR transition operators, can be written and evaluated in an algebraic language[7]. The key point for the construction of the Hamiltonian operator is that of determining a proper spectrum generating algebra (SGA). It has been shown[1,2] that the n-dimensional roto-vibrational problem for the single oscillator is correctly solved with an unitary algebra in n+1 dimensions, U(n+1). This means that the most general Hamiltonian operator is constructed as a function of elements of U(n+1), that is,

$$H = f(g_i), \qquad g_i \in U(n+1) \tag{1}$$

i.e. the Hamiltonian operator (with all the operators of interest) is in the enveloping algebra of U(n+1). A special case of such enveloping algebra is obtained when considering only invariant (Casimir) operators of U(n+1) and of its subalgebras. In this situation the Hamiltonian operator is expressed as

$$H = f(C_k) \tag{2}$$

where C_k is a generic Casimir operator of some subalgebra of U(n+1). The Hamiltonian (2) is now said to describe a dynamical symmetry of the physical system. The use of dynamical symmetries allows one to obtain particularly simple expressions for operators of interest. More specifically, analytical formulas for the eigenvalues of the operator at issue are available once a proper algebraic basis is chosen. In the following sections an overview of the two mostly used SGA in molecular spectroscopy, based on U(4) and U(2) algebras, will be given. Mathematical aspects of U(2)-based models are by far simpler than those using the U(4) algebra. Nonetheless, practical applications of the U(2) algebra to non-trivial physical systems are more recent than those making use of U(4). For this reason in the present paper U(4) will be discussed first.

U(4) AND THE VIBRON MODEL

Any diatomic molecule can be thought as a three-dimensional anharmonic oscillator. Its rotovibrational spectrum is thus readily reproduced by using as SGA the unitary algebra U(4). To do that, one must specify the representation within this algebra acts. The most convenient representation is [N,0,0,0], which is totally symmetric. To clarify the meaning of the quantum number N it is necessary to explicitly realize U(4) in terms of boson operators (vibrons)

$$b_i^+, b_i \quad i=1, \dots, 4 .$$ (3)

The generating operators of U(4) are thus written as

$$G_{ij} = b_i^+ b_j \quad i, j=1, \dots, 4 .$$ (4)

The most general Hamiltonian operator can then be written in the form

$$H = E_0 + \sum_{ij} \xi_{ij}^{(1)} G_{ij} + \frac{1}{2} \sum_{ijhk} \xi_{ijhk}^{(2)} G_{ij} G_{hk} + \cdots$$ (5)

This form can be greatly simplified by considering possible dynamical symmetries, that is, by using only Casimir operators of subalgebras of U(4). Before doing that, it is important to observe that only states with good angular momentum are considered in the physical description of the system. This means that *(i)* the rotation algebra O(3) must be included in the chain of subalgebras of U(4) and *(ii)* the boson operators (3) must be grouped in operators with well defined properties under rotations[3]. It is then possible to obtain two different chains (thus giving two distinct dynamical symmetries) given by

$$U(4) \supset O(4) \supset O(3) \supset O(2)$$ (6a)
$$U(4) \supset U(3) \supset O(3) \supset O(2) .$$ (6b)

These two chains correspond to dynamical symmetries suited respectively for rigid and soft diatomic molecules[3,4]. Starting from the total symmetric representation [N,0,0,0]≡[N] of the SGA one obtains two branching schemes for the chains (6a,b). As we are dealing only with rigid molecules, branching (6b) will not be considered anymore from now on. One then obtains the algebraic kets

$$\left| \begin{matrix} U(4) \supset O(4) \supset O(3) \supset O(2) \\ [N] \quad (\omega,0) \quad J \quad M_J \end{matrix} \right\rangle$$ (7)

where $(\omega,0)$ denotes a symmetric representation of O(4) and J, M_J have the usual meaning. The algebraic branching rules give for the quantum numbers defining kets (7) the following formulas:

$$\omega=N, N-2, \dots, 1 \text{ or } 0 \quad (N=\text{odd or even})$$
$$J=\omega, \omega-1, \dots, 0 \text{ and } -J \leq M_J \leq J .$$ (8)

It is now possible to write (up to quadratic order in the generators G_{ij}) the Hamiltonian corresponding to the dynamical symmetry (6a):

$$H = E_0 + A\, C(O(4)) + B\, C(O(3)),\qquad (9)$$

where $C(g)$ denotes a Casimir operator of g. The contribution coming from O(2) has not been added to eq.(9) since it is needed only in presence of an external field (lifting the 2J+1-fold degeneracy). Having at disposal analytical expressions for expectation values of invariant operators on the basis (7), one easily obtains for the spectrum of operator (9) the simple formula

$$E = E_0 + A\,\omega(\omega+2) + B\, J(J+1).\qquad (10)$$

This equation can be rewritten in a more familiar form by replacing the O(4) quantum number ω by $v = (N-\omega)/2$. With this substitution the spectrum (10) is now given by

$$E = \omega_e\,(v+1/2) - \omega_e x_e\,(v+1/2)^2 + B\, J(J+1)\qquad (11)$$

which, a part from very small corrections, gives the energies of a three-dimensional Morse oscillator[8]. The notation of eq.(11) is typical for a Dunham expansion stopped at the quadratic term for a Morse potential. The explicit function of co-ordinates of such potential is given by

$$V(r) = V_0\,[e^{-2\alpha(r-r_0)} - 2e^{-\alpha(r-r_0)}].\qquad (12)$$

This potential is one of the most widely used forms because of its accuracy in reproducing the experimental features of diatomic molecules. The results contained in eqs.(10) and (11) mean that the three-dimensional Morse interaction has a dynamical symmetry of the type (6a). The number N appears to be strictly connected with the total number of bound states supported by the potential well (or, equivalently, with its anharmonicity).

At this point it should be clear that the algebraic model for diatomic molecules can be generalized as an expansion in terms of Morse potentials. Given that the Morse potential is a good approximation for the real interatomic interaction, usually very few terms are needed to achieve spectroscopical resolution. This is obtained in practice by adding to the Hamiltonian (9) higher powers and/or products of Casimir operators.

The extension of the vibron model to polyatomic molecules is in principle straightforward. One replaces each interatomic bond by an U(4) algebraic structure. If the molecule contains n bonds then the SGA is given by

$$U_1(4) \oplus U_2(4) \oplus \ldots \oplus U_n(4).\qquad (13)$$

Again, the most general Hamiltonian operator in the enveloping algebra of (13) could be written. It is however extremely convenient to look for possible dynamical symmetries in order to reduce the complexity of the problem.

Starting for example with a three-atomic molecule, it has been found that, among others, the following algebraic lattice can be used[3,4]:

$$U_1(4)\oplus U_2(4) \quad\left\langle\begin{array}{c} O_1(4)\oplus O_2(4) \\[4pt] U_{12}(4) \end{array}\right\rangle\quad O_{12}(4)\supset O_{12}(3).\qquad (14)$$

To this very simple structure a direct deep physical significance can be attached. The chain containing the reduction $U_1(4)\oplus U_2(4)\supset O_1(4)\oplus O_2(4)$ describes the stretching motions of the two interatomic bonds. This is always true for the reduction $U_i(4)\supset O_i(4)$, as explained in the diatomic case. The reduction to $O_{12}(4)$ is directly connected with the anharmonic bending motion of the molecule. With the introduction of both quadratic Casimir operators of $O_{12}(4)$ (which can now include non-symmetrical representations) any geometrical configuration of the molecule - from linear to bent - can be described[3,4,7,9]. The final reduction to $O_{12}(3)$ allows one to label the rotational states of the system, as usual. In the vibron model an algebraic basis is chosen such that the first chains of subalgebras give a diagonal Hamiltonian operator. The kets are given by

$$\left| \begin{array}{cccccc} U_1(4)\oplus U_2(4) \supset O_1(4)\oplus O_2(4) \supset & O_{12}(4) \supset O_{12}(3) \\ {[N_1]} \quad [N_2] \quad\quad (\omega_1,0)\quad(\omega_2,0)\quad (\tau_1,\tau_2)\quad J \end{array} \right\rangle . \tag{15}$$

The Hamiltonian operator can thus be written as

$$H^{local}=H_0+A_1\,C(O_1(4))+A_2\,C(O_2(4))+A_{12}\,C(O_{12}(4))+B\,C(O_{12}(3)) . \tag{16}$$

Simple expressions for the eigenvalues of this Hamiltonian in terms of both algebraic and local vibrational quantum numbers are readily obtained[3]. It should be clear however that in a real molecule this "local" picture is only an approximation[10]. In particular one should expect non-diagonal contributions (mixing local modes) to this Hamiltonian. This is achieved by adding to (16) the Casimir operator associated with the second chain of the lattice (14), that is, a Majorana operator M_{12} related to $C(U_{12}(4))$ [3]. The second chain has again a simple physical interpretation. The reduction $U_1(4)\oplus U_2(4)\supset U_{12}(4)\supset O_{12}(4)$ is equivalent to consider the molecule as a "global" anharmonic oscillator in which local modes lose their own identity. The complete Hamiltonian operator then reads

$$H=H^{local}+\lambda_{12}M_{12} . \tag{17}$$

From an algebraic point of view this implies that the original dynamical symmetry (established by the first chain in (14)) is broken. Such symmetry breaking is proportional, in its effects, to the normal versus local behavior of the molecule. It should however pointed out that the $O_{12}(4)$ symmetry (as it can be clearly seen in the lattice (14)) is preserved. This means that a blocked form of the Hamiltonian matrix is to be expected, allowing for a great simplification of numerical computations[11].

The extension of the vibron model to larger molecules strictly follows the scheme explained above[12-14]. For a molecule containing n interatomic bonds, as a first step a local picture based on the algebraic reductions $U_i(4)\supset O_i(4)$ and $O_i(4)\oplus O_j(4)\supset O_{ij}(4)$ is considered. Here the algebraic quantum numbers of $O_i(4)$ and $O_{ij}(4)$ are respectively related to local stretching and bending vibrational quantum numbers. The single uncoupled $O_i(4)$ symmetries are then eventually broken to take into account non-diagonal contributions to the local Hamiltonian. This is achieved by adding Majorana operators M_{ij} anharmonically coupling local modes i and j.

As a general comment, it is worth saying that the real advantages of the vibron model come out when dealing with molecules with no less than three atoms. For a molecule with n atoms the number of parameters needed in a second order Dunham expansion is given in fact by $9(n-2)(n-1)/2$ while in the algebraic approach the needed parameters are $4n-7$.

As it has been explained for the diatomic case, higher order Hamiltonian operators can be constructed by adding products and/or powers of Casimir operators to the first order Hamiltonian (17). This procedure is again equivalent to consider the Morse potential as a good starting approximation for the anharmonic molecular interactions. It can happen however that the potential function strongly deviates from the Morse law (this is the case of some kind of torsional motions in complex molecules). Problems like this can be solved by introducing proper analytical functions of the Casimir operators[13].

U(2) AND THE ONE-DIMENSIONAL MODEL

The use of U(2) algebra in simple physical applications is a well-settled fact since many years[1,2,15-19]. It is however a very recent story that the U(2) algebra can also be widely and successfully used to attack non trivial problems of molecular spectroscopy[20-22].

In this section a brief description of the most important aspects of the U(2)-based dynamical symmetry will be given. The group theoretical approach to the one-dimensional Morse oscillator has been fully reviewed for example in ref.16 for both discrete and scattering states. Here only the bound states problem will be addressed.

The U(2) algebra is realized through the Schwinger[23] representation in terms of two boson creation and annihilation operators usually grouped as in

$$J_+ = a^+ b, \quad J_- = b^+ a, \quad J_z = (a^+ a - b^+ b)/2, \quad N = a^+ a + b^+ b . \tag{18}$$

In this way the isomorphism $U(2) = U(1) \oplus SU(2) \approx U(1) \oplus SO(3)$ is naturally emphasized. U(2) has in fact two Casimir operators: the first one, linear in the group generators, is $C^{(1)}(U(2)) \equiv N$ (the usual number operator); the second one, quadratic in the group generators, can be written in terms of the pseudo angular momentum operator S of SO(3) as $C^{(2)}(U(2)) \equiv 4S^2 - N = N^2 + N$. In order to obtain the most general Hamiltonian operator the following algebraic chains must be considered:

$$\begin{cases} U(2) \supset U(1) \\ U(2) \supset O(2) \end{cases} \tag{19}$$

with corresponding basis kets

$$\begin{cases} |[N], n\rangle \\ |[N], m\rangle \end{cases} \tag{20}$$

in which n, m label respectively eigenstates of $C^{(1)}(U(1))$ and $C^{(1)}(O(2))$. The branching rules are n = 0, 1, ... , N and m = ±N, ±(N-2), ... , ±1 or 0 (N even or odd). The most general Hamiltonian operator (Lipkin Hamiltonian[24]) is thus given by

$$H = E_0 + \alpha\, C^{(1)}(U(1)) + \beta\, [C^{(1)}(U(1))]^2 + A\, \{[C^{(1)}(O(2))]^2 - N^2\} . \tag{21}$$

Exactly solvable formulas are obtained when considering dynamical symmetries. In the U(2) case two possibilities arise:

U(1) dynamical symmetry:

$$H^{(1)} = E_0 + \alpha\, C^{(1)}(U(1)) + \beta\, [C^{(1)}(U(1))]^2 , \tag{22a}$$

O(2) dynamical symmetry:

$$H^{(2)} = E_0 + A\, \{[C^{(1)}(O(2))]^2 - N^2\} . \tag{22b}$$

The O(2) dynamical symmetry has an important physical counterpart obtained by writing down explicitly the spectrum of the Hamiltonian (22b):

$$E(m) = E_0 + A (m^2 - N^2) \qquad (23)$$

or, in terms of the vibrational quantum number $v \equiv (N-m)/2$,

$$E(v) = E_0 + 4A \, v \, (v-N), \quad v = 0, 1, \dots, N/2 \text{ or } (N-1)/2 \text{ (N even or odd)} \qquad (24)$$

in which only the positive branch of m has been considered. Equation (24) thus gives energy levels of the one-dimensional Morse oscillator whose physical parameters are easily put in correspondence with algebraic parameters E_0, A and N.

It is now important to extend this simple mathematical result to a system of several interacting one-dimensional oscillators. The simplest case is that of two oscillators. The SGA is then $U_1(2) \oplus U_2(2)$. Two possible dynamical symmetries arise, given by the algebraic lattice

$$U_1(2) \oplus U_2(2) \quad \Big\langle \begin{array}{c} O_1(2) \oplus O_2(2) \\ \\ U_{12}(2) \end{array} \Big\rangle \quad O_{12}(2) \qquad (25)$$

Starting from the first path in the lattice (25) a "local" basis ket can be defined as in

$$\left| \begin{array}{ccccc} U_1(2) \oplus U_2(2) \supset O_1(2) \oplus O_2(2) \supset O_{12}(2) \\ N_1 & N_2 & m_1 & m_2 & m_{12} = m_1 + m_2 \end{array} \right\rangle. \qquad (26)$$

The following Hamiltonian operator is then readily constructed

$$H = E_0 + A_1 \, C(O_1(2)) + A_2 \, C(O_2(2)) + A_{12} \, C(O_{12}(2)), \qquad (27)$$

in which $C(O_i(2)) \equiv [C^{(1)}(O_i(2))]^2$, i=1,2 and 12. The coupled operator $C(O_{12}(2))$ has matrix elements on the basis (26) given by

$$\langle N_1 m'_1, N_2 m'_2 | C(O_{12}(2)) | N_1 m_1 N_2 m_2 \rangle = ((m_1 + m_2)^2 - (N_1 + N_2)^2) \, \delta_{m_1, m'_1} \delta_{m_2, m'_2} \qquad (28a)$$

or, equivalently, after the introduction of vibrational quantum numbers $v_i = (N_i - m_i)/2$, i=1,2,

$$\langle N_1 v'_1, N_2 v'_2 | C(O_{12}(2)) | N_1 v_1 N_2 v_2 \rangle = 4(v_1 + v_2)(v_1 + v_2 - (N_1 + N_2)) \, \delta_{v_1, v'_1} \delta_{v_2, v'_2}, \qquad (28b)$$

The algebraic lattice (25) allows one to consider another interbond coupling on the local basis (26). Such coupling is given by the Majorana operator M_{12} related to the Casimir operator $C^{(2)}(U_{12}(2))$ through the following relation (with $C(U_i(2)) \equiv C^{(2)}(U_i(2))$):

$$M_{12} = -\frac{1}{2} \{ C(U_{12}(2)) - C(U_1(2)) - C(U_2(2)) - N_1 \cdot N_2 \} =$$
$$= \frac{N_1 \cdot N_2}{2} - 2S_1 \cdot S_2 \; . \qquad (29)$$

This operator has both diagonal ($N_1 \cdot N_2$) and non-diagonal ($S_1 \cdot S_2$) contributions to the spectrum of the Hamiltonian

$$H = E_0 + A_1\, C(O_1(2)) + A_2\, C(O_2(2)) + A_{12}\, C(O_{12}(2)) + \lambda_{12}\, M_{12}. \tag{30}$$

Matrix elements of the Majorana operator are given by

$$
\begin{aligned}
\langle N_1 v_1' N_2 v_2' | M_{12} | N_1 v_1 N_2 v_2 \rangle &= (N_1 v_1 + N_2 v_2 - 2 v_1 v_2)\, \delta_{v_1 v_1'}\, \delta_{v_2 v_2'} + \\
&- \sqrt{(N_1 - v_1)(v_1 + 1) v_2 (N_2 - v_2 + 1)}\, \delta_{v_1 + 1 v_1'}\, \delta_{v_2 - 1 v_2'} + \\
&- \sqrt{v_1 (N_1 - v_1 + 1)(v_2 + 1)(N_2 - v_2)}\, \delta_{v_1 - 1 v_1'}\, \delta_{v_2 + 1 v_2'}\,.
\end{aligned}
\tag{31}
$$

To summarize, a situation completely analogous to the three-dimensional case (see previous section) is obtained in the U(2) framework. In particular, an algebraic Hamiltonian (30) for two interacting anharmonic oscillators is constructed. Its diagonalization on a local algebraic basis (26) is easily performed by using equations (24), (28a) and (31). Two different kinds of coupling between bonds arise. These are *(i)* an anharmonic diagonal coupling C_{12} preserving the original uncoupled dynamical symmetries of the single oscillators and *(ii)* an anharmonic non-diagonal coupling M_{12} mixing local oscillators (thus leaving only $m_{12} = m_1 + m_2$ or $v_{12} = v_1 + v_2$ as good quantum numbers for the entire system).

This scheme can be easily extended in other directions which are the following:

(i) an higher order algebraic Hamiltonian is obtained by adding powers and/or products of Casimir operators to operator (30);

(ii) higher order Majorana operators (practically ladder operators, see next section) allowing for changes in the m_{12} or v_{12} global quantum numbers can be introduced[14]. This is equivalent to consider more general and complex interactions between single oscillators (breaking the $O_{12}(2)$ symmetry);

(iii) the extension of this model to a very large number of oscillators is a straightforward procedure;

(iv) the single oscillator degree of freedom can be interpreted in different physical ways, that is, not only stretching motions but bending vibrations as well are described within this framework. This point is the most important one for practical applications. The fact that not only stretching motions can be described by using U(2) is explained by observing that in one dimension both the Morse and the Pöschl-Teller potentials have the same discrete spectrum[16]. The symmetric shape about the origin of the Pöschl-Teller potential is particularly well-suited for describing non-degenerate molecular bending motions.

RECENT DEVELOPMENTS

In this section a brief description of novel aspects of algebraic approaches to molecular spectroscopy will be given. In these last years a large number of different problems have been attacked in this field. It would be simply impossible to describe with great detail all these problems. For this reason only a list of the most important results will be now presented.

Vibrational Analysis of Three- and Four-Atomic Molecules in the U(4) Model

Following the original formulation of the vibron model, its possibilities and limits have been explored by performing vibrational analysis of many different molecules with three[7,9,11] and four[12-14] atoms. In particular it has been found that a spectroscopical quality of fits is achieved by using second order algebraic Hamiltonian operators. This result represents nonetheless a considerable saving of parameters when compared with a Dunham expansion stopped at the same order. In the specific case of four-atomic linear molecules (HCCH, HCCF), the problem of describing in the algebraic framework combination bands of degenerate bending modes arises. Such combination bands constitute the basis of both l-splitting and l-doubling phenomena. It has been found[12] that suitable algebraic operators exist which take into account in a correct, quantitative way these effects. When dealing with molecules with four (or more) atoms, different possible geometrical configurations must be considered. One in fact can have linear symmetrical (HCCH), linear non-symmetrical (HCCF), quasi-linear (HCNO), bent planar (HNCO), non-chained planar (H_2CO) and non-planar (NH_3, HOOH) molecules. The problem of the correlation among rotational and vibrational quantum states characterizing the same molecule in different geometrical configurations (a trivial problem in the three-atomic case) is still open when internal hindered torsional motions have to be considered. This problem has been successfully solved with the help of the algebraic parametrization of the Hamiltonian operator[25].

Rovibrational Analysis in the U(4) Model

The problem of reproducing in the algebraic framework both energy shift and energy splitting caused by the coupling between rotations and vibrations has been attacked[14]. The simple case of vibrational energy shift has been easily solved by considering an effective rovibrational Hamiltonian whose form reads

$$H^{RV} = \sum_i A_i^{RV} C_i \, C(O(3)) . \tag{32}$$

In this expression the C_i denote Casimir operators reproducing vibrational terms affected by rotations. The O(3) invariant operator obviously describes molecular rotations for a given vibrational state. A more subtle problem arises when degenerate molecular bending modes are taken into account. In this case the rovibrational coupling lifts the vibrational angular momentum degeneracy[6]. A non-diagonal contribution to the rovibrational Hamiltonian operator is thus expected. Preliminary results for three-atomic molecules are now available[26]. A non-diagonal rovibrational algebraic operator has been introduced. This operator preserves the final O(3) molecular symmetry, but it also breaks the final O(4) symmetry in order to lift the associated vibrational angular momentum degeneracy. The full rovibrational operator is given by

$$H^{RV} = \sum_{i,j} A_{ij}^{RV} \left[\left[D_i \times D_j \right]^{(L)} \times [J \times J]^{(L)} \right]_0^{(0)} . \tag{33}$$

The sums i, j here go over the bond indices. In this expression the D operator is related to $C(O_{12}(4))$ through the relation $C(O_{12}(4))=D\cdot D+J\cdot J$ with $D\equiv\Sigma_i D_i$. Operators like (33) have the correct selection rules for reproducing the physical effect of interest. The computation of matrix elements of H^{RV} requires an extended use of Racah algebra and will not reported here. The final result (expressed on the local basis (15)) is

$$\left\langle \omega_1, \omega_2, (\tau_1, \tau_2), J, M \middle\| \left[\left[D_i \times D_j \right]^{(L)} \times [J \times J]^{(L)} \right]_0^{(0)} \middle| \omega_1', \omega_2', (\tau_1', \tau_2'), J', M' \right\rangle =$$

$$= \delta_{JJ'} \, \delta_{MM'} \, J(J+1)\sqrt{2J+1}\sqrt{2L+1} \begin{Bmatrix} 1 & 1 & L \\ J & J & J \end{Bmatrix} \sum_{\substack{l_1 l_2 \\ l_1' l_2'}} \left\langle \begin{matrix} \omega_1 & \omega_2 \\ l_1 & l_2 \end{matrix} \Big| \begin{matrix} (\tau_1, \tau_2) \\ J \end{matrix} \right\rangle \left\langle \begin{matrix} \omega_1' & \omega_2' \\ l_1' & l_2' \end{matrix} \Big| \begin{matrix} (\tau_1', \tau_2') \\ J \end{matrix} \right\rangle \cdot \tag{34}$$

$$\begin{Bmatrix} l_1 & l_2 & J \\ l_1' & l_2' & J \\ 1 & 1 & L \end{Bmatrix} \left\langle \omega_1 l_1 \| D_1 \| \omega_1' l_1' \right\rangle \left\langle \omega_2 l_2 \| D_2 \| \omega_2' l_2' \right\rangle$$

in which isoscalar factors for the reduction O(4)⊃O(3) following the Racah factorization lemma and reduced matrix elements of the operator D have been used[3,4].

U(2) Dynamical Symmetry and Point Groups

As it is well known, computation of symmetry adapted vibrational wave functions of molecules - especially for complicated point groups and/or in case of highly excited levels - is by no means an easy task[27]. As a matter of fact, one-dimensional algebraic techniques can give a precious help in solving this problem. It has been shown in fact that the U(2) Hamiltonian operator (30) for n-interacting bonds automatically provides states carrying irreducible representations of the appropriate molecular point group[20]. As it has been recently pointed out[28], this result has a general validity. This can be explained by establishing the isomorphism between the discrete group of the physical system and a subgroup of the permutation group (as imposed by Cayley's theorem[29]). The Hamiltonian operator is characterized by U(2) invariant operators (Majorana operators and their linear combinations) and the representations of unitary groups are related to those of the symmetric group. It can be shown that the diagonalization of appropriate linear combinations of Majorana operators M_{ij} gives states transforming in the proper way. As a consequence of this important mathematical result, vibrational analyses of complex molecules like XF_6 [20] and C_6H_6 [21,22] (and its deuterated forms) have been carried out with minimum computational effort.

High Resolution Analysis of Very Complex Molecules in the U(2) Model

Recent experimental techniques allow for very high resolution spectroscopy of excited vibrational modes of molecules whose structure is often beyond the actual possibilities of theoretical interpretation. A key point of this fact is the existence of resonances (called Fermi resonances[30]) perturbing several molecular states. These resonances, if on one side characterize in an unique way the IR spectrum, on the other side make really challenging - if not impossible at all - any theoretical attack to this situation. The one-dimensional algebraic approach seems again to offer an alternative or even necessary way to the solution of this problem. In the U(2) model it is an easy task to reproduce any kind of interbond resonance. As already mentioned in the previous section, higher order (Majorana-like) operators breaking the final O(2) symmetry are constructed[14]. These operators can be thought as shift operators changing both single oscillator and total vibrational quantum numbers. For a system of two interacting oscillators one can construct, for example, an operator mixing states whose total quantum number changes by one in this simple way:

$$\langle v_1, v_2 | S_1^+ S_2^{-2} | v_1 - 1, v_2 + 2 \rangle = \sqrt{v_1 (N_1 - v_1 + 1)(N_2 - v_2)(N_2 - v_2 + 1)(v_2 + 1)(v_2 + 2)} \; . \tag{35}$$

Operators like this have been successfully applied to a detailed analysis of the vibrational spectrum of the HCCF molecule[14], whose Fermi pattern is extremely complex due to the accidental degeneracy between fundamental and low-energy combination modes[31]. A more recent application deals with overtone spectroscopy of CH stretching modes in the benzene molecule[21,22]. In this case the huge density of vibrational levels (thousands per cm^{-1}) makes extremely difficult and often unsatisfactory to use traditional methods. Preliminary results show an excellent agreement with the experiments.

CONCLUSIONS

A brief review of the most important features of dynamical symmetries applied to molecular spectroscopy has been presented. The essential mathematical aspects of the vibron model in both its three- and one-dimensional versions have been discussed. It has been pointed out in particular the quite surprising possibilities of the simple U(2) model. Its advantages, if compared with the U(4) model, are to be found in the mathematical simplicity and flexibility to attack a large number of different physical problems. On the other side, the more exacting U(4) model (also from the computational point of view) is still indispensable when rotations and their coupling to other modes have to be taken into account.

ACKNOWLEDGMENTS

I think that it should be pointed out the unique role of Franco Iachello in the development of these techniques. In particular, I want to thank him for his personal guidance in these years, during which I have been able to aknowledge his skill and wish to share his many ideas with other people.

REFERENCES

1. F.Iachello, Chem.Phys.Lett. **78**, 581 (1981).
2. F.Iachello and R.D.Levine, J.Chem.Phys. **77**, 3046 (1982).
3. O.S.van Roosmalen, A.E.L.Dieperink and F.Iachello, Chem.Phys.Lett. **85**, 32 (1982).
4. O.S.van Roosmalen, F.Iachello, R.D.Levine and A.E.L.Dieperink, J.Chem.Phys. **79**, 2515 (1983).
5. A.Arima and F.Iachello, Phys.Rev.Lett. **35**, 1069 (1975).
6. G.Herzberg, "Infrared and Raman Spectra of Polyatomic Molecules," Van Nostrand, New York (1950).
7. F.Iachello and S.Oss, J.Mol.Spectrosc. **142**, 85 (1990).
8. P.M.Morse, Phys.Rev. **34**, 57 (1929).
9. F.Iachello, S.Oss and R.Lemus, J.Mol.Spectrosc. **146**, 56 (1991).
10. I.M.Mills and A.G.Robiette, Mol.Phys. **56**, 743 (1985).
11. S.Oss, N.Manini and R.Lemus, Comput.Phys.Commun., in print (1992).
12. F.Iachello, S.Oss and R.Lemus, J.Mol.Spectrosc. **149**, 132 (1991).
13. F.Iachello, N.Manini and S.Oss, J.Mol.Spectrosc., in print (1992).
14. F.Iachello, S.Oss and L.Viola, Mol.Phys., in print (1992).
15. Y.Alhassid, F.Gürsey and F.Iachello, Phys.Rev.Lett. **50**, 873 (1983).
16. Y.Alhassid, F.Gürsey and F.Iachello, Ann.Phys. (N.Y.) **148**, 346 (1983).
17. R.D.Levine and C.E.Wulfman, Chem.Phys.Lett. **60**, 372 (1979).
18. R.D.Levine, Chem.Phys.Lett. **95**, 87 (1983).
19. H.Berrondo and A.Palma, J.Phys.A **13**, 773 (1980).
20. F.Iachello and S.Oss, Phys.Rev.Lett. **66**, 2976 (1991).
21. F.Iachello and S.Oss, Chem.Phys.Lett. **187**, 500 (1991).

22. F.Iachello and S.Oss, J.Mol.Spectrosc. **153**, 225 (1992).

23. J.Schwinger, "Quantum Theory of Angular Momentum," Academic Press, New York (1965).

24. H.J.Lipkin, N.Meshkov and A.J.Glick, Nucl.Phys. **62**, 188 (1965).

25. S.Oss and N.Manini, in preparation.

26. S.Oss and L.Viola, in preparation.

27. L.Halonen and M.S.Child, J.Chem.Phys. **79**, 559 (1983).

28. A.Frank and R.Lemus, Phys.Rev.Lett. **68**, 413 (1992).

29. M.Hamermesh, "Group Theory and Its Applications to Physical Problems," Addison Wesley, Reading (1964)

30. E.Fermi, Z.Physik, **72**, 250 (1931).

31. J.K.Holland, D.A.Newnham, I.M.Mills and M.Herman, J.Mol.Spectrosc. **151**, 346 (1992).

MICROSCOPIC PICTURE OF INTERACTING BOSON MODEL

Takaharu Otsuka

Department of Physics, University of Tokyo, Hongo, Bunkyo-ku, Tokyo, 113, Japan

1. INTRODUCTION

The Interacting Boson Model (IBM) is discussed from two viewpoints related to the microscopic nuclear structure. One is the derivation of the IBM using the OAI mapping method. The results of recent microscopic calculations along this line are presented for Te-Xe-Ba isotopes. The IBM-2 Hamiltonian is derived from a nucleon Hamiltonian comprised of single-particle energy, pairing and quadrupole-quadrupole interactions, by carrying out mapping from multi-nucleon systems to interacting-boson systems. The microscopic origin of the O(6) dynamical symmetry or γ-unstable deformation is discussed. The second major subject is a novel interpretation of nuclear rotation. It is shown that, both in IBM and nucleonic calcuations, the proton and neutron deformed ellipsoids are rotating in the opposite directions near the ground state. The rotational energy can then be interpreted in terms of inertial parameter against tilting of the proton rotaion axis towards the neutron axis. This paper thus consists of the following sections.
1. Introduction
2. Brief review on the microscopic basis of the IBM
3. Microscopic derivation of IBM-2 Hamiltonian and the O(6) dynamical symmetry
4. Nuclear rotation seen in the laboratory frame
5. Summary

2. MICROSCOPIC BASIS OF THE INTERACTING BOSON MODEL

2.1. The Outline

The Interacting Boson Model (IBM) has been introduced by Francesco Iachello with Akito Arima nearly two decades ago, and has made tremendous success in a unified description of quadrupole collective states[1]. The description is made in terms of a scalar boson of positive parity, called s, and a quadrupole boson of positive parity, called d. There are two characteristic features in IBM; (i) the total boson number is conserved for a given nucleus, (ii) the boson Hamiltonian consists of single particle energy and two-body interaction terms. By choosing specific values for the parameters in a the Hamiltonian, one obtains limiting cases where the solutions can be derived analytically. Those limiting situations are referred to as dynamical symmetries, and are actually comprised of three cases, (i) U(5), (ii) SU(3), and (iii) O(6). It is widely recognized that Francesco Iachello has made significant and indispensable contributions to finding, developing and establishing these dynamical symmetries. The U(5) limit corresponds, in geometrical terms, to anharmonic quadrupole vibrations, the SU(3) to axially symmetric rotation, while the O(6) corresponds to γ-unstable deformation[1].

The underlying microscopic picture of the IBM is given in terms of collective pairs of nucleons[2-5]. First, collective pairs of valence nucleons of $J^{\pi}=0^+$ (called S) and $J^{\pi}=2^+$ (called D) are introduced. These pairs are nothing but generalized or extended Cooper pairs. In fact, the S pair in spherical nuclei is exactly the Cooper pair of valence nucleons in the BCS ground state. The D pair absorbs much of the quadrupole strength when the quadrupole (moment) operator is acting on the S pair. The S and D pairs, thus, should be strongly connected through the quadrupole collectivity. I shall discuss more in detail on the structure of the S and D pairs in subsect. 2.3. The microscopic basis of the IBM can be given, at least for nuclei without strong static deformation, by the Otsuka-Arima-Iachello (OAI) mapping[4], where these pairs correspond intuitively to the s and d bosons as,

$$S \text{ pair} \rightarrow s \text{ boson} \qquad \text{and} \qquad D \text{ pair} \rightarrow d \text{ boson}. \tag{1}$$

The number of bosons is hence given by the number of pairs. The number of s bosons, N_s, is equal to the number of the S pairs, while the number of d bosons, N_d, is equal to the number of the D pairs. Since these pairs are comprised of valence particles, the total number of bosons is determined by the number of valence particles, as discussed more in detail in the next subsection.

The states constructed by the S and D pairs form a subspace truncated from the entire shell-model space. This subspace is called SD (pair) subspace. The states in the SD subspace are mapped onto the IBM states as

$$|S^N; 0^+\rangle \qquad \rightarrow |s^N; 0^+),$$

$$|S^{N-1}D; 2^+\rangle \qquad \rightarrow |s^{N-1}d; 2^+), \qquad \ldots \tag{2}$$

The boson Hamiltonian, H^B, is then determined so that the matrix elements of the Hamiltonian H in the SD subspace can be reproduced to a good approximation by the corresponding sd boson matrix elements as

$$\langle \psi | H | \psi' \rangle \approx (\varphi | H^B | \varphi'), \tag{3}$$

where ψ and ψ' are SD states, and φ and φ' are the corresponding sd boson states obtained by the mapping in eq. (2). This process can be generalized to more general operators, for instance, the quadrupole operator, as will be shown. For an introductory description of the mapping, one can refer to a recent overview[5].

2.2. Proton-Neutron IBM or IBM-2

Since the nucleus consists of protons and neutrons, SD pairs of protons, called S_{π} and D_{π}, and SD pairs of neutrons, called S_{ν} and D_{ν} are introduced based on the above discussion. Consequently, the proton bosons, denoted as s_{π} and d_{π}, and neutron bosons denoted as s_{ν} and d_{ν} are introduced. The IBM with the proton bosons and the neutron bosons is referred to as the Proton-Neutron IBM or IBM-2 [2-4]. The mapping process in eqs. (2)-(3) can be generalized to the IBM-2, by mapping the proton states (operators) on to proton boson states (operators), and likewise neutron states (operators) on to neutron boson states (operators). The number of proton bosons is denoted as N_{π}, while the value of N_{π} is given by half of the number of valence protons, if the proton system is in the first half of the major shell. If it is in the second half of the major shell, pairs of proton holes are mapped on to bosons, and N_π is given by half of the number of proton holes. A similar rule holds for neutrons, and the number of neutron bosons is denoted as N_v [4]. Likewise, the suffix π (v) indicates hereafter quantities referring to protons (neutrons), although this suffix may be omitted for brevity if no confusion occurs.

The original IBM is sometimes called the IBM-1 in order to distinguish it from the IBM-2. Evidently, there is no distinction between proton boson and neutron boson in the IBM-1.

2.3. Structure of the S and D pairs

The structure of the S and D pairs is discussed first. The type of nucleon in the following discussions should be kept to either proton or neutron, although not mentioned explicitly.

The S pair creation operator is written as

$$S^\dagger = \Sigma_j \; \alpha_j \; \frac{1}{\sqrt{2}} \; [a_j^\dagger a_j^\dagger]^{(0)} , \tag{4}$$

where $a_j^\dagger$ denotes the creation operator of a nucleon on the orbit j, and α_j means the amplitude. Similarly the D pair is written with the β_{ij} amplitude as,

$$D^\dagger = \Sigma_{i \le j} \; \beta_{ij} \; \frac{1}{\sqrt{1+\delta_{ij}}} [a_i^\dagger a_j^\dagger]^{(2)} . \tag{5}$$

The S and D pairs have been discussed often in terms of the seniority scheme of shell model [2,6]. I present here a slightly different formulation based on the many-body approach such as BCS or Hartree-Fock-Bogoljubov (HFB), but the basic physics remains the same. In general, the amplitudes α_j and β_{ij} should be determined so that the S and D pairs produce the most optimum description of low-lying quadrupole collective states. Basically the α_j amplitude of the S pair is determined by the competition between the pairing correlation and the one-body mean field (*i.e.*, single-particle energy). The β_{ij} amplitude of the D pair can be more sensitive to the dynamical and/or static quadrupole deformation.

In the present method, the Hartree-Fock-Bogoljubov (HFB) ground state is obtained first by the HFB calculation. This HFB (and BCS) ground state is generally written as

$$|\Psi\rangle = \Pi_{m>0} \; (u_m + v_m a_m^\dagger \, a_{\bar{m}}^\dagger \;) \; |0\rangle \tag{6}$$

where the u's (v's) are the usual u (v) factors, m stands for the magnetic quantum number of single-particle state, and $\bar{m}$ is the time-reversed state of m. Introducing the particle number projector, one projects $|\Psi\rangle$ onto a good particle number 2N as

$$P_N |\Psi\rangle \propto [\Sigma_{m>0} \; (v_m \, / u_m \;) \; a_m^\dagger \, a_{\bar{m}}^\dagger \;]^N |0\rangle. \tag{7}$$

This is nothing but the N-pair condensate of the Cooper pair. The Cooper pair is denoted as the Λ pair hereafter, and its creation operator is written as

$$\Lambda^\dagger \equiv N \; \Sigma_{m>0} \; (v_m \, / u_m \;) \; a_m^\dagger \, a_{\bar{m}}^\dagger \tag{8}$$

where N denotes the normalization constant. Equation (7) can then be rewritten as

$$P_N |\Psi\rangle \propto (\Lambda^\dagger)^N |0\rangle. \tag{9}$$

479

We calculate the u and v factors in eq. (6) with the particle number conservation by the method of Ref. 7, and hence what is actually determined is $P_N |\Psi>$ rather than $|\Psi>$.

If the nucleus being considered is spherical, u_m and v_m become independent of the magnetic quantum number m, while they are still dependent on the angular momentum j of the single-particle orbit. In this case, the Λ pair is scalar, and carries $J^\pi=0^+$. We then identify the Λ pair as the S pair as,

$$\Lambda \text{ pair } \rightarrow \text{ S pair,} \tag{10}$$

and thereby

$$| S^N ; \ 0^+ > \ \propto \ P_N |\Psi>. \tag{11}$$

The seniority (v) of this state is v=0 in the generalized seniority scheme [8].

In order to introduce the D pair in such spherical cases, one can use the broken-pair formalism [9]. The broken pair state is created, in general, by changing one of the Λ pairs in eq. (9) into a pair of nucleons, called a broken pair, as

$$| N, \rho\sigma > \ \propto \ a_\rho^\dagger a_\sigma^\dagger \ (\Lambda^\dagger)^{N-1} |0>, \tag{12}$$

where ρ and σ stand for single-particle states (or their linear combinations) including magnetic quantum numbers. The ρ-σ pair thus introduced is called a broken pair, because it is produced by destroying one of the condensed Cooper pairs. The states given by eq. (12) are referred to as one-broken-pair states, and are denoted also as (number-projected) TDA states. There are a number of broken pair states due to the many combinations of ρ and σ in eq. (12).

In the spherical cases (i.e., eq. (10)), the Λ pair is nothing but the S pair. The one-broken-pair states can be redefined by a proper linear combination among themselves, so as to have a good angular momentum, as

$$| N, ij; J^\pi > \ \propto \ \frac{1}{\sqrt{1+\delta_{ij}}} [a_i^\dagger a_j^\dagger]^{(J)} \ (S^\dagger)^{N-1} |0>, \tag{13}$$

where J^π stands for the angular momentum and parity of the broken pair, and i and j refer to the single particle orbits (i ≤ j). The states in eq. (13) form an orthonormal set of basis states, except that one needs to reorthogonalize for $J^\pi=0^+$. In order to determine the structure of the D pair, we calculate matrix elements of the Hamiltonian with respect to such an orthonormal basis for $J^\pi=2^+$,

$$< N, ij; 2^+ | \ H \ | N, kl; 2^+ >, \tag{14}$$

where the bra and ket states are defined in eq. (13). The lowest eigensolution of this matrix can be identified as an SD state,

$$| D \ S^{N-1} ; \ 2^+ > \ \propto \ D^\dagger \ (S^\dagger)^{N-1} |0>. \tag{15}$$

Thus, the β_{ij} amplitudes in eq. (5) are determined [9-11]. In the diagonalization of the matrix in eq. (14), the proton-neutron interaction can be included in the Hamiltonian, and then the proton D pair and the neutron D pair are determined simultaneously. This is the outline of how one can determine the structure of S and D pairs in spherical nuclei. Note that the one-broken-pair states in spherical cases are v=2 states in the generalized seniority. An actual example of spherical calculations is found with Cd isotopes [12].

If the HFB ground state is deformed, the Λ pair is no longer scalar, and contains components of various angular momenta. One can then extract the S and D pairs directly from the Λ pair. We decompose the Λ pair into such components as

$$\Lambda^\dagger = x_0 \ S^\dagger + x_2 \ D^\dagger + x_4 \ G^\dagger + ..., \tag{16}$$

where the x's are amplitudes, and S, D, G, *etc.* denote the positive-parity pairs of angular momenta 0, 2, 4, *etc.* projected from the Λ pair. The S and D pairs thus fixed can be rewritten in the forms of eqs. (4) and (5), and we use the same notation as in the spherical cases discussed above. In the intermediate situation, however, the mixing of the $J^\pi=2^+$ pair in the Λ pair is not sufficiently large, and hence the D pair may not be well determined by eq. (16). In such cases, the S pair is determined by eq. (16), whereas the D pair is obtained from the broken-pair calculation in eqs. (14)-(15). This is the case I am going to discuss in this talk, showing examples of Te-Xe-Ba isotopes.

The values of α_j and β_{ij} thus calculated vary gradually as functions of the neutron and proton numbers within a major shell. This property has been mentioned in Refs. 6, 9-11 and 13. In earliest papers of the IBM, it was suggested that the α_j and β_{ij} amplitudes are determined by solving two-body problems, *i.e.*, N=1 in the above discussion. This is true only in the limit that the changes of α_j and β_{ij} are negligible as functions of nucleon number, although this is approximately the case in quite a few situations.

I have presented a basic method to determine the α_j and β_{ij} amplitudes, starting from the BCS/HFB ground state. The HFB method has been used in Ref. 14 also. There can be several variations in the calculation of α_j and β_{ij}, for instance a shell-model approach by Nakada *et al.* [15].

2.4. The OAI Mapping

Once the S and D pairs are fixed, one can construct SD (pair) states. We normally construct them in the order that the number of D pairs increases. Non-orthogonality may occur, and the Schmidt orthogonalization is carried out to construct orthogonal basis vectors of the SD subspace. With the orthogonalization, each SD basis state is normalized. The SD basis vectors are then mapped onto the sd boson basis vectors. Some examples of this mapping are shown below,

$$| S^N ; 0^+ > \qquad \rightarrow \qquad | s^N ; 0^+), \qquad\qquad (17a)$$

$$| D\, S^{N-1} ; 2^+ > \qquad \rightarrow \qquad | d\, s^{N-1} ; 2^+), \qquad\qquad (17b)$$

$$| D^2\, S^{N-2} ; J^+ > \qquad \rightarrow \qquad | d^2\, s^{N-2} ; J^+), \qquad J = 0, 2, 4, \quad (17c)$$

$$\vdots \qquad\qquad\qquad\qquad \vdots$$

I would like to emphasize that the SD pair states and sd boson states in eq. (17) are all normalized. Equation (17) presents the state-to-state mapping which constitutes the first step of the OAI mapping [4].

The next step of the OAI mapping is the mapping of the Hamiltonian and other operators. The *boson image* of this mapping is obtained, as shown schematically in eq. (3), by requiring that SD pair matrix elements of a given operator can be well reproduced by the corresponding sd boson matrix elements [4]. The correspondence here means the one given by eq. (17).

I would like to illustrate this process by an example. The boson image of the proton Hamiltonian can be written as

$$H_\pi^B = E_\pi + \varepsilon_\pi \hat{N}_{d\pi} + \dots \qquad\qquad (18)$$

where E_π and ε_π are coefficients to be calculated, and $\hat{N}_{d\pi}$ denotes the number operator of the proton d bosons. Clearly, E_π is a c-number dependent on the total number of proton bosons, N_π, and contributes to the binding energy only. These two coefficients are determined by equations for the proton Hamiltonian H_π ;

$$< S^N; 0^+ \,|\, H_\pi \,|\, S^N; 0^+ > \quad = \quad (s^N; 0^+ \,|\, H_\pi^B \,|\, s^N; 0^+) \; = \; E_\pi , \qquad (19a)$$

and

$$< D\, S^{N-1}; 2^+ \,|\, H_\pi \,|\, D\, S^{N-1}; 2^+ > \; = \; (d\, s^{N-1}; 2^+ \,|\, H_\pi^B \,|\, d\, s^{N-1}; 2^+)$$

$$= \; E_\pi + \varepsilon_\pi . \qquad (19b)$$

Other terms of the boson Hamiltonian are determined similarly.

The quadrupole operator plays a crucial role in the description of quadrupole collective states. The quadrupole operator of either protons or neutrons is written in the second quantized form as

$$Q^F \;=\; \Sigma_{k,m} \; <k\,|\, r^2\, Y^{(2)} \,|\, m> \; a_k{}^\dagger a_m , \qquad (20)$$

with k and m being single-particle states. The boson image of Q^F is written as

$$Q^B \;=\; q\,\{ \; d^\dagger s + s^\dagger \tilde{d} + \chi\,[\,d^\dagger \tilde{d}\,]^{(2)} \; \}, \qquad (21)$$

where q and χ are coefficients. The values of q and χ are determined by the equations

$$< D\, S^{N-1}; 2^+ \,\|\, Q^F \,\|\, S^N; 0^+ > \quad = \quad (d\, s^{N-1}; 2^+ \,\|\, Q^B \,\|\, s^N; 0^+)$$

$$= \; q\,\sqrt{5}\,\sqrt{N}. \qquad (22)$$

and

$$< D\, S^{N-1}; 2^+ \,\|\, Q^F \,\|\, D\, S^{N-1}; 2^+ > \; = \; (d\, s^{N-1}; 2^+ \,\|\, Q^B \,\|\, d\, s^{N-1}; 2^+)$$

$$= \; q\,\chi\,\sqrt{5}. \qquad (23)$$

The parameters for the proton system, denoted as q_π and χ_π, and those for the neutron system, denoted as q_ν and χ_ν, are determined separately by considering proton matrix elements and neutron matrix elements, respectively. The parameters χ's play crucial roles in determining shapes of the deformation in the IBM, i.e., prolate, oblate, triaxial, etc. The parameter χ will then be referred to as "*shape*" parameter hereafter.

For systems comprised of single orbit or degenerate many orbits, one can take advantage of the quasi-spin scheme and express the q and χ parameters in a more analytic way. Following the prescription of Otsuka et al.[4], one obtains

$$q \;=\; \sqrt{\frac{\Omega - N}{\Omega - 1}} \; < D; 2^+ \,\|\, Q^F \,\|\, S; 0^+ > \qquad (24)$$

and

$$\chi \;=\; \frac{\Omega - 2N}{\Omega - 1} \; \sqrt{\frac{\Omega - 1}{\Omega - N}} \; \frac{< D; 2^+ \,\|\, Q^F \,\|\, D; 2^+ >}{< D; 2^+ \,\|\, Q^F \,\|\, S; 0^+ >}, \qquad (25)$$

where N stands for the number of SD pairs, and Ω means the total degeneracy; $\Omega \equiv \Sigma_j (j + 1/2)$. The N dependencies in eqs. (24) and (25) are clearly due to the Pauli principle, and indicate that the IBM includes Pauli effects. In realistic cases with non-degenerate multi-orbits, one cannot use formulas in eqs. (24) and (25), and then has to compute numerically N-pair SD matrix elements, for instance, those in eq. (22) and (23). This is not a trivial or easy task, but we have recently succeeded in making a computer program for this purpose. I shall show, in this talk, some of their first results.

The proton quadrupole operator, Q_π^F, and the neutron quadrupole operator, Q_ν^F are mapped respectively onto the proton boson quadrupole operator, $q_\pi Q_\pi(\chi_\pi)$, and the neutron boson quadrupole operator, $q_\nu Q_\nu(\chi_\nu)$, where the Q operator is defined as

$$Q(\chi) = d^\dagger s + s^\dagger \tilde{d} + \chi [d^\dagger \tilde{d}]^{(2)}, \tag{26}$$

where subscript π or ν is omitted. The proton-neutron quadrupole-quadrupole interaction is then mapped as

$$V_{\pi\nu} = -f (Q_\pi^F \cdot Q_\nu^F) \quad \rightarrow \quad V_{\pi\nu}^B = -\kappa (Q_\pi(\chi_\pi) \cdot Q_\nu(\chi_\nu)) \tag{27}$$

with the parameter κ defined as

$$\kappa \equiv -f q_\pi q_\nu. \tag{28}$$

The E2 transition operator, T(E2), is mapped similarly,

$$T(E2) \equiv e_\pi Q_\pi^F + e_\nu Q_\nu^F \quad \rightarrow \quad T(E2)^B = e_\pi^B Q_\pi(\chi_\pi) + e_\nu^B Q_\nu(\chi_\nu), \tag{29}$$

where e_π and e_ν stand for shell-model effective charges of proton and neutron respectively, and e_π^B and e_ν^B are boson (effective) charges defined as

$$e_\pi^B \equiv e_\pi q_\pi \quad \text{and} \quad e_\nu^B \equiv e_\nu q_\nu. \tag{30}$$

2.5. F-spin

The symmetry associated with the proton-neutron degree of freedom in the IBM-2 is represented in terms of F-spin [2,4]. F spin is an isospin-like concept for bosons, and has similarities and differences compared to isospin. In the isospin scheme, each nucleon carries isospin T=1/2, and protons and neutrons are treated as the T_z=1/2 and T_z=-1/2 states, respectively. Similarly, each boson carries F-spin F=1/2. The proton boson is the F_z=1/2 state, while the neutron boson the F_z=-1/2 state. This assignment is tabulated in Table 1.

The F spin obeys the angular momentum algebra, as the isospin does. Because of the vector addition rule of the angular momentum algebra, a system with N_π proton bosons and N_ν neutron bosons has $F_z=(N_\pi-N_\nu)/2$, and can have F value, $|N_\pi-N_\nu|/2 \leq F \leq (N_\pi+N_\nu)/2$. The F spin characterizes proton-neutron symmetry in IBM-2 wave functions, and this symmetry property can be seen most clearly in two-boson systems. Table 2 presents F spin classification of two-boson states. First of all, the states with N_π=2 and N_ν=0 (i.e., F_z=1) and the states with N_π=0 and N_ν=2 (i.e., F_z=-1) are IBM-1 states because they contain either proton bosons or neutron bosons. Note that these states are in F=1 multiplets. Most of the states with N_π=N_ν=1 in Table 2 are symmetric with respect to the interchange between the proton boson and the neutron boson. These symmetric states have F=1 and F_z=0, and can be transformed to F_z=±1 (i.e., IBM-1) states by F-spin raising and lowering operators, which are introduced analogously to the angular-momentum or isospin raising and lowering operators. The states with the maximum F value, $F=(N_\pi+N_\nu)/2$, can be comprised only of F=1 pairs of bosons which are all symmetric as mentioned above. They therefore possess maximal proton-neutron symmetry, and are referred to as totally symmetric states.

Table 1 F spin quantum numbers

		F	F_z
proton boson	s_π and d_π	1/2	1/2
neutron boson	s_ν and d_ν	1/2	-1/2

The totally symmetric states can be transformed, by F-spin raising or lowering operator, to the states of $F_z=\pm(N_\pi+N_\nu)/2$ (i.e., purely proton-boson or neutron-boson state) with the same F value. Because of these properties, the totally symmetric states are equivalent to the corresponding IBM-1 states of $F_z=\pm(N_\pi+N_\nu)/2$. As explained below, lowest collective states are dominated by components comprised of the totally symmetric states, and one can describe such lowest states in a good approximation by IBM-1 with an appropriately chosen Hamiltonian. Such an IBM-1 Hamiltonian is obtained essentially by extracting the symmetric part of the original IBM-2 Hamiltonian with respect to the proton and neutron bosons. I shall mention this point in subsect. 3.2.

Table 2 F spin classification of two-boson states

	F	F_Z
$\lvert s_\pi^2 ; 0^+ \rangle$	1	1
$\lvert s_\pi s_\nu ; 0^+ \rangle$	1	0
$\lvert s_\nu^2 ; 0^+ \rangle$	1	-1
$\lvert d_\pi s_\pi ; 2^+ \rangle$	1	1
$\frac{1}{\sqrt{2}}\{\lvert d_\pi s_\nu ; 2^+ \rangle + \lvert d_\nu s_\pi ; 2^+ \rangle\}$	1	0
$\lvert d_\nu s_\nu ; 2^+ \rangle$	1	-1
$\frac{1}{\sqrt{2}}\{\lvert d_\pi s_\nu ; 2^+ \rangle - \lvert d_\nu s_\pi ; 2^+ \rangle\}$	0	0
$\lvert d_\pi^2 ; L^+ \rangle$ (L=0, 2, 4)	1	1
$\lvert d_\pi d_\nu ; L^+ \rangle$ (L=0, 2, 4)	1	0
$\lvert d_\nu^2 ; L^+ \rangle$ (L=0, 2, 4)	1	-1
$\lvert d_\pi d_\nu ; L^+ \rangle$ (L=1, 3)	0	0

It may be worth noting that the total wave functions should be antisymmetric for nucleons, and, by making the isospin part antisymmetric, the rest of the wave function becomes symmetric and can gain binding energy. The antisymmetric isospin wave function means lower isospin. The situation is reversed in boson systems where the total wave function should be symmetric; by keeping the F spin part symmetric, the rest of the wave function remains symmetric, coming down in energy. This is the basic reason why lowest states are dominated usually by components of largest F spin in the IBM-2, while the isospin takes minimum value in lowest states. IBM-2 states containing F=0 pair of proton boson and neutron boson are called mixed-symmetry states.

3. MICROSCOPIC DERIVATION OF IBM-2 HAMILTONIAN AND THE O(6) DYNAMICAL SYMMETRY IN THE XE-BA REGION

3.1. Determination of the parameters of the nucleon Hamiltonian

I shall now present results for Te, Xe and Ba isotopes. The doubly-closed shell of Z=50 and N=82 is assumed. Since the neutron number of nuclei to be considered is between 66 and 82, neutrons are treated as holes, whereas protons are (valence) particles. The nucleon Hamiltonian is assumed to consist of the single-particle energy, the monopole and quadrupole pairing interactions between identical nucleons, and the quadrupole-quadrupole interaction between identical nucleons and that between proton and neutron.

All parameters such as single-particle energies and interaction strengths are determined by the spectra of single-closed nuclei except the strength of the proton-neutron interaction. The single-particle energies are determined so as to fit observed levels of one-quasi-particle states of relevant single-closed nuclei, for instance Sn isotopes for neutrons. The proton (neutron) single-particle energies are basically constant, but include small variation (less than 1 MeV) as a function of the proton (neutron) number.

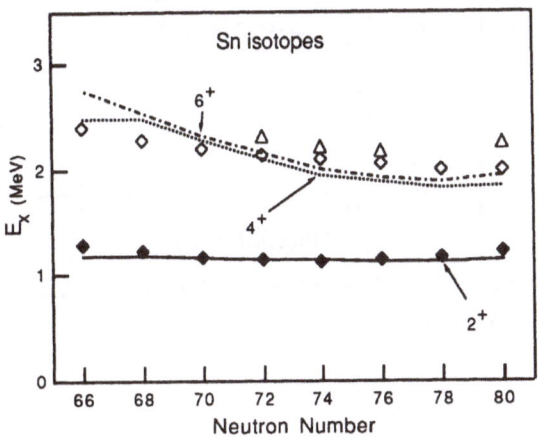

Fig. 1 Energy levels of 2^+_1, 4^+_1 and 6^+_1 for Sn isotopes as a function of the neutron number. Lines are calculations. Points are experiments; closed diamonds, open diamonds and triangles indicate, respectively, the 2^+_1, 4^+_1 and 6^+_1 levels.

The strengths of the interactions between identical nucleons are determined so that levels of 2^+_1 and 4^+_1 of single-closed nuclei are reproduced well. Those levels are calculated in terms of one-broken pair approximation, where the states are written as a linear combination of components of the form in eq. (13). Figure 1 shows the 2^+_1, 4^+_1 and 6^+_1 levels for Sn isotopes, while Fig. 2 for N=82 isotones. One finds reasonable agreement between the present calculation and experiments.

The proton-neutron interaction strength thus remains as only adjustable parameter, and it is determined so as to reproduce the levels of ^{132}Te which is the minimal proton-neutron open-shell nucleus; two valence protons and two neutron holes.

3.2. Calculation of the *"shape"* parameters χ_π and χ_ν.

We show, in Fig. 3, the values of the *"shape"* parameters χ_π and χ_ν defined in eq. (23) for Te, Xe and Ba isotopes. The parameters χ_π and χ_ν have opposite signs with similar magnitudes for most of the Xe and Ba isotopes shown in Fig. 3. This is very important

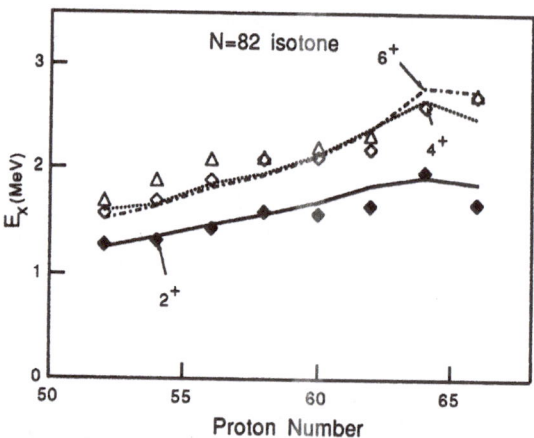

Fig. 2   Energy levels of 2^+_1, 4^+_1 and 6^+_1 for N=82 isotones as a function of the proton number. See the caption of Fig. 1.

feature to understand the appearance of the O(6) dynamical symmetry. I shall turn to this problem.

The O(6) dynamical symmetry can be obtained by an IBM-1 Hamiltonian

$$H^B_{O(6)} = - \kappa \ (Q \ (\chi=0) \cdot Q \ (\chi=0)), \tag{31}$$

where the $Q \ (\chi)$ operator is defined in eq. (26). As pointed out in subsect. 2.5, an IBM-2 Hamiltonian can be projected into an IBM-1 Hamiltonian by taking its proton-neutron symmetric part. For $V^B_{\pi\nu}$ in eq. (27) with $\chi_\pi = -\chi_\nu$, the projected term is proportional to eq. (31) except for a term ($[d^\dagger \tilde{d}]^{(2)} \cdot [d^\dagger \tilde{d}]^{(2)}$) which has much minor effects. Thus, for totally symmetric states, $V^B_{\pi\nu}$ in eq. (27) with $\chi_\pi = -\chi_\nu$ works effectively as an O(6) Casimir operator, and eigenstates show properties very close to the O(6) limit.

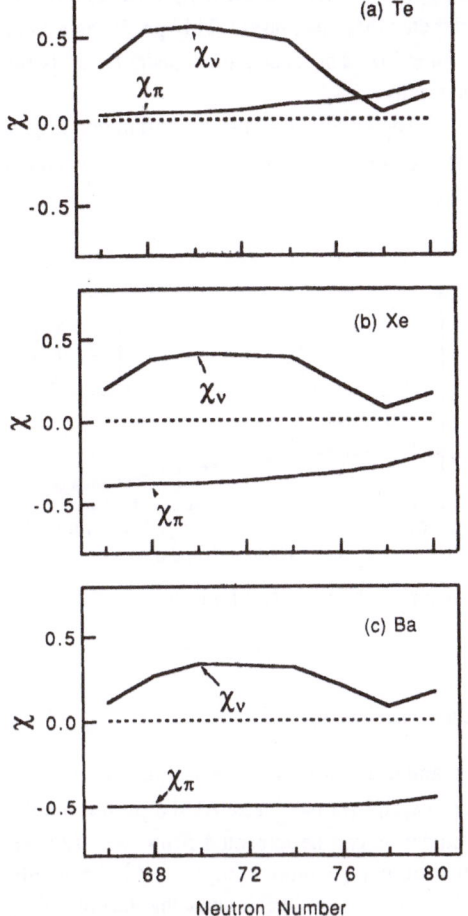

Fig. 3 Parameters χ_π and χ_ν for (a) Te, (b) Xe and (c) Ba isotopes.

The relation $\chi_\pi = -\chi_\nu$ holds to a good extent for many Xe and Ba isotopes. In fact, this relation holds much better in the present calculation than in a phenomenological fitting by Puddu *et al.* in Ref. 16. It has been discussed by Casten and von Brentano[17] that the O(6) region can be extended in the Xe-Ba region rather widely. The present calculation clearly shows that this empirical observation is quite reasonable from the microscopic viewpoint. The value of the χ parameter is determined by shell effects and Pauli effects. In order to understand this, we consider an extreme case that there is a sub-shell closure due to the separation of the $h_{11/2}$ single-orbit shell and the shell of the other natural-parity orbits, and that neutrons (*i.e.*, neutron valence particles) occupy the natural-parity shell first and the $h_{11/2}$ shell next. The neutron number dependence of χ_ν can be seen from eq. (25). Up to the neutron number 70, the natural-parity shell is being occupied, and χ_ν increases because of the sign of the ratio of two-nucleon matrix elements on the right-hand side of eq. (25)

(see Ref. 5). After the neutron number 70, the $h_{11/2}$ shell is being occupied, and the sign is reversed. The χ_v parameter then decreases. This produces a bump in the positive side around the neutron number 70. This schematic consideration is far from reality, but still explains what occurs basically.

The results in Fig. 3 shows also that the χ_π parameter is positive for Te isotopes. Hence the relation $\chi_\pi = -\chi_v$ does not hold even approximately and the O(6) structure is not anticipated for Te isotopes.

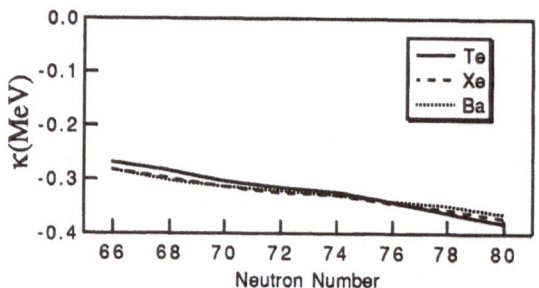

Fig. 4 The parameter κ for Te, Xe and Ba isotopes. The legend is in the insert.

3.3. Other parameters and energy levels

We calculate all parameters in the IBM-2 Hamiltonian. For instance, the value of κ is shown in Fig. 9 for Te, Xe and Ba isotopes. As the proton number increases, κ should decrease in a simple model as can be expected from eqs. (24) and (28). However, in realistic cases, the situation can be more complex. As the number of valence protons increases, the deformation becomes stronger, enhancing the collectivity of the S and D pairs. The parameter κ can then be increased as a function of the proton number, for a fixed neutron number. This is the case for lighter part of Fig. 9.

The energy levels and E2 transition matrix elements are being calculated, and will be published elsewhere.

4. NUCLEAR ROTATION SEEN IN THE LABORATORY FRAME

4.1. Rotation in the IBM as observed in the laboratory frame in the IBM

I now turn to the next major subject. The IBM has been very successful in describing rotational motion of medium-mass and heavy-mass even-even nuclei at lower rotational frequency. Among the three dynamical limits of the IBM, the SU(3) dynamical symmetry limit is suitable for the rotational motion [1]. The intrinsic state of rotational band is obtained in the SU(3) limit, and its relation to the intrinsic state of the geometrical description has been discussed [18-20]. However, the physical picture of the SU(3) description of the nuclear rotation has not been discussed in the laboratory frame. In this section, we shall consider this subject, presenting a novel result.

We shall analyze the wave functions in the SU(3) limit of the Proton-Neutron IBM, *i.e.*, IBM-2. The system with $N_\pi = N_\nu = 6$ is taken as an example. We assume the prolate deformation where both the proton system and the neutron system are deformed in prolate ellipsoid shapes. We then calculate the matrix elements of the operator

$$F^B_{\pi\nu} = (J^B_\pi \cdot J^B_\nu) \tag{32}$$

where the symbol ($\cdot$) means a scalar product, and J^B_π and J^B_ν are axial vectors denoting the proton-boson and neutron-boson angular momentum operators, resepectively. The proton-boson angular momentum J^B_π represents the rotation of the proton ellipsoid about an axis perpendicular to the symmetry axis of this ellipsoid. The neutron-boson angular momentum J^B_ν is explained similarly. We first show, in Fig. 5, the expectation value of this scalar

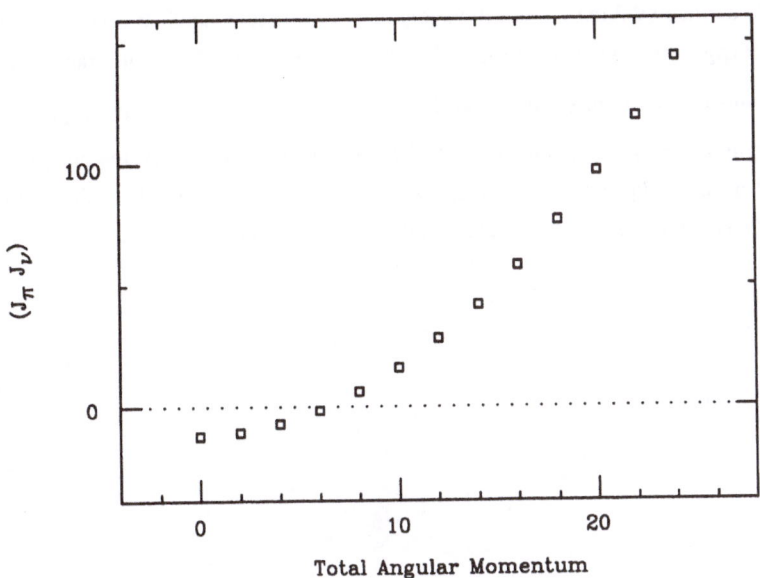

Fig. 5   Scalar product of the proton-boson and neutron-boson angular momentum operators for the SU(3) limit of the IBM-2 with $N_\pi = N_\nu = 6$.

product in eq. (32) for states in the ground state band. In the usual view based on the F-spin symmetry argument, it has been believed without confirmation that, in low-lying states, the proton ellipsoid and the neutron ellipsoid rotate together but very slowly. In this view, the expectation value of $F^B_{\pi\nu}$ should be a small positive number. The most striking feature in Fig. 5 is that the expectation value of $F^B_{\pi\nu}$ in eq. (32) is negative and not necessarily small for the states of J=0 ~ 8 where J stands for the total angular momentum. In order to see the physical meaning of this tendency, we calculate the angle θ defined by

$$\cos\theta = \frac{<(J_\pi^B \cdot J_\nu^B)>}{\sqrt{<(J_\pi^B \cdot J_\pi^B)><(J_\nu^B \cdot J_\nu^B)>}} \qquad (33)$$

where the symbol < > means the expectation value with respect to the ground-band state of the total angular momentum J. The implication of the angle θ is clear; the angle between the axial vectors J_π^B and J_ν^B. If it is negative, the proton ellipsoid and the neutron ellipsoid rotate in the opposite ways. This is the case here, and is illustrated for the ground state schematically in Fig. 6. Figure 7 presents the classical picture of the axial vectors J_π^B and J_ν^B for all the states in the ground-state band of the present case. In Fig. 7, the magnitudes of J_π^B and J_ν^B are represented by $\sqrt{<(J_\pi^B \cdot J_\pi^B)>}$ and $\sqrt{<(J_\nu^B \cdot J_\nu^B)>}$, while their open angle is defined by θ in eq. (33). It is very clear now that the angle θ becomes smaller as the total angular momentum increases. In lowest states, J_π^B and J_ν^B are of sizable magnitude with almost opposite directions, and, in high angular momentum states, J_π^B and J_ν^B are aligned to nearly the same directions. In other words, in lowest states, the proton ellipsoid and the neutron ellipsoid rotate inversely so as to yield nearly vanished angular momentum, while these two ellipsoids rotate in the almost same ways in high spin states.

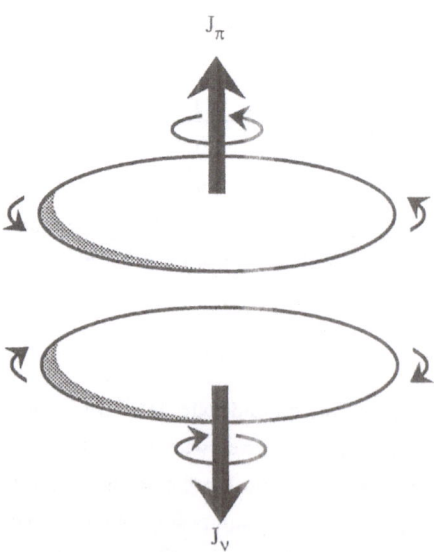

Fig. 6   Schematic picture of the rotation of the proton ellipsoid (upper) and that of the neutron ellipsoid (lower) in the ground state.

4.2. Rotation produced from the Nilsson wave function

We thus obtain a rather novel picture of the nuclear rotation. It is then of great interest whether this is characteristic to the IBM or not. We then carry out a similar calculation for multi-nucleon systems using the usual Nilsson wave functions. The nucleus we take is ^{156}Sm. Since this nucleus is described with $N_\pi = N_\nu = 6$ in the IBM, this nucleus has direct

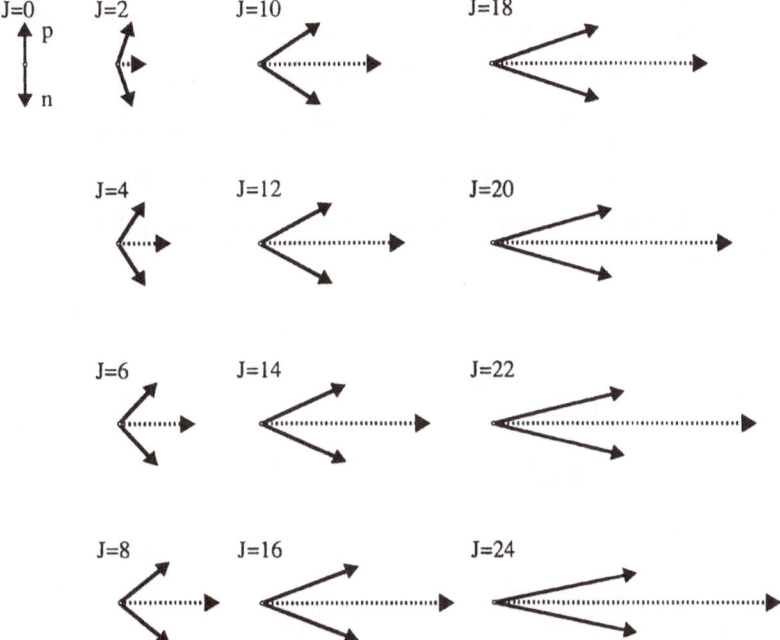

Fig. 7 Classical picture of the angular momenta of the proton ellipsoid rotation and the neutron ellipsoid rotation for the states of the total angular momentum J=0~24. The SU(3) limit of the IBM-2 is considered with $N_\pi = N_\nu = 6$. Upper solid arrows mean the proton angular momentum axial vectors, whereas lower arrows indicate the neutron angular momentum axial vectors. Dotted arrows are the total angular momentum axial vectors, the magnitudes of which are equal to $\sqrt{J(J+1)}$.

correspondence to the above example within the IBM. However, the following discussions are fully nucleonic. We first obtain the Nilsson wave function with the deformation parameter $\delta = 0.30$ and the pairing gap $\Delta \approx 0.8$ (MeV). These values are realistic, but detailed values of these parameters are irrelevant to the following arguments. Since we are concerned about low-frequency phenomena, the cranked term is irrelevant. The calculation is carried out with the particle number conservation, and the pairing gap is calculated in the definition of Ref. 7. We truncate the configuration space to the major shells Z=50-82 and N=82-126 for simplicity of the numerical calculation. We next project out this Nilsson wave function onto a good angular momentum J. Using such projected wave functions, we calculate matrix elements of the operator

$$F_{\pi\nu}^{F} = (\, J_{\pi}^{F} \cdot J_{\nu}^{F}\,) \tag{34}$$

where J_{π}^{F} and J_{ν}^{F} are axial vectors denoting the proton and neutron angular momentum operators, resepectively. The proton angular momentum J_{π}^{F} represents the rotation of the proton ellipsoid produced by the Nilsson wave function. The neutron angular momentum J_{ν}^{F} is explained similarly. Figure 8 shows the expectation value of $F_{\pi\nu}^{F}$ for J = 0 ~ 24 in the ground state band of ^{156}Sm. One finds quite similar pattern to Fig. 5. Figure 9 shows J_{π}^{F} and J_{ν}^{F} in a classical manner similar to Fig. 7. It is thus confirmed that the picture being proposed for the nuclear rotation is a general one and is indeed valid also for the Nilsson wave function.

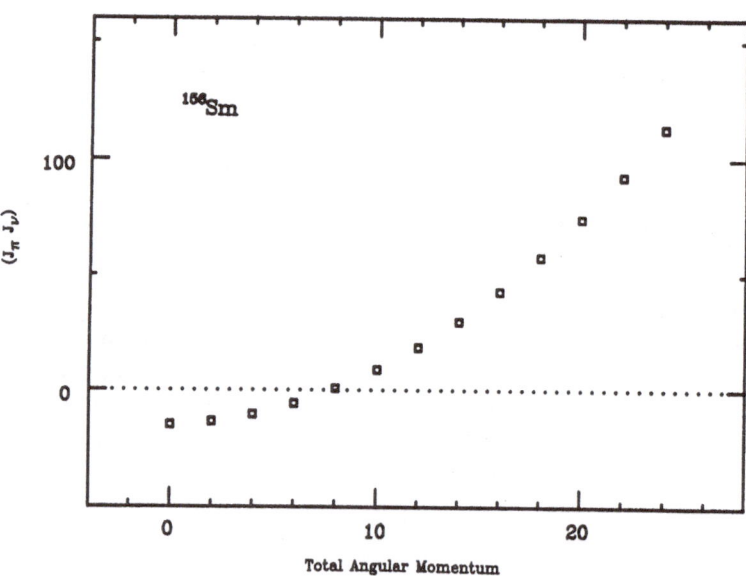

Fig. 8   Scalar product of the proton and neutron angular momentum operators for ^{156}Sm. The wave functions are obtained by the angular momentum projection of the Nilsson wave function.

4.3. Interpretation of the moment of inertia

After having the picture of the nuclear rotation in the laboratory frame, we should study what can be stated for the moment of inertia. For this purpose, the study is still qualitative. In the ground state, the wave function of "rotating" proton ellipsoid and that of "rotating" neutron ellipsoid have the same axis. Since the rotation occurs in the precisely opposite directions about the same axis, the wave function of proton ellipsoid and that of neutron ellipsoid have maximum spacial overlap. This situation is similar to the pairing interaction.

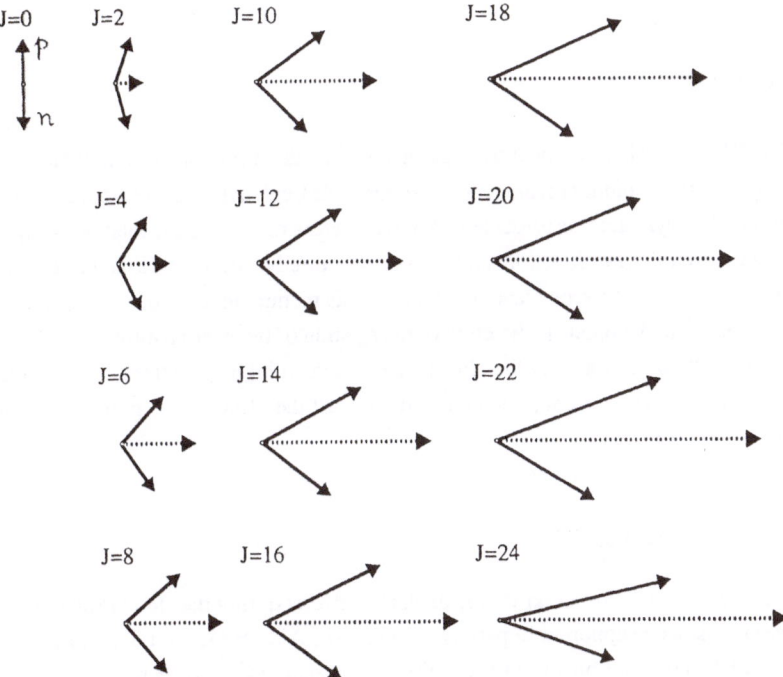

Fig. 9   Classical picture of the proton and neutron angular momenta in the ground state rotational band of ^{156}Sm. Upper solid arrows mean the proton angular momentum axial vectors, whereas lower arrows indicate the neutron angular momentum axial vectors. Dotted arrows are the total angular momentum axial vectors, the magnitudes of which are equal to $\sqrt{J(J+1)}$.

This maximum overlap explains why the ground state should be the 0^+ state in rotational nuclei. In the 2^+ state, the J^π axis is not completely opposite to the J^ν axis, where we do not distinguish bosons and nucleons because they can be discussed in the same way. Thus, the axis of J^π is tilted somewhat to the axis of J^ν. Here we introduce the rotation disks on which the proton or neutron ellipsoid rotates. The proton rotation disk and the neutron rotation disk are completely identical in the ground state because J^π and J^ν are just opposite.

In the 2⁺ state, due to the tilting mentioned above, the rotating disk of the proton ellipsoid is also tilted, and hence the wave function of the rotating proton ellipsoid has somewhat less overlap with that of neutrons in comparison to this overlap in the ground state. The less overlap means in general less gain in the binding energy. Thus, by tilting J^π towards J^ν (or vice versa), the total system loses some energy, and this energy loss seems to be one of the sources of the rotational energy, at least for lowest angular momentum states. In other words, the moment of inertia can be considered partly as the inertial parameter against the tilting of the proton and neutron rotation axes. This is a very intriguing new aspect, and more quantitative studies are being made.

5. SUMMARY

The IBM-2 has been derived from the standard nucleon (*i.e.*, shell-model) Hamiltonian by using the OAI mapping method. The microscopic explanation of the appearance of the O(6) dynamical symmetry is presented. A novel picture of the nuclear rotation is suggested by using the SU(3) wave function, and the same picture has been confirmed in the angular momentum projected Nilsson wave function. In this picture, in the ground state, the proton ellipsoid rotates in the opposite direction of the rotation of the neutron ellipsoid. The proton and neutron ellipsoids rotate in the almost same ways in high spin states. The moment of inertia can be viewed as the inertial parameter against the tilting of the proton rotation axis towards the neutron one.

ACKNOWLEDGMENTS

I acknowledge T. Mizusaki for his dedicated efforts during the study shown in sect. 3. This work has been supported in part by the Grant-in-Aid for General Scientific Research (no. 01540231) by the Ministry of Education, Science, and Culture as well as by Research Center of Nuclear Physics, Osaka University. The numerical calculations have been carried out partly by HITAC S-820/80 and M880 at Computer Centre, University of Tokyo, and partly by VAX 6640 at the Meson Science Laboratory, University of Tokyo.

REFERENCES

1. F. Iachello and A. Arima, *The Interacting Boson Model* (Cambridge Univ. Press, New York, 1987), and references therein.
2. A. Arima, T. Otsuka, F. Iachello and I. Talmi, Phys. Lett. **66B** (1977) 205.
3. T. Otsuka, A. Arima, F. Iachello and I. Talmi, Phys. Lett. **76B** (1978) 139.
4. T. Otsuka, A. Arima and F. Iachello, Nucl. Phys. **A309** (1978) 1.
5. T. Otsuka, *Algebraic Approaches to Nuclear Structure: Interacting Boson and Fermion Models*, edited by R.F. Casten, (Gordon and Breach, New York, 1993), Chapter 4.

494

6. C.H. Druce *et al.*, Ann. Phys. **176** (1987) 114, and references therein.
7. J.L. Egido and P. Ring, Nucl. Phys. **A383** (1982) 189.
8. I. Talmi, Nucl. Phys. **A172** (1971) 1.
9. K. Allaart *et al.*, Phys. Rep. **169** (1988) 209.
10. O. Scholten, Phys. Rev. **C28** (1983) 1783.
11. A. van Egmond and K. Allaart, Nucl. Phys. **A425** (1984) 275.
12. M. Deleze *et al.*, Nucl. Phys. **A551** (1993) 269.
13. W. Pannert, P. Ring and Y.K. Gambhir, Nucl. Phys. **A443** (1985) 189.
14. L.C. de Winter *et al.*, Phys. Lett. **179B** (1986) 322.
15. H. Nakada, T. Otsuka and T. Sebe, Phys. Rev. Lett. **67** (1991) 1086;
    T. Otsuka, in Understanding the Variety of Nuclear Excitations, edited by A. Covello
    (World Scientific, Singapore, 1991), p. 321;
    H. Nakada and T. Otsuka, to be published.
16. G. Puddu, O. Scholten and T. Otsuka, Nucl. Phys. **A348** (1980) 109.
17. R.F. Casten and P. von Brentano, Phys. Lett. **152B** (1985) 22.
18. A. Bohr and B.R. Mottelson, Physica Scripta **22** (1980) 468.
19. J.N. Ginocchio and M.W. Kirson, Nucl. Phys. **A350** (1980) 31.
20. A.E.L. Dieperink, O. Scholten and F. Iachello, Phys. Rev. Lett. **44** (1980) 1747.

A PAIRING EFFECT IN γ-SOFT NUCLEI

Xing-Wang Pan [1,2], Takaharu Otsuka [1] and Akito Arima [1]

[1]Department of Physics, University of Tokyo
Hongo, Bunkyo-ku, Tokyo 113, Japan
[2]Department of Physics, Drexel University
Philadelphia, PA 19104, U.S.A.

1. INTRODUCTION

For the spectra of axially-symmetric nuclei, there is a well known phenomenon that the experimental energy levels with high angular momenta lie below what is expected from the $J(J+1)$ rule.[1] To be more precise, the level spacings of high spin states are compressed because of the increase of the moment of inertia with spin, of which the driving force is the reduction of the pairing correlation due to faster rotation, *i.e.*, the Mottelson-Valatin effect.[2] This kind of deviation has been explained by various models. For instance, the deviated rotational spectra for strongly deformed nuclei can be described by the $SU(3)$ limit of the Interacting Boson Model (IBM) and the Fermion Dynamical Symmetry Model (FDSM), where the reduction of pairing correlation in high-spin states can be incorporated perturbatively.[3] The deviation in the rotational spectra has also been explained in terms of the Geometric Model of Bohr and Mottelson,[4] such as the Rotational-Vibrational Model (RVM) [5] or the Variable Moment-of-Inertia Model (VMI),[6] showing success in describing the yrast states before backbending.

In the case of γ-soft nuclei, the low-lying states are classified under the dynamical symmetry of $O(6) \supset O(5) \supset O(3)$ in the IBM.[7] A beautiful level pattern according to the $O(5)$ quantum number τ has been established in a large number of nuclei.[7-10] Recently, It is claimed that similar type of deviation as in the rotational spectrum also appears in the spectra of γ-soft nuclei.[11] whereas the deviation in the spectra of γ-soft nuclei is associated with the level spacings of $O(5)$ scheme. In other words, the experimental energy levels of higher $O(5)$ multiplets lie below what is expected from the $\tau(\tau+3)$ rule, *i.e.*, the eigenvalue of $C_{2O(5)}$ Casimir operator in the $O(6)$ dynamical symmetry. We shall refer to this sort of deviation as τ-compression (of the energy levels) hereafter, because the spacings between higher multiplets with different value of τ are compressed depending almost solely on τ without destroying the relative level structure among the same τ multiplets. Thus, the τ –compression effect for the *γ-soft* nuclei is introduced as a uniform downward shift of the energy levels in the same $O(5)$ multiplets. It will be shown that the pairing correlation still plays a crucial role to the specific deviation in γ-soft nuclei. It will also be shown that the τ –compression phenomenon can be interpreted as a manifestation of the $O(6) \supset O(5) \supset O(3)$ limit in γ-soft nuclei, because of its specific way of the symmetry breaking.

2. τ–COMPRESSION PHENOMENON

In the $O(6)$ limit of IBM-1, the Hamiltonian is defined as [7]

$$H = E_0 + A\hat{C}_6 + B\hat{C}_5 + C\hat{C}_3 , \tag{1}$$

where E_0, A, B and C are constants, and $\hat{C}_6$, $\hat{C}_5$ and $\hat{C}_3$ denote the $O(6)$, $O(5)$ and $O(3)$ two-body Casimir operators. The eigenvalue of eq.(1) is given by [7]

$$E(N,\sigma,\tau,n_\Delta, J,M_J) = E_0 + A\sigma(\sigma+4) + B\tau(\tau+3) + CJ\,(J+1) , \tag{2}$$

where σ and τ are quantum numbers to denote the irrep. of the $O(6) \supset O(5)$ chain, while n_Δ is an additional quantum number. N is the boson number (*i.e.*, half of the valence particle).

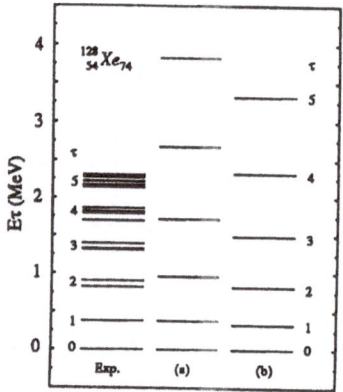

Figure 1. The O(5) energy levels of ^{128}Xe . For column (a); the parameters are taken as B=95.8 (KeV) and C=10(KeV) fitted by the 2_1^+ state. For column (b); the parameters are taken as B=83 (KeV) and C=10 (KeV) fitted by the 2_2^+ and 4_1^+ states

A good $O(6)$ symmetry in A ≈130 region is claimed by Casten *et al.* in ref.8, where general level patterns are presented for the above region, but without being fitted to each individual nucleus. However, If we fit the $O(5)$ parameter B by the lower $O(5)$ multiplets of γ-soft nuclei, for instance, $\tau \le 2$ states, then a specific deviation from $O(6)$ spectrum will show up clearly. We present the experimental levels and the pure $O(6)$ spectrum (for $\sigma=N$ band) of ^{128}Xe in Fig.1 as an example, where the experimental E_τ is obtained by extracting the $O(3)$ term $(CJ\,(J+1)\,)$ from the total excitation energy. *i.e.*, $E_\tau = B\tau(\tau+3)$. In column

(a) of Fig.1, we fixed the parameters B by the 2_1^+ level, while column (b) is fitted by $\tau=2$ multiplets. From Fig.1, first, we can see that there are relatively large energy gaps between neighboring $O(5)$ multiplets, which evidently indicate the goodness of the classification of low-lying states in the $O(5)$ scheme (More precisely speaking, those $O(5)$ multiplets are not exactly degenerate, which means the breaking of the $O(5)$ symmetry. Comparing the gaps between different τ values, we see that such a violation of the $O(5)$ degeneracy is small). Second, in Fig.1, the $O(5)$ spectrum manifests that the $O(5)$ gaps become larger with higher τ values, in contrast, the experimental spectrum does not display such increasing of the $O(5)$ spacings with τ. In other words, it shows the energy compression of higher $O(5)$ multiplets in the experimental levels. We emphasize that here the energy compressions are not dependent on the choice of the $O(3)$ parameter C, which should be determined by the relative level spacing among the same $O(5)$ multiplets, because the dominant contributions to the excitation energies of eq.(2) are from the $O(5)$ term. Thus if one takes a larger value of C, the energy compressions of higher $O(5)$ multiplets will not be eliminated, though the magnitude of the compression may be somehow reduced. However, as a price, the $O(5)$ scheme will be destroyed.

Note that in Fig.1 the 0_2^+ and 2_3^+ are not contained because these two states always locate above their other multiplets. It may be due to other mechanisms. For instance, the two-level mixing between 0_1^+ and 0_2^+ (also 2_1^+ with 2_3^+) shifts up the excitation energy of 0_2^+ (2_3^+) state.

Differing from the rotational case, where the compression of level spacings is related to spin J, while the energy compression is dependent on the $O(5)$ quantum number τ in γ-soft nuclei. What is more, the ground-state band, quasi-γ band and other bands are able to be treated in an unified way since low-lying states are classified in terms of the $O(5)$ multiplet in the $O(6)$ limit,

3. A MODIFIED $O(6)$ DESCRIPTION FOR γ-SOFT NUCLEI IN IBM-1

In order to explain the τ-compression effect in γ-soft nuclei, we introduce the pairing interaction to the $O(6)$ Hamiltonian.[11] The pairing Hamiltonian of nucleons is defined as,

$$H_{pair} \equiv G_0 S^\dagger \cdot S ,\qquad(3)$$

where G_0 is a negative constant , and

$$S^\dagger = \frac{1}{2} \sum_{jm} (-1)^{j-m} a_{jm}^\dagger a_{j-m}^\dagger ,$$
$$\qquad(4)$$

with $a_{jm}^\dagger$ being the usual nucleon creation operator.

From the OAI mapping procedure,[12] the corresponding boson image of the pairing term can be generally constructed as;

$$H_{pair} \equiv G_0 S^\dagger \cdot S \Rightarrow H_{pair}^B \equiv \alpha s^\dagger . s + \beta (s^\dagger \cdot s)^2, \qquad(5a)$$

$$= E_p + \varepsilon n_d + \beta \sum_{(L=0,2,4)} ([d^\dagger d^\dagger]^{(L)} \cdot [\tilde{d}\tilde{d}]^{(L)}), \qquad(5b)$$

where α, β and ε are parameters, and E_p is a function of boson number. For in single-orbit or degenerate multi-orbit systems, we have

$$E_p = G_0 N(\Omega+1-N), \quad \varepsilon = -\Omega G_0 \quad \text{and} \quad \beta = G_0, \tag{6}$$

where Ω stands for the total degeneracy of the fermion system, i.e., $\Omega = \Sigma_j (j+\frac{1}{2})$. We point out that H_{pair}^B can be written only in terms of the s-boson operators in eq.(5a). This is a characteristic feature of the boson image of the monopole pairing interaction between nucleons. We mention that Eq.(5) is obtained under the framework of the IBM-2, i.e., the proton-neutron IBM.[12] To get the IBM-1 H_{pair}^B, one need to project the IBM-2 H_{pair}^B onto the corresponding IBM-1 operator by using the projection technique. In ref.11, It is shown that the resultant IBM-1 H_{pair}^B has an expression similar to eq.(5)

Since H_{pair}^B can be written only in terms of the s-boson operators in eq.(5a), the boson image of the pairing term is a scalar for the group $O(5)$. This means that its matrix elements are independent of the $O(3)$ quantum numbers, i.e., the angular momentum within a subset belonging to a given $O(5)$ quantum number τ.

For the $\sigma=N$ band, the expectation values of the imaged pairing term between $O(6)$ basis states are derived as[13]

$$\alpha <s^\dagger \cdot s> + \beta <(s^\dagger \cdot s)^2>$$

$$= \left[\frac{\alpha}{2(N+1)} + \frac{(N^2+3N-2)\beta}{4N(N+1)} \right] N(N+3) - \left[\frac{\alpha}{2(N+1)} + \frac{(N^2+3N-1)\beta}{2N(N+1)} \right] \tau(\tau+3)$$

$$+ \beta \frac{1}{4N(N+1)} \, [\tau(\tau+3)]^2 . \tag{7}$$

Note that the angular momentum J does not appear on the right hand side of eq.(7). The first two terms on the right hand side of eq.(7) preserve the $O(6)$ pattern of energy levels. It is the last term that produces deviations from the $O(6)$ values. In fact, due to $\beta < 0$ (β is directly related to the pairing strength, for example, as eq.(6)), the higher the $O(5)$ quantum number τ, the stronger the lowering of the excitation energy from the $O(6)$ value. By adding the expectation value eq.(7) as a first-order perturbation to the Hamiltonian of eq.(2), we obtain the pairing-perturbed $O(6)$ spectra for the $\sigma=N$ states,

$$E_x = B' \frac{1}{6}\tau(\tau+3) + \lambda \, [\tau(\tau+3)]^2 + CJ \, (J+1) , \tag{8}$$

where E_x stands for the excitation energy, the renormalized $O(5)$ parameter B' is

$$B' = B - \left[\frac{\alpha}{2(N+1)} + \frac{(2N^2+2N-1)\beta}{2N(N+1)} \right] \quad \text{and} \quad \lambda = \beta \frac{1}{4N(N+1)} . \tag{9}$$

Up to now we only considered the first-order correction due to H_{pair}^B of eq.(5a). In order to treat H_{pair}^B more accurately, the total Hamiltonian,

$$H = E_0 + A \, \hat{C}_6 + B\hat{C}_5 + C\hat{C}_3 + H_{pair}^B , \tag{10}$$

is to be diagonalized numerically. To carry out the numerical calculation, we use the form of eq.(5b) for H_{pair}^B and obtain

$$H = E_0' + A \, \hat{C}_6 + B\hat{C}_5 + C\hat{C}_3 + \varepsilon n_d + \beta \sum_{(L=0,2,4)} ([d^\dagger d^\dagger]^{(L)} \cdot [\tilde{d}\tilde{d}]^{(L)}) . \tag{11}$$

We comment that, when one adopts sizable values of ε and β to fit the observed energy levels, the resultant value of A is varied from the value adjusted by the formula eq.(2). This is because ε contains a similar effect to the parameter A.

Since both the aforementioned deviation and $O(6)$ character are claimed for ^{128}Xe and its neighboring isotopes, we shall apply the $O(6)$+Pairing Hamiltonian to study the nuclei ^{126}Xe - ^{128}Xe. In Figs. 2-3, which correspond to ^{126}Xe and ^{128}Xe respectively, we show (a) the experimental levels, (b) the pure $O(6)$ spectrum according to eq.(2), (c) the perturbative results of eq.(7) and (d) the exact calculation of eq.(11).

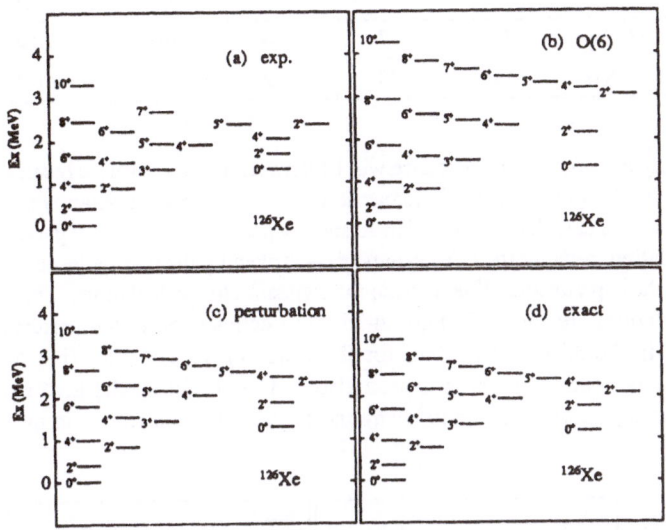

Figure 2. Energy levels of ^{126}Xe with $\sigma=N$. (a) Experimental levels [14-15]. (b) $O(6)$ spectrum of eq.(2). (c) Perturbed spectrum of $O(6)$+Pairing of eq.(8). (d) Exact spectrum by the $O(6)$+Pairing Hamiltonian of eq.(11).

For the pure $O(6)$ spectra, we keep the parameters as constants for ^{126}Xe and ^{128}Xe; $B = 74$ (KeV) and $C=12$ (KeV). These values are close to the values adopted in ref.[8]. Since only $\sigma=N$ states are shown in Figs.(2b-3b), the value of A is irrelevant. Comparing the experimental spectra (Figs.2a-3a) with the pure $O(6)$ ones (Figs.2b-3b), it is clear that the experimental energy levels of higher τ multiplets are much below those of $O(6)$ if the parameters in the Hamiltonian are fixed by fitting the lower-lying levels. A more careful comparison manifests a new regularity in the pattern of this deviation. To be more precise, the experimental energy levels are lower than the corresponding $O(6)$ ones, and the magnitude of the energy lowering is nearly equal among the states of the same $O(5)$ multiplets, though their spins are different. The magnitude of the downward shift is increased with higher τ, and therefore the spacing between neighboring τ multiplets is compressed as τ increases. As we have pointed out, this effect is denoted as τ-compression.

We first discuss case where the pairing interaction is included as a first-order perturbation correction to the $O(6)$ Hamiltonian. The energy levels can then be calculated by eq. (8). The parameters B and C are kept the same as in the $O(6)$ limit (see Figs. 2b-3b), because the $O(5)$ and $O(3)$ symmetries are not broken by the pairing terms. The values of all parameters are tabulated in Table 1. In this perturbative calculation, B' and λ are related to α and β by eq.(9), where $\alpha = -\varepsilon - (2N-1)\beta$. From Figs.2c-3c, We can see that the perturbative inclusion of the pairing interaction in the O(6) Hamiltonian indeed improves the agreement significantly, in particular for high τ states. It is remarkable that, with the addition of a single term $[\tau(\tau+3)]^2$, the compression of the level spacings can be reproduced up to τ =5 states.

Table 1. Parameters for ^{126}Xe and ^{128}Xe isotopes

Nuclei	B (KcV)	C (KeV)	A' (KeV)	ε (MeV)	β (MeV)
^{126}Xe	74	12	- 74	0.62	- 0.150
^{128}Xe	74	12	- 88	0.64	- 0.172

One may ask how much difference will be between the perturbative result and the exact solution of the O(6)+pairing Hamiltonian in eq.(11). The Hamiltonian is taken to be the same in these two calculations. One has to specify the value of the O(6) parameter A in the exact calculation, because the O(6) symmetry is broken by the pairing terms. This value is adjusted by the experimental 0_3^+ level as in the standard O(6) calculations. The results of the exact calculations are shown in Figs.2d-3d. The comparison to the perturbative results shown in Figs.2c-3c indicates that, for the states up to τ=5, the first-order pairing perturbation is quite accurate. It is seen also that the exact results give rise to a good agreement with experiments, reproducing the τ-compression effect as in the perturbative calculation.

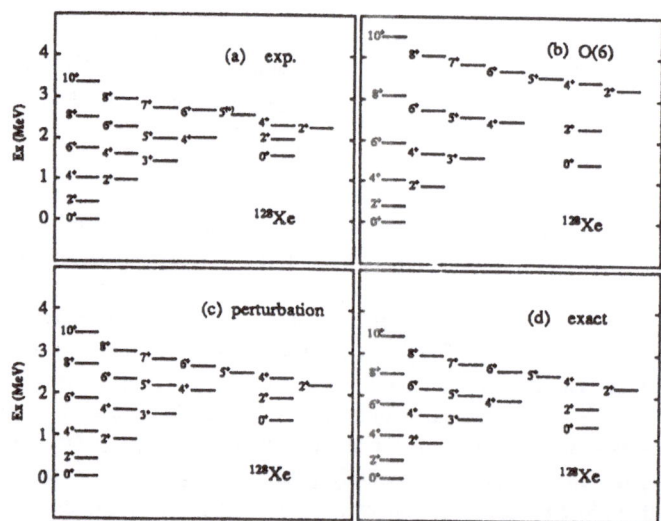

Figure 3. Energy levels of ^{128}Xe with σ=N (See the caption of Fig.2). Experimental levels are taken from [14,16].

From Figs.2-3, one can clearly see that the energy gap between 0_2^+ and 2_3^+ is characteristically large in the O(6) spectra in Figs.2b-3b compared to the corresponding experimental gap in Figs.2a-3a. This gap is nothing but the energy spacing between higher $O(5)$ multiplets with different τ values ($\tau=3$ and 4) in the $O(6)$ limit. However, the experimental gap is relatively small. This discrepancy can be explained as a natural consequence of the τ–compression. As shown in Figs.2c-3c and Figs.2d-3d, the large gap in the $O(6)$ spectra is indeed reduced when the pairing interaction is taken into account.

For the typical spectrum of the $O(6)$ limit, the states belonging to different $O(5)$ multiplets are split by a large energy gap (mainly by the $O(5)$ gaps), and the states within the same O(5) multiplets are closely spaced. This pattern is referred to as staggering for the quasi-γ band; the quasi-γ band is built up by pairs of almost degenerate levels as $(3_1^+, 4_2^+)$, $(5_1^+, 6_2^+)$, $(7_1^+, 8_2^+)$, ..., as seen in the O(6) limit.[8] In contrast, these states are nearly equally spaced in the experimental spectra of some nuclei in A≈130 region, indicating weakened or no staggering. The parameters of the $\tau(\tau+3)$ and $J(J+1)$ terms in the $O(6)$ spectrum are determined by the lower-lying levels, describe the lower-lying states well but produce always strong staggering. Thus, the weakened level staggering in the quasi-γ band of some $O(6)$ nuclei cannot be explained in the $O(6)$ limit. From the $O(6)$ spectrum of Fig.1, we see that two facts cause the level staggering in the quasi-γ band, one is the large $O(5)$ gaps between neighboring $O(5)$ multiplets, which is produced by the $\tau(\tau+3)$ term, the other is the $O(5)$ degeneracy among the same $O(5)$ multiplets. The present $O(6)$+pairing Hamiltonian in eq.(11) changes the position of energy levels within the quasi-γ band from the $O(6)$ limit. In other words, the level spacings between the neighboring $O(5)$ multiplets can be drastically reduced by the pairing term, hence the amplitude of the staggering is weakened to some extent. While the spacing between 3_1^+ and 4_2^+ (also 5_1^+ and 6_2^+) is not improved by the additional pairing interaction since it is the scalar term of $O(5)$ group, and some $O(5)$ symmetry breaking term is required to split the spacings among the $O(5)$ multiplets. The aforementioned staggering in the γ–band definitely characterize the $O(5)$ level scheme. On the other hand, the ARM spectrum does show rather large splitting between 3_1^+ and 4_2^+ (also 5_1^+ and 6_2^+...).[13] This fact implies that the deviation of the level staggering phenomenon in the γ-soft nuclei may be related to the (rigid) triaxiality. Recently, many efforts are paid to the introduction of the triaxiality to the $O(6)$ symmetry. For instance, Casten et al. introduced a cubic term to incorporate the triaxiality to the $O(6)$ like nuclei to explain the γ-band staggering.[8] Another approach is by using the $O(6)+ SU(3)^*$ model.[17] At some aspect, these two methods reproduce the properties of triaxiality successfully. In the following part of this paper, however, we will try to investigate the mentioned effect in γ-soft nuclei in terms of the $O(5)$ symmetry breaking under the framework of IBM-2, providing a different explanation of the deviation of the staggering effect in γ-soft nuclei.

4. THE IBM-2 DESCRIPTION FOR THE LEVEL STRUCTURE OF γ-SOFT NUCLEI

When proton and neutron are distinguished, the $O_{\pi+\nu}(6)$ symmetry can be realized by the following group chain,

$$U_\pi(6) \times U_\nu(6) \supset U_{\pi+\nu}(6) \supset O_{\pi+\nu}(6) \supset O_{\pi+\nu}(5) \supset O_{\pi+\nu}(3) , \qquad (12)$$

$$[N_\pi] \quad [N_\nu] \quad [N\text{-}ff] \quad < \sigma_1, \sigma_2 > \quad (\tau_1, \tau_2) \quad J$$

where N is the total boson number, $i.e., N = N_\pi + N_\nu$. The value of f can take 0, or 1, 2...to min.(N_π, N_ν), labeling the symmetric or mixed-symmetry irrep. of $U(6)_{\pi+\nu}$. The $U(6)_{\pi+\nu}$ components are also denoted by the F-spin values $(F = N/2-f)$.[11]

Then the $O(6)$ Hamiltonian in eq.(1) is extended to

$$H = E_0 + A\hat{C}_{6\pi+\nu} + B\hat{C}_{5\pi+\nu} + C\hat{C}_{3\pi+\nu} + M, \tag{13}$$

where M is the total Majorana interaction, which contains three components $(i.e., M = M_1 + M_2 + M_3)$

$$M_2 = \xi_2 \left[s_\nu^\dagger d_\pi^\dagger - s_\pi^\dagger d_\nu^\dagger \right] \cdot \left[s_\nu \tilde{d}_\pi - s_\pi \tilde{d}_\nu \right],$$

$$M_k = -2\xi_k \left[d_\nu^\dagger d_\pi^\dagger \right]^{(k)} \cdot \left[\tilde{d}_\nu \tilde{d}_\pi \right]^{(k)}, \quad k=1,3. \tag{14}$$

Taking $\xi_1 = \xi_2 = \xi_3 = \xi$, the Hamiltonian of eq.(13) has the eigenvalues

$$E = E_0 + A[\sigma_1(\sigma_1 +4) + \sigma_2(\sigma_2 +2)] + B[\tau_1(\tau_1 +3) + \tau_2(\tau_2 +1)]$$

$$+ CJ(J+1) + \xi \left(\frac{N}{2} - F \right)\left(\frac{N}{2} - F + 1 \right). \tag{15}$$

For the [N,0] symmetry component, the $O(6)_{\pi+\nu}$ spectrum is identical with that of the IBM-1. The Majorana term (also each individual term M_i $(i=1,2,3)$ in eq.(14)) makes no contribution to the excitation energy, whereas, for the [N-1,1] mixed-symmetry component, the total Majorana term with $\xi_1 = \xi_2 = \xi_3 = \xi$ lifts up the [N-1,1] mixed-symmetry states by ξN. Therefore, in the $\xi_1 = \xi_2 = \xi_3$ case, the total Majorana interaction merely affects the levels of the [N-1,1] mixed-symmetry states by a constant shift upward without changing the $O(5)_{\pi+\nu}$ level structure.

By playing with each individual terms of eq.(14) $(i.e.,$ without the $\xi_1 = \xi_2 = \xi_3$ condition), we can see the resultant change in the $O(5)_{\pi+\nu}$ level structure. Each M_i affects the energy levels in different manners. We will discuss yrast mixed-symmetry states due to the major contribution of the yrast mixed-symmetry states to the F-spin mixing in low-lying states (Here we define the yrast mixed-symmetry states as the lowest mixed-symmetry state for a given spin). To preserve the $O(5)_{\pi+\nu}$ symmetry, we add the $\Delta(M_1 + M_3)$ term (here Δ is just to make difference from the total Majorana interaction in eq.(13)) to the Hamiltonian due to $[M, \Delta(M_1 + M_3)] = 0$. Note that $2(M_1 + M_3) = \hat{n}_d (\hat{n}_d + 4) - C_{U(5)}$. Thus the $(M_1 + M_3)$ term maintains the $O(5)_{\pi+\nu}$ level structure for the mixed-symmetry states. On right of Fig.4, we plot the energy variation of the being discussed states against this particular set of the Majorana interaction, which indicates that the energy contribution from the $\Delta(M_1 + M_3)$ term depends on the irrep. $(\tau_1,1)$ of $O(5)_{\pi+\nu}$ (under irrep. <N-1,1> of $O(6)_{\pi+\nu}$). Comparing with the neighboring even-spin states $(i.e.,$ with the $(\tau,0)$ irrep.), the energies of the odd-spin states $(i.e.,$ with $(\tau_1,1)$ irrep.) are sharply increased as the strength of the $\Delta(M_1 + M_3)$ term is increased, showing a evident transition of level pattern from $(2^+, 3^+)$, $(4^+, 5^+)$, $(6^+, 7^+)$...type to 2^+, $(3^+, 4^+)$, $(5^+, 6^+)$, $(7^+, 8^+)$... type $(i.e., O(5)$ staggering). The former has the form which is analogous to the ARM-type level structure. Comparing the above property with the compression of the $O(5)$ level spacing of the symmetry states by the pairing interaction discussed in last section, we get another sort of variation in the $O(5)$ level structure. However, it is for mixed-symmetry states and caused by the Majorana term.

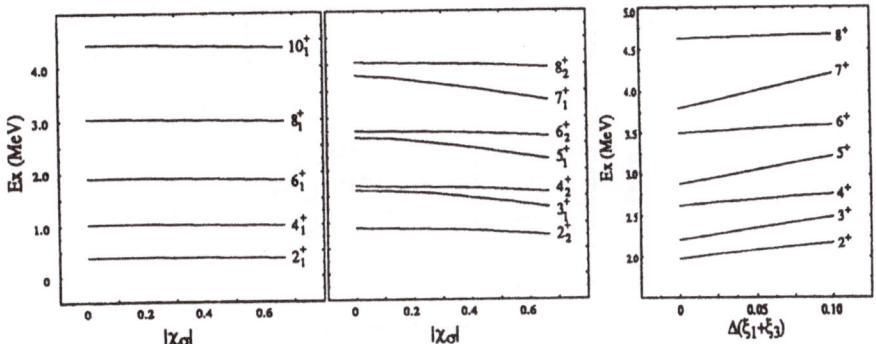

Figure 4. The right: The energy variation of the mixed-symmetry states against the $\Delta(M_1+M_3)$ Majorana term. The calculation is carried out by the Hamiltonian of eq.(13) with the parameters $A=-67$ (KeV), $B=83$ (KeV), $C=10$ KeV) and $\xi=130$ (KeV), and the additional $\Delta(M_1+M_3)$ term. **The left:** The energy variation of states in the ground-state band, γ-band against the parameter $|\chi_\sigma|$.

Note that the transition property of the energy levels with the $\Delta(M_1+M_3)$ Majorana term is for the mixed-symmetry states. On the other hand, considering a realistic situation, the states in the quasi-γ band of γ-soft nuclei are not completely symmetric. In ref.13, a systematic calculation with a realistic IBM-2 Hamiltonian over a wide range of Xe isotopes shows that mixed-symmetry components are contained in the low-lying states of quasi-γ band. In fact, the systematic calculations about the percentage of the F-spin mixing indicate that the odd-spin states in the quasi-γ band have more mixed-symmetry components than even-spin states. Naturally, it leads us to consider the role of the F-spin mixing in the staggering phenomenon.

In IBM-2, We write the $O(6)$+Pairing Hamiltonian as

$$H = \varepsilon(n_{d\pi}+n_{d\nu}) + \kappa Q_\pi Q_\nu + \kappa/2(Q_\pi Q_\pi + Q_\nu Q_\nu) + g_J J \cdot J + M$$

$$+ \sum_{(\sigma,L=0,2,4)} \lambda_L [d^\dagger d^\dagger]_\sigma^{(L)} \cdot [\tilde{d}\tilde{d}]_\sigma^{(L)} , \tag{16}$$

with

$$Q_\sigma = (d^\dagger s + \tilde{d}s^\dagger)_\sigma^{(2)} + \chi_\sigma (d^\dagger \tilde{d})_\sigma^{(2)} , \qquad (\sigma=\pi,\nu)$$

and

$$M = \xi_2(s_\pi^\dagger d_\nu^\dagger - s_\nu^\dagger d_\pi^\dagger) \cdot (\tilde{d}_\pi s_\nu - \tilde{d}_\nu s_\pi) - 2\sum_{(L=1,3)} \xi_L [d_\pi^\dagger d_\nu^\dagger]^{(L)} \cdot [\tilde{d}_\nu \tilde{d}_\pi]^{(L)} .$$

In the framework of IBM-2, the $O(6)_{\pi+\nu}$ limit can also be realized when the parameters ε, λ and χ in the Hamiltonian of eq.(16) are set to zero. According to what we have discussed in the last section, the ε and λ terms represent the pairing interaction if appropriate values are chosen. Usually, in numerical calculations of the IBM-2, the one-body d-boson term is already included and the parameter χ is non-zero.[18-19] According to the systematic calculation in ref.18, the first two terms in eq.(16) play a crucial role in the discussion of the nuclear global properties. In the detailed discussion of the $O(6)$ symmetry properties for the specific region (i.e., A ≈130), the quadrupole interactions between identical bosons are included to maintain $O(6)$ structure as more as possible. Therefore, in this work, the

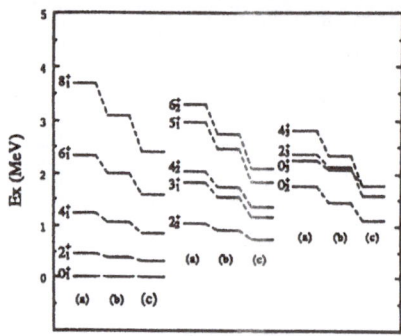

Figure 5. The perturbed $SO(6)$-like spectra by the Hamiltonian of eq.(16) with $\kappa=-0.158$ (MeV), $\varepsilon=0.3$ (MeV), $g_J=0.012$ (MeV), $\xi=0.3$ (MeV) and $(\chi_\pi,\chi_\nu)=(-0.511,0.425)$. Column (a), (b) and (c) are for the parameter $\lambda=0$, -0.175 and -0.25 (MeV) cases.

$O(6)$+Pairing IBM-2 Hamiltonian is defined as eq.(16). Just for simplicity, we choose the same strengths in the two-body d-boson interactions (the λ term in eq.(16)) for proton and neutron.

First, the effect of the two-body d-boson interaction on the spectrum of the $O(6)$-like nuclei can be seen by changing the parameter λ and keeping all the other parameters in the eq.(16) as constants. Fig.5 shows the variations of energy level in the $O(6)$-like spectrum with the increase of the parameter λ. This figure clearly shows that the higher the $O(5)$ multiplets, the larger the amplitude of the compression of energy. We point out that the effect of aforementioned staggering phenomenon in the $O(6)$ spectrum is due to the comparison of the level spacings between the $O(5)$ gaps (i.e., $\tau(\tau+3)$ term) and the relative spacings among the same $O(5)$ multiplets. In other words, if those $O(5)$ gaps were originally small, no evident staggering phenomenon would show up. Thus from Fig.5, we can say that the level staggering effect in the quasi-γ band is weakened to some extend by the pairing terms.

From the group theoretical point of view, to split $O(5)$ multiplets, some $O(5)$ symmetry breaking term is required. For instance, the Hamiltonian of eq.(16) (in the case of $\chi\neq0$) breaks the $O(5)$ symmetry. The energy variations in the ground-state band, γ-band against $|\chi_\sigma|$ (i.e., take $\chi_\pi=-\chi_\nu$) are plotted in the right figure of Fig.4. From this figure, we see that changing the parameters χ_σ in eq.(16) will give rise to a level transition in the quasi-γ band, from the ARM-type level scheme to the γ-unstable-type staggering. Now that the states in the quasi-γ-band contain components of mixed-symmetry due to the F-spin mixing, and such mixing is larger in odd-spin states. Therefore, the splitting between the 3_1^+ and 4_2^+ states (also 5_1^+ and 6_2^+... etc.) can be further produced by decreasing the(M_1+M_3) term in eq.(16).

In ref.17, Sevrin *et al.* discussed the staggering effect in γ-soft nuclei, using the $SU(3)^*$ perturbation to the $O(6)$ symmetry to introduce the triaxiality. The relative energies of the levels in the quasi-γ band such as 3_1^+, 4_2^+, 5_1^+ and 6_2^+ are indeed improved. However, the absolute energies do not fit experimental results so well. For example, the excitation energies for the lower-lying states as 2_1^+, 2_2^+ and 4_1^+ are underestimated, while overestimations appear among higher states such as 5_1^+, 6_2^+, 8_1^+, 10_1^+ etc. This shows necessity to include the pairing interaction explicitly in the $O(6)+SU(3)^*$ calculation to obtian a quantitative description of the level structure of γ-soft nuclei.

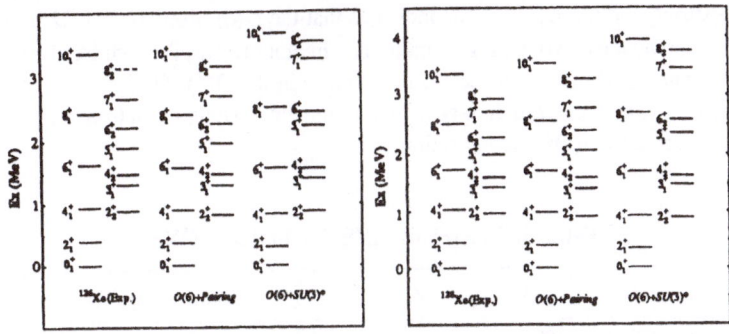

Figure 6. The comparison between experimental states (in ground-state band and γ-band) and calculated energy levels from the O(6)+pairing Hamiltonian and O(6)+SU(3)* one for ^{126}Xe and ^{128}Xe.

In Fig.6, ground-state band and quasi-γ band calculated by the Hamiltonian in eq.(16) are compared with experimentally observed levels of ^{126}Xe and ^{128}Xe. The theoretical spectra by $O(6)+SU(3)*$ calculation are also presented to compare with the $O(6)+Pairing$ calculation calculations. In our calculation, we have the same χ_σ values as in the systematic calculations in ref.18. For the $O(6)+SU(3)*$ case, Sevrin *et al.* take $\chi_\pi = -\chi_\nu$ to incorporate $SU(3)*$ characters. The parameters in our calculations are listed in Table 2. The parameters $\xi_2 = 0.10$ (MeV) and ξ_1 are fixed at 0.23 (MeV) to put 1_1^+ state above 2.5 (MeV). Comparing the levels in the quasi-γ band obtained by the $O(6)+pairing$ Hamiltonian with the results of $O(6)+SU(3)*$, one can observe that the calculations reproduce reasonable splittings among the same $O(5)$ multiplets (now the $O(5)$ scheme is not maintained exactly, but we still use the scheme to classify states for clarification). The large $O(5)$ gaps still sustained in the $O(6)+SU(3)*$ case because the pairing interaction is not taken into account. In the $O(6)+pairing$ calculation, the ground-state band and quasi-γ band are reproduced in an unified way. The levels in the quasi-γ band are well located. The agreement in Fig.6 for ^{126}Xe is particularly good. We comment that in the $O(6)+SU(3)*$ model, the triaxiality character in the γ-soft nuclei is introduced in terms of the $SU(3)*$ (*i.e.*, $\chi_\pi = -\chi_\nu = -\sqrt{7}/2$). In the $O(6)+pairing$ case, the triaxiality is incorporated by the $O(5)$ symmetry breaking and the F-Spin mixing. Both cases indicate the requirement of the pairing interaction to explain level spacings in γ-soft nuclei.

We know that the most stringent way to test whether a Hamiltonian is properly defined or not is by calculating the corresponding E2 transitions and making a comparison with the observed E2 data. Recently, rich relative B(E2) data for ^{126}Xe are available in ref.15, where a comparison between the experimental ratios and the $O(6)_\chi$ results (with the parameter $\chi \neq 0$ in the E2 operator of the $O(6)$ limit) is made. The comparison indicates that ^{126}Xe possess good $O(6)$ structure in terms of the E2 transition. Thus the data of E2 transition enable us to test the $O(6)+pairing$ Hamiltonian to see whether this Hamiltonian still can keep the $O(6)$ character with the pairing included. In the calculation for ^{126}Xe, the E2 operator is defined the same as the quadrupole operator in eq.(16), and the effective boson charges are microscopically determined by $e_\sigma = e_\sigma^0 \sqrt{(\Omega_\sigma - N_\sigma)/(\Omega_\sigma - 1)}$ with $e_\pi^0 = 0.132$ (eb) and $e_\nu^0 = 0.110$(eb).[18] In table3, we list the relative E2 results of experiment and the calculations by the $O(6)+pairing$ and $O(6)+SU(3)*$. It shows a nice consistency between experimental ratios and the $O(6)+pairing$ calculations, also presents the closeness between the results of the $O(6)$+pairing and the $O(6)_\chi$. This indicates that the $O(6)$ properties of E2 transition is well maintained by the $O(6)$+pairing Hamiltonian. In other words, the additional pairing term does not change the E2 transition so much. whereas in the $O(6)+SU(3)^*$ case, most of the relative B(E2) values are rather large, indicating the serious deviation from the $O(6)$ symmetry due to the $SU(3)^*$ incorporation.

Table 3. The relative B(E2) values for ^{126}Xe

transition	exp.	O(6)	O(6)χ	SO(6)+pairing	SO(6)+SU(3)*
$2_2^+ \to 2_1^+$	100	100	100	100	100
$\to 0_1^+$	1.5 ± 0.4	0	1.5	1.3	11
$3_1^+ \to 2_2^+$	100	100	100	100	100
$\to 4_1^+$	34^{+10}_{-34}	40	40	38	42
$\to 2_1^+$	$2.0^{+0.6}_{-1.7}$	0	2.0	2.9	20
$4_2^+ \to 2_2^+$	100	100	100	100	100
$\to 4_1^+$	76 ± 22	91	91	89	84
$\to 2_1^+$	0.4 ± 0.1	0	2.0	0.4	1.6
$0_2^+ \to 2_2^+$	100	100	100	100	100
$\to 2_1^+$	7.7 ± 2.2	0	2.0	1.94	1.5
$2_3^+ \to 0_2^+$	100	100	100	100	100
$\to 3_1^+$	67 ± 25	125	125	118	97
$\to 4_1^+$	2.0 ± 0.8	0	1.1	0.6	0
$\to 2_2^+$	2.2 ± 1.0	0	3.5	0	6.4
$\to 2_1^+$	0.14 ± 0.06	0	0	0	1.2
$\to 0_1^+$	0.13 ± 0.04	0	0	0.17	7.5
$4_3^+ \to 4_2^+$	100	100	100	100	100
$\to 3_1^+$	43 ± 13	115	115	105	98
$\to 4_1^+$	4.5 ± 1.4	0	0.4	0.3	5.6
$\to 2_2^+$	2.8 ± 0.9	0	2.5	1.6	11.6
$4_4^+ \to 2_3^+$	100	100	100	100	100
$\to 2_1^+$	0.9 ± 0.4	0	0	0.78	15.8
$\to 4_1^+$	7.9 ± 3.4	0	0	0	4.8
$5_1^+ \to 3_1^+$	100	100	100	100	100
$\to 6_1^+$	75 ± 23	45	45	46	52
$\to 4_2^+$	76 ± 21	45	45	51	47
$\to 4_1^+$	2.9 ± 0.8	0	2.0	1.6	16
$5_2^+ \to 5_1^+$	100	100	100	100	100
$\to 4_3^+$	100 ± 45	286	286	303	301
$\to 6_1^+$	2.2 ± 1.8	0	0.03	0	0
$\to 4_2^+$	1.9 ± 0.9	0	1.5	1.8	14.6
$\to 3_1^+$	4.2 ± 1.6	0	3.2	2.5	26.3
$6_2^+ \to 4_2^+$	100	100	100	100	100
$\to 6_1^+$	34^{+15}_{-25}	47	47	43	47
$\to 4_1^+$	0.49 ± 0.15	0	1.1	0.15	0.2
$7_1^+ \to 5_1^+$	100	100	100	100	100
$\to 6_2^+$	40 ± 26	18	18	20	19

To summarize this contribution, we suggest a new regularity in the $O(5)$ level structure of γ-soft nuclei, namely the τ-compression effect, *i.e.*, it is the uniform lowering of the levels within the same τ multiplets. The origin of τ-compression is found to be the pairing interaction between nucleons. we found that the additional pairing interaction provides a particular way to correct the $O(5)$ level spacings, meanwhile the $O(5)$ level scheme is still preserved due to the scalar property of this additional interaction in $O(5)$ group. When the pairing effect is added perturbatively to the $O(6)$ dynamical symmetry, the explicit formulas for the spectra are derived in the framework of IBM-1, which gives rise to a very simple but effective explanation of the τ-compression, *i.e.*, the $[\tau(\tau+3)]^2$ correction to the levels. Good agreement between the theoretical calculation based on the $O(6)$+pairing Hamiltonian and the experimental results is obtained for [126- 128]Xe , presenting a good example of the τ-compression. With the extension of $O(6)$+pairing Hamiltonian to IBM-2, the staggering effect in quasi-γ band of [126- 128]Xe isotopes is explained in terms of $O(5)$ breaking and F-spin mixing. The calculation of E2 transitions consistently shows the character of $O(5)$ scheme in the $O(6)$+pairing Hamiltonian. It also indicates the validity of O(6) symmetry in γ-soft nuclei, because the τ-compression effect can be explained very naturally under this symmetry but probably not without it.

REFERENCES

1. D. Bonatsos and A. Klein, *At. Data Nucl. Data Tables*, 30:27 (1984) , and references therein.
2. B.R. Mottelson and J.G. Valatin, Phys. Rev. Lett. 5: 511 (1960).
3. N. Yoshida, H. Sagawa, T. Otsuka and A. Arima, Phys. Lett. B256:129 (1991).J.L. Pin, J.Q. Chen, C.L. Wu and D.H. Feng, Phys.Rev. C43: 2224 (1991).
4. A. Bohr and B. R. Mottelson, Mat. Fys. Medd. Dan. Vid. Selsk. 16: 27(1953); A. Bohr and B.R. *Mottelson, Nuclear Structure Vol. II* (Benjamin, New York, 1975).
5. A. Faessler, W. Greiner and R.K. Sheline, Nucl. Phys. 70:33 (1965) .
6. M.A..J. Mariscotti, G. Scharff-Goldhaber and B.Buck, Phys. Rev. 178:1864 (1969).
7. A. Arima and F. Iachello, Ann.Phys.123:468 (1979) .
8. R. F. Casten, and P. Von Brentano, Phys. Lett. B152:22 (1985).
9. P. Von Brentano, A. Gelberg, S. Harissopulos and R.F. Casten, Phys. Rev. C38:2386 (1988) .
10. N.V. Zamfir and R.F. Casten, Phys.Lett. B260: 265 (1991).
11. X.W. Pan, T. Otsuka, J.Q. Chen and A. Arima, Phys. Lett. B287:1 (1991).
12. T. Otsuka, A. Arima, F. Iachello and I. Talmi, Phys. Lett. B66:20 (1977). T. Otsuka, A. Arima and F. Iachello, Nucl. Phys. A309:1 (1979).
13. X.W. Pan, Ph.D. Thesis (The University of Tokyo 1992).
14. M. Sakai, At. Data. Nucl. Data Tables 31:399 (1984).
15. W. Lieberz, A. Dewald, W. Frank, A. Gelberg, W. Krips, D. Lieberz, R. Wirowski and P. Von Brentano, Phys. Lett. B240: 38(1990).
16. R. Reinhardt, A. Dewald, A. Gelberg, W. Lieberz, K. Schiffer, K. P. Schmittgen, K. O. Zell and P. Von Brentano, Z.Phys. A329: 507 (1988).
17. A. Sevrin, K. Heyde and J. Jolie, Phys.Rev. C36:2621, 2631 (1987).
18. T. Otsuka, X.W. Pan and A. Arima, Phys.Lett. B247:191(1991).
19. G. Puddu, O. Scholten and T. Otsuka, Nucl. Phys. A348:109 (1980).

"DEFORMED U(Gl(3))" FROM $SO_q(3)$

A. Sciarrino

Università di Napoli "Federico II"
Dipartimento di Scienze Fisiche
and I.N.F.N. - Sezione di Napoli
I-80125 Napoli - Italy

1. Introduction

The quantum groups or quantum algebras, introduced in the study of quantum integrable systems[1], are presently well defined mathematical structures[2,3,4,5].

The quantum algebras are indeed deformation of enveloping Lie algebras. This field is now under very active investigation both from mathematical and physical point of view.

The main, neither unique nor conclusive, physical applications are in

- Conformal field theories

- Squeezed states in quantum optics

- Rotational spectra of nuclei and molecules

- 1-dim solid state physics and statistical mechanics systems.

These applications make use essentially of $SU_q(2)$. Recently the chain

$$SU_q(3) \supset SU_q(2) \supset SO_q(2)$$

has been applied in nuclear physics[6]. The underlying idea in many applications of q-algebras is to use a q-deformed algebra instead of a Lie algebra to realize a *dynamical symmetry*. A different approach has also been proposed, introducing the concept of invariance for q-group of motion for physical systems with a fundamental length scale[7]. In this approach q is a variable with dimension, no more a c-number.

It is well known that the dynamical symmetry in many models of nuclear, hadronic, molecular and chemical physics is displayed through embedding chain of algebras of the type

$$G_0 \supset G_1 \supset \ldots \supset SO(3) \supset SO(2) \tag{1}$$

Symmetries in Science VII, Edited by B. Gruber
and T. Otsuka, Plenum Press, New York, 1994

where $SO(3)$ describes the angular momentum and the Lie algebras are realized in terms of bosonic creation-annihilation operators.

The contribution in this field of F. Iachello has been extremely relevant: the well kmown Arima-Iachello[8] model in nuclear physics, the vibron model[9] in molecular physics, the "generalised" dynamical symmetry approach to scattering problem[10]. So it is particularly appropriate to present this contribution in this Symposium.

An essential step to carry forward the program of application of q-algebras as *dynamical symmetry* is to dispose on a formalism which allows to build up chains analogous to eq.(1) replacing the Lie algebras by the deformed ones.

The simplest, not trivial, chain is the q-analogous of the embedding chain of Elliott model[11],i.e.

$$SU_q(3) \supset SO_q(3)$$

In the Elliott model $SO(3)$ is the 3-dim principal subalgebra[12] of $SU(3)$. Quite recently Van der Jeugt[13] has investigated the existence of 3-dim principal q-subalgebra for $Gl_q(n+1)$, showing that such a subalgebra exists only for n = 2 when the algebraic relations are restricted to the symmetric representations (see below for definition). This restriction is not really a strong limitation for physical applications when the algebras are written in terms of bosonic oscillators. The 3-dim principal q-subalgebra is defined by the following equation:

$$[L_+, L_-] = [2L_0]_q \qquad [L_0, L_+] = L_+ \qquad [L_0, L_-] = -L_- \tag{2}$$

where

$$[x]_q = \frac{q^x - q^{-x}}{q - q^{-1}} \tag{3}$$

and by the requirement that in the limit $q \to 1$ it reduces to the 3-dim principal subalgebra. However the coproduct of $Gl_q(3)$ **does not induce** a coproduct in the 3-dim principal subalgebra.

Let us recall the definition of G_q associated with a simple Lie algebra G of rank r defined by the Cartan matrix (a_{ij}). G_q, deformation of the universal enveloping algebra of G which should be more precisely denoted by $U_q(G)$, is generated by $3r$ elements E_i, F_i and H_i which satisfy ($i, j = 1 = 1, \ldots, r$)

$$[E_i, F_j] = \delta_{ij}[H_i]_{q_i} \qquad\qquad [H_i, H_j] = 0$$

$$[H_i, E_j] = a_{ij} E_j \qquad\qquad [H_i, F_j] = -a_{ij} F_j \tag{4}$$

where $q_i = q^{d_i}$, d_i being non-zero integers with greatest common divisor equal to one such that $d_i a_{ij} = d_j a_{ji}$. Further the generators have to satisfy the Serre relations:

$$\sum_{0 \le n \le 1 - a_{ij}} (-1) \begin{bmatrix} 1 - a_{ij} \\ n \end{bmatrix}_{q_i} (E_i)^{1-a_{ij}-n} E_j (E_i)^n = 0 \tag{5}$$

where

$$\begin{bmatrix} m \\ n \end{bmatrix}_q = \frac{[m]_q!}{[m-n]_q![n]_q} \tag{6}$$

$$[n]_q! = [1]_q [2]_q \ldots [n]_q$$

Analogous equations hold replacing E_i by F_i.

G_q is endowed with a Hopf algebra structure. The action of the coproduct Δ, antipode S and co-unit ε on the generators is as follows:

$$\Delta(H_i) = H_i \otimes 1 + 1 \otimes H_i$$

$$\Delta(E_i) = E_i \otimes q_i^{H_i} + q_i^{-H_i} \otimes E_i$$

$$\Delta(F_i) = F_i \otimes q_i^{H_i} + q_i^{-H_i} \otimes F_i$$

$$S(H_i) = -H_i$$

$$S(E_i) = -q_i^{-a_{ii}/2} E_i \qquad S(F_i) = -q_i^{a_{ii}/2} F_i$$

$$\varepsilon(H_i) = \varepsilon(E_i) = \varepsilon(F_i) = 0 \qquad \varepsilon(1) = 1$$

(7)

The definition of the coproduct is essential to define the tensor product of spaces, i.e. to allow the composition of the (q-analogous) angular momentum in the case of $SO_q(3)$.

$Gl_q(n+1)$ is defined by $3n+1$ generators (E_i, F_i, N_i, N_0) satisfying:

$$(i, j = 1, \ldots, n)$$

$$[E_i, F_i] = [N_{i-1} - N_i]_q \qquad [N_i, N_j] = 0$$

(8)

$$[N_i, E_j] = (\delta_{i,j-1} - \delta_{ij}) E_j \qquad [N_i, F_j] = -(\delta_{i,j-1} - \delta_{ij}) F_j$$

and the Serre relations can be written replacing

$$a_{ij} \implies -(\delta_{i,j-1} + \delta_{i-1,j})$$

2. q-Boson Oscillators

For $Gl_q(n+1)$ there exists a realization in terms of q-boson operators[14] which are defined by the following relations (in the following we shall omit to write explicitly the q-dependence):

$$b_i b_j^+ - q\, b_j^+ b_i = \delta_{ij} q^{-N_i}$$

$$[N_i, b_j^+] = \delta_{ij} b_j^+ \qquad [N_i, b_j] = -\delta_{ij} b_j$$

(9)

$$[N_i, N_j] = [b_i, b_j] = [b_i^+, b_j^+] = 0$$

Analogous equations hold replacing q by q^{-1}.

The realization of $Gl_q(3)$ in terms of q-bosons can be written:

$$E_1 = b_1^+ b_0 \qquad F_1 = b_0^+ b_1 = E_1^+$$

$$E_2 = b_0^+ b_{-1} \qquad F_2 = b_{-1}^+ b_0 = E_2^+$$

(10)

where we have used the labels $(1, 0, -1)$ instead of $(1, 2, 3)$. The Serre relation, e.g. for $a_{12} = -1$, can be written:

$$[E_1, [E_1, E_2]_{q^{-1}}]_q = 0$$

where

$$[A, B]_q = AB - q\, BA$$

The states of symmetric representations are realized in the Fock space of the q-bosons and are defined by:

$$|n_1, n_0, n_1 \rangle = \frac{(b_1^+)^{n_1} \, (b_0^+)^{n_0} \, (b_{-1}^+)^{n_{-1}} \, |0\rangle}{\sqrt{[n_1]_q! \, [n_0]_q! \, [n_{-1}]_q!}} \tag{11}$$

These states are eigenstates of the N_μ operators

$$N_\mu |n_1, n_0, n_{-1} \rangle = n_\mu |n_1, n_0, n_{-1} \rangle \tag{12}$$

The q-boson operator acts on the states in the following way:

$$b_\mu^+ |n_\mu, \cdot, \cdot \rangle = [n_\mu + 1]_q^{1/2} |n_\mu + 1, \cdot, \cdot \rangle$$

$$b_\mu |n_\mu, \cdot, \cdot \rangle = [n_\mu]_q^{1/2} |n_\mu - 1, \cdot, \cdot \rangle \tag{13}$$

In this realization the $SO_q(3)$ can be written[13]:

$$L_0 = N_1 - N_{-1} \tag{14}$$

$$L_+ = q^{N_{-1}} q^{-1/2N_0} \sqrt{q^{N_1} + q^{-N_1}} \, b_1^+ b_0$$
$$+ \, b_0^+ b_{-1} q^{N_1} q^{-1/2N_0} \sqrt{q^{N_{-1}} + q^{-N_{-1}}} \tag{15}$$

$$L_- = b_0^+ b_1 q^{N_{-1}} q^{-1/2N_0} \sqrt{q^{N_1} + q^{-N_1}}$$
$$+ \, q^{N_1} q^{-1/2N_0} \sqrt{q^{N_{-1}} + q^{-N_{-1}}} \, b_{-1}^+ b_0 \tag{16}$$

The dimension of the representations of $SO_q(3)$ in the Fock space of the q-boson oscillators is given by $2N + 1$ ($N = n_1 + n_0 + n_{-1}$) therefore it is always **odd**. So this is a realization of $SO_q(3)$ and not of $SU_q(2)$. Moreover for $q = 1$ we obtain the usual $SO(3)$ realization in terms of boson oscillators.

Let us make a few remarks on eqs.(14-16):
1) L_+, L_- are **not invariant** for $q \to q^{-1}$ while L_0 is invariant;
2) L_+, L_- really belong to $Gl_q(3)$ not to $SU_q(3)$, as only $N_1 - N_0$ and $N_0 - N_{-1}$ belong to $SU_q(3)$;
3) The coproduct defined on $Gl_q(3)$, i.e. on E_i (F_i) (i =1,2) **does not induce a** coproduct on L_+ (L_-) defined by eqs.(15,16).

It is natural to wonder if one can solve the problem the other way around: to define a true $SO_q(3)$, i.e. a deformed $SO(3)$ in which a coproduct on the generators is defined and to build up a deformed structure of the type $Gl_q(3)$ or $SU_q(3)$. In order to better define the problem let us go back to the non-deformed case and let us review how it is possible to build up $SU(3)$ from $SO(3)$.

3. From $SO(3)$ to $SU(3)$

Let us start from the bosonic oscillators realization of $SO(3)$. We shall use the same notations for bosons and q-bosons, hoping that no confusion arises. Let b_μ^+ (b_μ)

($\mu = 1, 0, -1$) be 3 standard creation (annihilation) operators. Then the $SO(3)$ algebra can be realized as:

$$L_+ = \sqrt{2}\,(b_1^+ b_0 + b_0^+ b_{-1}) \tag{17}$$

$$L_- = \sqrt{2}\,(b_0^+ b_1 + b_{-1}^+ b_0) \tag{18}$$

$$L_0 = b_1^+ b_1 - b_{-1}^+ b_{-1} = N_1 - N_{-1} \tag{19}$$

The commutation relations are:

$$[L_+, L_-] = 2L_0 \qquad [L_0, L_\pm] = \pm L_\pm \tag{20}$$

Let us add to the generators of $SO(3)$ the highest weight of a symmetric tensor operator of rank 2 T_2^2 which in terms of the b-bosons can be written as:

$$T_2^2 = b_1^+ b_{-1} \tag{21}$$

By the action of L_-

$$[L_-, T_\alpha^2] = \sqrt{[2 + \alpha]\,[2 - \alpha + 1]}\, T_{\alpha-1}^2 \tag{22}$$

we get all the other operators:

$$T_1^2 = -\frac{1}{\sqrt{2}}(b_1^+ b_0 - b_0^+ b_{-1}) \qquad T_0^2 = -\frac{1}{\sqrt{6}}(2b_0^+ b_0 - b_1^+ b_1 - b_{-1}^+ b_{-1}) \tag{23}$$

$$(T_\alpha^2)^+ = (-1)^\alpha\, T_{-\alpha}^2 \qquad (\alpha = 2, 1, 0, \text{-1}, \text{-2})$$

We can then compute the commutation relations between the operators T_α^2 and we obtain:

$$[T_\alpha^2, T_{-\alpha}^2] = (-1)^\alpha\, \alpha/2\, L_0$$

$$[T_2^2, T_1^2] = [T_2^2, T_0^2] = 0 \tag{24}$$

$$[T_1^2, T_0^2] = \frac{\sqrt{3}}{2\sqrt{2}}\, L_+$$

The above equations and the hermitian coniugate ones show that we have obtained a 8-dim., rank 2 algebra which has to be identified with $SU(3)$.

The following expression is a scalar for $SO(3)$:

$$I = \sum (-1)^\alpha T_\alpha^2 T_{-\alpha}^2 \tag{25}$$

and the 2nd order Casimir operator of $SU(3)$ can be written as

$$C_2 = I + \frac{1}{4}(L_+ L_- + L_- L_+ + 2L_0 L_0) \tag{26}$$

4. From the realization of $SO_q(3)$ to a deformed algebra

There is a procedure analogous to what we have done in Sec. 3 in the case of $SO_q(3)$? To try to answer this question let us start from a realization of $SO_q(3)$ in terms of q-bosons. We can use eqs.(14-16), but now, **forgetting** about $Gl_q(3)$, we define a coproduct, antipode and counit on the generator $L_\pm$ and L_0, i.e.

$$\Delta(L_0) = L_0 \otimes \mathbf{1} + \mathbf{1} \otimes L_0$$

$$\Delta(L_+) = L_+ \otimes q^{L_0} + q^{-L_0} \otimes L_+$$

$$\Delta(L_-) = L_- \otimes q^{L_0} + q^{-L_0} \otimes L_- \tag{27}$$

$$S(L_0) = -L_0 \qquad S(L_+) = -q^{-1}L_+ \qquad S(L_-) = -q\,L_-$$

$$\varepsilon(L_0) = \varepsilon(L_+) = \varepsilon(L_-) = 0 \qquad \varepsilon(\mathbf{1}) = 1$$

Once more let us emphasize that we identify this deformed structure with $SO_q(3)$ and not with $SU_q(2)$ as its representations, in the Fock space of the q-bosons, are of odd dimension.

In analogy with the not deformed case, we add to the generators of $SO_q(3)$ the highest weight of q-tensor operator[15] of rank 2 of $SO_q(3)$ or equivalently of $SU_q(2)$. From the definition of coproduct Biedenharn and Tarlini[15] have shown that the q-tensor operators have to satisfy the following equations ($-k \le m \le k$):

$$(L_\pm T_m^k(q) \; - \; T_m^k(q)\, L_\pm\, q^{-m})q^{-L_0} \; = \; \sqrt{[k \mp m]_q\, [k \pm m + 1]_q}\, T_{m\pm1}^k(q) \tag{28}$$

$$[L_0, T_m^k(q)] \; = \; m\, T_m^k(q) \tag{29}$$

$T_2^2(q)$ can be written in terms of q-bosons as (in the following the q-dependence is omitted):

$$T_2^2 = \sqrt{q^{N_1} + q^{-N_1}}\; b_1^+ b_{-1} \sqrt{q^{N_{-1}} + q^{-N_{-1}}}\; \exp\left(c_1 N_1 + c_0 N_0 + c_{-1} N_{-1}\right) \tag{30}$$

where

$$c_1 - c_0 = 1 \qquad c_0 - c_{-1} = 3$$

We compute T_1^2

$$[L_-, T_2^2]_{q^{-2}} = [4]_q\, T_1^2\, q^{L_0} \tag{31}$$

and then

$$[T_2^2, T_1^2] \neq 0 \tag{32}$$

The corresponding commutator in the not deformed case eq.(24) is vanishing. It seems reasonable to require that the vanishing relations are preserved in the deformation procedure, so we try another choice.

Let us look for an operator $X_2(q)$ ($X_{-2} = X_2^+$) such that:

$$[L_+, X_2] = 0 \qquad [L_0, X_2] = 2\,X_2 \tag{33}$$

X_2 can be written in terms of q-bosons as:

$$X_2 = b_1^+ b_{-1}\, q^{-X_0} \sqrt{(q^{N_1+1} + q^{-N_1-1})(q^{N_{-1}} + q^{-N_{-1}})} \tag{34}$$

$$X_0 = N_1 + N_{-1} \tag{35}$$

Then we can compute the action of L_- on X_2, the commutator of X_2 with X_{-2} and so on. We find:

$$[X_2, X_{-2}] = [2L_0]_q (q + q^{-1}) q^{-2X_0} \tag{36}$$

$$[X_0, L_\pm] = \pm X_{\pm 1} \qquad [L_0, X_0] = [X_0, X_\pm] = 0 \tag{37}$$

where

$$\begin{aligned} X_+ &= q^{N_{-1}} q^{-1/2N_0} \sqrt{q^{N_1} + q^{-N_1}} \, b_1^+ b_0 \\ &\quad - b_0^+ b_{-1} \, q^{N_1} \, q^{-1/2N_0} \sqrt{q^{N_{-1}} + q^{-N_{-1}}} \end{aligned} \tag{38}$$

$$[X_{+1}, X_{-1}] = [2L_0]_q \tag{39}$$

$$[L_0, X_{\pm 1}] = \pm X_{\pm 1} \tag{40}$$

$$[X_0, X_{\pm 1}] = \mp L_\pm \tag{41}$$

$$[X_{+1}, X_2] = 0 \tag{42}$$

$$\begin{aligned} [L_-, X_2] &= (2q)^{-1} (q + q^{-1}) \\ &\quad \times \{(L_+ - X_{+1}) q^{-2L_0} - (L_+ + X_{+1}) q^{2L_0}\} q^{-2X_0} \end{aligned} \tag{43}$$

$$[L_-, X_{+1}] = 2q(q - q^{-1})^{-1} \times \{q^{2X_0}[q - (q + q^{-1})q^{-2N_0}] + q^{2L_0} + q^{-2L_0}\} \tag{44}$$

$$[L_+, X_{+1}] = 2X_2 q^{-N_0} \times \{q^{N_0+1} + q^{-N_0}(q + q^{-1})\} \tag{45}$$

$$\begin{aligned} [X_{-1}, X_2] &= (2q)^{-1} (q + q^{-1}) \\ &\quad \times \{(L_+ - X_{+1})q^{-2(X_0+L_0)} + (L_+ + X_{+1})q^{-2(X_0-L_0)}\} \end{aligned} \tag{46}$$

$$[N_0, X_{\pm 2}] = 0 \qquad [N_0, L_\pm] = \mp X_{\pm 1} \qquad [N_0, X_{\pm 1}] = \pm L_\pm \tag{47}$$

The above equations show that the set

$$L_\pm \oplus L_0 \oplus X_{\pm 2} \oplus X_{\pm 1} \oplus X_0 \oplus N_0 \tag{48}$$

closes in the sense of the enveloping algebra. However in order to state that we have build up $Gl_q(3)$ we should be able to recover the Hopf structure; while all the generators of eq.(48) belong to $Gl_q(3)$ in the q-boson realization of Sec. 2 the Hopf structure of the generators of $Gl_q(3)$ eq.(10) does not induce an analogous structure on the $\{X\}$. So we have kept the Hopf structure on $SO_q(3)$ but not in the larger algebraic structure we have build up, which we call to avoid confusion with already established notation "deformed $U(Gl(3))$". Nevertheless let us remark that we can write invariant operators for this *deformed* $U(Gl(3))$ as for $Gl_q(3)$[16], e.g. the following expression which reduces in the limit q=1 to the 2nd order Casimir operator for $SU(3)$ is invariant under the action (commutator) of the operators of eq.(48)

$$
\begin{aligned}
(q+q^{-1})\,C_2(q) \; = \; & (q-q^{-1})^{-2} \times [q^2 q^{2/3(2L_0-N_0)} + q^{-2} q^{-2/3(2L_0-N_0)} + q^{-2/3(X_0-2N_0)} - 3] \\
& + qq^{1/3(3/2L_0-1/2X_0+N_0)} b_0^+ b_0 b_1 b_1^+ \\
& + q^{-1} q^{-1/3(3/2L_0-1/2X_0+N_0)} b_{-1}^+ b_{-1} b_0 b_0^+ \\
& - q^1/3(X_0 - 2N_0)[b_0^+ b_1,\ b_{-1}^+ b_0]_q\,[b_1^+ b_0,\ b_0^+ b_{-1}]_{q^{-1}} \\
& + \quad (q \to q^{-1})
\end{aligned}
\tag{49}
$$

Using operators of eq.(48) it is possible obtain invariants for $SO_q(3)$:
(C $\equiv$ q-Casimir operator)

$$
C = L_+ L_- + [L_0]_q\,[L_0 + 1]_q
\tag{50}
$$

$$
I_0 = q^{X_0 + N_0}
\tag{51}
$$

$$
I = f(X_2 X_{-2},\ X_{+1} X_{-1},\ q^{\pm X_0},\ q^{\pm N_0})
\tag{52}
$$

Let us finally remark that in the limit q = 1 the set $\{X\}$ reduces to the tensor operator $\{T^2\}$, up to sign and numerical normalization factors.

5. Alternative realization of $SO_q(3)$

It is natural to wonder if and in which way the construction of Sec. 4 depends on the explicit realization of $SO_q(3)$ in terms of q-boson we have used.

Let us introduce a new type of deformed boson operator, which we shall denote by B_0^+ (B_0) which is defined by the following equations:

$$
q\,B_0(q)B_0^+(q) - q^{-1} B_0^+(q)B_0(q) = \frac{q^{2M_0+1} + q^{-2M_0-1}}{q + q^{-1}}
\tag{53}
$$

$$
[M_0, B_0^+(q)] = B_0^+(q) \qquad [M_0, B_0(q)] = -B_0(q) \qquad M_0 = M_0^+
\tag{54}
$$

A solution of eq.(53) is:

$$
B_0^+(q)B_0(q)\,q = \frac{q^{2M_0} - q^{-2M_0}}{q^2 + q^{-2}}
\tag{55}
$$

$$
B_0(q)B_0^+(q)\,q^{-1} = \frac{q^{2M_0+2} - q^{-2M_0-2}}{q^2 + q^{-2}}
\tag{56}
$$

Moreover we require that B_0^+, B_0 and M_0 commute with the q-bosons $b_{\pm 1}^+$, $b_{\pm 1}$ and with $N_{\pm 1}$ of Sec. 2.

Let us remark

- the deformed boson $B_0^+(q)$ is not invariant for $q \to q^{-1}$.

- in the limit q = 1 eq.(53) reduces to the relation defining a bosonic oscillator.

Then we can write the following realization of $SO_q(3)$:

$$L_+ = \sqrt{q^{-1}(q+q^{-1})} \times \{b_1^+ B_0 q^{M_0-N_{-1}} - q^{M_0-N_1} B_0^+ b_{-1}\} \tag{57}$$

$$L_- = \sqrt{q^{-1}(q+q^{-1})} \times \{q^{M_0-N_{-1}} B_0^+ b_1 - b_{-1}^+ B_0 q^{M_0-N_1}\} \tag{58}$$

$$L_0 = N_1 - N_{-1} \tag{59}$$

Let us remark that in this realization too L_+, L_- are **not invariant** for $q \to q^{-1}$ while L_0 is invariant.

Let us look for an operator $X_2(q)$ $(X_{-2} = X_2^+)$ such that:

$$[L_+, X_2] = 0 \qquad [L_0, X_2] = 2 X_2 \tag{60}$$

X_2 can be written in terms of q-bosons as:

$$X_2 = b_1^+ b_{-1} q^{X_0} \tag{61}$$

where X_0 is the same operator of Sec. 4.

Then we can compute the action of L_- on X_2, the commutator of X_2 with X_{-2} and so on, as in Sec. 4. We find:

$$[X_2, X_{-2}] = [2L_0]_q (q+q^{-1})^{-1} q^{2X_0} \tag{62}$$

$$[X_0, L_\pm] = \pm X_{\pm 1} \qquad [L_0, X_0] = [X_0, X_{\pm 2}] = 0 \tag{63}$$

where

$$X_+ = \sqrt{q^{-1}(q+q^{-1})} \times \{b_1^+ B_0 q^{M_0-N_{-1}} + q^{M_0-N_1} B_0^+ b_{-1}\} \tag{64}$$

$$[X_{+1}, X_{-1}] = [2L_0]_q \tag{65}$$

$$[L_0, X_{\pm 1}] = \pm X_{\pm 1} \tag{66}$$

$$[X_0, X_{\pm 1}] = \mp L_\pm \tag{67}$$

$$[X_{+1}, X_2] = 0 \tag{68}$$

$$[L_-, X_2] = q/2 \times \{(X_{+1} - L_+) q^{2(X_0+L_0)} + (X_{+1} + L_+) q^{2(X_0-L_0)}\} \tag{69}$$

$$[L_-, X_{+1}] = [2M_0 - X_0 + L_0]_q q^{2M_0-X_0+L_0} - [X_0 - L_0 - 2M_0]_q q^{2M_0-X_0-L_0} \tag{70}$$

$$[L_{-1}, X_{-1}] = -2q(q+q^{-1}) \times X_{-2} q^{2(2M_0-X_0)} \tag{71}$$

$$[M_0, X_{\pm 2}] = 0 \qquad [M_0, L_\pm] = \mp X_{\pm 1} \qquad [M_0, X_{\pm 1}] = \pm L_\pm \tag{72}$$

$$[X_{-1}, X_2] = q/2 \times \{(X_{+1} - L_+) q^{2L_0} - (X_{+1} + L_+) q^{-2L_0}\} q^{2X_0} \tag{73}$$

From a comparison of the above equations with the analogous ones of Sec. 4 we see that the detailed form of the commutation relations is depending of the particular realization of $SO_q(3)$ we have used, but, however, the general structure is essentially the same. We have a 9-dim algebraic structure which *closes* in the sense of an enveloping algebra, as in the right side of the commutators product of the generators times q to the power of the diagonalizable elements (X_0, L_0, M_0) appear. The above equations look simpler than the equivalent ones of Sec. 4, but let us remark that by means of $(B_0, b_{\pm 1})$ it is not possible to write a simple realization of $Gl_q(3)$ as eq.(10). For q = 1 also in this case the set $\{X\}$ reduces to the tensor operator $\{T^2\}$, up to numerical factors.

6. Conclusions

The problem to build up a formalism suitable to discuss embedding chain for q-algebra is really an hard one. We have shown that if we impose the Hopf structure on $Gl_q(3)$ we loose this structure on the principal 3-dim subalgebra while if we start from $SO_q(3)$ it is not clear how to impose the Hopf structure in the obtained deformed $U(Gl(3))$. It should be the physical problem one is dealing with to suggest which structure one should choice.

The deep root of the difficulties is on the fact that G_q is well mathematically defined only in the Chevalley-Cartan basis, see Sec. 1, and this basis is not suitable to discuss embedding of subalgebras except the trivial ones.

More generally, let G be a Lie algebra and $H \subset G$ a singular subalgebra, then it follows that deforming and endowing the generators of G with an Hopf structure G_q does not induce an analogous structure on H, loosely speaking H_q is not contained in G_q. Also in this general case, in analogy with the construction presented in this paper, one could start from H_q and then adding a suitable set of operators $\{X\}$ to get a deformed $U(G)$.

Acknowledgements

It is a pleasure to thank F. Iachello for stimulating discussions which have drawn my interest on the problem.

References

1. L.D. Faddeev, *"Integrable models in 1+1 dimensional quantum field theory"*, Les Houches Lecture 1982, Eds. J. Zuber and R. Stora (Elsevier, Amsterdam, 1984)

2. P.P. Kulish and N.Yu. Reshetikhin, *Zapiski Nauchn. Semin. LOMI* **95**, 129 (1980), English translation *J.Soviet.Math.* **23**, 2435 (1983)

3. E.K. Sylvanin, *Funct.Anal.Appl.* **16**, 263 (1982)

4. V.G. Drinfeld, *Sov.Math.Dokl.* **32**, 254 (1985); Proc. ICM-86, Berkeley (1987)

5. M.J. Jimbo, *Lett.Math.Phys.* **10**, 63 (1985); **11**, 247 (1986)

6. D. Bonatsos, A. Faessler, P.P. Raychec, R.P. Roussev and Yu. F. Smirnov, *J.Phys.* A **25**, L267 (1992)

7. E. Celeghini, R. Giachetti, E. Sorace and M. Tarlini, *Phys. Lett.* B **280**, 180 (1992) and refs. therein;
   F. Bonechi, E. Celeghini, R. Giachetti, E. Sorace and M. Tarlini, *Heisenberg XXZ model and quantum Galilei group*, DFF 160//04/92 Firenze and refs. therein

8. A. Arima and F. Iachello, *Ann.Phys.* **99**, 253 (1976); **111**, 201 (1978); **123**, 468 (1979)

9. F. Iachello, *Chem.Phys.Lett.* **78**, 581 (1981);
   F. Iachello and R.D. Levine, *J.Chem.Phys.* **77**, 3046 (1982);
   O.S. van Roosmalen, F. Iachello, R.D. Levine and A.E.L. Dieperink, *J.Chem.Phys.* **79**, 2515 (1983)

10. Y. Alhassid, F. Gürsey and F. Iachello, *Phys.Rev.Lett.* **50** , 873 (1983); *Ann.Phys.* **148**, 346 (1983); **167**, 181 (1986);
    J. Wu, F. Iachello and Y. Alhassid, *Ann.Phys.* **173**, 68 (1987)

11. J.P. Elliott, *Proc.R.Soc.* **A 245**, 128, 562 (1958)

12. B. Kostant, *Am.J.Math.* **81**, 973 (1959)

13. J. Van der Jeugt, *J.Phys.* **A 25**, L213 (1992)

14. L.C. Biedenharn, *J. Phys.* **A 22**, L873 (1989).
    A.J. MacFarlane, *J. Phys.* **A 22**, 4581 (1989).

15. L.C. Biedenharn and M. Tarlini, *Lett.Math.Phys.* **20**, 272 (1990)

16. V. Pasquier and H. Sauler, *Nucl.Phys.* **B330**, 523 (1990)

MEAN FIELD APPROACH TO THE ALGEBRAIC TREATMENT OF MOLECULES

Bin Shao, Niels R. Walet and R. D. Amado

Department of Physics, University of Pennsylvania
Philadelphia, PA 19104, U.S.A.

Abstract

We study recently proposed algebraic models for diatomic and simple poly-atomic molecules in the mean field approximation. In this approach, we recover the harmonic molecular spectra efficiently and accurately and more importantly, we are able to explore the geometrical meaning of boson operators with an eye to suggesting boson Hamiltonians that have close links to molecular geometries.

INTRODUCTION

Recently Iachello and co-workers[1] have demonstrated the great power of algebraic approaches to molecular dynamics. These methods have been applied with success to the spectra and transition intensities in diatomic molecules,[2] bent and linear triatomic molecules,[3,4] linear four atom molecules,[5] and even complex polyatomic molecules.[6] The algebraic method has been shown to be a rich starting point for studies of electron scattering from molecules,[7,8] and promises to be a fertile starting point for other reaction studies.[9] The principle disadvantage of the vibron model, as this approach is called, is that the molecular dynamics is expressed in terms of boson operators that are not easily interpreted in terms of the atomic position coordinates of traditional molecular physics. The standard way of extracting geometric meaning from boson models is through the use of coherent-state mean-field techniques.[10-17]

In this paper we show that by applying mean-field methods to the algebraic description, we not only greatly simplify the algebra, but also make a connection between the algebraic degrees of freedom and the coordinate degrees of freedom. The mean field approximation is basically an expansion in $1/N$ and is particularly appropriate in a discussion of the algebraic treatment of molecular dynamics since the number of bosons (N) is normally quite large. In this limit we have both the simplification of the mean field approximation and the fact that the operators approach classical values. It is these two features that we exploit.

We will show how the MFA comes out of the vibron model and how the correct dynamical and zero modes emerge. We will show how the molecular equilibrium shape (in the intrinsic frame) comes out and how it depends on the input parameters. We will stress the connection between the equilibrium shape, the dynamical modes and the zero modes. We will compare our results with those obtained from the full vibron model. We will see that the spectra are not quite so accurately given as in the full model, since we omit both $1/N$ corrections and anharmonicities. These can be included, but that is not the point of the approach. For many purposes, such as electron scattering, reactions and the description of complex molecules that are beyond the reach of the vibron model, the accuracy we obtain is sufficient, and the simplicity of the description a boon. It is that description and its direct contact to molecular shape that is the strongest appeal of the method. In this paper we demonstrate the use of the MFA for diatomic molecules, linear and bent triatomic molelcules and for linear four atom molecules, and we compare with full vibron calculations.

We begin with a discussion of the MFA for the $U(4)$ vibron model appropriate to diatomic molecules. This section is primarily pedagogic and establishes our notation. We then consider triatomic molecules, described in the vibron model by $U(4) \times U(4)$. We treat first linear molecules and then bent triatomic molecules. We close with a brief discussion of linear four-atom molecules.

$U(4)$ VIBRON MODEL - DIATOMIC MOLECULES

In the vibron model the appropriate group for a diatomic molecule is $U(4)$. It is realized in terms of four bosons, p_x, p_y, p_z and s, where the three components of $\vec{p}$ form a vector. A general Hamiltonian that conserves angular momentum and is at most quadratic in the generators of the $u(4)$ algebra has two possible group chains where we can solve for the eigenvalues analytically.[1] One is the so-called $U(3)$ limit with the group chain $U(4) \supset U(3) \supset O(3)$. This describes a harmonic oscillator. The other one is the $O(4)$ limit with the group chain $U(4) \supset O(4) \supset O(3)$ and can be shown to correspond to the Morse potential.[1] Most diatomic molecules are well described by the $O(4)$ limit and hereafter we shall concentrate on this case.

We express the Hamiltonian in terms of the Casimir invariants in the group chain $U(4) \supset O(4) \supset O(3)$ by

$$H = -AC_2(O(4)) + BC_2(O(3)). \tag{1}$$

where, of course, $C_2(O(3))$ is the square of the angular momentum. In terms of the bosons the angular momentum is given by $\vec{L} = -i\vec{p}^\dagger \times \vec{p}$. The Casimir invariant for $O(4)$, $C_2(O(4))$, is given in terms of the dipole operator $\vec{D}$ and the angular momentum operator by

$$C_2(O(4)) = \vec{D} \cdot \vec{D} + \vec{L} \cdot \vec{L}, \tag{2}$$

where

$$\vec{D} = s^\dagger \vec{p} + \vec{p}^\dagger s. \tag{3}$$

The Hamiltonian commutes with the total boson number operator $\hat{N} = s^\dagger s + \vec{p}^\dagger \cdot \vec{p}$, which is the Casimir invariant of $U(4)$.

Using group theoretical methods we find for the eigenvalues of H

$$E(v, L) = -A\sigma(\sigma + 2) + BL(L + 1), \tag{4}$$

where σ can be written in terms of the vibrational quantum number v as $\sigma = N - 2v$. When $v \ll N$, the vibrational energy reduces to a harmonic spectrum $E(v) \approx \omega v$ with frequency $\omega = 4AN$. This approximation is good if N is very large, and v is far from N. We now show how one can obtain the same large N result using the MFA. In this method one takes as variational *ansatz* a mean field or boson condensate state. In this state there are N identical bosons which are formed from a particular linear combination of the dynamical bosons. The correct combination is found variationally. The state associated with this combination is then treated as a classical mean field, the condensate, and we study quantum fluctuations around it. Because the mean field is a variational solution to the dynamics, the fluctuations must be orthogonal to that field. The variational state can be written

$$|N, \vec{r}> = \frac{1}{\sqrt{N!}}(b_c^\dagger)^N |0>.$$ (5)

We parameterize the condensate boson as

$$b_c = \frac{1}{\sqrt{1+r^2}}(s + \vec{r} \cdot \vec{p}),$$ (6)

where $\vec{r}$ is the variational parameter. Minimizing H in the condensate state gives $r = 1$, but does not fix the direction of $\vec{r}$. We are free to pick that direction, so we take $\vec{r} = \hat{z}$. This breaking of the rotational invariance of the Hamiltonian by the intrinsic state is familiar in many branches of physics. For our case it corresponds to the fact that the dipolar molecule has an intrinsic orientation, which we take here to be along z. The optimal condensate boson is then

$$b_c^\dagger = \frac{1}{\sqrt{2}}(s^\dagger + p_z^\dagger).$$ (7)

The orthogonal fluctuation bosons (FBs) are then $b_z^\dagger = \frac{1}{\sqrt{2}}(-s^\dagger + p_z^\dagger)$, $b_x^\dagger = p_x^\dagger$ and $b_y^\dagger = p_y^\dagger$. We rewrite the Hamiltonian in terms of the condensate and fluctuation bosons. We then make the following replacement as often as possible:

$$b_c^\dagger b_c \rightarrow N - b_z^\dagger b_z - b_x^\dagger b_x - b_y^\dagger b_y.$$ (8)

The remaining terms that still contain condensate bosons are then replaced by $\sqrt{N}$,

$$b_c^\dagger \rightarrow \sqrt{N}, \qquad b_c \rightarrow \sqrt{N}.$$ (9)

These rules are an algebraic equivalent of deriving the standard RPA hamiltonian in the large N-limit.[15]

Neglecting terms which contain more than two FBs as well as constant terms, we obtain for the Hamiltonian in the MFA

$$H \approx 4AN b_z^\dagger b_z - \frac{BN}{2}[(b_x^\dagger - b_x)^2 + (b_y^\dagger - b_y)^2].$$ (10)

This shows that we have a harmonic term in the b_z boson and zero modes in the b_x and b_y bosons. If we replace the boson operators by the standard combination of coordinates and momenta, the last two terms are clearly pure momentum operators. They can be shown to correspond to the mean-field limit of L_x^2 and L_y^2. The angular momentum term (the second term in (10)) has expectation value proportional to N in the condensate for any $\vec{r}$, and in the large N limit does not contribute to the energy of the condensate. It contributes only to the kinetic energy of the molecule. Its energy is in the two zero modes (Goldstone modes) associated with the breaking of rotational invariance. We note that the MFA with the prescription given in (8) and (9) gives the correct number of modes, one stretching vibration and two rotational zero modes. The four original bosons lead to only three modes because the condensate boson has been promoted to a classical variable and given its classical expectation value.[16]

The simple condensate with $r = 1$ presented here reflects our choice of a Hamiltonian in the $U(4) \supset O(4) \supset O(3)$ group chain. An alternate choice would be the dynamical symmetry group chain $U(4) \supset U(3) \supset O(3)$. The Casimir invariant of $U(3)$, $C_2(U(3))$, equals $\hat{n}_p(\hat{n}_p + 2)$, where $\hat{n}_p$ is the number operator of p-bosons. A Hamiltonian with $C_2(U(3))$ alone would lead, in MFA, to $r = 0$. Clearly a Hamiltonian containing terms in both $C_2(O(4))$ and $C_2(U(3))$ would, in MFA, lead to some r between zero and one. The eigenvalues of such a Hamiltonian cannot be derived in closed form using group theoretic methods, but are easily found in MFA. Since most diatomic molecules, although close to the $O(4)$ limit, do not have the exact dynamical symmetry, realistic Hamiltonians can have minima at r slightly different from 1. The ease of solution in MFA of the mixed Hamiltonian should make this physically attractive solution more accessible.

We have presented this treatment of diatomic molecules in the MFA largely for pedagodic purposes. (For more pedagogic detail see ref. 18.) There are many accurate and well understood methods for treating diatomic molecules to which the MFA adds little. Our interest is in extending the lessons of the MFA in the diatomic case to more complex molecular systmes.

U(4) × U(4) VIBRON MODEL: TRIATOMIC MOLECULES

Linear molecules

A triatomic molecule is described in the vibron model by the combined groups $U(4) \times U(4)$. The picture is that each $U(4)$ describes the properties of a single bond, and that terms involving operators drawn from both $U(4)$'s describe the coupling between the bonds. We first consider parameter choices for this description appropriate to linear triatomic molecules in the $U(4) \times U(4) \supset O(4) \times O(4) \supset O_{12}(4) \supset O_{12}(3)$ group chain. We will return to bent molecules below. The $U(4)$'s are realized with two sets of commuting bosons, $s_1, \vec{p}_1$ and $s_2, \vec{p}_2$. The Hamiltonian for the triatomic molecule is taken to be

$$
\begin{aligned}
H = {} & -A_1 C_2(O_1(4)) + B_1 C_2(O_1(3)) - A_2 C_2(O_2(4)) + B_2 C_2(O_1(3)) - \\
& A_{12} C_2(O_{12}(4)) + \lambda \mathcal{M}_{12} + B C_2(O_{12}(3)),
\end{aligned} \tag{11}
$$

where the Casimir invariants of the combined $O(4)$ and $O(3)$ are given in terms of the combined dipole operator $\vec{D}_{12} = \vec{D}_1 + \vec{D}_2$ and total angular momentum operator $\vec{L}_{12} = \vec{L}_1 + \vec{L}_2$. In terms of the bosons, the Majorana operator $\mathcal{M}_{12}$ is given by

$$
\mathcal{M}_{12} = (\vec{p}_1^\dagger s_2^\dagger - s_1^\dagger \vec{p}_2^\dagger) \cdot (\vec{p}_1 s_2 - s_1 \vec{p}_2) + (\vec{p}_1^\dagger \times \vec{p}_2^\dagger) \cdot (\vec{p}_1 \times \vec{p}_2). \tag{12}
$$

Note that the Hamiltonian commutes separately with the number operator for each boson type, $\hat{N}_i = s_i^\dagger s_i + \vec{p}_i^\dagger \cdot \vec{p}_i$. We prefer to rewrite the $O(3)$ parts of the Hamiltonian as

$$
\begin{aligned}
H = {} & -A_1 C_2(O_1(4)) - A_2 C_2(O_2(4)) - A_{12} C_2(O_{12}(4)) + \lambda \mathcal{M}_{12} + \\
& B_L L^2 + B_\Delta \Delta^2 + B_{L\Delta} \vec{L} \cdot \vec{\Delta},
\end{aligned} \tag{13}
$$

where $\vec{L} = \vec{L}_1 + \vec{L}_2$ and $\vec{\Delta} = \sqrt{\frac{N_2}{N_1}} \vec{L}_1 - \sqrt{\frac{N_1}{N_2}} \vec{L}_2$. The operator created by $\vec{\Delta}$ is not a spin, but rather is the kinetic part of the bending mode. It is analogous to a similar operator in the description of the scissors mode in nuclear physics.[19]

We now solve this Hamiltonian using the MFA. Because each boson number is separately conserved, we introduce a condensate boson for each boson type by

$$b_{ci} = \frac{1}{\sqrt{1 + r_i^2}}(s_i + \vec{r}_i \cdot \vec{p}_i), \tag{14}$$

and use a product of mean field states as variational estimate

$$|N_1, \vec{r}_1; N_2, \vec{r}_2\rangle = \frac{1}{\sqrt{N_1! N_2!}}((b_{c1}^\dagger)^{N_1})((b_{c2}^\dagger)^{N_2})|0\rangle, \tag{15}$$

where $\vec{r}_1$ and $\vec{r}_2$ are the variational parameters. We take the expectation value of the Hamiltonian in the condensate and find (noting that the angular momentum operators always have zero expectation value in the condensate)

$$\langle N_1, \vec{r}_1; N_2, \vec{r}_2 | H | N_1, \vec{r}_1; N_2, \vec{r}_2 \rangle =$$
$$-(A_1 + A_{12})\frac{4N_1^2 r_1^2}{(1 + r_1^2)^2} - (A_2 + A_{12})\frac{4N_2^2 r_2^2}{(1 + r_2^2)^2} - 8A_{12}\frac{N_1 N_2 \vec{r}_1 \cdot \vec{r}_2}{(1 + r_1^2)(1 + r_2^2)}$$
$$+\lambda\frac{N_1 N_2}{(1 + r_1^2)(1 + r_2^2)}((\vec{r}_1 - \vec{r}_2)^2 + (\vec{r}_1 \times \vec{r}_2)^2). \tag{16}$$

The expectation value of the Hamiltonian in the condensate depends on the angle between $\vec{r}_1$ and $\vec{r}_2$, which we call θ. Solving for the minimum energy with respect to that angle we find two roots, $\sin\theta = 0$ and $\cos\theta = -(4A_{12} + \lambda)/\lambda r_1 r_2$. For a linear molecule one typically uses[4] parameters where $|-(4A_{12} + \lambda)/\lambda r_1 r_2| > 1$, which shows that we should only have a solution $\theta = 0$ or $\theta = \pi$, which corresponds to a linear molecule. One can also show that for the parameters considered here, the mean field minimum comes at $r_1 = r_2 = 1$ and $\theta = \pi$. To make the r's different requires some combination of a large λ (strong Majorana term) and additional terms proportional to $C_2(U(3))$ in one or both of the bosons. We do not introduce such terms at this point.

To study quantum fluctuations around the condensate, we introduce the orthogonal fluctuation bosons. As before we take the molecular symmetry axis along z. For the $r = 1$ case, the FBs are just as in the single $U(4)$ case. Explicitly, $b_{zi} = \frac{1}{\sqrt{2}}(-s_i + p_{zi})$, $b_{xi} = p_{xi}$ and $b_{yi} = p_{yi}$. We rewrite the Hamiltonian (up to constant terms) in terms of the condensate bosons and the FBs and make the substitutions of (8) and (9) for each boson type to find

$$\begin{aligned} H = {} & [4(A_1 + A_{12})N_1 + (4A_{12} + \lambda)N_2] \, b_{z1}^\dagger b_{z1} + [4(A_2 + A_{12})N_2 + (4A_{12} + \lambda)N_1] \, b_{z2}^\dagger b_{z2} \\ & - \lambda\sqrt{N_1 N_2}(b_{z1}^\dagger b_{z2} + b_{z2}^\dagger b_{z1}) \\ & + (2A_{12} + \lambda)(N_1 + N_2) \left[b_{xB}^\dagger b_{xB} + b_{yB}^\dagger b_{yB} \right] + B_\Delta \left[\Delta_x^2 + \Delta_y^2 \right] \\ & + B_L \left[L_x^2 + L_y^2 \right] + B_{L\Delta} \left[L_x \Delta_x + L_y \Delta_y \right], \end{aligned} \tag{17}$$

where we have introduced the bosons that create and destroy the bending modes. In terms of the b_{xi} they are given by

$$b_{xB}^\dagger = \frac{\sqrt{N_2} b_{x1}^\dagger - \sqrt{N_1} b_{x2}^\dagger}{\sqrt{N_1 + N_2}}, \tag{18}$$

with a similar expression for the y bending mode. These are closely related to Δ:

$$\Delta_x = \frac{i}{\sqrt{2}} \left(b_{xB}^\dagger - b_{xB} \right). \tag{19}$$

Finally we also introduce the intrinsic components of the angular momentum

$$L_x = \frac{i}{\sqrt{2}} \left[\sqrt{N_1}(b_{x1}^\dagger - b_{x1}) + \sqrt{N_2}(b_{x2}^\dagger - b_{x2}) \right], \tag{20}$$

and similarly for L_y. Even though the rotational and the vibrational modes are not decoupled in the Hamiltonian (17), we can do so by making a canonical transformation.[18] Clearly there are two degenerate bending modes, one in the yz-plane, generated by $b_{xB}^\dagger$ and another in the xz plane generated by $b_{yB}^\dagger$. The explicit form of the bending modes is completely determined from the geometry. The bending modes have frequency $(2A_{12} + \lambda)(N_1 + N_2)$, so that they are generated from a combination of the mixed $O(4)$ term (A_{12}) and the Majorana term. In the same part of the Hamiltonian we also find the rotational zero modes, corresponding to a rotation around the x or y axis. There is a dynamical coupling between the bending and rotational modes, $B_{L\Delta}$. This term is well known in the mean-field limit of the nuclear analog of the vibron model.[19] We also see that there are two stretching modes carried by the b_z bosons. Only the Majorana term mixes the stretching modes of the two types. The resulting two by two matrix can be diagonalized. Thus in total we have six harmonic modes of a linear triatomic molecule.

In order to show the power of the discussion given above let us compare our results for linear triatomic molecules with those obtained in ref. 4. The Hamiltonian used in that paper contains terms quadratic in the Casimir operators, beyond the terms used in (13). Fortunately these terms are constructed in such a way that the higher order terms vanish in the harmonic limit.

In Table 1 we compare the harmonic energies with the exact vibrational levels using the parameters in Table I from Fit III in ref. 4. Our boson expansion results[18] agree very well with the exact calculation. To obtain this agreement we have included the effects of Fermi resosnances in the two isotopic CO_2 molecules. We use the coupling defined in Eq. (2.14) of ref. 4, The agreement between the exact and MFA results is at least as good as one would expect for a $1/N$ expansion, and the method is easily extended to include transition strengths.

Bent molecules

We want to construct a geometrically correct vibron model Hamiltonian for bent triatomic molecules. The existence of a bent condensate requires three rotational zero modes. We go one step further, and require that we produce a condensate state with the correct bond angle. A convenient way to do this is to include the Casimir and Majorana operators of $U(3)$ subgroups: $U_1(3), U_2(3)$ and $U_{12}(3)$. They are effectively represented by $\hat{n}_{p1}^2, \hat{n}_{p2}^2, \hat{n}_{p1}\hat{n}_{p2}$ and $(\vec{p}_1^\dagger \times \vec{p}_2^\dagger) \cdot (\vec{p}_1 \times \vec{p}_2)$. We also include rotational terms involving only the angular momentum operators: $\vec{L}_1^2, \vec{L}_2^2$ and $\vec{L}_1 \cdot \vec{L}_2$.

Since we are only interested in the MFA or the harmonic approximation of the vibron model, we limit the expansion in terms of the generators of the group $U_1(4) \times U_2(4)$ here to quadratic order. From the considerations outlined above, we take the Hamiltonian for a bent triatomic molecule to be:

$$\begin{aligned} H =\ & A_1(\vec{L}_1 \cdot \vec{L}_1 + \vec{D}_1 \cdot \vec{D}_1) + A_2(\vec{L}_2 \cdot \vec{L}_2 + \vec{D}_2 \cdot \vec{D}_2) + A_{12}(\vec{L}_1 \cdot \vec{L}_2 + \vec{D}_1 \cdot \vec{D}_2) \\ & + \lambda(s_1^\dagger \vec{p}_2^\dagger - s_2^\dagger \vec{p}_1^\dagger) \cdot (s_1 \vec{p}_2 - s_2 \vec{p}_1) + \mu(\vec{p}_1^\dagger \times \vec{p}_2^\dagger) \cdot (\vec{p}_1 \times \vec{p}_2) \\ & + h_1 n_{p1}^2 + h_2 n_{p2}^2 + h_{12} n_{p1} n_{p2} + B_1 L_1^2 + B_2 L_2^2 + B_{12} \vec{L}_1 \cdot \vec{L}_2. \end{aligned} \tag{21}$$

The ground state condensate is still of the form (15). The condensate bosons allow us to study the geometry of a bent molecule since the vectors $\vec{r}_1, \vec{r}_2$ are assumed to give the directions of the two bonds carrying the $U(4)$ groups. We take the angle between

Table 1. A comparison of the exact vibron model calculation to the harmonic approximation including Fermi resonances: linear triatomic molecules.

HCN ($N_1 = 140, N_2 = 47$)			
$v_1 v_2^{l_2} v_3$	exact	MFA	error
$01^1 0$	712.4	707.8	0.65%
$10^0 0$	2095.8	2092.1	0.18%
$00^0 1$	3312.6	3318.0	0.16%
OCS ($N_1 = 190, N_2 = 159$)			
$v_1 v_2^{l_2} v_3$	exact	MFA	error
$01^1 0$	519.4	518.3	0.21%
$10^0 0$	860.2	861.8	0.19%
$00^0 1$	2062.2	2064.9	0.13%
$v_1 v_2^{l_2} v_3$	exact	MFA	error
$01^1 0$	588.9	586.9	0.34%
$10^0 0$	1284.9	1270.7	1.1%
$00^0 1$	2223.6	2224.9	0.06%
$C^{12} O_2$ ($N_1 = N_2 = 153$)			
$v_1 v_2^{l_2} v_3$	exact	MFA	error
$01^1 0$	666.8	664.0	0.42%
$10^0 0$	1388.4	1384.0	0.32%
$00^0 1$	2348.3	2350.7	0.10%
$C^{13} O_2$ ($N_1 = N_2 = 154$)			
$v_1 v_2^{l_2} v_3$	exact	MFA	error
$01^1 0$	648.6	645.0	0.56%
$10^0 0$	1370.1	1362.7	0.53%
$00^0 1$	2283.2	2284.7	0.07%

the two bonds of the molecule to be equal to the angle between $\vec{r}_1$ and $\vec{r}_2$. Due to the rotational invariance of the Hamiltonian, its expectation value in the condensate state only depends on the lengths of $\vec{r}_1$ and $\vec{r}_2$ and the angle θ between them.

It is much easier to analyze the MFA Hamiltonian expressed in terms of the fluctuation bosons than the original Hamiltonian since the fluctuation bosons become just a set of coupled harmonic oscillators in the mean field limit. Normal mode frequencies can be found by a simple diagonalization of the RPA matrices which can be read off from the MFA Hamiltonian. To further simplify the discussion, we specialize to the study of bent $X_2 Y$ molecules for which we can obtain analytical solutions in the MFA with the aid of the C_{2v} symmetry of these molecules.[20,21] We choose the identical XY bonds to be the carriers of the $U(4)$ groups and because of the point group symmetry we can set $A_1 = A_2 = A, h_1 = h_2 = h, B_1 = B_2 = B, N_1 = N_2 = N$ and $r_1 = r_2 = r$ in the following.

For bent $X_2 Y$ molecules, the extremum conditions become

$$[4A + 2A_{12} \cos \theta + \lambda(1 - \cos \theta)](1 - r^2) + 2hr^2 + h_{12}r^2 + \mu r^2 \sin^2 \theta = 0, \qquad (22)$$

and

$$\sin \theta(-2A_{12} + \lambda + \mu r^2 \cos \theta) = 0. \qquad (23)$$

From these equations, we can solve for the condensate parameters r and θ. As we can see, if $h = h_{12} = 0$, there is a solution with $\theta = 0$ (or π) and $r = 1$ which corresponds to a linear molecule. Therefore, the $U(3)$ terms ($h \neq 0, h_{12} \neq 0$) are essential for bent geometry.

The MFA Hamiltonian for a bent X_2Y molecule is long, but straightforward. All terms are quadratic in the fluctuation bosons and hence the full expression simply describes a set of coupled oscillators. The mode couplings correspond to symmetric stretching and bending modes, an antisymmetric stretching mode coupled to rotations about the normal to the molecular plane and two rotational modes about axis in the molecular plane. The coupled modes are easily diagonalized to yield the three normal modes; symmetric stretching, antisymmetric stretching and bending, and the three rotational zero modes.

We have used this picture to describe the water molecule.[21] There are more than enough (one extra) parameters to allow a fit of the molecule bend angle and the frequencies of the low lying vibrational and rotational states of the molecule. Note this means fitting both vibrational frequencies and moments of inertia. If we include the low lying transition strengths, the fit is over constrained. Nontheless a fit is possible, indicating that the picture is sensible. The quality of this fit and its comparison, where possible, with a full vibron calculation[3] is shown in Table 2.

Table 2. Comparison of fits to the spectrum and transitions of water.

Physical quantity	units	experiment[22]	present	Iachello & Oss[3]
ω_b	cm^{-1}	1595	1595	1595
ω_s	cm^{-1}	3657	3657	3657
ω_a	cm^{-1}	3756	3756	3756
I_x	g cm^{-1}	1.928[-40]	1.928[-40]	
I_y	g cm^{-1}	3.017[-40]	3.017[-40]	
I_z	g cm^{-1}	1.009[-40]	1.009[-40]	
D_0	e Å	0.38	0.38	
$I_{(000)\to(010)}$	cm/molecule	1040	1059	1040
$I_{(000)\to(100)}$	cm/molecule	49.5	292	51.8
$I_{(000)\to(001)}$	cm/molecule	720	709	732.5

$U(4) \times U(4) \times U(4)$ VIBRON MODEL: LINEAR FOUR ATOM MOLECULES

For four atom molecules the appropriate group structure is $U(4) \times U(4) \times U(4)$. Again the picture is that each $U(4)$ describes the properties of a single bond, and that bond coupling is realized in terms of operators drawn from different $U(4)$'s. The $U(4)$'s are realized in terms of three sets of commuting bosons, $s_i, \vec{p}_i$. In this work we will only consider parameters appropriate to linear four atom molecules of the structure ABBA, e.g., acetylene (HCCH). In terms of the boson labels we take $i = 1, 2$ for the two AB bonds and $i = 3$ for the BB bond. Our Hamiltonian will commute with the number operator for each boson type, and the symmetry of the molecule requires $N_1 = N_2$. We take for our Hamiltonian

$$
\begin{aligned}
H = & -A_1 C_2(O_1(4)) - A_2 C_2(O_2(4)) - A_3 C_2(O_3(4)) - A_{12} C_2(O_{12}(4)) \\
& -A_{123} C_2(O_{123}(4)) + \lambda M_{12} + B C_2(O_{123}(3)),
\end{aligned} \tag{24}
$$

where the symmetry of the molecule will require that $A_1 = A_2$ and where $C_2(O_{123}(3)) = L^2$ with $\vec{L} = \vec{L}_1 + \vec{L}_2 + \vec{L}_3$. We take parameters such that the molecule is linear and

Table 3. A comparison of the exact vibron model calculation to the harmonic approximation:linear four atom molecules.

HCCH ($N_1 = N_2 = 43, N_3 = 137$)			
$v_1 v_2 v_3; v_4^{l_4} v_5^{l_5}$	exact	MFA	error
$100; 0^0 0^0$	3286.93	3286.73	0.00%
$001; 0^0 0^0$	3366.88	3368.42	0.05%
$010; 0^0 0^0$	1975.80	1983.68	0.40%
$000; 1^1 0^0$	617.12	611.20	0.96%
$000; 0^0 1^1$	724.50	717.21	1.00%
DCCD ($N_1 = N_2 = 61, N_3 = 137$)			
$v_1 v_2 v_3; v_4^{l_4} v_5^{l_5}$	exact	MFA	error
$100; 0^0 0^0$	2435.58	2430.44	0.21%
$001; 0^0 0^0$	2706.35	2704.44	0.07%
$010; 0^0 0^0$	1769.57	1769.18	0.02%
$000; 1^1 0^0$	513.55	510.80	0.05%
$000; 0^0 1^1$	534.08	528.79	0.99%

the condensate has $r_1 = r_2 = r_3 = 1$. Thus we have not included any Casimirs from $U(3)$ in our Hamiltonian. As usual we take the axis of the molecule in the z direction. We introduce the orthogonal fluctuation bosons and find

$$C_2(O_i(4)) \rightarrow N_i^2 - 4N_i b_{zi}^\dagger b_{zi}, \tag{25}$$

$$
\begin{aligned}
C_2(O_{12}(4)) \rightarrow{} & C_2(O_1(4)) + C_2(O_2(4)) \\
& -(4N_1 + 2N_2)b_{z1}^\dagger b_{z1} - (4N_2 + 2N_1)b_{z2}^\dagger b_{z2} \\
& -\sqrt{N_1 + N_2}(b_{xB}^\dagger b_{xB} + b_{yB}^\dagger b_{yB}),
\end{aligned} \tag{26}
$$

(in terms of the bending bosons of (18)),

$$
\begin{aligned}
\mathcal{M}_{12} \rightarrow{} & N_2 b_{z1}^\dagger b_{z1} - N_1 b_{z2}^\dagger b_{z2} - \sqrt{N_1 N_2}(b_{z1}^\dagger b_{z2} + b_{z2}^\dagger b_{z1}) \\
& +(N_1 + N_2)(b_{xB}^\dagger b_{xB} + b_{yB}^\dagger B_{yB}),
\end{aligned} \tag{27}
$$

and finally for $C_2(O_{123}(4))$ in what is an obvious notation

$$C_{123} = C_{12} + C_{23} + C_{13} - C_1 - C_2 - C_3, \tag{28}$$

where C_{23} (C_{13}) can be obtained from C_{12} by making the proper replacements. For the rotational energy we have

$$C_2(O_{123}(3)) \rightarrow L_x^2 + L_y^2, \tag{29}$$

with

$$L_x = \frac{i}{\sqrt{2}}\left[\sqrt{N_1}(b_{x1}^\dagger - b_{x1}) + \sqrt{N_2}(b_{x2}^\dagger - b_{x2}) + \sqrt{N_3}(b_{x3}^\dagger - b_{x3})\right]. \tag{30}$$

and similarly for L_y.

From the MFA, we find three stretching modes, two doubly degenerate bending modes and two zero modes. In Table 3, we compare our results[18] with the exact ones given in ref. 5 (the parameters are taken from Table II in ref. 5). Again we find excellent agreement.

DISCUSSION AND SUMMARY

Mean field methods have been employed to study algebraic models by a number of groups.[10-17] There are many complementary routes for obtaining classical geometric insight from these algebraic methods. In this work we apply the mean field approximation (MFA) to the vibron model, an algebraic model for molecular dynamics. We show that the MFA gives a very accurate description of the molecular spectra and provides a closer geometrical pictures of the molecular modes. The success of the MFA is ensured by the large N values in vibron models (on the order of 100).

The MFA in the large N limit gives promise of applications to many problems, from electron molecule scattering to the description of large complex molecules for which the geometry is known but for which the full algebraic method is prohibatively complex.

ACKNOWLEDGEMENTS

This work is supported in part by grants from the National Science Foundation and by the Department of Energy.

REFERENCES

1. F. Iachello, Chem. Phys Lett. **78**, 581 (1981); F. Iachello and R.D. Levine, J. Chem. Phys. **77**, 3046 (1982).

2. O. S. van Roosmalen, A. E. L. Dieperink and F. Iachello, Chem. Phys. Lett. **85**, 32 (1985); O. S. van Roosmalen, F. Iachello, R. D. Levine and A. E. L. Dieperink, J. Chem. Phys. **79**, 2515 (1983).

3. F. Iachello and S. Oss, J. Mol. Spectrosc. **142**, 85 (1990).

4. F. Iachello and S. Oss and R. Lemus, J. Mol. Spectrosc. **146**, 56 (1991).

5. F. Iachello and S. Oss, J. Mol. Spectrosc. **149**, 132 (1991).

6. F. Iachello and S. Oss, Phys. Rev. Lett. **66**, 2976 (1991).

7. R. Bijker and R.D. Amado, Phys. Rev. **A46**, 1388 (1992); R. Bijker and R.D. Amado, Phys. Rev. **A34**, 71 (1986); R. Bijker, R.D. Amado and D.A. Sparrow, Phys. Rev. **A33**, 871 (1986).

8. A. Mengoni and T. Shirai, Phys. Rev. **A44**, 7258 (1991); Y. Alhassid and B. Shao, Phys. Rev. **A** (to be published), 1992.

9. R. D. Amado, J. A. McNeil and D. A. Sparrow, Phys. Rev. **C25**, 13 (1982).

10. R. Gilmore and D. H. Feng, Nucl. Phys. **A301**, 189 (1978); J. N. Ginocchio and M. W. Kirson, Phys. Rev. Lett. **44**, 1744 (1980), Nucl. Phys. **A350**, 31 (1980); A. Dieperink, O. Scholten and F. Iachello, Phys. Rev. Lett. **44**, 1747 (1980).

11. S. Levit and U. Smilansky, Nucl. Phys. **A389**, 56 (1982).

12. O. S. van Roosmalen and A. E. L. Dieperink, Ann. Phys. **139**, 198 (1982).

13. I. Benjamin and R. D. Levine, Chem. Phys. Lett. **117**, 314 (1985).

14. O. S. van Roosmalen, I. Benjamin and R. D. Levine, J. Chem. Phys. **81**, 5986 (1984).

15. J. Dukelsky, G. G. Dussel, R. P. J. Perazzo, S. L. Reich and H. M. Sofía, Nucl. Phys. **A425**, 93 (1984).

16. A. Leviatan and M. Kirson, Ann. Phys. **188**, 142 (1988); A. Leviatan, J. Chem. Phys. **91**, 1706 (1989).

17. F. Iachello, A. Leviatan and A. Mengoni, J. Chem. Phys. **95**, 1449 (1991).

18. B. Shao, N. R. Walet and R. D. Amado, Phys. Rev. A, in press.

19. N. R. Walet, P. J. Brussaard and A. E. L. Dieperink, Phys. Lett. **B163**, 1 (1985).

20. G. Herzberg, Molecular Spectra and Molecular Structure I. (Van Nostrand, New York, 1950).

21. B. Shao, N. R. Walet and R. D. Amado, submitted to Phys. Rev. A, (1992).

22. R. T. Lawton and M. S. Child, Mol. Phys. **40**,773 (1980).

SPECTRUM GENERATING q-ALGEBRAS FOR ANYONS [1]

Allan I. Solomon
Faculty of Mathematics
The Open University
Milton Keynes MK7 6AA, UK

and

Joseph L. Birman
Department of Physics
City College of the City University of New York
New York 10031, USA

To Franco Iachello, who has shown us that experiment is the motive force of theory.

Abstract

We review some of the uses of the spectrum generating algebra approach using conventional Lie algebras applied to exactly solvable many-body models. We compare these to the possible applications of spectrum generating q-algebras. We describe a generalized deformation of the canonical commmutation relations (CCR), which reduce both to the CCR for standard q-bosons, and to the Canonical Anti-commutation Relations (CAR) for fermions in appropriate limits. By a choice of the deformation, we show that the new "qummutation relations" define bosonic-type particles having finite number occupancy (anyons) and relate these to standard descriptions of anyons, and use them to construct a model hamiltonian for interacting anyons which has the quantum group $SU_q(2)$ as dynamical group.

INTRODUCTION

The discovery of the spectrum generating algebra (SGA) in the Sixties [1] provided another theoretical tool for the analysis of models in quantum mechanics, in particular those of condensed matter physics. The original SGAs were the Lie algebras of non-compact Lie groups; their infinite-dimensional unitary representations gave an elegant description of the infinite spectrum of models such as the hydrogen atom [2], and superfluid helium [3]. Subsequently, compact groups have been used in many-electron

---

[1]Talk contributed to SYMMETRIES IN SCIENCE VII: Spectrum Generating Algebras and Dynamic Symmetries in Physics, on the occasion of the fiftieth birthday of Professor Franco Iachello.

physics [4], especially models involving several coexisting phases [5]. Here, the infinite character of the spectrum is provided by the appropriate algebra being an infinite sum of (identical) compact algebras. The Interacting Boson Model of Iachello and associates provided an important application of these algebras to nuclear physics [6]. More recently, these methods have been developed beyond the Lie algebra framework to that of Lie superalgebras, enabling both bosons and fermions to be incorporated in the same model, or perhaps more relevantly, allowing both single fermions and fermion pairs to be treated within the same model. This approach has been used to advantage in nuclear physics[7], and in superconductivity, treating extensions of the conventional BCS theory[8] and the Hubbard model [9].

The most recent development in this field has been the advent of the so-called quantum groups [10]. These mathematical systems, although not groups, may be thought of as "the next best thing"; they are essentially deformations of Lie groups which retain some group properties (product and inverse) and reduce in a well-defined limit to the classical Lie groups. Interesting in their own right as complex mathematical systems, they also allow us to solve models which are extensions of previously solved systems. One may mention applications to statistical mechanics [11], spin models [12], nuclear [13] and molecular [14] physics. It is as yet not clear that these "applications" have very much to do with reality; however, it is always possible that the symmetry of the real world is not exactly that of Lie group theory, and it is at the very least an interesting intellectual exercise to evaluate the deviations obtained from assuming a less exact symmetry.

Since the pioneering work of Drinfeld and others [10, 15, 16] the theory of quantum groups has been rapidly developing. The introduction of bosonic realizations (q-bosons) [18, 19] has permitted applications to be made to generalized boson systemsincluding the Jaynes-Cummings model [21], and other models in quantum optics [22, 23].

An alternative, and apparently unrelated, development in condensed matter physics, came with the earlier realization that elementary excitations - quasiparticles - with fractional statistics can occur. These "anyons" [25] have been identified as existing in the fractional quantum Hall effect (FQHE) [24] following Laughlin's work [26] and in certain soliton theories based on non-linear σ models with Hopf terms, as well as in $(2+1)$ dimensional relativistic field theories. One characteristic property of these anyons is that of finite occupancy; that is, like fermions only a finite number of states are available for each mode, this number being greater than two, associated with fermion occupancy.

In this note we shall relate a generalized deformation of the boson operator to the usual description of anyons.

1. STANDARD q BOSONS

We first note that the naive adaptation of the quantum boson equations, as has been proposed, is inadequate to give a non-trivial quantum fermionic structure.

Quantum-deformed single boson equations have been presented in two standard fashions.

The first examples of q-deformed boson operators were given by Arik and Coon [17]. They introduced creation (*resp* annihilation) operators $a^\dagger(a)$ obeying:

$$aa^\dagger - qa^\dagger a = 1. \tag{1}$$

A putative modification to describe "q-fermions" would naively be:

$$aa^\dagger + qa^\dagger a = 1. \tag{2}$$

Here q is the deformation parameter, often taken as $q \equiv e^s$, so that when $s \to 0$ then $q \to 1$, and the usual CAR are recovered.

We may introduce a number operator N such that $[N, a] = -a$. As usual in physical applications we define a normalized vacuum state by $a|0> = 0$. Number states $|n>$, obtained by the repeated action of $a^\dagger$ on the vacuum (and normalizing) may be shown to satisfy $N|n> = n|n>$. $N(a, a^\dagger)$ is not yet specified. The constraints on q are that it should be real, which is easily seen by taking the hermitian conjugate of Equation (2) and subtracting; and that $|q| < 1$ which arises from the normalizability of $|2>$. Thus changing the sign of q does not introduce an essentially different statistics and the fermionic property of finite occupancy is not a consequence of the equation with negative q.

The more recent proposal [18, 19] introduced creation (*resp* annihilation) operators $a^\dagger(a)$ satisfying:

$$aa^\dagger - q\, a^\dagger a = q^{-N}. \tag{3}$$

A modification suggested to describe "q-fermions" was:

$$aa^\dagger + qa^\dagger a = q^N \tag{4}$$

A "one" particle state is $|1> = a^\dagger|0>$, and is already normalized. Suppose we wish to create the state $|2>$. We take $\lambda_2|2> = a^\dagger|1>$, where λ_2 is a normalization constant. Then:

$$
\begin{aligned}
|\lambda_2|^2 < 2|2> &= <1|aa^\dagger|1> \\
&= <1|q^N - qa^\dagger a|1> \\
&= <1|q^N - q|1> \\
&= (q - q) < 1|1> = 0
\end{aligned}
\tag{5}
$$

where the third equality follows by repeatedly using the QDCAR (21). It follows that in this case only $\{0, 1\}$ eigenstates of N exist, just as in the conventional case.

Conversely, it is easy to show that conventional fermions obeying the usual CAR

$$aa^\dagger + a^\dagger a = 1, \quad N \equiv a^\dagger a$$

satisfy Equation (4) for any q; this follows by expanding $\exp sN$ and using the idempotency of N. Thus Equation (4) is equivalent to the usual CAR.

It is clear that we may circumvent the vanishing norm of Equation (5) by choosing

$$aa^\dagger + qa^\dagger a = q^{-N} \tag{6}$$

as our q-fermion equation [20]. This however does not lead to finite occupancy (for real q).

To avoid these difficulties we propose a more general q-deformed algebra (i.e. QDCAR) with a non-trivial Fock space, and non-standard statistics.

In order to emphasize the advantage of the algebraic approach, we recall some consequences of the algebraic formalism in the conventional Lie algebra setting.

2. CONVENTIONAL SPECTRUM GENERATING ALGEBRAS

A general approach to the use of an SGA is to assume that we have a representation for which the hamiltonian is an element of a Lie algebra $\mathcal{L}$. This arises in exactly-solvable models in quantum mechanics when, for example, the hamiltonian is expressed as a quadratic form in boson or fermion operators. Examples include:

(i)	Hydrogen Atom	$\mathcal{L} = so(4, 2)$	[27]
(ii)	Harmonic Oscillator (3D)	$\mathcal{L} = su(3, 1)$	[28]
(iii)	Superfluid Helium Four	$\mathcal{L} = su(1, 1)$	[3]

(iv)	Superfluid Helium Three	$\mathcal{L} = so(5)$	[4]
(v)	Superconductivity	$\mathcal{L} = su(2)$	[29]
(vi)	X-Y Model	$\mathcal{L} = so(n)$	[30]
(vii)	Charge Density Waves	$\mathcal{L} = u(2)$	[31]
(viii)	Spin Density Waves	$\mathcal{L} = u(1) \oplus so(5)$	[32]
(ix)	Quantum Optics (coherent states)	$\mathcal{L} = hw$ (Heisenberg-Weyl algebra)	
(ix)	Quantum Optics (squeezed states)	$\mathcal{L} = su(1,1) + hw$ (semi-direct sum).	

The advantage of the SGA approach is threefold:

(1) The *spectrum* is obtained by the process of diagonalization, which in the many-particle context corresponds to a Bogoliubov transformation. This is implemented by an automorphism of the algebra, usually inner and calculated by an adjoint action within the algebra; but sometimes outer. Schematically, we have

$$H \mapsto U H U^\dagger \equiv H_D$$

where U is a unitary representative of an element of the Lie group of the algebra.

(2) The eigenstates of H are given by $\psi = U^\dagger \psi_0$ where ψ_0 is an eigenstate of H_D. These are the *coherent states* of the system. A paradigm would be the coherent states of the Heisenberg-Weyl Group (the usual coherent states of quantum optics) where

$$U = \exp(\xi a - \xi^* a^\dagger). \tag{7}$$

We may apply the Campbell-Baker-Hausdorff Theorem or, equivalently, use normal ordering, to obtain a normalized state purely in terms of the creation operator.

(3) For a condensed many-body system, the *order parameters* are given by the non-Cartan elements $[b_\pm, b_\pm]$ of a Cartan basis ($b_\pm$ are the Borel subalgebras), chosen such that the Cartan elements h are conserved operators of the non-condensed phase [33].

One of the goals of the application of the so-called quantum algebras as SGA's should be to see to what extent the above features carry over to the q-algebra case.

(Q1) The most problematic is the first. In fact, one can show that it is impossible to implement a direct analogue of the Bogoliubov transformation in the case of $SU_q(1,1)$ over the complex numbers. However, one can obtain an analogue by introducing quantum plane variables *á la Manin*. Consider two elements α and γ as defined by Woronowicz [34], satisfying the following commutation relations:

$$\begin{aligned}
\alpha\gamma &= \mu\gamma\alpha \\
\alpha\gamma^* &= \mu\gamma^*\alpha \\
\gamma\gamma^* &= \gamma^*\gamma \\
\alpha^*\alpha - \gamma^*\gamma &= 1 \\
\alpha\alpha^* - \mu^2\gamma^*\gamma &= 1.
\end{aligned} \tag{8}$$

We now introduce a conjugation $A \mapsto \check{A}$ defined by its effect on

- c-numbers $c \mapsto c^*$, (complex conjugation)

- q-numbers (quantum plane) $\tilde{\alpha} = \alpha^*$ $\tilde{\alpha^*} = \alpha$ $\tilde{\gamma} = \mu\gamma^*$ $\tilde{\gamma^*} = \frac{1}{\mu}\gamma$

- operators $\check{A} = q^{\frac{1}{2}(N^2-N)} A^\dagger q^{-\frac{1}{2}(N^2-N)}$ (q real).

Under this transformation, $\check{\check{A}} = A$, $\check{AB} = \check{B}\check{A}$; and the boson a satisfying

$$aa^\dagger - qa^\dagger a = q^{-N}$$

maps to $\tilde{a}$, with the pair $a, \tilde{a}$ satisfying

$$a\tilde{a} - \mu\tilde{a}a = 1. \tag{9}$$

with $\mu = q^2$. The two-dimensional fundamental representation of $SU_\mu(1,1)$ is given by

$$u = \begin{bmatrix} \alpha & \mu\gamma^* \\ \gamma & \alpha^* \end{bmatrix} \tag{10}$$

and u satisfies $uJ\tilde{u} = J$ where

$$J = \begin{bmatrix} 1 & 0 \\ 0 & -\mu \end{bmatrix}$$

The transformation

$$[a, \tilde{a}] \mapsto [a, \tilde{a}]u \tag{11}$$

is an automorphism which preserves Equation (9), and thus gives a direct analogue of the Bogoliubov transformation [35].

(Q2) The classical q-analogue of the exponential function is given by [36]

$$E_q(x) = \sum_{n=0}^{\infty} \frac{x^r}{[r]_q!}. \tag{12}$$

The symbol $[r]_q!$ is defined by $[r]_q! = [r]_q[r-1]_q[r-2]_q \cdots [1]_q$ where

$$[r]_q = (q^r - 1)/(q - 1) \tag{13}$$

Using the classical result of Jackson[36]

$$E_q(x)E_{1/q}(-x) = 1 \tag{14}$$

it is possible to define an analogue of the unitary property of (2) above. Aternatively, a q-analogue of the Campbell-Baker-Hausdorff Theorem [37] is also available in this case, as well as an appropriate form of "normal-ordering".

(Q3) The third advantage of the SGA approach carries over without difficulty, since the essential equations required to ensure that the order parameter properties hold

$$[h, X_i{}^\pm] = \pm\alpha_i(h)X_i{}^\pm \tag{15}$$

remain true in the q-algebra case [10].

3. GENERAL q-DEFORMED BOSONS: One-Mode Case

We modify Equation 1 as follows:

$$aa^\dagger - f_q(N)a^\dagger a = 1. \tag{16}$$

Here $f_q(N)$ is a real operator-valued function of the number operator N which also depends on the q-deformation parameter with the property that $f_q(N) \to 1$ as $q \to 1$. As usual, we first define a vacuum state $|0>$ by $a|0>= 0$, and the one-particle state by $a^\dagger|0>= |1>$. Both $|0>$ and $|1>$ are normalized. Dropping the q-suffix on the function f, whose dependence on q is now implicit, we find generally that the normalized n-particle state is given by

$$|n>= \frac{(a^\dagger)^n|0>}{(F(n)!)^{1/2}} \tag{17}$$

where the function $F(N)$ satisfies

$$F(n) = 1 + f(n-1)F(n-1) \tag{18}$$

and is defined recursively with $F(0) = 0, F(1) = 1; F(2) = 1 + f(1)$, etc. We also define $F(n)! \equiv F(0)F(1)\ldots F(n)$.

Equation (16) recovers many previous forms of "qummutator" relation:

(i) $f(N) = 1;$ $F(N) = n$, [conventional bosons]

(ii) $f(N) = -1;$ $F(N) = \{0, 1\}$, [conventional fermions]

(iii) $f(N) = q;$ $F(N) = \frac{q^N - 1}{q - 1}$, ["Maths" qummutator (1)]

(iv) $f(N) = -q;$ a "fermionic" version of (iii)

               which does not lead to finite occupancy, Equation (2)

(v) $f(N) = \frac{q^{N+2}+1}{q(q^N+1)};$ $F(N) = \frac{q^N - q^{-N}}{q - q^{-1}}$, ["Physics" qummutator Equation (3)]

(vi) $f(N) = \frac{1 - q^N - q^{N+2} + (-)^n q^{2N+2}}{q - (-)^n q^{2N+1}};$

$$F(N) = \frac{q^{-N} - (-q)^N}{q + q^{-1}}, \text{ a "fermionic" version of (v), Equation (6)}$$

The form which we shall consider here is that corresponding to (v), with $q = \exp is$. This leads to $F(N) = \frac{\sin Ns}{\sin s}$, which by suitable choice of s leads to finite occupancy ($q^n = 1$ implies $0, \ldots, (n-1)$ occupancy). Of course, this choice is well-known; however, our interest here is in the comparison with the anyon formulation by means of Equation 16, which leads us to a consideration of the two-mode case.

4. GENERAL q-DEFORMED BOSONS: Two-Mode Case

We now extend to two particles. Let i and j be the labels for two distinct states We assume the following system:

$$
\begin{aligned}
a_i a_j^\dagger - g(N_i, N_j) a_j^\dagger a_i &= \delta_{ij} & \{i, j = 1, 2\} \\
[N_i, a_i] &= -a_i & \{i = 1, 2\} \\
a_1 a_2 &= e^{i\delta} a_2 a_1.
\end{aligned}
\tag{19}
$$

As above, $g(N_i, N_j)$ is an operator-valued function of the number operators N_1, N_2 and also depends on the q-deformation parameter with the property that $g \to 1$ as $q \to 1$.

One solution for the function g consistent with hermicity and the one-mode case above is $g(N_1, N_2) = f(\frac{1}{2}(N_1 + N_2))$ which recovers the one-mode case of the previous section.

Another solution is $g(N_1, N_2) = \exp i\pi(N_1 - N_2)$ for $N_1 \neq N_2$.

Note that Equation (19) is consistent with the work of Fradkin [38], and others.

The Equations 19 do not violate the "no-go" theorems of Greenberg [40] and Fivel [41] which give constraints on the possibility of defining commutation relations for a_1, a_2. It may be shown that they imply

$$[N_i, a_j] = -\delta_{ij}, \quad [N_1, N_2] = 0.$$

A choice of basis number states has to be made; for example, if we choose

$$|m, n> \equiv (a_1^\dagger)^m (a_2^\dagger)^n |0, 0 >$$

appropriately normalized, other states with the same numbers of modes, such as

$$(a_2^\dagger)^n (a_1^\dagger)^m |0, 0 >$$

will only differ by phase factors.

In order to examine the physical consequences of our new QDCAR, we shall consider an "anyon" problem.

540

5. RELATION WITH THE ANYON PROBLEM

Field theoretic descriptions of anyons have been given given by several authors including Read, Zhang et al, Fradkin, and Semenoff [42, 43, 38, 44]. In most cases a first quantized theory was presented although Read and Fradkin presented second quantized versions of the theory. Of particular interest to us is the work of Fradkin [38], who gave a Jordan-Wigner Transformation for 2-D spin-one-half systems on a lattice. The relevant particle operators, denoted by $\tilde{a}(x)$ and $\tilde{a}^\dagger(x)$ are related to the operators for spinless fermions $a(x), a^\dagger(x)$ by expressions:

$$\tilde{a}(x) = e^{i\phi(x)} a(x); \quad \tilde{a}^\dagger(x) = a^\dagger(x)e^{-i\phi(x)} \tag{20}$$

where $e^{i\phi(x)}$ is a "disorder operator" and the gradient of the scalar operator $\phi(x)$ gives the vector potential. . The equal-time generalized canonical anti-commutation relation (CAR) for these operators is:

$$\tilde{a}(x)\tilde{a}^\dagger(x') = \delta_{x,x'} - e^{i\Delta(x,x')}\tilde{a}^\dagger(x')\tilde{a}(x). \tag{21}$$

The structure of Equation 21, suggests a q-deformed canonical anticommunication relation (QDCAR) for the field operators of the form of the two-mode q-boson Equation (19). . As representations of q-deformed algebras such as $su_q(2)$ and $su_q(1,1)$ [18, 19] and others are available in terms of q-bosons; one may therefore describe physical systems whose hamiltonians, defined in terms of these anyon-like operators, belong to spectrum generating q-algebras.

For example, we may obtain one simple model by assuming that the anyon Hamiltonian is represented by an element of the dynamical symmetry group $su_q(2)$. Anyon operators are a_i with commutation relations defined by Equation (19). We take

$$H = \epsilon J_3 + \gamma J_+ + \bar{\gamma} J_- \tag{22}$$

where γ is a complex coupling constant, and we make the usual identification $J_+ \equiv a_1{}^\dagger a_2 = J_-{}^\dagger$, and $J_3 \equiv \frac{1}{2}(N_1 - N_2)$. The operators $J_\pm$ are two of the generators of $su_q(2)$. Further, if we use the expression for f given in (v) above, leading to the conventional quantum group form

$$F(N) \equiv [N] = (q^N - q^{-N})/(q - q^{-1})$$

then $[J_+, J_-] = [2 \, J_3]$ where $[A, B]$ is the usual commutator and "box " is as defined above. Writing $q \equiv e^{is}$, then

$$f(n) = 1 + (s^2/2)(n - 1) + O(s^4) + \dots .$$

This expression for $f(n)$ provides a number density analogue of the potential function $\phi(x)$ of Equation (20) and $\Delta(x, x')$ of Equation (21).

REFERENCES

[1] Y. Dothan, M. Gell-Mann and Y. Neeman, Phys.Rev.Lett. **17**,148 (1965)

[2] A.O. Barut and A. Bohm, Phys.Rev.**139**,B107 (1965)

[3] A.I. Solomon, J.Math. Phys., **12**,390, (1971)

[4] A.I. Solomon ,J.Phys.A: Math.Gen., **14**, 2177, (1981)

[5] A.I. Solomon and J.L. Birman, J.Math. Phys. **28**,1526 (987)

[6] F. Iachello and A. Arima , *The Interacting Boson Model*, Cambridge University Press (1987)

[7] F. Iachello, Phys. Rev. Lett. **44**, 772 (1980)

[8] A.Montorsi, M. Rasetti and A. I.Solomon, Phys.Rev.Lett.**59**,2243, (1987)

[9] A.Montorsi, M. Rasetti and A. I.Solomon, Intl. J.Mod. Phys. B,**2**, 247 (1989)

[10] V.G.Drinfeld, in *Proc.Int.Cong.Math.*, ed.A.M. Gleason,Univ.Calif.Press,798-820(1986)

[11] For a review see H.J. de Vega, Nucl. Phys. B (Proc. Suppl.) **18A**,229 (1990)

[12] S. Meljanac, M. Milekovic and S. Pallua, J.Phys.A.:Math.Gen. **24**,581(1991)

[13] E. Celeghini, R. Giachetti, E. Sorace and M. Tarlini, J. Math. Phys. **32**, 1159 (1991)

[14] D. Bonatsos, P.P. Raychev and A. Faessler, Chem. Phys. Lett. **178**,221 (1991)

[15] M. Jimbo, Comm. Math. Phys. **102**, 537 (1986)

[16] S. Woronowicz, Comm. Math. Phys. **111**, 613 (1987)

[17] M Arik and DD Coon, J.Math.Phys. **17**,524 (1976)

[18] A.J.MacFarlane, J.Phys.A:Math.Gen.**22**,4581- 4588 (1989)

[19] L.C.Biedenharn, J.Phys.A:Math.Gen.**22**,L873-878(1989)

[20] K.S. Viswanathan, R. Parthasarathy and R.Jagganathan, J.Math.Phys. (1992)

[21] M. Chaichian, D. Ellinas and P. Kulish, Phys. Rev. Lett **65**,980 (1990)

[22] A.I.Solomon and J.Katriel. J.Phys.A:Lett,**23**,L1209(1990)

[23] J. Katriel and A.I.Solomon. J.Phys.A:Math.Gen.,**24**,2093(1991)

[24] For an experimental review, see D.C. Tsui, Mod. Phys. Lett. **B4**,301 (1990)

[25] F. Wilczek, Phys. Rev. Lett. **49**, 957 (1982)

[26] R.B. Laughlin, Phys. Rev. Lett. **50**, 1395 (1983)

[27] A.O. Barut and G. L. Bornzin, J.Math.Phys. **12**,841 (1971)

[28] R.C. Hwa and J. Nuyts, Phys. Rev. **145**,1188 (1966)

[29] This is implicit in the approach of P.W. Anderson, Phys. Rev. **112**,1900 (1958)

[30] A. I. Solomon and William Montgomery, Journal of Physics A:Math.Gen., **11**,1633 (1978)

[31] A. I. Solomon and Joseph L. Birman, Physics Letters, **88A**, 413(1982)

[32] A. I. Solomon and Joseph L. Birman, Physics Letters, **104A**235,(1984)

[33] A. I. Solomon, Annals of the New York Academy of Sciences, **410**, 63, (1983)

[34] SL Woronowicz, Commun. Math. Phys. **136** , 399(1991)

[35] A.I. Solomon in *Workshop on Harmonic Oscillators*, University of Maryland at College Park (1992)

[36] FH Jackson, Quart.J.Pure Appl.Math., **41**,193(1910)

[37] J. Katriel and A.I.Solomon, J.Phys.A:Math.Gen.,**24**, L1139, (1991)

[38] E. Fradkin, Phys. Rev. Lett. 63, 322 (1990)

[39] E. Fradkin, Intl. Jrnl. Mod. Phys. **3**,239(1989)

[40] O.W. Greenberg, Phys. Rev. D **43**, 4111 (1991)

[41] D.I. Fivel, Phys. Rev. Lett. **65**,3361 (1990)

[42] N. Read, Phys. Rev. Lett., **62**,86 (1989)

[43] L.C. Zhang, T.H. Hansson and S. KivelsonPhys. Rev. Lett., **62**,(1989)

[44] G.W. Semenoff, Phys. Rev. Lett. **61**,517(1988)

[45] A. Bincer, J.Phys.A.:Math.Gen **24**,L1133(1991)

INTERACTING BOSON MODEL
AND 3⁻ NUCLEAR STATES IN EVEN–EVEN NUCLEI

Stanisław Szpikowski and Krystyna Zając

Institute of Physics
Department of Theoretical Physics
University of M. Curie–Skłodowska
pl. M. Curie–Skłodowskiej 1, Lublin, Poland

1. INTRODUCTION

It is a pleasure to make a contribution to the volume of Franco Iachello 50th birthday. It is also almost exactly 20 years after the Interacting Boson Model have been invented by Iachello and Arima. Let us also mention that, due to our opinion, there are at present three nuclear models of a great power of prediction and description of nuclear data. There are Nuclear Shell Model of Goeppert-Mayer and Jensen, Collective Model of Bohr and Mottelson and Interacting Boson Model of Arima and Iachello. It is also worthy to say that first two models had been honoured by the Nobel Prize not immediately but after around 20 years of invention.

We come now to our work in which we apply the IBM to a systematics of the experimental data of the first excited 3⁻ levels for even-even nuclei in the regions $A \approx 100$ and $A \approx 150$. Such systematics of different measured observables had been and have been made not very seldom. In the relation to the problem discussed in this paper we can mention the systematics of B(E2) transition probabilities made more then 25 years ago by Hamamoto (1965) and more recently by Raman et al. (1985). The interesting point of the discussion is that the experimental values were displayed versus $N_p N_n$, the product of valence protons and neutrons. The product $N_p N_n$ has been considered as a measure of the neutron-proton interaction. The more detailed discussion, Zhang et al. (1989), has shown the validity of such interpretation. Casten et al. (1985) not only completed the experimental energies of the first 2⁺ and 4⁺ nuclear levels but also provided an interpretation within the Interacting Boson Model. The leading idea for the interpretation has come from the observation that experimental points versus $N_p N_n$ followed a regular curve. Then one of the IBM parameter instead of being changed from nucleus to nucleus has been constructed as a function of $N_p N_n$. The IBM calculations could then in a satisfactory way describe the experimental data.

Recently, we have made a similar observation concerning the first excited 3⁻ states of the even–even nuclei in the $A \approx 100$ and $A \approx 150$ regions (Lederer and

Shirley 1978). For the $A \approx 100$ region the energies of the first excited 3^- levels behave quite regularly versus $N_p N_n$ product (Fig.1).

The energy points for the Cd isotopes deviate from the curve but there is an reasonable interpretation of a such deviation given by the IBM. Namely, in Cd isotopes there is only two proton-holes in the $Z = 50$ magic shell, however near the closed proton shell the IBM is not quite applicable. There is quite a regular position of experimental points as compare to the same points versus the atomic number A (Fig.1) even with the Cd isotopes.

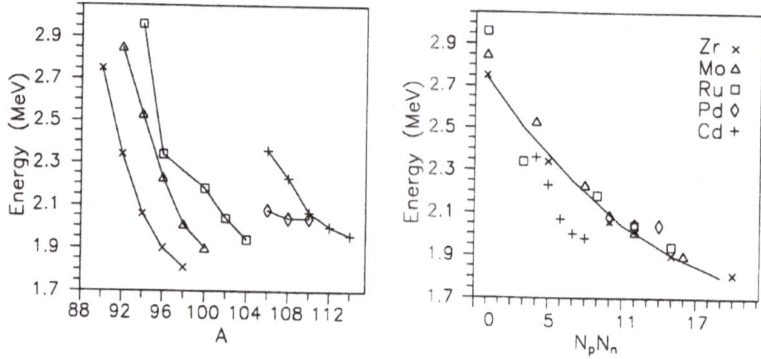

Fig.1. The experimental (Lederer, Shirley 1978) energies of the first excited 3^- states versus A-number and versus the $N_p N_n$ product of valence proton- and neutron-bosons.

The same analysis of the nuclei in the $A \approx 150$ region gives bigger splitting of energy points. Still, the experimental points deviate not so much as in the case of A-dependence (Fig.2) but it is not so striking regularity as in the $A \approx 150$ region. It should be mentioned that the valence protons and neutrons to form valence bosons were taken in Fig.1 and Fig.2 due to the magic numbers 28–50–82–126 and, as usual, in the second half of the magic shell there were taken holes rather then particles. We add that the systematics of the 3^- first excited levels has been also done by Cottle and Bromley (1986) who have made a microscopical qualitative interpretation of the behaviour of excited 3^- levels. Such a compilation and interpretation of the 3^- levels has been recently done by Zamfir et al. (1989). On the basis of the discussion the authors introduced the general phenomenological formula

$$E(3^-) = 19A^{-1/3} - 0.5\sqrt{N_t}$$

where besides of atomic number A there is also the number N_t of valence neutrons and protons. The formula describes well experimental data except of the region $A \approx 150$.

In our IBM interpretation of the first excited 3^- levels (Szpikowski, Zając and Król 1988) we have introduced one f-boson besides of usual s-d bosons. It has been also suggested the p-boson role in the description of negative parity excitations (Engel and Iachello 1987) due to the possible octupole deformations (Kusnezov 1988, Kusnezov and Iachello 1988). It also means that the p-boson is an essential one to describe the E1 branching rates. However, available data in the rare–earth region can be reproduced equally well with or without octupole deformations (Kusnezov and Iachello

1988, Cottle 1986) hence the model IBM1 + f-boson should work reasonably well (Barfield et al 1988).

In our first results (Szpikowski, Zając and Król 1988) we have got for couple of nuclei and their isotopes from $A \approx 100$ and $A \approx 150$ regions quite a good agreement with experimental data with the same IBM parameters but with the $N_p N_n$ dependence. However, trying to extend the calculations on the rest of even–even nuclei in

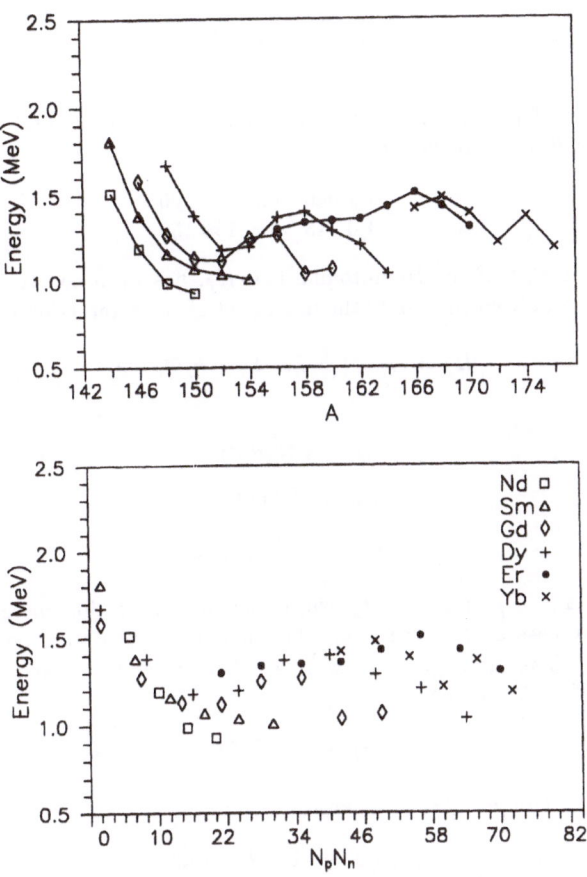

Fig.2. The same as in Fig.1 but for even-even nuclei of the $A \approx 150$ region.

the $A \approx 150$ region we have faced the difficulties with numerical adjustment of calculated and experimental data while taking the valence nucleons between the magic shells. Before further discussion of the results we introduce now our version of the Interacting Boson Model.

2. THE IBM HAMILTONIAN AND STATE-VECTORS

The IBM boson Hamiltonian consists of the s-d boson part H_{sd}, as well as the H_f part coming from the presence of the f-boson :

$$H = H_{sd} + H_f, \tag{1}$$

where H_{sd} is the same as in Casten et al. (1985), i.e.

$$H_{sd} = \epsilon_d n_d + k Q_d \cdot Q_d, \tag{2}$$

where n_d is the d-boson number operator,

$$Q_{dm} = (s^\dagger \tilde{d} + d^\dagger \tilde{s})_m^{(2)} + \frac{1}{\sqrt{5}} \chi (d^\dagger \tilde{d})_m^{(2)}, \tag{3}$$

$$\epsilon_d = \alpha_d e^{-\theta N_p N_n} \tag{4}$$

and $N_p(N_n)$ is the number of valence proton (neutron) bosons.
The parameters of the H_{sd} are taken also from Casten et al. (1985), namely $k = -0.0919, \chi = 0.90, \theta = 0.0405$ and

$$\alpha_d = \begin{cases} 1.080 & \text{for } A \approx 100 \\ 0.443 & \text{for } A \approx 150 \end{cases}. \tag{5}$$

The second part of the Hamiltonian (1), H_f, has been constructed also as an approximation and it consists only the main part of the interaction (sd)–(f), namely

$$H_f = \epsilon_f n_f + k_1 L_d \cdot L_f + k_2 Q_d \cdot Q_f \tag{6}$$

where

$$L_{dm} = \sqrt{10}(d^\dagger \tilde{d})_m^{(1)}, \tag{7}$$

$$L_{fm} = \sqrt{7}(f^\dagger \tilde{f})_m^{(1)}, \tag{8}$$

$$Q_{fm} = -2\sqrt{7}(f^\dagger \tilde{f})_m^{(2)}. \tag{9}$$

The operators $L_d \cdot L_f$ and $Q_d \cdot Q_f$ are needed in general for energy calculations, however, in the case of 3^- excitations their matrix elements are equal to zero (or almost to zero). The single particle energy ϵ_f of the f-boson is taken, similarly to ϵ_d, as a function of $N_p N_n$, namely

$$\epsilon_f = \alpha e^{-\beta N_p N_n} \tag{10}$$

with adjusted parameters $\beta = 0.022$ and

$$\alpha = \begin{cases} 2.56 & \text{for } A \approx 100 \\ 1.40 & \text{for } A \approx 150 \end{cases}. \tag{11}$$

The IBM calculations have been performed in the complete s-d boson basis coupled to the f-boson part in the angular momentum space. The complete s-d basis has been taken from the work (Szpikowski and Góźdź 1980) and the vectors read

$$|N n_d v x L M> = \frac{1}{\sqrt{(N - n_d)!}} (s^\dagger)^{N-n_d} |n_d v x L M> \tag{12}$$

where
$|n_d v x L M>$ is the complete basis for d-bosons
N - the number of s-d bosons,
n_d - the number of d-bosons,
v - the seniority number of d-bosons,
L, M - angular momentum quantum numbers,

x - an additional number for complete classification of states which has a meaning of a maximum number of d-boson triplets $(d^\dagger d^\dagger d^\dagger)^{L=0}$, coupled to zero angular momentum.

The complete quadrupole basis $|n_d v x LM>$ was obtained with the aid of the $SU(1,1)$ quasi–spin generator ,namely $K_+ = \frac{\sqrt{5}}{2}(d^\dagger d^\dagger)^{(0)}$, and with recoupling coefficients $W(n_{-2}\cdots n_2|vxLM)$ (Szpikowski and Góźdź 1980) :

$$|n_d v x LM >= \mathcal{N}(K_+)^{\frac{n_d-v}{2}}. \tag{13}$$

$$\times \sum_{n_{-2}\cdots n_2} W(n_{-2}\cdots n_2|vxLM)(d^\dagger_{-2})^{n_{-2}}\cdots(d^\dagger_2)^{n_2}|0 > .$$

Here $\mathcal{N}$ is the normalizing constant and the summation is limited by $\sum_{i=-2}^{i=2} n_i = v$ and $\sum_{i=-2}^{i=2} i n_i = M$.

The Hamiltonian (2) can be rewritten in terms of $SU(1,1)$ and $SO(3)$ tensors and, consequently, all needed matrix elements can be expressed (Zając and Szpikowski 1986) by Wigner coefficients of both groups and recoupling coefficients contained in (13). For example the three d-boson $SO(3)$ scalar which appears in the Hamiltonian (1) has the rank 3/2 in $SU(1,1)$ group :

$$(d^\dagger(\tilde{d}\tilde{d})^{(2)})^{(0)} = \frac{1}{\sqrt{3}}T^{(3/2,0)}_{-1/2} \tag{14}$$

and the quadrupole operator Q_{dm} is an $SO(3)$ tensor operator of a rank 2 and can be expressed in terms of the $SU(1,1)$ tensors of definite ranks :

$$Q_{dm} = s^\dagger T^{(1/2,2)}_{-1/2,m} + T^{(1/2,2)}_{1/2,m} s + \frac{\chi}{\sqrt{10}}T^{(1,2)}_{0,m}. \tag{15}$$

Reduced matrix elements of these tensors have been given by Zając and Szpikowski (1986).

The lowest negative parity states are interpreted in the IBM as states with only one f-boson (one pair of nucleons coupled to $l = 3$). Hence, the f-part of the vector state with only one f-boson is

$$|n_f L_f M > \tag{16}$$

with $n_f = 1$ and $L_f = 3$.

The coupled state-vectors of s-d-f bosons reads

$$|N n_d v x L_d n_f L_f; LM >= (|N - n_d, n_d v x L_d > \otimes |n_f L_f >)^{(L)}_M \tag{17}$$

where now $N = n_s + n_d + n_f$.

The matrix elements of the H_f part of (1) are then calculated in the basis (19). The first term, $L_d \cdot L_f$ is diagonal in the basis with an eigenvalue $\frac{1}{2}[L(L+1)-L_d(L_d+1) - L_f(L_f + 1)]$.

The non-zero matrix elements of the second term, $Q_d \cdot Q_f$ are more complicated

$$< N n'_d v' x' L'_d n'_f L'_f; LM |Q_d \cdot Q_f| N n_d v x L_d n_f L_f; LM >= \tag{18}$$

$$= (-1)^{L+L_d+L_f} \left\{ \begin{matrix} L'_d & L_d & 2 \\ L_f & L'_f & L \end{matrix} \right\} < N - n'_f, n'_d v' x' L'_d \|Q_d\| N - n_f, n_d v x L_d >$$

$$\times < n'_f L'_f \|Q_f\| n_f L_f > .$$

The last matrix element of (18) with $n_f = 1$ and $L_f = 3$ can be obtained by simply straightforward calculations

$$< n_f L_f \| Q_f \| n_f L_f >= -2\sqrt{35}. \qquad (19)$$

The results of the calculation for 3^- energy levels for several nuclei and their comparison with experimental data (Lederer and Shirley 1978, Bielenkij and Grigoriev 1987) are presented in Fig. 3–6.

3. DISCUSSION

In the beginning let us stress the following. We have adopted, without further adjustment, the part H_{sd} of the Hamiltonian exactly from Casten et al. (1985). The new part of H, namely H_f , possesses two new parameters α and β (10–11). However, β has been taken the same for all of the considered nuclei and α takes on only two different values for $A \approx 100$ and $A \approx 150$. In contrast, in the usual IBM calculation the

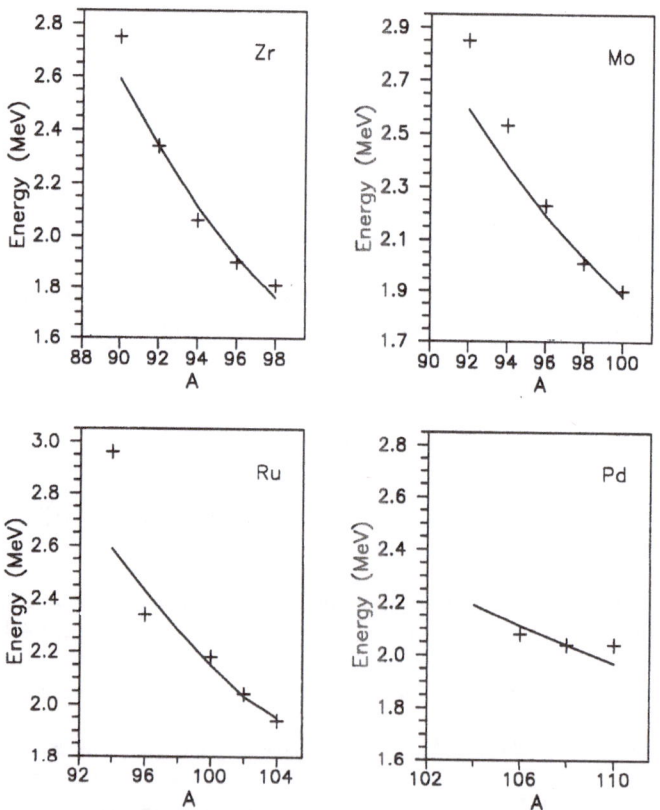

Fig.3. Comparison of experimental (+) and calculated energies of the first 3^- levels in the region of $A \approx 100$ nuclei.

model parameters are adopted from nuclei to nuclei showing not seldom non smooth behaviour. The most interesting thing which Fig. 3–6 show is not the very good agreement of experimental and theoretical values for the region $A \approx 100$ and partly for the region $A \approx 150$, but completely wrong theoretical prediction for $A > 152(N > 90)$.

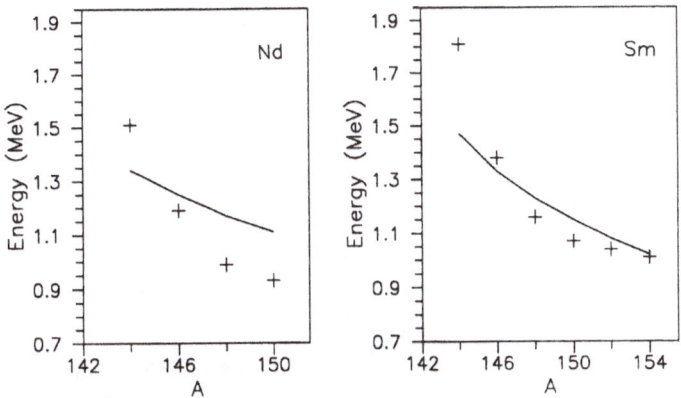

Fig.4. The experimental $(+)$ and calculated 3_1^- energies for Nd and Sm isotopes.

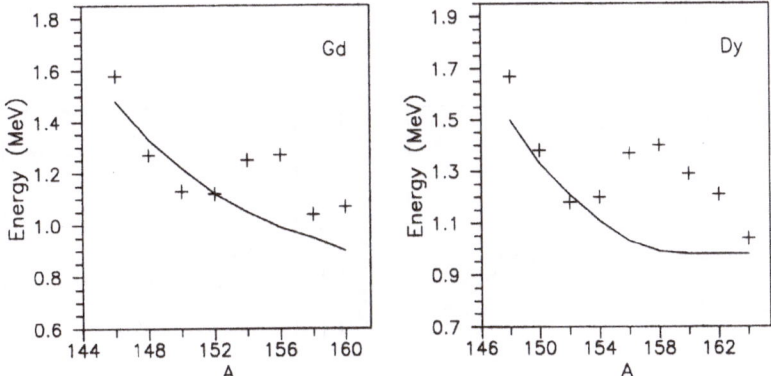

Fig.5. The experimental $(+)$ and calculated 3_1^- energies for Gd and Dy isotopes.

If there would not be another sign of such a splitting of nuclei under consideration, we would say, only on the basis of our calculations, that nuclei with $A > 152(N > 90)$ should possess different structures from those with $A < 152(N < 90)$. But there are well established facts that from $N = 90$ begins the static quadrupole deformation of nuclei and hence, the IBM without a proper correction which would take into account that deformation, could not be applied which is shown just in Fig. 5–6. We have come to the conclusion that correction should be taken in numbers of acting bosons. Namely, if we examine the Nilsson diagrams (Cottle 1986, Soloviov 1971) then we find that there are energy gaps for $N = 88$ and 98 and for $Z = 76$ in the region of nuclei with the quadrupole deformation $0.2 < \epsilon < 0.3$ (Dudek et al. 1980, de Voigt et al. 1983, Cottle 1986).

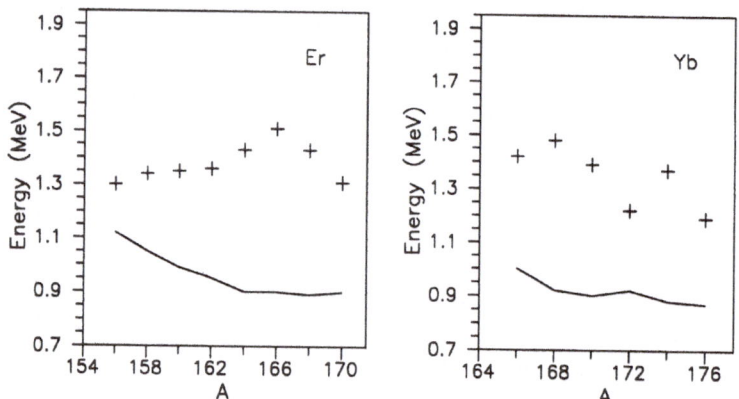

Fig.6. The experimental (+) and calculated 3_1^- energies for Er and Yb isotopes.

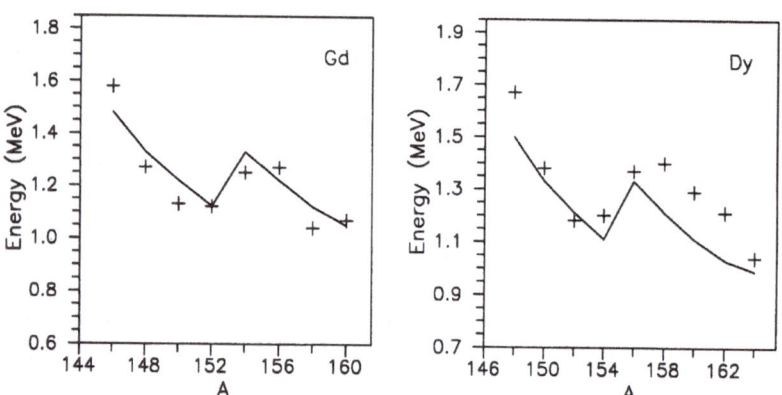

Fig.7. Comparison of experimental (+) and calculated energies of the first excited 3^- levels for Gd and Dy isotopes. The semi-magic numbers for the definition of valence bosons are given in text.

Those are the deformations of nuclei in our considerations for $A > 152$. In the spirit of any boson model one is free to take the number of active bosons and hence let us take the active bosons (or boson-holes) as valence bosons due to the magic or semi-magic shells:

for nuclei with $N < 90$ — magic numbers 50-82-126 for protons and neutrons,

for nuclei with $N > 90$ — in addition : $Z = 76, N = 88$ and

for nuclei with $Z > 68$ — in addition : $N = 98$

which follow the Nillson diagrams.

Keeping the same parameters as before (Fig. 3–6) we repeat the calculation for Gd, Dy, Er and Yb isotopes using the above definition of active bosons (boson-holes). The result of the calculation is given in Fig. 7–8. There is not only a quantitative improvement in much higher calculated energy values but also the qualitative staggering theoretical lines as compared to Fig. 5–6.

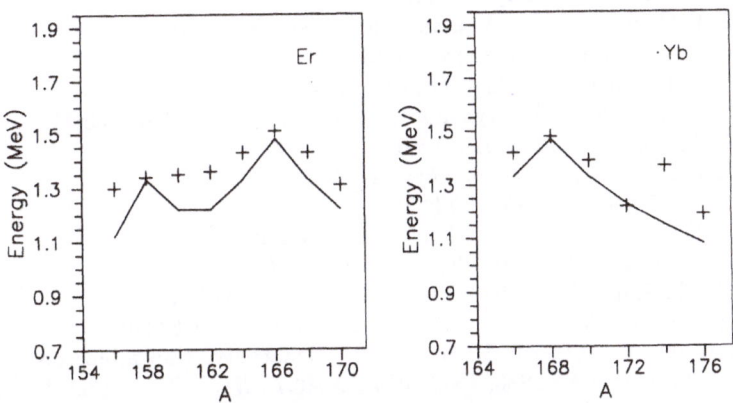

Fig.8. Comparison of experimental (+) and calculated energies of the first excited 3^- levels for Er and Yb isotopes. The semi-magic shells leading to the new proton- and neutron-boson numbers are the same as for Gd and Dy isotopes (Fig. 7).

Although the new quasi-magic numbers have been practically our definition of active bosons, we make a comment concerning spherical bosons in non-spherical Nilsson diagrams. The main success, in our opinion, of the IBM has a source in its phenomenology which makes it possible to comprise in a boson both: single particle (pair of nucleons) and collective degrees of freedom (d-bosons as a quantisation of Bohr quadrupole parameters). The single-particle part of a boson can be understood in the non-spherical (internal) frame of reference and a collective part of a boson, also deformed, can be then coupled to its single particle part to form a spherical boson. If such a program could be worked out (if any) it would provide the microscopical interpretation of the IBM but, for sure, the IBM would lost its beautiful simplicity. Hence, let us better repeat that the quasi-magic numbers merely define here the valence bosons (boson-holes).

In the end we mention that such "magic numbers" in the similar context has been also taken by Zamfir et al. (1990) who adopted the following "magic shells" N = 106, 148 and Z = 40, 64, 70, 100 in order to parametrize the energies of the first excited 2^+_γ states.

ACKNOWLEDGEMENTS

A part of the presented work has been prepared during the stay of the first author in the Technische Universität, München and the Max Planck Institute für Kernphysik, Heidelberg, Germany. The author is very grateful to Professor Klaus Dietrich and Professor Hans Weidenmüller for a kind invitation and also to the Deutsche Forschungsgemeinschaft and to the Max Planck Institute for financial support during the stay.

The work is also partly supported by grant No PB 2 1201 9102.

REFERENCES

Barfield,A.F., Barret,B.R., Wood,J.L., and Scholten,O., 1988, *Ann. of Phys.* 182:344

Bielenkij,W.M., Grigoriev,E.P.,1987,"Struktura Tschetnych Jadier", Energoatomizdat,Moscow

Casten,R.F., 1985, *Nucl. Phys.* A443:1

Casten,R.F., Frank,W., Von Brentano, 1985, *Nucl. Phys.* A444:133

Cottle,P.D., 1986, Ph.D.Thesis, Yale University

Cottle,P.D., Bromley,D.A., 1986, *Phys. Lett.* B182:129

De Voigt,M.J.A., Dudek,J., Szymanski,Z., 1983, *Rev. Mod. Phys.* 55:949

Dudek,J., Majhofer,A., Skalski,J., 1980, *J. Phys.* G6:447

Engel,J., Iachello,F., 1987, *Nucl. Phys.* A472:61

Hamamoto,I., 1965, *Nucl. Phys.* 73:225

Kusnezov,D., 1988, Ph.D.Thesis, Princeton University

Kusnezov,D., Iachello,F., 1988, *Phys. Lett* B209:420

Lederer,C.M., Shirley,V.S., 1978, "Tables of Isotopes",New York

Raman,S., Nestor,C.W.Jr., Bhatt,K.H., 1988, *Phys. Rev.* C37:805

Soloviov,W.G., 1971,"Teorija Sloznych Jadier",Izdatelstvo Nauka,Moscow

Szpikowski,S., Góźdź,A., 1980, *Nucl. Phys.* A340:7 ; *Nucl. Phys.* A399:359

Szpikowski,S., Zając,K.,and Król,D., 1988, "Thesis of 2-nd International Spring Seminar on Nuclear Physics", Capri, 513

Zając,K., Szpikowski,S., 1986, *Acta Phys. Pol.* B17:1109

PARITY DOUBLET LEVELS AT SUPERDEFORMATION

K. Sugawara-Tanabe[1] and A. Arima[2]

[1] School of Social Information Studies,Otsuma Women's University
Tokyo 206 Japan
[2] Department of Physics, University of Tokyo, Tokyo 113 Japan

INTRODUCTION

Recently many yrare superdeformed bands and subsequently the quantization of alignment[1] are observed in the superdeformed region of $A \sim 190$ (Hg, Os)[2] and $A \sim 150$ (Dy, Tb, Gd)[3], but they are not found in the $A \sim 130$ region (Ce, Nd)[4]. It is our motivation to start this work to find the origin to produce this difference. When we look at the simple Nilsson diagram[5,6], we find the degeneracy of $[880]\frac{1}{2}$ and $[761]\frac{1}{2}$ in ν shell and $[770]\frac{1}{2}$ and $[651]\frac{1}{2}$ in π shell around $\beta \sim 0.6$ and just over the fermi surface of $N = 112$ and $Z = 80$. This kind of degeneracy is always observed among the two levels with asymptotic quantum numbers $[N, n_z = N, \Lambda = 0]\frac{1}{2}$ and $[N-1, n_z = N-2, \Lambda = 1]\frac{1}{2}$ at large deformation. The former level belongs to the unique parity levels with $\Omega = \frac{1}{2}$, and the latter level to the pseudo-spin family as $[\tilde{N} - 2, \tilde{N} - 2, \tilde{0}]\frac{1}{2}$ in the pseudo-spin representation[7,8], which has no pseudo-spin partner. We will name this different parity-pair levels as parity-doublet (P-D) levels from now on. The P-D levels become nearly degenerate around the fermi surface in superdeformed shape for $A \sim 150$ and $A \sim 190$ regions, but not for $A \sim 130$ region. We propose the idea that these P-D levels are responsible for the quantization of alignment.

SPIN-ORBIT FORCE IN P-D LEVELS

At first we will comment on the energy levels of P-D levels. We adopt the conventional Nilsson Hamiltonian[9] in an axially symmetric deformed oscillator potential with a rational ratio $a : b$ between the frequencies $\omega_\perp$ and ω_z.

$$
\begin{aligned}
H &= -\frac{\hbar^2}{2M}\mathbf{P}^2 + \frac{M}{2}(\omega_\perp^2(x^2 + y^2) + \omega_z^2 z^2) + \xi_{ll}(\mathbf{l}^2 - <\mathbf{l}^2 >_N) + \xi_{ls}\mathbf{ls}, \\
&\equiv H_0 + \xi_{ll}(\mathbf{l}^2 - <\mathbf{l}^2 >_N) + \xi_{ls}\mathbf{ls}.
\end{aligned}
\tag{1}
$$

When we neglect the $\mathbf{l}^2$ and $\mathbf{ls}$ interactions in (1), the degeneracy of the P-D levels is easily seen by applying the cylindrical coordinate,

$$
x_\pm = x \pm iy = \pm i(\frac{\hbar}{M\omega_\perp})^{\frac{1}{2}}(c_\pm^\dagger + c_\mp), \quad z = -i(\frac{\hbar}{2M\omega_z})^{\frac{1}{2}}(c_z^\dagger - c_z),
\tag{2}
$$

$$
p_\pm = p_x \pm ip_y = \mp(\hbar M\omega_\perp)^{\frac{1}{2}}(c_\pm^\dagger - c_\mp), \quad p_z = (\frac{\hbar M\omega_z}{2})^{\frac{1}{2}}(c_z^\dagger + c_z),
\tag{3}
$$

to H_0 in (1). Then the energy eigenvalue of H_0 is described by the shell quantum number $N_{sh} = an_\perp + bn_z$ and its eigenfunction corresponds to the asymptotic wave function $|n_z, n_\perp, \Lambda >$.

$$
H_0|n_z, n_\perp, \Lambda >= \hbar\omega_{sh}(N_{sh} + a + \frac{b}{2})|n_z, n_\perp, \Lambda >,
\tag{4}
$$

where $n_\perp = n_+ + n_-$, $\Lambda = n_+ - n_-$, $\omega_{sh} = \omega_0(a^2 b)^{-1/3}$ and n_i is the eigenvalue of $c_i^\dagger c_i$ ($i = +, -$ or z). The deformation parameter δ is given by $3(a - b)/(2a + b)$. In the superdeformed shape where $a : b = 2 : 1$, the P-D levels belong to the same $N_{sh} = 2n_\perp + n_z = N$ with the same shell energy $\hbar\omega_{sh}(N_{sh} + 5/2)$, which proves the P-D levels are almost degenerate in the energy diagram. Including the residual $\mathbf{l}^2$ and $\mathbf{ls}$ interactions in (1) up to first order perturbation (1st o.p.) and up to second order perturbation (2nd o.p.), we can calculate the total energy in the same approximation used by ref.[10] (see eqs. (6-501) and (6-506) in ref.[10]). Then the energy difference between the P-D levels becomes 0.0191 MeV up to 2nd o.p. with parameters of $\xi_{ll}/\hbar\omega_0 = -0.019$ and $\xi_{ls}/\hbar\omega_0 = -0.127$ ($\hbar\omega_0 = 41A^{-1/3}$). In (1) the most important term for high spin states is neglected, i.e. the Coriolis coupling term $\hbar\omega_{rot}J_x$. If one assumes $\hbar\omega_{rot}J_x = 0.3$ MeV, the orbital part of the Coriolis force up to 2nd o.p. change the energy difference to 0.0589 MeV. Thus the orbital part of the Coriolis force does not much destroy the energy degeneracy in the P-D levels, but only its spin part becomes important as shown in Fig. 1. For the P-D levels, levels connected by residual interactions are far from them in superdeformation and so the degeneracy of the energy levels is not much destroyed.

When we diagonalize the total H in(1), the situation in Fig.1 becomes much clear. We show in Fig.2 the eigenvalue $E(\sigma, \Omega)$ obtained by the diagonalization of total H at fixed δ,

$$
H|\sigma, \Omega >= E(\sigma, \Omega)|\sigma, \Omega >,
\tag{5}
$$

where Ω corresponds to the eigenvalue of J_z (= $\frac{1}{2}$ for the P-D levels) and σ denotes all the other quantum numbers except for Ω. The degeneracy of the P-D levels around $\delta \sim 0.6$ is shown for ^{192}Hg, which are obtained numerically by the exact diagonalization of total H in the space $\Delta N = 0$ with the same parameters as in Fig.1. All the levels in Fig.2 are denoted by the asymptotic quantum numbers $[N, n_z, \Lambda]\Omega$.

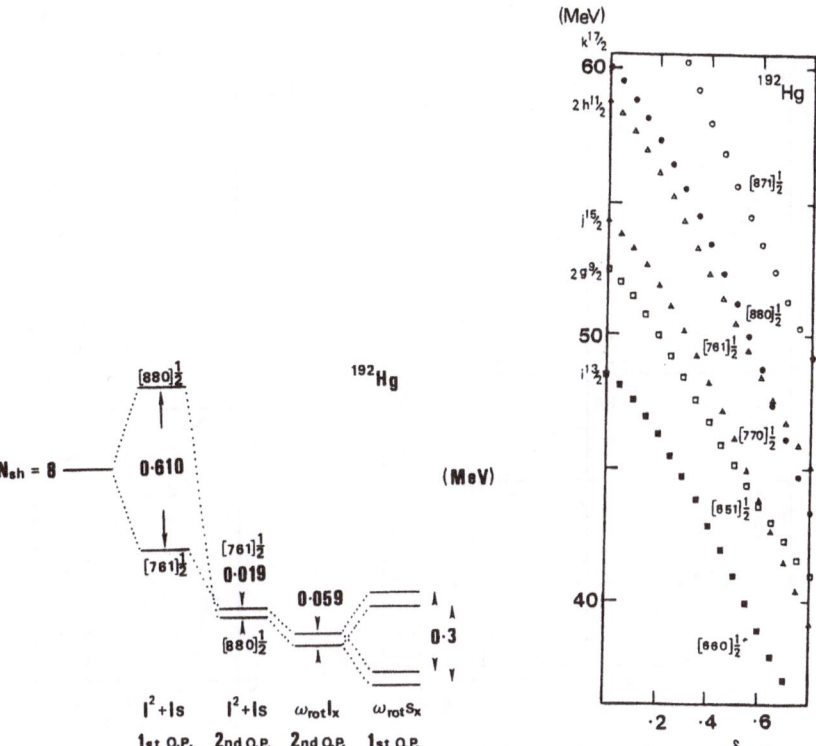

Fig.1. Perturbation calculation of energies.

Fig.2. $E(\sigma, \Omega)$ as a function of δ.

Now we will evaluate the expectation values of the ls interaction using the eigenvector obtained from (5), i.e. $< \sigma, \Omega |ls| \sigma, \Omega >$. Fig.3 shows these values for the P-D levels both for $A = 192$ and 152 cases. We see the decrease of $< \sigma, \Omega |ls| \sigma, \Omega >$ according with the increasing deformation for the P-D levels both for the Dy and Hg cases. It is easily seen that $< \sigma, \Omega |ls| \sigma, \Omega >$ reaches to 0 for $\Lambda = 0$ and $-\frac{1}{2}$ for $\Lambda = 1$. It is remarkable that those two values correspond to the values calculated by the asymptotic wave function without spin-orbit interaction (4) from the formula $< n_z, n_\perp, \Lambda, \Sigma |ls| n_z, n_\perp, \Lambda, \Sigma >= \Lambda\Sigma$, where $\Omega = \Lambda + \Sigma$ and Σ is the eigenvalue of S_z. This result indicates $L - S$ coupling scheme is recovered in the P-D levels. As for the $[660]\frac{1}{2}$ and $[541]\frac{1}{2}$ in $A = 152$ case, which locate not over but far under the fermi surface of $Z = 66$ in contrast to the other cases, they reach $\frac{1}{2}$ and -1 respectively, indicating that $L - S$ scheme is still not recovered for these levels. This feature of the small expectation value of the ls interaction in superdeformation is characteristic not only for the P-D levels, but also for some other single-particle levels in large deformation. For the case of $\Omega = \frac{3}{2}$ the asymptotic value of $\Lambda\Sigma$ is 0.5 or -1. Some levels with the asymptotic quantum numbers $[701]\frac{3}{2}$, $[732]\frac{3}{2}$ and $[752]\frac{3}{2}$ show a good tendency to the asymptotic values. This is because of the pseudo-spin mechanism, and explained later.

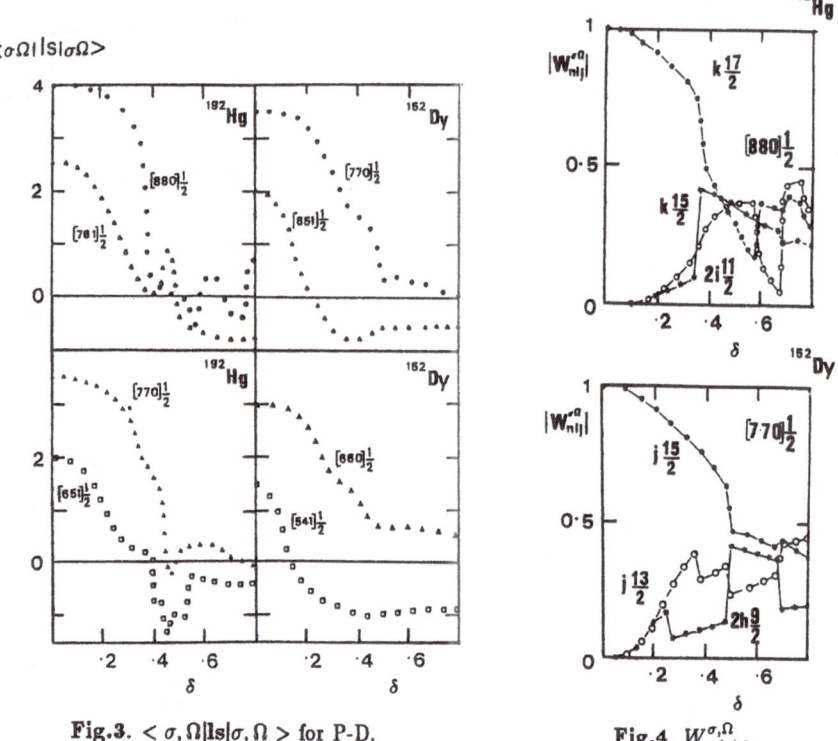

$<\sigma\Omega|s|\sigma\Omega>$

Fig.3. $<\sigma,\Omega|ls|\sigma,\Omega>$ for P-D.

Fig.4. $W^{\sigma,\Omega}_{n,l,j,\Omega}$.

The state $|\sigma,\Omega>$ in (5) is expanded by the spherical basis $|n,l,j,\Omega)$ through the formula $|\sigma,\Omega> = \sum W^{\sigma,\Omega}_{n,l,j,\Omega}|n,l,j,\Omega)$. To make our observation much clear we show the absolute values of the dominant components of the P-D levels, i.e. $|W^{\sigma,\frac{1}{2}}_{n,l,j,\frac{1}{2}}|$ as functions of δ in Fig.4. Here the dominant spherical level $|n,l,j=l+\frac{1}{2},\frac{1}{2})$ fulfills 100 % of $|\sigma,\Omega=\frac{1}{2}>$ at $\delta=0$. In the same figure we show the coefficients of the spin-orbit partner of the dominant component, i.e. $|W^{\sigma,\frac{1}{2}}_{n,l,j'=j-1,\frac{1}{2}}|$, and also that for the spherical basis $|n,l'=l-2,j'=j-3,\frac{1}{2})$ or $|n,l'=l+2,j'=j+1,\frac{1}{2})$ which corresponds to the spin-orbit partner of the next dominant component around $\delta=0.1\sim0.3$. The dominant component decreases with increasing δ, and instead the spin-orbit partner's components increase. Finally both have a tendency to reach nearly to the same amount around $\delta\sim0.6$. For example in the state with asymptotic quantum number $[880]\frac{1}{2}$, the coefficients of $k17/2$ come to the same amount with those of $k15/2$ and $2i11/2$, and in the state $[770]\frac{1}{2}$ $j15/2$ with $j13/2$ and $2h9/2$. What Figs.3 and 4 indicate is that the $L-S$ coupling scheme works well in the P-D levels.

As the axially symmetric deformed field is proportional to δY_{20}, these phenomena shown in the figures are simply explained from the nature of the matrix element of spherical harmonics. The matrix element of $(j'=l-2\pm\frac{1}{2}|Y_{20}|j=l\pm\frac{1}{2})$ is proportional to $l\sqrt{(l-1)(l+1)}/((2l-1)(2l+1))$ which is a comparable order to the diagonal element $(j' = l \pm \frac{1}{2}|Y_{20}|j = l \pm \frac{1}{2}) \propto (\frac{3}{4} - j(j+1))^2/(j(j+1)(2j+3)(2j-1))$. On the other hand the matrix element between spin-orbit partners $(j' = l - \frac{1}{2}|Y_{20}|j = l + \frac{1}{2})$ is proportional to $\sqrt{(2j+1)(2j'+1)}/(jj'(j+1))$, which is usually much smaller than the diagonal element. Thus the coupling among $j = l + \frac{1}{2}$ family ($s1/2$, $d5/2$, $g9/2$, ..., or $p3/2$, $f5/2$, $h11/2$,....) or that among $j = l - \frac{1}{2}$ ($d3/2$, $g7/2$,..., or $p1/2$, $f5/2$, $h9/2$,...) are strongly enhanced. However the matrix element between spin-orbit partners of the spherical basis with small value of l, i.e. between $s1/2$ and $d3/2$ or $p1/2$ and $p3/2$ becomes non-negligible order but almost comparable order to the diagonal element of Y_{20}. This causes a strong mixture between the spin-orbit partners with large l value, and the small expectation values of the ls interaction for the P-D levels at $\delta \sim 0.6$ shown in Figs.3 and 4. As for the levels with $\Omega \geq \frac{3}{2}$ the pseudo-spin mechanism works. The non-diagonal element between the pseudo-spin pair states $p3/2$ and $f5/2$ becomes larger at superdeformation than the energy difference between their diagonal elements, i.e. pseudo-spin degeneracy. This causes a large mixing of spin-orbit partners with larger l similar to the case of $\Omega = \frac{1}{2}$. Here we want to mark that in contrast to the case of pseudo-spin scheme where the $\tilde{l}\tilde{s}$ interaction is nearly zero, the ls interaction remains large in the P-D levels, but the $L - S$ coupling scheme is recovered because of the large deformed field.

The calculation with the Coriolis term does not change these results around superdeformation. In Fig.5 we show the effect of the Coriolis coupling term in the exact diagonalization within the space of $\Delta N = 0$. The orbital part of the Coriolis term mixes $\Delta N_{sh} = \pm 1$ and ± 3. For example within $\Delta N_{sh} = \pm 1$, $[770]\frac{1}{2}$ is mixed with $[761]\frac{3}{2}$ and the time reversed level of $[761]\frac{1}{2}$, and $[761]\frac{1}{2}$ with $[752]\frac{3}{2}$ and time reversed levels of $[750]\frac{1}{2}$ and $[770]\frac{1}{2}$. These levels which give influence on the P-D levels are locating near the P-D levels at small δ, but moving away from them with increasing δ. In Fig.5 we show the results obtained by the diagonalization of Coriolis term for $[770]\frac{1}{2}$ (squares) and $[761]\frac{1}{2}$ (triangles) together with those without the Coriolis term (solid lines) and $[761]\frac{3}{2}$ and $[752]\frac{3}{2}$ levels (dashed lines). The closed symbols in both squares and triangles correspond to the time-reversed levels of the open symbols respectively, and $\hbar\omega_{rot} = 0.2$ MeV. We see the difference between the energies with and without Coriolis term decreases according with increasing δ. We show the values of $< \sigma|ls|\sigma >$ and $<< \sigma|J_x|\sigma >>$ with Coliolis term as a function of δ in Fig.6. Here $|\sigma >$ is the eigenfunction of $H - \hbar\omega_{rot}J_x$ within the space of $\Delta N = 0$, which converges to $|\sigma, \frac{1}{2} >$ at the limit of $\hbar\omega_{rot} = 0$. With these $|\sigma >$ states, $|\sigma >>$ state is defined as $|\sigma >>= 1/\sqrt{2}[|\sigma > +T|\sigma >]$ with time reversal operator T. The solid lines are without Coriolis term, closed symbols correspond to the time reversal of the open symbols. The squares (triangles) correspond to $[770]\frac{1}{2}$ ($[761]\frac{1}{2}$) at $\hbar\omega_{rot} = 0$ limit, respectively. Again we see the difference between the values with and without Coriolis term becomes small at superdeformation.

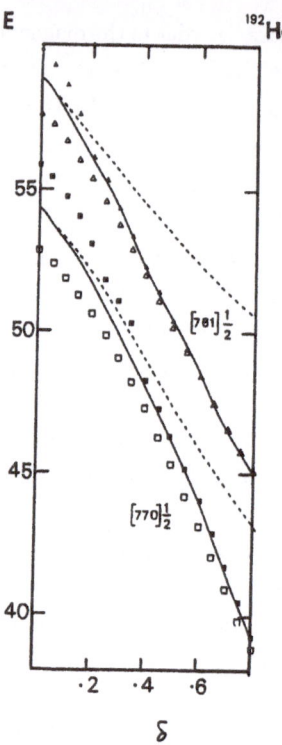

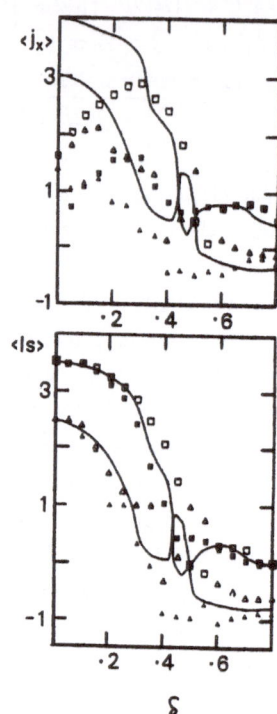

Fig.5. Energies with Coriolis.

Fig.6. ls and J_x with Coriolis.

Using the same approximation as in Fig.1, we can estimate the value of $< \sigma|l_x|\sigma >$ in perturbation approximation. Up to second order perturbation of $l^2 + ls$ interactions, the values of $< \sigma|l_x|\sigma >$ are zero both for the states which correspond to $[880]\frac{1}{2}$ and $[761]\frac{1}{2}$ at the limit of $\hbar\omega_{rot} = 0$. If we include Coriolis term with $\hbar\omega_{rot} = 0.2$ MeV in addition to the $l^2 + ls$ interaction, $< \sigma|l_x|\sigma >$ becomes -0.467 up to second order perturbation of the Coriolis force for the state with asymptotic $[880]\frac{1}{2}$. This value should be compared with the value of 8 at spherical limit, i.e. it is reduced to -6%. As for the state with asymptotic $[761]\frac{1}{2}$, $< \sigma|l_x|\sigma >$ become -0.295, which is reduced to -4% compared with the value of 7 at spherical limit.

SU(3) generator for the P-D levels

As the residual ls interaction is small for the P-D levels, we can limit our discussion to H_0 in (1) for the P-D levels. As an analogy to Holstein-Primakoff transformation in spin operators, we can construct new boson operators d and $d^\dagger$ from the harmonic oscillator bosons, c_i and $c_i^\dagger$ in (2) and (3).

$$d = \frac{1}{\sqrt{2 + 2c_z^\dagger c_z}} c_z c_z, \quad d^\dagger = c_z^\dagger c_z^\dagger \frac{1}{\sqrt{2 + 2c_z^\dagger c_z}}, \tag{6}$$

or

$$d = \frac{1}{\sqrt{4 + 2c_z^\dagger c_z}} c_z c_z, \quad d^\dagger = c_z^\dagger c_z^\dagger \frac{1}{\sqrt{4 + 2c_z^\dagger c_z}}. \tag{7}$$

Since we are costructing new bosons from the product of original bosons, there is the degeneracy of the vacuum, i.e. both $|0 > (c_z|0 >= 0)$ and $c_z^\dagger|0 >$ become the vacuum of b, subsequently we have two kinds of bosons. Eq.(6) corresponds to $N_{sh} =$ even and $d^\dagger d = c_z^\dagger c_z/2$, while (7) to $N_{sh} =$ odd and $d^\dagger d = (c_z^\dagger c_z - 1)/2$. For $N_{sh} =$ even (odd) case, $N_{sh}/2 - n_d =$ even ($(N_{sh} - 1)/2 - n_d =$ odd) corresponds to positive (negative) parity level and odd (even) to negative (positive) parity level, respectively. Here n_d corresponds to the eigenvalue of $\hat{n}_d = d^\dagger d$. So long as we are treating a fixed N_{sh}, both of (6) and (7) do not mix each other. Eqs. (6) and (7) are the same as the ladder operator introduced by ref.[11]. Analogous to the spherical case the following 9 operators are all commutable with Hamiltonian and the commutation relations among them are closed inside of them.

$$\begin{pmatrix} c_x^\dagger c_x & c_y^\dagger c_x & d^\dagger c_x \\ c_x^\dagger c_y & c_y^\dagger c_y & d^\dagger c_y \\ c_x^\dagger d & c_y^\dagger d & d^\dagger d \end{pmatrix} \tag{8}$$

The sum of these biproduct operators are also commutable with H_0.

Elliot[12] has found the quadrupole group operator which is commutable with the spherical Hamiltonian through the formula,

$$Q_q = \sqrt{\frac{4\pi}{5}} (r^2 Y_{2q} + b^4 p^2 Y_{2q})/b^2, \tag{9}$$

where $b^4 = 1/(M^2\omega_0^2)$. His definition of quadrupole operator corresponds to the approximation to cut $\pm 2\hbar\omega_0$ excitation, i.e. (9) is equivalent to a exact quadrupole operator within a shell. According to this recipe developed in the spherical case, we replace

$$x \rightarrow x\sqrt{M\omega_\perp}, \; y \rightarrow y\sqrt{M\omega_\perp}, \; z \rightarrow z\sqrt{M\omega_z},$$
$$p_x \rightarrow \frac{p_x}{\sqrt{M\omega_\perp}}, \; p_y \rightarrow \frac{p_y}{\sqrt{M\omega_\perp}}, \; p_z \rightarrow \frac{p_z}{\sqrt{M\omega_\perp}}, \tag{10}$$

in the equation

$$Q_q = \sqrt{4\pi/5}(r^2 Y_{2q} + p^2 Y_{2q}). \tag{11}$$

This replacement of (10) in (11) is equivalent to get (9) when we replace $r \rightarrow r\sqrt{M\omega_0}$ and $p \rightarrow p/\sqrt{M\omega_0}$ in (11). After inserting (10) in (11), then replacing $c_z, c_z^\dagger$ by $d, d^\dagger$, we get the new set of group operators $\tilde{Q}$ and $\tilde{\ell}$ for our new SU(3) group.

$$\tilde{Q}_0 = 2d^\dagger d - c_+^\dagger c_+ - c_-^\dagger c_-,$$
$$\tilde{Q}_{\pm 1} = -\sqrt{3}(d^\dagger c_\mp - dc_\pm^\dagger),$$

$$\tilde{Q}_{\pm 2} = -\sqrt{6}c_{\pm}^{\dagger}c_{\mp},$$
$$\tilde{\ell}_{\pm} = d^{\dagger}c_{\mp} + dc_{\pm}^{\dagger},$$
$$\tilde{\ell}_z = c_{+}^{\dagger}c_{+} - c_{-}^{\dagger}c_{-}. \tag{12}$$

The oscillator Hamiltonian H_0 is invariant with respect to the group SU(3) described by this eight infinitesimal group operators $\tilde{\ell}_q$ and $\tilde{Q}_q$. The commutation relations among these operators are closed within them. In contrast to Elliot's case which corresponds to cut $\pm 2\hbar\omega$ excitations, our SU(3) operators neglect $\pm 2\hbar\omega_{sh}$ excitations. From these generators we find the Casimir operator as,

$$C = \frac{1}{2}(\tilde{Q}\cdot\tilde{Q} + 3\tilde{\ell}\cdot\tilde{\ell}) = 2(\hat{n}_d + \hat{n}_{\perp})(\hat{n}_d + \hat{n}_{\perp} + 3), \tag{13}$$

where $\hat{n}_{\perp} = c_{+}^{\dagger}c_{+} + c_{-}^{\dagger}c_{-}$. In the reduction of the SU(3) group to $R_3 \times U_1$ subgroups, the group U_1 is described by $\tilde{Q}_0$, while representation of R_3 are labelled by an integral or half integral number $n_{\perp}/2$ like an angular momentum.

$$v^2 = \frac{1}{4}\tilde{\ell}_0^2 + \frac{1}{12}(\tilde{Q}_2\tilde{Q}_{-2} + \tilde{Q}_{-2}\tilde{Q}_2) = \frac{\hat{n}_{\perp}}{2}(\frac{\hat{n}_{\perp}}{2} + 1) \tag{14}$$

The linear combination of $\tilde{Q}_{\pm 1}$ and $\tilde{\ell}_{\pm 1}$, i.e. $\xi_{\pm} = \tilde{Q}_{\pm 1} + \sqrt{3}\tilde{\ell}_{\pm}$ ($\eta_{\pm} = \tilde{Q}_{\pm 1} - \sqrt{3}\tilde{\ell}_{\pm}$) becomes the lowering (raising) operator of the value of $< \tilde{Q}_0 >$ to $< \tilde{Q}_0 > \mp 3$. The operator $\tilde{Q}_0$ describes the excess of quanta in z-direction to those in the $x - y$ plane, while ξ and η are operators which shift an oscillator quantum from the z-direction into the $x - y$ plane by unit 3 and vice versa. Then we can summarize the classification of the single particle states in a superdeformed oscillator of $\tilde{Q}_0$ and $\tilde{\ell}_0$ for $N_{sh} \leq 5$ in Table 1. Here the half integer values of $n_{\perp}/2$ (in other words odd integer values of Λ) correspond to the different parity levels in N_{sh}.

We see the difference between the Elliot's SU(3) case and the superdeformed SU(3) case: (1) For the fixed value of Casimir operator, there correspond two superdeformed shells, N_{sh} and $N_{sh} + 1$ (in other words SU(3) $\times$ SU(3)). (2) One N_{sh} contains both parity levels corresponding to the values of $n_{\perp}/2$ or the values of Λ. (3) Because of the replacement of $c_z^{\dagger}$ and c_z by $b^{\dagger}$ and b and also the approximation of (10) in deriving (12), the tilded operators except for $\tilde{\ell}_z$ are not realistic operators. When we want to get the rotational energy, we must take care of the difference between the realistic operators and (12).

$$Q\cdot Q = 2C - 3\ell\cdot\ell + 2\Delta C, \tag{15}$$

where Q and ℓ are realistic operators and ℓ^2 gives the rotational energy proportional to $\ell(\ell + 1)$. The difference ΔC contains nondiagonal operators and its expectation value by the state $|n_d, n_{\perp} >$ becomes

$$< n_d, n_{\perp}|\Delta C|n_d, n_{\perp} > = 2(n_z^2 - n_d^2) + n_{\perp}(5n_z - 4n_d) + 7n_z - 6n_d$$
$$+ \frac{1}{8}(9n_{\perp}^2 + 18n_+n_- + 31n_{\perp} + 22). \tag{16}$$

Table 1. The single particle eigenfunction of $\tilde{Q}_0$ and $\tilde{\ell}_0$ for $N_{sh} \leq 5$.

N_{sh}	$n_\perp/2$	$<\tilde{Q}_0>$	n_d	C	N_{sh}	$n_\perp/2$	$<\tilde{Q}_0>$	n_d	C
0	0	0	0	0	1	0	0	0	0
2	0	2	1	8	3	0	2	1	8
	$\frac{1}{2}$	-1	0	8		$\frac{1}{2}$	-1	0	8
4	0	4	2	20	5	0	4	2	20
	$\frac{1}{2}$	1	1	20		$\frac{1}{2}$	1	1	20
	1	-2	0	20		1	-2	0	20

This is the third difference of the SU(3) in superdeformation from Elliot's case.

Conclusion

Although our numerical analysis is based on a simple Nilsson calculation in $\Delta N = 0$ space, we can conclude the following results. The P-D levels are almost degenerate at superdeformation because the levels mixed with them through the residual interactions are locating far from them at superdeformation. For these P-D levels the $L-S$ coupling scheme is recovered with increasing deformation, which is caused from the nature of the large quadrupole deformed field δY_{20}. Then we can use "$L-S$ coupling" scheme instead of "$j-j$ coupling" scheme for the P-D levels. Since these P-D levels are locating near the fermi surface of ^{192}Hg and ^{152}Dy (i.e. reference system in the discussion of the quantization of alignment), we can use real spin " s " instead of the pseudo-spin " $\tilde{s}$ ". If the pseudo-spin is responsible for the quantization of alignment, the degeneracy of pseudo-spin multiplet is always found in any deformation, which contradicts the fact that the quantization of alignment is not observed in Ce region. Thus we propose this P-D levels are the candidate for the excited superdeformed bands and the quantization of alignment.

We can construct new boson operators from the harmonic oscillator bosons only for these P-D levels, and eight generators of the SU(3) group using the recipe developed by Elliot. Using the Casimir operator made from eight generators and the representation of the subgroup R_3, we find the difference between Elliot's SU(3) and superdeformed SU(3). Our further work is the many body wave function based on this SU(3) representation, which will be published anywhere.

REFERENCES

1) F.S. Stephens et al, Phys. Rev. Lett. 64:2623(1990); 65:301 (1990).

2) E.F. Moore et al, Phys. Rev. Lett. 63:360(1989).

3) P.J. Twin et al, Phys. Rev. Lett. 57:811(1986).

4) P.J. Nolan et al, J. Phys. G. 11:L17(1985).

5) C.Gustafson, I.L. Lamm, B. Nilsson and S.G. Nilsson, Arkiv. Fys. 36:613(1967).

6) P. Ring and P. Schuck, "Nuclear Many-Body Problem" Springer-Verlag,(1980).

7) A. Arima, M. Harvey and K. Shimizu, Phys. Lett. 30B:517(1969).

8) K.T. Hecht and A. Adler, Nucl. Phys. A137:129(1969).

9) S.G. Nilsson, Dan Mat. Fys. Medd. 29:no.16(1955).

10) A. Bohr and B.R. Mottelson, "Nuclear Structure" 2 Benjamin,(1975).

11) W. Nazarewitch and J. Dobaczewski, Phys. Rev. Lett. 68:154(1992).

12) J.P. Elliot, Proc. Roy. Soc. A245:128; 562(1958).

COEFFICIENTS OF FRACTIONAL PARENTAGE OF THE BOSON SYSTEM WITH SINGLE l AND F-SPIN 1/2[1]

Hong-zhou Sun[a,b], Qi-zhi Han[c], and Yu-xin Liu[d,b]

[a] Department of Physics, Tsinghua University, Beijing 100084, China

[b] Institute of Theoretical Physics, Academia Sinica
Beijing 100080, China

[c] Department of Physics, Peking University, Beijing 100871,China

[d] CCAST(World Laboratory), P. O. Box 8730, Beijing 100080, China

Abstract

A system of identical bosons with single angular momentum l and F-spin 1/2 is under consideration. The classification of it's wave functions and the factorization of it's coefficients of fractional parentage are discussed by using Lie group theory. The branching rules of the group reductions and an approach to evaluate the CFP of the system are given.

I Introduction

Interacting boson model(IBM) has been successful in describing the nuclear structure[1]. In its original version(IBM1), which does not distinguish the proton boson from neutron boson, the three kind nuclear collective motions can be described quite well. With the IBM1 being developed, the version of IBM2 which treats the proton boson and neutron boson separately was put forward. Then both the theory and the agreement between the prediction and experimetal data were made a remarkable progress[2,3]. Thereafter with the g-boson being introduced, the well deformed nuclei and the higher excited states can be described[4,5]. And the nuclear octupole deformation is described by means of considering the f- and p-bosons[6,7].

---

[1]The project supported by Doctoral Programm Foundation of Institution of Higher Education of China and National Natural Science Foundation of China

In recent years, the nuclear high spin states and the superdeformed states have greatly been paid attention to[8]. On the basis of IBM, considering the mechanism of pair broken, we can discuss the properties of high spin states and the backbending[9]. With the core excitation being taken into account, the superdeformed nuclei can be described by the normal deformed bosons and superdeformed bosons. The nuclear superdeformed state and the shape coexsistance can then be described well[9,10].

It is known that the coefficients of fractional parentage(CFP)[11−14] method is one of the most efficient techneque for constructing the interacing boson model wave functions. Even though many works have been done to study nuclear physics, the discussion on the CFP of IBM is relatively meager. As a consequence, the work load of calculation of some computer codes in the framework of IBM is tremendously heavy. Moreover some of them can not be performed in the case that the number of bosons in the nucleus is relatively larger. These defects have limitted the application region of IBM. It is fortunate that Sun and their colaborators have recently put forward a simple formula to calculate the CFP of IBM1[15]. And a computer code basing on the formula has also been set up to determine the CFP of IBM1[16]. With the code the CFP's of the system including 42 d-bosons can be obtained. It provides us convenience to describe the nuclear superdeformed state and chaotic behavier of many boson system.

In fact, as the nuclear physics being discussed in the framework of IBM, it is usually recongnised that the IBM2 is the most efficient version to be used. It is certain that the proton boson and the neutron boson can be treated separately in this case. However the configuration space is enlarged remarkablely. It increases a great amount of the work of calculation, so that the application region is limitted. In the other hand the physical meaning is not transparent enugh. To solve this problem, a formalism to determine the CFP of IBM2, which is just the CFP of boson system with single angular momentum l and F-spin 1/2, is proposed in the light of Lie group theory. In this paper we discuss the classification of the wave function of the boson system with single l and F-spin 1/2 and the factorization of the CFP, and provide an approach to evaluate the factors.

2 Classification of Wave Functions

The wave functions of the system with many bosons, each with angular momentum l and F-spin 1/2, are classified according to the group chain

$$U(2N) \supset (U(N) \supset O(N) \supset O(3)) \otimes SU(2) \qquad (N = 2l + 1). \qquad (2.1)$$

Supposing

$$b_{m\sigma}^\dagger = b_{lm1/2\sigma}^\dagger,$$
$$b_{m\sigma} = b_{lm1/2\sigma}, \qquad (2.2)$$

are the creation and annihilation operators of boson with angular momentum l, F-spin $1/2$ and z component m and σ. And let

$$\tilde{d}_{m\sigma} = (-1)^{l+m+1/2+\sigma} b_{l-m1/2-\sigma}$$

be the irreducible tensor corresponding to $b_{m\sigma}$. The generators, Casimir operators, and the labels of IRRPs for each subgroup in the group chain (2.1) are given in Table 1.

Table 1. The generators, Casimir operators and IRRP labels
for groups in chain (2.1)

Group	Generators	Casimir Operators	IRRP Labels
$U(2N)$	$B_{qu}^{kf} = (b^\dagger \tilde{b})_{qu}^{kf}$	$C_{1U(2N)} = \sum\limits_{m,\sigma} b_{m\sigma}^\dagger b_{m\sigma} = \hat{N}_b$	$[n]$
		$C_{2U(2N)} = \sum\limits_{k,f} B^{kf} \cdot B^{kf}$	
		$= \hat{N}_b(\hat{N}_b + 2N + 1)$	
$U(N)$	$P_q^k = \sqrt{2}\, B_{q0}^{k0}$	$C_{2U(N)} = \sum\limits_k P^k \cdot P^k$	$[n_1, n_2]$
		$= \dfrac{\hat{N}_b(\hat{N}_b + 2N - 4)}{2} + 2\hat{F} \cdot \hat{F}$	
$O(N)$	$P^k \quad k = odd$	$C_{2O(N)} = \sum\limits_{k=odd} P^k \cdot P^k$	(σ_1, σ_2)
		$= C_{2U(N)} - \hat{N}_b - 2\wp^\dagger \cdot \tilde{\wp}$	
$O(3)$	$\hat{L}_q = \sqrt{l(l+1)N/3}\, P_q^1$	$C_{2O(3)} = \hat{L} \cdot \hat{L}$	L
$SU(2)$	$\hat{F}_u = \sqrt{N/2}\, B_{0u}^{01}$	$C_{2SU(2)} = \hat{F} \cdot \hat{F}$	F

in which $\wp_u^\dagger = \sqrt{\frac{N}{2}}(b^\dagger b^\dagger)_{0u}^{01}$, $\tilde{\wp}_u = \sqrt{\frac{N}{2}}(\tilde{b}\tilde{b})_{0u}^{01}$ is the creation and annihilation operator of boson pair respectively. They are the invariant quantities of group $O(N)$.

The wave functions of the boson system can then be written as

$$|[n]_{2N} \quad [n_1\, n_2]_N \quad (\sigma_1\sigma_2) \ \alpha \ L \quad \beta \ F \,\rangle,$$

$$U(2N) \quad U(N) \quad O(N) \quad O(3) \ SU(2) \tag{2.3}$$

where α and β are additional quantum numbers. The reason for including these additional quantum number is that the reductions of $U(N) \supset O(N)$ and $O(N) \supset O(3)$ are not simple reducible.

The eq.(2.3) can be rewritten as

$$|n(sf)\alpha L\beta F\rangle = |[n]_{2N} [n_1\, n_2]_N (\sigma_1\sigma_2)\, \alpha\, L\, \beta\, F\rangle, \tag{2.4}$$

where

$$s = \sigma_1 + \sigma_2, \qquad f = \frac{\sigma_1 - \sigma_2}{2}.$$

From eq.(2.3) we know that it requires 7 parameters to label the wave function completely. These 7 parameters can be chosen as the ones in eq.(2.4), i.e., $n, s, f, \alpha, L, \beta, F$. According to the definition of eq.(2.4), $|n(sf)\alpha L\beta F\rangle$ satisfies the following relations,

$$
\begin{bmatrix}
C_{2U(2N)} \\
C_{2U(N)} \\
C_{2O(N)} \\
C_{2O(3)} \\
C_{2SU(2)}
\end{bmatrix}
|n(sf)\alpha L\beta F\rangle =
\begin{bmatrix}
n(n + 2N - 1) \\
\dfrac{n(n + 2N - 4)}{2} + 2F(F + 1) \\
\dfrac{n(n + 2N - 6)}{2} + 2f(f + 1) \\
L(L + 1) \\
F(F + 1)
\end{bmatrix}
|n(sf)\alpha L\beta F\rangle. \quad (2.5)
$$

Because $\wp_u^\dagger, \tilde{\wp}_u$ is the invariant quantities of group $O(N)$, $|n(sf)\alpha L\beta F\rangle$ can be obtained by means of acting the operator $\wp_u^\dagger$ on $|(sf)\alpha L\rangle = |s(sf)\alpha Lf\rangle$ step by step, i.e.,

$$
|n(sf)\alpha L\beta F\rangle = C\{(\wp_u^\dagger)^\rho|(sf)\alpha L\rangle\}^{\beta F}. \quad (2.6)
$$

where $\rho = \frac{n-s}{2}$, C is the normalizing constant. $|(sf)\alpha L\rangle$ satisfies the restrictions shown in the following

$$
\begin{bmatrix}
\hat{N}_b \\
\hat{L}^2 \\
\hat{F}^2 \\
\tilde{\wp}_u
\end{bmatrix}
|(sf)\alpha L\rangle =
\begin{bmatrix}
s \\
L(L + 1) \\
f(f + 1) \\
0
\end{bmatrix}
|(sf)\alpha L\rangle. \quad (2.7)
$$

The discussion above shows that s is the seniority and f can be regarded as the reduced F-spin.

3 The Branching Rule of the Reduction: $U(2N) \supset (U(N) \supset O(N) \supset O(3)) \otimes SU(2)$

(a) The Reduction of $U(2N) \supset U(N) \otimes SU(2)$
The branching rule for this reduction is quite simple. It can be expressed as

$$
[n]_{2N} = \sum_{n_1 n_2} [n_1 \, n_2]_N \otimes F, \quad (3.1)
$$

where

$$
n = n_1 + n_2, \qquad F = \frac{n_1 - n_2}{2}.
$$

(b) The Reduction of $U(N) \supset O(N)$

By considering the kronecker product of the IRRPs of group $U(N)$

$$[n_1]_N \otimes [n_2]_N = [n_1 + n_2]_N \oplus [n_1 + n_2 - 1, 1]_N \oplus \cdots\cdots \oplus [n_1, n_2]_N, \qquad (3.2)$$

where

$$n_1 \geq n_2,$$

we get

$$[n_1, n_2]_N = [n_1]_N \otimes [n_2]_N \ominus [n_1 + 1]_N \otimes [n_2 - 1]_N. \qquad (3.3)$$

By the same way we know that the relation between the kronecker product of IRRPs of group $O(N)$ is[17]

$$\langle \sigma \rangle \otimes \langle \sigma' \rangle = \langle \sigma + 1 \rangle \otimes \langle \sigma' - 1 \rangle \oplus \sum_{\alpha=0}^{\sigma'} \langle \sigma - \sigma' + \alpha, \alpha \rangle. \qquad (3.4)$$

Taking the following branching rule for the totlely symmetric IRRP of group $U(N)$ into account

$$[n]_N = \langle n \rangle \oplus \langle n - 2 \rangle \oplus \langle n - 4 \rangle \oplus \cdots\cdots, \qquad (3.5)$$

we get the following recurrent relations of the branching rule of the reduction $U(N) \supset O(N)$

$$[n, n]_N = [n - 2, n - 2]_N + F(n, n) - F(n - 1, n - 1);$$

$$[n, n - 1]_N = F(n, n - 1); \qquad (3.6)$$

$$[n_1, n_2]_N = [n_1 - 2, n_2]_N + F(n_1, n_2),$$

where

$$F(n_1, n_2) = \sum_{\beta \, \alpha} \langle n_1 - \beta + \alpha, \alpha \rangle;$$

$$\beta = n_2, n_2 - 2, n_2 - 4, \cdots\cdots, \geq 0; \qquad (3.7)$$

$$\alpha = \beta, \beta - 1, \beta - 2, \cdots\cdots, 0.$$

With the eqs.(3.6) and (3.7), all of the branching rules of this reduction can be obtained. The branching rules for the IRRPs $[n]_N$ and $[n, 1]_N, \cdots\cdots$, are given in table 2.

Table 2. Some Branching Rule of the Reduction $U(N) \supset O(N)$

(a) $[n]_N$

$n = even$	$< n >$	$< n - 2 >$	$< n - 4 >$	$\cdots\cdots$	$< 0 >$
$n = odd$	$< n >$	$< n - 2 >$	$< n - 4 >$	$\cdots\cdots$	$< 1 >$

(b) $[n, 1]_N$

$n = even$	$< n - 1 >$	$< n - 3 >$	$\cdots\cdots$	$< 3 >$	$< 1 >$
	$< n, 1 >$	$< n - 2, 1 >$	$\cdots\cdots$	$< 4, 1 >$	$< 2, 1 >$
$n = odd$	$< n - 1 >$	$< n - 3 >$	$\cdots\cdots$	$< 2 >$	
	$< n, 1 >$	$< n - 2, 1 >$	$\cdots\cdots$	$< 3, 1 >$	$< 1, 1 >$

The corresponding wave functions labelled by these IRRPs can then be expressed as

$$|n\,(s\,f)\,L_{max}\,F\rangle = |[n]_{2N}\,[n]_N\langle\sigma\rangle L_{max}F\rangle$$

$$= \sqrt{\frac{(N+2\sigma-2)!!}{\rho!(N+2\sigma+2\rho-2)!!}}\{\wp^{\dagger\rho}|\langle\sigma\rangle L_{max}f\rangle\}^F, \tag{3.8}$$

with $\qquad \rho = \dfrac{n-\sigma}{2},\ F = \rho + f.$

$$|n\,(s\,f)\,L_{max}\,F\rangle = |[n]_{2N}\,[n-1,1]_N\langle\sigma,1\rangle L_{max}F\rangle$$

$$= \sqrt{\frac{(N+2\sigma-2)!!}{\rho!(N+2\sigma+2\rho-2)!!}}\{\wp^{\dagger\rho}|\langle\sigma,1\rangle L_{max}f\rangle\}^F, \tag{3.9}$$

with $\qquad \rho = \dfrac{n-\sigma-1}{2},\ F = \rho + f.$

$$|n\,(s\,f)\,L_{max}\,F\rangle = |[n]_{2N}\,[n-1,1]_N\langle\sigma\rangle L_{max}F\rangle$$

$$= \sqrt{\frac{(N+2\sigma-2)}{\rho!(N+2\sigma+2\rho-4)!!(N+\sigma-1)}}\{\wp^{\dagger\rho}|\langle\sigma\rangle L_{max}f\rangle\}^F, \tag{3.10}$$

with $\qquad \rho = \dfrac{n-\sigma}{2},\ F = \rho + f - 1.$

$$\cdots\cdots\cdots,$$

where

$$|(s\,f)L_{max} = |\langle\sigma_1,\sigma_2\rangle L_{max}f\rangle = \sqrt{\frac{(\sigma_1-\sigma_2+1)!}{(\sigma_1+1)!\sigma_2!}}B^{\dagger\sigma_2}b_{l\,1/2}^{\dagger(\sigma_1-\sigma_2)}|0\rangle;$$

$$B^\dagger = b_{l\,1/2}^\dagger b_{l-1\,-1/2}^\dagger - b_{l-1\,1/2}^\dagger b_{l\,-1/2}^\dagger.$$

(c) The Reduction of $O(N) \supset O(3)$

The method to get the branching rules for this reduction analytically has been proposed by Wang et. al.[18]. With the computer codes[18,19] all the branching rules can be obtained.

It is well known that reducing the IRRPs of group is rather complicated, so that whether the reduction is correct should be checked. One of the efficient checking measures is comparing the dimension of the IRRPs. The dimension of the IRRPs of group $U(2N)$, $U(N)$ and $O(N)$ can be found with the following formulae

$$d([n]_{2N}) = \begin{pmatrix} n + 2N - 1 \\ n \end{pmatrix};$$

$$d([n_1, n_2]_N) = \frac{n_1 - n_2 + 1}{n_1 + 1} \begin{pmatrix} n_1 + N - 1 \\ n_1 \end{pmatrix} \begin{pmatrix} n_2 + N - 2 \\ n_2 \end{pmatrix}; \tag{3.11}$$

$$d(\langle \sigma_1 \sigma_2 \rangle) = \frac{(\sigma_1 + N - 4)!(\sigma_2 + N - 5)!}{(\sigma_1 + 1)!\sigma_2!(N - 2)!(N - 4)!}(2\sigma_1 + N - 2)(2\sigma_2 + N - 4)$$

$$(\sigma_1 + \sigma_2 + N - 3)(\sigma_1 - \sigma_2 + 1).$$

4 The Coefficient of Fractional Parentage

The coefficient of fractional parentage for the system whose wave founction can be labelled by the group chain (2.1) can be expressed in the second quantization representation as

$$\langle n(sf)\alpha L\beta F\{|n - 1(s'f')\alpha'L'\beta'F'\rangle = \sqrt{\frac{1}{n}}\langle n(sf)\alpha L\beta F\|b^\dagger\|n - 1(s'f')\alpha'L'\beta'F'\rangle, \tag{4.1}$$

where $\langle \cdots \|b^\dagger\| \cdots \rangle$ is the reduced matrix element of the irreducible tensor $b^\dagger_{m\sigma}$. It is easy to show that $b^\dagger_{m\sigma}$ is the irreducible tensor with rank $[1]_{2N}$ under the group chain (2.1), i.e.,

$$b^\dagger_{m\sigma} = b^\dagger([1]_{2N} \quad [1]_N \quad \langle 1, 0 \rangle \quad lm \quad 1/2\sigma). \tag{4.2}$$
$$\phantom{b^\dagger_{m\sigma} = b^\dagger(}U(2N)\, U(N) \quad O(N) \quad O(3) \quad SU(2)$$

Then according to the generalized Wigner-Eckart theorem the reduced matrix element of $b^\dagger_{m\,\sigma}$ can be factorized, with respect to the group chain (2.1), as the following

$$\langle n\,(s\,f)\,\alpha\,L\beta\,F\|b^\dagger\|n - 1\,(s'\,f')\,\alpha'\,L'\beta'\,F'\rangle$$

$$= \sqrt{n} \begin{bmatrix} [1]_{2N} & [n - 1]_{2N} & [n]_{2N} \\ [1]_N & [n_1'\,n_2']_N & [n_1\,n_2]_N \end{bmatrix} \begin{bmatrix} [1]_N & [n_1'\,n_2']_N & [n_1\,n_2]_N \\ <1> & \beta' < \sigma_1'\,\sigma_2' > & \beta < \sigma_1\,\sigma_2 > \end{bmatrix} \tag{4.3}$$

$$\begin{bmatrix} <1> < \sigma_1'\,\sigma_2' > & <\sigma_1\,\sigma_2 > \\ l & \alpha'L' & \alpha L \end{bmatrix},$$

where

$$\begin{bmatrix} [1]_{2N} & [n-1]_{2N} & [\tilde{n}_1\,\tilde{n}_2]_{2N} \\ [1]_N & [n'_1\,n'_2]_N & [n_1\,n_2]_N \end{bmatrix} \quad \text{is the isoscalar factor of } U(2N) \supset U(N).$$

$$\begin{bmatrix} [1]_N & [n'_1\,n'_2]_N & [n_1\,n_2]_N \\ <1> & \beta' <\sigma'_1\,\sigma'_2> & \beta <\sigma_1\,\sigma_2> \end{bmatrix} \quad \text{is the isoscalar factor of } U(N) \supset O(N).$$

$$\begin{bmatrix} <1> & <\sigma'_1\,\sigma'_2> & <\sigma_1\,\sigma_2> \\ l & \alpha'L' & \alpha\,L \end{bmatrix} \quad \text{is the isoscalar factor of } O(N) \supset O(3).$$

The isoscalar factor of $U(2N) \supset U(N)$ can be easily obtained by using the highest weight state of the IRRP $[n_1\,n_2]_N$ of group $U(N)$. The results are shown in table 3.

Table 3. The Isoscalar Factor of $U(2N) \supset O(N)$

$\begin{matrix}\tilde{n}_1\ \tilde{n}_2 \\ n'_1\ n'_2\end{matrix}$	$n \quad 0$	$n-1 \quad 1$
$n_1 - 1\ n_2$	$\sqrt{\dfrac{(n_1+1)(n_1-n_2)}{(n_1+n_2)(n_1-n_2+1)}}$	$\sqrt{\dfrac{n_2(n_1-n_2+2)}{(n_1+n_2)(n_1-n_2+1)}}$
$n_1\ n_2 - 1$	$\sqrt{\dfrac{n_2(n_1-n_2+2)}{(n_1+n_2)(n_1-n_2+1)}}$	$-\sqrt{\dfrac{(n_1+n_2)(n_1-n_2)}{(n_1+n_2)(n_1-n_2+1)}}$

The CFP as shown in eq.(4.1) can then be rewritten as

$$\langle n(sf)\alpha L\beta F\{|n-1(s'f')\alpha'L'\beta'F'\rangle$$
$$= \langle n(sf)\beta F\{|n-1(s'f')\beta'F'\rangle\langle(sf)\alpha L\{|(s'f')\alpha'L'\rangle, \tag{4.4}$$

in which

$$\langle n(sf)\beta F\{|n-1(s'f')\beta'F'\rangle$$

$$= \begin{bmatrix} [1]_{2N} & [n-1]_{2N} & [n]_{2N} \\ [1]_N & [n'_1\,n'_2]_N & [n_1\,n_2]_N \end{bmatrix} \begin{bmatrix} [1]_N & [n'_1\,n'_2]_N & [n_1\,n_2]_N \\ <1> & \beta' <\sigma'_1\,\sigma'_2> & \beta <\sigma_1\,\sigma_2> \end{bmatrix}, \tag{4.5}$$

can be called as the F-spin part of CFP.

$$\langle(s\,f)\alpha\,L\{|(s'\,f')\alpha'\,L'\rangle = \begin{bmatrix} (1\,1/2)\,(s'\,f') & (s\,f) \\ l & \alpha'L' & \alpha\,L \end{bmatrix}$$
$$= \begin{bmatrix} <1> & <\sigma'_1\,\sigma'_2> & <\sigma_1\,\sigma_2> \\ l & \alpha'L' & \alpha\,L \end{bmatrix}, \tag{4.6}$$

can be referred as the orbital part of CFP.

$\langle n(sf)\beta F\{|n-1(s'f')\beta'F'\rangle$ can be calculated using $|n(sf)L_{max}\beta F\rangle$ (see eq.(3.8 – 3.10)). For $\langle (s\ f)\alpha\ L\{|(s'\ f')\alpha'\ L'\rangle$, we find the following simple formula to evaluate it

$$\begin{bmatrix} (1\ 1/2)\ (s-1,f') & (s,f) \\ l & \alpha'L' & (\alpha_1'\ L_1'\ f_1')\ L \end{bmatrix} = C\ \langle (s\ f)(\alpha_1'\ L_1'\ f_1')L\|b^\dagger\|(s-1\ f')\alpha'L'\rangle,$$

(4.7)

where C is the normalizing constant,

$$\langle (s\ f)(\alpha_1'\ L_1'\ f_1')L\|b^\dagger\|(s-1\ f')\alpha'L'\rangle = \frac{P(\alpha_1'\ L_1'\ f_1'\ \alpha'\ L'\ f')}{\sqrt{P(\alpha_1'\ L_1'\ f_1'\ \alpha_1'\ L_1'\ f_1')}},$$

(4.8)

where

$$P(\alpha_1'\ L_1'\ f_1'\ \alpha'\ L'\ f') = \delta(\alpha_1',\alpha')\ \delta(L_1',L')\delta(f_1',f')$$

$$- \sum_{f''\ \alpha''L''} (-1)^{f_1'+f'+L_1'+L'}\sqrt{(2f_1'+1)(2f'+1)(2L_1'+1)(2L'+1)}$$

$$\left[\begin{Bmatrix} 1/2\ f''\ f_1' \\ 1/2\ f\ f' \end{Bmatrix} \begin{Bmatrix} l\ L''\ L_1' \\ l\ L\ L' \end{Bmatrix} + \right.$$

$$\left. \frac{(-1)^{f_1'+f'}6\delta(L'',L)}{(2L+1)((N+s-4)-f''(f''+1)+f(f+1))} \begin{Bmatrix} 1/2\ f''\ f_1' \\ f\ 1/2\ 1 \end{Bmatrix} \begin{Bmatrix} 1/2\ f''\ f' \\ f\ 1/2\ 1 \end{Bmatrix} \right]$$

$$\langle (s-1\ f_1')\alpha_1'\ L_1'\|b^\dagger\|(s-2\ f'')\alpha''L''\rangle\langle (s-1\ f')\alpha'\ L'\|b^\dagger\|(s-2\ f'')\alpha''L''\rangle,$$

(4.9)

and we have the reciprocal relation

$$\begin{bmatrix} (1\ 1/2) & (s,f) & (s-1,f') \\ l & (\alpha_1'\ L_1'\ f_1')L & \alpha'\ L' \end{bmatrix} = (-1)^{l+L'+L}\sqrt{\frac{d(s-1,f')(2L+1)}{d(s,f)(2L'+1)}}$$

$$\begin{bmatrix} (1\ 1/2)\ (s-1,f') & (s,f) \\ l & \alpha'L' & (\alpha_1'\ L_1'\ f_1')\ L \end{bmatrix},$$

(4.10)

where $d(s,f)$ is the dimension of the $IRRP\ (s,f)$ of group $O(N)$.

In this paper the classification of the wave function of the boson system with single angular momentum l and F-spin 1/2 and the branching rules of the IRRP reductions in the group chain $U(2N) \supset (U(N) \supset O(N) \supset O(3)) \otimes SU(2)$ are discussed. The coefficient of fractional parentage for the boson system is factorized into the F-spin part $\langle n(sf)\beta F\{|n-1(s'f')\beta'F'\rangle$ and the orbital part $\langle (sf)\alpha L\{|(s'f')\alpha'L'\rangle$. A recurrent formula to evaluate the CFP with well-defined F-spin and seniority is given. We hope these discussion will improve the IBM calculations, and then promote the discussion on superdeformed nuclei and the quantum chaotic hehavier of many boson system.

References

[1] F. Iachello and A. Arima, *The Interacting Boson Model* (Cambridge; Cambridge University Press, 1987)

[2] A. Arima, T. Otsuka, F. Iachello and I. Talmi, Phys. Lett. **66B**(1977), 205

[3] G. L. Long, Ph. D. Thesis(Tsinghua University, 1987)

[4] H. Z. Sun, M. Moshinsky, A. Frank and F. Isacker, Kinam **5** (1983), 135

[5] T. Otsuka, Phys. Rev. Lett. **46**(1981), 710; ibid, **48** (1982), 387

[6] J. Engel and F. Iachello, Phys. Rev. Lett. **54**(1985), 1126; ibid, Nucl. Phys. **A472**(1987), 61

[7] H. Z. Sun, M. Zhang and Q.Z. Han, Chinese J. Nucl. Phys., **13**(1991), 121

[8] P. J. Twin, Nucl. Phys. **A520**(1990), 17c; ibid; **A522** (1991), 13c

[9] F. Iachello and D. Vretener, Phys. Rev. **C43**(1991), R945; F. Iachello, Nucl. Phys. **A522**(1991), 83c

[10] T. Otsuka and M. Honma, Phys. Lett., **268B**(1991), 305

[11] R. F. Bacher and S. Goudsmit, Phys. Rev. **46**(1934)948

[12] G. Racah, Phys. Rew. **63**(1943)367

[13] P. J. Redmond, Proc. Roy. Soc. London **A222**(1954)84.

[14] A. de-Shalit and I. Talmi, *Nuclear shell Theory*(Academic, New York 1963)

[15] H. Z. Sun, M. Zhang, Q. Z. Han ang G. L. Long, J. Phys. **A22**(1989), 4769; ibid, **A23**(1990), 1957

[16] Y. X. Liu, H. Z. Sun and E. G. Zhao, Comput. Phys. Commun. **70**(1992), 154

[17] Q. Z. Han and H. Z. Sun, *Group Theory*(Peking University, 1987)

[18] J. J. Wang and H. Z. Sun, High Energy Phys. Nucl. Phys., **14** (1990), 842; H. Z. Sun, J. J. Wang and Y. X. Liu, to be published

[19] Y. X. Liu, Chinese J. Comput. Phys. **9**(1992), 163

GIANT DIPOLE RESONANCES IN THE SU(3)⊗SU(2) LIMIT
OF THE INTERACTING BOSON-FERMION MODEL

Alberto Ventura[1], Giuseppe Maino[1], and Lina Zuffi[2]

[1]Ente Nuove Tecnologie, Energia e Ambiente, Bologna, Italy, and Istituto Nazionale di Fisica Nucleare, Sezione di Firenze, Firenze, Italy
[2]Dipartimento di Fisica dell' Universita' di Milano, and Istituto Nazionale di Fisica Nucleare, Sezione di Milano, Milano, Italy

INTRODUCTION

The scope of the present work is to describe the excitation of high-energy collective states, such as the giant dipole resonance (GDR), in deformed odd-mass nuclei, on the assumption that the low-energy spectrum belongs to an irreducible representation (irrep) of a limit symmetry of the interacting boson-fermion model[1] (IBFM), namely the Bose-Fermi symmetry SU(3)⊗SU(2), where SU(3) is the group of pseudo-orbital angular momentum and SU(2) the group of pseudospin.

As is known, the strong spin-orbit coupling in the nuclear shell model prevents total orbital momentum, L, and total spin, S, from being good quantum numbers. However, it is empirically observed that, with a realistic choice of shell model parameters, the major shells of interest to medium-heavy nuclei are made up with single-particle doublets, $\{ l_j, (l+2)_{j+1} \}$, of the same parity, the so-called normal parity, up to a maximum angular momentum, j_{max}, to which a single-particle state is added with opposite (abnormal) parity and $j = j_{max} + 2$. It becomes thus natural to introduce a pseudo-shell quantum number, $\tilde{n} = j_{max} - 1/2 = n - 1$, n being the major shell number, a pseudo-orbital angular momentum, $\tilde{l} = \tilde{n}, \tilde{n} - 2, \ldots 0$, or 1, according to whether $\tilde{n}$ is even, or odd, and a pseudospin $\tilde{s} = 1/2$ [2,3].

An exact pseudospin symmetry would correspond to the degeneracy of the members of the pseudospin doublets, $\{ \tilde{l}_{\tilde{l}-1/2}, \tilde{l}_{\tilde{l}+1/2} \}$, belonging to irreps of SU(2), while pseudo-orbital symmetry would lead to the classification of normal parity states in a major shell according to the unitary group U(N), where $\tilde{N} = \Sigma_{\tilde{l}} (2\tilde{l} + 1) = (\tilde{n} + 1)(\tilde{n} + 2)/2$ is the total pseudoshell degeneracy. If, in addition, the nuclear states are characterized by large quadrupole deformations, they can be further classified according to the group reduction chain U(N) ⊃ SU(3), where the harmonic oscillator group SU(3) is generated by the five components of the pseudo-quadrupole operator, Q, and the three components of the pseudo-orbital momentum, L, which, in turn, generate an SO(3) subgroup.

Such a classification holds for protons and neutrons separately, but, if they are strongly coupled, i. e. coupled at the SU(3) level, the low-energy nuclear states made up with single-particle states of normal parity will belong to irreps of a global SU(3)⊗SU(2) group

Fermion models with the above dynamic symmetry have been developed, in particular, by Draayer and collaborators[4,5,6,7] and applied with good success to deformed nuclei in the lanthanide and actinide regions.

Empirical basis of this approach is the Nilsson scheme of levels, labelled with their asymptotic pseudo-quantum numbers, $[\tilde{n}\tilde{n}_z\tilde{\Lambda}]\tilde{\Omega}$, where $\tilde{n}$ has already been defined, $\tilde{n}_z$ is the number of oscillator quanta along the z-axis, taken to be the symmetry axis, $\tilde{\Lambda}$ is the projection of the pseudo- orbital momentum, $\tilde{l}$, and $\Omega = j_z$ the projection of the total angular momentum on the same axis. Filling up Nilsson orbitals of an even nucleon system with pairs of valence particles creates a quadrupole moment $<Q_0> \propto \Sigma(2\tilde{n}_z - \tilde{n}_x - \tilde{n}_y) = 2\lambda + \mu$, where (λ, μ) are Elliott's quantum numbers of the SU(3) irrep to which the many-nucleon state belongs[8,9]. The leading irrep of the low-lying spectrum is expected to have the highest possible value of $\varepsilon = 2\lambda + \mu$, and, once ε is fixed, the highest possible $\mu = \Sigma(\tilde{n}_x - \tilde{n}_y)$. If the leading irreps for protons and neutrons are (λ_π, μ_π) and (λ_v, μ_v), respectively, the nuclear ground state will belong to $(\lambda_\pi + \lambda_v, \mu_\pi + \mu_v)$, as a consequence of the strong coupling between protons and neutrons. In this way, the majority of deformed even-even nuclei have leading irreps with $\mu \geq 2$, containing both the (K=0) ground-state band and the (K=2) γ band.

Similar group-theory techniques, but different physical assumptions are employed in the interacting boson model[1] of deformed nuclei, where the angular momentum coupling of valence pairs is restricted to L=0,2; the fermion pairs are thus approximated by s and d bosons and the basic unitary group has $N = 6$. The low-lying spectrum is associated with the totally symmetric U(6) irrep $[N, 0, 0, 0, 0, 0]$, where $N = N_\pi + N_v$ is the effective boson number. N_π (N_v) is usually taken to be the smaller of the numbers of particle pairs and hole pairs in the proton (neutron) valence shell. Differently from Draayer, the SU(3) irrep containing the ground-state band of an even- even nucleus with axial symmetry, originated from the decomposition of boson representations of the U(6)⊃SU(3) chain, is always symmetric, (2N,0), while the γ band belongs to $(2N - 4,2)$.

THE IBFM FOR DEFORMED NUCLEI

The limit symmetry of the IBFM suitable to the description of low energy spectra of odd-mass nuclei has been discussed in detail in refs.[1,10]. Therefore, only a few points, necessary to the extension to high-energy modes, such as the GDR, will be recalled here.

We assume that the fermion pairs in the valence shells of the even-even core are bosonized and their states classified according to the $U_B(6) \supset SU_B(3)$ chain; moreover, the unpaired fermion occupies a Nilsson level of the $\tilde{n}$-th pseudoshell and the reduction $U_F(N) \otimes SU(2) \supset SU_F(3) \otimes SU(2)$ is meaningful; it is then possible to combine the two chains into a common Bose-Fermi symmetry

$$U_B(6) \otimes U_F(\tilde{N}) \otimes SU(2) \supset SU_B(3) \otimes SU_F(3) \otimes SU(2)$$

$$\supset SU_{BF}(3) \otimes SU(2) \supset SO_{BF}(3) \otimes SU(2) \supset Spin(3) \tag{1}$$

In that case, the IBFM Hamiltonian, restricted for simplicity's sake to a sum of one-body and two-body terms, can be written as a combination of linear and quadratic Casimir operators made up with the generators of the groups appearing in the reduction chain (1):

$$\hat{H} = \hat{H}_1 + \beta_1 \hat{C}_2(SU_B(3)) + \beta_2 \hat{C}_2(SU_{BF}(3)) + \gamma_1 \hat{C}_2(SO_{BF}(3)) + \gamma_2 \hat{C}_2(Spin(3)). \quad (2)$$

Here, the one-boson and one-fermion terms contribute only to the binding energy and are included in $\hat{H}_1$, while the quadratic Casimir operators, $\hat{C}_2$, give rise to the excitation spectrum and are written as follows, together with their eigenvalues:

$$\hat{C}_2(SO(3)) = \hat{L} \cdot \hat{L} \Rightarrow <\hat{C}_2> = \tilde{L}(\tilde{L} + 1), \quad (3)$$

where $\hat{L}$ is the pseudo-orbital momentum;

$$\hat{C}_2(Spin(3)) = \hat{J} \cdot \hat{J} \Rightarrow <\hat{C}_2> = J(J + 1), \quad (4)$$

where J is the total angular momentum;

$$\hat{C}_2(SU(3)) = 2\hat{Q} \cdot \hat{Q} + \frac{3}{4} \hat{L} \cdot \hat{L} \Rightarrow <\hat{C}_2> = \lambda^2 + \mu^2 + \lambda\mu + 3(\lambda + \mu). \quad (5)$$

At Bose level, $\hat{L}$ and $\hat{Q}$ of formula (5) have well known expressions

$$\hat{L}_{sd} = \sqrt{10} (d^+ \times \tilde{d})^{(1)}, \quad (6)$$

$$\hat{Q}_{sd} = (d^+ \times \tilde{s} + s^+ \times \tilde{d})^{(2)} \pm \frac{\sqrt{7}}{2} (d^+ \times \tilde{d})^{(2)}, \quad (7)$$

where the - sign is used for prolate nuclei, the + sign for oblate nuclei.

At Bose-Fermi level, $\hat{Q}$ and $\hat{L}$ are sums of boson and fermion terms:

$$\hat{L} = \hat{L}_{sd} + \hat{L}_F, \quad (8)$$

$$\hat{Q} = \hat{Q}_{sd} \pm \hat{Q}_F, \quad (9)$$

The expressions of $\hat{L}_F$ and $\hat{Q}_F$ are given in refs[1,10]; the + sign on the r. h. s. of formula (9) is adopted if the unpaired fermion is a particle, the - sign if it is a hole.

The SU(3) irreps relevant to the low-lying spectrum are the following: $(\lambda_B, \mu_B) = (2N, 0)$, for the ground-state band of the even-even core, $(\lambda_F, \mu_F) = (\tilde{n}, 0)$, or $(0, \tilde{n})$, if the unpaired fermion is a particle, or a hole, respectively. The irreps containing the states of the odd-mass nucleus are thus obtained by decomposition of the product of representations:

$$(2N, 0) \otimes (\tilde{n}, 0) = \oplus \sum_{r=0}^{r_{max}} (2N + \tilde{n} - 2r, r), \tag{10}$$

in the case of particle coupling, and

$$(2N, 0) \otimes (0, \tilde{n}) = \oplus \sum_{r=0}^{r_{max}} (2N - r, \tilde{n} - r), \tag{11}$$

in the case of hole coupling; in both formulae, $r_{max} = \min\{2N, \tilde{n}\}$.

In order to determine the irrep on the right-hand side of eq. (10), or (11) containing the nuclear ground state, we can exploit the one-to-one correspondence, in the large N limit, between Nilsson wavefunctions, $|[Nn_z\Lambda]\Omega>$, and IBFM wavefunctions, $|(\lambda, \mu)K_{\tilde{L}}K>$, written in a basis labelled with pseudo-L projection, $K_{\tilde{L}}$, and total angular momentum projection, K, on the symmetry axis; as a consequence:

$$\mu = N - 1 - n_z = \tilde{n} - \tilde{n}_z, \tag{12}$$

$$K = \Omega, \tag{13}$$

$$K_{\tilde{L}} = 2\Omega - \Lambda, \tag{14}$$

Since the Nilsson quantum numbers for the ground state of an odd-mass nucleus are known, the relevant irrep on the r. h. s. of eq. (10), or (11), is the one with μ given by eq. (12).

As it stands, the Hamiltonian of eq. (2) presents the drawback that the boson-fermion interaction is of pure quadrupole type, while microscopic interpretation of the IBFM and subsequent realistic calculations have stressed the importance of an exchange term, restoring in part the effects of the Pauli principle when the odd fermion interacts with a boson, which is, in reality, a pair of correlated fermions. While an exchange interaction breaks the SU(3) symmetry, in general, it has been shown that, in the asymptotic coupling regime[10], where both the coefficients of the quadrupole and exchange interactions, Γ_0 and Λ_0, respectively, go to infinity, but $|\Gamma_0/\Lambda_0| < 1$, both interactions become diagonal in K, in the large N limit. The intrinsic energies depend on Γ_0, Λ_0, $\tilde{n}$, μ, and the occupation probabilities, v^2, of the single-particle levels accessible to the odd fermion. In the asymptotic coupling regime, a simple IBFM Hamiltonian including the exchange term makes it possible to reproduce various experimental bands of odd-mass nuclei, without breaking the SU(3) symmetry.

THE GIANT DIPOLE RESONANCE

The SU(3)⊗SU(2) limit of the IBFM lends itself to a natural extension to high-lying collective modes, or giant resonances. While s and d bosons simulate $0\hbar\omega$ transitions in valence shells, of the particle-particle, or hole-hole type, high-energy bosons, S, D, P, ..., with spin-parity $L^\pi = 0^+, 2^+, 1^-, ...$, can be introduced into the model in order to simulate collective particle-hole transitions across shell closures. If we limit ourselves to the one-phonon approximation, the excitation of giant dipole resonances is produced by one P boson, whose three components span an (1,0) irrep of SU(3), while giant monopole and quadrupole modes are represented by S and D bosons, belonging to a (2,0) irrep[1,11]. The present work will be focused on P bosons, or GDR's.

The interaction of a P boson with the low-energy s, and d modes, and one unpaired fermion can be introduced into the Hamiltonian in a form that conserves the SU(3) symmetry, if one defines the P-boson contributions to the pseudo-orbital momentum, $\hat{L}_P$, and to the quadrupole operator, $\hat{Q}_P$, in such a way that $\hat{L}_P$ and $\hat{Q}_P$ generate an SU(3) algebra with the same commutation rules as $\hat{L}_{s,d}$ and $\hat{Q}_{s,d}$, defined in formulae (6) and (7); the result is

$$\hat{L}_P = \sqrt{2}\,(P^+ \times \tilde{P})^{(1)}, \tag{15}$$

$$\hat{Q}_P = \pm\frac{\sqrt{3}}{2}\,(P^+ \times \tilde{P})^{(2)}, \tag{16}$$

where the sign on the r. h. s. of (16) is chosen equal to that of the d-conserving part of $\hat{Q}_{s,d}$ in (7). The Casimir operators on the r. h. s. of formula (2) should be modified by replacing everywhere $\hat{L}_{BF}$ and $\hat{Q}_{BF}$ with $\hat{L}_{BF} + \hat{L}_P$ and $\hat{Q}_{BF} + \hat{Q}_P$, respectively. The new reduction chain containing P bosons turns out to be[1,12]

$$SU_{BF}(3)\otimes SU_P(3)\otimes SU(2) \supset SU_{BFP}(3)\otimes SU(2)$$
$$\supset SO_{BFP}(3)\otimes SU(2) \supset Spin(3), \tag{17}$$

where $SU_P(3)$ is generated by the eight operators of formulae (15) and (16), and $SU_{BFP}(3)$ by $\hat{L}_{BF} + \hat{L}_P$ and $\hat{Q}_{BF} + \hat{Q}_P$. However, since we are mainly interested in the energy splitting of the GDR in an odd-mass nucleus, we give up the rigorous symmetry of the Hamiltonian and keep only the leading term of the interaction, which can be written as a quadrupole-quadrupole coupling[12]

$$\hat{H}_{int.} = \alpha\hat{Q}_{BF} \cdot \hat{Q}_P, \tag{18}$$

where α can be adjusted, for instance, on the experimental GDR splitting in a photoabsorption process, while, in a strict SU(3) symmetry, α would be connected to β_2 of formula (2) by the relation $\alpha = 2\beta_2$.

By expressing the quadrupole-quadrupole coupling (18) in terms of Casimir operators (3) and (5), $\hat{H}_{int.}$ can be recast in the form

$$\hat{H}_{int.} = (\alpha/4)[\hat{C}_2(SU_{BFP}(3)) - \hat{C}_2(SU_{BF}(3)) - \hat{C}_2(SU_P(3))]$$
$$- 3(\alpha/16)[\hat{C}_2(SO_{BFP}(3)) - \hat{C}_2(SO_{BF}(3)) - \hat{C}_2(SO_P(3))] \tag{19}$$

The last two Casimir operators are not diagonal in the coupled BFP basis, but, in the large N limit, their eigenvalues are always much smaller than those of the SU(3) Casimir operators; therefore, we neglect them and obtain an interaction Hamiltonian diagonal in the BFP basis. Moreover, the absolute values of the coupling constants β_i and γ_i ($i = 1, 2$) in formula (2) are expected to be much smaller than that of α, since the former are responsible for the energy splitting of the low-lying states, the latter for the the splitting of the GDR. In adiabatic approximation, one obtains, for the total Hamiltonian

$$\hat{H} \simeq E_P \hat{n}_P + \hat{H}_{int.} \,,$$

(20)

where E_P is the P-boson energy, and $<\hat{n}_P> = 0$, or 1, in the one-phonon approximation adopted in the present work.

In order to determine the approximate values of the energies of GDR states, we only need the $SU_{BFP}(3)$ irreps they belong to; if the nuclear ground state is contained in the $SU_{BF}(3)$ irrep (λ, μ), the GDR states belong to the irreps obtained by decomposing SU $(3) \otimes SU_P(3)$ according to the rule[10]

$$(\lambda, \mu) \otimes (1, 0) = \oplus \sum_{r=0}^{1} \sum_{s=0}^{s_{max}} (\lambda + 1 - 2r - s, \mu + r - s) \,,$$

(21)

where $s_{max} = \min\{1 - r, \mu\}$. If $\mu = 0$, the above decomposition yields two terms, $(\lambda + 1, 0)$ and $(\lambda - 1, 1)$, if $\mu > 0$, three terms, $(\lambda + 1, \mu)$, $(\lambda - 1, \mu + 1)$ and $(\lambda, \mu - 1)$. In the latter case, computing the eigenvalues of the Casimir operators on the r. h. s. of formula (20) yields the following approximate energies

$$E_{GDR}((\lambda + 1, \mu), \tilde{L}) \simeq E_P + \frac{\alpha}{4} (2\lambda + \mu) - \frac{3}{16} \alpha \tilde{L}(\tilde{L} + 1) \,,$$

(22)

$$E_{GDR}((\lambda - 1, \mu - 1), \tilde{L}) \simeq E_P - \frac{\alpha}{4} (\lambda - \mu + 3) - \frac{3}{16} \alpha \tilde{L}(\tilde{L} + 1) \,,$$

(23)

$$E_{GDR}((\lambda, \mu - 1), \tilde{L}) \simeq E_P - \frac{\alpha}{4} (\lambda + 2\mu + 6) - \frac{3}{16} \alpha \tilde{L}(\tilde{L} + 1) \,.$$

(24)

If $\alpha < 0$, as expected for a quadrupole-quadrupole interaction, the states of formula (22) are the lowest, those of formula (24) the highest in energy. The $\tilde{L}$ values contained in the generic irrep, $(\tilde{\lambda}, \tilde{\mu})$, are[1,10]

$$\tilde{L} = K_{\tilde{L}}, K_{\tilde{L}} + 1, \ldots, K_{\tilde{L}} + \max\{\tilde{\lambda}, \tilde{\mu}\} \,,$$

(25)

$$K_{\tilde{L}} = \min\{\tilde{\lambda}, \tilde{\mu}\}, \min\{\tilde{\lambda}, \tilde{\mu}\} - 2, \ldots, 1, \text{ or } 0 \,,$$

(26)

with the exception of $K_{\tilde{L}} = 0$, for which

$$\tilde{L} = \max\{\tilde{\lambda}, \tilde{\mu}\}, \max\{\tilde{\lambda}, \tilde{\mu}\} - 2, \dots, 1 \text{ or } 0. \tag{27}$$

In the above formulae, the spin splitting of the GDR's has been neglected in comparison with the L splitting. In a more general formulation, the spin splitting can be obtained by inserting in the Hamiltonian a quadratic Casimir operator of Spin(3), analogous to the last term on the r. h. s. of formula (2).

The E1 transitions between low-lying states and GDR states are described by the dipole operator

$$\hat{D}^{(1)} = D_0(P^+ + \tilde{P}), \tag{28}$$

where D_0 is an adjustable parameter; $\hat{D}^{(1)}$ does not conserve the number of P bosons, because they represent collective particle-hole transitions across shell closures. Since $\hat{D}^{(1)}$ acts only on the $\tilde{L}$-dependent part of the wavefunctions, not on the spin part, its reduced matrix elements can be factorized according to a general reduction rule[1], and, in the SU(3) limit, the spin independent factors taken to be proportional to the corresponding Wigner coefficients of the SU(3) $\supset$ SO(3) reduction chain[8,9]

$$\begin{aligned}
&< (\lambda_f, \mu_f), K_{\tilde{L}_f}, \tilde{L}_f \frac{1}{2}, J_f || \hat{D}^{(1)} || (\lambda_i, \mu_i), K_{\tilde{L}_i}, \tilde{L}_i \frac{1}{2}, J_i > = \\
&D \hat{J}_i \hat{J}_f \, W(\tilde{L}_f \frac{1}{2}, 1, J_i, J_f, \tilde{L}_i) < (\lambda_i, \mu_i), K_{\tilde{L}_i}, \tilde{L}_i, (1,0), 0, 1 || (\lambda_f, \mu_f), K_{\tilde{L}_f}, \tilde{L}_f > .
\end{aligned} \tag{29}$$

Here, the initial state is a low-lying one, the final state a GDR component, W is a standard Racah coefficient, D a new adjustable parameter, and $\hat{J} \equiv \sqrt{2J+1}$.

The formalism worked out till now is sufficient for schematic calculations of GDR energy splitting and E1 transition strengths, as done in ref.[12] for the case of an odd fermion in the $\tilde{n} = 1$ pseudoshell; realistic calculations of photonuclear reactions in the GDR energy region require, however, introduction of the GDR decay, which lies beyond the limits of the present model, but can be phenomenologically treated by means of a power-law dependence of the decay width on excitation energy[1]

$$\Gamma(E) = \gamma_0 (E/E_P)^\delta, \tag{30}$$

where γ_0 and δ are adjustable parameters.

The energies, E_k, of the GDR states, their widths, $\Gamma_k \equiv \Gamma(E_k)$, and the reduced matrix elements (29) of the dipole operator are then introduced in the generalized Kramers-Heisenberg amplitudes for photon scattering[13]

$$\begin{aligned}
A_j = C_j \sum_k &< J_f || \hat{D}^{(1)} || J_k > < J_k || \hat{D}^{(1)} || J_0 > W(J_0, J_f, 1, 1, j, J_k) \\
&\times \left[\frac{1}{E_k - E - i\Gamma_k/2} + \frac{(-1)^j}{E_k + E' + i\Gamma_k/2} \right] + \sqrt{3} \, A_T \delta_{j0} \delta_{0f},
\end{aligned} \tag{31}$$

where

$$C_j = (-1)^j \left(\frac{e}{\hbar c}\right)^2 \left(\frac{E'}{E}\right)^{1/2} EE' \frac{\hat{j}}{\hat{J_0}},$$ (32)

and

$$A_T = -\frac{Ze^2}{AMc^2}$$ (33)

is the classical Thomson amplitude. J_0, J_f, and J_k are the total angular momenta of the initial, final and intermediate (GDR) state, respectively; j is the transferred angular momentum, with possible values 0, 1 and 2. Finally, E is the incident photon energy, E' the scattered energy.

The total photoabsorption cross section is given by the optical theorem

$$\sigma_{abs.}(E) = 4\pi \frac{\hbar c}{E} Im\left(\frac{A_0}{\sqrt{3}}\right),$$ (34)

and the scattering cross sections for unpolarized radiation are

$$\frac{d\sigma}{d\Omega} = \sum_{j=0}^{2} \frac{|A_j|^2}{2j+1} g_j(\theta),$$ (35)

where

$$g_0(\theta) = \frac{1}{6}(1 + \cos^2\theta),$$ (36)

$$g_1(\theta) = \frac{1}{4}(2 + \sin^2\theta),$$ (37)

$$g_2(\theta) = \frac{1}{12}(13 + \cos^2\theta).$$ (38)

RESULTS AND COMMENTS

The formalism of the previous section has already been applied to the analysis of experimental photon absorption and scattering by even-even deformed nuclei[14], with one main simplification, that the reduced matrix elements of the dipole operator do not contain a spin dependent factor, like in formula (29), but are simply proportional to the Wigner coefficients of the SU(3)⊃SO(3) reduction

582

chain, which have been evaluated by means of a modified version of the computer programs of ref.[9]. The agreement between theory and experiment turns out to be remarkably good, in spite of the extreme simplicity of the model, for the axially symmetric nucleus ^{238}U ($\mu = 0$), and the triaxial nucleus ^{188}Os ($\mu > 0$).

In the present section we show preliminary results for the odd-mass nucleus ^{169}Tm ($Z = 69$), whose low-lying spectrum of positive parity has already been analysed in the SU(3)$\otimes$SU(2) limit of the IBFM[10]. Its even-even core, ^{170}Yb ($Z = 70$), has 9 pairs of neutron particles in the 82-126 shell and 6 pairs of proton holes in the 50-82 shell, so that the effective boson number is taken to be $N = 15$, and the ground-state band belongs to a (30,0) irrep of SU(3). The ground state of ^{169}Tm, with $J^\pi = 1/2^+$, is described in terms of a proton hole in the Nilsson level with asymptotic quantum numbers [411]1/2, i. e. pseudo-quantum numbers [310]1/2. Since the hole coupling is appropriate to this case, formula (11) predicts low-lying bands belonging to the representations $(30, 0)\otimes(0, 3) = (30, 3)\oplus(29, 2)\oplus(28, 1)\oplus(27, 0)$, and formulae (12-14) a ground-state band with $\mu_{\sim} = 2$, $K_{\bar{L}} = 0$. The corresponding irrep is thus (29,2), and the ground state has $L = 1, J = 1/2$. The experimental ground-state band is well reproduced in the asymptotic coupling regime of the SU(3)$\otimes$SU(2) model[10]; however, other low-lying levels of the (29,2) irrep, with $K_{\bar{L}} = 2$, are not experimentally observed, whence the necessity to break the SU(3) symmetry, by introducing $K_{\bar{L}}$ -dependent terms in the Hamiltonian.

If one introduces a P mode, the resulting high-energy excitations built on the ground state belong to $(29, 2)\otimes(1, 0) = (30, 2)\oplus(28, 3)\oplus(29, 1)$, and their quantum numbers are easily determined by means of formulae (25-27). In particular, (30,2) contains 3 GDR states with $(K_{\bar{L}}, L, J^\pi) = (0, 0, 1/2^-)$, $(0, 2, 3/2^-)$, and $(2, 2, 3/2^-)$; (28,3) contains $(1, 1, 1/2^-)$, $(1, 1, 3/2^-)$, $(1, 2, 3/2^-)$; finally, (29,1) contains $(1, 1, 1/2^-)$, $(1, 1, 3/2^-)$, and $(1, 2, 3/2^-)$.

The corresponding distribution of E1 strengths, defined by the formula

$$B(E1, J_0 \rightarrow J_{GDR}) = \frac{1}{2J_0 + 1} \mid < J_{GDR} \mid\mid \hat{D}^{(1)} \mid\mid J_0 > \mid^2, \tag{39}$$

is shown in Fig. 1, with superimposition of the second and third state of (30,2), since both have $L = 2$, the first and second state of (28,3), since spin splitting has been neglected, and the first and second state of (29,1), for the same reason. The parameters of the P-boson Hamiltonian (20) and of the decay width (30) used for computing the photoabsorption cross section (34), shown in the same figure, have been determined on the basis of experimental systematics in the lanthanide region[15], and have the following values: $E_P = 14.37$ MeV, $\alpha = -0.150$ MeV, $\gamma_0 = 3.92$ MeV, $\delta = 1.71$, and $D = 15.2$ fm.

Finally, Fig. 2 shows elastic and inelastic scattering cross sections (35), at scattering angle $\theta = 120°$, where the Delbrück contribution to elastic scattering is expected to be negligible. In particular, the dashed line represents the sum of coherent ($j = 0$) and incoherent ($j > 0$) elastic scattering, the solid curve is the sum of elastic scattering and inelastic excitation of the $(3/2)^+_1$ and $(5/2)^+_1$ states, hardly distinguishable from the ground state in a low-resolution experiment.

Acknowledgment

We are particularly pleased to dedicate the present work to Professor Francesco Iachello on the occasion of his 50-th birthday, since our study of giant resonances in the frame of the IBM and IBFM, started in collaboration with him, has always benefited from his valuable comments and encouragement.

583

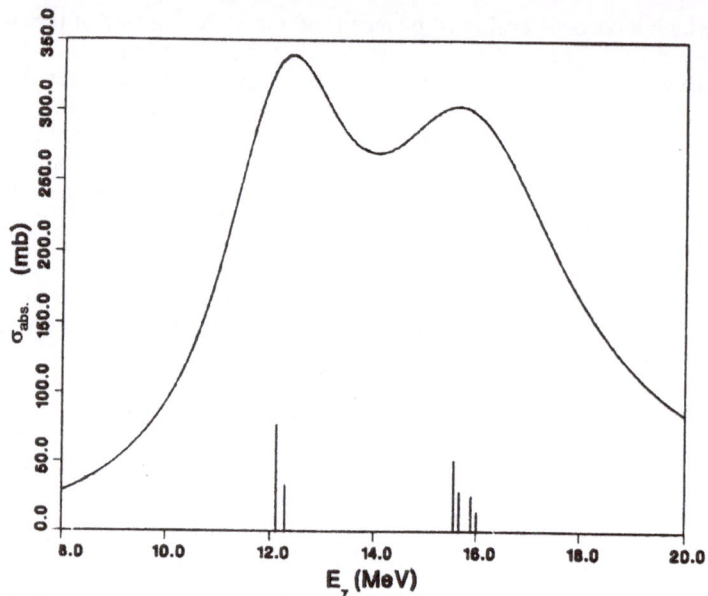

Figure 1. Photoabsorption cross section of ^{169}Tm (mb) vs incident photon energy (MeV). The vertical bars are E1 transition strengths, eq. (39), in arbitrary units.

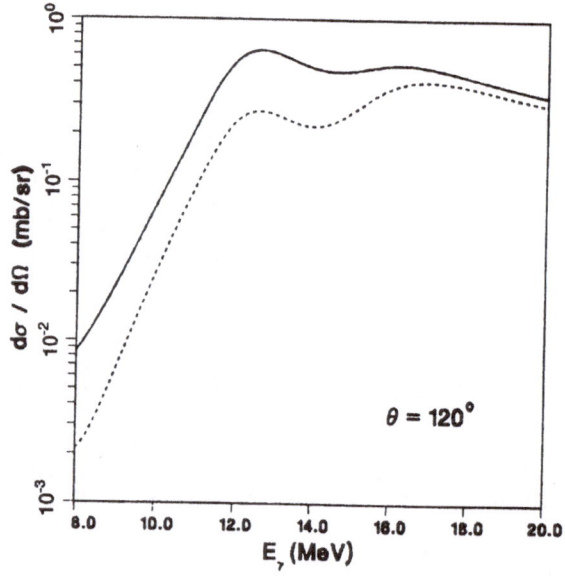

Figure 2. Scattering cross sections (mb/sr) for ^{169}Tm at $\theta = 120°$ vs incident photon energy (MeV). Dashed curve: elastic scattering; solid curve: elastic plus inelastic scattering to the $(3/2)^+_1$ and $(5/2)^+_1$ states.

REFERENCES

1. F. Iachello and P. Van Isacker, "The Interacting Boson-Fermion Model", Cambridge University Press, Cambridge (1991); see, in particular, Chapters 3 (Bose-Fermi symmetries) and 12 (giant resonances).
2. A. Arima, M. Harvey, and K. Shimizu, Pseudo LS coupling and pseudo SU(3) coupling schemes, *Phys. Lett. B* 30:517(1969).
3. K. T. Hecht and A. Adler, Generalized seniority for favored $J \neq 0$ pairs in mixed configurations, *Nucl. Phys. A* 137:129(1969).
4. R. D. Ratna Raju, J. P. Draayer, and K. T. Hecht, Search for a coupling scheme in heavy deformed nuclei: the pseudo SU(3) coupling, *Nucl. Phys. A* 202:433(1973).
5. J. P. Draayer and K. J. Weeks, Towards a shell model description of the low energy structure of deformed nuclei: I. Even-even systems, *Ann. Phys. (N.Y.)* 156:41(1984).
6. O. Castaños, J. P. Draayer, and Y. Leschber, Towards a shell model ...: II. Electromagnetic properties of collective M1 bands, *Ann. Phys. (N.Y.)* 180:290(1987).
7. O. Castaños, P. O. Hess, J. P. Draayer, and P. Rochford, Pseudo-symplectic model for strongly deformed heavy nuclei, *Nucl. Phys. A* 524:469(1991).
8. J. P. Draayer and Y. Akiyama, Wigner and Racah coefficients for SU(3), *J. Math. Phys.* 14:1904(1973).
9. J. P. Draayer and Y. Akiyama, A user's guide to FORTRAN programs for Wigner and Racah coefficients of SU(3), *Comput. Phys. Commun.* 5:405(1973).
10. R. Bijker and V. K. B. Kota, Interacting boson-fermion model of collective states: IV. The SU(3)⊗U(2) limit, *Ann. Phys. (N.Y.)* 187:148(1988).
11. D. J. Rowe and F. Iachello, Group theoretical models of giant resonance splittings in deformed nuclei, *Phys. Lett. B* 130:231(1983).
12. G. Maino, Possible Bose-Fermi symmetries in the giant resonance fragmentation of deformed odd-even nuclei, *Phys. Rev. C* 40:988(1989).
13. E. Hayward, Photonuclear reactions, *in*: "Nuclear Structure and Electromagnetic Interactions", N. Mac Donald, ed., Oliver and Boyd Ltd., Edinburgh (1965).
14. L. Zuffi, P. Van Isacker, G. Maino, and A. Ventura, Gamma-ray scattering by deformed nuclei, *Nucl. Instr. Meth. Phys. Res. A* 255:46(1987).
15. G. M. Gurevich, L. E. Lazareva, V. M. Mazur, S. Yu. Merkulov, G. V. Solodukhov, and V. A. Tyutin, Total nuclear photoabsorption cross sections in the region $150 < A < 190$, *Nucl. Phys. A* 351:257(1981).

COULOMB EXCITATION OF OCTUPOLE
DEFORMED ROTORS

Andrea Vitturi

Dipartimento di Fisica and INFN
Padova, Italy

We study the Coulomb excitation patterns in the case of nuclear systems displaying static octupole deformations. The specific symmetries associated with these reflection-asymmetric systems leads to characteristic features in the final spin population. The most relevant is the occurrence of two rainbow-like edges, which arise from the combined effects of the quadrupole and octupole components of the field. Q-value effects associated with the relative displacement of the positive- and negative-parity branches of the rotational band lead to clear-cut differences between the cases of static and dynamic octupole deformation.

The occurrence of permanent deformations in nuclei far from closed shells underscores the importance of shell effects in defining the properties of the self-consistent nuclear mean-field. The precise characteristics of the equilibrium deformation depend, on a microscopic basis, on the single-particle levels available for the valence particles and on the values of the associated matrix-elements of the interaction. For this reason the deformation parameters defining the macroscopic nuclear shape in the multipole expansion

$$R(\theta, \varphi) = R_0 \left[1 + \sum_{\lambda\mu} \beta_{\lambda\mu} Y_{\lambda\mu}(\theta, \varphi) \right]$$

are expected to vary with the number of valence particles due to the change of the Fermi surface or, for instance, with the angular momentum due to the rearrangement of single-particle energies arising from the modification of the average field.

Characteristic of the nuclear coupling is that most of the stable deformed nuclei in their ground state exhibit prolate axially-symmetric quadrupole deformations. Evidence for non-vanishing hexadecapole deformations – predicted on the basis of the hexadecapole moments of the orbits near the Fermi surface – has also been obtained, for example from inelastic excitation of the 4+ states in the rotational bands, as well as from processes which are particularly sensitive to the nuclear density profile such as heavy-ion fusion at subbarrier energies. The possibility of exploring new dimensions in both angular momentum and mass has further opened the way to the systematic study of shape transitions and to the discovery of situations characterized by more extreme deformations, such as the quadrupole ones characterized by axis ratio 2:1 (superdeformation) or 3:1 (hyperdeformation).

Within this line an item which has been lengthily debated is the possible occurrence of reflection-asymmetric components (odd λ's) in the multipole expansion of the nuclear shape, to begin with the octupole term [1]. Conditions for the presence of permanent axially-symmetric octupole deformation are essentially the availability, around the Fermi surface, of close single-particle levels which can be connected by the octupole operator (proportional to $Y_{3\mu}$). This condition is easily fulfilled in the case of stable systems with a very large numbers of interacting fermions [2] (as for example metallic clusters), but difficult to obtain in the nuclear case where the production of very heavy nuclei is prevented by the severe instability agaist fission. In the standard nuclear range all predicted regions lie somewhat far from the stability line. Best candidates [3] are the already partially studied actinide region around ^{222}Ra ($Z \approx 88$, $N \approx 132$), the light lanthanides as Ba or Ce ($Z \approx 58$, $N \approx 68$), as well as the further regions around the neutron-deficient nuclei with $Z \approx 34$, $N \approx 34$ ($\approx^{64}$Ge) and the neutron-rich with and $Z \approx 34$, $N \approx 56$ ($\approx^{88}$Ge). Other favourable conditions, not limited to the axially symmetric one but also involving more exotic shapes as those associated with the Y_{31} (*banana shapes*), are expected in connection with quadrupole superdeformation at medium and high spins, where the stability of octupole deformations is expected to increase due to the Coriolis force.

The formalisms used to described these octupolly-deformed systems are the natural extensions of those introduced in connection with the quadrupole degree of freedom. So, for example, single particle energies in the now more general quadrupole + octupole axially-deformed (and eventually rotating) potential are used to construct the energy surfaces in the $\beta_2 - \beta_3$ plane. Similarly self-consistent HFB calculations have been performed [4], allowing for equilibrium solutions which break the reflection symmetry in addition to the spherical symmetry. Other approaches, as those derived by the Interacting Boson Model, use group-theoretical techniques, exploiting the dynamical symmetries of the hamiltonian associated with these systems. The inclusion of the octupole degree of freedom will in general enlarge the dimensions and the complexity of the corresponding symmetry groups. Within a boson representation of these algebras, this is obtained by adding to the standard interacting s and d bosons associated with the quadrupole degree of freedom, odd-parity bosons carrying angular momentum l=1 (p boson) and l=3 (f boson) [5]. Different choices of the form of the hamiltonian describing the interacting bosons leads to rather different behaviours, ranging from sphericity in the presence of octupole vibrations to stable quadrupole deformation with octupole vibrations, with the extreme case of stable quadrupole + octupole deformation.

From both theoretical and experimental points of view, it is important to single-out unambiguous signatures for this last kind of static reflection-asymmetric deformations, as opposed for example to the ones associated with quadrupole-deformed systems in the presence of low-lying $K = 0^-$ octupole bands of vibrational character. Those well known in the case of molecular physics are of spectrocopic nature, e.g. the identification of parity doublets in odd-A nuclei or of rotational bands with alternating parity levels connected by strongly enhanced El transitions, due to the intrinsic dipole moment. Other signatures for permanent octupole deformation, on the other hand, involve dynamical excitation and deexcitation characteristics of the system via the electromagnetic or nuclear fields. In fact, specific transition and selection rules are expected to hold in the case of a rotor characterized by quadrupole and octupole deformations, leading to population patterns of the members of the band which should differ significantly from those obtained in the case of a quadrupole rotor coupled to an octupole vibration.

Coulomb excitation processes induced by heavy-ion collisions offer the simplest case where these ideas can be tested. Experimental data are gradually becoming available in the relevant regions and further progress is expected following the use of new accelerating facilities and detection techniques. The problem of Coulomb excitation of a rigid rotor has been extensively studied within the quadrupole degree of freedom for axially symmetric [6], non-axially symmetric and γ-unstable systems [7]. Large projectile charges and bombarding energies can be exploited to reach very high-lying members of the rotational bands. In this limit the fundamental – but rather involved – coupled-channel treatment of the collision process has been successfully complemented by approaches where the relative motion and/or the internal degrees of freedom are described with classical concepts. Complex excitation patterns have thus found intuitive interpretation in terms of interfering contributions associated with specific initial conditions. One should also note that, under simplifying assumptions to describe the relative motion, the group theoretical structure underlying a quadrupole rotor leads to analytical close solutions.

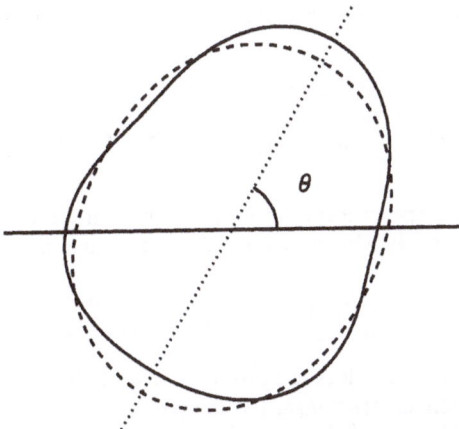

Fig. 1 - *Schematic picture of the nuclear surface for a quadrupole plus octupole deformed axially-symmetric system. The angle θ characterizes the orientation of the symmetry axis with respect to the beam direction.*

In order to point out the specific features associated with the octupole degree of freedom let us consider an even-even system with mass A_T and charge Z_T characterized by static axially-symmetric quadrupole and octupole deformations, with parameters $\beta_{20} \equiv \beta_2$ and $\beta_{30} \equiv \beta_3$. A sketch of a representative equilibrium nuclear profile is given in fig. 1. As a consequence these nuclei will display a ground-state rotational band (with $K = 0^{\pm}$), characterized by integer spins I and alternating parities $\pi_I = (-)^I$. Such rotational bands can be populated as a consequence of the action of an external field. In particular we consider a head-on collision induced by a (structureless) spherical projectile of mass A_p and charge Z_p at energies below the Coulomb barrier. The relevant coupling V is given by the multipole expansion of the Coulomb interaction with the deformed quadrupole+octupole target charge distribution. The lowest multipole terms read

$$V(t) = eZ_p \left(\frac{4\pi}{3} \frac{q_1 Y_{1,0}(\theta, 0)}{r(t)^2} + \frac{4\pi}{5} \frac{q_2 Y_{2,0}(\theta, 0)}{r(t)^3} + \frac{4\pi}{7} \frac{q_3 Y_{3,0}(\theta, 0)}{r(t)^4} \right) . \tag{1}$$

The quadrupole and octupole moments q_2 and q_3 are in leading order related to the deformation parameters in the form $q_\lambda = (3/4\pi)\beta_\lambda Z_T e R_T^\lambda$, $(\lambda = 2, 3)$ The simultaneous presence of both quadrupole and octupole deformations introduces also a dipole moment q_1 which can be approximated [8] as $\sqrt{(3/4\pi)}c_{LD} A_T Z_T e \beta_2 \beta_3$, with the liquid drop estimate for the constant $c_{LD} = 0.00069$ fm.

If we restrict ourselves to conditions where the sudden limit applies, i.e. cases in which the intrinsic motion is slow compared to the collision time, the amplitude for populating the generic state n starting from the ground state is simply given by

$$a_n = \langle \Psi_n | \exp\left[-\frac{i}{\hbar} \int_{-\infty}^{+\infty} V(t')\, dt' \right] | \Psi_0 \rangle , \tag{2}$$

In a head-on collision one can take the z-direction as that of the beam, in which case only the substates with $M = 0$ can be populated. The amplitude associated with the population of the generic state I of the ground-state rotational band is then

$$a_I \equiv a_{I,0} = (\pi)^{-1/2} \int_0^\pi \sin\theta\, d\theta\; Y_{I,0}(\theta, 0)\, e^{-i\kappa_1 Y_1(\theta,0) - i\kappa_2 Y_2(\theta,0) - i\kappa_3 Y_3(\theta,0)} . \tag{3}$$

The quantities κ's incorporate the integrals of the different terms of the coupling interaction (1) over the Rutherford trajectory and are given by

$$\kappa_\lambda = \frac{4\pi Z_p e q_1}{3\hbar} \int_{-\infty}^{+\infty} \frac{dt}{r(t)^{\lambda+1}} = \frac{8\pi Z_p e q_1}{\hbar v_0 r_0^\lambda (2\lambda + 1)!!} , \tag{4}$$

in terms of the distance of closest approach r_0 and the asymptotic relative velocity v_0. These dimensionless parameters give a measure of the strength of the different terms of the interaction coupling and thus control the extent of the population of the target states for a given bombarding energy. Given the different dependence of the κ's on the bombarding energy, the relative importance of the different multipole moments in (3) varies with the choice of kinematical conditions.

As an illustrative example the probabilities for the excitation of the different members of the rotational band, calculated according to (3), are displayed in fig. 2 for the case of the excitation of ^{224}Ra by a ^{208}Pb projectile at a bombarding center-of-mass energy E_{cm} equal to 650 MeV. The multipole deformation parameters β_2 and β_3 in ^{224}Ra have been taken equal to 0.13 and 0.09, respectively. Even-spin (open circles) and odd-spin states (solid circles) have been separately identified. Fig. 2c gives the results obtained when both quadrupole and

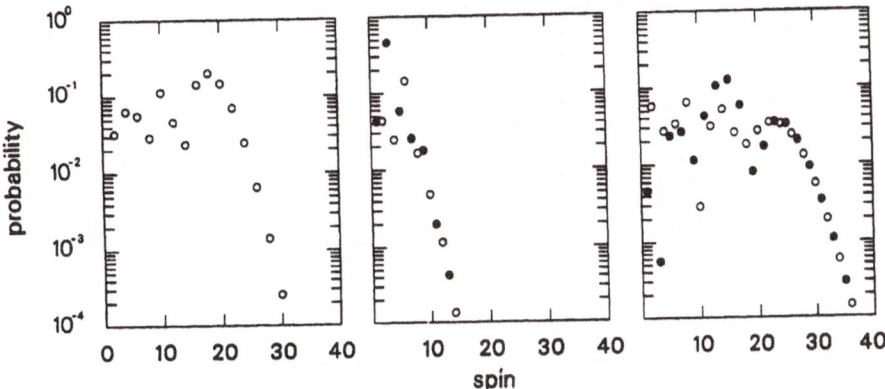

Fig. 2 - *Probability for the Coulomb excitation of the different members of the ground-state rotational band, characterized by their spins, calculated in the sudden limit (cf. eq. 3). The considered reaction is the excitation of ^{224}Ra by a ^{208}Pb projectile at a bombarding center-of-mass energy E_{cm} equal to 650 MeV. The multipole deformation parameters β_2 and β_3 in ^{224}Ra have been taken equal to 0.13 and 0.09, respectively. Even-spin (open circles) and odd-spin states (solid circles) are separately identified. Fig. 2c gives the full results obtained when both quadrupole and octupole deformations are included, while figs. 2a and 2b give the excitation probabilities obtained when only the quadrupole or the octupole couplings are active.*

octupole deformations are included but, for illustration, we also show the excitation probabilities obtained when only the quadrupole (fig. 2a) or the octupole couplings (fig. 2b) are active.

Parity conservation in the case of pure quadrupole coupling prevents the population of the odd-spin states. Both positive and negative parity states can, on the other hand, be populated when the octupole coupling is switched on. In such case the excitation pattern is rather similar as a function of the spin, displaying oscillations followed by a sharp decrease beyond some critical value. The overall behaviour is, however, less smooth than in the pure quadrupole case. Only at high values of the spin the curves associated with even- and odd-spin states merge into a common pattern.

To gain an intuitive understanding of the behaviour displayed in the figures, let us examine the classical probability of population of a state at a given angular momentum. To this end we can first evaluate the spin acquired by the target (as a consequence of a head-on collision) as a function of the initial orientation angle θ_0 between the symmetry axis of the system and the beam axis (cf. fig. 1). In a classical description of the scattering process the final angular momentum $I \equiv p_\theta(t = +\infty)$ can be obtained by integrating the corresponding hamilton equation. In the sudden limit, where one assumes that the rotation is slow as compared to the interaction time, one can freeze the orientation angle to its initial value θ_0, so that

$$I(\theta_0) = \int_{-\infty}^{+\infty} \frac{\partial V(r(t), \theta_0)}{\partial \theta} \, dt =$$

$$-\sqrt{\frac{3}{4\pi}} \kappa_1 \sin(\theta_0) + 3\sqrt{\frac{5}{4\pi}} \kappa_2 \sin(2\theta_0) + \sqrt{\frac{7}{4\pi}} \kappa_3 (15 \cos^2(\theta_0) \sin(\theta_0) - 3 \sin(\theta_0)). \quad (5)$$

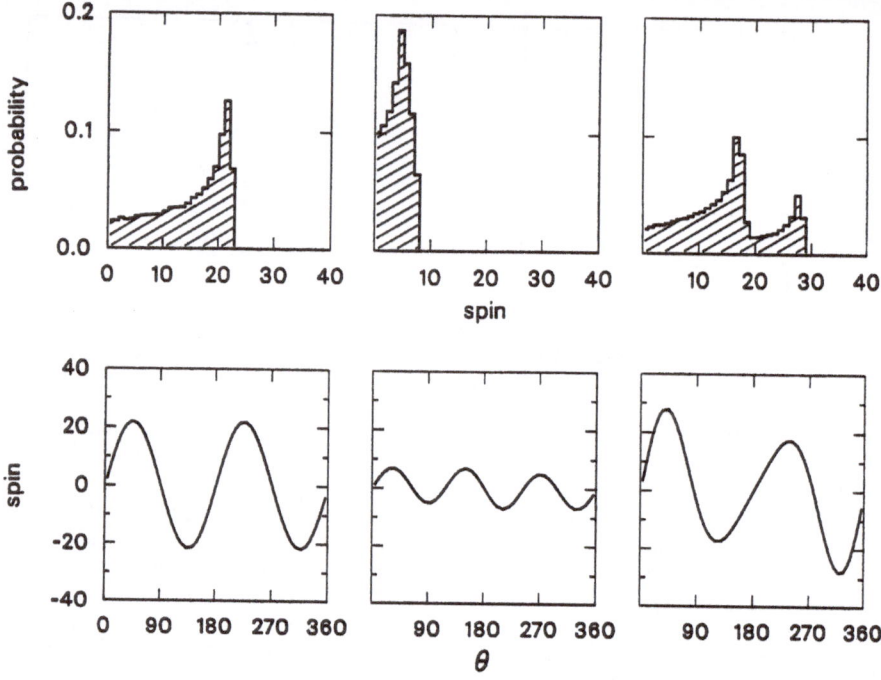

Fig. 3 - *Classical probability for the Coulomb excitation of the different members of the ground-state rotational band, characterized by their spins, for the reaction* $^{208}Pb + ^{224}Ra$ *at* E_{cm} *equal to 650 MeV (upper part). For the multipole deformation parameters* β_2 *and* β_3 *in* ^{224}Ra *cf. caption to fig. 2. As in fig. 2 the figure on the right-hand side gives the full results obtained when both quadrupole and octupole deformations are included, while figures on the left-hand side and at the center give the excitation probabilities obtained when only the quadrupole or the octupole couplings are active. In the lower part of the figure we display the corresponding final spin transferred classically to the target in the sudden limit during the reaction as a function of the initial orientation of the system.*

Although abscribing this characteristic pattern to the octupole degree of freedom, one may enquire whether a similar behaviour might also arise in the case of an octupole vibration strongly coupled to a permanent quadrupole deformation. Under these conditions the low-lying spectrum displays two parellel bands with opposite parity, approximately shifted by the energy of the octupole vibration, e.g. of the order of 1-2 MeV. One expects that the parity-changing intraband transitions will be partly depressed by the adiabaticity conditions, due to the relatively large Q-values. To verify this point, we have performed two parallel calculations in the rotational and in the vibrational case, within a full coupled channel formalism which releases the sudden-limit approximation assumed in (3). The results are shown in Figs. 4b and 4c, compared with the sudden-limit predictions displayed in Fig. 4a. The final Q-value is clearly responsible for the quenching of the maximum spin transferred into the system.

The resulting final angular momenta are displayed in the lower part of Fig. 3 for the previously considered cases, with the corresponding classical probabilities shown in the upper part. To compare with the quantal calculation, the values have been plotted as histograms in integer steps. As it is known, a pure quadrupole coupling leads to a $\sin(2\theta_0)$ dependence with a maximum allowed value of the spin I_{max} (acquired at $\pi/4$) and to a final spin distribution characterized by the presence of a classical rainbow at I_{max}. The period of oscillations in the final spin as a function of the initial orientation moves from π when the octupole component (or any reflection-asymmetric term) is included. The pure octupole case leads – although not exactly – to a $\sin(3\theta_0)$-like dependence with three maxima close in value and thus, again, to the presence of a single rainbow in the spin distribution. When both quadrupole and octupole coupling are active the combination of the different periods produces a curve with two relative maxima and, as a result, the spin distribution displays now a characteristic behaviour with a double rainbow.

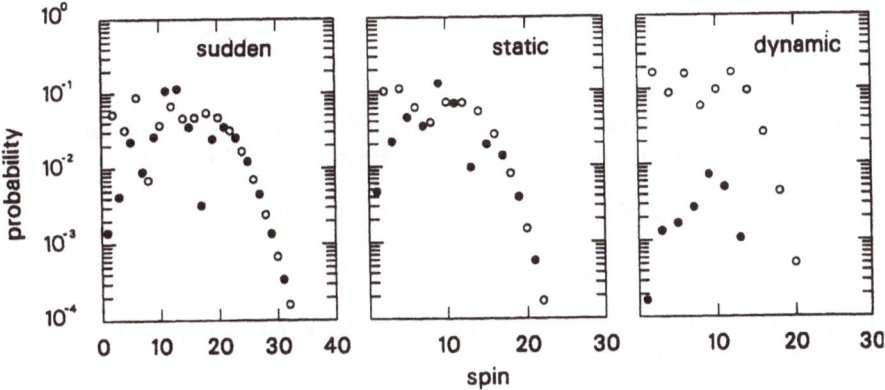

Fig. 4 - *Probability for the Coulomb excitation of the different members of the ground-state rotational band in the reaction* $^{208}Pb + {}^{224}Ra$ *at* $E_{cm} = 600$ *MeV. The results in the figure on the left-hand-side has been obtained in the sudden limit assuming both quadrupole and octupole permanent deformation. The central figure corresponds to the same case with the coupled-channel code COULEX. The energies of the states of the band were obtained assuming a moment of inertia equal to* $100 \, \hbar^2/MeV$. *The last figure refers to the case of a quadrupole-deformed band coupled to a negative-parity vibrational octupole band. The value of the dynamical deformation parameter* β_3 *is taken equal to the previous static value. The energy of the octupole vibration is taken equal to 2 MeV.*

The qualitative features arising from the classical description are reflected in the quantal results. The sharp decrease for high values of the spin corresponds to the classical forbidden region, while the oscillations in the allowed region are a quantal effect arising from the interference of contributions from different orientations leading to the same classical final spin. The double rainbow expected in the case od quadrupole plus octupole defermation is clearly evident in the quantal result shown in Fig. 2c, and therefore assumed as a clear evidence for octupole deformation. But more striking is, in the case of the vibrational octupole band, the strongly reduced population of the negative parity states. This is at variance with the case of the static deformation, where the spin population shows a uniform behavour irrespectively of the parity of the states.

We therefore conclude by stating that Coulomb excitation processes induced by heavy ions offer an interesting tool for evidence stable octupole deformation in nuclei. The excitation probabilities of the different members of the rotational band display in fact a characterisic pattern, which is rather different from those expected in the case of pure quadrupole deformation as well as in the case of octupole vibrational state coupled to the quadrupole deformation.

REFERENCES

1. G.A. Leander, Static octupole deformation, in: "Nuclear Structure 1985", R.A. Broglia et al, eds, North Holland, Amsterdam (1985)

2. I. Hamamoto, B. Mottelson, H. Xie and X. Zhang, Zeit. Physik D21:163 (1991)

3. S. Aberg, H. Flocard and W. Nazarewicz, Nuclear shapes in mean field theory, in: "Annual Review of nuclear science", 40:439 (1990)

4. J.L. Egido and L.M. Robledo, Nucl. Phys. A545:589 (192)

5. J. Engel and F. Iachello, Phys. Rev. Lett. 54:1126 (1985)

6. K. Alder and A. Winther, "Electromagnetic Excitation", North Holland, Amsterdam (1975)

7. C.H. Dasso and A. Vitturi, Nucl. Phys. A536:179 (1992)

8. G.A. Leander et al, Nucl. Phys. A453:58 (1986)

A POSSIBLE WAY TO GENERALIZE THE O-A-I METHOD
TO NON-DEGENERATE j SHELLS

Yang Ze-sen[1], Qi Hui[1], Li Xian-hui[1],
Akito Arima[2] and Takaharu Otsuka[2]

[1]Department of Physics, Peking University, Beijing 100871, China
[2]Department of Physics, University of Tokyo, Tokyo, Japan

I. INTRODUCTION

The phenomenological success of the interacting boson model[1] implies that it should be possible to construct a subspace V_F which is formed by SDG pairs and is approximately invariant under the Hamiltonian H if the creation operators of these correlated pairs are appropriately defined. Once a 1-to-1 mapping between V_F and a boson subspace V_B of the sdg state space is chosen, a boson description is set up and V_B is the physical boson state subspace in the method. Such a kind of boson descriptions can be included in a framework of "Modified boson mapping" in the sense that the mapping is only applied to a subspace of fermion states which is or is assumed to be approximately invariant under H. Special choices of the subspace and of the mapping can yield various methods. For example, the OAI method[2,3], the Suzuki-Matsuyamagi method[4,5] and the modified Morumori boson expansion method[6,7] are of this type.

We will introduce a new type of fermion operators that can not couple to S-pair. The creation operators of D and G pairs formed by these fermion operators can ensure that two states with different number of S-pairs are always orthogonal. Keeping in mind the advantages of the Dyson image of the Hamiltonian and considering the importance of the explicit construction of the physical boson subspace V_B, we will use a couple of transformation operators which are somewhat similar to the Usui operators in Dyson mapping to define V_B and construct our mapping. We will also study a kind of boson operators that replace Dyson images. An approximation method will be described for determining the boson image of the Hamiltonian and other quantities. If equation of motion is emploied within the subspace V_B, the influnces of the unphysical boson states can be avoided. We will also find a way to use the eigen-equation of the non-hermitean Hamiltonian in the entire sdg subspace. It will be shown that the OAI procedure of constructing the correlated D-pair and the low-order expression of the boson images of physical quantities is a special case of the procedure of the present work.

II. A NEW TYPE OF FERMION OPERATORS

Denote by $a^\dagger_{jm}$ and a_{jm} the usual creation and annihilation operator of a nucleon in state (jm) where jm stand for the single-particle quantum numbers $nljm$. We have for each kind of nucleons:

$$a^\dagger_{j_1 m_1} a^\dagger_{j_2 m_2} + a^\dagger_{j_2 m_2} a^\dagger_{j_1 m_1} = 0 \,, \tag{2.1}$$

$$a_{j_1 m_1} a^\dagger_{j_2 m_2} + a^\dagger_{j_2 m_2} a_{j_1 m_1} = \delta_{j_1 j_2} \delta_{m_1 m_2} \,, \tag{2.2}$$

$$a_{jm} \,|\, 0\rangle = 0 \,, \tag{2.3}$$

where $|\, 0\rangle$ stands for the vacuum state.

We now introduce a new type of fermion operators defined by

$$C^\dagger_{jm} = A^\dagger_{jm} \Gamma(\Omega_j - \hat{n}_j) \,, \tag{2.4}$$

$$C_{jm} = \Gamma(\Omega_j - \hat{n}_j) A_{jm} \,, \tag{2.5}$$

where

$$A^\dagger_{jm} = a^\dagger_{jm} + \sqrt{\frac{\Omega_j}{2}} \left(a^\dagger_j a^\dagger_j\right)_0 \frac{1}{\Omega_j - \hat{n}_j} \hat{a}_{jm} \,, \tag{2.6}$$

$$\hat{a}_{jm} = (-1)^{j+m} a_{j-m} \,, \tag{2.7}$$

$$\hat{n}_j = \sum_m a^\dagger_{jm} a_{jm} \,, \tag{2.8}$$

$$\Omega_j = j + \frac{1}{2} \,, \tag{2.9}$$

$$\left(a^\dagger_j a^\dagger_j\right)_0 = \sqrt{\frac{1}{2\Omega_j}} \sum_m (-1)^{j-m} a^\dagger_{jm} a^\dagger_{j-m} \,, \tag{2.10}$$

$$\Gamma(s) = \begin{cases} 1 \text{ for } s = 1, 2 \cdots \\ 0 \text{ for } s = 0, -1, -2 \cdots \end{cases} . \tag{2.11}$$

Obviously,

$$\Gamma(s)\Gamma(s+1) = \Gamma(s) \,, \tag{2.12}$$

$$\Gamma(s+1-\hat{n}_j) - \Gamma(s-\hat{n}_j) = \delta_{s \hat{n}_j} \,, \tag{2.13}$$

$$\Gamma(\Omega_j - \hat{n}_j) a^\dagger_{jm} = a^\dagger_{jm} \Gamma(\Omega_j - 1 - \hat{n}_j) \,, \tag{2.14}$$

$$\Gamma(\Omega_j - \hat{n}_j) a_{jm} = a_{jm} \Gamma(\Omega_j + 1 - \hat{n}_j) \,, \tag{2.15}$$

$$C^\dagger_{jm_1} \cdots C^\dagger_{jm_k} = A^\dagger_{jm_1} \cdots A^\dagger_{jm_k} \Gamma(\Omega_j + 1 - k - \hat{n}_j) \,. \tag{2.16}$$

$$C^\dagger_{j_1 m_1} C^\dagger_{j_2 m_2} = A^\dagger_{j_1 m_1} A^\dagger_{j_2 m_2} \Gamma(\Omega_{j_1} - \delta_{j_1 j_2} - \hat{n}_{j_1})\Gamma(\Omega_{j_2} - \hat{n}_{j_2}) \,, \tag{2.17}$$

$$C^\dagger_{j_1 m_1} C_{j_2 m_2} = A^\dagger_{j_1 m_1} A_{j_2 m_2} \Gamma(\Omega_{j_1} + \delta_{j_1 j_2} - \hat{n}_{j_1})\Gamma(\Omega_{j_2} + 1 - \hat{n}_{j_2}) \,. \tag{2.18}$$

The operators $C^\dagger_{j_1 m_1}$, $C^\dagger_{j_2 m_2}$ for arbitrary $j_1 j_2$ still satisfy the usual anticommutation relation:

$$C^\dagger_{j_1 m_1} C^\dagger_{j_2 m_2} + C^\dagger_{j_2 m_2} C^\dagger_{j_1 m_1} = 0 \,, \tag{2.19}$$

which also implies

$$C_{j_1 m_1} C_{j_2 m_2} + C_{j_2 m_2} C_{j_1 m_1} = 0 \,. \tag{2.20}$$

For the later convenience we express the operators $(A^\dagger_j A^\dagger_j)_0$ and $(A^\dagger_j \hat{A}_j)_0$ as:

$$\left(A^\dagger_j A^\dagger_j\right)_0 = \frac{\Omega_j}{2} \left(a^\dagger_j a^\dagger_j\right)_0 \left(a^\dagger_j a^\dagger_j\right)_0 (\hat{a}_j \hat{a}_j)_0 \frac{1}{\Omega_j - \hat{n}_j} \frac{1}{\Omega_j + 1 - \hat{n}_j} \,, \tag{2.21}$$

$$\sqrt{2\Omega_j} \left(A^\dagger_j \hat{A}_j\right)_0$$
$$= \hat{n}_j + \left[\frac{3}{2}\left(\frac{\Omega_j}{\Omega_j + 2 - \hat{n}_j}\right) - \frac{1}{2}\left(\frac{\Omega_j}{\Omega_j + 2 - \hat{n}_j}\right)^2\right] \left(a^\dagger_j a^\dagger_j\right)_0 (\hat{a}_j \hat{a}_j)_0 \,. \tag{2.22}$$

We next consider the operation of the $C^\dagger$-operators on a special kind of states ψ_0 specified by

$$\left(\hat{a}_j\hat{a}_j\right)_0\psi_0 = 0 \ , \tag{2.23}$$

which also implies that

$$\left(C_j^\dagger C_j^\dagger\right)_0\psi_0 = 0 \ , \tag{2.24}$$

$$\left(C_j^\dagger C_j^\dagger\right)_0 C_{j_1 m_1}^\dagger \cdots C_{j_k m_k}^\dagger\psi_0 = 0 \ . \tag{2.25}$$

Operating with $\left(\hat{a}_j\hat{a}_j\right)_0$ on $C_{jm}^\dagger\psi_0$ one has

$$\left(\hat{a}_j\hat{a}_j\right)_0 C_{jm}^\dagger\psi_0$$
$$= \left[\left(\hat{a}_j\hat{a}_j\right)_0 a_{jm}^\dagger + \sqrt{\frac{\Omega_j}{2}}\frac{1}{\Omega_j - \hat{n}_j}\left(\hat{a}_j\hat{a}_j\right)_0\left(a_j^\dagger a_j^\dagger\right)_0 \hat{a}_{jm}\right]\Gamma(\Omega_j - \hat{n}_j)\psi_0 \ .$$

Using

$$\left(\hat{a}_j\hat{a}_j\right)_0 \Gamma(\Omega_j - \hat{n}_j)\psi_0 = 0 \ ,$$

$$\left[\left(\hat{a}_j\hat{a}_j\right) \ , \ a_{jm}^\dagger\right] = \sqrt{\frac{2}{\Omega_j}}\hat{a}_{jm} \ ,$$

$$\left[\left(\hat{a}_j\hat{a}_j\right)_0 \ , \ \left(a_j^\dagger a_j^\dagger\right)_0\right] = \frac{2}{\Omega_j}\left(\hat{n}_j - \Omega_j\right) \ ,$$

one gets

$$\left(\hat{a}_j\hat{a}_j\right)_0 C_{jm}^\dagger\psi_0 = 0 \ . \tag{2.26}$$

Therefore

$$\left(\hat{a}_j\hat{a}_j\right)_0 C_{jm_1}^\dagger C_{jm_2}^\dagger \cdots C_{jm_k}^\dagger\psi_0 = 0 \ , \tag{2.27}$$

$$\left(\hat{a}_j\hat{a}_j\right)_0 C_{j_1 m_1}^\dagger C_{j_2 m_2}^\dagger \cdots C_{j_k m_k}^\dagger\psi_0 = 0 \ . \tag{2.28}$$

In order to find the last relation, the following obvious equation is needed:

$$\left[\left(\hat{a}_j\hat{a}_j\right)_0 \ , \ C_{j'm'}^\dagger\right] = 0 \text{ for } j' \neq j \ . \tag{2.29}$$

One sees that the operation of $C_{jm}^\dagger$ on a state with no $\left(a_j^\dagger a_j^\dagger\right)_0$-pairs yields a state of the same kind and increases by 1 the number of the unpaired nucleons in orbit j unless the result vanishes. The projecting operator Γ appearing in the definition of the C-operators ensures that it is impossible to form a state with more than Ω_j unpaired protons or neutrons in orbit j by operating with $C_{jm_1}^\dagger \cdots C_{jm_k}^\dagger$ on ψ_0. It is also worthwhile to notice that:

$$\left(\sum_m C_{jm}^\dagger C_{jm}\right)\psi_0 = \hat{n}_j\Gamma(\Omega_j + 1 - \hat{n}_j)\psi_0 \ , \tag{2.30}$$

$$\left(\sum_m C_{jm}^\dagger C_{jm}\right) C_{jm_1}^\dagger \cdots C_{jm_k}^\dagger\psi_0$$
$$= C_{jm_1}^\dagger \cdots C_{jm_k}^\dagger (\hat{n}_j + k)\,\Gamma(\Omega_j + 1 - k - \hat{n}_j)\psi_0 \ , \tag{2.31}$$

which show that for a kind of states satisfying condition (2.30), the operator $\sum_m C_{jm}^\dagger C_{jm}$ represents the number operator $\hat{n}_j$.

III. CONSTRUCTION OF SUBSPACE V_F, V_B AND MAPPING

For convenience we consider a system with $2N$ active identical nucleons moving in many j-orbits which are in general non-degenerate. The generalization of our method to the case with protons and neutrons will be straightforward. As usual the creation operator of the correlated S-pair will be expressed as a linear combination of the pairs of $a_{jm}^\dagger$-operators:

$$S^\dagger = \sum_{\alpha\beta} \chi_{\alpha\beta}(0,0) a_\alpha^\dagger a_\beta^\dagger = \frac{1}{2\sqrt{\Omega}} \sum_j v_j \sum_m (-1)^{j-m} a_{jm}^\dagger a_{j-m}^\dagger .$$ (3.1)

The creation operators of the correlated DG pairs can be formed by either $a_{jm}^\dagger$ or $C_{jm}^\dagger$ -operators. For the sake of definiteness and of convenience in generalizing the OAI method, we will use the $C_{jm}^\dagger$ -operators and define the creation operators of the DG pairs as

$$D_\mu^\dagger = \sum_{\alpha\beta} \chi_{\alpha\beta}(2\mu) C_\alpha^\dagger C_\beta^\dagger, \quad (\chi_{\alpha\beta}(2\mu) = -\chi_{\beta\alpha}(2\mu)) ,$$ (3.2)

$$G_\mu^\dagger = \sum_{\alpha\beta} \chi_{\alpha\beta}(4\mu) C_\alpha^\dagger C_\beta^\dagger, \quad (\chi_{\alpha\beta}(4\mu) = -\chi_{\beta\alpha}(4\mu)) .$$ (3.3)

In relations (3.1)–(3.3), α or β stands for the whole set of the single-particle quantum numbers (jm). The coefficients $\chi_{\alpha\beta}(J\mu)$ are to be determined under the conditions that they are real numbers, yield the parity and angular momentums of the SDG pairs and satisfy

$$\langle 0 \mid S\, S^\dagger \mid 0 \rangle = \langle 0 \mid D_\mu D_\mu^\dagger \mid 0 \rangle = \langle 0 \mid G_\mu G_\mu^\dagger \mid 0 \rangle = 1 .$$ (3.4)

From relation (2.25), one gets

$$S\, C_{j_1 m_1}^\dagger C_{j_2 m_2}^\dagger \cdots C_{j_k m_k}^\dagger \mid 0 \rangle = 0 ,$$ (3.5)

and therefore

$$S\, f(D^\dagger, G^\dagger) \mid 0 \rangle = 0 ,$$ (3.6)

where $f(D^\dagger, G^\dagger) \mid 0 \rangle$ stands for an arbitrary state formed by DG pairs. This relation implies that in a state containing only the DG degrees of freedom, no S-pairs can be found. Furthermore, since for a state $f(C^\dagger) \mid 0 \rangle$ formed by operating with $C^\dagger$ operators on $\mid 0 \rangle$, $S^k S^{\dagger k} f(C^\dagger) \mid 0 \rangle$ is a sum of states of the same kind and (3.5) implies that

$$S\, S^k\, S^{\dagger k} f(C^\dagger) \mid 0 \rangle = 0 .$$ (3.7)

In particular one has

$$S\, S^k\, S^{\dagger k} f(D^\dagger, G^\dagger) \mid 0 \rangle = 0 ,$$ (3.8)

and

$$\langle 0 \mid f(D^\dagger, G^\dagger)^\dagger S^k\, S^{\dagger k'} f'(D^\dagger, G^\dagger) \mid 0 \rangle = 0 \quad \text{for } k' \neq k .$$ (3.9)

Problem of how to determine the coefficients $\chi_{\alpha\beta}(I\mu)$ $\left(\text{or } \chi'_{\alpha\beta}(I\mu)\right)$ will not be treated in the present paper, and we assume here that the operators $S^\dagger$, $D_\mu^\dagger$ and $G_\mu^\dagger$ are known and that all the linear combinations of the N pair states of the form

$$f_1(S^\dagger) f_2(D^\dagger, G^\dagger) \mid 0 \rangle$$ (3.10)

construct an approximately invariant subspace of H. This subspace is what we mean by V_F and is assumed to be invariance under H.

The subspace V_B and our mapping will be constructed with the help of the following transformation operators:

$$U = \langle 0 \mid \exp\{\sum_\mu d_\mu^\dagger D_\mu + \sum_\mu g_\mu^\dagger G_\mu\}\exp\{s^\dagger S\} \mid 0 \rangle ,$$ (3.11)

$$U^\dagger = \langle 0 \mid \exp\{S^\dagger s\}\exp\{\sum_\mu D_\mu^\dagger d_\mu + \sum_\mu G_\mu^\dagger g_\mu\} \mid 0 \rangle .$$ (3.12)

where $s^\dagger, d_\mu^\dagger, g_\nu^\dagger$ are the creation operators of the s, d, g bosons respectively. $\mid 0)$ stands for the boson vacuum. They satisfy the usual boson commutation relations and

$$s \mid 0 \,) = d_\mu \mid 0 \,) = g_\mu \mid 0 \,) = 0 \ . \tag{3.13}$$

The operator U has a structure somewhat similar to the Usui's transformation operator[8,9] in Dyson mapping. However the former contains only the SDG degrees of freedom. Besides, operators D_μ or G_ν do not commute with S.

It is easy to see

$$U \mid 0 \rangle = \mid 0 \,) \ , \tag{3.14}$$

$$U^\dagger \mid 0 \,) = \mid 0 \rangle, \tag{3.15}$$

$$U^\dagger s^\dagger = S^\dagger U^\dagger \ , \tag{3.16}$$

$$U^\dagger f_1(s^\dagger) f_2(d^\dagger, g^\dagger) \mid 0 \,) = f_1(S^\dagger) f_2(D^\dagger, G^\dagger) \mid 0 \rangle \ . \tag{3.17}$$

Obviously, the subspace V_{sdg} formed by all the N-body sdg boson states will be mapped onto V_F under $U^\dagger$:

$$U^\dagger V_{sdg} = V_F \ . \tag{3.18}$$

However, there can be different sdg states which, under $U^\dagger$, are mapped into the same state in V_F. We now define a subspace V_0 of V_{sdg} formed by all the states Φ_0 which satisfies

$$U^\dagger \Phi_0 = 0 \ (\Phi_0 \in V_{sdg}) \ . \tag{3.19}$$

Since this condition is equivalent to

$$UU^\dagger \Phi_0 = 0 \ (\Phi_0 \in V_{sdg}) \ , \tag{3.20}$$

V_0 can also be regarded as the set of all the N-body sdg states with a zero eigenvalue of the operator $UU^\dagger$. It is clear that in the subspace formed by all the N-body sdg states orthogonal to V_0, no different states can be mapped into the same fermion state under $U^\dagger$. This subspace will be what we mean by V_B, and our mapping can be expressed as

$$\psi = U^\dagger \Gamma_B \Phi \ (\Phi \in V_B) \ , \tag{3.21}$$

where Γ_B is the projection operator to the subspace V_B. Needless to say, (3.21) expresses a 1-to-1 correspondence between V_B and V_F. V_B and V_0 are the completely physical and unphysical boson subspace respectively.

Since no physical states can have zero eigenvalue of the operator $UU^\dagger$ which is equal to $\Gamma_B UU^\dagger \Gamma_B$, the following relation expresses a reversible transformation from V_B onto itself:

$$\Phi' = \Gamma_B UU^\dagger \Gamma_B \Phi \ (\Phi \in V_B) \ . \tag{3.22}$$

This also shows that the operator $\Gamma_B U$ can be used to construct a 1-to-1 correspondence from V_F onto V_B:

$$\Phi' = \Gamma_B U \psi \ (\psi \in V_F) \ . \tag{3.23}$$

Moreover, one can define a hermitean operator $\left(UU^\dagger \right)_B^{-1/2}$ and get for an arbitrary $\phi \, (\in V_B)$:

$$\left(UU^\dagger \right)_B^{-1/2} \Gamma_B UU^\dagger \Gamma_B \left(UU^\dagger \right)_B^{-1/2} \phi = \phi \ . \tag{3.24}$$

Therefore $U^\dagger \Gamma_B \left(UU^\dagger \right)_B^{-1/2}$ and $\left(UU^\dagger \right)_B^{-1/2} \Gamma_B U$ can be regarded as "unitary" operators. Namely under each of the following mappings an orthogonal-normalized basis will be mapped into another:

$$\psi = U^\dagger \Gamma_B \left(UU^\dagger \right)_B^{-1/2} \phi \ (\phi \in V_B) \ , \tag{3.25}$$

$$\phi = \left(UU^\dagger \right)_B^{-1/2} \Gamma_B U \psi \ (\psi \in V_F) \ . \tag{3.26}$$

We will regard these as the rewritten form of (3.21) and (3.23) respectively. Correspondingly, ϕ and Φ apprearing in these formulae are related to each other through

$$\Phi = \left(UU^\dagger\right)_B^{-1/2}\phi .$$

(3.27)

The operators $\left(UU^\dagger\right)_B^{1/2}$ and $\left(UU^\dagger\right)_B^{-1/2}$ can be defined as follows. Each of $\left(UU^\dagger\right)_B^{1/2}\Phi_0$ and $\left(UU^\dagger\right)_B^{-1/2}\Phi_0$ vanishes for $\Phi_0 \in V_0$. For a given basis of V_B, the matrix of $\left(UU^\dagger\right)_B^{1/2}$ is the real and positive-definite square root of the matrix of $\Gamma_B UU^\dagger\Gamma_B$, and the matrix of $\left(UU^\dagger\right)_B^{-1/2}$ is the inverse of that of $\left(UU^\dagger\right)_B^{1/2}$. We will also denote $\left\{\left(UU^\dagger\right)_B^{1/2}\right\}^2$ and $\left\{\left(UU^\dagger\right)_B^{-1/2}\right\}^2$ by $\left(UU^\dagger\right)_B$ and $\left(UU^\dagger\right)_B^{-1}$ respectively.

IV. BOSON IMAGES OF HAMILTONIAN AND OTHER OPERATORS

Since the subspace V_F of Fermion states formed by the SDG pairs is assumed to be invariant under H and the correspondence given in (3.14) is 1-to-1, the equation of motion of the collective states can be written as:

$$HU^\dagger\Gamma_B\Phi_\lambda = E_\lambda U^\dagger\Gamma_B\Phi_\lambda,\ (\Phi_\lambda \in V_B) ,$$

(4.1)

which is equivalent to

$$\Gamma_B UHU^\dagger\Gamma_B\Phi_\lambda = E_\lambda \left(UU^\dagger\right)_B \Phi_\lambda ,$$

(4.2)

or

$$H_B\phi_\lambda = E_\lambda\phi_\lambda ,$$

(4.3)

with

$$\phi_\lambda = \left(UU^\dagger\right)_B^{1/2}\Phi_\lambda ,$$

(4.4)

and

$$H_B = \left(UU^\dagger\right)_B^{-1/2}\Gamma_B UHU^\dagger\Gamma_B\left(UU^\dagger\right)_B^{-1/2} .$$

(4.5)

It is the operator H_B which is regarded as the IBM Hamiltonian in our method.

Subspace V_F will not be invariant under an arbitrary operator F even when it is exactly invariant under H. However, when we only describe the collective states, the Boson image F_B can still be derived from the formula given in the introduction section and written as:

$$F_B = \left(UU^\dagger\right)_B^{-1/2}\Gamma_B UFU^\dagger\Gamma_B\left(UU^\dagger\right)_B^{-1/2} .$$

(4.6)

It should be noticed that the product $F_B^{(1)}F_B^{(2)}$ formed by the Boson images of two operators $F^{(1)}$ and $F^{(2)}$ is usually impossible to be regarded as the Boson image of $F^{(1)}F^{(2)}$. This is also a drastic difference between the present method and the usual Boson mapping.

We next introduce operators h_1, h_2 and f_1, f_2 according to the following conditions:

$$UHU^\dagger = h_1 UU^\dagger = UU^\dagger h_2 ,$$

(4.7)

$$UFU^\dagger = f_1 UU^\dagger = UU^\dagger f_2 .$$

(4.8)

The operators h_1, f_1 replace the usual Dyson images H_D, F_D which satisfy

$$U_s HU_s^\dagger = H_D U_s U_s^\dagger ,$$

(4.9)

$$U_s F U_s^\dagger = F_D U_s U_s^\dagger \,, \tag{4.10}$$

with

$$U_s = \langle 0 \mid \exp\{\frac{1}{2}\sum_{\alpha\beta} b_{\alpha\beta}^\dagger a_\beta a_\alpha\} \mid 0 \rangle \tag{4.11}$$

where U_s is the Usui operator[8,9] and $\{b_{\alpha\beta}^\dagger\}$ are the creation operators of so-called ideal bosons. The operators h_1, f_1 (or h_2, f_2) are in general a series, while their counterparts in Dyson mapping are finite. We expect that, they can be approximated by taking the low order terms for the usual H and F. The method of finding the expressions for these operators will be given in the next section.

From (4.5)–(4.8), one can express H_B and F_B as:

$$H_B = \left(UU^\dagger\right)_B^{-1/2} \Gamma_B h_1 \Gamma_B \left(UU^\dagger\right)_B^{1/2} = \left(UU^\dagger\right)_B^{1/2} \Gamma_B h_2 \Gamma_B \left(UU^\dagger\right)_B^{-1/2} \,, \tag{4.12}$$

$$F_B = \left(UU^\dagger\right)_B^{-1/2} \Gamma_B f_1 \Gamma_B \left(UU^\dagger\right)_B^{1/2} = \left(UU^\dagger\right)_B^{1/2} \Gamma_B f_2 \Gamma_B \left(UU^\dagger\right)_B^{-1/2} \,. \tag{4.13}$$

Moreover the equation of motion can be transformed into

$$\left(\Gamma_B h_2 \Gamma_B\right) \Phi_\lambda = E_\lambda \Phi_\lambda \,, \ (\Phi_\lambda \in V_B) \tag{4.14}$$

or

$$h_1 \Phi_\lambda' = E_\lambda \Phi_\lambda' \,. \qquad \left(\Phi_\lambda' = UU^\dagger \Phi_\lambda\right) \tag{4.15}$$

Usually, h_1 and h_2 are non-hermitean. In some cases they can be hermitean and then commute with $UU^\dagger$ and therefore H_B is equal to $\Gamma_B h_1 \Gamma_B$ or $\Gamma_B h_2 \Gamma_B$.

While various forms of the equation of motion as written in (4.3), (4.14) and (4.15) can only be emploied within the physical subspace, the eigen-equation of h_2 and h_1 can be used in the entire V_{sdg} with the help of the following theorems:

Theorem 1. If $\Phi \in V_{sdg}$ is an eigenstate of h_2 belonging to the eiganvalue E and satisfies $U^\dagger \Phi \neq 0$, then $U^\dagger \Phi$ is an eigenstate of H belonging to the same eigenvalue.

Theorem 2. If $\Phi \in V_{sdg}$ and $U^\dagger \Phi(\neq 0)$ satisfies equation $HU^\dagger \Phi = EU^\dagger \Phi$, then $UU^\dagger \Phi$ is a completely physical eigenstate of h_1 belonging to eigenvalue E.

V. LOW ORDER APPROXIMATION TO THE BOSON HAMILTONIAN AND OTHER BOSON OPERATORS

A. f_2 and f

We will cosider the E2 transition operator $F(2\mu)$ and Hamiltonian H with a single-particle part and a two-body interaction for a 2N-body fermion system. The lowest order expression for $f_2(2\mu)$ can be written in the following form with real coefficients $\alpha_2 \ \alpha_2' \ \beta' \ \gamma_2 \ \gamma_2'$ and δ':

$$f_2^{(0)}(2\mu) = \alpha_2 s^\dagger \hat{d}_\mu + \alpha_2' d_\mu^\dagger s + \beta' \left(d^\dagger \hat{d}\right)_{2\mu}$$
$$+ \gamma_2 \left(d^\dagger \hat{g}\right)_{2\mu} + \gamma_2' \left(g^\dagger \hat{d}\right)_{2\mu} + \delta' \left(g^\dagger \hat{g}\right)_{2\mu} \,. \tag{5.1}$$

We only use the boson states with $2N \leq \sum_j \Omega_j$. If the particle number is larger than $\sum_j \Omega_j$ we will deal with holes. We also assume that $\sum_j \Omega_j$ is large enough and the following states are physical:

$$s^{\dagger N} \mid 0 \rangle, \ s^{\dagger N-1} d_\mu^\dagger \mid 0 \rangle, \ s^{\dagger N-1} g_\mu^\dagger \mid 0 \rangle \,. \tag{5.2}$$

These are all the eigenstates of $UU^\dagger$. Thus, by evaluating with these states the matrix elements of the operators in the two sides of the equation

$$UF(2\mu)U^\dagger \approx UU^\dagger f_2^{(0)}(2\mu) \,, \tag{5.3}$$

one can find the formula for the coefficients appearing in (5.1). One of the usages of $f_2^{(0)}(2\mu)$ is that one can express the approximate boson image $F_B(2\mu)$ of $F(2\mu)$ as

$$F_B(2\mu) \approx \left(UU^\dagger\right)_B^{1/2} f_2^{(0)}(2\mu) \left(UU^\dagger\right)_B^{-1/2} . \tag{5.4}$$

A more rough procedure of finding the approximate formula for $F_B(2\mu)$ is to expressed itself as

$$F_B(2\mu) \approx \Gamma_B f^{(0)}(2\mu)\Gamma_B , \tag{5.5}$$

where $f^{(0)}(2\mu)$ is of the form with real cofficients:

$$f^{(0)}(2\mu) = \alpha\left(s^\dagger \hat{d}_\mu + d_\mu^\dagger s\right) + \beta\left(d^\dagger \hat{d}\right)_{2\mu}$$
$$+ \gamma\left\{\left(d^\dagger \hat{g}\right)_{2\mu} + \left(g^\dagger \hat{d}\right)_{2\mu}\right\} + \delta\left(g^\dagger \hat{g}\right)_{2\mu} . \tag{5.6}$$

By evaluating with the states listing in (5.2) the matrix elements of the two sides of the equation

$$\Gamma_B\left(UU^\dagger\right)_B^{-1/2} UF(2\mu) U^\dagger\left(UU^\dagger\right)_B^{-1/2}\Gamma_B \approx \Gamma_B f^{(0)}(2\mu)\Gamma_B , \tag{5.7}$$

one can obtain the formula for α, β, γ and δ and find that

$$\beta = \beta', \qquad \delta = \delta',$$
$$\alpha^2 = \alpha_2\,\alpha_2', \gamma^2 = \gamma_2\gamma_2' .$$

B. h_2 and h

We write the lowest order expression of h_2 in the following form with real coefficients

$$h_2^{(0)} = H_0 + h_0 + \epsilon_d \hat{N}_d + \epsilon_g \hat{N}_g$$
$$+ \sum_I \sqrt{2I+1}Y_I(b_1 b_2, b_3 b_4)\left[\left(b_1{}^\dagger b_2{}^\dagger\right)_I \left(\hat{b}_3\hat{b}_4\right)_I\right]_0 , \tag{5.8}$$

where $H_0 \equiv \langle 0 \mid H \mid 0\rangle$, $\left(b_1^\dagger b_2^\dagger\right)$ stands for $s^\dagger s^\dagger$, $\left(d^\dagger d^\dagger\right)_I$, $\left(g^\dagger g^\dagger\right)_I$, $\left(s^\dagger d^\dagger\right)_I$, $\left(s^\dagger g^\dagger\right)_I$ and $\left(d^\dagger g^\dagger\right)_I$ and $\left(\hat{b}_3\hat{b}_4\right)_I$ for ss, $\left(\hat{d}\hat{d}\right)_I$, $(\hat{g}\hat{g})_I$, $\left(\hat{d}s\right)_I$, $(\hat{g}s)_I$ and $\left(\hat{g}\hat{d}\right)_I$, $\sqrt{2I+1}Y_I(b_1 b_2, b_3 b_4)$ is the coefficient of the term $\left[\left(b^\dagger b^\dagger\right)_I \left(\hat{b}_3\hat{b}_4\right)_I\right]_0$. The summation includes all the independent terms with the understanding that a term containing $s^\dagger s$ as a factor is not independent because it can be replaced by $\left(N - \hat{N}_d - \hat{N}_g\right)$.

The formulae for the coefficients of the independent terms can be found by evaluating the matrix elements of the two sides of the equation

$$U(H - H_0) U^\dagger \approx UU^\dagger\left(h_2^{(0)} - H_0\right) , \tag{5.9}$$

with the following states

$$s^N \mid 0) , \qquad s^{N-1}d_\mu^\dagger \mid 0) , \qquad s^\dagger{}^{N-1}g_\mu^\dagger \mid 0),$$
$$s^\dagger{}^{N-2}\left(d^\dagger d^\dagger\right)_{I\mu} \mid 0) , \; s^\dagger{}^{N-2}\left(d^\dagger g^\dagger\right)_{I\mu} \mid 0) , \; s^\dagger{}^{N-2}\left(d^\dagger g^\dagger\right)_{I\mu} \mid 0) . \tag{5.10}$$

According to section 4, $h_2^{(0)}$ can be used as a non-hermitean boson Hamiltonian or to form an expression for H_B, and the latter takes the form

$$H_B \approx \left(UU^\dagger\right)_B^{1/2} h_2^{(0)} \left(UU^\dagger\right)_B^{-1/2} = \left(UU^\dagger\right)_B^{-1/2} h_2^{(0)\dagger} \left(UU^\dagger\right)_B^{1/2} . \tag{5.11}$$

On the other hand we can also follow the rough procedure mentioned above and write H_B as

$$H_B \approx \Gamma_B h^{(0)}\Gamma_B , \tag{5.12}$$

where $h^{(0)}$ is a hermitean operator of the from with real cofficients

$$h^{(0)} = H_0 + h_0 + \epsilon_d \hat{N}_d + \epsilon_g \hat{N}_g$$
$$= + \sum_I \sqrt{2I + 1} W_I (b_1 b_2, b_3 b_4) \left[\left(b_1^\dagger b_2^\dagger \right)_I \left(\hat{b}_3 \hat{b}_4 \right)_I \right]_0 . \tag{5.13}$$

The summation includes all the independent terms. The coefficients can be found by evaluating with the states in (5.10) the matrix elements of the two sides of equation

$$\Gamma_B \left(U U^\dagger \right)_B^{-1/2} U (H - H_0) U^\dagger \left(U U^\dagger \right)_B^{-1/2} \Gamma_B \approx \Gamma_B \left(h^{(0)} - H_0 \right) \Gamma_B . \tag{5.14}$$

Since the low-order formulae for boson operators are derived with the states listed in (5.10) each of which contains a very low number of DG pairs, these formulae can be used for performing calculations by employing the method as adopted in ref. 10. Of course, it is also worthwhile to develop approximation procedures.

VI. SINGLE j-ORBIT

We now apply the present method to a single j-orbit and compare it with the OAI method. From relations (2.4) and (2.6) one has

$$\left(c_j^\dagger c_j^\dagger \right)_{I\mu} = \left(A_j^\dagger A_j^\dagger \right)_{I\mu} \Gamma \left(\Omega_j - 1 - \hat{n}_j \right) , \tag{6.1}$$

where

$$\left(A_j^\dagger A_j^\dagger \right)_{I\mu} = \left(a_j^\dagger a_j^\dagger \right)_{I\mu} + \frac{\Omega_j}{2} \frac{1}{\Omega_j + 2 - \hat{n}_j} \frac{1}{\Omega_j + 3 - \hat{n}_j} \left(a_j^\dagger a_j^\dagger \right)_0 \left(a_j^\dagger a_j^\dagger \right)_0 (\hat{a}_j \hat{a}_j)_{I\mu}$$
$$+ \sqrt{2\Omega_j} \frac{1}{\Omega_j + 2 - \hat{n}_j} \left(a_j^\dagger a_j^\dagger \right)_0 \left(a_j^\dagger \hat{a}_j \right)_{I\mu} - \Omega_j \frac{1}{\Omega_j + 2 - \hat{n}_j} \left(a_j^\dagger a_j^\dagger \right)_{I\mu} \delta_{I0} . \tag{6.2}$$

For $I \neq 0$, one can use

$$\left(a_j^\dagger \hat{a}_j \right)_{I\mu} = \frac{\sqrt{2\Omega_j}}{4} \left[(\hat{a}_j \hat{a}_j)_0 , \left(a_j^\dagger a_j^\dagger \right)_{I\mu} \right] ,$$
$$(\hat{a}_j \hat{a}_j)_{I\mu} = \frac{\Omega_j}{4} \left[(\hat{a}_j \hat{a}_j)_0 , \left[(\hat{a}_j \hat{a}_j)_0 , \left(a_j^\dagger a_j^\dagger \right)_{I\mu} \right] \right] .$$

On operating with $\left(c_j^\dagger c_j^\dagger \right)_{I\mu}$ on ψ_0 which satisfies condition (2.23), one gets

$$\left(c_j^\dagger c_j^\dagger \right)_{I\mu} \psi_0 = P \left(a_j^\dagger a_j^\dagger \right)_{I\mu} \Gamma(\Omega_j - 1 - \hat{n}_j) \psi_0 \text{ for } I \neq 0 , \tag{6.3}$$

where

$$P = 1 + \frac{\Omega_j}{2} \frac{1}{\Omega_j + 2 - \hat{n}_j} \left(a_j^\dagger a_j^\dagger \right)_0 (\hat{a}_j \hat{a}_j)_0$$
$$+ \frac{\Omega_j^2}{8} \frac{1}{\Omega_j + 2 - \hat{n}_j} \frac{1}{\Omega_j + 3 - \hat{n}_j} \left(a_j^\dagger a_j^\dagger \right)_0 \left(a_j^\dagger a_j^\dagger \right)_0 (\hat{a}_j \hat{a}_j)_0 (\hat{a}_j \hat{a}_j)_0 . \tag{6.4}$$

Relation (6.3) also implies that

$$\left(c_j^\dagger c_j^\dagger \right)_{I_1 \mu_1} \left(c_j^\dagger c_j^\dagger \right)_{I_2 \mu_2} \cdots \left(c_j^\dagger c_j^\dagger \right)_{I_k \mu_k} \psi_0$$
$$= \left(P \left(a_j^\dagger a_j^\dagger \right)_{I_1 \mu_1} \right) \left(P \left(a_j^\dagger a_j^\dagger \right)_{I_2 \mu_2} \right) \cdots \left(P \left(a_j^\dagger a_j^\dagger \right)_{I_k \mu_k} \right)$$
$$\times \Gamma(\Omega_j + 1 - 2k + \hat{n}_j) \psi_0 \text{ for } I_1, I_2 \cdots \neq 0 . \tag{6.5}$$

In particular,

$$\left(c_j^\dagger c_j^\dagger \right)_{I_1 \mu_1} \left(c_j^\dagger c_j^\dagger \right)_{I_2 \mu_2} \cdots \left(c_j^\dagger c_j^\dagger \right)_{I_k \mu_k} | 0 \rangle$$
$$= \left(P \left(a_j^\dagger a_j^\dagger \right)_{I_1 \mu_1} \right) \left(P \left(a_j^\dagger a_j^\dagger \right)_{I_2 \mu_2} \right) \cdots \left(P \left(a_j^\dagger a_j^\dagger \right)_{I_k \mu_k} \right) | 0 \rangle$$
$$\times \Gamma(\Omega_j + 1 - 2k) . \tag{6.6}$$

The operator P given in (6.4) is just the operator used by OAI to insure that the operation of $P\left(a_j^\dagger a_j^\dagger\right)_{2\mu}$ on a highest seniority state yields a state of the same kind. It follows that the OAI procedure of defining the D-pair is a special case of the present procedure.

Moreover, in the OAI method and the special form of the present method for a V_F containing only the SD degrees of freedom the low-order approximation boson image of the Hamiltonian and other physical quantities are determined with the help of those SD states which can be completely specified by the total pair number N, D-pair number n_d and angular momentum (IM). Besides, under the restriction of our mapping to the case of a single j-orbit with a V_F containing only the SD degrees of freedom, the boson image of a SD state which is completely specified by (N n_d I M) is the same as the OAI boson image. It follows that in the case of single j-orbit, the low-order expression of the sd boson image of physical quantities will be the same for the OAI method and the restriction form of the present method.

VII. CONCLUDING REMARKS

By constructing and using the basic tools, namely, operators $C_{jm}, C_{jm}^\dagger$, subspaces V_F, V_B and mapping $\{U^\dagger \Gamma_B : V_B \longrightarrow V_F\}$, we have developed a generalized OAI method for calculating low-lying spectra in medium-to-heavy even-even nuclei and expounding the microscopic foundation of the interacting boson model. When the expression for H_B obtained through $h^{(0)}$ is found to be poor, another kind of expression obtained through $h_2^{(0)}$ may show some advantages because the factors $\left(UU^\dagger\right)_B^{1/2}$ and $\left(UU^\dagger\right)_B^{-1/2}$ are explicitly retained in the later kind of expression. One can also apply the method of section 5 to include the higher order terms in $h^{(0)}, h_2^{(0)}$ and $f^{(0)}, f_2^{(0)}$ by adding to (5.10) some completely physical states with more d-g bosons. If some of these states are not completely physical then the unphysical components should be excluded beforehand. In principle our method can be generalized to include directly the single-particle degrees of freedom besides that of SDG pairs.

ACKNOWLEDGEMENTS

This work is a part of the program of the joint research between the University of Tokyo and Peking University, under the sponsorship of the Ministry of Education, Culture and Science of Japan. The Project is also supported in part by Doctoral Program Foundation of Institution of Higher Education and by National Natural Science Foundation of China.

REFERENCES

1. A. Arima and F. Iachello, Phys. Rev. Lett. 35:1069 (1975).
2. A. Arima, T. Otsuka, F. Iachello and I. Talmi, Phys. Lett. 66B:205 (1977).
3. T. Otsuka, A. Arima and F. Iachello, Nucl. Phys. A309:1 (1978).
4. T. Suzuki and K. Matsuyanagi, Prog. Theor. Phys. 56:1156 (1976).
5. T. Suzuki, M. Fuyuki and K. Matsuyanagi, Prog. Theor. Phys. 61:1082 (1979).
6. T. Marumori, M. Yamamura and A. Tokunaga, Prog. Theor. Phys. 31:1009 (1964).
7. S. Y. Li, R. M. Dreizler and A. Klein, Phys. Rev. C4:1571 (1971).
8. T. Usui, Prog. Theor. Phys. 23:787 (1960).
9. D. Jansen, F. Donau, S. Frauendorf and R. V. Jolos, Nucl. Phys. A172:145 (1971).
10. S. Pittel, P. D. Duval and B. R. Barrett, Phys. Rev. C25:2843 (1982); Ann. Phys. (NY) 144:168 (1982).

DYNAMICAL SYMMETRY AND CHAOS

IN NUCLEAR COLLECTIVE MODELS

Naotaka Yoshinaga

Department of Physics, College of Liberal Arts, Saitama University
Urawa-shi, Saitama 338, Japan

INTRODUCTION

The Interacting Boson Model [1] (**IBM** for short) was introduced by Franco Iachello and Akito Arima almost 20 years ago. Since then it has served as one of the basic models to understand nuclear collective motion. In this talk I would like to show that the IBM still serves as a basic model for a new research of understanding nature.

My talk is divided into two parts. In the first part the Interacting Boson Model of Arima and Iachello is investigated from a view point of energy level statistics. In the second part the Ginocchio Model [2] is investigated from the same point of view. Since the first part has been already reported in the reference [3] in detail, I shall only briefly mention our main result its importance for the first part and concentrate on the latter half part.

One of our purposes in this work is to find a generic feature of the nuclear collective motion from a view point of energy level statistics. We study the nuclear collective motion from the view point how and to what extent the nuclear collective motion becomes regular to chaos.

In this talk I consider only the nearest-neighbour level-spacing distribution : $P(s)$, probability that the spacing of a pair of neibouring (in energy) levels lies between s and $s + ds$. A classical integrable system, when quantized, gives a poisson distribution for $P(s)$. This is proved by Berry and Tabor [4], while it is believed that a strongly chaotic system gives a Wigner distribution. This is only demonstrated by using Random Matrix Theory [5].

THE INTERACTING BOSON MODEL

I shall talk about an analysis of the IBM from a view point of energy level statistics. There are some interesting features in the IBM. First the IBM is a realistic model to investigate chaos; up to now the study of chaos is carried out only by simple, but unrealistic models, such as the kicked rotator or billiards. Second due to a change of parameters the system of the IBM moves from integral to non-integral and returns to integral again. Thus the system falls on the class of "a soft chaos system", the word coined by Gutswiller [6]. Third we would like to investigate a relation between regular motion and dynamical symmetry. Finally we are interested in the statistics in between Poisson and Wigner. It is not a settled problem what kind of distribution appears in the intermediate situation between regular and chaos.

The IBM has the group symmetry U(6) which has three dynamical symmetries as its subgroups: U(5), SU(3) and O(6). The U(5) symmetry corresponds to the vibrational nuclei, SU(3) the rotational nuclei, and O(6) the gamma-unstable nuclei. Varying two parameters is sufficient to connect the three dynamical symmetries. In this sense our model hamiltonian is essentially the same which was considered by Alhassid et al. [7] The model Hamiltonian we study is written as

$$H = (1-\kappa)\left(\sqrt{2}\, n_d^2 + \sqrt{3}U \cdot U\right) - \kappa Q(\chi) \cdot Q(\chi) \tag{1}$$

where d-boson number n_d, the octupole operator U, and the quadrupole operator $Q(\chi)$ are defined respectively by

$$n_d = d^\dagger \cdot \tilde{d} \, , \qquad\qquad U = [d^\dagger \tilde{d}]^{(3)} \tag{2}$$

and

$$Q(\chi) = \left([s^\dagger \tilde{d}]^{(2)} + [d^\dagger \tilde{s}]^{(2)}\right) - \frac{\sqrt{7}}{2} \chi \, [d^\dagger \tilde{d}]^{(2)} \tag{3}$$

To define the chaoticity we use M_n the moments of the distribution $P(s)$ defined by

$$M_n = \int_0^\infty ds\, s^n\, P(s) \tag{4}$$

as measures of the level statistics. For "chaotic" Wigner distribution, M_n take the values

$M_2 = 1.273..,\qquad M_3 = 1.909..,\qquad M_4 = 3.242.., \ldots$,

and for "integrable" Poisson ditribution, they are

$M_2 = 2,\qquad M_3 = 6,\qquad M_4 = 24, \ldots$.

They interpolate in between these two values for intermediate statistics.

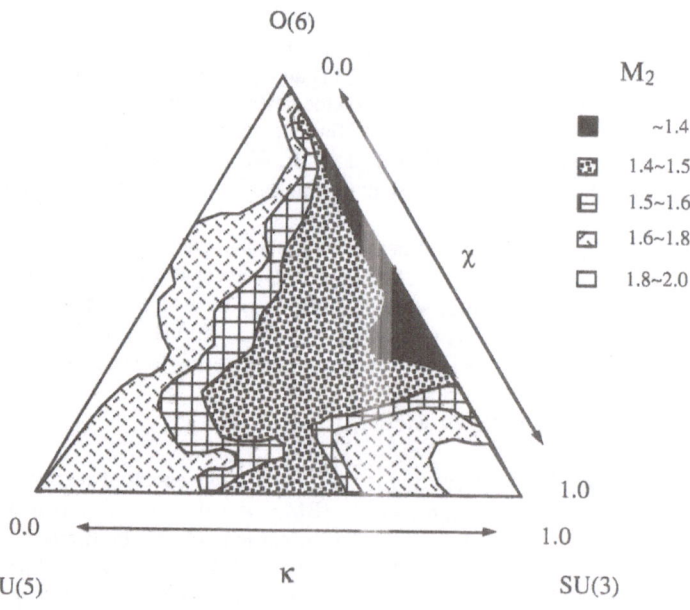

Figure 1. Map of M_2 on the Casten triangle. Equilateral lines of parameters κ and χ are shown.

Our primary result is shown in figure 1, a contour map of the value of M_2 on the symmetry triangle of parameter space (κ, χ). Denser hatch corresponds to the smaller value of M_2 indicating more Wigner like level spacing distribution. Two Poissonian limits SU(3) and O(6) are separated by the chaotic region where the $P(s)$ becomes almost completely Wigner-like. The chaotic domain persists to the central region of the triangle away from all symmetry limits. A remarkable feature of this map is the existence of an unexpected strip of Poissonian statistics along the line connecting the two symmetry limits U(5) and O(6). This curious behavior of the nearest-neighbor level-spacing distribution can be understood in terms of the unbroken O(5) symmetry along that line which corresponds to $\chi = 0$. Because of the conserved quantum number associated with the O(5) symmetry (v and n_Δ) Hamiltonian matrix possesses the block-diagonal structure.

What we have learned from this result is that (1) even without fermionic degree of freedom, we have chaotic motion and (2) the regularity is closely connected to the dynamical symmetry.

THE GINOCCHIO MODEL

Our motivation why we take the Ginocchio model [2,8] for the study of chaos is as follows. First it is a pure fermion model. We can explicitly incorporate the Pauli effect in the model. While in a boson model the Pauli effect is only incorporated into the coefficients of the hamiltonian. Second the model has a close relation to the IBM. The nucleon pairs in the Ginocchio model approach to the corresponding s and d bosons in the IBM for large nuclear degeneracy. That is why the Ginocchio model has been used to test the IBM [9]. Third the Ginocchio model has no dynamical symmetry in the deformed limit, which is in contrast to the SU(3) limit of the IBM. It should be noted that realistic models such as the nuclear shell model or the Hartree Fock Bogoliubov have no dynamical symmetry in their rotational limits. Thus it is quite interesting to investigate the Ginocchio model in the above respects.

We start from a single j-shell model in which a nucleus is described as a system of fermions with pseudo spin i filling a shell specified by its pseudo orbital angular momentum k and total angular momentum $j = k + i$. In the Ginocchio model the intrinsic spin is given by $i = 3/2$ in place of the nucleon's one-half. We refer to this fictitious spin as the pseudo spin. In the L-S coupling scheme, the two-particle state is specified by its total orbital and spin angular momenta $K = k_1 + k_2$ and $I = i_1 + i_2$. The identity of the two fermions prohibits the odd values for the total spin, namely only $I = 0$ and $I = 2$ are allowed. The unique feature of the Ginocchio model is in the fact that the operators made of two fermions coupled to $K = 0$ form a closed algebra SO(8) whose 28 generators are given by

$$S^\dagger = \sqrt{2k+1}\,[\, b_{ki}^\dagger b_{ki}^\dagger \,]^{(0,0)} \,, \qquad\qquad S = \sqrt{2k+1}\,[\, \tilde{b}_{ki} \tilde{b}_{ki} \,]^{(0,0)} \,, \qquad (5a)$$

$$D^\dagger = \sqrt{2k+1}\,[\, b_{ki}^\dagger b_{ki}^\dagger \,]^{(0,2)} \,, \qquad\qquad \tilde{D} = \sqrt{2k+1}\,[\, \tilde{b}_{ki} \tilde{b}_{ki} \,]^{(0,2)} \,, \qquad (5b)$$

$$R^{(L)} = \sqrt{2k+1}\,[\, b_{ki}^\dagger \tilde{b}_{ki} \,]^{(0,L)} \qquad\qquad L = 0,\,1,\,2,\,3, \qquad (5c)$$

where $b_{ki}^\dagger$ and b_{ki} are the fermion creation and annihilation operators. The notation $(0,L)$ means that the pseudo orbital angular momenta are coupled to 0 and the pseudo spins to L. The number operator of fermions n is given by $n = 2R^{(0)}$. We also define $N = n/2$ which represents the number of pairs.

The SO(8) symmetry of this model has three dynamical sub-symmetries SO(5)×SU(2), SO(6) and SO(7), all sharing the common subgroup SO(5) in addition to the rotational symmetry SO(3), namely,

$$SO(8) \supset \begin{cases} SO(5) \times SU(2) \\ SO(6) \\ SO(7) \end{cases} \supset SO(5) \supset SO(3) \quad .$$

The common subgroup SO(5) poses a problem in the actual application of the model, since no such universal symmetry is known among real nuclei. This difficulty has been solved in the extended model developed by Arima, Ginocchio and Yoshida [8] where protons and neutrons are distinguished. In our current work, we adopt a simplified version of this extended model, whose hamiltonian is given by

$$H = -g_\pi S_\pi^\dagger S_\pi - g_\nu S_\nu^\dagger S_\nu - \kappa Q_\nu^{(2)} \cdot Q_\pi^{(2)} \quad , \tag{6}$$

where quadrupole operator $Q^{(2)}$ is defined by

$$Q^{(2)} = R^{(2)} - \frac{\sqrt{7}}{2\Omega} [\, D^\dagger \, \tilde{D} \,]^{(2)} \quad . \tag{7}$$

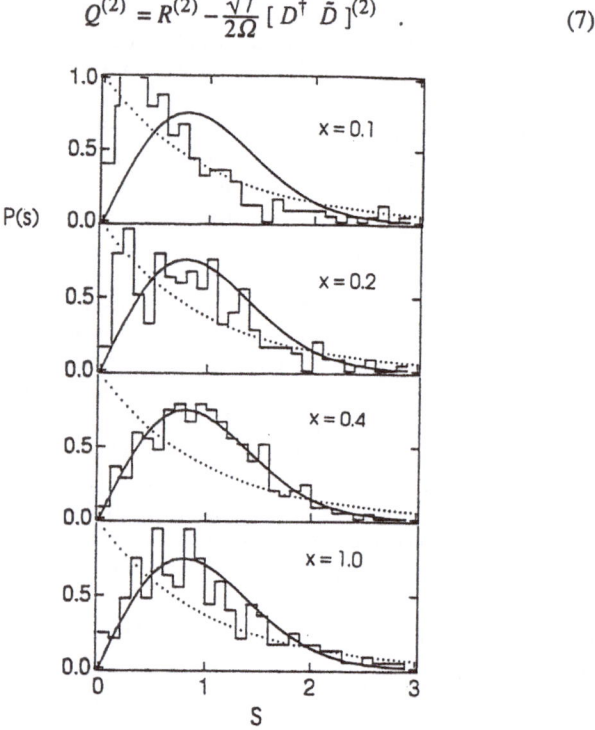

Figure 2.   The level spacing distribution $P(s)$ of the Ginocchio model with the various values of x, with Wigner (solid lines) and Poisson (dotted lines) formulae.

The subscripts π and ν in eq. (6) refer to the proton and neutron respectively. The first two terms in eq. (6) represent the paring interactions between like particles, while the last term represents the quadrupole interaction between different kind of particles. Thus this hamiltonian has the essential feature of a "Paring plus QQ" hamiltonian. The QQ interaction explicitly breaks the unwanted sub-symmetry SO(5). The shell degeneracies Ω_π and Ω_ν appearing in eq. (7) are defined in terms of the values of orbital angular momenta k_π and k_ν

as $\Omega_\pi = 2 (2k_\pi+1)$ and $\Omega_\nu = 2 (2k_\nu+1)$. With this definition the shell degeneracy stands for half of the maximum number of fermions in each shell. It is known that the Ginocchio model approaches to the IBM when Ω_π and Ω_ν approach infinity [8]. Without losing the essential feature, we simplify the hamiltonian by introducing a single parameter x and putting the constraints among the three parameters g_π, g_ν and κ as

$$g_\pi = g_\nu = 1-x , \qquad \kappa = 5x . \qquad (8)$$

When the parameter x is varied from 0 to 1, we can simulate nuclei from pairing-vibrational to rotational limits. The factor 5 in the second of eq. (8) is chosen just for convenience, and it can be re-scaled to any value without changing the physical content. The SO(5) dynamical symmetry exists in the pairing-vibrational limit. On the contrary, no dynamical symmetry is present in the rotational limit of the Ginocchio model, except in the limiting case of $\Omega_\pi = \Omega_\nu = \infty$ where the rotational limit becomes the SU(3) limit of the IBM [8]. In physical term, the collectivity realized at $\Omega = \infty$ is hindered by the Pauli-blocking effect when Ω is small.

We look at the change in the level statistical of the Ginocchio model along the vibrational and rotational limits as we vary the value of x from zero to one. In figure 2, the nearest level spacing distributions $P(s)$ at $x = 0.1, 0.2, 0.4$ and 1 are displayed. Throughout, the shell degeneracies and the numbers of the pairs are kept to be $\Omega_\pi = 11$, $\Omega_\nu = 22$, and $N_\pi = 5$, $N_\nu = 4$, which corresponds to ^{150}Nd. One observes that the $P(s)$ starts out as Poisson-like around the pairing limit, then becomes Wigner-like at intermediate region $x = 0.4$. Further to the rotational limit, it stays Wigner-like up to $x = 1$. The result in figure 2 is interpreted that the dynamics of actual rotational nuclei can be more chaotic than the previous studies based on the IBM indicate. In hindsight, this is a natural consequence of the absence of dynamical symmetry in the rotational limit of the Ginocchio model.

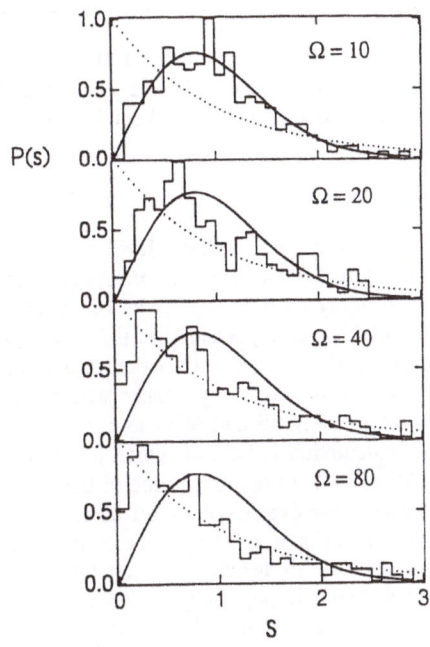

Figure 3. $P(s)$ for various Ω at the rotational limit $x = 1$.

The figure 3 shows how large Ω one needs in order to reach the SU(3) limit of the IBM in terms of the level statistics. Here, x is fixed to be one and $\Omega \equiv \Omega_\pi = \Omega_\nu$ is varied as 10, 20, 40, 80. N_π and N_ν are again fixed to be 5 and 4. One observes that the Poisson statistics as in the boson limit is certainly recovered, but only at very large value of Ω.

The result shown in figure 3 also has an interesting implication to the global trend across the different nuclear mass region. Within the framework of single j-model, larger mass number corresponds to the lager value of Ω for the valence shell. Therefore, the transition to Poisson-like statistics in figure 3 seems to offer a natural explanation for the experimentally observed global shift toward the regularity in the spectra of heavier nuclei [10]. In order to make a simple estimate, we assume that most nuclei are deformed, and most levels are rotational. We also assume the spectral statistics of the whole range of nuclei in a major shell is represented by a single typical nucleus. These assumptions may appear quite drastic, but we argue that they still keep the underlying physics intact. For the first point, it can be pointed out that even spherical nuclei display rotational bands in their excited states whose number tends to dominate over the sparsely found single particle excitations. For the second point, we may recall that a fixed parameter set has successfully described the entire samarium isotopes [8]. Also, we have checked in the actual calculation that the results are insensitive to the modest variation of the parameters.

Table 1. The Brody parameter β at various mass region. Here $\beta = 0$ corresponds to the Poisson (or regular) statistics, and $\beta = 1$ to the Wigner (or chaotic) statistics. The β (calc) means calculated Brody parameter in the present model. The β (exp) E indicates the experimentally compiled Brody parameter for even-even nuclei, while β (exp) A , for all nuclei.

Ω_ν / Ω_π	11 / 11	16 / 16	22 / 16	29 / 22	
β (calc)	0.74	0.60	0.51	0.37	
A	50 - 100	100 - 150	150 - 180	180 - 210	210 -
β (exp) E	----------	0.62 ± 16	0.26 ± 11	0.30 ± 18	0.27 ± 32
β (exp) A	0.88 ± 41	0.55 ± 11	0.33 ± 07	0.43 ± 17	0.24 ± 11

In Table **1**, the Brody parameters[11] of the level spacing statistics in the various nuclear mass regions are tabulated. Four columns correspond to $(\Omega_\pi, \Omega_\nu)=(11,11)$, (16,16), (16,22) and (22,29), each roughly representing nuclei of the mass regions of $50<A<100$, $100<A<150$, $150<A<210$ and $210<A$, respectively. In the second line, theoretical estimates of the Brody parameter β are shown. They are calculated from the second moment of $P(s)$. The parameter x is fixed to be one in accordance with our assumption. The numbers of the fermion pairs are set to be $N_\pi=5$ and $N_\nu=4$ as a typical example. The number of the levels included in the calculation is 900, of which 80 are of $L=0$ levels, 250 of $L=2$ levels, 250 of $L=3$ and 320 of $L=4$. In the fourth line of the Table (1), the "experimental" values of the Brody parameter for even-even nuclei taken from the paper of Shriner et al.[10] are tabulated. The agreement is as good as one can expect from our simple estimate and marginal statistics. Based on this result, we assert that the experimentally observed trend in the nuclear level statistics is physically understood within a schematic single j-shell model: namely, the chaotic spectrum in lighter nuclei is the result of the incomplete realization of dynamical symmetry due to the Pauli-blocking effect.

CONCLUSIONS

We have investigated the energy levels of the IBM and the Ginocchio model to understand general features of collective motion from the view point of chaos.

As for the IBM, we have seen that (a) strong chaotic spectra are seen in between SU(3) and O(6) symmetries, (b) persistence of Poisson statistics is found along the line connecting U(5) and O(6) symmetries, (c) Brody distribution is valid except regions close to the integrable limits and (d) a new " regular region" not related to any of the three dynamical symmetries has been confirmed which was first confirmed by Alhassid et al.

As for the Ginocchio Model, we have found that (a) generally collective motion in the Ginocchio model is "chaotic" even in the rotational limit due to the lack of dynamical symmetry , (b) when degeneracies are made larger, the system restores its collectivity and also the regularity and (c) the above result may explain the mass dependence of chaoticity.

ACKNOWLEDGEMENT

I am grateful for the support by the Grand-in-Aid for General Science Research (No. 03640260) by the Ministry of Education, Science and Culture. This work has been done in collaboration with T. Cheon, T. Shigehara and T. Mizusaki.

REFERENCES

1. A. Arima and F. Iachello, Ann. of Phys. (N.Y.), **99**:253 (1976); **111**:201 (1978); **123**:468 (1979),"The Interacting Boson Model", (Cambridge University Press, Cambridge, 1987).
2. J. N. Ginocchio, Ann. of Phys. **126**:234 (1980).
3. T. Mizusaki, N. Yoshinaga, T. Shigehara and T. Cheon, Phys. Lett. **B269**: 6 (1991).
4. M.V. Berry and M. Tabor, Proc. R. Soc. London, **A400**:229(1985).
5. M. L. Mehta, *Random Matrices 2nd edition*, (Academic, New York, 1990).
6. M. C. Gutzwiller, *Chaos in Classical and Quantum Mechanics* (Springer, Berlin, 1990).
7. Y. Alhassid and N. Whelan, Phys. Rev. Lett. **67**:816 (1991); Y. Alhassid and A. Noveselsky, Phys. Rev. **C45**:1677 (1992).
8. A. Arima, N. Yoshida and J.N. Ginocchio, **B101**:209(1981).
9. N. Yoshida, A. Arima and T. Otsuka, Phys. Lett. **B114**:86 (1982).
10. J. F. Shriner, Jr , G. E. Mitchell, T. von Egidy, Z. Phys. **A338**:309 (1991).
11. T. A. Brody, Lett. Nuovo Cimento **7**: 482 (1973).

ANISOTROPICNESS IN LATTICE GAUGE THEORY

Zheng-Kun Zhu, Sachiko Takeuchi[1] and Da Hsuan Feng

Department of Physics and Atmospheric Science, Drexel University
Philadelphia, PA 19104, USA

1. INTRODUCTION

Symmetry is a fundamental concept in nuclear physics and no where can it be more important and prevalent than in quantum chromodynamics (QCD). Although QCD is a very difficult problem, some important understandings have been made since the pioneering work of Wilson[1] and Polyakov[2] in lattice gauge theory (LGT). For example, the pure SU(2) and SU(3) gauge fields were shown to have the important feature of confinement[3] and with quark fields, a chiral symmetry restoration [4]. With the rapid development of massively parallel machines, one would expect that lattice QCD will turn into a tool for practical numerical predications in the future , and perhaps near. Still, current simulations have been limited by computational power and the slow pace of alogrithmic improvement. Therefore it is still necessary to study the type of lattice in order to improve calculations.

Thus far, nearly all the calculations were carried out on an isotropic lattice. Here, "isotropic" means that the lattice spacing in the space direction is the same as that in the time direction. If not, then it is "anisotropic". Of course, the physics in question cannot depend on the choice of lattice, but the approach to carry out the calculations can be. Indeed, it has been shown that anisotropic lattice may provide a quick approach to the continuum limit[5]. We therefore feel that the study of the LGT on an anisotropic lattice deserves a closer look, either numerically or analytically. In this paper, we have used an analytical technique to carry out such an investigation.

In the past decade, a number of analytical techniques have appeared in the market. For example, methods like strong coupling [1, 6, 7], weak coupling[7], saddle-point[8] and cumulant expansions[9]. All of these methods have their individual advantages and disadvantages. The first two were able to illuminate the nature of QCD in the strong or weak coupling regions while the latter two have the advantage to study the LGT for arbitrary coupling strengths. However, for the latter, when the coupling constant is in the intermediate range, higher order corrections are shown to be mandatory. On the other hand, it was shown[10] that to develop a systematic procedure for LGT to

---

[1]Present address: Department of Physics, Tokyo Institute of Technology, Tokyo, Japan

Symmetries in Science VII, Edited by B. Gruber
and T. Otsuka, Plenum Press, New York, 1994

compute higher order expansion terms is nontrival. A prerequisite rests on whether one can find a small expansion parameter. As it turns out, because one has the freedom to choose any kind of lattices[11, 12] since at the continuum limit, physics should be independent of the choice [7], such a parameter can indeed be found. Thus, the inverse of the anisotropic parameter (defined as the ratio between space and time spacing), which can be adjusted arbitarily small, can serve as an expansion parameter and with it we will have the possibility of an effective expansion method. In this paper, we will explore this possibility.

This idea is applied to the transverse Ising model, and the physics around the continuum limit is investigated. It is shown that our approach is rather effective: for the free energy in the symmetric phase, the zeroth order approximation seems to provide already highly accurate results while the critical coupling constant can be reproduced fairly accurately with corrections up to second-order. Also, the critical coupling constant seems to be insensitive to the lattice anisotropicness for small time spacing. This suggests that there is a region where physical observables are independent of the choice of lattice. This nontrival feature suggests that our method can be used to verify the independence of LGT on the choice of the lattice.

We have organized the paper as follows. In section 2, the pure $SU(N)$ gauge theory is discussed. The temporal gauge fixing is first employed to simplify the calculation. Then the characteristics of the lattice dynamics of the pure LGT on an anisotropic lattice is investigated. Due to a particular property of the lattice dynamics, the action is reduced to a one-dimensional system in a mean field. In section 3, The lattice dynamics of a ϕ^4 field and the transverse Ising model are also discussed on the anisotropic lattice. We have shown that they both can be reduced to one-dimensional problems. A general argument of dimensional reduction on lattice boson field is then given. In section 4, we introduced the variational cumulant expansion technique to treat the proposed problem discussed in sections 2 and 3. In order to examine the validity and practicability of our method, we have applied it to the critical property of transverse Ising model. This is given in section 5. Using this model, the expansion convergence is investigated as well as the continuum limit. Finally, the conclusions are given in section 6.

2. DIMENSIONAL REDUCTION OF PURE LGT

We begin by studying the zero temperature pure $SU(N)$ LGT, which can be formulated on an anisotropic $(d+1)$-dimensional hypercubical lattice $N_s^d \times N_\tau$ with space and time spacing a_s and a_τ, respectively. The zero temperature is obtained by letting $N_\tau a_\tau$ to be infinite, and the partition function is

$$Q = \sum e^S, \tag{1}$$

where $\sum$ represents the summation over the entire configuration space, S is the Wilson action.

$$S = \beta_s \sum_x \sum_{\mu < \nu < 4} P_x^{\mu\nu} + \beta_\tau \sum_x \sum_{\mu < 4} P_x^{\mu 4}, \tag{2}$$

where

$$P_x^{\mu\nu} = \frac{1}{N} ReTr(U_x^\mu U_{x+\mu}^\nu U_{x+\nu}^{\mu+} U_x^{\nu+}). \tag{3}$$

Here U_x^μ is a SU(N) group element and x and μ are the coordinate and the direction of the link, respectively. β_s and β_τ are

$$\beta_s = \xi^{-1}(\beta_E + c_s(\xi)) \tag{4}$$

and

$$\beta_\tau = \xi(\beta_E + c_\tau(\xi)). \tag{5}$$

where $c_s(\xi)$ and $c_\tau(\xi)$ can be obtained by background field calculations[13]. The anisotropicness of the lattice ξ is:

$$\xi = \frac{a_s}{a_\tau}. \tag{6}$$

It is well known that in a full gauge fixing formulation, the mean field approximation can be used without violating Elitzur's theorem. In addition, it can simplify the calculations as well. Therefore, we will first introduce a temporal gauge fixing, which means that all the time-like link variables chosen as identity matrices. In order to keep the theory gauge invariant, these fixed links must not produce loops[14], and the time direction has an open boundary. Thus the action S becomes

$$S = S_{space} + S_{time} \tag{7}$$

where

$$S_{space} = \beta_s \sum_x \sum_{\mu < \nu < 4} P_x^{\mu\nu} \tag{8}$$

$$S_{time} = \beta_\tau \sum_x \sum_{\mu < 4} \frac{1}{N} Re Tr(U_x^\mu U_{x+\hat{\tau}}^{\mu^+}). \tag{9}$$

We recognize that S_{time} resembles the hamiltonian of a set of one-dimensional nearest-neighbor coupling spin chains. To illustrate this formalism, let us take the four-dimensional pure SU(2) LGT as an example.

For convenience, the four dimensional vector x is written explicitly as $(\vec{x}, \tau)$. The link variables in equation (9) are now the SU(2) group elements, and each can be expressed as

$$U_x^\mu = a_0^\tau(\vec{x}, \mu) + i\vec{a}^\tau(\vec{x}, \mu) \cdot \vec{\sigma} \tag{10}$$

$$U_{x+\hat{\tau}}^{\mu^+} = a_0^{\tau+1}(\vec{x}, \mu) - i\vec{a}^{\tau+1}(\vec{x}, \mu) \cdot \vec{\sigma}, \tag{11}$$

where $\vec{\sigma}$ is the Pauli matrix and

$$(a_0^\tau(\vec{x}, \mu))^2 + (\vec{a}^\tau(\vec{x}, \mu))^2 = 1. \tag{12}$$

To show the similarity more succinctly, we rewrite $(a_0^\tau(\vec{x}, \mu), \vec{a}^\tau(\vec{x}, \mu))$ as a unitary four-dimensional vector, $\vec{S}_{\vec{x},\mu}^\tau$. Then S_{time} becomes

$$S_{time} = \beta_\tau \sum_{\vec{x},\mu < 4} \sum_\tau \vec{S}_{\vec{x},\mu}^\tau \cdot \vec{S}_{\vec{x},\mu}^{\tau+1}, \tag{13}$$

which is a one-dimensional spin model with a 4-dimensional internal space.

Obviously, if S_{space} is ignored, S_{time} can be solved. Then the effect of S_{space} can be introduced by an expansion method. Before we discuss this in detail, we will first discuss the lattice dynamics for various regions of ξ.

First, we will consider the case of $0 < \xi < 1$, or $\beta_s > \beta_\tau$. When ξ approaches zero, i.e., $\beta_s \gg \beta_\tau$, the space spacing goes to zero and LGT will return to the continuous field theory. Obviously this region will not be feasible for further exploration.

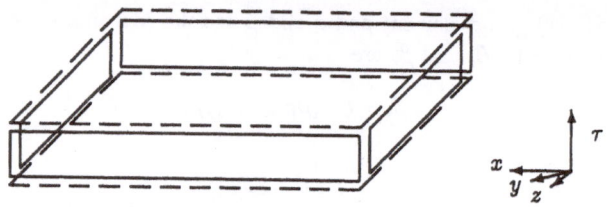

Figure 1. Pure LGT on the anisotropic lattice. The full (dashed) line plaquettes indicate that the coupling between links is strong (weak).

The second is $\xi = 1$, *i.e.*, the isotropic case. Most LGT studies made in the past were carried out on such a lattice. As was mentioned in the introduction, in the intermediate-coupling region, it is difficult to analytically study the LGT with such a lattice.

The last is $1 < \xi$. For this, ξ can be chosen in such a way that β_s and β_τ can be adjusted far from the intermediate coupling region, and then we can perform an expansion around the smaller coupling constant, β_s. The focus of this paper is in this region.

We should note that according to equations (4) and (5), for fixed β_E, an increase of ξ will decrease β_s while increasing β_τ. In particular, when $\xi \to \infty$, which corresponds to the hamiltonian approach of LGT,

$$\beta_s \to 0 \quad \text{and} \quad \beta_\tau \to \infty, \tag{14}$$

while the product of β_s and β_τ remains finite:

$$\beta_s \beta_\tau = C. \tag{15}$$

where C is a finite value.

On the very anisotropic lattice, we have a small β_s. This allow us to approximate S_{space} and include its effect by an expansion technique. In the zeroth order, the action S is approximated as S_0:

$$S_0 = \beta_\tau \sum_{x,\hat\tau} \sum_{\mu < 4} \frac{1}{N} ReTr(U_x^\mu U_{x+\hat\tau}^{\mu^+}) + \sum_x \sum_{\mu < 4} Tr(J U_x^\mu + U_x^{\mu^+} J^+), \tag{16}$$

where $\{J\}$ is a mean field parameter. We see from this that the study of pure SU(N) LGT becomes the study of one-dimensional spin model (intra-chain interaction) with residual effects of S_{space} (inter-chain interaction). That is why we have referred such an this approximate system as a quasi one-dimensional system, and succinctly this idea is graphically displayed in Fig.1.

3. DIMENSIONAL REDUCTION OF LATTICE BOSON FIELD ON THE ANISOTROPIC LATTICE

It should be emphasized that the dynamics of pure $SU(N)$ LGT on the anisotropic lattice is by no means a special case. In fact, the dynamics of the lattice ϕ^4 field and the transverse Ising model on the anisotropic lattice are both similar to that and can be reduced to quasi one-dimensional problems. In this section, we will further argue that such a reduction is suitable for any boson fields on the anisotropic lattice.

3.1 LATTICE DYNAMICS OF ϕ^4 FIELD

In this subsection, we will show how a ϕ^4 filed can be reduced to a one-dimensional system.

The Lagrangian of a $(d+1)$-dimensional self-coupled scalar field [7] is

$$L = \int d^d x \left\{ \frac{1}{2} \left(\partial \phi / \partial t \right)^2 - \frac{1}{2} \left(\nabla \phi \right)^2 - \frac{1}{2} \mu_0^2 \phi^2 - \lambda_0 \phi^4 \right\}. \tag{17}$$

In order to formulate it on the anisotropic lattice with a_s and a_τ as it space and time spacing respectively, its Euclidean action is[19]

$$-S = \sum_{\vec{n},\tau} \Big[K_\tau \{ \phi(\vec{n}, \tau+1) - \phi(\vec{n}, \tau) \}^2$$
$$+ K \sum_{\vec{k}} \{ \phi(\vec{n}+\vec{k}, \tau) - \phi(\vec{n}, \tau) \}^2 + b_0 \phi(\vec{n}, \tau)^2 + u_0 \phi(\vec{n}, \tau)^4 \Big], \tag{18}$$

where

$$K_\tau = \frac{a_s^d}{2a_\tau}$$

$$K = \frac{1}{2} a_s^{d-2} a_\tau$$

$$b_0 = \frac{1}{2} a_s^d a_\tau \mu_0^2$$

$$u_0 = a_s^d a_\tau \lambda_0. \tag{19}$$

To facilitate the study of the (3+1)-dimensional lattice ϕ^4 field on the anisotropic lattice, the following definitions are made:

$$\phi_{latt}(\vec{n}, \tau) = \phi(\vec{n}, \tau) a_s \tag{20}$$

and

$$\xi = \frac{a_s}{a_\tau}. \tag{21}$$

Using eqs. (20) and (21), eq.(18) can be rewritten as

$$-S = \sum_{\vec{n},\tau} \Big[K_\tau^l \{ \phi_{latt}(\vec{n}, \tau+1) - \phi_{latt}(\vec{n}, \tau) \}^2$$
$$+ K^l \sum_{\vec{k}} \{ \phi_{latt}(\vec{n}+\vec{k}, \tau) - \phi_{latt}(\vec{n}, \tau) \}^2 + b_0^l \phi_{latt}(\vec{n}, \tau)^2 + u_0^l \phi_{latt}(\vec{n}, \tau)^4 \Big], \tag{22}$$

where

$$K_\tau^l = \frac{\xi}{2} a_s^{d-3}$$

$$K^l = \frac{1}{2\xi} a_s^{d-3}$$

$$b_0^l = \frac{1}{2\xi} a_s^{d-1} \mu_0^2$$

$$u_0^l = \xi^{-1} a_s^{d-3} \lambda_0. \tag{23}$$

Again we will write S as the sum of S_{space} and S_{time}

$$-S_{time} = \sum_{\vec{n},\tau} K_\tau^l \{\phi_{latt}(\vec{n},\tau+1) - \phi_{latt}(\vec{n},\tau)\}^2$$

$$+ \sum_{\vec{n},\tau} \left[K^l 2d \{\phi_{latt}(\vec{n},\tau)\}^2 + b_0^l \phi_{latt}(\vec{n},\tau)^2 + u_0^l \phi_{latt}(\vec{n},\tau)^4 \right] \quad (24)$$

$$-S_{space} = -2K^l \sum_{\vec{n},\vec{k}} \phi_{latt}(\vec{n}+\vec{k},\tau)\phi_{latt}(\vec{n},\tau). \quad (25)$$

For an extremely anisotropic lattice ($\xi \gg 1$) with a fixed value of a_s, K^l is much smaller than K_τ^l while their product remains fixed:

$$K^l K_\tau^l = \frac{a_s^{2d-6}}{4}. \quad (26)$$

With a small coupling K^l on the very anisotropic lattice, S can be approximated as S_0:

$$S_0 = S_{time} + \alpha \sum_{\vec{n},\tau} \phi_{latt}(\vec{n},\tau), \quad (27)$$

where α is a mean field parameter. The effect of S_{space} can then be introduced by an expansion technique. Thus, we see that the ϕ^4 field also becomes one-dimensional.

3.2 THE TRANSVERSE ISING MODEL

The next example we will investigate is the transverse Ising model. The Hamiltonian of a d-dimensional transverse Ising model is

$$\hat{H} = -\sum_i \hat{\sigma}_i^x - \lambda \sum_{<j,k>} \hat{\sigma}_j^z \hat{\sigma}_k^z, \quad (28)$$

where λ is the strength of the spin-spin coupling between nearest-neighbor lattice sites j and k, and $\hat{\sigma}^x$ and $\hat{\sigma}^z$ are the Pauli matrices. In the path integral form with N_τ time sites, the system at temperature T can be regarded as a $(d+1)$-dimensional Ising model with action[19]

$$S = \sum_{(\vec{n},n_\tau)} (\beta_s \sum_{\vec{i}} \sigma_{\vec{n},n_\tau} \sigma_{\vec{n}+\vec{i},n_\tau} + \beta_\tau \sigma_{\vec{n},n_\tau} \sigma_{\vec{n},n_\tau+1}), \quad (29)$$

where $(\vec{n},n_\tau)$ represents the coordinate of a site in the $(d+1)$-dimensional lattice, $\vec{i}$ the d-dimensional unit vector, and β_s and β_τ are the two coupling constants in the space and time directions, respectively. $\sigma_{\vec{n},n_\tau}$ represents spin at site $(\vec{n},n_\tau)$. β_s and β_τ are

$$\beta_s = \lambda \frac{\beta}{N_\tau}$$

$$\beta_\tau = \frac{1}{2} \ln(\coth(\frac{\beta}{N_\tau})) \quad (30)$$

with $\beta = \frac{1}{kT}$. The spacing in the time direction a_τ is

$$a_\tau = \frac{\beta}{N_\tau}. \quad (31)$$

Therefore, we can rewrite equation (30) as

$$\beta_s = \lambda a_\tau \qquad (32)$$

$$\beta_\tau = \frac{1}{2}\ln(\coth a_\tau). \qquad (33)$$

With a fixed value of λ, as $a_\tau \to 0$, which is the continuum limit, β_τ approaches infinity while β_s will approach zero. However, the relation between β_s and β_τ remains fixed:

$$\frac{e^{2\beta_\tau}}{\coth(\frac{\beta_s}{\lambda})} = 1. \qquad (34)$$

To study the dynamics explicitly, we define the anisotropic parameter ξ as

$$\xi = \sqrt{\frac{\beta_\tau}{\beta_s}} \qquad (35)$$

Again S is written as the sum of S_{space} and S_{time}.

$$S_{space} = \sum_{(\vec{n},n_\tau)} \beta_s \sum_i \sigma_{\vec{n},n_\tau}\sigma_{\vec{n}+\vec{i},n_\tau}$$

$$S_{time} = \beta_\tau \sum_{(\vec{n},n_\tau)} \sigma_{\vec{n},n_\tau}\sigma_{\vec{n},n_\tau+1}. \qquad (36)$$

On the extremely anisotropic lattice, β_τ is much larger than β_s. Therefore, we can approximate S_{space} as the "mean field action":

$$S_0 = \beta_\tau \sum_{(\vec{n},n_\tau)} \sigma_{\vec{n},n_\tau}\sigma_{\vec{n},n_\tau+1} + \alpha \sum_{(\vec{n},n_\tau)} \sigma_{\vec{n},n_\tau}, \qquad (37)$$

where α is a mean field parameter. Hence, S_0 becomes a one-dimensional Ising model which is exactly solvable, and the zero temperature lattice for the transverse Ising model can be obtained by choosing $N_\tau a_\tau \to \infty$, just as the $SU(N)$ and ϕ^4 case.

3.3 THE LATTICE BOSON FIELD

In the previous two subsections, we have reduced an arbitrary dimensional pure lattice gauge theory, the lattice ϕ^4 field and the transverse Ising model to be one-dimensional problems. We argue that this can be applied to any lattice boson field with nearest-neighbor interactions in the time direction. We will first give a general definition of an "anisotropic parameter" ξ:

Definition The "anisotropic parameter" ξ is defined by the square root of the ratio of coupling constant β_τ in the time direction and β_s in the space direction or the ratio of space and time spacing.

We now will make a remark about the bosonic field lattice dynamics.

Remark On the very anisotropic lattice ($\xi \gg 1$), the coupling constant (β_s) in the space direction becomes much weaker than that (β_τ) in the time direction. While ξ goes to infinite, β_s goes to zero and β_τ goes to infinite but a relation of β_s and β_τ remains fixed:

$$f(\beta_s, \beta_\tau, \lambda) = 0 \qquad (38)$$

Here $f(\beta_s, \beta_\tau, \lambda)$ is a function of β_s, β_τ and λ, where λ is a characteristic constant given by the field.

For a pure lattice gauge theory, the lattice ϕ^4 field and the transverse Ising model, eq.(38) represents eqs.(15) (26) and (34), respectively.

Now we see that the various lattice boson fields have a generic dynamical structure, and can be reduced to a quasi one-dimensional problem. Let us emphasize the importance of the constraint condition (38). When ξ is infinite, the system is reduced to a trivial one-dimensional problem[17]. It is different for the gauge field. Due to the eq. (38), the LGT remains nontrivial and require us to introduce the residual effects of S_{space} to the system.

4. MEAN FIELD CORRECTION

The mean field approximation to LGT discussed above is a zeroth order approximation. The results can be improved by including higher order correction terms. Various groups have found that the variational cumulant expansion (VCE) technique is a powerful method to study the nonperturbative problems[9]. We will also employ this method to carry out our investigation.

With the "mean field action" S_0 as a trial action, the residual effects of S_{space} are introduced by higher order terms. The partition function of the system with the original action, Q, can be written as:

$$
\begin{aligned}
Q &= \sum \exp(S_0)\exp(S - S_0) \\
&= Q_0 \langle \exp(S - S_0) \rangle_0 \\
&= Q_0 \exp(\sum_{n=1}^{\infty} \frac{1}{n!} K_n),
\end{aligned}
$$
(39)

where

$$
Q_0 = \sum \exp(S_0)
$$
(40)

$$
\langle \cdots \rangle_0 = \frac{1}{Q_0} \sum \exp(S_0)(\cdots)
$$
(41)

and

$$
\begin{aligned}
K_1 &= \langle S - S_0 \rangle_0 \\
K_2 &= \langle (S - S_0)^2 \rangle_0 - \langle S - S_0 \rangle_0^2 \\
K_3 &= \langle (S - S_0)^3 \rangle_0 - 3\langle S - S_0 \rangle_0 \langle (S - S_0)^2 \rangle_0 + 2\langle S - S_0 \rangle_0^3.
\end{aligned}
$$
(42)
$$
\vdots
$$

Then, the free energy F is

$$
F = -L^{-1} \ln Q,
$$
(43)

L is the number of lattice sites, i.e., $N_\tau \times N_s^d$. The mean field parameters can be determined by minimizing the free energy with respect to the mean field parameters. Once these parameters are known, the expectation value of any observable G can be computed in the cumulant expansion fashion as follows:

$$
\begin{aligned}
\langle G \rangle &= \frac{1}{Q} \sum_{\{U\}} G\{U\} \times \exp(S\{U\}) \\
&= \sum_{m=1}^{\infty} \frac{1}{m!} L_m,
\end{aligned}
$$
(44)

where

$$
\begin{aligned}
L_1 &= \langle G \rangle_0 \\
L_2 &= 2\langle G(S - S_0) \rangle_0 - 2\langle G \rangle_0 \langle (S - S_0) \rangle_0 \\
L_3 &= 3\langle G(S - S_0)^2 \rangle_0 - 6\langle G(S - S_0) \rangle_0 \langle (S - S_0) \rangle_0 \\
&\quad + 6\langle G \rangle_0 \langle (S - S_0) \rangle_0^2 - 3\langle G \rangle_0 \langle (S - S_0)^2 \rangle_0.
\end{aligned}
\tag{45}
$$
$$\vdots$$

and $\{L_n\}$ are the cumulant expansion terms. Thus, we have a way to obtain any physical observables.

5. APPLICATION

Once we have an expansion technique, it is of course important to determine the validity and practicability of this method. To show that, the transverse Ising model will be employed to demonstrate its usefulness. This is also as a guide to study the LGT.

Our calculations will be performed on $N_s^d \times N_\tau$ lattice where N_s and N_τ are the number of the sites of the lattice in the space and time directions, respectively. In VCE, N_s is assumed to be infinite. With the "mean field action" S_0, the free energy F per site is expressed as:

$$
\begin{aligned}
F &= -\frac{1}{L} \ln Q \\
&= -\frac{1}{L} \ln Q_0 - \frac{1}{L} K_1 - \frac{1}{2L} K_2 + \cdots
\end{aligned}
\tag{46}
$$

We will compute only the first three terms of the rhs of eq.(46). They are

$$
Q_0 = \sum_{\{\sigma\}} \exp(S_0\{\sigma\}),
\tag{47}
$$

$$
K_1 = \langle \sum_{\vec{n},n_\tau} \beta_s \sum_{\vec{i}} \sigma_{\vec{n},n_\tau} \sigma_{\vec{n}+\vec{i},n_\tau} - \alpha \sum_{\vec{n},n_\tau} \sigma_{\vec{n},n_\tau} \rangle_0
\tag{48}
$$

and

$$
\begin{aligned}
K_2 &= \langle (\sum_{\vec{n},n_\tau} \beta_s \sum_{\vec{i}} \sigma_{\vec{n},n_\tau} \sigma_{\vec{n}+\vec{i},n_\tau} - \alpha \sum_{\vec{n},n_\tau} \sigma_{\vec{n},n_\tau})^2 \rangle_0 \\
&\quad - \langle \sum_{\vec{n},n_\tau} \beta_s \sum_{\vec{i}} \sigma_{\vec{n},n_\tau} \sigma_{\vec{n}+\vec{i},n_\tau} - \alpha \sum_{\vec{n},n_\tau} \sigma_{\vec{n},n_\tau} \rangle_0^2
\end{aligned}
\tag{49}
$$

These calculations can be performed with the aid of graphical expression, and the results are given in Appendix A.

5.1 THE CONVERGENCE OF EXPANSION IN THE SYMMETRY PHASE

It is of course vital to know whether the expansion converges, and if so, whether it converges quickly. We will now discuss the situation in the symmetry phase.

In this phase, both the average spin in the mean field action, $\langle\sigma\rangle_0$, and the variational parameter α vanish. Therefore, K_1 must vanish as well. To calculate K_2, we will consider the $N_\tau \to \infty$ case. The two spin correlation is given by

$$\langle\sigma_i\sigma_j\rangle_0 = (\tanh\beta_\tau)^{|j-i|}. \tag{50}$$

and K_2 is

$$K_2 = L(\beta_s)^2\{d + d\sum_{\substack{|i-j|=1\,(i\,\text{fixed})}}^{N_\tau-1}\langle\sigma_i\sigma_j\rangle_0^2\}. \tag{51}$$

After substituting eq.(50) into eq.(51), we have

$$\frac{K_2}{L} = d\beta_s^2\frac{1 + (\tanh\beta_\tau)^2}{1 - (\tanh\beta_\tau)^2}. \tag{52}$$

With eq.(35), one sees that in the time continuum limit, K_2 vanishes as well. In fact, explicit calculations of K_3 and K_4 which will not be given here also show that they vanish at the continuum limit as well. Let us emphasize this aspect by comparing our results with the Onsager solution for the $(1+1)$-dimensional Ising model (Appendix B) [20]. Remarkably, at the continuum limit in the time direction, the partition function given by the zeroth order term in our expansion, Q_0, is equal to the partition function of the Onsager solution. This suggests that the higher order terms simply do not contribute here.

This example shows that even if the coupling constant is very close to the critical value, as long as it is in the symmetric phase, the system seems to be well approximated by the one-dimensional model, and the correlation in the space direction can be neglected. It may also imply rapid convergence, even in the broken symmetric phase.

5.2 THE CRITICAL PROPERTY OF THE TRANSVERSE ISING MODEL

The phase transition from the symmetric to the broken symmetric phase is known to be second order. So in this case α is non-zero and very small when λ is very close to the critical point, because α is zero in the symmetric phase. Hence if we differentiate the free energy, F, with respect to α near the critical point from the broken symmetric phase side, the derivative can be expressed as:

$$\frac{\partial F}{\partial\alpha} = P\alpha + O(\alpha^2). \tag{53}$$

The condition of a minimum value of F is

$$\frac{\partial F}{\partial\alpha} = 0, \tag{54}$$

which means that

$$P = 0. \tag{55}$$

Thus, using equations (61) and (60) in Appendix A, the critical value equation of the transverse Ising model, which corresponds to eq.(55), is shown to be

$$\{\beta_s e^{2\beta_\tau} \frac{(\lambda_+)^{N_\tau} - (\lambda_-)^{N_\tau}}{(\lambda_+)^{N_\tau} + (\lambda_-)^{N_\tau}} (2d-1) - 1\}$$
$$+ \frac{1}{4}\beta_s \frac{\frac{\partial}{\partial\alpha}(\sum_{|i-j|=1, (i\,\text{fixed})}^{N_\tau-1} \langle\sigma_i\sigma_j\rangle_0^2)}{(\sum_{|i-j|=1, (i\,\text{fixed})}^{N_\tau-1} \langle\sigma_i\sigma_j\rangle_0 + 1)\frac{\partial}{\partial\alpha}(\langle\sigma\rangle_0)\alpha} = 0. \tag{56}$$

With $N_\tau \to \infty$ for zero temperature, the last equation is reduced to

$$\{\beta_s e^{2\beta_\tau}(2d - \frac{5}{8}) - 1\} - \frac{\beta_s e^{-2\beta_\tau}}{4} - \frac{\beta_s e^{-6\beta_\tau}}{8} = 0. \tag{57}$$

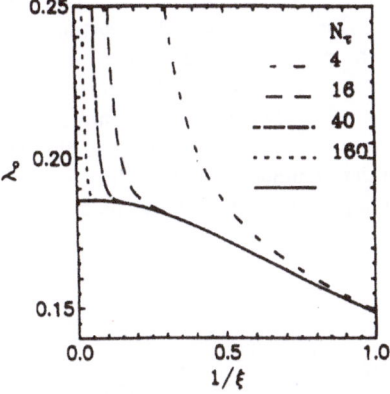

Figure 2. Critical coupling λ_c of the transverse Ising model with respect to the inverse anisotropic parameter $\frac{1}{\xi}$.

The time continuum limit can be obtained by letting $a_\tau \to 0$, which means that $\beta_s \to 0$ and $\beta_\tau \to \infty$. Therefore this limit can be obtained simply by choosing ξ as infinity. Using eq.(35), the above equation in the continuum limit becomes

$$(2d - \frac{5}{8})\lambda - 1 = 0. \tag{58}$$

In the $(3+1)$-dimensional space, the following physical result on the very anisotropic lattice is obtained. The critical value, λ_c, according to the second-order approximation is

$$\lambda_c^{(2)} = 0.186. \tag{59}$$

In order to study the convergence of the expansion in the broken phase, we will also give the critical value of $(3 + 1)$-dimensional space up to first-order: $\lambda_c^{(1)} = 0.167$. It is known from previous calculation that the result with up to tenth order corrections (by a high temperature expansion method) is 0.197 ± 0.001 [22]. Hence, our expansion on the anisotropic lattice converges rapidly even near critical point. Certainly, the result can be further improved by introducing higher order terms.

In the following, we will discuss the behavior of the system as it approaches the continuum limit in the time direction. Various types of lattice with finite N_τ ($= 4$, 16, 40, 160, ∞) are considered and the ξ dependence of $\lambda_c^{(2)}$, with eq.(56), are numerically evaluated. The results are plotted in Fig.2. As can be seen from the figure, as long as N_τ is finite, the critical value for each N_τ will eventually diverge as $\xi \to \infty$, i.e., $a_\tau \to 0$. This phenomenon of course originates from the finite size effect. We will also point out that the divergence of the critical value is in agreement with the result of ref.[12], which claims that large ξ at fixed N_τ and N_s will worsen the lattice approximation. In this method, we can avoid this problem.

For $N_\tau = 40$ and 160, we see that a plateau exists for the critical value curve which will widen as N_τ increases. The critical value at the plateau is approximately 0.186, corresponding to the zero-temperature value at the continuum limit, eq.(59). This shows that we can still obtain the physical value by extrapolation although the system with finite N_τ cannot go to the continuum limit. So we conclude that by changing ξ from 1 to infinity for the fixed large N_τ, one can study the continuum limit of the zero temperature LGT. This is particularly important for problems which are not exactly solvable even after the reduction to the one-dimensional problems. Moreover, the existence of the plateau implies that if a_τ is sufficiently small, the observable is not sensitive to the variation of ξ. Therefore we should expect that there is a region where the observables does not depend on the choice of the lattice, and this certainly agrees with the intuitive understanding of LGT. In fact, our method seems to provide a procedure to verify the independence of LGT on the choice of the type of lattice.

6. CONCLUSION

Through several examples, we have shown that it is possible to reduce an arbitrary dimensional lattice boson field to a quasi one-dimensional problem approximately on the extremely anisotropic lattice.

Our study of the transverse Ising model shows that the method is indeed effective. In the symmetric phase, the mean field result (zeroth order approximation) is found to be in excellent agreement with exact results: For the free energy, the correction terms up to fourth order (the highest order we have calculated) vanish. For the $(1+1)$-dimensional case, the mean field result for free energy is exact. For the critical coupling constant, our method with the correction up to the second order already gives very close value to the exact one. The results will be further improved by introducing higher-order terms, which are feasible to calculate.

We also found that there exists a plateau for sufficiently large but finite N_τ lattice which corresponds to the continuum limit. This is a nontrival feature and suggests that we can expect a region where physical observables are independent of the choice of the lattice. This also suggests that our method can be used to verify the independence of LGT on the choice of the type of lattice. Furthermore this feature confirms the validity of our method. We therefore conclude that to go to the continuum limit, the following procedure can be taken: First we fix N_τ at a large but finite value, and then change the anisotropic parameter from 1 to a very large value. To obtain the result at the continuum limit, we need to study the behavior of the physical observable

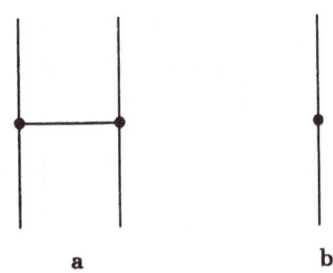

a   b

Figure 3. Contribution of cumulant term K_1. The vertical line represents the one-dimensional chain in the time direction and • represents spin σ_i.

carefully as the anisotropic parameter approachs infinity. We can expect that when N_τ is sufficiently large, but before the emergence of finite size effect, the observable has a stationary value from which the result at the continuum limit can be obtained by extrapolation.

Our investigation shows that the study on the anisotropic lattice is indeed nontrival and it may help us to understand some other nontrival features. We also see that it is possible to apply our approach to study other problems, the approximate one-dimension system has to be solvable at least for finite N_τ. The boson fields we have investigated here, i.e., pure $SU(N)$ gauge field, ϕ^4 field, and transverse Ising model, seem to satisfy this condition. We then need to enumerate the cumulant terms to include the residual correlations in the space direction by using graphical forms. We found out that the graphs are identical to the usual cumulant expansion results, which suggests that our method can be an approach for nonperturbative study.

ACKNOWLEDGEMENT: Many discussions with Hai-Cang Ren, Xiangdong Ji, Yong-Shi Wu and Lay Nam Chang are acknowledged. One of us (DHF) is particularly grateful to Professor Franco Iachello who showed him the great importance of symmetry in nuclear physics. This work was supported by the National Science Foundation of the United States.

APPENDIX A

In this appendix, we want to present the calculation of K_1 and K_2 for the transverse Ising model.

(1) The calculation of K_1

The calculation of K_1 can be represented by Fig.3 a) and b). Each graph corresponds to the first and second terms in the following result, respectively.

$$K_1 = L(d\beta_s\langle\sigma\rangle_0^2 - \alpha\langle\sigma\rangle_0). \tag{60}$$

Here $\langle\sigma\rangle_0$ is the average spin by the action S_0.

(2) The calculation of K_2 The calculation of K_2 can be represented by Fig.4. The terms

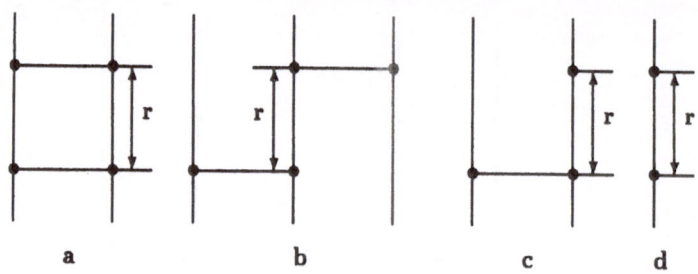

Figure 4. Contribution of cumulant term K_2. The vertical line represents the one-dimensional chain in the time direction and • represents spin σ_i. (•——•) represents $\sigma_i\sigma_j$ with r, the distance in the time direction between two spins.

do not vanish only when two (or more) spins are on the same spatial point. The result is:

$$
\begin{aligned}
K_2 &= L\beta_s^2 d \sum_{\substack{|i-j|=0,\,(i\text{ fixed})}}^{N_\tau-1} (\langle\sigma_i\sigma_j\rangle_0^2 - \langle\sigma\rangle_0^4) \\
&+ L\beta_s^2 2d(2d-1) \sum_{\substack{|i-j|=0,\,(i\text{ fixed})}}^{N_\tau-1} (\langle\sigma_i\sigma_j\rangle_0\langle\sigma\rangle_0^2 - \langle\sigma\rangle_0^4) \\
&- L2\alpha\beta_s 2d \sum_{\substack{|i-j|=0,\,(i\text{ fixed})}}^{N_\tau-1} (\langle\sigma_i\sigma_j\rangle_0\langle\sigma\rangle_0 - \langle\sigma\rangle_0^3) \\
&+ L\alpha^2 \sum_{\substack{|i-j|=0,\,(i\text{ fixed})}}^{N_\tau-1} (\langle\sigma_i\sigma_j\rangle_0 - \langle\sigma\rangle_0^2).
\end{aligned} \tag{61}
$$

Here $\langle\sigma_i\sigma_j\rangle_0$ is the two-spin correlation in the time direction. Each term in the above equation corresponds to Fig.4 a)–d), respectively.

By means of matrix diagonalizing technique[21], it is easy to obtain

$$
Q_0 = ((\lambda_+)^{N_\tau} + (\lambda_-)^{N_\tau})^{N_s^d} \tag{62}
$$

$$
\langle\sigma\rangle_0 = \left\{ \frac{(\lambda_+)^{N_\tau} - (\lambda_-)^{N_\tau}}{(\lambda_+)^{N_\tau} + (\lambda_-)^{N_\tau}} \right\} \frac{\sinh\alpha}{\sqrt{(\sinh\alpha)^2 + e^{-4\beta_\tau}}} \tag{63}
$$

$$
\langle\sigma_i\sigma_j\rangle_0 = \frac{(\sinh\alpha)^2}{(\sinh\alpha)^2 + e^{-4\beta_\tau}}
$$
$$
+ \frac{e^{-4\beta_\tau}}{(\sinh\alpha)^2 + e^{-4\beta_\tau}} \left\{ \frac{(\frac{\lambda_-}{\lambda_+})^r}{1 + (\frac{\lambda_-}{\lambda_+})^{N_\tau}} + \frac{(\frac{\lambda_-}{\lambda_+})^{N_\tau-r}}{1 + (\frac{\lambda_-}{\lambda_+})^{N_\tau}} \right\}, \tag{64}
$$

where $|i-j| = r$, and $\lambda_\pm$ are given by

$$
\lambda_\pm = e^{\beta_\tau}\left(\cosh\alpha \pm \sqrt{(\sinh\alpha)^2 + e^{-4\beta_\tau}} \right). \tag{65}
$$

APPENDIX B

The partition function of two dimensional Onsager solution of each point, Q, is given by

$$\ln Q = \frac{1}{2}\ln(2\sinh(2\beta_\tau)) + \frac{1}{2\pi}\int_0^\pi \gamma(\omega)d\omega, \tag{66}$$

where

$$\cosh(\gamma(\omega)) = \cosh(2\beta_s)\cosh(2\beta_\tau^*) - \sinh(2\beta_s)\sinh(2\beta_\tau^*)\cos\omega \tag{67}$$

and

$$\beta_\tau^* = \frac{1}{2}\ln(\coth\beta_\tau). \tag{68}$$

When $\beta_\tau \to \infty$, both β_τ^* and β_s vanish. Therefore, at this limit,

$$\ln Q = \beta_\tau. \tag{69}$$

It is easy to see that Q here is exactly the same as what we have obtained for the zeroth-order term Q_0/L in equation (47).

REFERENCES

[1] K.G.Wilson, Phys. Rev. D14(1974)2455.

[2] A.M.Polyakov, Phys. Lett. B59(1975)79; Phys. Lett. B59(1975)82.

[3] M.Creutz, Phys. Rev. D21(1980)2308; Phys. Rev. Lett. 45(1980)313.

[4] H.Hamber and G.Parisi, Phys. Rev. Lett. 47(1981)1792,
E.G.Marinari, G.Parisi and C.Rebbi, Phys. Rev. Lett. 47(1981)1795,
J.Kogut, M.Stone, H.W.Wyld, J.Shigemitsu, S.H.Shenker and D.K.Sinclair, Phys. Rev. Lett. 48(1982)1140.

[5] S.Das and J.B.Kogut, Nucl. Phys. B265(1986)303 and references therein,
Z. K. Zhu, X. Zhang, X. Jin and B. Sa, Chinese Physics, 10(1990)601.

[6] J.M.Drouffe and J.B.Zuber, Phys. Rep. 102(1983)1 and references therein.

[7] J.Kogut, Rev. Mod. Phys. D51(1979)659
J.Kogut, Rev. Mod. Phys. D55(1983)775 and references therein.

[8] J.M.Drouffe, Nucl. Phys. B170(1980)211,
E.Brezin and J.M.Drouffe, Nucl. Phys. B200(1982)93.

[9] X.H.He, T.C.Hsien and Y.X.Song, Phys. Lett. B153(1985)417,
X.T.Zheng, Z.G.Tan and J.Wang, Nucl. Phys. B287(1987)171.

[10] W.Kerler, Phys. Rev. Lett. 60(1988)1906.

[11] B.McCoy and T.T.Wu, Nucl. Phys. B190(1981)519,
H.B.Nielsen and Y.K.Fu, Nucl. Phys. B236(1984)167.

[12] J.Engels, F.Karsch and H.Satz, Nucl. Phys. **B205**(1982)239,
J.Engels, F.Karsch, H.Satz and I.Montvay, Nucl. Phys. **B205**(1982)545.

[13] F. Karsch, Nucl. Phys. **B205**(1982)285.

[14] M.Creutz, Phys. Rev. **D15**(1977)1128.

[15] D.P.Landau and R.H.Swenden, Phys. Rev. **B30**(1984)2787.

[16] J.Kogut and L.Susskind, Phys. Rev. **D11**(1975)395.

[17] L.L.Liu and H.E.Stanley, Phys. Rev. Lett. **29**(1972)927, Phys. Rev. **B8**(1973)2279,
L.J.de Jongh and H.E.Stanley, Phys. Rev. Lett. **36**(1976)817
T. Draim and D.P.Landau, Phys. Rev. **B24**(1981)5156.

[18] M.Suzuki, Prog. Theor. Phys. **56**(1976)1454.

[19] M.Kato, Prog. Theor. Phys. **73**(1985)426.

[20] L.Onsager, Phys. Rev. **65**(1944)117.

[21] See, for example, K.Huang, *Statistical Mechanics*, John Wiley&Sons Inc. 1967.

[22] J.Oitmaa and M.Plischke, Physica **B86-88**(1977)577.

INDEX